Volcano–Ice Interactions on Earth and Mars

Special Publication reviewing procedures

The Society makes every effort to ensure that the scientific and production quality of its books matches that of its journals. Since 1997, all book proposals have been refereed by specialist reviewers as well as by the Society's Books Editorial Committee. If the referees identify weaknesses in the proposal, these must be addressed before the proposal is accepted.

Once the book is accepted, the Society has a team of Book Editors (listed above) who ensure that the volume editors follow strict guidelines on refereeing and quality control. We insist that individual papers can only be accepted after satisfactory review by two independent referees. The questions on the review forms are similar to those for *Journal of the Geological Society*. The referees' forms and comments must be available to the Society's Book Editors on request.

Although many of the books result from meetings, the editors are expected to commission papers that were not presented at the meeting to ensure that the book provides a balanced coverage of the subject. Being accepted for presentation at the meeting does not guarantee inclusion in the book.

Geological Society Special Publications are included in the ISI Science Citation Index, but they do not have an impact factor, the latter being applicable only to journals.

More information about submitting a proposal and producing a Special Publication can be found on the Society's web site: www.geolsoc.org.uk.

GEOLOGICAL SOCIETY SPECIAL PUBLICATION NO. 202

Volcano–Ice Interaction on Earth and Mars

EDITED BY

J. L. SMELLIE
British Antarctic Survey, UK

M. G. CHAPMAN
US Geological Survey, USA

2002

Published by

The Geological Society

London

THE GEOLOGICAL SOCIETY

The Geological Society of London (GSL) was founded in 1807. It is the oldest national geological society in the world and the largest in Europe. It was incorporated under Royal Charter in 1825 and is Registered Charity 210161.

The Society is the UK national learned and professional society for geology with a worldwide Fellowship (FGS) of 9000. The Society has the power to confer Chartered status on suitably qualified Fellows, and about 2000 of the Fellowship carry the title (CGeol). Chartered Geologists may also obtain the equivalent European title, European Geologist (EurGeol). One fifth of the Society's fellowship resides outside the UK. To find out more about the Society, log on to www.geolsoc.org.uk.

The Geological Society Publishing House (Bath, UK) produces the Society's international journals and books, and acts as European distributor for selected publications of the American Association of Petroleum Geologists (AAPG), the American Geological Institute (AGI), the Indonesian Petroleum Association (IPA), the Geological Society of America (GSA), the Society for Sedimentary Geology (SEPM) and the Geologists' Association (GA). Joint marketing agreements ensure that GSL Fellows may purchase these societies' publications at a discount. The Society's online bookshop (accessible from www.geolsoc.org.uk) offers secure book purchasing with your credit or debit card.

To find out about joining the Society and benefiting from substantial discounts on publications of GSL and other societies world-wide, consult www.geolsoc.org.uk, or contact the Fellowship Department at: The Geological Society, Burlington House, Piccadilly, London W1J 0BG: Tel. +44 (0)20 7434 9944; Fax +44 (0)20 7439 8975; Email: enquiries@geolsoc.org.uk.

Published by The Geological Society from:
The Geological Society Publishing House
Unit 7, Brassmill Enterprise Centre
Brassmill Lane
Bath BA1 3JN, UK

(*Orders*: Tel. +44 (0)1225 445046
Fax +44 (0)1225 442836)
Online bookshop: http://bookshop.geolsoc.org.uk

British Library Cataloguing in Publication Data
A catalogue record for this book is available from the British Library.

ISBN 1-86239-121-1
ISSN 0305-8719

Distributors
USA
AAPG Bookstore
PO Box 979
Tulsa
OK 74101-0979
USA
Orders: Tel. +1 918 584-2555
Fax +1 918 560-2652
E-mail: *bookstore@aapg.org*

India
Affiliated East-West Press PVT Ltd
G-1/16 Ansari Road, Daryaganj,
New Delhi 110 002
India
Orders: Tel. +91 11 327-9113
Fax +91 11 326-0538
E-mail: *affiliat@nda.vsnl.net.in*

Japan
Kanda Book Trading Co.
Cityhouse Tama 204
Tsurumaki 1-3-10
Tama-shi
Tokyo 206-0034
Japan
Orders: Tel. +81 (0)423 57-7650
Fax +81 (0)423 57-7651

Typeset by Aarontype Ltd, Bristol, UK

Printed by Cromwell Press, Trowbridge, UK

Contents

It is recommended that reference to all or part of this book should be made in one of the following ways:

SMELLIE, J. L. & CHAPMAN, M. G. (eds) 2002. *Volcano–Ice Interaction on Earth and Mars.* Geological Society, London, Special Publications, **202**.

EDWARDS, B. R. & RUSSELL, J. K. 2002. Glacial influences on morphology and eruptive products of Hoodoo Mountain volcano, Canada. *In*: SMELLIE, J. L. & CHAPMAN, M. G. (eds) *Volcano–Ice Interaction on Earth and Mars*. Geological Society, London, Special Publications, **202**, 179–194.

Referees

The Editors are very grateful to the following people who, through their insightful comments, provided the additional motivation that materially improved the scientific quality of the papers submitted to this Special Publication.

Philip Kyle, New Mexico Institute of Mining and Technology, Socorro, USA
Magnus Gudmundsson, University of Iceland, Reykjavik, Iceland
Jamie Allan, Appalachian State University, North Carolina, USA
Sally Newman, California Institute of Technology, Pasadena, USA
Cathie Hickson, Geological Survey of Canada, Vancouver, Canada
Thom Wilch, Albion College, Michigan, USA
John Behrendt, University of Colorado at Boulder, USA
David Lescinsky, University of Western Ontario, London, Canada
Olivier Bourgeois, Centre National de la Recherche Scientifique, Vandoeuvre-lès-Nancy, France
Wes LeMasurier, University of Colorado at Denver, USA
John Stix, McGill University, Montreal, Canada
Jim Moore, U.S. Geological Survey, Menlo Park, California, USA
Steve Self, The Open University, Milton Keynes, UK
Kristján Sæmundsson, Orkufstofnun – National Energy Authority, Reykjavik, Iceland
Dave McGarvie, The Open University, Leeds, UK
Ian Skilling, University of Southern Mississippi, Hattiesburg, USA
Pietro Armienti, Univesità degli Studi di Pisa, Italy
Bill McIntosh, New Mexico Geochronology Research Laboratory, Socorro, USA
Gudmundur Sigvaldason, Nordic Volcanological Institute, Reykjavik, Iceland
James White, University of Otago, Dunedin, New Zealand
Jennie Gilbert, Lancaster University, UK
Carl Allen, NASA Johnson Space Center, Houston, Texas, USA
Bruce Houghton, University of Hawaii, Honolulu, USA
Ben Edwards, Grand Valley State University, Allendale, Michigan, USA
Armann Hoskuldsson, South Iceland Institute of Natural History, Vestmannaeyjar, Iceland
Hugh Tuffen, Lancaster University, UK
Sue Loughlin, British Geological Survey, Edinburgh, UK
Lothar Viereck, Friedrich-Schiller Universitat, Germany
Shaun Fitzgerald, BP Institute, University of Cambridge, UK
Jim Head, Brown University, Providence, Rhode Island, USA
Alfred McEwen, Lunar and Planetary Lab, University of Arizona, Tucson, USA
Tracy Gregg, The University at Buffalo, New York, USA.
Thor Thordarson, University of Hawaii at Manoa, USA
Andy Woods, BP Institute, University of Cambridge, UK
Jeff Johnson, U.S. Geological Survey, Flagstaff, Arizona, USA
Peter Schiffman, University of California at Davis, USA
Jeff Kargel, U.S. Geological Survey, Flagstaff, Arizona, USA
Matt Staid, U.S. Geological Survey, Flagstaff, Arizona, USA
Wendy Calvin, University of Nevada at Reno, USA
Norm Sleep, Stanford University, Stanford, California, USA
Jack Farmer, Arizona State University, Tempe, USA
Sveinn Jakobsson, Icelandic Institute of Natural History, Reykjavik, Iceland
Sarah Fagents, University of Hawaii at Manoa, USA
Ken Herkenhoff, U.S. Geological Survey, Flagstaff, Arizona, USA
Janice Bishop, SETI Institute/NASA Ames Research Center, Moffett Field, California, USA
Susan Sakimoto, GEST at NASA Goddard Space Flight Center, Greenbelt, Maryland, USA

Introduction: volcano–ice interaction on Earth and Mars

JOHN L. SMELLIE[1] & MARY G. CHAPMAN[2]

[1] *British Antarctic Survey, High Cross, Madingley Road, Cambridge CB3 0ET, UK (e-mail: JLSM@bas.ac.uk)*

[2] *US Geological Survey, 2255 N. Gemini Drive, Flagstaff, Arizona 86001, USA (e-mail: mchapman@usgs.gov)*

The theme of this volume was conceived during discussions between the editors and many colleagues, particularly Ian Skilling, Magnus Gudmundsson, Virginia Gulick and Sveinn Jacobsson, in response to a burgeoning growth of interest in volcano–ice systems by geologists working on terrestrial and putative martian examples. Both communities of geologists have been travelling essentially parallel paths in pursuit of their science, but using very different tools: principally remote sensing (satellite data) for Mars; mainly outcrop geology for Earth studies. At present, there are no publications that span the divide that artificially exists between the terrestrial and martian investigations, and, thus, the concept for this volume was borne. Isolated papers have addressed volcano–ice topics but this is the first attempt to assemble a thematic group of contributions addressing the diverse range of interactions known or thought to occur on both planets.

The broad focus of this volume is the interaction between magmas and cryospheres, whether on Earth or Mars. On Earth, snow and ice are found in extensive polar ice caps and on the summits of mountains even down to tropical latitudes, and ice sheets were much more widespread in the geological past. The exploration of Mars, by satellite and instrumental lander, has also revealed abundant examples of water and ice: in polar ice caps today and formerly elsewhere on the surface, in the crust and in the megaregolith, and the planet may even have sustained frozen oceans early in its history. Very different eruptive environments are implied, however, with Mars experiencing about a quarter of Earth's gravity and a much thinner atmosphere. These are physical properties that significantly affect the basic principles of magma ascent and eruption, leading to large differences in eruptive styles on both planets. The question arises: to what extent do terrestrial volcano–ice interactions provide a plausible analogue for putative martian examples? There is no simple answer to that question, but the papers in this volume are an important step forward. The purpose of the volume is to provide a snapshot of current research in volcano–ice interactions. Until we understand the similarities and differences between processes that occur on Earth and Mars, it is unwise to transfer blindly Earth-based knowledge to interpreting remote martian systems.

Eruptive, hydrological and glacial dynamics, and tephra chronology of subglacial eruptions

The first two contributions are complementary theoretical syntheses of magma–ice interactions on Earth and Mars. For Earth, **Wilson & Head** show how, at the very high strain rates anticipated, an advancing dyke tip will propagate at least 20–30% into the thickness of an overlying ice sheet before collapsing to form a basal rubble. Other intrusions will spread sideways to form a sill at the ice–substrate interface. Their analysis also suggests that the course of an eruption is determined by the fate of any meltwater formed. If it drains, explosive fragmentation may be triggered leading to further enhanced ice melting. These results have important implications for subglacial lava effusion and flow morphology, and some sills may be intruded largely as hyaloclastite breccias. **Head & Wilson** review the distribution of water and ice on Mars. They show how the wide range of eruption styles differs from those on earth because of modulation by the martian environment. A major difference is the presence on Mars of a several km-thick global permafrost layer in the upper crust. In particular, generation of mega-lahars may be a unique martian eruptive phenomenon. Latitudinal variations in the availability of cryospheric water with time may also have caused

From: SMELLIE, J. L. & CHAPMAN, M. G. (eds) 2002. *Volcano–Ice Interaction on Earth and Mars*. Geological Society, London, Special Publications, **202**, 1–4. 0305-8719/02/$15.00

evolutionary changes in volcanic morphology and eruption styles, which can be used as clues to ancient palaeolatitudes. There are few well-observed and documented subglacial eruptions on Earth, that of Gjálp (Iceland) in 1996 being the best example. However, **Smellie** describes an earlier eruption of an Antarctic volcano, in 1969, which crossed a thin ice cap on Deception Island and was also well documented. Unusually, the eruption was associated with widespread supraglacial sheet flooding as meltwater overflowed from ice fissures and chimneys above the erupting vents. He invokes gas-driven melting of an essentially 'dry' cavity roof to explain the enigmatic speed with which subglacial eruptions of relatively fluid magma can melt through substantial thicknesses of overlying glacier ice. **Larsen** completes the section with a statistical summary of 11 centuries of eruptions from Icelandic volcanoes. She shows that a minimum of 60% were from partly ice-covered and ice-capped volcanic systems.

Reconstruction of sub-ice volcanoes and ice sheet thicknesses from geomorphological and lithofacies analysis and volatile compositions

In this section, **Skilling** reviews and describes the lithofacies, depositional processes and architecture of basaltic lava-fed deltas. It is the most comprehensive published account of the construction of these morphologically distinctive and common subglacial volcanic features. Despite superficial similarities with alluvial deltas, there are important contrasts, particularly the absence of any effluent force and presence of hot clasts in the volcanic systems. **LeMasurier** describes basaltic and trachytic lava-fed deltas in Marie Byrd Land (Antarctica) and focuses particularly on documenting superbly exposed post-depositional structures. He also highlights an anomalous apparently province-wide lack of pillow lava cores in Marie Byrd Land subglacial volcanoes, and suggests that the presence of abundant microlites in the erupted lavas might have altered the magma rheology and inhibited pillow formation. **Loughlin** describes the very wide range of lithofacies involved in the construction of a single, large, long-lived alkali basalt stratovolcano (Eyjafjallajökull, Iceland). The volcano is dominated by subaqueous lithofacies emplaced under relatively thin ice (<150 m), in associations bounded by glacial unconformities representing significant time gaps formed during glacial advances. Most deposits were formed beneath valley-confined glaciers, others beneath an extensive ice cap, and the volcano probably grew mainly during periods of deglaciation. By contrast, from studies by **Edwards & Russell**, **Kelman**, **Russell & Hickson** and **Tuffen**, **McGarvie**, **Gilbert & Pinkerton**, it is now apparent that subglacial eruptions of more evolved magmas show significant differences in volcano construction compared with those at basaltic centres. Centres in British Columbia (Hoodoo Mountain: phonolites–trachytes; Garibaldi Volcanic Belt: andesites–dacites), and in Iceland (Rauðufossafjöll: rhyolites) are lava flow-dominated. Early-formed clastic products are not well known but, exceptionally, vigorous phreatomagmatic explosions were a feature of the Rauðufossafjöll edifice and built a pile of unbedded ash up to 300 m thick. Later phases in all these volcanoes involved compound lava flows emplaced within essentially dry ice cauldrons. Flow and edifice shapes were strongly influenced by ponding against surrounding thick masses of ice. It is postulated that the differences between subglacial rhyolitic and basaltic eruptions are principally caused by contrasting hydrological patterns. Subglacially erupted lithofacies at two stratovolcanoes in Marie Byrd Land (Antarctica) are interpreted by **Wilch & McIntosh** and provide evidence for the existence of a widespread mid-Miocene ice sheet, and Miocene–Pleistocene ice-level changes in West Antarctica. Together with extensive $^{40}Ar/^{39}Ar$ dating, the sequences provide a uniquely detailed proxy record and chronological framework for the glacial history of the region. They urge caution in interpreting former ice levels from volcanic sequences unless features such as coastal proximity and local palaeotopography are also taken into account, and they speculate about a link between inception of a dynamic West Antarctic Ice Sheet and increased volcanism in Marie Byrd Land. In a unique study in this volume, **Dixon**, **Filiberto**, **Moore & Hickson** measured dissolved volatile concentrations (H_2O, CO_2, S and Cl) in tholeiitic and alkali basaltic glasses from Tanzilla Mountain, a subglacial volcano from British Columbia. The edifice was erupted entirely subglacially and the vapour saturation pressures suggest former ice thicknesses within the range 300–900 m consistent with eruption during the waning phases of the coeval Fraser glaciation.

Remote sensing of terrestrial and martian subglacial features

Data from instruments on the currently orbiting Mars Global Surveyor spacecraft indicate the existence of widespread layered, massive and thin-bedded sediments on Mars. Using the spatial associations of the material in the geological units, suggested spectral compositions, possible palaeowater/ice localities and geomorphological attributes, **Chapman** suggests that these outcrops may be (1) widespread tephra layers formed from eruptions following explosive magma–ground-ice/water interaction, and (2) interior-deposit tuyas (subglacial table mountain volcanoes) which may have formed beneath confined ice in the chasmata. New high-resolution Mars Orbiter Camera images are used by **Fagents**, **Lanagan & Greeley** to provide a description and statistical synthesis of data for putative rootless cones on Mars and a comparison with terrestrial examples. They propose a new model of the dynamics of cone formation, involving only very modest amounts of water ice, consistent with the likely low availability of water ice in the martian regolith. They also surmise that many martian cones may have very young ages (<10–100 Ma), and their distribution can be used as a proxy for mapping ground ice on Mars, which is of key significance in understanding the evolution of the martian climate. **Gudmundsson**, **Pálsson**, **Björnsson & Högnadóttir** use remote-sensing geophysical techniques to continue the documentation and interpretation of the 1996 eruption of Gjálp (Iceland). Their data suggest strongly that the morphology of the subglacial Gjálp volcano resembles many Pleistocene hyaloclastite ridges in Iceland, and that much of the eruption comprised fragmented volcanic glass rather than pillow lava, which is critically important for understanding possible heat-exchange processes acting in the englacial vault. **Behrendt**, **Blankenship**, **Morse**, **Finn & Bell** review aeromagnetic and radar ice soundings in central West Antarctica to penetrate the 1–2 km-thick West Antarctic Ice Sheet (WAIS) and affirm the presence there of widespread subglacially erupted hyaloclastite edifices. Their modelling, and comparisons with the deglaciated formerly subvolcanic landscape in Iceland, suggest that many edifices beneath the WAIS have been glacially removed beneath a dynamic ice sheet with a divide that migrated through time. Subaerial eruption at the presently subglacial Sinuous Ridge might have provided a nucleus for early (late Miocene?) glaciation there and may have forced the advance of the WAIS, although the timing of these postulated events is unknown.

Hydrothermal evolution, and mineralogical and biological formation of palagonite

The compositional and spectroscopic characteristics of hydrothermal ferrihydrite from Iceland are described by **Bishop & Murad**. They speculate that volcanic activity on Mars may have been associated with hydrothermal springs and ferrihydrite formation, and that dehydrated ferrihydrite may have contributed to the widespread ferric oxide-rich surface material postulated on Mars. Palagonitization is a common process that greatly modifies the physical and chemical properties of glassy basaltic tephra deposited in subaquatic (including subglacial) environments on Earth and perhaps Mars. Using samples of subglacially-erupted, altered pillow basalts and hyalotuffs from Iceland, **Bishop**, **Schiffman & Southard** characterize the properties that distinguish palagonitization from other forms of low temperature alteration in the Icelandic environment. They suggest that the Icelandic palagonite samples may be similar to the altered basaltic surface fines on Mars and note that they share spectral characteristics similar to the bright martian soils measured by Pathfinder and martian dust measured by the Mariner missions. **Schiffman**, **Southard**, **Eberl & Bishop** suggest a definition and criteria by which hydrothermally and pedogenetically altered palagonite may be distinguished, noting that the two processes have very different results. They conclude that palagonitization is accomplished principally by a short hydrothermal process rather than by longer-term pedogenesis and show that JSC Mars-1, a terrestrial sample that NASA uses as a Martian soil simulant, is not undergoing palagonitization despite weathering for several thousand years. Palagonitization may also be biotically mediated as well as abiotic. The biotic influence on palagonite formation in altered basaltic glass is investigated by **Furnes**, **Thorseth**, **Torsvik**, **Muehlenbachs**, **Staudigel & Tumyr**. They describe a variety of indicators, including bio-generated textures, filamentous organic remains, C and N DNA and ribosomal RNA, and bio-fractionated ^{12}C and ^{13}C isotopes, which indicate that bio-alteration dominates over abiotic alteration in the upper 300 m of oceanic crust. The alteration apparently takes place at any depth where temperatures permit life to exist, a conclusion that seems set to influence strongly the continuing search for life in extreme environments.

Heat transfer and melting in subglacial basaltic volcanic eruptions: implications for volcanic deposit morphology and meltwater volumes

LIONEL WILSON[1,2] & JAMES W. HEAD, III[2]

[1] *Department of Environmental Science, Lancaster University, Lancaster LA1 4YQ, UK (e-mail: L.Wilson@Lancaster.ac.uk)*

[2] *Department of Geological Sciences, Brown University, Providence, RI 02912, USA*

Abstract: Subglacial volcanic eruptions can generate large volumes of meltwater that is stored and transported beneath glaciers and released catastrophically in jökulhlaups. At typical basaltic dyke propagation speeds, the high strain rate at a dyke tip causes ice to behave as a brittle solid; dykes can overshoot a rock–ice interface to intrude through 20–30% of the thickness of the overlying ice. The very large surface area of the dyke sides causes rapid melting of ice and subsequent collapse of the dyke to form a basal rubble pile. Magma can also be intruded at the substrate–ice interface as a sill, spreading sideways more efficiently than a subaerial flow, and also producing efficient and widespread heat transfer. Both intrusion mechanisms may lead to the early abundance of meltwater sometimes observed in Icelandic subglacial eruptions. If meltwater is retained above a sill, continuous melting of adjacent and overlying ice by hot convecting meltwater occurs. At typical sill pressures under more than 300 m ice thickness, magmatic CO_2 gas bubbles form *c.* 25 vol% of the pressurized magma. If water drains and contact with the atmosphere is established, the pressure decreases dramatically unless the overlying ice subsides rapidly into the vacated space. If it does not, further CO_2 exsolution plus the onset of H_2O exsolution has the potential to cause explosive fragmentation, i.e. a fire-fountain that forms at the dyke–sill connection, enhancing melting and creating another candidate pulse of meltwater. The now effectively subaerial magma body becomes thicker, narrower, and flows faster so that marginal meltwater drainage channels become available. If the ice overburden thickness is much less than *c.* 300 m the entire sill injection process may involve explosive magma fragmentation. Thus, there should be major differences between subglacial eruptions under local or alpine glaciers compared with those under continental-scale glaciers.

Subglacial volcanic eruptions have been studied extensively in Iceland (Björnsson 1975; Allen 1980; Gudmundsson & Björnsson 1991; Gudmundsson *et al.* 1997; Johannesson & Saemundsson 1998) due to the ongoing nature of the process and the many beautifully exposed landforms and deposits. Of particular interest is the generation of large volumes of meltwater, its storage and transport below the glaciers, and the catastrophic meltwater release at glacial margins to produce jökulhlaups (Björnsson 1975, 1992). Documentation of the products and landforms resulting from these eruptions (Björnsson 1975; Allen 1980; Grönvold & Johannesson 1983; Gudmundsson *et al.* 1997; Johannesson & Saemundsson 1998) and continuing study of active examples (Gudmundsson *et al.* 1997), together with the development of qualitative and quantitative models of the processes (Einarsson 1966; Gudmundsson *et al.* 1997; Höskuldsson & Sparks 1997; Hickson 2000; Smellie 2000), has led to the recognition of candidates for these processes elsewhere on Earth (Mathews 1947; Skilling 1994; Smellie & Skilling 1994; Chapman *et al.* 2000) and on Mars (Allen 1979; Hodges & Moore 1994; Head & Wilson 2002).

Dykes represent the propagation, both laterally and vertically, of sub-vertical magma-filled cracks from crustal or subcrustal reservoirs into the surrounding area. Dykes may propagate to the surface to cause eruptions; may propagate to the near-surface to set up stress fields, which under suitable conditions result in graben (Mastin & Pollard 1988; Rubin 1992); or may stall and cool in the crust at depths too great to produce visible indications of their presence. The latter includes the possibility that they may cease vertical propagation at some relatively shallow depth and then spread sideways to produce sills. This process is encouraged if the least principal stress ceases to be horizontal and becomes vertical. The discontinuity in density and other material properties provided by the contact between a glacier or ice-cap and the underlying

From: SMELLIE, J. L. & CHAPMAN, M. G. (eds) 2002. *Volcano–Ice Interaction on Earth and Mars*. Geological Society, London, Special Publications, **202**, 5–26. 0305-8719/02/$15.00

rocks may also be a trigger for such activity, and subglacial eruptions are likely to begin with the intrusion of a sill at the rock–ice boundary.

Commonly, a subaerial basaltic eruption is initially manifested as a curtain of fire along a fissure tens to hundreds of metres long which marks the surface trace of the dyke. Cooling along the narrow parts of the dyke (Wilson & Head 1988) causes localization of extrusion within a few hours to a few days, and transition to a centralized vent eruption (Head & Wilson 1987; Bruce & Huppert 1989). In submarine (Head *et al.* 1996) and subglacial basaltic eruptions, a classical initial curtain of fire does not generally occur because of the inhibition of gas exsolution due to the pressure of the overlying water or ice. In submarine eruptions, the suppression of gas release continues throughout the eruption, but in subglacial eruptions the situation may become much more complex. Melting of the ice overlying the initial sill may form a cavity. As long as the overlying ice does not deform too quickly, the pressure in the cavity may be less than the lithostatic load which acted on the sill during the early stages of the intrusion process, and this may lead to an increase in gas exsolution and magma vesiculation, possibly resulting in magma fragmentation and some form of explosive activity. The overlying ice cover may be completely removed, exposing magma to the pressure of the atmosphere and leading to more vigorous explosive activity.

With suitable additions, existing physical models for the ascent and eruption of magma (Wilson & Head 1981, 1983) can be applied to subglacial environments. Here we develop some simple physical principles for the intrusion of magma into a glacial cover and assess the implications for eruption behaviour and the nature of the resulting volcanic deposits and meltwater release processes. We discuss the conditions under which hyaloclastites and lava breccias form, and show how either lava flow units or sill-like bodies can form at the base of the ice, depending on the melting rate and behaviour of the ice dictated by its thickness.

Subglacial and englacial dyke emplacement

Mafic dykes sourced in crustal magma reservoirs are driven upward by magma buoyancy, by the presence of an excess pressure in the reservoir, or by a combination of the two. We shall show in later sections that typical mafic magmas have bulk densities smaller than those of their host rocks by $\Delta\rho = c.\ 200\ \mathrm{kg\,m^{-3}}$, so that the buoyancy pressure gradient acting on them ($g\,\Delta\rho =$) is $c.\ 2000\ \mathrm{Pa\,m^{-1}}$. Excess pressures in crustal mafic magma reservoirs are typically $c.\ 3$ MPa (Parfitt 1991) and for reservoir depths of a few kilometres these correspond to similar pressure gradients of $c.\ 1000\ \mathrm{Pa\,m^{-1}}$. The consequence is that the magma in mafic dykes with typical widths of $c.\ 1$ m propagates upward at speeds of $c.\ 1\ \mathrm{m\,s^{-1}}$ (Wilson & Head 1981). The strain rates near the dyke tips implied by these speeds are $c.\ 1\ \mathrm{s^{-1}}$, about seven orders of magnitude larger than the strain rates at which the surrounding ice can flow plastically given the rheological models (a pseudo-plastic power-law fluid with a yield strength) proposed by Glen (1952), Nye (1953) and Paterson (1994). Thus a dyke can easily overshoot an ice–rock interface because the ice appears to the propagating crack as a brittle, low-density rock with elastic properties similar to those of the basalt substrate. We show that the amount of ice melting which takes place on the timescale of dyke emplacement may be small enough for the emplacement process to be stable, though subsequent, more extensive ice melting may lead to collapse of the dyke.

The pressure distribution in a dyke propagating through an elastic medium is dictated by several requirements that must be met simultaneously. Most fundamental is that the distribution of stress across the dyke wall (dictated by both the internal pressure distribution and the external stress distribution) must be such as to hold open the sides of the fracture into which magma is moving. There must also be a vertical pressure gradient in the magma to support the static weight of the magma, and an additional pressure gradient in the direction of magma travel to drive the motion against wall friction. To maximize the flow speed, and hence the mass and volume fluxes through a dyke of a given shape, the pressure in the propagating tip of the dyke, P_{pt}, must decrease to a low value. The theoretical ideal tip pressure is zero, but Rubin (1993) suggested that tip pressure would in fact be no smaller than the pressure at which the most soluble volatile species which the magma contains, commonly water, becomes saturated. The argument is that if the pressure falls slightly below the value at which the magma is saturated in this volatile, more of the volatile exsolves. The solubility function for water in basalt (Wilson & Head 1981) is:

$$n_w = K_w P^{0.7} \tag{1}$$

where the constant K_w is 6.8×10^{-8} if n_w is expressed as a mass fraction and P is the pressure in Pascals. If the magma contains 0.25 mass% water, a plausible value for a mafic magma (Gerlach 1986), $n = 0.0025$ and the saturation pressure, and hence the propagating dyke tip pressure, is close to 3.3 MPa; we use this value in many subsequent calculations. We note, however, that reducing the assumed water content by a factor of two would imply a pressure of 1.2 MPa whereas increasing it by a factor of two would imply 9.0 MPa. We comment on the implications of this later.

When a vertically-propagating dyke comes to rest with its tip at some point below the surface, the pressure gradient due to the motion will by definition have vanished. In general, any excess pressure originally present in the magma reservoir and driving the intrusion will also have vanished, though residual pressure gradients may still be present if the magma in any part of the dyke system has a non-Newtonian rheology involving a finite yield strength (Parfitt & Wilson 1994). Figure 1 shows the configuration of such a dyke propagating from a reservoir at depth z below the rock–ice interface. A layer of ice of thickness y exists at this location and the tip of the dyke comes to rest at a depth x below the ice surface. The density ρ_i of the ice is c. 917 kg m^{-3}. The density ρ_r of the crustal rocks is controlled by their likely origin as a mixture of vesicular lavas and possibly poorly packed pyroclastics which have undergone various kinds of weathering and alteration: we assume a value of 2300 kg m^{-3}, close to that implied by the inversion of seismic data (Hill 1969; Zucca *et al.* 1982; Gudmundsson 1987; Head & Wilson 1992). To estimate the average magma density between the reservoir and the trapped tip we recall that the tip pressure is likely to be buffered by H_2O exsolution so that the only exsolved volatile phase will be CO_2, present as bubbles of gas or supercritical fluid in the magma. We assume that z is likely to be in the range 1 to 3 km, based on the depths of shallow magma reservoirs in Iceland (Björnsson *et al.* 1977), and that y will lie in the range 500 to at most 2000 m based on ice cap thicknesses under current (up to c. 900 m Sigmundsson & Einarsson 1992; Einarsson 1994) and glacial (1000–1500 m Einarsson & Albertsson 1988; Geirsdottir & Ericksson 1994; Bourgeois *et al.* 1998) conditions. The lithostatic pressure P_r at the top of the relaxed magma reservoir will then lie within the extremes of 27 and 86 MPa. The solubility n_c of CO_2 in basaltic magmas is given by (Harris 1981)

$$n_c = J_c + K_c P \qquad (2)$$

where J_c equals 3.4×10^{-6} and K_c equals 6×10^{-12} Pa^{-1} when n_c is expressed as a mass fraction. Assuming a plausible basaltic magma content n_t of this volatile, say 0.2 mass% i.e. 0.002 mass fraction, the mass fraction exsolved at 27 MPa is 0.00183 and at 86 MPa is 0.00148. At the dyke tip, where the pressure is likely to be no less than the value during propagation ($P_{pt} = c$. 3.3 MPa), the amount of CO_2 exsolved will be $n_e = (n_t - n_c) = c$. 0.00198. The bulk density β of the magma is given by

$$1/\beta = [n_e/\rho_c] + [(1 - n_e)/\rho_m] \qquad (3)$$

where ρ_c is the density of the CO_2 given to an adequate approximation by the ideal gas law:

$$\rho_c = (m_c P)/(Q T_m). \qquad (4)$$

Here m_c is the molecular mass of CO_2, 43.99 kg kmol^{-1}, Q is the universal gas constant, 8.314 kJ kmol^{-1} K^{-1}, T_m is the magma temperature, 1473 K (1200°C), and ρ_m is the density of the basaltic magmatic liquid, say 2700 kg m^{-3}. Using these values, the magma bulk density varies from 1863 kg m^{-3} at 3.3 MPa to 2574 kg m^{-3} at 27 MPa to 2669 kg m^{-3} at 86 MPa. The mean bulk density, β_m, of the magma in the dyke between the pressure in the tip, P_{pt}, and at the reservoir roof, P_r, is evaluated from

$$\beta_m = (P_r - P_{pt})^{-1} \int_{P=P_{pt}}^{P=P_r} \beta \, dP \qquad (5a)$$

Using equations (2) and (3), and defining the convenient constants

$$a = [\rho_m Q T_m (n_t - J_c)],$$

$$b = [m_c(1 - n_t + J_c) - \rho_m Q T_m K_c],$$

$$c = [m_c K_c],$$

$$e = [\rho_m/(2K_c)],$$

$$h = [(b^2 - 4ac)^{1/2}],$$

and

$$f = [(\rho_m b)/(2K_c h)],$$

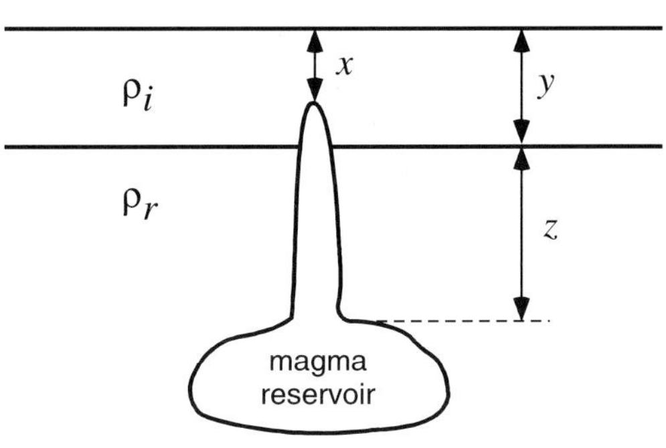

Fig. 1. Geometry of an englacial dyke extending to within a distance x of the upper surface of an ice layer of thickness y from a reservoir a distance z below the ice–rock interface.

we find:

$$\beta_m = (P_r - P_{pt})^{-1}$$
$$\times [e \ln\{(a + bP_r + cP_r^2)/(a + bP_{pt} + cP_{pt}^2)\}$$
$$- f \ln\{[(2cP_r + b - h)$$
$$\times (2cP_{pt} + b + h)]/[(2cP_r + b + h)$$
$$\times (2cP_{pt} + b - h)]\}]. \quad (5b)$$

If the roof of the magma reservoir is at the $P_r = 27$ MPa level, the mean density of magma in the dyke will be *c.* 2390 kg m^{-3} and if the reservoir roof pressure is 86 MPa the mean dyke magma density will be *c.* 2567 kg m^{-3}. Whatever the geometry, therefore, the mean density of the magma will lie within *c.* 4% of the value $\beta = 2480$ kg m^{-3}. This value will change if the assumed magma water content is changed, because the dyke tip pressure during dyke emplacement will be buffered at a different value. Equation (1) shows that varying the water content between 0.125% and 1% causes the propagating tip pressure P_{pt} to vary between 1.2 MPa and 24 MPa. Equation (2) shows that the exsolved amount of CO_2 would change by *c.* 7% as a result. The change in tip pressure is thus somewhat more important than the resulting change in CO_2 content of the magma. The mean magma densities correponding to $P_{pt} = 1.2$ and 24 MPa are 2361 and 2564 kg m^{-3}, respectively, if $P_r = 27$ MPa, and are 2553 and 2635 kg m^{-3}, respectively, if $P_r = 86$ MPa, typically a 4% variation for the smaller P_r and a 1.6% variation for the larger P_r.

As the tip of the dyke comes to rest, the pressure in the gas in the tip cavity will increase from the low, buffered value maintained during magma motion and will reach a final value P_t.

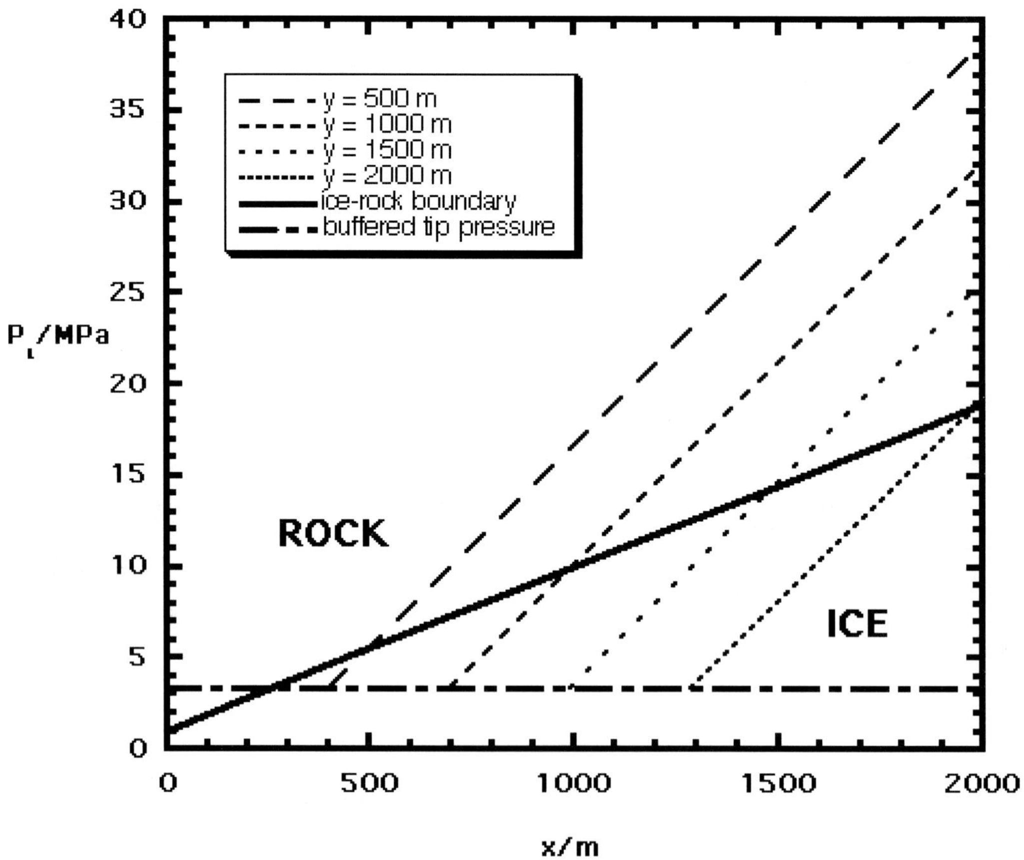

Fig. 2. The pressure, P_t, in the gas pocket in the tip of a dyke after it has been intruded into an ice layer (see Fig. 1) as a function of the depth, x of the tip below the ice surface and the thickness, y of the ice layer. The horizontal broken line indicates the smallest pressure, for the chosen magma volatile content (see text), likely to exist in the dyke tip while it is propagating. The inclined solid line shows the location of the ice–rock interface.

This pressure can be found by assuming that the hot rocks near the roof of the magma reservoir cannot support significant stresses (whereas the cold rocks or ice around the tip of the dyke can support stresses as large as their mechanical strengths). The balance between lithostatic ($g[z\rho_r + y\rho_i]$) and magma ($P_t + g\beta[z + \{y - x\}]$) stresses at the roof of the magma reservoir (see Fig. 1) implies, after collecting terms, that

$$P_t = g[\beta x - (\beta - \rho_i)y + (\rho_r - \beta)z]. \quad (6)$$

Using the densities adopted above, we find $(\beta - \rho_i) = 1333\ \mathrm{kg\,m^{-3}}$ and $(\rho_r - \beta) = 50\ \mathrm{kg\,m^{-3}}$. The relatively small value of $(\beta - \rho_r)$ means that P_t is only weakly dependent on the magma reservoir depth, z and is controlled mainly by the ice thickness, y and the depth of the dyke tip below the surface, x. Figure 2 shows how P_t varies with x for y = 500, 1000, 1500 and 2000 m. The horizontal line on this graph shows the water pressure of 3.3 MPa, the pressure in the dyke tip before it came to rest: any decrease in pressure below this value would lead to additional exsolution of water from the magma. This would lead to a decrease in magma density near the dyke tip but would not greatly change the mean magma density used in the calculation. The oblique line on the graph shows the boundary between dyke tips located within the ice layer (below the oblique line) and those located within the silicate rock crust (above the line). Clearly there is a wide range of conditions under which a dyke could penetrate, and stall within, the ice.

The pressure in the dyke tip in excess of the local lithostatic load of the overlying ice, P_e, is equal to $(P_t - g\rho_i x)$ and so using equation (6)

$$P_t = g[(\beta - \rho_i)(x - y) + (\rho_r - \beta)z]. \quad (7)$$

Figure 3 shows how P_e varies with x for the same set of values of y as Figure 2. Physically, P_e may be either positive or negative. The boundary between dyke tips in ice and dyke tips in

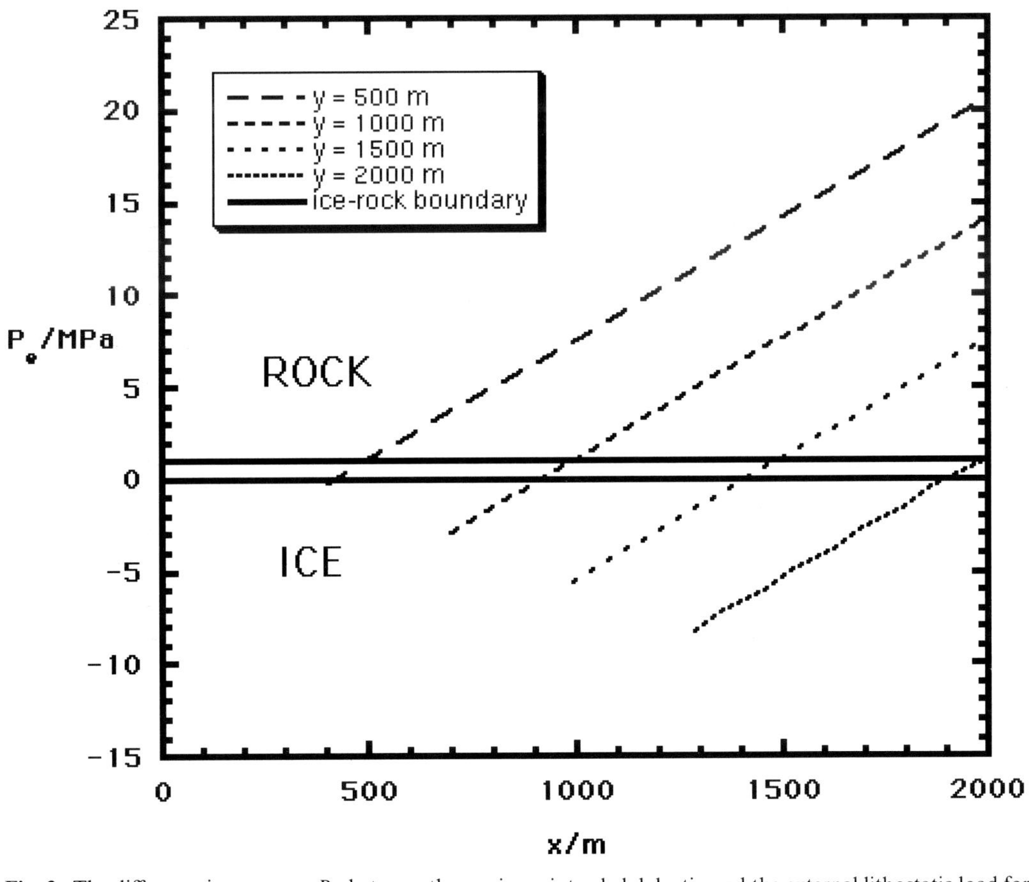

Fig. 3. The difference in pressure, P_e, between the gas in an intruded dyke tip and the external lithostatic load for the intrusion geometries corresponding to Figure 2. See text for discussion.

rock, which corresponds to setting x equals y in equation (7), is now a horizontal line. The line shown in Figure 3 corresponds to z = 2 km at $P_e = 0.98$ MPa; equation (7) shows that the corresponding values of P_e for z = 1 km and 3 km are 0.49 and 1.47 MPa, respectively. Thus, for all cases where the dyke tip penetrates into and stalls within the ice, the excess pressure in the tip can be positive but less than about 0.5–1.5 MPa, the exact value depending on z. The requirement that P_t be no less than c. 3.3 MPa leads to the truncation of the lines in Figure 3, and so the excess pressure P_e can also become negative by up to about −10 MPa.

None of the dykes modelled above (using plausible magma densities and volatile contents) are expected to break through to the upper surface of the ice. Thus magmatic eruptions at the surfaces of glaciers and ice-caps should not be a common occurrence even when dykes do penetrate into overlying ice. However, there are some potential consequences of the fact that the pressure differential, P_e, between the water vapour in a dyke tip and the surrounding ice could range from c. 1.5 MPa positive to c. 10 MPa negative. Positive pressure differentials this small will probably not lead to brittle failure of the surrounding ice, being less than the likely tensile strength of the ice, but large negative pressure differentials may lead to failure in tension or shear of the ice forming the dyke walls and collapse of blocks of ice into the gas cavity at the dyke tip. This process would be encouraged by the c. 8% volume decrease which occurs when ice melts to water. Progressive collapse might occur until a pressure path to the surface was formed, in which case the excess water vapour pressure in the dyke tip would be vented to the atmosphere and the consequent unloading of the magma would lead to further magma vesiculation and the onset of explosive activity. This activity would almost certainly be phreatomagmatic because of the intimate contact between magma, water and spalled blocks of ice. It would not be long-lived, however: even complete relaxation of the pressure at the top of the magma column to atmospheric pressure would not cause magma to rise to the surface of the ice, and so the magma at the top of the column would rapidly be chilled, causing explosive activity to cease.

We can obtain an idea of the timescale for the dyke emplacement process using the typical magma rise speed, c. 1 m s^{-1}, quoted earlier. Figure 1 shows that dykes will penetrate a distance (y–x) into the ice layer. Figure 2 shows that, as y takes the values 500, 1000, 1500 and 2000 m, the value of x at which P_t is equal to the buffered value of c. 3.3 MPa takes the values c. 400, c. 700, c. 1000 and c. 1300 m respectively. Thus the penetration distances are [(y − x) =] about 100, 300, 500 and 700 m respectively. At a magma rise speed of 1 m s^{-1}, the corresponding dyke emplacement times would range from about 100 to 700 s and in these time intervals any temperature changes caused solely by thermal conduction would penetrate a distance (d) of order $(\kappa t)^{1/2}$ where κ is the thermal diffusivity of the ice or chilling dyke magma. Thermal diffusivities of both ice and basalt are c. 10^{-6} m^2 s^{-1} and so d would be at most a few centimetres. Thus englacial dykes could well be emplaced in the initial phase of an eruption (Fig. 1).

Soon after their emplacement, dykes intruded into ice would provide relatively efficient ice melting because of the formation of two broad and extensive surface areas (the sides of the dyke) in contact with the ice. Anticipating calculations given below for heat loss from a sill, typical average heat transfer rates during the first 10 seconds after emplacement exceed 3 MW m^{-2}, and this could be a factor in the rapid initial production of meltwater reported in some Icelandic eruptions (Gudmundsson *et al.* 1997). Over the subsequent few tens of hours, solidification of the magma and formation of cooling cracks, together with melting of adjacent ice, would almost certainly cause the magma column to lose coherence and collapse to form a 'dyke rubble pile'. If the dyke were c. 200 m high and c. 1 metre wide (200 m^2 cross-sectional area), then its eventual collapse could produce a rubble pile at least c. 15 m wide by 15 m high even with minimal bulking (or more likely c. 20 m wide by c. 10 m high if it eventually attained angle of rest slopes). The cores of eruptive structures beginning with this type of event might contain a breccia pile with morphology diagnostic of its dyke-induced origin.

We now turn our attention to the consequences of magma intruding at the ice–rock interface instead of propagating as a dyke into the ice.

Sill intrusion at the ice–basalt substrate interface

The conditions that determine where the tip of an initially vertically propagating dyke ceases to move upwards, and instead initiates a fracture propagating sideways to allow the intrusion of a sill, are complex. Lister (1990), in modelling the rise of mafic magmas from deep levels, has argued that lateral intrusion will be favoured at the level of magma neutral buoyancy, and this is

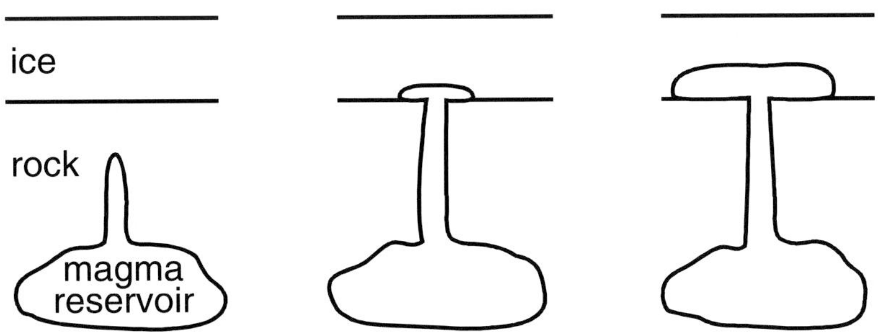

Fig. 4. Successive stages in the intrusion of a sill at the base of an ice layer. The thickness of the sill relative to its horizontal extent is exaggerated for clarity.

an attractive model for the origin of crustal magma reservoirs (e.g. Ryan 1987). However, at shallower levels in general, and especially when the tip of a dyke is nearing the shallowest level to which the stresses controlling it will allow it to penetrate, it seems inevitable that local variations in host rock properties will also play a part. We infer that dykes capable of penetrating a significant distance into overlying ice will not be excessively sensitive to the presence of the large density contrast at the rock–ice interface, whereas those which would otherwise have stalled just above the interface will initiate sill intrusion even if part of the magma rises into a dyke somewhat overshooting the interface.

When magma rises in a dyke and then intrudes as a sill, there must be a finite vertical pressure gradient in the sub-vertical feeder dyke due to the weight of the magma; if the sill is intruded horizontally, there is of course no pressure gradient in the sill due to magma weight. However, a pressure gradient required to overcome wall friction associated with magma flow must exist in both the vertical and the horizontal parts of the system. Figure 4 shows the geometry at various stages during the sill injection. The injection pressure P_i in the magma at the ice–rock interface is equal to P_t at the moment sill injection begins and increases thereafter, but must always be less (because of the pressure difference required to drive magma motion against wall friction losses) than the pressure in a static column of magma extending from the reservoir up to this point, which Figure 1 shows to be P_c given by

$$\begin{aligned} P_c &= P_a + g\,y\,\rho_i + g\,z\,\rho_r - g\,z\,\beta \\ &= P_a + g\,y\,\rho_i + g\,z\,(\rho_r - \beta) \qquad (8) \end{aligned}$$

where P_a is the atmospheric pressure, *c.* 0.1 MPa. Earlier we specified that y would probably lie in the range 500 to 2000 m and that z would lie in the range 1000 to 3000 m. Also, we found that the bulk magma density averaged over the vertical extent of the feeder dyke would lie within *c.* 10% of 2250 kg m^{-3}. Thus if ρ_r equals 2300 kg m^{-3}, the value of $(\rho_r - \beta)$ will lie between about +75 and −175 kg m^{-3}. Then since ρ_i equals 917 kg m^{-3}, P_c is dominated by the first term in equation (8) and is only a little greater than the weight of the overlying lithostatic (cryostatic) load. Thus as the sill grows, P_i increases from P_t towards P_c, just attaining this value when sill growth ceases and the pressure gradients due to magma motion vanish. Furthermore, Rubin (1993) showed that most of the pressure decrease used to overcome friction will occur over a disproportionately short distance near the dyke tip. As a result, the pressure in nearly all of the sill will be quite close to P_c for most of the duration of its emplacement after the brief initial period when most of the sill consists of 'tip'.

We use this fact in Table 1 to illustrate conditions in a typical mafic magma intruded at the interface between glaciers of various thicknesses, y and the underlying silicate surface. The magma reservoir depth is assumed to be just greater than 1 km so that $[g\,z\,(\rho_r - \beta)] = 0.5$ MPa, and so the sill pressure P_c exceeds the overlying load by this amount. At great depths the magma contains 0.25 mass% H_2O and 0.2 mass% CO_2 as before; the table shows the amounts of these volatile phases exsolved, the bulk magma density, and the volume proportion of the magma consisting of gas bubbles for a sill intruded under various ice thicknesses (y) from 50 m to 2000 m. The entry in Table 1 for y = 303 m corresponds to a sill pressure of 3.3 MPa, which is the water saturation pressure for the assumed water content of 0.25 mass%. The fact that under shallower ice thicknesses the sill inlet pressure must be less than this value inevitably implies that excessive amounts of water vapour would have to be exsolved in the dyke tip during such intrusions. Indeed, it calls into question the advisability of ever assuming that the pressure in

Table 1. *Illustration of the amounts of CO_2 and H_2O exsolved from a mafic magma intruding beneath glacial ice layers of various thicknesses and the consequences for the bulk density of the magma and the volume fraction of the magma that consists of gas bubbles*

Glacial ice thickness (m)	Pressure in most of sill (MPa)	Exsolved CO_2 amount (mass%)	Exsolved H_2O amount (mass%)	Magma bulk density (kg/m^3)	Exsolved gas proportion (volume%)
50	1.049	0.19903	0.13856	557	79.4
100	1.499	0.19876	0.10692	817	69.8
250	2.847	0.19795	0.02583	1600	40.9
303	3.327	0.19766	0.0	1869	30.9
500	5.093	0.19660	0.0	2096	22.5
1000	9.587	0.19391	0.0	2348	13.2
1500	14.080	0.19121	0.0	2454	9.3
2000	18.573	0.18852	0.0	2513	7.1

The total volatile content of the magma prior to any gas exsolution is 0.2 mass% CO_2 and 0.25 mass% H_2O. See text for discussion.

the tip of a propagating dyke is exactly buffered by the saturation pressure of the most soluble magma volatile, especially if that volatile is present in large amounts. For the case shown in Table 1 any magma containing more than 0.25 mass% water would exsolve a significant amount of that water if it were intruded under an ice thickness less than 300 m. For ice thicknesses less than about 100 m the gas volume fraction in much of the sill would be greater than 70%, and spontaneous magma fragmentation would be expected. Enhanced interaction between magma clots, rapidly chilled by intimate contact with the water being produced, would lead to the formation of hyaloclastite breccia. It seems quite possible that, under these circumstances, sill formation would not occur; instead, continuing explosive instability would lead to the formation of a hyaloclastite ridge, or series of cones, along the feeder dyke.

The ice thickness under which magma disruption of this kind will occur will be a function of the magma water content. Table 2 shows the water saturation pressures corresponding to a range of magma water contents and the minimum ice thicknesses needed to suppress magma disruption during intrusion. We stress that these are only approximate depths, because an exact calculation would have to take more detailed account of the pressure distribution and density structure of the magma throughout the dyke–sill system. However, this important influence of ice thickness may explain some of the differences observed between the subglacial deposits of local or alpine glaciers and continental-scale glaciers (Smellie & Skilling 1994).

In all of the cases shown in Table 1 which intrude as undisrupted magmatic liquids, the gas bubbles will drift upward through the liquid to become more concentrated at the top of the sill, where bubbles will eventually burst to release gas into a continuous pocket. Formation of a continuous gas phase facilitates gas loss through any fractures which exist or subsequently form in the overlying ice. However, the timescale for gas loss will be determined by the rise speed of the gas bubbles and the thickness of the sill. A typical timescale can be illustrated by considering the sill intruded under 303 m of ice in Table 1. With the assumed carbon dioxide content of 0.2 mass% gas bubbles will have nucleated at a pressure of 338 MPa, deep in the magma source region, and they are expected to have initial diameters of *c.* 20 microns (Sparks 1978). By the time that they have decompressed to the sill pressure of 3.3 MPa they will have expanded to diameters of *c.* 93 microns (assuming that the emplacement time will have been short enough that little diffusion of gas into the bubbles occurs; Sparks 1978). The bubbles will by this stage be rising through the magma at a speed u determined by the balance between their

Table 2. *Examples of the minimum ice thicknesses needed to suppress spontaneous magma disruption during sill injection as a function of the magma water content*

Magma water content (mass%)	Water saturation pressure (MPa)	Required minimum ice thickness (m)
1.0	24.0	563
0.5	9.0	245
0.25	3.3	98
0.125	1.2	28

The magma is assumed to contain 0.2 mass% CO_2 in addition to the amount of H_2O shown.

buoyancy (given by $[4/3]\pi r^3[\beta - \sigma_g]g$ where r is the bubble radius, β is the magma density, σ_g is the gas density and g is the acceleration due to gravity) and the drag force acting on them ($6\pi\eta_m ru$ where η_m is the magma viscosity). Using $\beta = 2250\,\mathrm{kg\,m^{-3}}$, $g = 9.8\,\mathrm{m\,s^{-2}}$ and $\eta_m = 100\,\mathrm{Pa\,s}$ for a mafic magma, and neglecting σ_g because it is approximately 100 times smaller than β, the rise speed of a 93 micron diameter carbon dioxide bubble is $0.42\,\mu\mathrm{m\,s^{-1}}$. At this speed, about four weeks are needed to segregate all of the bubbles from a sill one metre thick; proportionally greater times are needed for thicker sills, and sills intruded under thicker ice layers will have smaller bubbles with longer rise times. These timescales are much longer than the emplacement times of the sills (at most a few hours given the typical rise speeds of mafic magmas) and so gas loss can be ignored in all cases.

Magma intruding into a sill spreads sideways and, if the ice–rock interface is inclined, preferentially downslope. Magma in a sill probably always forms a thinner and more widespread layer than lava in a surface flow with the same mass flux and hence causes a more geometrically efficient transfer of heat to the ice; this may be a second explanation for the initial abundance of meltwater that is observed in some Icelandic subglacial eruptions (Gudmundsson *et al.* 1997). We base this assertion on the following series of arguments. The thicknesses of subaerial flows are determined by the bulk density, viscosity and effective yield strength of the magma, the acceleration due to gravity, and the surface slope: the requirements are that in the levees the stress at the base (the product of levee thickness, gravity and ground slope) is equal to the effective magma yield strength, and that in the central channels the product of channel width, magma depth and magma flow speed (the speed being in turn controlled by flow depth, magma viscocity and ground slope) must equal the volume flux from the vent (e.g. Pinkerton & Wilson 1994). The same is true for pahoehoe toes with the complication that a yield-strength-like component of the magma rheology, in addition to the other factors, influences the 'central channel' depth. In contrast to this, a subglacial flow or sill has no free upper surface. The thicknesses of the 'levees' and the 'central channel' are controlled only by the stress distribution in the host rocks. On the largest spatial scales, that same stress distribution prevents the sill from thickening locally into a series of lava flow-like fingers in the same way that vertically propagating dykes travel upward as sheets of finite lateral extent, not as a series of nearby tubes. On smaller spatial scales, especially in the early stages of growth of a sill while it is still thin, there may be the possibility of minor instabilities causing the front of the sill to grow initially as a series of pahoehoe-like toes; as the sill extends and thickens, however, we expect any such toes to be overidden by the more nearly sheet-like intrusion.

To quantify some of these considerations, consider a basaltic shield volcano having a magma reservoir within which an excess pressure of 1 MPa causes a dyke to propagate to the surface. The length of the dyke, A, is equal to the depth of the roof of the reservoir, say 2 km (e.g. Gudmundsson 1987; Ryan 1987). The mean width of the dyke, W, will be given by

$$W = [(1 - \nu/\mu](\pi/2)PA, \qquad (9)$$

where ν and μ are the Poisson's ratio and shear modulus for the crustal rocks, *c.* 0.25 and 3 GPa, respectively (Rubin 1993), so that W equals 0.8 m. The excess pressure drives magma with viscosity η_m upward through the dyke at an average speed U_M where, if the magma motion is laminar,

$$U_M = (W^2 P)/(8\eta_m A). \qquad (10)$$

If $\eta_m = 100\,\mathrm{Pa\,s}$, $U_M = c.\ 0.4\,\mathrm{m\,s^{-1}}$. The Reynolds number for the magma motion is

$$\mathrm{Re} = (2WU_M\beta)/\eta_m, \qquad (11)$$

where β is the magma density, *c.* $2200\,\mathrm{kg\,m^{-3}}$, in which case Re = *c.* 14, confirming that the magma motion is laminar. The total volume flux, V through the dyke is the product of the magma speed U_M, the dyke width W and the horizontal extent of the dyke, L. Assuming that L is of the same order as A, say 1 km, we find V equals $320\,\mathrm{m^3\,s^{-1}}$. For comparison, this value is quite similar to the *c.* $200\,\mathrm{m^3\,s^{-1}}$ eruption rates typical of recent basaltic activity on the East Rift Zone of Kilauea volcano, Hawai'i (Wolfe *et al.* 1987; Parfitt & Wilson 1994).

Assume first that this dyke feeds a subaerial basaltic lava flow which has a thickness D, density ρ, effective yield strength Y, viscosity η_L, and is moving down a slope α. Heslop *et al.* (1989) analysed the fluid mechanics of the proximal parts of a flow on the south edge of the summit caldera of Kilauea volcano for which typically D is *c.* 2 m, ρ is *c.* $1000\,\mathrm{kg\,m^{-3}}$, Y is *c.* 700 Pa, η_L is *c.* 50 Pa s and α is 2^0. The mean advance speed, U_L, of such a flow is given by

$$U_L = (\rho g \sin\alpha D^2)/(3\eta_L), \qquad (12)$$

so that in this case U_L is *c.* $9\,\mathrm{m\,s^{-1}}$. To accommodate the total estimated magma volume flux of $V = 320\,\mathrm{m^3\,s^{-1}}$, the width of the flow must

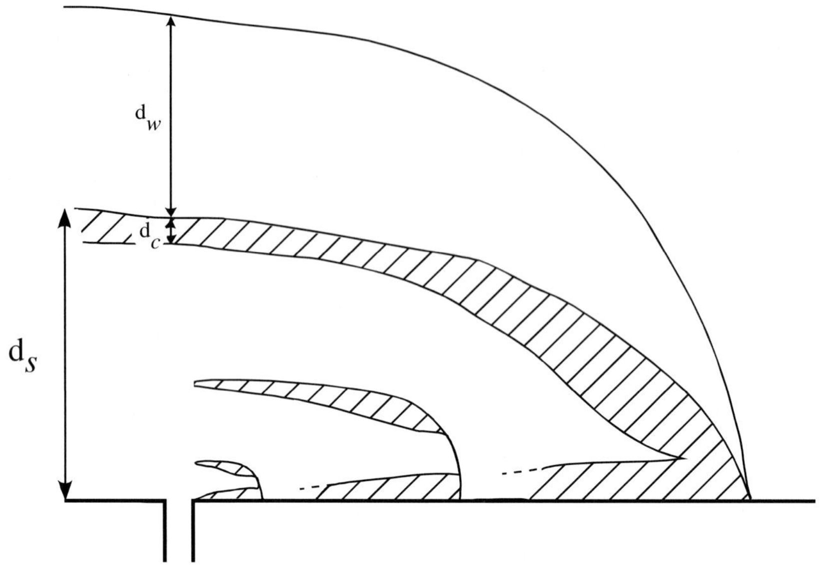

Fig. 5. Details of the development of a chilled crust, thickness d_c, and an overlying meltwater lens, thickness d_w, during the progressive intrusion of a subglacial sill of thickness d_s.

then be about 18 m, in good agreement with the observed width.

Now assume that, instead of erupting, the dyke magma ceases to propagate upward when it encounters the base of an ice layer 500 m thick (so that the magma density is close to $2100\,kg\,m^{-3}$, see Table 1) and intrudes as a horizontal sill. Initially the sill will extend along the entire 1 km horizontal extent of the dyke (L) and will be growing laterally away from it on both sides (Figs 4 & 5). Let the proximal sill thickness be d_s (Fig. 5) and the magma flow speed be U_S. The total volume flux must be the same as that in the dyke and so d_s and U_S are related via

$$V = 2d_sLU_S. \quad (13)$$

However, the sill grows by deforming the host materials (rock below, ice above) in an elastic manner, and the elastic properties of ice are not grossly different from those of rock (Hobbs 1974). Let the horizontal extent of the sill on either side of the dyke be E at time t and let the magma pressure at the point of injection be P_s. Then by analogy with equation (9),

$$d_s = [(1 - \nu)/\mu](\pi/2)P_sE. \quad (14)$$

By eliminating d_s between equations (13) and (14) and noting that by definition U_S equals (dE/dt), we find the relationship V equals $2[(1 - \nu)/\mu] \times (\pi/2)P_sEL(dE/dt)$, which integrates to give E as a function of time:

$$E = [V/\{[(1 - \nu)/\mu](\pi/2)P_sL\}]^{1/2}t^{1/2}, \quad (15)$$

from which d_s can be found as a function of time by substituting equation (15) for E in equation (14):

$$d_s = \{[(1 - \nu)/\mu](\pi/2)P_s\}^{1/2}(V/L)^{1/2}t^{1/2}. \quad (16)$$

Finally, from $U_S = (dE/dt)$ we have

$$U_S = 0.5[V/\{[(1 - \nu)/\mu](\pi/2)P_sL\}]^{1/2}t^{-1/2}. \quad (17)$$

Still using V equal to $320\,m^3\,s^{-1}$ and L equal to 1 km, and taking P_s as 5 MPa, a suitable value for a sill intruded under about 500 m of ice (see Table 1), we find the values of E, d_s and U_s as a function of time shown in Table 3. After the first second the sill has grown horizontally to 13 m on

Table 3. *Variation with time,* t *of the extent,* E, *thickness,* d_s, *and magma inflow speed,* U_s, *for a sill driven by an injection pressure of* $P_s = 5$ *MPa from a* c. *0.8 m thick dyke 1 km in horizontal extent when the volume flux* (V) *is* $320\,m^3\,s^{-1}$

t (s)	E (m)	d_s (m)	U_s ($m\,s^{-1}$)
1	13	0.025	6.38
3	22	0.043	3.69
10	40	0.079	2.02
30	70	0.14	1.17
100	128	0.25	0.64
300	221	0.43	0.37
1 000	404	0.79	0.20
3 000	699	1.37	0.12
10 000	1277	2.51	0.064
30 000	2211	4.34	0.037

either side of the dyke; its advance speed of c. 6 m s^{-1} is already smaller than the c. 9 m s^{-1} advance rate of the surface flow described above. Although the sill is only c. 25 mm thick, its surface area in contact with overlying ice by this time is 26 000 m^2, whereas the surface area of the 2 m thick, 18 m wide lava flow after it has advanced 13 m is only 234 m^2. Only if the sill grows to a distance of 6.4 km from the dyke will it have a mean thickness as large as the 2 m thickness of the lava flow; the magma injection speed of the sill will then be c. 0.1 m s^{-1}, two orders of magnitude less than the lava flow advance rate. Note that, even as early as one second after the start of sill injection, the thickness of the chilled skin on the sill magma, c. $(\kappa_m t)^{1/2}$ equals c. 1 mm, where κ_m is the thermal diffusivity of basalt, is much less than the c. 25 mm thickness of the sill, so heat transfer to the overlying ice does not hinder sill injection in this example.

Admittedly, the above comparison has various deficiencies. For example, by the time the sill extends horizontally for a distance comparable to the along-strike length of the dyke (c. 1 km in the earlier examples), it will be spreading sideways, i.e. its horizontal growth will be taking place parallel to the strike of the dyke as well as normal to it, and by the time it has extended to approximately twice this distance magma will be flowing more nearly radially away from the source region, and so the continuity relationship used above will overestimate the sill thickness and advance rate. Also, it has been tacitly assumed that the excess pressure in the magma is preserved throughout the emplacement event, whereas in fact that pressure is likely to decrease steadily as the magma reservoir at depth is deflated by magma removal. Further, the stress distribution in the feeder dyke has been assumed to remain constant, whereas in fact the growth of the sill will have a feedback effect on the overall geometry of the dyke–sill system, changing the magma volume flux to some extent. Even so, the comparison serves to support the assertion that sills generally have a larger contact area with adjacent ice than equivalent surface lava flows.

Some of the consequences of the injection of magma beneath an ice layer were investigated by Höskuldsson & Sparks (1997), who evaluated the variations with time of the thickness of the chilled crust on the magma and also the heat loss rate from the magma, and hence the thickness of ice melted. They did not, however, deal explicitly with the rate of thickening of the magma layer, instead introducing an efficiency factor which represented the fraction of the heat available from the magma that was actually transferred to the ice. Their analysis also effectively assumed that the overlying ice and underlying rock behave in a rigid fashion. The fact that water is denser than ice, leading to a volume decrease on melting, potentially provides some of the volume needed to accommodate the magma. Additionally, if the pressure in the water increases, some magma volume is accommodated by the small but finite amount of compression of the water produced (water is much more compressible than the overlying ice, the magma, or the underlying rock). If the water pressure becomes large enough to support the weight of the entire overlying ice layer, then sudden and large-scale (but short-lived) escape of the water along the margins of the ice–rock contact becomes possible.

We are not convinced that this is how the system behaves. The injection of magma into a sill fed by a dyke explicitly requires some deformation and local compression of the adjacent host materials as typified, for example, by the shear modulus and Poisson's ratio in equation (9). The fact that the host material overlying the sill is ice rather than rock does not change this. Any water created by ice melting is a Newtonian fluid and transmits stresses isotropically (as does the unchilled part of the magma as long as its properties are near-Newtonian), so it is not appropriate to consider pressure changes in the water independently of the pressure in the rest of the fluid system. Indeed, any potential pressure increase in the water (possibly caused, for example, by the very rapid conversion of a thin film of ice directly to supercritical vapour at the magma–ice contact) would first be accommodated by the compression of the bubbles of exsolved carbon dioxide in the adjacent magma. In our view, the melting of ice into water during the intrusion process, and the consequent reduction in volume of the H_2O component (due to liquid water being denser than ice), simply makes it possible to inject a greater volume of magma for a given set of magma pressure conditions.

We do, however, agree with the analysis of Höskuldsson & Sparks (1997) as regards the rate of cooling of the injected magma and melting of the overlying ice, and now develop these ideas to illustrate the importance of the magma injection rate and the ultimate consequences of the intrusion process. Figure 5 shows schematically the thickening of the sill, its chilled crust and the overlying water layer, and defines the total thickness of the sill, d_s, and the sill crust thickness, d_c, near the sill injection point. The corresponding depth of ice melted is d_i. Using treatments based on those developed by Carslaw & Jaeger (1947), Höskuldsson & Sparks (1997) give the crust thickness, d_c, and the heat loss rate per unit area of magmaice contact, q, as a function of time, t, as

$$d_c = 2\lambda(\kappa_m t)^{1/2}, \quad (18)$$

$$q = [k_m(T_m - T_w)]/[\mathrm{erf}(\lambda)(\pi\kappa_m t)^{1/2}], \quad (19)$$

where T_m is the temperature of the uncooled sill magma, T_w is the temperature of the meltwater above the crust, κ_m and k_m are the thermal diffusivity and thermal conductivity, respectively, of the solidified magma, and λ is a constant given by the solution of

$$[\lambda^{-1}\exp(-\lambda^2)]/[\mathrm{erf}(\lambda)] = [\pi^{1/2}L_m]/[c_m(T_m - T_w)] \quad (20)$$

where L_m and c_m are the latent heat of fusion and the specific heat, respectively, of solidified magma. Taking L_m as $2.09 \times 10^5\,\mathrm{J\,kg^{-1}}$, c_m as $1200\,\mathrm{J\,kg^{-1}\,K^{-1}}$, T_m as 1473 K (1200°C), and T_w as 277 K (i.e. close to the melting point and just above the temperature at which water has its maximum density) we find λ equals 1.1514 and erf(λ) equals 0.8968. We note that Höskuldsson & Sparks (1997) found erf(k) = 0.84, and suspect that they inadvertently used the latent heat of fusion of ice, rather than that of magma, in solving equation (20), but this does not lead to any major differences between their results and ours. We now integrate equation (19) to find the total amount of heat absorbed by the ice and the resulting water as a function of time, $H(t)$:

$$H(t) = \int_0^t q(t')\,dt' = \{[2k_m(T_m - T_w)]/\mathrm{erf}(\lambda)\}[t/(\pi\kappa_m)]^{1/2} \quad (21)$$

and equate this to the amount of heat needed to melt the thickness d_i of ice, $d_i = H/(\rho_i L_i)$, where ρ_i and L_i are the density and latent heat of fusion of ice, respectively, giving

$$d_i = \{[2k_m(T_m - T_w)]/[\rho_i L_i\,\mathrm{erf}(\lambda)]\} \times [t/(\pi\kappa_m)]^{1/2} \quad (22)$$

Using these results, the first five columns of Table 4 show how q, H, d_i and d_c are expected to change with time. Also shown is the thickness of the water layer produced by the ice melting, d_w, given by $d_w = d_i(\rho_i/\rho_w) = c.\,0.917 d_i$.

It has tacitly been assumed in the above analysis that the sill is injected fast enough that the total sill thickness at the vent, d_s, is greater than the chilled crust thickness d_c; in other words, there is some uncooled magma in the core of the sill. The calculations given above for the increase in the crust thickness with time therefore imply a minimum magma injection rate into the sill. In the example of sill injection calculated earlier for comparison with an equivalent volume–flux lava flow, we saw that the sill was easily able to avoid excess cooling. To establish the minimum magma volume flux to allow sill injection to be thermally viable, we note that the essential requirement is that the sill thickness d_s given by equation (16) must exceed the chilled crust thickness d_c given by equation (18). Both have the same time dependence, and so the requirement is simply

$$\{[(1 - \nu)/\mu](\pi/2)P_s\}^{1/2}(V/L)^{1/2} \gg 2\lambda\kappa_m^{1/2} \quad (23)$$

which, since λ equals 1.1514, is more conveniently written

$$[(1 - \nu)/\mu](\pi/2)P_s(V/L) \gg 5.3\kappa_m \quad (24)$$

We saw in Table 1 that P_s probably lies between 3 and 18 MPa; κ_m is $c.\ 0.8 \times 10^{-6}\,\mathrm{m^2\,s^{-1}}$, and we have $\nu = c.\,0.25$ and $\mu = c.\,3$ GPa. Thus the requirement is essentially that (V/L) should be greater than a critical value which lies between 6×10^{-4} and $36 \times 10^{-4}\,\mathrm{m^2\,s^{-1}}$. Some values of (V/L) observed in, or deduced for, subaerial eruptions include $c.\,3\,\mathrm{m^2\,s^{-1}}$ for the 1961 fissure eruption at Askja, Iceland (Thorarinsson & Sigvaldason 1962), $c.\,0.6\,\mathrm{m^2\,s^{-1}}$ for the 1783 Lakagigar eruption in Iceland (Thorarinsson 1969), $c.\,7\,\mathrm{m^2\,s^{-1}}$ for the July, 1974 summit eruption of Kilauea, Hawai'i (Heslop *et al.* 1989) and $12\,\mathrm{m^2\,s^{-1}}$ for the Yakima member of the Columbia River Basalt series (Swanson *et al.* 1975). These are all orders of magnitude greater than the minimum flux required, and so it seems likely that sill injection beneath ice should be a common occurrence, uninhibited by cooling problems, when the stress regime favours it.

Further stages of activity

When cooling does not limit sill injection at an ice–rock interface, magma injection will continue until one of two possible events happens: (1) the supply of magma from the source feeding the eruption ceases because the stresses driving the magma have been relaxed; (2) the sill spreads far enough laterally that the stresses at the propagating tip of the sill cause the precursor fracture (recall that the tip of the sill will contain pressurized water vapour and not magma) to reach the edge of the ice pile so that a connection is made to the atmosphere. We now consider the consequences of these events in turn.

Magma supply ceases

If the magma supply is cut off, growth of the sill immediately ceases, but melting of overlying ice continues as heat is convected upward through the water layer in contact with the sill. Because of the volume reduction as ice melts to water, the first consequence of this is that any residual excess pressure which may be present in the magma, together with any residual non-hydrostatic vertical compressive stress which may be present in the overlying ice, is quickly relaxed. Subsequently, if the overlying ice cannot deform downward fast enough, or alternatively if expansion of gas bubbles in the as-yet unsolidified part of the sill magma cannot crack the cooled crust and expand the sill sufficiently, a gap filled with water vapour will exist between the water and the overlying ice. Assuming that the temperature in the convecting water remains at only a few K above the melting point, the absolute pressure in the water would have decreased to *c.* 10^3 Pa, i.e. 10^{-2} atmospheres, by the time that a vapour phase appeared. An absolute pressure this low would cause an enormous stress gradient in the overlying ice and an equally impressive pressure gradient across the chilled margin of the underlying sill, and so probably in practice no vapour layer ever forms. However, if it did so it would form a good insulator: the vapour density would be *c.* 10^{-2} kg m^{-3} and so, although the specific heat of the vapour is only a factor of approximately two smaller than that of liquid water, the thermal capacity per unit volume of the vapour would be *c.* 2×10^5 times smaller. Presumably in practice an equilibrium will be reached between ice deformation, sill inflation, ice melting and heat transfer in which an appropriately narrow vapour space exists (if it exists at all).

This process will continue until the available sill magma heat content is exhausted. An upper limit on the thickness of ice which can be melted by a given thickness of magma can be found by assuming that heat transfer through water and water vapour continues to cause ice melting until all of the magma has cooled to 274 K, the temperature at which water has its maximum density, at which point convective heat transfer ceases. On this basis, and assuming no net lateral transfer of heat, each one metre thickness of sill magma could melt $(\rho_m[L_m + c_m(T_m - T_w)]/(\rho_i L_i) = 14.5$ m thickness of ice, to form a water layer $14.5 \times (\rho_i/\rho_w) = 13.3$ metres deep. The $(14.5 - 13.3 =)$ 1.2 metres of space thus created by the time ice melting ceases more than accommodates the one metre thickness of magma intruded, and so although the initial intrusion of the sill must have caused some small net uplift of the ice mass, subsidence of the overlying ice ultimately occurs. There seems no reason why any activity should be seen at the surface other than slow ice subsidence, greatest over the vent, to form an ice cauldron (Fig. 6). Only if the accumulated water escapes, either by slow seepage or by sudden release in a jökulhlaup, will there be more complex short term topographical changes (Björnsson 1992). Of course, if water does not escape, it will eventually freeze again, and its expansion as this happens will induce new stresses in the ice layer. However, the freezing process will be so slow (the timescale for conductive heat loss from under 100 m of ice is about 300 years) that any required ice deformation will probably be by plastic creep.

Finally, we note that the heat sharing calculation just employed assumes that all of the heat lost from the sill causes ice melting. This may not be the case. The temperature of the water between the sill and the ice is by definition higher than the ice melting point, whereas the ice itself must have a temperature at or below the melting point. If the ice temperature is even infinitesimally below the melting point, some heat is conducted into the ice ahead of the melting front and is not available to supply latent heat to melt ice. However, this is not a large effect. Consider the $d_s = 4.3$ m thick sill intruded on a timescale of 3×10^4 s $= c.$ 8 hours illustrated in Table 3. According to the above calculation this sill could generate a $4.3 \times 13.3 = c.$ 57 m deep water layer. The timescale for cooling the sill is $c. [d_s^2/\kappa_m] = c.\ 1.8 \times 10^7$ s $= c.$ 200 days. On this time scale a thermal wave would penetrate a comparable *c.* 4 m distance into the ice ahead of the melting front. Assume that the ice was as much as 10 K below the freezing point. Then the average amount of ice heating would be *c.* 5 K and the amount of heat leaked into the ice per unit area would be *c.* $4\ \text{m} \times 5\ \text{K} \times 2100\ \text{J kg}^{-1}\ \text{K}^{-1} \times 917\ \text{kg m}^{-3} = 3.9 \times 10^7\ \text{J m}^{-2}$. The amount of heat contained in the *c.* 57 m thick layer of water (heated to 4 K above the melting point) would be *c.* $57\ \text{m} \times 4\ \text{K} \times 4200\ \text{J kg}^{-1}\ \text{K}^{-1} \times 1000\ \text{kg m}^{-3} = 9.6 \times 10^8\ \text{J m}^{-2}$. This suggests that the heat transfer to the water is more than 95% efficient. In contrast, the heat-sharing calculation employed earlier shows that the efficiency of the process would have to be less than 83% before there was no net subsidence of the ice.

There is a potentially useful diagnostic consequence of activity in which the intruded sill is never exposed to atmospheric pressure. With ice overburdens of several hundred metres, basaltic magmas should typically exsolve most of their CO_2 but little of their H_2O. Thus, as pointed out

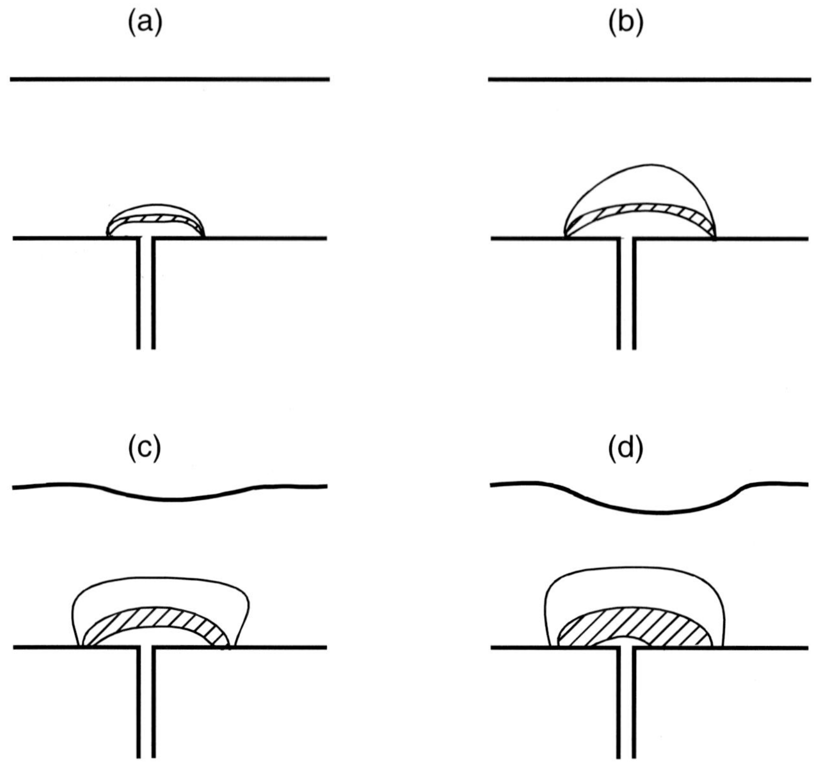

Fig. 6. Successive events during and after the intrusion of a sill at the base of an ice layer when the sill does not reach the edge of the ice sheet. (**a**) Early stage of intrusion; (**b**) sill has grown in all directions, chilled crust and overlying water lens are both thicker; (**c**) sill growth has ceased due to termination of magma supply, chilled crust and water lens have both thickened, and some subsidence of the surface of the ice has begun because the ice-to-water volume decrease has more than compensated for the sill thickness; (**d**) all available heat has been extracted from sill and vertical extents of water lens and subsidence have reached their maximum values.

by Dixon *et al.* (2002), analysis of the residual CO_2 and H_2O contents of eruption products should help distinguish between magma that has been emplaced under an ice overburden and that which has been erupted subaerially.

A pathway to the edge of the ice forms

As soon as a growing sill (Fig. 7a) approaches close enough to the edge of the ice cover that a direct connection between the intruded materials and the atmosphere is made (Fig. 7b), the pressure in the sill tip will decrease to that of the atmosphere as the pressurized water vapour escapes. The elastic constraints on the aspect ratio of the sill will then decay very quickly as the water which has already been produced above the sill begins to leak out onto the surrounding surface. For a short time, the pressure acting at the magma–water interface will become equal to the hydrostatic weight of the overlying ice; we showed earlier that the pressure in the sill is always fairly close to this value, so no major change in the overall magma flow rate through and into the sill will occur at this stage.

However, as soon as a significant amount of water has drained from above the sill, the pressure in this region will start to decrease toward atmospheric pressure, because the water can be replaced by atmospheric air leaking in. Only if the overlying ice can deform on a short enough timescale to replace the water will ice automatically stay in close proximity to the top of the sill magma. For a set of conditions similar to that envisaged here, Höskuldsson & Sparks (1997) calculated an ice deformation rate of order 1 mm s^{-1}, so if the rate of thinning of the water layer exceeds this value, the pressure will inevitably start to decrease.

Any pressure reduction in the sill will lead to an increase in the pressure difference between the

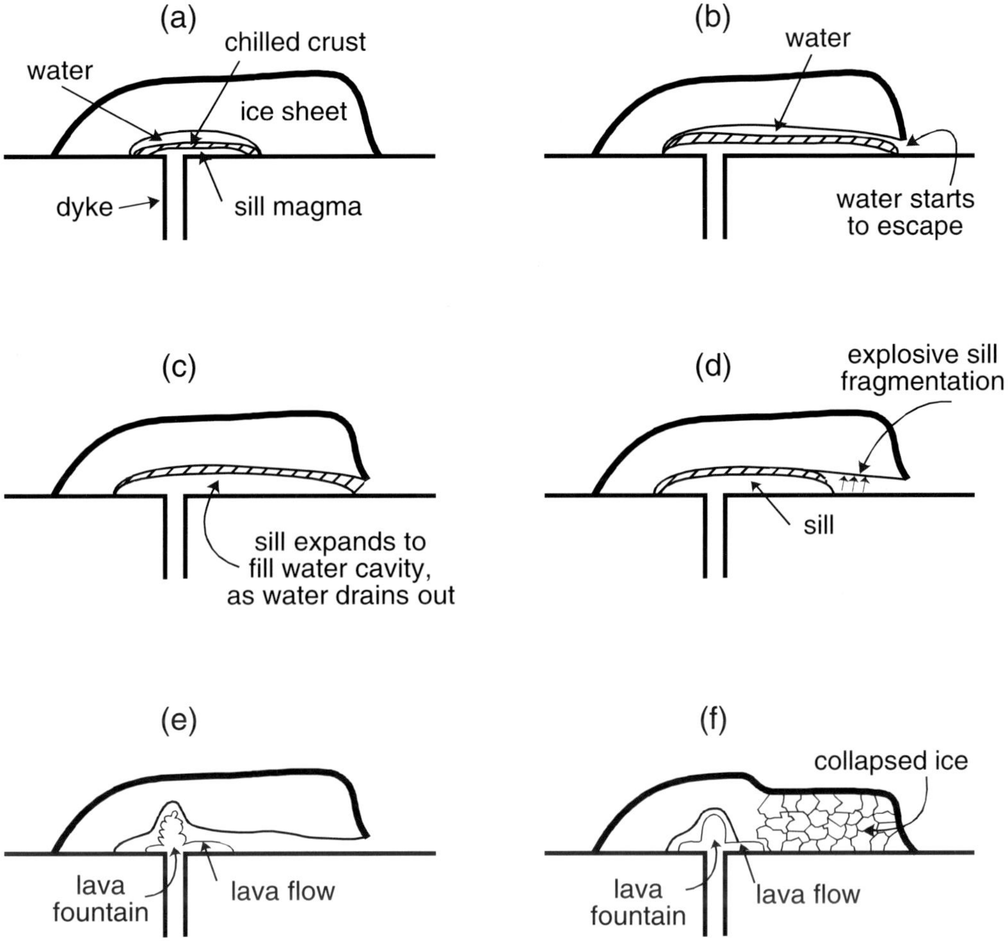

Fig. 7. Successive events during and after the intrusion of a sill at the base of an ice layer when the sill extends as far as the edge of the ice sheet. (**a**) Early stage of intrusion; (**b**) sill has reached edge of ice sheet and some water leakage begins; (**c**) much of the water generated earlier has drained out from beneath the ice and reduced interface pressure has allowed additional magma vesiculation; (**d**) pressure has become low enough near the drainage point for sill magma fragmentation to begin, enhancing the heat transfer to overlying ice; (**e**) all of the sill has been disrupted and a lava fountain exists at the outlet of the feeder dyke, rapidly eroding the overlying ice and feeding a subglacial lava flow; (**f**) the equivalent of stage (e) when the overlying ice has collapsed, greatly increasing the efficiency of thermal contact between the lava flow and ice.

bottom and the top of the dyke, and hence an increase, albeit probably small, in the magma flow rate through the dyke system. It may also have dramatic consequences, because it will lead to gas exsolution from the sill magma beneath the chilled crust. In the initial stages, the magma will simply vesiculate: existing carbon dioxide bubbles will expand and new bubbles of both CO_2 and H_2O will form at a rate which causes the magma to stay in physical contact with the overlying ice (Fig. 7c). This will lead to continued water production, and the system will tend toward a new equilibrium in which the pressure in the water is greatest near the dyke and least at the edge of the ice sheet, the resulting pressure gradient driving the water toward the exit. However, this state of affairs cannot persist for long. It seems inevitable that, as water drainage becomes more efficient, and the flowing water itself begins to melt and erode overlying ice, the pressure at the sill–ice contact will decrease to approach the atmospheric pressure. The key issue is then whether or not the magma contains enough volatiles so that at atmospheric pressure the volume fraction of gas bubbles in the magma becomes so great that magma fragmentation begins to occur. For the plausible volatile mixture used earlier (0.2 mass% CO_2, 0.25 mass% H_2O), magma fragmentation would begin at about 1.2 MPa, i.e. 12 bars, and so we assume that such fragmentation is common.

Since the lowest pressure in the system must always be at the distal end of the sill closest to the connection to the atmosphere, it is in this region that magma fragmentation will begin. As the pressure in the space above the chilled magma crust decreases, the crust will initially prevent any response from the underlying magma. However, due to the presence of cooling cracks in its outermost parts, the crust is unlikely to have great strength. Once the pressure difference across the crust exceeds this strength it will fail, and an expansion wave will propagate vertically downward into the sill. The speed of the wave will be some fraction of the speed of sound in the vesicular magma, at most *c*. 100 m s^{-1} (Kieffer 1977; Wilson & Head 1981). Thus for a sill a few metres thick (Table 3) the timescale will be only a few hundredths of a second. Passage of the expansion wave will fragment the magma, and expansion of the released gas through a pressure difference equal to the effective crustal strength will accelerate disrupted magma clots to impact the overlying ice (Fig. 7d). As an illustration, formulae given by Wilson (1980) for transient explosions show that if the effective strength were 1 MPa, then under a 500 m thick ice layer where the sill pressure was *c*. 5 MPa (see Table 1), expansion of the *c*. 0.2 mass% of CO_2 from *c*. 5 MPa to *c*. 4 MPa would generate speeds in the hot vesiculated pyroclasts up to *c*. 30 m s^{-1}. This should result in locally enhanced ice melting and magma chilling, and might be enough to trigger a sustained violent fuel-coolant type of interaction (Wohletz & McQueen 1984; Zimanowski *et al.* 1991). The products of the explosive mixing would be directed toward the exit to the atmosphere, and the wave of pressure reduction, vesiculation and fragmentation would also propagate from the distal end of the sill toward the feeder dyke. In this case the propagation speed would be a balance between the speed of the wave front into the unaffected sill magma (again some fraction of the local speed of sound) and the speed at which water and fragmented magma could be expelled from the discharge region. The rate of escape will be influenced mainly by the lateral extent of the sill; we saw earlier that the rate at which the overlying ice can deform downwards is not likely to be more than a few mm s^{-1}. This explosive fragmentation process is, of course, an excellent candidate for the origin of sudden jökulhlaup production. It is not clear what fraction of the fragmented magma would be washed out with the escaping water and what fraction would be left behind to form a hyaloclastite deposit.

A major change occurs when the wave of magma disruption reaches the feeder dyke. During the fragmentation process magma is still flowing up the dyke and being injected into the sill. However, the fragmentation process greatly reduces the frictional energy losses associated with magma motion in the sill and so as soon as all of the sill magma has been fragmented, the flow rate up the dyke will inevitably increase somewhat. The pressure at the dyke outlet will now be very close to atmospheric, and so the system will behave just as it would have done if the eruption had started subaerially. A chain of lava fountains will form along the dyke and magma clots falling from the fountains will begin to form lava flows (Head & Wilson 1989). The lava fountains will impinge on the overlying ice, greatly increasing the ice melting rate above the dyke (Fig. 7e). The resulting cavity 'drilled' into the overlying ice will grow upward until the subaerial height of the lava fountain is reached (Head & Wilson 1987), after which heat will only be transferred to the ice by radiation from pyroclasts in the fountain. From this time onward a new balance between ice subsidence and melting will be established but, if the eruption continues for long enough, it is clear that the explosive activity may eventually emerge through the ice; interaction with the water being produced will cause the activity observed to be phreatomagmatic. This scenario would be complicated somewhat if the ice layer above the now fragmented sill residue underwent fracturing and collapse rather than slow plastic deformation (Fig. 7f). In this case the pressure acting at the exit from the dyke would still be very close to atmospheric as long as there was a reasonably high porosity and permeability in the collapsed ice block pile, but the interaction between the magma and the ice would be more vigorous because of the tendency of ice blocks to settle as their bases were melted.

There is a second possible consequence of efficient water drainage once a pressure pathway to the atmosphere is established, one which is particularly applicable to magmas that do not have a large volatile content. As soon as the elastic constraint on the shape of the magma–ice contact is removed, the cross-sectional shape of the magma body is free to evolve under more local forces; specifically, magma should begin to concentrate into one or more structures resembling subaerial lava flows (Fig. 8a, b). The change will happen because the energy losses due to

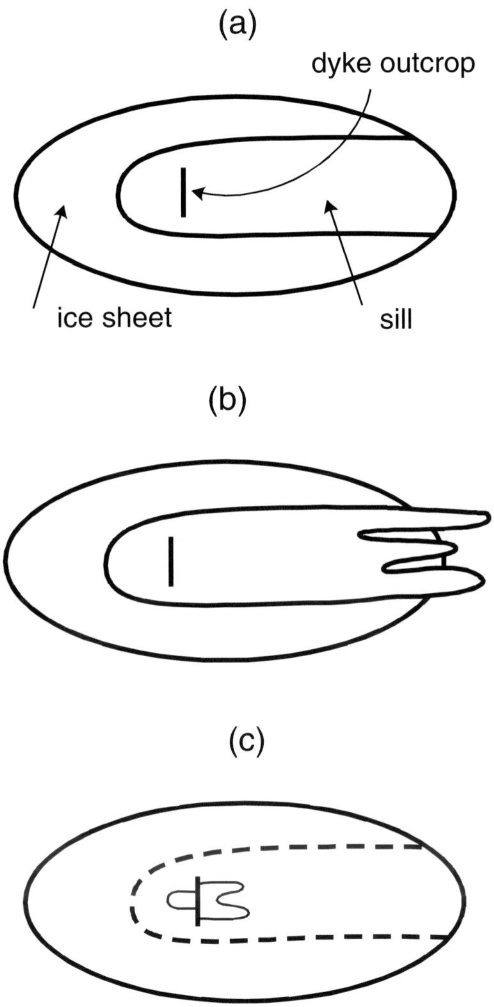

Fig. 8. The development of subglacial lava flow structures. (**a**) Sill reaches edge of ice sheet and elastic constraints are relaxed but no explosive fragmentation of sill magma occurs; (**b**) water escapes and flow regime evolves to resemble that of subaerial flows.
(**c**) Alternative source of subglacial lava flows formed when sill is explosively fragmented, dyke exit experiences atmospheric pressure, and flows are generated from a lava fountain over the vent (see Fig. 7e, f).

friction decrease as the cross-sectional shape of a moving fluid body becomes more equant. Consider the comparison made earlier between the advance of a sill and a surface flow with the same volume flux. The flow was 2 m thick and 18 m wide, thus having a total perimeter at right angles to the direction of travel of 40 m of which the 18 m at the base is in contact with the ground. The corresponding sill was initially *c.* 1 km wide (the same horizontal length as the feeder dyke) and increased in thickness as it grew, the mean thickness reaching several metres after about 10 hours. Thus the thickness is largely irrelevant and the sill perimeter is somewhere between 1 and 2 km. The uncertainty arises because the base is always in contact with a stationary rock surface but the top has a layer of water between it and the stationary ice, thus making the frictional slip conditions more complicated. The same issue would apply to a subglacial lava flow, because even if its top were in contact with the overlying ice, there would be a layer of low-viscosity water, however thin, at the interface. Again this hardly matters, however, because even the conservative sill perimeter of *c.* 1 km is vastly greater than the worst case $(2 \times 18 =)$ 36 m friction-generating perimeter of the flow. Changing the assumed slope down which the flow-like structure moves would change its cross-sectional shape somewhat (note the presence of $\sin \alpha$ in equation (12)), but again not enough to change the fact that any small instability which causes the advance of the magma to become concentrated into one or more flow-like structures will be favoured. Water generated by heat transfer into the ice will tend to be channelled along the side(s) of the flow(s), and the system will only remain stable as long as the pressure in the water is maintained high enough to suppress magma vesiculation to the point of fragmentation. If this occurs, one or more discrete lava flows will emerge from beneath the ice (Fig. 8b). However if instead fragmentation occurs, then the factors already discussed relating to subglacial explosive activity come into play (Fig. 7e, f), and new lava flow lobes will grow away from the dyke (Fig. 8c).

Summary

(1) With appropriate modifications, the principles used to analyze subaerial eruptions and intrusions (both dyke- and sill-like) in silicate rocks can be applied to eruptions under, into and through ice sheets, as illustrated in Figures 9 and 10. The geometries of dyke and sill emplacement and subsequent behaviour (decompression, transition to phreatomagmatic behaviour, etc.) are very efficient at delivering heat to the surrounding ice and creating high volumes of meltwater early in the eruptions, perhaps accounting for the production of major initial pulses of meltwater sometimes observed in Icelandic eruptions (e.g. Björnsson 1992).

(2) Typical basaltic magma densities and volatile contents are such that dykes which

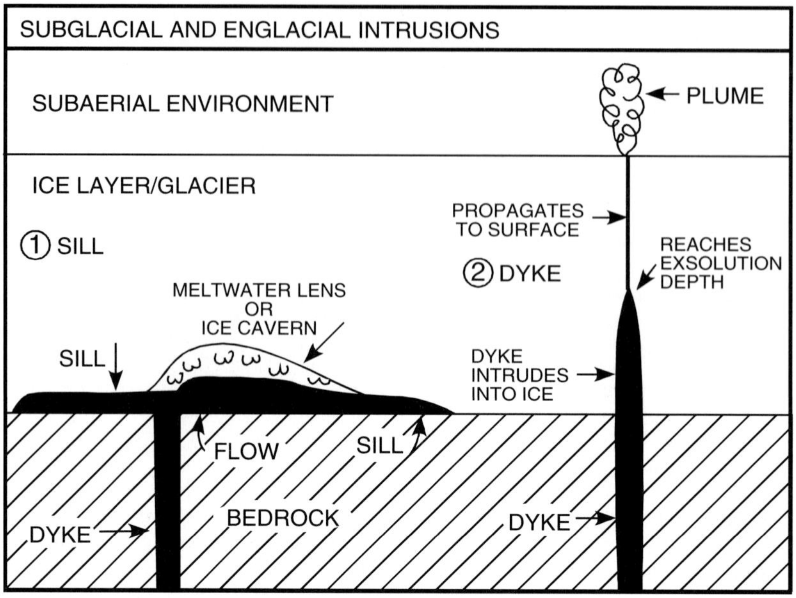

Fig. 9. Diagrammatic representation of subglacial and englacial intrusions. At (1) the dyke may become a sill at the bedrock–ice interface, and subsequent heating of the ice can lead to meltwater production or, if drainage occurs, an ice cavern and transition to a flow. In (2) the dyke propagates a significant distance into the overlying ice, which appears rheologically similar to the underlying silicates at these strain rates; if enough volatile exsolution occurs, propagation to the surface may occur and an eruption plume could be produced. Heating and ice melting at the dyke margin causes it to lose coherence and collapse to form a rubble pile. Such rubble piles could lie at the cores of hyaloclastite ridges.

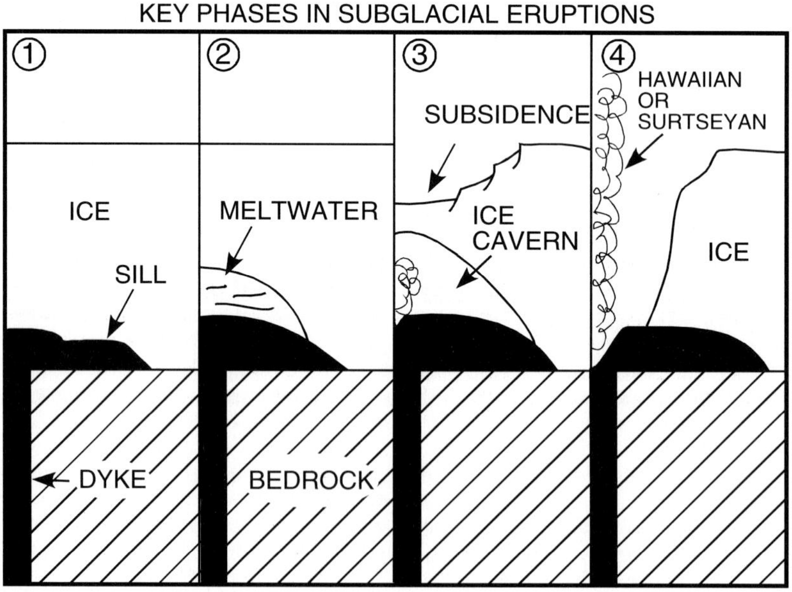

Fig. 10. Diagrammatic representation of key phases of subglacial eruptions. At (1) dyke intrusion leads to sill formation at the bedrock–dyke interface; at (2) heating produces a meltwater lens. If meltwater is drained and ambient atmospheric pressure is reached, phreatomagmatic eruptions will occur, accompanied by subsidence and ice cauldron formation; At (3) collapse of the ice surface can lead to Hawaiian or Surtseyan eruptions, depending on the involvement of meltwater in the vent, at (4).

Table 4. *Variation with time,* t *of the heat flux,* q *and the total amount of heat released so far,* H, *from a sill intruded under ice. Also given are the thicknesses of the cooled crust on the sill,* d_c, *the ice layer melted,* d_i, *and the layer of water produced,* d_w

t (s)	q (kW m^{-2})	H (MJ m^{-2})	d_c (m)	d_i (m)	d_w (m)
1	2395	4.8	0.0021	0.0156	0.0143
3	1383	8.3	0.0036	0.0270	0.0248
10	757	15.1	0.0066	0.0493	0.0452
30	437	26.2	0.0115	0.0854	0.0783
100	240	47.9	0.0210	0.1559	0.1430
300	138	83.0	0.0363	0.2710	0.2477
1 000	76	151.5	0.0663	0.4931	0.4522
3 000	44	262.4	0.1149	0.8541	0.7832
10 000	24	479.0	0.2098	1.5593	1.4299
30 000	14	830.0	0.3630	2.7100	2.4770

would have reached the surface and erupted subaerially can, if they reach the surface under an ice sheet, penetrate 20 to 30% of the way through the ice and stall as dyke-like intrusions (Figs 1–3). In most cases there would be no surface manifestation of these events other than possible minor subsidence, but in some cases gas venting, surface disturbance, and even minor phreatomagmatic activity might be observable. Subsequent ice melting will render these intrusions unstable and they will collapse to form characteristic fragmental deposits at the base of the ice.

(3) Sills can form at the bases of ice sheets (Fig. 4). The pressures in the magmas in these sills will typically be *c.* 0.5 MPa higher than the lithostatic pressure of the overlying ice (Table 1) and at low magma water contents exsolution of mainly CO_2 will cause the sills to have vesicularities typically ranging from 10% (up to 2 km ice cover) to 30% (a few hundred metres ice cover). Under shallower ice depths and with high magma water contents (Table 2), enough water exsolution may occur that spontaneous magma fragmentation takes place and sills may be intruded largely as hyaloclastite deposits. Such intrusions can reach lateral extents of *c.* 1 km and thicknesses of 1–2 metres in *c.* 1 hour (Table 3).

(4) Comparison of the typical rates of increase of thickness of subglacial sills (Table 3) with the rate of growth of chilled crust as they interact with overlying ice (Table 4, Fig. 5) shows that cooling will almost never inhibit their emplacement; intrusion will continue until either the magma supply ceases or the sill reaches the edge of the ice sheet.

(5) If magma supply ceases before the sill magma reaches the edge of the ice sheet, all of the available heat is extracted from the magma over a long time scale and subsidence of overlying ice occurs to form an ice cauldron (Fig. 6).

(6) If magma supply continues after the sill magma reaches the edge of the ice sheet, the release of confining pressure can have several consequences (Figs 7a–c & 8a–c). Rapid water release (jökulhlaup formation) can occur, exacerbated by the explosive decompression of sill magma and enhanced heating of the overlying ice (Fig. 7d). A subglacial lava fountain will form over the feeder dyke, locally greatly increasing the ice melting rate (Fig. 7e), and a new subglacial lava flow or group of flows (Figs 7e, f & 8c) will form, the ice-melting efficiency of which will be enhanced if overlying ice collapses into the cavity vacated by disruption of the initial sill (Fig. 7f). Alternatively, if explosive decompression of the sill does not occur, the shape of the subglacial sill may evolve into that of one or more lava flow-like structures (Fig. 8c).

(7) A wide array of volcanic landforms has been observed on Mars (Hodges & Moore 1994). Application of the principles developed here to Mars provides criteria to assess possible examples of intrusion and eruption below polar deposits, ice fields, and glaciers (Garvin *et al.* 2000; Ghatan & Head 2001; Head & Wilson 2002).

Discussions in the field with Magnus Gudmundsson, Snorri Snorrason, Elsa Vilmundardottir, Sveinn Jakobsson, J. Smellie and I. Skilling are gratefully acknowledged. Comments on the manuscript by J. Smellie and two anonymous reviewers helped us to clarify a number of issues. We thank A. Côté for help in drafting. This paper is based on an invited presentation given at the Volcano/Ice Interaction meeting in Reykjavik, Iceland, in August, 2000. We gratefully acknowledge financial support from NASA through the Planetary Geology and Geophysics Program and the Mars Data Analysis Program, and from PPARC through grant PPA/G/S/2000/00521.

Appendix

A vertical extent of dyke (m)
D thickness of subaerial lava flow (m)
E horizontal extent of sill on either side of feeder dyke (m)
H amount of heat released by magma per unit area of ice contact ($J\ m^{-2}$)
J_c constant in CO_2 solubility law, equal to 3.4×10^{-6} (dimensionless)
K_c constant in CO_2 solubility law, equal to 6×10^{-12} (Pa^{-1})
K_w constant in water solubility law, equal to 6.8×10^{-8} ($Pa^{-0.7}$)
L horizontal extent of dyke (m)
L_i latent heat of fusion of ice, equal to 3.3×10^5 ($J\,kg^{-1}$)
L_m latent heat of fusion of magma, equal to 2.09×10^5 ($J\,kg^{-1}$)
P ambient pressure (Pa)
P_a atmospheric pressure, equal to *c.* 10^5 (Pa)
P_c pressure in static magma column extending from reservoir to ice–rock interface (Pa)
P_e pressure in dyke tip in excess of local lithostatic load (Pa)
P_i magma pressure at ice–rock interface (Pa)
P_{pt} pressure in dyke tip while dyke is propagating (Pa)
P_r pressure in magma at roof of magma reservoir (Pa)
P_s magma pressure at sill inlet from feeder dyke (Pa)
P_t residual pressure in dyke tip after it comes to rest (Pa)
Q universal gas constant, equal to 8.314 ($kJ\,kmol^{-1}\,K^{-1}$)
T_m magma temperature, equal to 1473 (K)
T_w temperature of meltwater above chilled sill crust, equal to 277 (K)
U_L flow speed of subaerial lava ($m\,s^{-1}$)
U_M rise speed of magma in dyke ($m\,s^{-1}$)
U_S speed of magma flowing into sill ($m\,s^{-1}$)
V volume flux of magma flowing though dyke ($m^3\,s^{-1}$)
W mean width of dyke (m)
Y yield strength of subaerial lava, equal to 700 (Pa)
a constant used in equation (5b), equal to $[\rho_m Q T_m (n_t - J_c)]$ ($kg^2\ m^{-1}\ s^{-2}\ mol^{-1}$)
b constant used in equation (5b), equal to $[m_c(1 - n_t + J_c) - \rho_m Q T_m K_c]$ ($kg\,mol^{-1}$)
c constant used in equation (5b), equal to $[m_c K_c]$ ($m\,s^2\,mol^{-1}$)
c_m specific heat of solidified magma, equal to 1200 ($J\,kg^{-1}\,K^{-1}$)
d distance penetrated by thermal changes due to conduction (m)
d_c thickness of chilled crust on sill (m)
d_i thickness of ice melted adjacent to sill (m)
d_s thickness of sill near feeder dyke (m)
d_w thickness of water layer produced by ice melting (m)
e constant used in equation (5b), equal to $[\rho_m/(2K_c)]$ ($kg^2\,m^{-4}\,s^{-2}$)
f constant used in equation (5b), equal to $[(\rho_m b)/(2K_c h)]$ ($kg^2\,m^{-4}\,s^{-2}$)
g acceleration due to gravity, equal to 9.8 ($m\,s^{-2}$)
h constant used in equation (5b), equal to $[(b^2 - 4ac)^{1/2}]$ ($kg\ mol^{-1}$)
k_m thermal conductivity of solidified magma, equal to 3.1 ($W\,m^{-1}\,K^{-1}$)
m_c molecular weight of CO_2, equal to 43.99 ($kg\,kmol^{-1}$)
n weight fraction of water dissolved in basalt (dimensionless)
n_c solubility of CO_2 in basalt (dimensionless)
n_e weight fraction of CO_2 exsolved from magma (dimensionless)
n_t total CO_2 content of magma (dimensionless)
n_w solubility of water in basalt (dimensionless)
q heat loss rate per unit area of magma–ice contact ($W\ m^{-2}$)
r radius of gas bubble (m)
t time (s)
u rise speed of gas bubbles through magma ($m\,s^{-1}$)
x depth of upper dyke tip below ice surface (m)
y thickness of surface ice layer (m)
z depth of magma reservoir below rock–ice interface (m)
α slope of ground under subaerial lava flow (degrees)
β bulk density of magma ($kg\,m^{-3}$)
β_m mean bulk density of magma in dyke ($kg\,m^{-3}$)
η_m viscosity of magma in dyke, equal to 100 (Pa s)
η_L viscosity of magma in subaerial flow, equal to 50 (Pa s)
κ_i thermal diffusivity of ice, equal to *c.* 10^{-6} ($m^2\,s^{-1}$)
κ_m thermal diffusivity of solidified magma, equal to *c.* 10^{-6} ($m^2\,s^{-1}$)
λ constant in heat transfer equation, equal to 1.1514 (dimensionless)
μ shear modulus of crustal rocks, equal to 3×10^9 (Pa)
ν Poisson's ratio of crustal rocks, equal to 0.25 (dimensionless)
ρ density of subaerial lava flow, equal to 1000 ($kg\ m^{-3}$)
ρ_c density of CO_2 gas ($kg\,m^{-3}$)
ρ_i density of ice, equal to 917 ($kg\,m^{-3}$)

ρ_m density of basaltic magmatic liquid, equal to 2700 ($kg\,m^{-3}$)
ρ_r density of crustal rocks, equal to 2300 ($kg\,m^{-3}$)
σ_g gas density in bubble ($kg\,m^{-3}$)

References

ALLEN, C. C. 1979. Volcano–ice interactions on Mars. *Journal of Geophysical Research*, **84**, 8048–8059.

ALLEN, C. C. 1980. Icelandic subglacial volcanism: thermal and physical studies. *Geology*, **88**, 108–117.

BJÖRNSSON, H. 1975. Subglacial water reservoirs, jökulhlaups and volcanic eruptions. *Jökull*, **25**, 1–11.

BJÖRNSSON, H. 1992. Jokulhlaups in Iceland: prediction, characteristics and simulation. *Annals of Glaciology*, **16,** 95–106.

BJÖRNSSON, A., SAEMUNDSSON, K., EINARSSON, P., TRYGGVASON, E. & GRONVOLD, K. 1977. Current rifting episode in North Iceland. *Nature*, **266**, 318–323.

BOURGEOIS, O., DAUTEUIL, O. & VAN VLIET-LANOE, B. 1998. Pleistocene subglacial volcanism in Iceland: Tectonic implications. *Earth and Planetary Science Letters*, **164**, 165–178.

BRUCE, P. M. & HUPPERT, H. E. 1989. Thermal control of basaltic fissure eruptions, *Nature*, **342**, 665–667.

CARSLAW, H. S. & JAEGER, J. C. 1947. *Conduction of Heat in Solids*. Clarendon Press, Oxford.

CHAPMAN, M., ALLEN, C. C., GUDMUNDSSON, M. T., GULICK, V. C., JAKOBSSON, S. P., LUCCHITTA, B. K., SKILLING, I. P. & WAITT, R. B. 2000. 'Fire and ice': volcanism and ice interactions on Earth and Mars. *In*: ZIMBELMAN, J. R. & GREGG, T. K. P. (eds) *Environmental Effects on Volcanic Eruptions: From Deep Oceans to Deep Space*. Kluwer Publishing, New York, 39–73.

DIXON, J. E., FILIBERTO, J. R., MOORE, J. G. & HICKSON, C. J. 2002. Volatiles in basaltic glasses from a subglacial volcano in northern British Columbia (Canada): implications for ice sheet thickness and mantle volatiles. *In*: SMELLIE, J. L. & CHAPMAN, M. G. (eds) *Volcano–Ice Interaction on Earth and Mars*. Geological Society, London, **202**, 255–271.

EINARSSON, T. 1966. Physical aspects of sub-glacial eruptions. *Jökull*, **16**, 167–174.

EINARSSON, T. 1994. *Geology of Iceland. Rocks and Landscape*. Málog Menning publishing company, Reykjavik.

EINARSSON, T. & ALBERTSSON, K. J. 1988. The glacial history of Iceland during the past three million years. *Philosophical Transactions of the Royal Society London Series A*, **318**, 637–644.

GARVIN, J. B., SAKIMOTO, S. E. H., FRAWLEY, J. J., SCHNETZLER, C. C. & WRIGHT, H. M. 2000. Topographic evidence for geologically recent near-polar volcanism on Mars. *Icarus*, **145**, 648–652.

GEIRSDOTTIR, A. & ERICKSSON, J. 1994. Growth of intermittent ice sheet in Iceland during the Late Pliocene and Early Pleistocene. *Quaternary Research*, **42**, 115–130.

GERLACH, T. M. 1986. Exsolution of H_2O, CO_2, and S during eruptive episodes at Kilauea Volcano, Hawaii. *Journal of Geophysical Research*, **91**, 2177–2185.

GHATAN, G. J. & HEAD, J. W. 2001. Candidate subglacial volcanoes in the south polar region of Mars. *Lunar and Planetary Science*, **32**, #1039 (CD ROM).

GLEN, J. W. 1952. Experiments on the deformation of ice. *Journal of Glaciology*, **2**, 111–114.

GRÖNVOLD, K. & JOHANNESSON, H. 1983. Eruption in Grímsvötn 1983, course of events and chemical studies of the tephra. *Jökull*, **34**, 1–11.

GUDMUNDSSON, A. 1987. Lateral magma flow, caldera collapse, and a mechanism of large eruptions in Iceland. *Journal of Volcanology and Geothermal Research*, **34**, 65–78.

GUDMUNDSSON, M. T. & BJÖRNSSON, H. 1991. Eruptions in Grímsvötn, Vatnajökull, Iceland 1934–1991. *Jökull*, **41**, 21–45.

GUDMUNDSSON, M. T., SIGMUNDSSON, F. & BJORNSSON, H. 1997. Ice-volcano interaction of the 1996 Gjalp subglacial eruption, Vatnajökull, Iceland. *Nature*, **389**, 954–957.

HARRIS, D. M. 1981. The concentration of CO_2 in submarine tholeiitic basalts. *Journal of Geology*, **89**, 689–701.

HEAD, J. W. & WILSON, L. 1987. Lava fountain heights at Pu'u 'O'o, Kilauea, Hawai'i: indicators of amount and variations of exsolved magma volatiles. *Journal of Geophysical Research*, **92**, 13715–13719.

HEAD, J. W. & WILSON, L. 1989. Basaltic pyroclastic eruptions: Influence of gas-release patterns and volume fluxes on fountain structure and the formation of cinder cones, spatter cones, rootless flows, lava ponds and lava flows. *Journal of Volcanology and Geothermal Research*, **37**, 261–271.

HEAD, J. W. & WILSON, L. 1992. Magma reservoirs and neutral buoyancy zones on Venus: Implications for the formation and evolution of volcanic landforms. *Journal of Geophysical Research*, **97**, 3877–3903.

HEAD, J. W. & WILSON, L. 2002. Mars: A review and synthesis of general environments and geological settings of magma/H_2O interactions. *In*: SMELLIE, J. L. & CHAPMAN, M. G. (eds) *Volcano–Ice Interaction on Earth and Mars*. Geological Society, London, Special Publications, **202**, 27–57.

HEAD, J. W., WILSON, L. & Smith, D. K. 1996. Mid-ocean ridge eruptive vent morphology and structure: evidence for dyke widths, eruption rates, and evolution of eruptions and axial volcanic ridges. *Journal of Geophysical Research*, **101**, 28265–28280.

HESLOP, S. E., WILSON, L., PINKERTON, H. & HEAD, J. W. 1989. Dynamics of a confined lava flow on Kilauea volcano, Hawai'i. *Bulletin of Volcanology*, **51**, 415–432.

HICKSON, C. J. 2000. Physical controls and resulting morphologic forms of Quaternary ice-contact volcanoes in western Canada. *Geomorphology*, **32**, 239–261.

HILL, D. P. 1969. Crustal structure of the island of Hawaii from seismic-reflection measurements.

Bulletin of the Seismological Society of America, **59**, 101–130.

HOBBS, P. V. 1974. *Ice Physics*. Clarendon Press, Oxford.

HODGES, C. A. & MOORE, H. J. 1994. *Atlas of Volcanic Landforms on Mars*., U.S. Geological Survey Professional Paper, **1534**.

HÖSKULDSSON, A. & SPARKS, R. S. J. 1997. Thermodynamics and fluid dynamics of effusive subglacial eruptions. *Bulletin of Volcanology*, **59**, 219–230.

JAUPART, C. & VERGNIOLLE, S. 1989. The generation and collapse of a foam layer at the roof of a basaltic magma chamber, *Journal of Fluid Mechanics*, **203**, 347–380.

JOHANNESSON, H. & SAEMUNDSSON, K. 1998. *Geological Map of Iceland, Bedrock Geology*, 1:500000, Icelandic Institute of Natural History, Reykjavik (2nd edition).

KIEFFER, S. W. 1977. Sound speed in liquid-gas mixtures: water-air and water-steam. *Journal of Geophysical Research*, **82**, 2895–2904.

LISTER, J. R. 1990. Buoyancy-driven fluid fracture: similarity solutions for the horizontal and vertical propagation of fluid-filled cracks. *Journal of Fluid Mechanics* **217**, 213–239.

MASTIN, L. G. & POLLARD, D. D. 1988. Surface deformation and shallow dyke intrusion processes at Inyo Craters, Long Valley, California. *Journal of Geophysical Research*, **93**, 13 221–13 235.

MATHEWS, W. H. 1947. 'Tuyas,' flat-topped volcanoes in Northern British Columbia. *American Journal of Science*, **245**, 560–570.

MCBIRNEY, A. R. & MURASE, T. 1984. Rheological properties of magmas. *Annual Review of Earth and Planetary Science*, **12**, 337–357.

NYE, J. F. 1953. The flow law of ice from measurements in glacier tunnels, laboratory experiments, and the Jungfraufirn borehole expedition. *Proceedings of the Royal Society Series A*, **219**, 477–489.

PARFITT, E. A. 1991. The role of rift zone storage in controlling the site and timing of eruptions and intrusions of Kilauea volcano, Hawai'i. *Journal of Geophysical Research*, **96**, 10 101–10 112.

PARFITT, E. A. & WILSON, L. 1994. The 1983–86 Pu'u 'O'o eruption of Kilauea volcano, Hawaii: a study of dyke geometry and eruption mechanisms for a long-lived eruption. *Journal of Volcanology and Geothermal Research*, **59**, 179–205.

PATERSON, W. S. B. 1994. *The physics of glaciers*. 3rd. edn. Pergamon Press, Oxford.

PINKERTON, H. & WILSON, L. 1994. Factors controlling the lengths of channel-fed lava flows. *Bulletin of Volcanology* **56**, 108–120.

RUBIN, A. M. 1992. Dyke-induced faulting and graben subsidence in volcanic rift zones. *Journal of Geophysical Research*, **97**, 1839–1858.

RUBIN, A. M. 1993. Dykes vs. diapirs in viscoelastic rock. *Earth and Planetary Science Letters*, **119**, 641–659.

RYAN, M. P. 1987. Neutral buoyancy and the mechanical evolution of magmatic systems. *In*: MYSEN, B. O. (ed.) *Magmatic Processes: Physico-chemical Principles. Geochemical Society Special Publication*, **1** , 259–287.

SHAW, H. 1969. Rheology of basalt in the melting range. *Journal of Petrology*, **10**, 510–535.

SIGMUNDSSON, F. & EINARSSON, P. 1992. Glacio-isostatic crustal movements caused by historical volume change of the Vatnajökull ice cap, Iceland. *Geophysical Research Letters*, **19**, 2123–2126.

SKILLING, I. P. 1994. Evolution of an englacial volcano: Brown Bluff, Antarctica. *Bulletin of Volcanology*, **56**, 573–591.

SMELLIE, J. L. 2000. Subglacial eruptions. *In*: SIGURDSSON, H. (ed.). *Encyclopedia of Volcanoes*. Academic Press, San Diego, 403–418.

SMELLIE, J. & SKILLING, I. P. 1994. Products of subglacial volcanic eruptions under different ice thickness: Two examples from Antarctica. *Sedimentary Geology*, **91**, 115–129.

SPARKS, R. S. J. 1978. The dynamics of bubble formation and growth in magmas: a review and analysis. *Journal of Volcanology and Geothermal Research*, **3**, 1–37.

SWANSON, D. W., WRIGHT, T. L. & HELZ, R. T. 1975. Linear vent systems and estimated rates of magma production and eruption for the Yakima basalt on the Columbia Plateau. *American Journal of Science*, **275**, 877–905.

THORARINSSON, S. 1969. The Lakagígar eruption of 1783. *Bulletin of Volcanology*, **33**, 910–929.

THORARINSSON, S. & SIGVALDASON, G. E. 1962. The eruption in Askja 1961: a preliminary report. *American Journal of Science*, **260**, 641–651.

WILSON, L. 1980. Relationships between pressure, volatile content and ejecta velocity in three types of volcanic explosion. *Journal of Volcanology and Geothermal Research*, **8**, 297–313.

WILSON, L. & HEAD, J. W. 1981. Ascent and eruption of basaltic magma on the Earth and Moon. *Journal of Geophysical Research*, **86**, 2971–3001.

WILSON, L. & HEAD, J. W. 1983. A comparison of volcanic eruption processes on Earth, Moon, Mars, Io and Venus. *Nature*, **302**, 663–669.

WILSON, L. & HEAD, J. W. 1988. Nature of local magma storage zones and geometry of conduit systems below basaltic eruption sites: the Pu'u 'O'o, Kilauea East Rift, Hawaii example. *Journal of Geophysical Research*, **93**, 14 785–14 792.

WOHLETZ, K. H., & MCQUEEN, R. G. 1984. Experimental studies of hydromagmatic volcanism. *In*: *Explosive Volcanism: Inception, Evolution, and Hazards*, Studies in Geophysics, National Academy Press, Washington, 158–169.

WOLFE, E. W., GARCIA, M. O., JACKSON, D. B., KOYANAGI, R. Y., NEAL, C. A. & OKAMURA, A. T. 1987. The Puu Oo eruption of Kilauea volcano, episodes 1–20, January 3 1983, to June 8 1984. *United States Geological Survey Professional Paper*, **1350**, 471–508.

ZIMANOWSKI, B., FROHLICH, G. & LORENZ, V. 1991. Quantitative experiments on phreatomagmatic explosions. *Journal of Volcanology and Geothermal Research*, **48**, 341–358.

ZUCCA, J. J., HILL, D. P. & KOVACH, R. L. 1982. Crustal structure of Mauna Loa volcano, Hawaii, from seismic refraction and gravity data. *Bulletin of the Seismological Society of America*, **72**, 1535–1550.

Mars: a review and synthesis of general environments and geological settings of magma–H_2O interactions

JAMES W. HEAD, III[1] & LIONEL WILSON[2]

[1] *Department of Geological Sciences, Brown University, Providence, RI 02912, USA (e-mail: james_head@brown.edu)*

[2] *Environmental Sciences Department, Lancaster University, Lancaster LA14YQ, UK (e-mail: L.Wilson@lancaster.ac.uk)*

Abstract: The advent of a global cryosphere likely occurred very early in the history of Mars, and much of the available water and related volatiles (CO_2, clathrates, etc.) were sequestered within and below the cryosphere. This means that magmatism (plutonism and volcanism) as a geological process throughout the history of Mars cannot be fully understood without accounting for the interaction of magma and water (and related species) in both solid and liquid form. We review and outline the probable configuration of water and ice deposits in the history of Mars, describe environments and modes of magma–H_2O interaction, and provide specific examples from the geological record of Mars. Magma and water–ice interactions have been interpreted to have formed: (1) massive pyroclastic deposits; (2) large-scale ground collapse and chaotic terrain; (3) major outflow channels; (4) mega-lahars dwarfing terrestrial examples; (5) sub-ice-sheet eruptions and edifices; (6) pseudocraters; (7) landslides on volcanic edifice flanks; and (8) hydrothermal sites. The global nature of the cryosphere, its longevity, and the diversity of environments means that Mars is an excellent laboratory for the study of magma–H_2O interactions and the role of related volatile species.

Water has occurred in a variety of environments on Mars in solid and liquid states (Fig. 1) (e.g. Clifford 1993; Carr 1996*a*, *b*, 2000), and has interacted with the ascent and eruption of magma in a host of different ways (Figs 2–7) (e.g. Allen 1979; Mouginis-Mark 1985; Squyres *et al.* 1987; Greeley & Crown 1990; Crown & Greeley 1993; Wilson & Head 1994; Gulick 1998; Chapman *et al.* 2000; Wilson & Head 2000; Head & Wilson 2001; Keszthelyi & McEwen 2001). The reader is referred to Chapman *et al.* (2000) for an excellent review of volcanism and ice interactions on Earth and Mars in which terrestrial volcano–ice environments are subdivided into two types and discussed: (1) alpine, beneath mountain snow or summit and valley glaciers, and (2) continental scale glaciers and ice sheets. The Chapman *et al.* (2000) summary sets the stage for discussion in this paper of these environments in the context of the dominant global permafrost layer.

The work of Carr (1996*a*) represents a fundamental and comprehensive treatment of water on Mars in all its forms and environments, and the reader is referred to this as an additional starting point. Furthermore, CO_2, clathrates and hydrates are likely to have been very important in magma interaction, and reference should be made to Miller (1985), Kargel & Strom (1998), Hoffman (2000), Longhi (2000), Hess & Longhi (2001) and Baker (2001) for additional discussion of these species and their relation to water. This contribution relies heavily on previous work, which largely considers an H_2O-dominated framework (e.g. Clifford 1993; Carr 1996*a*), but future work should carefully consider the implications of more complex combinations of volatiles and the role of CO_2 in surface and subsurface volatile reservoirs and magmatic interactions.

Mars Global Surveyor results, including Mars Orbiter Camera (MOC), Mars Orbiter Laser Altimeter (MOLA) and Thermal Emission Spectrometer (TES) data, are greatly contributing to the detailed documentation and understanding of present polar ice deposits (e.g. Zuber *et al.* 1998), past polar volatile-rich deposits (e.g. Malin & Edgett 2000*a*; Head & Pratt 2001), deposits from possible ancient standing bodies of water perhaps even at the oceanic scale (e.g. Malin & Edgett 1999, 2000*a*; Head *et al.* 1999), outflow channel timing and extent (e.g. Ivanov & Head 2001; Jakosky & Phillips 2001), and evidence for groundwater reservoirs (e.g. Baker 2001). MOLA and MOC data also reveal the

From: SMELLIE, J. L. & CHAPMAN, M. G. (eds) 2002. *Volcano–Ice Interaction on Earth and Mars*. Geological Society, London, Special Publications, **202**, 27–57. 0305-8719/02/$15.00

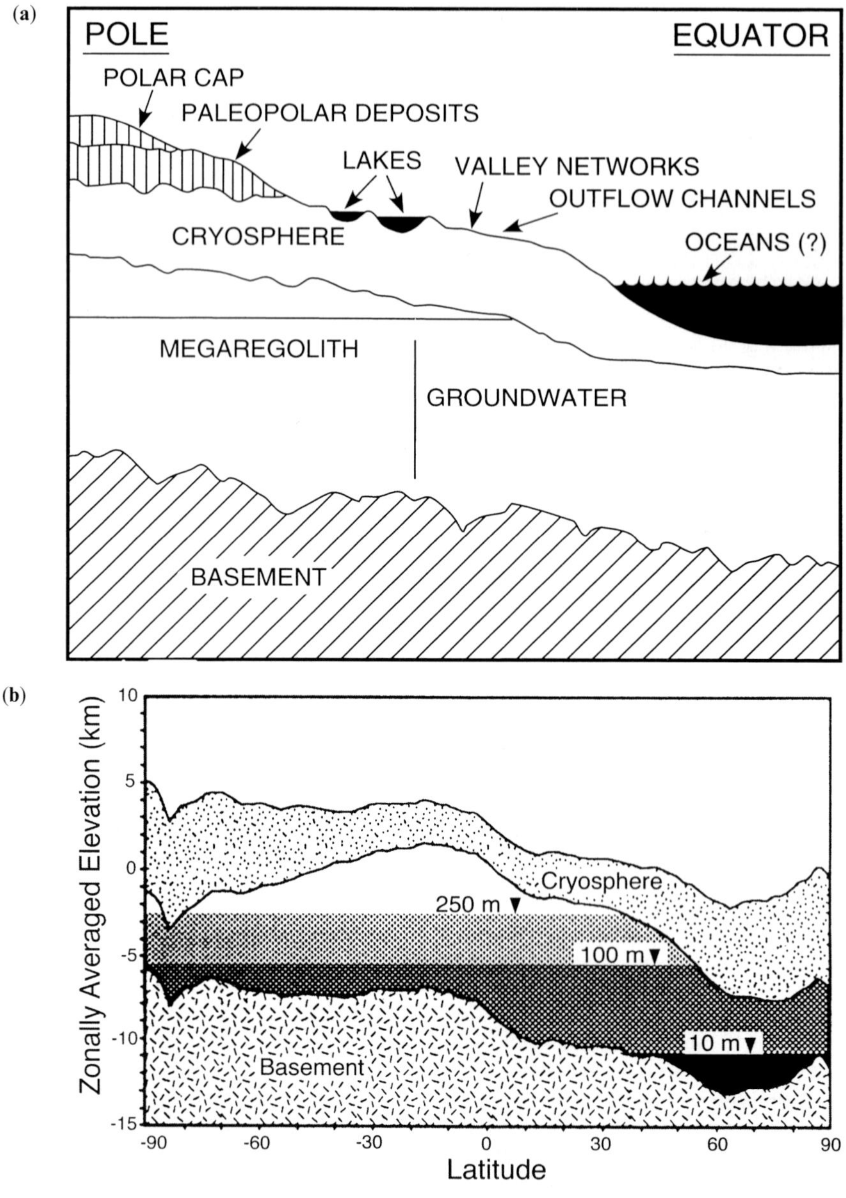

Fig. 1. (**a**) Environments of water and ice on the surface and in the crust of Mars. Surface deposits include polar and palaeopolar deposits, and evidence for ancient lakes, seas and perhaps oceans. Morphological evidence suggesting surface water flow includes valley networks and outflow channels. The cryosphere is thickest toward the poles and thins toward the equator (see **b**). Groundwater may exist in the porous megaregolith between the cryosphere and the basement, defined as the depth where cracks have closed. (**b**) Model of the cryosphere and underlying hydrosphere from Clifford (1993). The upper surface is the latitudinally averaged mean elevation, and the surface of the basement is 10 km below the surface of the ground. Water capacities are in globally averaged layer thicknesses (10, 100, 250 m) and are for a 50% surface porosity model. (**c**) Equilibrium permafrost thickness as a function of geothermal heat flow, with permafrost temperature shown for three mean surface temperatures. Equilibrium permafrost thickness is about 2 km. Heat flow above active hydrothermal systems may approach 3–5 W m^{-3}, indicating equilibrium thicknesses of less than several hundred metres (Gulick & Baker 1993; Gulick 1998). (**d**) Cryosphere dehydration in equatorial regions. Steady-state ice content of a porous rock regolith shown as a function of depth and latitude from pole to pole. Maximum ice content depends on the porosity (15%) and the density of ice (925 kg m^{-3}). From Mellon *et al.* (1997).

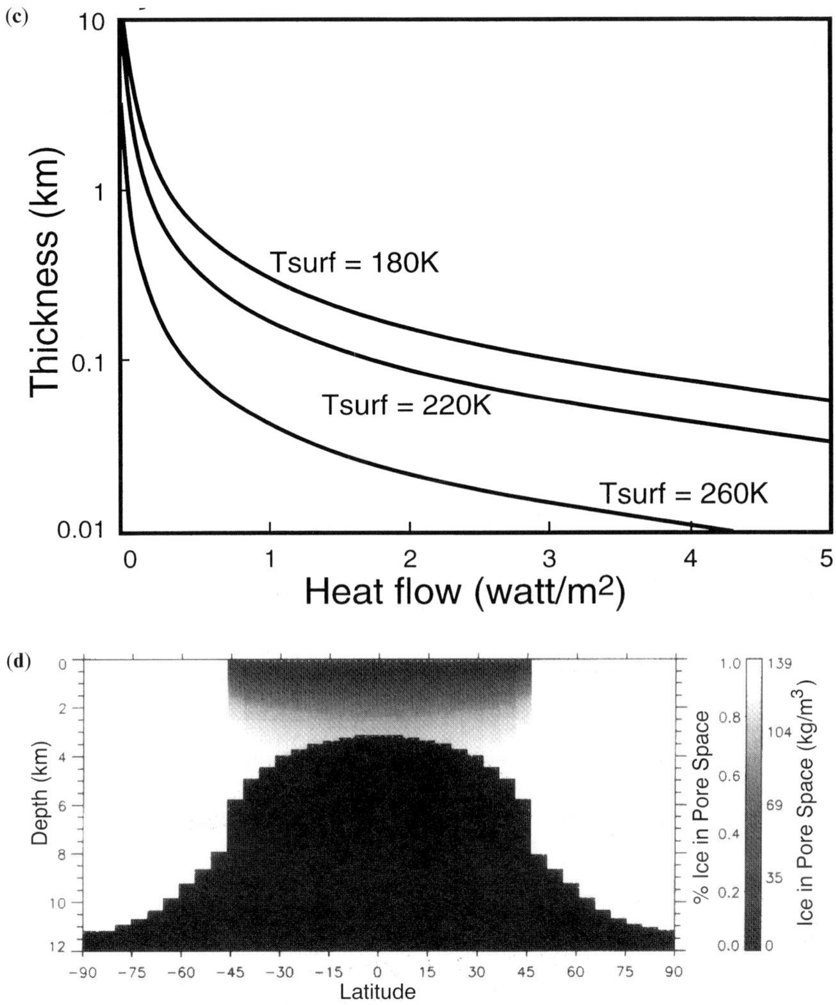

Fig. 1. (*continued*)

morphology and topography of volcanic edifices and deposits in a much more quantitative way (e.g. Hartmann *et al.* 1998; Hartmann 1999; Hartmann & Berman 2000). Quantitative measurements of flows (e.g. Keszthelyi *et al.* 2000; Sakimoto & Gregg 2001), surface manifestations of dykes (Wilson & Head 2000), regional slopes (Kreslavsky & Head 1999, 2000; Aharonson *et al.* 2001), and morphometry of edifices (e.g. Garvin *et al.* 2000; Sakimoto *et al.* 2001), together with improved topographic information revealing stratigraphic relationships, are greatly improving our knowledge of the volcanology of Mars. Together with much-improved mineralogical information (e.g. Bandfield *et al.* 2000; Hamilton *et al.* 2001; Wyatt *et al.* 2001), these new data have the potential to reveal a much more comprehensive picture of the ways in which ascending magma interacts with water and ice throughout the history of Mars. In this contribution we outline the framework of occurrence and possible interaction of magmatism and hydrospheric and cryospheric processes (e.g. Clifford 1987, 1993), and then assess several specific environments and their implications. In a separate contribution, we treat the theory of the interaction of ascending magma and ice deposits (Wilson & Head 2001*a*).

Water and ice on Mars

Geological evidence accumulated during the exploration of Mars has provided abundant

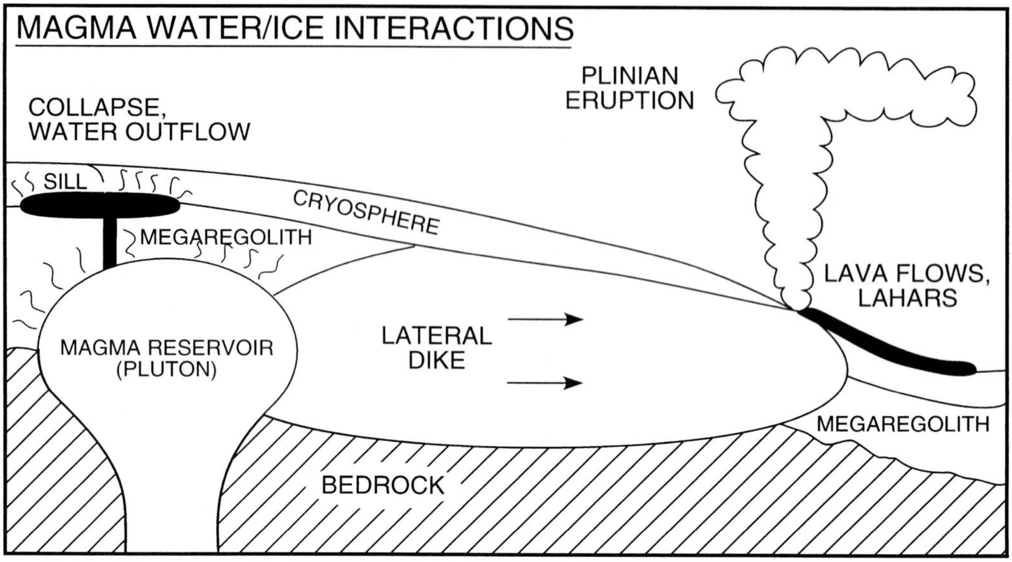

Fig. 2. Some environments of magma and water–ice interactions. Magma rising diapirically can stall at a neutral bouyancy zone, for example at a bedrock–megaregolith interface, and conduct heat into the surrounding region. Overpressurization can cause lateral dyke propagation and if the dyke penetrates the cryosphere, water under hydrostatic pressure can form eruptions and produce outflows of groundwater and magma–water–ice mixtures to produce mega-lahars. Evolution of the reservoir can cause vertical migration of magma and its intrusion along the base of the cryosphere to form a sill. Conduction of heat from the sill can cause melting and ground collapse.

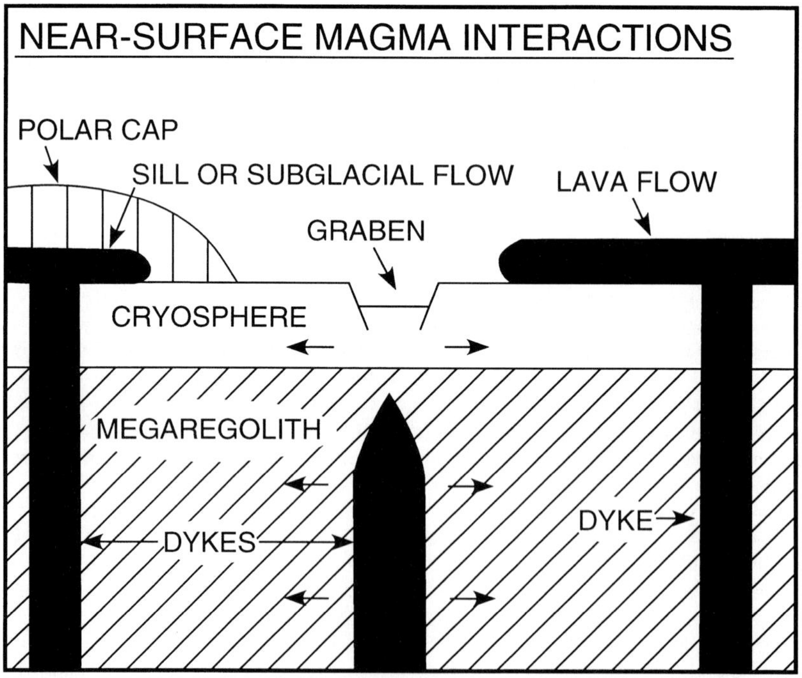

Fig. 3. Near-surface magma interactions. Magma-filled cracks (dykes) propagating from an overpressurized reservoir create near-surface extensional stress fields that can form graben along the trace of the dyke at shallow depth (middle). Dykes may also propagate to a density transition (the base of the cryosphere or the base of a polar ice cap) and intrude as a sill (left). In other cases they may extend to the surface to produce effusive flows (right).

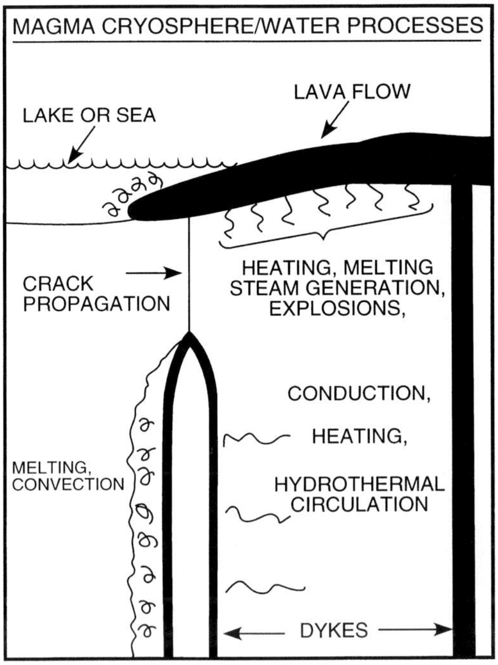

Fig. 4. Magma and cryosphere–water interactions. Dykes propagating to the surface can conduct heat into the cryosphere, causing melting and convection, and generating hydrothermal circulation (left). Convection is a much more efficient heat transfer mechanism than conduction, potentially causing enhanced cooling of the dyke. Magma reaching the surface forms lava flows that can heat and melt the substrate to generate steam and explosions, and can flow into standing bodies of water for further interactions (right).

examples of water and ice on the surface and in the crust and megaregolith (Fig. 1). Water vapour exists in the atmosphere, and a significant reservoir of ice exists at the present polar caps (Jakosky & Haberle 1992; Thomas *et al.* 1992). Evidence for ancient surface water (Carr 1996*a*) includes valley networks, outflow channels and possible lakes and oceans. Candidates for ancient volatile-rich deposits are abundant, including the south polar Dorsa Argentea Formation (Head & Pratt 2001) and the equatorial Medusae Fossae Formation (Sakimoto *et al.* 1999; Zimbelman *et al.* 1999; Bradley *et al.* 2000; Head 2001; Head & Kreslavsky 2001). The megaregolith clearly hosted groundwater and ground ice in the past and the cryosphere today is likely to hold abundant ice and contain groundwater at greater depths (Clifford 1987, 1993; Mellon & Jakosky 1995). Groundwater is an important component in many processes, but can be completely decoupled from the cryosphere (Clifford 1993; Fig. 1b) because permafrost is defined in terms of temperature (e.g. permanently frozen ground, soil or subsoil), and so it need not contain water/ice. Here the term cryosphere is used to indicate ice-rich deposits and permafrost that contains water/ice. The shallow crust is likely to be desiccated in its upper part from the equator up to about 50° latitude due to temperature-dependent sublimation (Fig. 1d; Mellon & Jakosky 1995; Mellon *et al.* 1997). At present, there appears to be much more limited subsurface–surface water exchange than there has been noted in the past, and there is clear evidence that the hydrological cycle has changed with time. Surface deposits, units and morphological features suggest that snow (Clow 1987) and ice (Baker *et al.*1991; Head & Pratt 2001) were abundant during past history, and periods of surface and cryospheric melting involved streams (sapping, drainage and outflow) and accumulation in standing bodies of water, perhaps ice-covered. At present, ice almost certainly exists in the subsurface (cryosphere) and not just on the surface and the present martian cryosphere is much more globally significant than that of present Earth.

Magmatism (plutonism and volcanism) as a process on Mars

The basic physical principles of ascent and eruption of magma can be used to determine the nature of magmatism on Earth and other planetary bodies and the way in which planetary variables (e.g. surface temperature, pressure, gravity, composition) can lead to important variations on the basic theme (e.g. Wilson & Head 1994). A wide range of eruption styles is seen on Mars (Hodges & Moore 1994; Chapman *et al.* 2000; Greeley *et al.* 2000); eruption styles are modulated by the Mars environment, with differences in gravity and the low density of the martian atmosphere being important (Wilson & Head 1994). The physical processes occurring in the vicinity of the vent on Mars (Fig. 5 expanded block) and variations in their relative importance, can result in a wide array of eruption styles. In Figure 5 at (1), essentially gas-free magma erupts to produce a low, liquid fountain feeding flows. At (2), gas rise rate exceeds magma rise rate; bubbles grow and coalesce, causing intermittent disruption of a lava lake surface and a strombolian eruption resulting in the near-vent accumulation of spatter. At (3), a steady explosive eruption creates a pyroclastic fire fountain

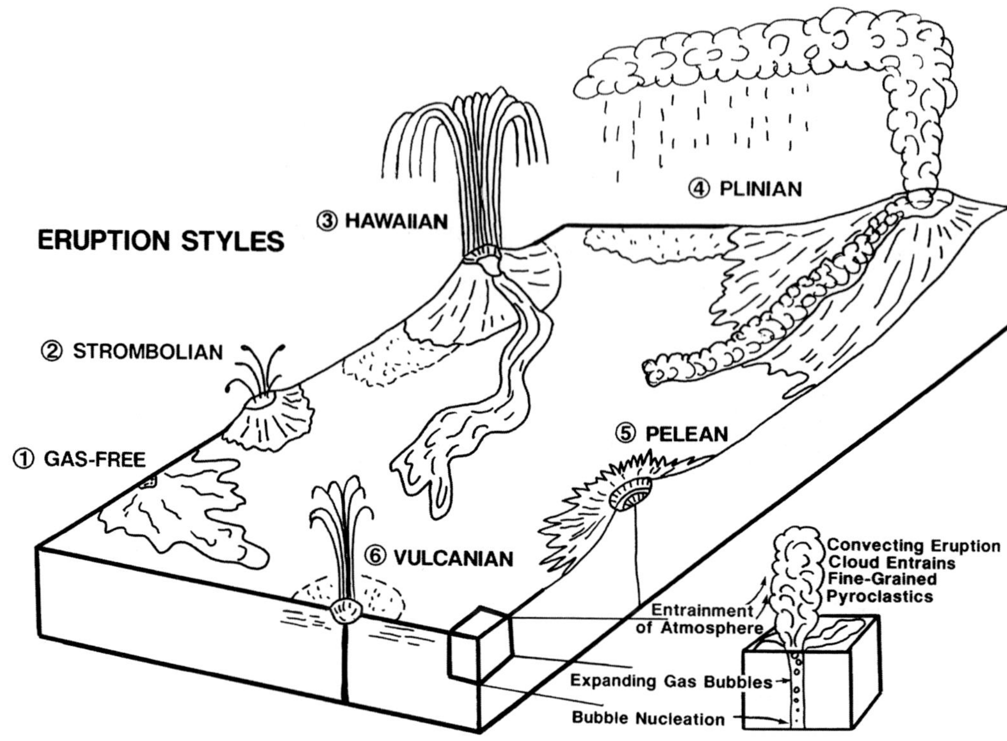

Fig. 5. The range of eruption styles seen or predicted on Mars. The most important physical processes occurring in the vicinity of the vent are shown in the expanded block, and variations in their relative significance can result in a wide array of eruption styles, as described in the text. From Wilson & Head (1994).

and an Hawaiian eruption; most clasts decouple from the gas and fall back to feed a flow. Some clasts cool and create scoria, building the cone; others are convected upward and create ash deposits. At (4), magma disrupts into small fragments that are locked to the gas stream resulting in a plinian eruption; the atmosphere is incorporated and heated and rises convectively to produce a plume, which can spread laterally and deposit tephra. If insufficient atmosphere is incorporated, the density of the column exceeds that of the atmosphere, collapse occurs, and pyroclastic flows (ignimbrites) stream down the slopes. At (5), low magma rise speed causes cooling and formation of a viscous dome with a solid carapace; catastrophic disruption of the carapace causes a pelean eruption and formation of a small convecting plume and pyroclastic flow. The aftermath, shown in Figure 5 is a crater often inside a caldera. At (6), solidification of magma and resultant gas build-up in a near-surface conduit causes explosive disruption of the solidified cap and adjacent country rock, producing a vulcanian eruption.

Mars differs from Earth in that it is a one-plate planet (Solomon 1978), has a stable lithosphere, and its upper crustal configuration is dominated by impact processes and the presence of a thick megaregolith which forms a reservoir for water and ice (e.g. Clifford 1993). The configuration of the crust of Mars represents the influence of endogenic, exogenic and surface processes, which shape the surface and modify the density structure. The southern highlands and northern lowlands are illustrated schematically in Figure 6 and impact cratering on the ancient crust produces a megaregolith composed of autochthonous (*in situ*) and allocthonous (transported) breccias. This outer fragmental layer is then reworked by mass wasting and eolian processes, and is further modified by the intrusion and extrusion of volcanic deposits. Ground ice and groundwater in the outer crust further add to the modification of the crust through chemical alteration, outflow channel erosion and deposit emplacement, and the formation of fractured ground. Increased pressure with depth causes fractured basement to be

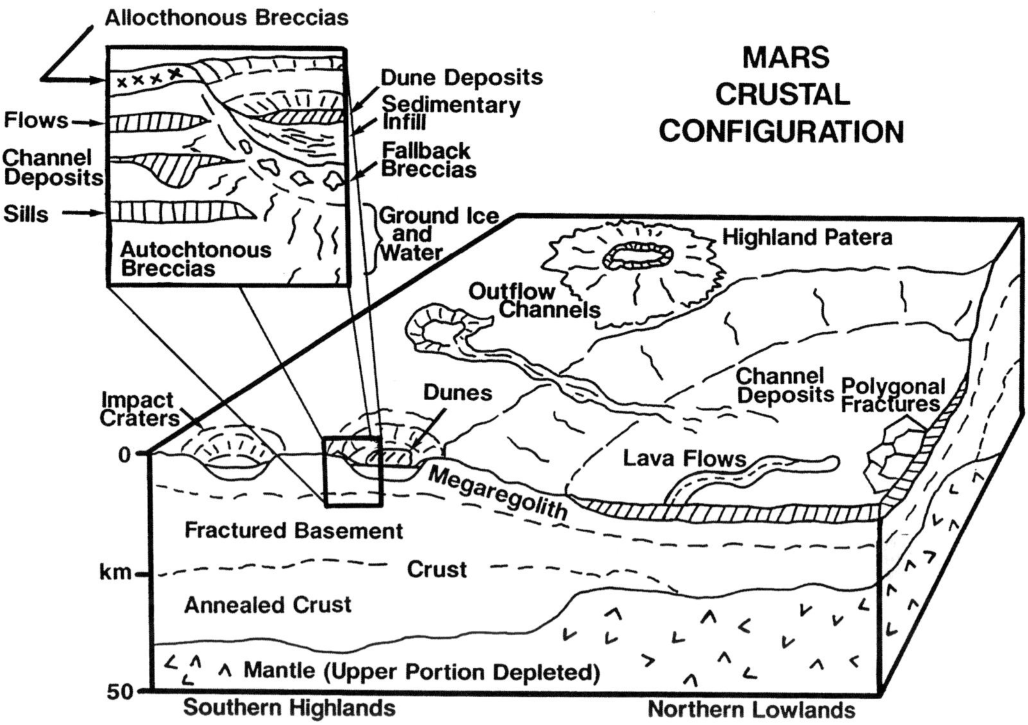

Fig. 6. Generalized configuration of the crust of Mars indicating the basic crustal structure and the influence of exogenic and surface processes in modifying its nature and density structure. The southern highlands and northern lowlands are illustrated schematically. Impact cratering on the ancient crust produces a megaregolith composed of autochthonous (*in situ*) and allochthonous (transported) breccias. This outer fragmental layer is then reworked by mass wasting and eolian processes, and is further modified by the intrusion and extrusion of volcanic deposits. Ground ice and groundwater in the outer crust further add to the modification of the crust through chemical alteration, outflow and channel erosion, and deposit emplacement. Increase in pressure with depth causes fractured basement to be annealed. The upper portion of the mantle is likely to be chemically depleted due to the derivation of the crust. Layer thicknesses are approximate. From Wilson & Head (1994).

annealed. The upper portion of the mantle is likely to be chemically depleted due to the initial derivation of the crust.

The relationship between ascending magma and the presence of groundwater and ice has many possible dimensions on Mars. In Figure 7 on the right, a pluton radiates heat to the surroundings and locally creates an aureole of alteration (A). Lava flows emplaced on the surface (D) can convert underlying ground ice into water to create melting and channel formation, or into steam to create local explosions and pseudocraters (shown as small cratered cones on the top of the flow). Sills emplaced at depth (B) can cause melting and release of groundwater at scarps to create outflow channels (C). Similar relations occur at shallow intrusions in crater floors (E). Active interactions result when magma rising in a dyke incorporates groundwater into the conduit (F) and draws down the local groundwater table in the course of the eruption, often at very high flow rates. This enhances the formation of plinian eruption plumes (G) which can yield fallout deposits or extensive pyroclastic flows, and the formation of highland paterae. These processes and environments have changed in magnitude and relative importance during the geological history of Mars.

The geological history of Mars

The geological history of Mars is divided into three periods, from oldest to youngest; Noachian, Hesperian and Amazonian (Scott & Tanaka 1986; Tanaka 1986; Greeley & Guest 1987; Tanaka & Scott 1987; Tanaka *et al.* 1988, 1992*a*). Relative ages of geological units are determined by impact crater size-frequency distributions, and absolute ages by comparisons to

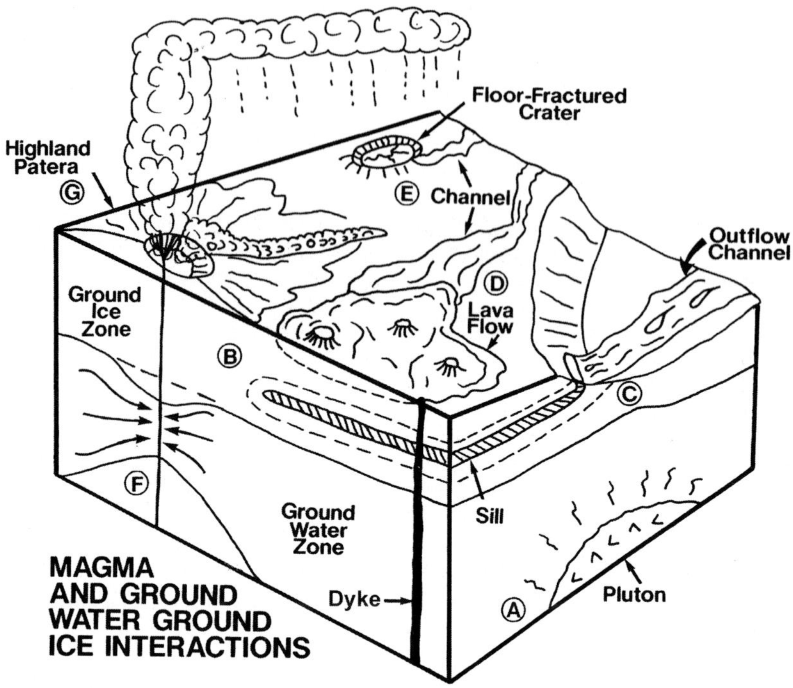

Fig. 7. Block diagram illustrating schematic relationships between ascending magma and the presence of groundwater and ice. On the right, passive interactions are illustrated, in which a pluton radiates heat to the surroundings and locally creates an aureole of alteration (**A**). Lava flows emplaced on the surface (**D**) can convert underlying ground ice into water to create melting and channel formation, or into steam to create local explosions and pseudocraters (shown as small cratered cones on the top of the flow). Sills emplaced at depth (**B**) can cause melting and release of ground water at scarps to create outflow channels (**C**). Similar relations occur at shallow intrusions in crater floors (**E**). Active interactions result when magma rising in a dyke incorporates groundwater into the conduit (**F**) and draws down the local groundwater table in the course of the eruption, often at very high flow rates. This enhances the formation of plinian eruption plumes (**G**) which can yield fallout deposits or extensive pyroclastic flows, and the formation of highland paterae. From Wilson & Head (1994).

projectile flux estimates. Recent work on the impact flux suggests that the Noachian period ended between 3.7 and 3.5 billion years ago, and the Amazonian Period began between 3.3 and 2.9 billion years ago (Hartmann & Neukum 2001; Ivanov 2001). This makes the duration of the Noachian about 0.8–1.0 billion years, the Hesperian about 0.4–0.6 billion years, and the Amazonian about 2.9–3.3 billion years.

The Noachian Period is dominated by a high, but declining impact flux, the formation of the northern lowlands, formation of abundant impact basins including Hellas and Argyre, emplacement of abundant highland and intercrater plains, dissection and erosion of these surfaces and valley network formation, and early, but very significant tectonic and volcanic activity in the Tharsis region. Significant magmatic and volcanic activity was certainly involved in the formation of the crust of Mars, but its nature and style are obscured by the accompanying and subsequent impact flux. A significant part of Tharsis (centred on Syria Planum) had been constructed by the end of the Noachian (e.g. Tanaka & Davis 1988; Phillips *et al.* 2001; Webb *et al.* 2001). Was the surface and atmosphere at this time 'warm and wet', and if so, when was the transition to conditions more like the present? The main evidence for an early warm and wet period is the presence of valley networks, thought by some to represent pluvial processes (see discussion in Sharp & Malin 1975), and significantly higher erosion rates than those seen later (see review in Carr 1996*a*, and discussions in Jakosky & Phillips 2001, and Baker 2001).

The mechanism(s) of formation of the valley networks and their implications for climate history and the hydrological cycle are keys to understanding the distribution of groundwater and ice in the Late Noachian. Initially, valley networks were thought to represent rainfall and runoff process in a warmer, wetter period of Mars' history (e.g. Masursky 1973; Gulick 2001), but other studies showed that their characteristics were unlike regional pluvial (due to the action of rain) processes, and that they were much more likely to represent groundwater sap ping and lateral transport of released water (Pieri 1980; Carr 1995; Williams & Phillips 2001). Sapping scenarios require active groundwater in the shallow crust, often located at the highest elevations, such as crater rims. This, in turn, appears to require significant groundwater recharge processes (e.g. Grant 2000).

Hydrothermal activity is one mechanism proposed to explain these characteristics. In this scenario, magmatic intrusions into regolith containing ground ice cause melting and mobilization and lateral transport of water-saturated sediment (Brackenridge *et al.* 1985; Gulick & Baker 1993). In other scenarios, the widespread, but patchy distribution of valley networks is related to precipitation of snow and its subsequent basal melting (Clow 1987; Head & Kreslavsky 2001). Many workers concluded that by the Late Noachian, the surface temperatures and pressures were more similar to those of today than they were to those required for a warm, wet Mars (e.g. Melosh & Vickery 1989; Clifford 1993; Carr 1996*c*; Brain & Jakosky 1998). In this case, a global cryosphere would have existed by the Late Noachian and its average thickness would be determined by the global thermal gradient and latitude-dependent insolation (Fig. 1c), thickening with time, and locally thinning due to variations and perturbations related to magmatic intrusions. Even at times of early higher heat flux (Zuber *et al.* 2000), the equilibrium cryosphere thickness would likely be at least several hundreds of meters (Fig. 1c), and would increase substantially toward the poles (Fig. 1b). This global cryosphere would have effectively precluded volumetrically significant volatile exchange between the atmosphere and surface above, and the subsurface groundwater below, except in certain circumstances such as at the poles (Clifford 1993). The presence of the global cryosphere (Fig. 1a, b) also meant that from the beginning of this time, any magma ascending toward the surface would have to pass through a shallow crustal region (approximately the upper 10 km) characterized by groundwater (whose configuration depended on the total amount of water) and an overlying cryosphere a few kilmometres thick (Figs 1b, c & 7). Such a Noachian configuration would have effectively set the stage for magma–H_2O interaction for the rest of the history of Mars.

The Hesperian Period is an important phase in the evolution of the major processes operating during the history of Mars. During the Hesperian Period, globally distributed ridged plains of volcanic origin were emplaced and deformed (Tanaka 1986; Tanaka *et al.* 1992*a*), and significant regional volcanism took place in Tharsis, Syria (Tanaka & Davis 1988) and Elysium. Radial and concentric deformation occurred around the Tharsis region, and Valles Marineris formed and was filled. Extensive south polar ice-rich deposits formed and were highly modified (Head & Pratt 2001), the majority of the outflow channels formed (Carr 1996*a*), and extensive resurfacing and infilling of the northern lowlands took place (Scott & Tanaka 1986; Greeley & Guest 1987; Tanaka & Scott 1987; Tanaka *et al.* 1992*a*; Frey *et al.* 2001; Head *et al.* 2002). During the Hesperian Period, Tharsis and Elysium took their final shape, contraction of the lithosphere from regional to global proportions may have occurred (e.g. Chicarro *et al.* 1985; Watters 1988, 1993), there was a major redistribution of water from the subsurface to the surface, and volcanism and tectonism evolved from global (Early Hesperian ridged plains) to regional (primarily concentrated at Tharsis and Elysium).

Volcanic centres in the Early Hesperian are characterized by an unusual morphology (paterae) that differs from the later edifices of the Tharsis Montes. Relatively low-lying edifices are surrounded by extensive volcanic plains in Malea and Hesperia Planum, and Syrtis Major Planitia (e.g. Greeley & Crown 1990; Crown & Greeley 1993; Hodges & Moore 1994; Greeley *et al.* 2000). Previous workers (e.g. Greeley & Crown 1990; Crown & Greeley 1993) attributed this morphology to the interaction of rising magma with groundwater and the generation of explosive eruptions, which dispersed the ejecta widely, forming low broad edifices that contrast to the later more steeply-sided lava-flow-dominated structures of Tharsis. New Mars Global Surveyor (MGS) data have emphasized the role of water in Amazonian-aged eruptions in Elysium and Utopia Planitia (e.g. Hartmann & Berman 2000; Keszthelyi *et al.* 2000; Russell & Head 2001). Although the Amazonian-aged deposits differ in morphology from those of the Hesperian, no hypothesis at present successfully accounts for both the involvement of water and the morphological differences between the two periods.

Four key trends in the Hesperian are important in the context of magma–H_2O interactions. The first is the change in distribution of volcanism from global to regional, and the related decrease in magmatic flux (Greeley & Schneid

1991). MGS observations of potentially more voluminous early volcanism (e.g. McEwen *et al.* 1999) and Late Amazonian volcanism (e.g. Hartmann *et al.* 1998), will likely change the exact flux, but the relative amounts and the trends are very likely to remain the same. This meant that there were fewer occurrences of magma–H_2O interaction, and that magmatism was concentrated in a small number of specific areas (Tharsis and Elysium). The second trend is that the global decrease in magmatic flux was almost certainly accompanied by a decrease in heat flux from the interior and a corresponding increase in the thickness of the cryosphere. A corollary is that the regional heat flux was higher in Tharsis and Elysium then the global average, which is decreasing with time, and the cryosphere will always be thinner there than the global average, which is increasing in thickness with time. The third trend is the large-scale water migration that took place during this time. The formation of the outflow channels (Milton 1973; Masursky *et al.* 1977) represents a breaching of the cryosphere and the large-scale transfer of water from the subsurface to the surface. Because most of the outflow channels have their sources in the circum-Tharsis region, magma–H_2O interactions have often been implicated in their formation (e.g. Baker *et al.* 1991; Baker 2001). The water clearly flowed into the northern lowlands, where it may have formed significant standing bodies of water (e.g. Parker *et al.* 1989, 1993). The residence time of the outflow channel effluent in the northern lowlands depends largely on the nature of the atmosphere and the thickness and extent of any sedimentary cover on top of ice (e.g. Carr 1996*a*). Under the wide range of scenarios envisioned for the post-outflow channel history of the northern lowlands, ascending magma may have encountered large standing bodies of water (Parker *et al.* 1993), thick ice sheets (Kargel *et al.* 1995; Chapman 1994), or thin sedimentary residues remaining from frozen and sublimated outflow channel effluents (Kreslavsky & Head 2002).

A further consequence of the formation of the outflow channels is the implication of a general depletion of water from the groundwater reservoir (e.g. Clifford 1993; Carr 1996*a*). The change in volcanic style from patera-type activity in the Hesperian to edifice-building activity in the Amazonian could be related to this depletion of upper crustal groundwater (Greeley & Spudis 1981), producing a trend toward more effusive and less explosive volcanism. Further testing of this idea must take into account the evidence cited previously for closely related Amazonian-aged volcanism and groundwater.

A fourth trend is the apparent loss of polar volatiles. An example is the proposed melt-back of the Dorsa Argentea Formation, an extensive volatile-rich deposit surrounding the south polar region. This deposit appears to have been emplaced in the earliest Hesperian (e.g. Plaut *et al.* 1988) and to have undergone extensive melt-back and drainage in the later Hesperian (e.g. Head & Pratt 2001). The timing of the formation and melt-back of this deposit is not well constrained, nor are the reasons for its emplacement and melt-back understood. Evidence for subglacial-like edifice-building eruptions (Ghatan & Head 2001, 2002) within the Dorsa Argentea Formation suggests that basal or internal heating and melting may have been important.

The Amazonian Period is characterized by waning volcanism and tectonism in the Tharsis and Elysium regions, late phases of channel formation, unusual lobate debris flow-like deposits in Elysium and Utopia, emplacement of the Medusae Fossae Formation, eolian modification of the surface, and late in the Amazonian, emplacement of the present polar deposits (Scott & Tanaka 1986; Greeley & Guest 1987; Tanaka & Scott 1987; Tanaka *et al.* 1992*a*). Most workers have agreed that during the Amazonian the cryosphere is relatively thick and similar to that at the present time (Fig. 1). The implications are that during the Amazonian, magma–H_2O interactions are generally limited due to the significantly lower magmatic flux, and that during the ascent of magma, the upper part of the crust contains less groundwater than during earlier periods, except perhaps locally. Lateral dyke emplacement, penetration of the cryosphere, and release of sediment-charged groundwater to produce lahar-like deposits and floods (Fig. 2) apparently occurred in Elysium during the Amazonian (e.g. Christiansen 1989; Tanaka *et al.* 1992*b*; Fuller *et al.* 2001; Lanagan *et al.* 2001; Russell & Head 2001). The presence of surface water in the form of ice-rich deposits (for example at the poles), creates an additional environment for magma interaction during the Amazonian (Figs 1a & 3), and unusual landforms at the margins of the north polar cap have been interpreted to be due to recent volcanism (e.g. Garvin *et al.* 2000; Sakimoto *et al.* 2000).

Magma/lava and water/ice environments

This brief review of the history of volcanism and water on Mars provides the basis for examining specific environments of magma–H_2O interactions as examples and case studies. Interaction between volcanism and hydrospheric/cryospheric bodies and their processes can take many forms. The general environment of water and ice on the surface and in the upper crust is summarized in Figure 1a. Magma rising from the mantle into the crust (Fig. 2) may form plutons that perturb the regional thermal gradient and that stall at neutral buoyancy zones. Plutons can be intruded into water-saturated crust and megaregolith and can conduct heat into the cryosphere above. Overpressurization of the reservoir by additional magma may cause lateral propagation of dykes, and their intersection with the cryosphere and the surface. Evolution of magma within the reservoir may cause further magma ascent, dyke and sill formation, and eruption. Sills above magma reservoirs which heat and melt ice in the cryosphere can cause mobilization and collapse of material (Squyres *et al.* 1987). Emplacement of dykes to shallow levels (Fig. 3) can set up near-surface stress fields that form graben over the dykes, and such deformation can crack the crust and cryosphere. Dykes can be emplaced laterally for thousands of kilometres (Ernst *et al.* 1995; Mege & Masson 1996; Ernst & Buchan 1997; Wilson & Head 2000, 2001*b*), encountering perched aquifers and ground ice, and cracking hydrostatic seals, perhaps leading to lahars and outflow channels.

Intruded dykes will conduct heat into the surrounding region, potentially causing melting and convection, and possibly setting up hydrothermal circulation patterns (Farmer 1996; Gulick 1998). The repetitiveness of dyke intrusions and the longevity of the dyke system are important in determining the scale and duration of hydrothermal systems. Penetration of dykes to the contact between rock and overlying ice can result in intrusion into polar deposits (Fig. 3 left), sill emplacement due to the density differences between the polar deposits and the substrate, and melting of the ice. Surface extrusions can occur at a rock–ice interface (e.g. underneath a body of ice such as a polar cap; actually forming a sill, see Wilson & Head 2001*a*), and may produce palagonite. Subaerial emplacement of lava flows (Figs 3 (right) & 4) can cause heating, melting and steam explosions in water and ice-rich substrates, forming rootless cones and pseudocraters (e.g. Lanagan *et al.* 2001). If lavas emerge into small lakes or larger seas (Fig. 4), quenching and explosive eruptions may occur and glasses and hyaloclastites may form. Should such lakes and seas be frozen, then unusual textures involving partial melting and collapse may form.

Terrestrial examples of many of these environments and features are known, such as pseudocraters (resulting from flows over water-rich substrates; Thorarinsson 1953), maars (resulting from shallow intrusions; Lorenz 1973), and situations where large quantities of groundwater are cycled through the substrate to the vicinity of the magmatic plumbing system to produce a variety of explosive eruption styles (the basaltic plinian eruptions of Tarawera; Walker *et al.* 1984; the eruptions of Heimaey and Surtsey; Thorarinsson 1967). The theory of large-scale volcano–ice interaction for a number of situations on Mars (Figs 1–4) has been treated (Squyres *et al.* 1987; Wilson & Head 1994), and a variety of martian features thought to involve groundwater or ground ice interactions have been studied (Frey *et al.* 1979; Mouginis-Mark *et al.* 1982, 1988; Mouginis-Mark 1985; Greeley & Crown 1990; Lanagan *et al.* 2001).

In summary, processes associated with magma/lava and water/ice interaction (Figs 3–4) include heating, melting, hydrosphere and cryosphere perturbations, hydrothermal circulation and systems, fracturing, steam generation, explosive interactions, magma mixing, and chemical interaction and alteration. Associated phenomena and morphologic effects include up-bowing, faulting, subsidence and collapse, H_2O outflow, lahars, explosive eruptions, surface alteration, and pseudocrater formation. Magma/lava and water/ice interactions will vary in different environments, and will be characterized by different efficiencies, associated processes and morphological manifestations.

Magmatic–H_2O interaction is thus an extremely important process on Mars and many examples show that it is widely distributed in space and time. Although the analysis of terrestrial analogues is very important to further understanding of Mars, it also provides a laboratory for the study of these processes under a wide range of conditions. Magma–water interactions may also be very significant in the production of fine-grained sedimentary material and deposits in martian history, such as hyaloclastite and palagonite (Allen 1979; Wilson & Head 1998; Chapman *et al.* 2000). Long-term interaction of magma and subsurface water may also provide environments conducive to the origin and evolution of life. In the following sections, we assess several case studies and compare them to observed features on Mars.

Plutons

Plutons and large magma reservoirs represent large magma bodies rising diapirically into the

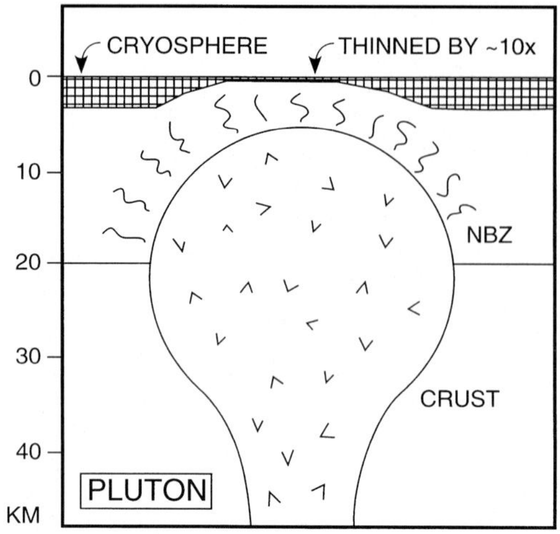

Fig. 8. Schematic diagram of a 30 km-diameter pluton intruded at a neutral buoyancy zone at 20 km depth, with its top at about 6 km depth (e.g. Wilson & Head 1994), near the base of the cryosphere. Heat conducted from the pluton will increase regional heat flow and melt portions of the base of the cryosphere, considerably thinning it above the pluton. The pluton will cool in only a few tens of millions of years unless resupplied.

crust and lithosphere (Fig. 8) and involve the advective transfer of heat to the intrusion zone and the conduction of heat into adjacent regions. Magmatic heat lost conductively from the reservoir can alter the local geothermal gradient, but not as efficiently as magma delivered in geometries with large surface areas/volume, such as dykes and sills. Calculations (Wilson & Head 1994) show that a static reservoir of diameter 10 km emplaced at a depth of about 10 km will perturb the geotherms at the surface by up to about 10% within about 12 km of the area above the reservoir, and that the reservoir will cool to ambient temperatures in about one million years. The early thermal pulse is sufficient to melt ground ice within a zone of about 25 km in diameter above the intrusion, and this might result in the formation of outflow channels, and of collapse features above the intrusion. In the specific example in Figure 8, a 30 km diameter pluton intruded to a typical neutral buoyancy zone (NBZ) depth of about 20 km (Wilson & Head 1994) would have its top at *c.* 6 km depth, below the base of the cryosphere. The advective heat transfer will initially increase heat flow through conduction to the surroundings, which will decrease cryosphere thickness by a factor of *c.* 10%. Unless resupplied, this pluton will cool significantly in only a few tens of millions of years. Post-emplacement evolution of the pluton (e.g. differentiation) may result in emergence of evolved melts above the NBZ as sills, dykes and eruptions. In their simplest form, plutons are most important for regional effects on heat flow.

Cryosphere thickness is critical to the effects of the pluton at the surface, and this particular example will differ depending on variations in cryosphere thickness in space and time. An intrusion such as that pictured in Figure 8 emplaced in the present Mars environment could melt the base of thick ice-rich deposits, such as those toward the poles of Mars (Fig. 1b, c), but a significantly shallower or larger pluton is required to completely melt the cryosphere under current conditions. In earlier history, when the cryosphere was likely to be thinner, a similar intrusion could have melted the cryosphere, particularly in near-equatorial regions (Fig. 1b).

Shallower magma reservoirs and plutons should characterize the major volcanic edifices seen on Mars (Wilson & Head 1994), such as those at Tharsis and Elysium. For example, Wilson & Head (1998) calculated the depths and sizes of reservoirs in the current Tharsis shields, and found that the reservoir centres should lie at depths of about 9 to 13 km below the summit. Assessment of this geometry (Fig. 9) shows that the reservoirs would be well within the edifice itself, and that the resulting enhanced thermal fluxes could cause near-surface melting and possible disruption of the cryosphere. Such effects may be responsible for slumping and mass movements associated with the western flanks of the Tharsis Montes and described by Zimbelman & Edgett (1992). In this case (Fig. 9), the slump deposits lie about 12–14 km below the summit of Arsia Mons, an elevation that is comparable to the expected midpoint of the reservoirs. Concentration of these effects on the western flanks of the edifices may be related to preferential deposition of H_2O toward the northern lowlands or to subsequent burial by flank eruptions on the east (Head 2001).

Plutons are excellent candidates for the production of hydrothermal systems (e.g. Farmer 1996; Gulick 1998). Heating of the groundwater and melting of the base of the cryosphere could significantly alter local and regional equipotential surfaces and hydrostatic relationships. Such perturbations could be responsible for the occurrence of young valley networks on the flanks of some volcanoes such as Alba Patera (Gulick & Baker 1990). In this situation, enhanced heat flux from the reservoir (or possibly dykes and sills) could melt ground ice, causing local sapping and runoff.

(a)

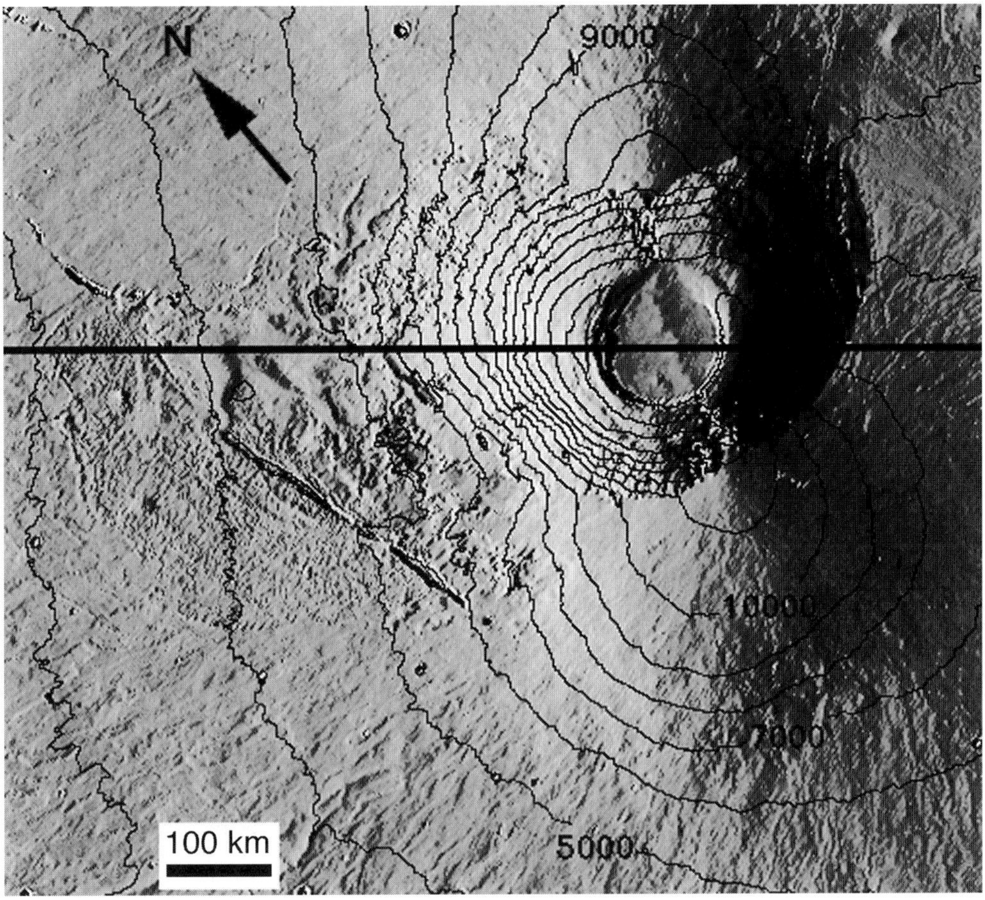

(b)

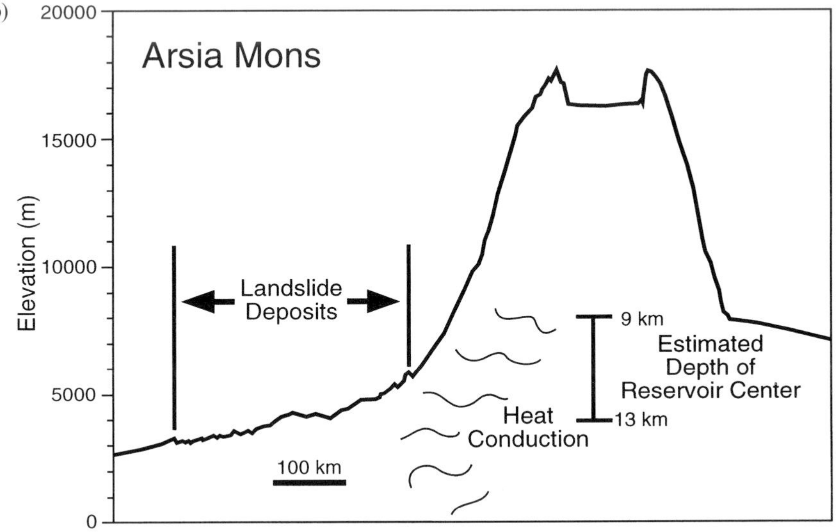

Fig. 9. Slump deposits on the western flanks of Arsia Mons. (**a**) MOLA topographic gradient map with one km contours. The summit caldera and flanking rift zones and flow aprons are readily seen. On the left flank, the extensive landslide deposit mapped by Zimbelman & Edgett (1992) is seen. The horizontal black solid line represents the location of the MOLA topographic profile shown in **b**. The estimated depth of the reservoir centre (Wilson & Head 1998) lies near the elevation of the landslide deposits (**b**). Heat conducted from the reservoir may have helped to melt the cryosphere, leading to instabilities, slumping, and landsliding.

Dykes

In contrast to the rather equidimensional nature of plutons, dykes represent the vertical and lateral emplacement of magma-filled cracks, often all the way to the surface to produce eruptions. Because of the brittle nature of the upper part of the crust, virtually all magma reaching the surface from sources at depth must be delivered through magma-filled cracks, or dykes. Such cracks can range in width from narrow structures (<1 m) propagating hundreds of metres to a few kilometres from a shallow reservoir (such as the East Rift Zone of Kilauea volcano (e.g. Wilson & Head 1988), up to much larger features and complexes, such as the Mackenzie dyke swarm, with dyke widths of tens of metres and lengths of several thousand

(a)

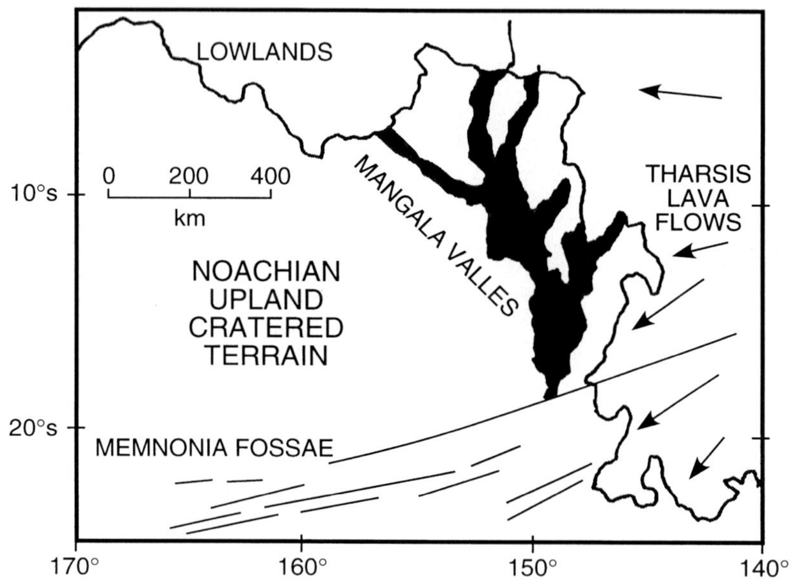

(b)

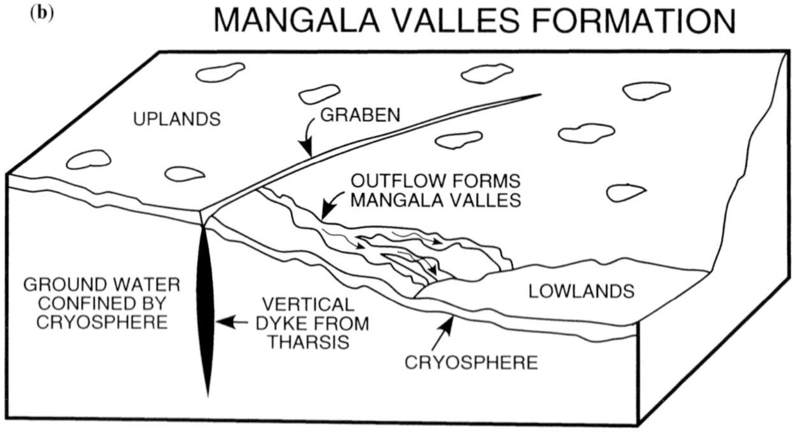

Fig. 10. Memnonia Fossae and Mangala Valles, showing the interpreted relation of graben, dykes and outflow channels. (**a**) Memnonia Fossae are graben radial to, and extending thousands of kilometres from, the centre of Tharsis. Mangala Valles appears to have its source at one of the Memnonia graben. (**b**) One explanation for this relationship is shown in the block diagram. Dykes propagating radially from overpressurized reservoirs in Tharsis approach the surface, creating near-surface stress fields that form graben, cracking the cryosphere and releasing groundwater confined by the cryosphere (e.g. Wilson & Head 2001*a*). Outflow continues until the cryosphere freezes and reseals, or hydrostatic equilibrium is reached.

kilometres (e.g. Ernst *et al.* 1995). On Mars, dyke widths are typically larger than Earth (Swanson *et al.* 1976; Wilson & Head 1994) and lengths may range up to as much as several thousands of kilometres (e.g. the radial graben to the SW of Tharsis; e.g. Wilson & Head 2000, 2001*b*).

Because of their geometry, dykes provide much more efficient heat transfer to the cryosphere than plutons. In addition, dyke intrusion to shallow depths can create a near-surface stress field, resulting in faulting of the overlying rocks, breaching of the cryosphere, and the opportunity for water under hydrostatic head to reach the surface through the cryosphere, potentially resulting in large-scale release of groundwater. A candidate for such a situation is Mangala Valles (Fig. 10; Masursky *et al.* 1986; Tanaka & Chapman 1990). In this case, graben extending radially from Tharsis are interpreted to be the surface manifestation of a dyke propagated laterally from magma reservoirs at Tharsis (Wilson & Head 2000, 2001*b*). The dykes penetrate into a region of Noachian upland cratered terrain on the northern slopes of the southern uplands, where groundwater is interpreted to be confined by a cryospheric seal. As the top of the dyke intrudes into the cryosphere, local melting, combined with the near-surface stress field associated with dyke emplacement, breaks the cryospheric seal, causing release of the groundwater under regional hydrostatic pressure, and formation of the outflow channels of Mangala Valles. Recurring emplacement of magma in the dyke due to additional reservoir overpressurization events prior to the complete solidification of the dyke could cause episodicity in outflow events. Temporary drawdown, or freezing and sealing of the cryosphere, might cause the flow to cease. The subsequent repressurization event might widen the dyke and lead to new cracking and outflow.

In addition, groundwater and melted ground ice can be incorporated into the magma to produce explosive eruptions and widening of the vent system (Wilson & Head 1994), such as the types of eruptions that are thought to be responsible for many of the paterae on Mars (Greeley & Crown 1990; Crown & Greeley 1993). For example, Tyrrhena Patera, a large low-relief volcano in the southern martian highlands, has been interpreted to be the site of pyroclastic flow deposits possibly generated by hydromagmatic explosive eruptions involving water volumes of *c.* 7.5×10^{16} kg and water flow rates of up to about 10^5–10^6 $m^3\,s^{-1}$ (Greeley & Crown 1990). Similar conditions are interpreted to have occurred at nearby Hadriaca Patera; the inferred high permeability of the crust of Mars apparently permitted large amounts of groundwater to be transported laterally into the region at flow rates estimated at 10^3–10^4 $m^3\,s^{-1}$ (Crown & Greeley 1993; Fig. 7 (G)).

In a variation on this theme, an environment may exist in which both effusive volcanic and lahar-like eruptions will occur. An example of such a situation where these two types of deposits exist is the Elysium region (Tanaka *et al.* 1992*b*, 2000; Fig. 11a). Recent work using new altimetry data has shown that the lava flows are erupted from topographically higher vents, while lahar-like flows (Christiansen 1989) are erupted from vents in a restricted range of lower elevations (Russell & Head 2001). One interpretation of this relationship (Fig. 11b) is that dykes emplaced laterally from reservoirs below Elysium propagate to the surface and penetrate the cryosphere to cause eruptions. At high elevations, effusive and explosive eruptions occur, while at low elevations, breaching of the cryospheric seal releases groundwater under hydrostatic head, which mixes with volcanic ash and other products to form mega-lahars (Russell & Head 2001). The volume of these deposits is considerably greater than lahars on Earth (e.g. Major & Newhall 1989; Lavigne & Thouret 2000), primarily because of the involvement of such large quantities of groundwater. The presence of much greater quantities of groundwater on Mars than Earth is apparently due primarily to the presence on Mars of the regional to global cryospheric seal, which is capable of trapping significant volumes of water under hydrostatic pressure (e.g. Carr 1996*c*).

In summary, in terms of the efficiency of heat transfer to the cryosphere, blade-like dykes are much more efficient than deeper, more circular plutons. Dykes can interact with the hydrosphere and cryosphere, pump groundwater into the eruptive system, break artesian seals, and produce major water outflows, lahars, and explosive eruptions. Dykes will also cause direct melting of ice out to several dyke widths solely by conduction; if convection occurs, the lateral influence can increase several fold (Fig. 4). Dykes are extremely important for lateral transport of heat, fracturing of the cryosphere, water release, and local heat generation in the vicinity of the dyke, which can also set up local hydrothermal circulation systems. But because the geometry of dykes is orthogonal to the cryosphere, the melting efficiency is focused at the top of the dyke. Obviously, greater efficiency for cryospheric melting is found in sills, where the geometry of the heat source is more parallel to the cryosphere.

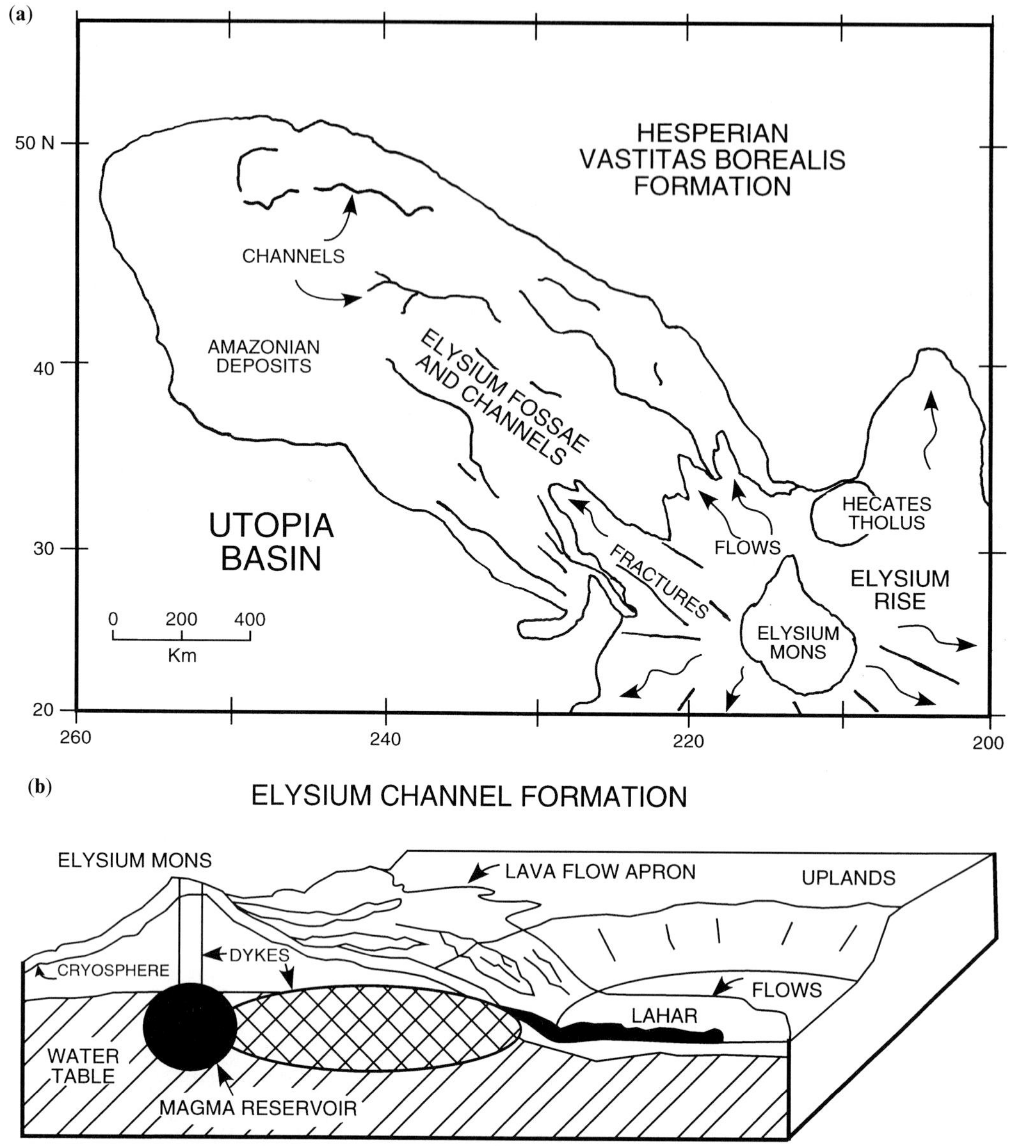

Fig. 11. Elysium Fossae, channels and Amazonian flow deposits, showing the interpreted relationship of magmatism, dyke emplacement and mega-lahar formation. (**a**) Extensive Amazonian-aged deposits occur in the Utopia Basin, overlying the Hesperian-aged Vastitas Borealis Formation. These deposits include lava flows, channels, debris flows, lahars and other features (e.g. Christiansen 1989; Tanaka *et al.* 1992*b*) that appear to be related to the Elysium rise to the south-east. (**b**) One explanation for this relationship is shown in the block diagram. Lateral dyke propagation from reservoirs underlying Elysium penetrate the cryospheric seal, releasing confined groundwater which mixes with the magmatic components to produce mega-lahars that flow into the adjacent Utopia Basin (e.g. Russell & Head 2001). Dykes propagated to higher levels are above the water table and form predominantly effusive flows.

Sills and surface flows

Sill geometry is best for delivery of heat to shallow crustal levels and thus sills are extremely efficient in melting near-surface ice and releasing groundwater (Squyres *et al.* 1987; Wilson & Head 1994). In addition, the presence of a low-density cryosphere can enhance density contrasts in the shallow crust and favour the intrusion of sills. Sill intrusion provides plausible models for formation of several outflow channels. For example, consider the intrusion of a sill about 150–200 km in diameter, 100 metres thick, and emplaced at the base of a hydrous cryosphere (Fig. 12a). Heat is lost primarily from the top and upward, not downward. Heat from the sill surface (surface T is *c.* 1000 K) is initially conducted into the surrounding ground ice; when this melts, heat is convected across to the receding ice surface. After about eight years, the sill is chilled to the middle and 250 m of ice is melted from the top. Loss of the rest of the heat from the sill would produce about a 500 m vertical column of ice melted over a 150–200 km diameter sill, for a total of *c.* 10^4 km^3 of water. The consequences are that sill intrusion can plausibly produce surface uplift, significant melting, collapse, chaos formation, and outflow channel formation (Fig. 12b, c). Sill intrusion thus provides plausible models for outflow channels, and may explain why some outbursts appear to be one-time events. If some of the water drains out, the heat transfer capacity is seriously influenced, and heat may then be simply deposited into the surroundings or lost inefficiently.

Squyres *et al.* (1987) considered the thermodynamics of the interactions of volcanic intrusions (sills) and extrusions (flows) with ground ice, and made quantitative estimates of the amounts and rates of liquid water and steam that might be generated or released under a variety of conditions. They then used these estimates

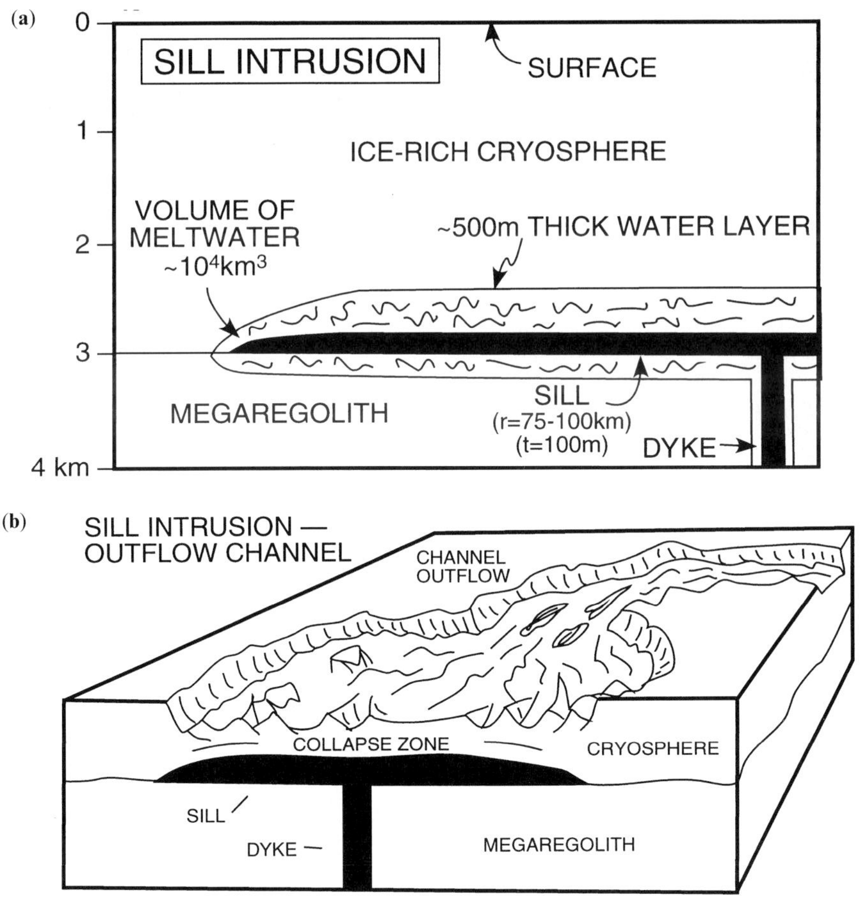

Fig. 12. Interaction of a sill intrusion to produce melting, chaos formation and outflow channels. In (**a**), a sill intrudes into the megaregolith at the base of the ice-rich cryosphere, and conduction and melting occur, producing about 10^4 km^3 of meltwater in only a few years (see text for details). (**b**) Sill intrusion can plausibly cause surface uplift and fracturing, cryosphere cracking, significant melting, collapse, chaos formation and outflow channels. (**c**) Region of chaos at the head of an outflow channel at 20°N, 33°W that might result from sill intrusion. Width of image is about 62 km, Viking Orbiter frame 742A12.

(c)

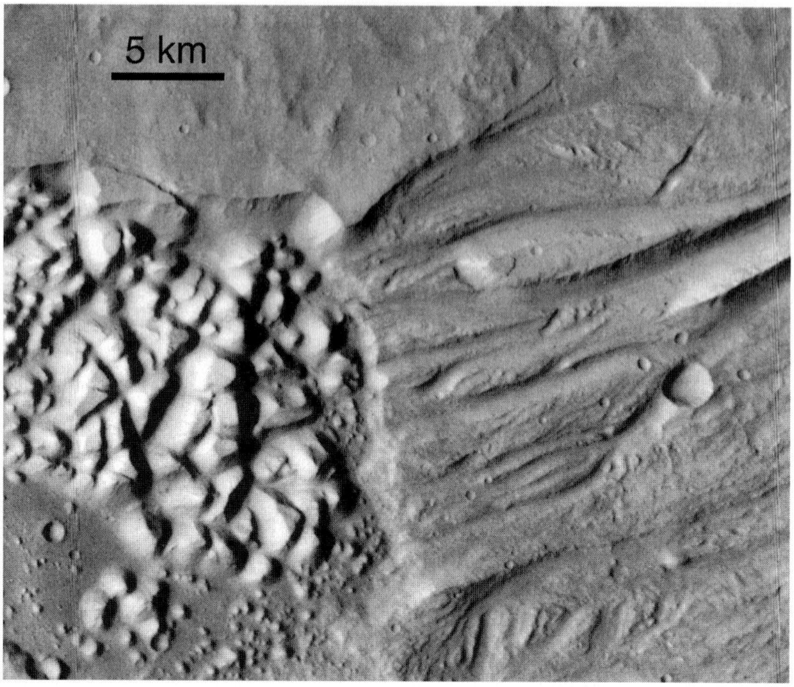

Fig. 12. (*continued*)

to investigate the relation to the amount of ground subsidence predicted to occur, and compared these with values inferred from fluvial channels and other features associated with volcanic complexes.

The simpler case is the eruption of lava over ice-rich permafrost; as heat is conducted downward, two phase-change boundaries exist, the melting front separating permafrost from wet soil, and the vaporization front separating wet soil from soil plus steam. The bulk of the heat is lost from the top of flow (Fig. 13a) but for a 10 m-thick flow over a substrate that is 25% ice, the maximum depth of dehydration is 3.0 m beneath the base of the flow. The total amount of ice melted or vaporized is equivalent to a column thickness of 3.7 m. As the lava flow thickness increases, the proportion of the column that is only liquid increases because of the higher vaporization temperature. In general, under a variety of conditions, the total column thickness of melted water is less than half the lava flow thickness.

Although not treated by Squyres *et al.* (1987), this same geometry can also lead to the explosive release of water vapour generated during substrate melting and the disruption of the lava flow to cause rootless pyroclastic deposits and cones (pseudocraters) on the flow surface (Thorarinsson 1953). Recent work on Icelandic rootless cones has further characterized these features and led to a better understanding of the processes of formation (Greeley & Fagents 2001). Rootless cones range from small spatter structures to larger tuff cones, which show alternating layers of lapilli scoria and red-baked lacustrine muds overlain by crudely bedded brick-red agglutinates. The upward trends are consistent with decreasing explosivity and decreasing amounts of incorporated lacustrine mud. Cone groups rest directly on the host lava but are undeformed. These observations were interpreted to require inflated tube-fed pahoehoe flows crossing wetlands, and creating phreatomagmatic eruptions fed by the lava tubes.

Pseudocrater-like features on Mars have been described in Acidalia Planitia (Frey *et al.* 1979; Allen 1980), Chryse Planitia (Greeley & Thelig 1978), and Deutronilus Mensae (Lucchitta 1978). Detailed analysis of these features shows that they are typically about 600 m in diameter (e.g. Frey & Jarosewich 1982), which is about twice the size of typical Icelandic examples. Other interpretations include degradational remnants of older cratered terrain, degraded ice-sheet remnants, and pingos.

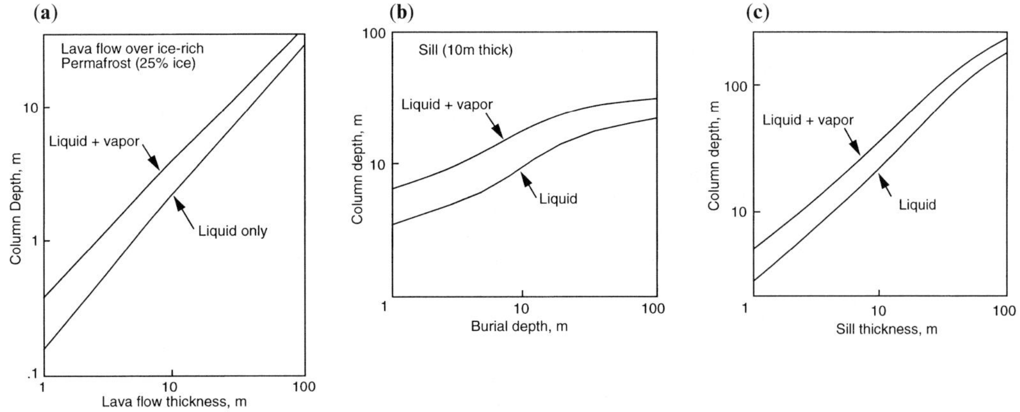

Fig. 13. Relationship between magma placement and effects for flows and sills. (**a**) Calculated column depth of liquid and liquid plus vapour generated by extrusion of a lava flow over an ice-rich (25% ice) permafrost, as a function of lava flow thickness, (**b**) for injection of a 10 metre thick sill, as a function of depth of burial, and (**c**) for injection of sills of various thicknesses at a depth of 100 m (from Squyres *et al.* 1987).

Mars Orbiter Camera images of parts of the Elysium Basin and Amazonis Planitia have revealed excellent high-resolution examples of features interpreted to be pseudocraters (Lanagan *et al.* 2001; Fig. 14). These features occur in areas of relatively recent (Late Amazonian) lava flows (Hartmann & Berman 2000; Keszthelyi *et al.* 2000) in small clusters independent of any obvious structural patterns. They are superimposed on these flows, are less than about 250 m in diameter, and show no obvious flows emerging from the cones. What were the conditions of formation of these features on Mars? Lanagan *et al.* (2001) pointed out that because the Elysium and Amazonis examples lie in an equatorial region, the substrate should be depleted of ground ice to depths of the order of 100–300 m (e.g. Fanale *et al.* 1986; Mellon & Jakosky 1995; Mellon *et al.* 1997). They attribute the formation of the pseudocrater-like features to the replenishment of ground ice by relatively recent Late Amazonian fluvial processes.

A significant amount of the volcanic activity on Mars occurs in equatorial and mid-latitudes, associated with Tharsis and Elysium. The general depletion of shallow ground ice in equatorial regions (e.g. Fanale *et al.* 1986; Mellon & Jakosky 1995; Mellon *et al.* 1997) may explain why these pseudocrater-like features are so rare, despite the presence of a thick global water-rich cryosphere throughout much of the history of Mars. As more and more high-resolution images of Mars become available, analysis of the nature and distribution of pseudocrater-like features may help to define regions which contained ice or water-saturated substrates in the past history of Mars. For example, the pseudocrater-like features described by Frey *et al.* (1979) and Frey & Jarosewich (1982) in Chryse and Acidalia occur in regions that are thought to have contained abundant water related to outflow channel effluents emplaced during the Hesperian.

Squyres *et al.* (1987) also treated the more complex case of injection of a sill into ice-rich permafrost. The area below the sill is identical to the surface lava flow case, while in the area above the sill, the boundaries propagate upward, and the water generated migrates downward and undergoes vaporization once it reaches regions where the temperature exceeds the boiling point. Therefore, liquid water is only generated once the temperatures everywhere above the sill are less than the boiling temperature. Because of the lack of a 'free' surface, cooling takes place much more slowly than for the flow case and significantly more liquid and vapour are generated by the sill. For a sill thickness of 10 m at a depth of 100 m in a substrate that is 25% ice the thickness of the water column is substantially greater than the thickness of the sill, by a factor of about 2–4 (Figure 13b). When the burial depth is less than the sill thickness, considerable heat is lost to the surface, reducing the total column thickness. Thus, sill injection into an ice-rich substrate can form net depressions (column height > sill thickness), while lava flows on the surface cannot (column height < flow thickness).

Schultz & Glicken (1979) suggested that many floor-fractured craters may be the sites of sills and there is evidence that these are also the locations of the release of groundwater. Squyres

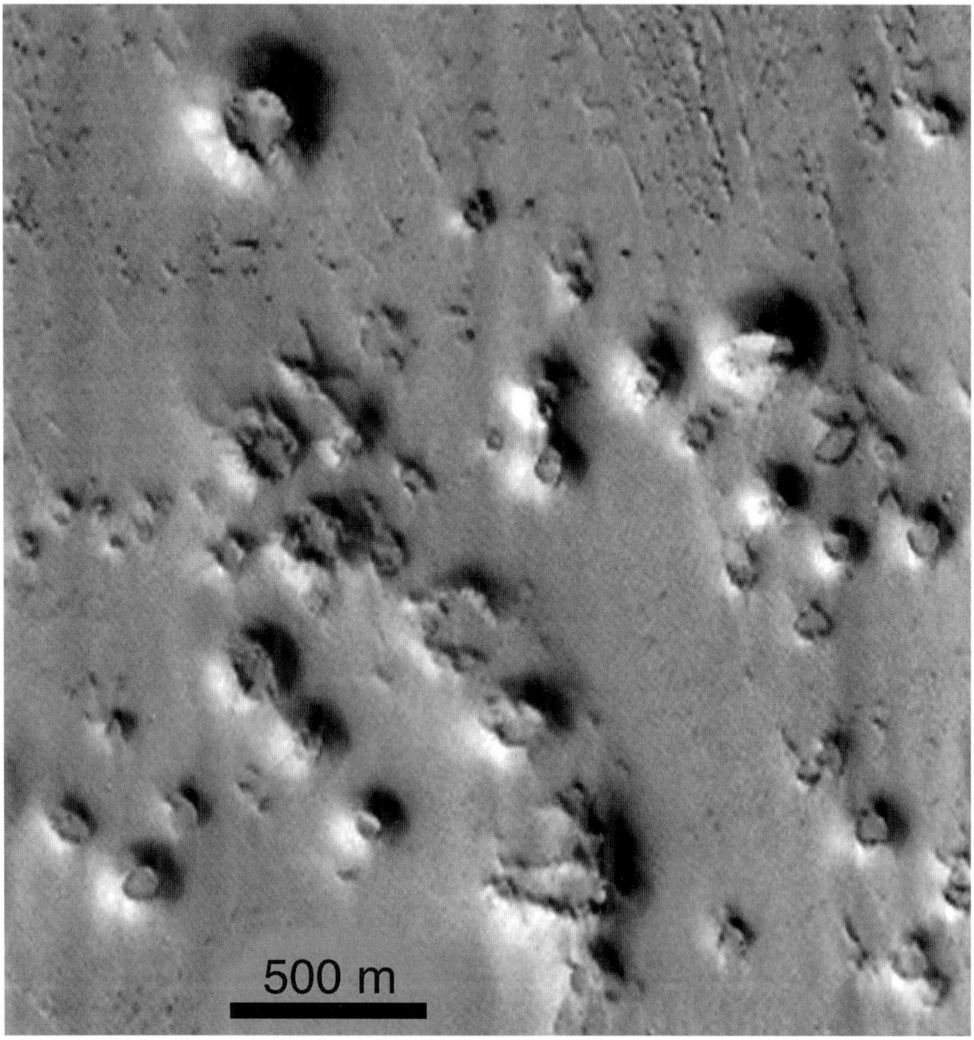

Fig. 14. Cluster of cones north of the Cerberus plains, Eastern Elysium. Cones lie on platy and ridged lava flows, interpreted to be rubbly pahoehoe (Lanagan *et al.* 2001). The size, density, overlapping nature and relation of the cones to the lava flow, suggested to Lanagan *et al.* (2001) that they are rootless cones or pseudocraters, very similar in appearance and scale to those in Iceland. Portion of MOC image M08/01962.

et al. (1987) outlined morphologic and stratigraphic evidence for a large flow lobe deposit in the Aeolis Mensae Region of Elysium Planitia adjacent to a sill-like unit interleaved between friable deposits and now exposed by erosion. Squyres *et al.* (1987) also presented evidence for a possible lahar generated by intrusion of a sill and, just south of Hadriaca Patera, a series of channels and deposits interpreted to be generated by sill and flow emplacement. Wilson & Mouginis-Mark (2001) describe a scenario for sill intrusion at the head of Hrad Valles in western Elysium Planitia in which they envision heat from sill intrusion into the ice-rich substrate melting the ice and boiling the meltwater to create phreatomagmatic explosions. Skilling *et al.* (2001) have described voluminous eruption-fed lahars associated with sill intrusions in flood basalt provinces on Earth. These deposits filled large collapse structures and generated large outflow sheets. In the Karoo examples, the comingling of magma and fluidized wet sediment occurred in association with the emplacement and replenishment of high-level sills, and Skilling *et al.* (2001) draw analogy to the collapse areas associated with the martian outflow channels (also see previous discussion).

Bergh & Sigvaldson (1991) have described an interesting combination of features in a basaltic hyaloclastite mass-flow deposit in South Iceland. Repeated voluminous extrusion of basaltic lava from sub-aquatic fissures on the Eastern Rift Zone produced a 700 m-thick sequence with basal columnar basalt underlying cube-jointed and pillow basalt, which in turn is overlain by thick unstructured hyaloclastite containing aligned basalt lobes, with the sequence capped by bedded hyaloclastite. Due to the extremely high effusion rates (these vents are in the same area as the Laki fissures) the effective water/melt ratio was low, precluding complete fragmentation of the melt. Thus, the melt sheet moved downslope at the base, below a heterogeneous mass of hyaloclastite and fluid melt, with vigorous interaction occurring at the boundary, producing increasingly vesicular hyaloclastite fragments above. This scenario may be common in high-effusion rate seafloor eruptions (see Head & Wilson 2002) and may have occurred on Mars (Russell & Head 2001).

In summary, a wide diversity of interactions of igneous intrusions and volcanic extrusions with permafrost and groundwater are possible on Mars (Figs 3 & 7) and the predicted landforms include collapse depressions and outflow channels, dykes and sills possibly exhumed by erosion of surface layers, and surface flows leading to melting of subsurface layers or formation of rootless explosion craters (pseudocraters).

Maars on Mars

On Earth, maars are broad, low-rimmed volcanic craters that typically result from phreatic or phreatomagmatic eruptions. Maars are explosive hydrovolcanic craters which form when rising magma in dykes interacts explosively with groundwater or surface water (Lorenz 1973, 1986). Maar deposits may contain little to no magmatic material (phreatic) or a mixture (phreatomagmatic), and they grade into tuff rings and tuff cones (Cas & Wright 1987). The presence of shallow ground ice and groundwater on Mars for virtually all of its history suggests that maar eruptions should be abundant. Mege & Masson (1996) interpret pits in some graben on Alba Patera to be maars, but features interpreted to be maars have not commonly been detected or reported in the literature.

The lack of identification of abundant maars on Mars may be due to several factors, including similarity to degraded impact craters and erosion of fragmental and poorly welded and widely dispersed deposits (Wilson & Head 1994). One additional important factor, however, may be the lack of shallow groundwater and standing bodies of water over much of martian history. If a cryosphere characterizes the upper several kilometres of the martian crust (Clifford 1993), then ascending magma will interact with ice, rather than groundwater, and this will reduce the probability of phreatic or phreatomagmatic eruptions operating continuously enough to cause maars. However, earlier conditions on Mars may have differed and maar deposits should be searched for wherever near-surface liquid water is suspected.

An additional factor may be the latitudinal depletion of shallow ground ice. Mellon & Jaksoky (1995) and Mellon *et al.* (1997) have modelled the migration and loss of shallow ground ice in the warmer equatorial regions of Mars (between about 50° north and south latitude), producing a dehydrated layer at depths of up to several hundred metres, depending on the type of material. Dykes reaching the near surface in this latitude range may thus encounter dehydrated material and where the dyke does interact with ground ice and groundwater at depth, overburden pressures might not be appropriate for the vigorous boiling and explosive disruption necessary to produce maars.

Additional environments and settings of magma–H_2O interactions on Mars

The floor of Valles Marineris

Extensive layered deposits have been documented on the floor of Valles Marineris and adjacent canyons (e.g. Nedell *et al.* 1987; Lucchitta *et al.* 1992, 1994) and have often been interpreted to be sedimentary in origin, perhaps related to explosive volcanic eruption products, either primary or deposited in standing bodies of water (Lucchitta 1990; Weitz 1999). Nedell *et al.* (1987) pointed out that subaqueous volcanism was an attractive hypothesis for the layered terrain on the valley floors, because such eruptions could supply volcanic material to the floor even if the standing bodies of water were covered with ice, and preferential deposition there would remove the need to deposit material on the surrounding uplands. Some of the landforms on the floors have been attributed to subaqueous and sub-ice eruptions (e.g. Lucchitta *et al.* 1994).

Recent work using MGS data has strengthened the case that some of the interior layered

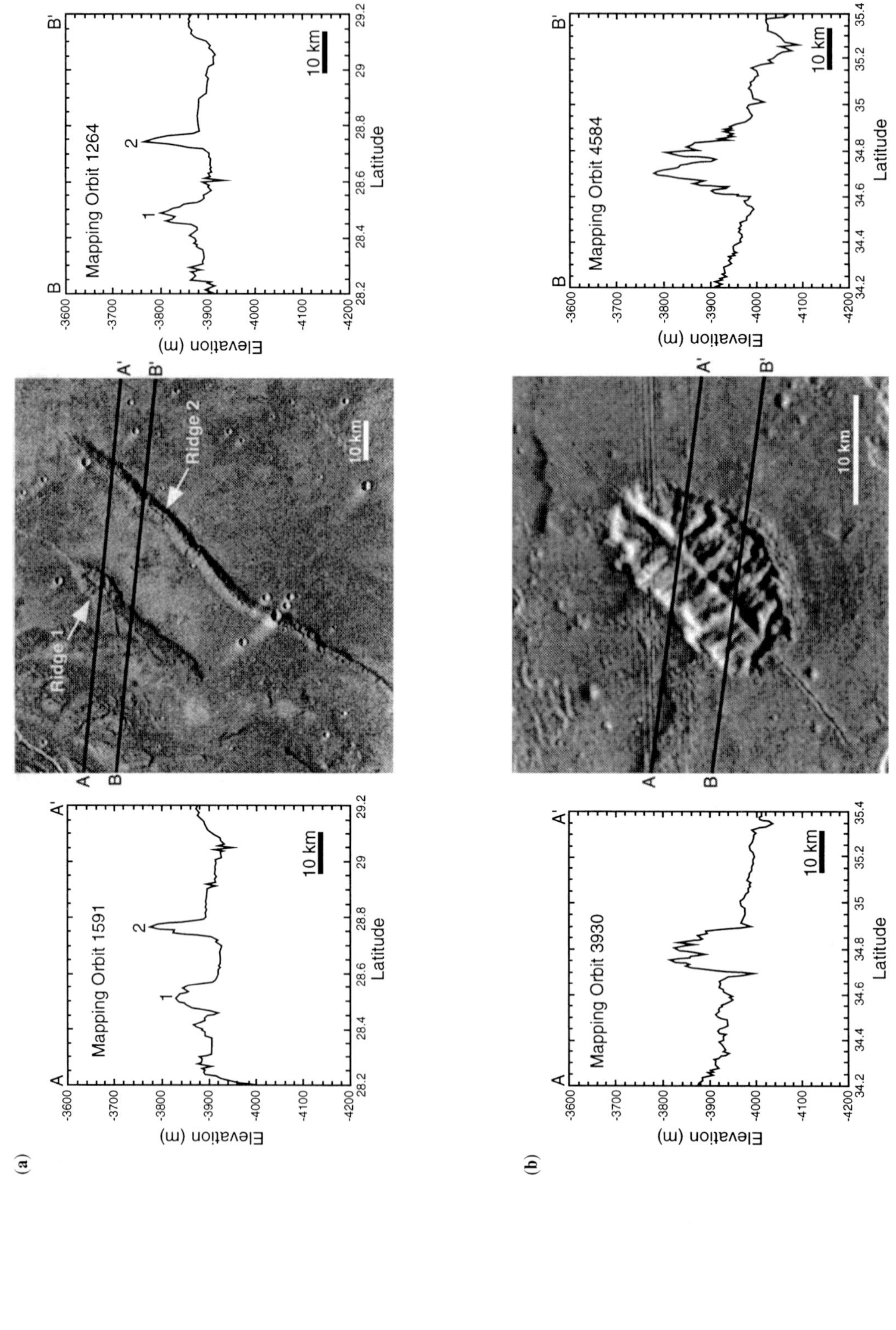
(a)
Mapping Orbit 1591
Mapping Orbit 1264
Ridge 1
Ridge 2
A
A'
B
B'
1
2
10 km
Elevation (m)
Latitude
(b)
Mapping Orbit 3930
Mapping Orbit 4584
10 km
Elevation (m)
Latitude

volatiles, perhaps dominated by CO_2, at the time of the associated igneous events. In their view, the martian crust was extraordinarily susceptible to erosion induced by voluminous magmatic activity. Shallow dykes or small sills may have produced collapse depressions, in some cases huge lahars and mass flows at major volcanic complexes.

In the Tanaka *et al.* (2001*a*, *b*) scenario, melting of water ice or release of groundwater alone is thought to be insufficient to explain the observations. In addition, the presence of CO_2 clathrate and dry ice (Hoffman 2000) may have had profound effects. CO_2 release could have fluidized an unconsolidated or fragmental regolith, leading to generation of huge debris flows and their deposition into adjacent basins. In addition, Tanaka *et al.* (2001*a*, *b*) suggested that such magma–subsurface volatile interactions could account for much of the nature of Valles Marineris and Noctis Labyrinthus. In their view, intrusions and eruptions in the west led to collapse and formation of Noctis Labyrinthus, to collapse of portions of Valles Marineris, and to the extensive outflows in the east, which, charged with CO_2, could travel thousands of kilometres to the north into the northern lowlands (Hoffman 2000). The role of CO_2 and possibly even methane (Max & Clifford 2001) may be very important, and has yet to be fully evaluated or modelled.

Conclusions

Environments of magma–H_2O interaction differ to a first order between the Earth and Mars due to the presence on Mars of a global permafrost layer in the upper crust throughout most of its observed history. This several kilometre thick cryosphere apparently contained water/ice throughout most of its history, except where its uppermost part underwent dehydration in equatorial and lower latitude regions. Between the base of the cryosphere and the depth at which cracks close, groundwater appears to have existed during much of the history of Mars, as evidenced by Hesperian and Amazonian outflow channels. Water/ice also exists on the surface in the form of polar and circumpolar deposits, and in the past, liquid water (perhaps ice-covered) flowed across the surface at relatively low rates (valley networks) and at high rates (outflow channels). Ponded water, perhaps ice-covered, appears to have existed on the surface as a result of outflow channel formation. Data on the stability of water under a variety of conditions likely to have existed on Mars in the last 3–4 billion years suggest that such ponded water underwent relatively rapid freezing and sublimation.

Magmatism (plutonism and volcanism) was a very important process in the history of Mars, representing the record of planetary heat loss and mantle convective processes in space and time. The geometry and extent of the cryosphere and hydrosphere in the history of Mars indicates that magma–H_2O interactions were varied and often very significant. The importance of magma–H_2O interactions was often determined by the geometry of the intrusion in relation to the cryosphere and hydrosphere (e.g. pluton, dyke or sill), and the amount of water in the hydrosphere at the time of the intrusion.

Clearly, Mars is an important laboratory for the study of magma–H_2O interactions. New insights into the geometry of interactions and their consequences from high-resolution images, radar sounding and surface exploration, and basic physical models (Clifford 1993; Wilson & Head 1994, 2001*b*), can help to decipher this record, and lead to further understanding of the history of Mars. For example, if the upper part of the cryosphere is dehydrated in equatorial regions, but hydrated at higher latitudes, then the presence of pseudocraters, maars and other indications of surface magma–H_2O interactions could potentially be used as clues to ancient paleolatitudes. If the martian paterae imply hydrovolcanic eruptions, the change in morphology and inferred eruption conditions and styles from paterae to edifices (e.g. Greeley &

Fig. 16. Candidate hyaloclastite ridges on Mars in Utopia Planitia interpreted to have formed during eruptions beneath an ice sheet (Chapman 1994; Chapman *et al.* 2000) and characterized here by MOLA altimetry data. (**a**) Ridge 1 is variable in width and up to about 3 km wide; it is about 100 m in height and both profiles show evidence for a 10–20 m deep summit depression. Ridge 2 is narrower and more consistent in width, ranging up to about 2.5 km; the ridge is 130–150 m high where the profiles cross it. Viking Orbiter frame 541A20. Centre is about 28.5°N, 224°W. (**b**) Broad oval-shaped, rough-textured ridge up to about 11 km wide lying along a narrow linear ridge; broad ridge is about 200 m high at the highest point of the profile. Structure has been embayed by later flows as evidenced by flow lobes at the top of the image and the asymmetry in topography along profile A–A′, with the surface there about 70 metres higher, and showing a moat typical of the margins of flows adjacent to a kipuka. Centre of figure is about 34.8°N, 217°W. Composite of Viking Orbiter frames 541A10 and 541A12.

Spudis 1981) could be used to map out evolutionary changes in the availability of groundwater. Other factors could also be responsible, including evolution in magma composition, changes in atmospheric pressure and density (resulting in variations in eruptive styles and morphology), or other factors. In an alternative scenario, Gregg & Williams (1996), citing the record of explosive mafic volcanoes on Earth, interpreted the deposits of the large central vent volcanoes Hadriaca and Tyrrhena Paterae to be constructed primarily of mafic pyroclastic rocks, rather than being the products of magma–water–ice interactions. Further assessment of the Greeley & Spudis (1981) hypothesis with new data and theoretical models will help to distinguish among these important alternatives.

New high-resolution altimetry, radar sounding and imaging data, as well as increased understanding of terrestrial analogues, offer the promise of developing criteria to identify more confidently subaqueous and sub-ice eruptions on Mars. This will contribute to the understanding of the residence time and fate of bodies of water and ice on the surface of Mars during its history. Furthermore, study of the interaction of high-eruption-rate effusive volcanic eruptions, the cryosphere and the groundwater table (e.g. Russell & Head 2001), has led to the recognition of mega-lahars, an eruption style that represents a unique signature of the magma–H_2O configuration on Mars. Finally, magma–H_2O interactions have significant implications for the production of fine-grained sedimentary material and layered deposits in martian history (Malin & Edgett 2000*a*). Tephra produced in hydromagmatic eruptions should be very widespread in early Mars history and hyaloclastite and palagonite may be globally abundant (e.g. Allen 1979, 1980), and locally concentrated, particularly where there is evidence of Icelandic-like, sub-ice sheet eruptions (e.g. Chapman *et al.* 2000). Hydrothermal systems and deposits should be common (Farmer 1996; Gulick 1998) and their mineralogy should be readily recognizable with high spatial and spectral resolution instruments. Further recognition and documentation of the range of magma–H_2O environments described here will lead to an important new insight into candidate conditions for the origin of life on Mars, and the evolution of plausible biological niches with time. A challenge in the next decade is to develop more sophisticated models of the distribution and nature of water, water/ice and CO_2 clathrates and hydrates, and other gaseous species such as methane. A more comprehensive understanding of magma and frozen volatiles in the subsurface and on the surface must await the development of a consensus on the presence and abundance of these species, and their behaviour with depth and with time.

Thanks are extended to A. Côté and S. Pratt for figure preparation, P. Neivert for figure reproduction, and E. Fuller for assistance with references. Thanks are also extended to K. Herkenhoff and an anonymous reviewer whose comments helped improve the quality of the paper. This paper is based on an invited presentation given at the Volcano–Ice Interaction meeting in Reykjavik, Iceland in August, 2000. We gratefully acknowledge financial assistance to JWH from NASA through the Planetary Geology and Geophysics Program and the Mars Data Analysis Program.

References

AHARONSON, O., ZUBER, M. T. & ROTHMAN, D. H. 2001. Statistics of Mars' topography from MOLA: Slopes, correlations and physical models. *Journal of Geophysical Research*, **106**, 20 527–20 546.

ALLEN, C. C. 1979. Volcano–ice interactions on Mars. *Journal of Geophysical Research*, **84**, 8048–8059.

ALLEN, C. C. 1980. Icelandic subglacial volcanism: Thermal and physical studies. *Geology*, **88**, 108–117.

BAKER, V. R. 2001. Water and the martian landscape. *Nature*, **412**, 228–236.

BAKER, V. R., STROM, R. G., GULICK, V. C., KARGEL, J. S., KOMATSU, G. & KALE, V. S. 1991. Ancient oceans, ice sheets, and the hydrological cycle on Mars. *Nature*, **352**, 589–594.

BANDFIELD, J. L., HAMILTON, V. E. & CHRISTENSEN, P. R. 2000. A global view of martian surface compositions from MGS-TES. *Science*, **287**, 1626–1630.

BERGH, S. G. & SIGVALDSON, G. E. 1991. Pleistocene mass-flow deposits of basaltic hyaloclastite on a shallow submarine shelf, South Iceland. *Bulletin of Volcanology*, **53**, 597–611.

BRADLEY, B. A., GROSFILS, E. B. & SAKIMOTO, S. E. H. 2000. Boundaries and stratigraphy of the Medusae Fossae Formation and Elysium Basin materials using Mars Orbiter Laser Altimeter (MOLA) data. *Lunar and Planetary Science*, **31**, No. 2055.

BRAIN, D. & JAKOSKY, B. 1998. Atmospheric loss since the onset of the Martian geologic record: Combined role of impact erosion and sputtering. *Journal of Geophysical Research*, **103**, 22 689–22 694.

BRAKENRIDGE, G. R., NEWSOM, H. E. & BAKER, V. R. 1985. Ancient hot springs on Mars: Origin and paleoenvironmental significance of small martian valleys. *Geology*, **13**, 859–862.

CARR, M. H. 1995. The martian drainage system and the origin of networks and fretted channels. *Journal of Geophysical Research*, **100**, 7479–7507.

CARR, M. H. 1996*a*. *Water on Mars*. Oxford University Press, Oxford.

CARR, M. H. 1996*b*. Water on early Mars. *In*: BOCK, G. R. & GOODE, J. A. (eds) *Evolution of Hydrothermal Ecosystems on Earth (and Mars?)*. Wiley, Chichester, 249–265.

CARR, M. H. 1996c. Channels and valleys on Mars: Cold climate features formed as a result of a thickening cryosphere. *Planetary and Space Science*, **44**, 1411–1423.

CARR, M. H. 2000. Martian oceans, valleys and climate. *Astronomy and Geophysics*, **41**, 320–326.

CAS, R. A. F. & WRIGHT, J. V. 1987. *Volcanic Successions: Modern and Ancient*. Allen and Unwin, London.

CHAPMAN, M. G. 1994. Evidence, age, and thickness of a frozen paleolake in Utopia Planitia, Mars. *Icarus*, **109**, 393–406.

CHAPMAN, M. G. & TANAKA, K. L. 2001. Interior trough deposits on Mars: Subice volcanoes? *Journal of Geophysical Research*, **106**, 10 087–10 100.

CHAPMAN, M. G. , ALLEN, C. C., GUDMUNDSSON, M. T., GULICK, V. C., JAKOBSSON, S. P., LUCCHITTA, B. K., SKILLING, I. P. & WAITT, R. B. 2000. 'Fire and ice': Volcanism and ice interactions on Earth and Mars. *In*: ZIMBELMAN, J. R. & GREGG, T. K. P. (eds) *Environmental Effects on Volcanic Eruptions: From Deep Oceans to Deep Space*. Kluwer Publishing, New York, 39–73.

CHICARRO, A. F., SCHULTZ, P. H. & MASSON, P. 1985. Global and regional ridge patterns on Mars. *Icarus*, **63**, 153–174.

CHRISTIANSEN, E. H. 1989. Lahars in the Elysium region of Mars. *Geology*, **17**, 203–206.

CLIFFORD, S. M. 1987. Polar basal melting on Mars. *Journal of Geophysical Research*, **92**, 9135–9152.

CLIFFORD, S. M. 1993. A model for the hydrologic and climatic behavior of water on Mars. *Journal of Geophysical Research*, **98**, 10 973–11 016.

CLOW, G. D. 1987. Generation of liquid water on Mars through the melting of a dusty snowpack. *Icarus*, **72**, 95–127.

CROWN, D. A. & GREELEY, R. 1993. Volcanic geology of Hadriaca Patera and the eastern Hellas region of Mars. *Journal of Geophysical Research*, **98**, 3431–3451.

EINARSSON, T. 1966, Physical aspects of sub-glacial eruptions. *Jokull*, **16**, 167–174.

ERNST, R. E. & BUCHAN, K. E. 1997. Giant radiating dyke swarms: Their use in identifying pre-Mesozoic large igneous provinces and mantle plumes. *In*: MAHONEY, J. J. & COFFIN, M. F. (eds) *Large Igneous Provinces: Continental, Oceanic, and Planetary Flood Volcanism*. American Geophysical Union, Washington DC, 297–333

ERNST, R. E., HEAD, J. W., PARFITT, E., GROSFILS, E. & WILSON, L. 1995. Giant radiating dyke swarms on Earth and Venus. *Earth Science Reviews*, **39**, 1–58.

FANALE, F. P., SALVAIL, J. R., ZENT, A. P. & POSTAWKO, S. E. 1986. Global distribution and migration of subsurface ice on Mars. *Icarus*, **67**, 1–18.

FARMER, J. D. 1996. Hydrothermal systems on Mars: an assessment of present evidence. *In*: BOCK, G. R. & GOODE, J. A. (eds) *Evolution of Hydrothermal Ecosystems on Earth (and Mars?)*. Wiley, Chichester, 273–295.

FREY, H. V. & JAROSEWICH, M. 1982. Subkilometer martian landscapes: Properties and possible terrestrial analogs. *Journal of Geophysical Research*, **87**, 9867–9879.

FREY, H. V., LOWRY, B. L. & CHASE, S. A. 1979. Pseudocraters on Mars. *Journal of Geophysical Research*, **84**, 8075–8086.

FREY, H. V., SCHICKEY, F. M., FREY, E.KL., ROARK, J. H. & SAKIMOTO, S. E. H. 2001. A Very Large Population of Likely Buried Impact Basins in the Northern Lowlands of Mars Revealed by MOLA Data. *Lunar and Planetary Science*, **32**, No. 1680.

FULLER, E. R., HEAD, J. W., KRESLAVSKY, M. A. & PRATT, S. 2001. Volcanism and sedimentation in Amazonis Planitia: The origin of the smoothest plains on Mars. *In*: *Eos, Transactions, AGU, 82(20), Spring Meeting Supplement*, Abstract P31A-07, S244.

GARVIN, J. B., SAKIMOTO, S. E. H., FRAWLEY, J. J., SCHNETZLER, C. C. & WRIGHT, H. M. 2000. Topographic evidence for geologically recent near-polar volcanism on Mars. *Icarus*, **145**, 648–652.

GHATAN, G. J. & HEAD, J. W. 2001. Candidate subglacial volcanoes in the south polar region of Mars. *Lunar and Planetary Science*, **32**, No. 1039.

GHATAN, G. J. & HEAD, J. W. 2002. Candidate subglacial volcanoes in the south polar region of Mars: Morphology, morphometry, and eruption conditions. *Journal of Geophysical Research*, **107**, 10.1029/2001JE001519.

GRANT, J. A. 2000. Valley formation in Margaritifer Sinus, Mars, by precipitation-recharged groundwater sapping. *Geology*, **28**, 223–226.

GREELEY, R. & CROWN, D. A. 1990. Volcanic geology of Tyrrhena Patera, Mars. *Journal of Geophysical Research*, **95**, 7133–7149.

GREELEY, R. & FAGENTS, S. 2001. Icelandic pseudocraters as analogs to some volcanic cones on Mars. *Journal of Geophysical Research*, **106**, 20 527–20 546.

GREELEY, R. & GUEST, J. E. 1987. *Geologic map of the eastern equatorial region of Mars, scale 1:15 000 000*. USGS Miscellaneous Investigations Series Map **I-1802-B**.

GREELEY, R. & SCHNEID, B. D. 1991. Magma generation on Mars: Amounts, rates, and comparisons with Earth, Moon, and Venus. *Science*, **254**, 996–998.

GREELEY, R. & SPUDIS, P. D. 1981. Volcanism on Mars. *Reviews of Geophysics – Space Physics*, **19**, 13–41.

GREELEY, R. & THELIG, E. 1978. Small volcanic constructs in the Chryse Planitia region of Mars, *NASA TM-79729*, **202**.

GREELEY, R., BRIDGES, N. T., CROWN, D. A., CRUMPLER, L., FAGENTS, S. A., MOUGINIS-MARK, P. J. & ZIMBELMAN, J. R. 2000. Volcanism on the Red Planet: Mars. *In*: ZIMBELMAN, J. R. & GREGG, T. K. P. (eds) *Environmental Effects on Volcanic Eruptions: From Deep Oceans to Deep Space*. Kluwer Publishing, New York, 75–112.

GREGG, T. K. P. & WILLIAMS, S. N. 1996. Explosive mafic volcanoes on Mars and Earth: Deep magma sources and rapid rise rate. *Icarus*, **122**, 397–405.

GULICK, V. C. 1998. Magmatic intrusions and a hydrothermal origin for fluvial valleys on Mars. *Journal of Geophysical Research*, **103**, 19 365–19 388.

GULICK, V. C. 2001. Origin of the valley networks on Mars: a hydrological perspective. *Geomorphology*, **37**, 241–268.
GULICK, V. C. & BAKER, V. R. 1990. Origin and evolution of valleys on martian volcanoes. *Journal of Geophysical Research*, **95**, 14 625–14 344.
GULICK, V. C. & BAKER, V. R. 1993. *Fluvial valleys in the heavily cratered terrains of Mars: Evidence for paleoclimate change?* LPI Technical Report, **93-03**, 12–13.
HAMILTON, V. E., WYATT, M. B., MCSWEEN, H. J. & CHRISTENSEN, P. R. 2001. Analysis of terrestrial and martian volcanic compositions using thermal emission spectroscopy, 2, Application to martian surface spectra from the Mars Global Surveyor Thermal Emission Spectrometer. *Journal of Geophysical Research*, **106**, 14 733–14 746.
HARTMANN, W. K. 1999. Martian cratering VI: Crater count isochrons and evidence for recent volcanism from Mars Global Surveyor. *Meteoritics and Planetary Science*, **34**, 167–177.
HARTMANN, W. K. & BERMAN, D. 2000. Elysium Planitia lava flows: Crater count chronology and geological implications. *Journal of Geophysical Research*, **105**, 15 011–15 025.
HARTMANN, W. K. & NEUKUM, G. 2001. Cratering chronology and evolution of Mars. *Space Science Reviews*, **95**, 167–196.
HARTMANN, W. K., MALIN, M., MCEWEN, A., CARR, M., SODERBLOM, L., THOMAS, P., DANIELSONS, E., JAMES, P. & VEVERKA, J. 1998. Evidence for recent volcanism on Mars from crater counts. *Nature*, **397**, 586–589.
HEAD, J. W. 2001. Medusae Fossae Formation as ancient polar deposits?: Tests and new data on stratigraphic relationships. *Lunar and Planetary Science*, **32**, No. 1394.
HEAD, J. W. & KRESLAVSKY, M. A. 2001. Medusae Fossae Formation as volatile-rich sediments deposited during high obliquity: An hypothesis and tests. *In*: *Conference on the Geophysical Detection of Subsurface Water on Mars*, No. 7053.
HEAD, J. W. & PRATT, S. 2001. Extensive Hesperian-aged South Polar ice sheet on Mars: Evidence for massive melting and retreat, and lateral flow and ponding. *Journal of Geophysical Research*, **106**, 12 275–12 299.
HEAD, J. W. & WILSON, L. 2001. Mars: General environments of magma/H_20 interaction. *Lunar and Planetary Science*, **32**, No. 1218.
HEAD, J. W. & WILSON, L. 2002. Deep submarine pyroclastic eruptions on seamounts: Theory and predicted landforms and deposits. *Journal of Volcanology and Geothermal Research*, in press.
HEAD, J. W., HIESINGER, H., IVANOV, M. A., KRESLAVSKY, M. A., PRATT, S. & THOMSON, B. J. 1999. Possible Oceans On Mars: Evidence from Mars Orbiter Laser Altimeter data. *Science*, **286**, 2134–2137.
HEAD, J. W. & KRESLAVSKY, M. A. & PRATT, S. 2001. Northern lowlands of Mars: Evidence for widespread volcanic flooding and tectonic deformation in the Hesperian Period. *Journal of Geophysical Research*, **107**, No. E1, 10.1029/2000JE001445.
HEAD, J. W., KRESLAVSKY, M. A. & PRATT, S. 2002. Northern lowlands of Mars: evidence for widespread volcanic flooding and tectonic deformation in the Hesperian period. *Journal of Geophysical Research*, **107**, 10.1029/2000JE001445.
HESS, P. C. & LONGHI, J. 2001. Capillary effects on the stability of ice in martian crust. *Lunar and Planetary Science*, **32**, No. 1702.
HODGES, C. A. & MOORE, H. J. 1994. *Atlas of volcanic landforms on Mars*, US Geological Survey Professional Paper **1534**.
HOFFMAN, N. 2000. White Mars: A new model for Mars' surface and atmosphere based on CO2. *Icarus*, **146**, 326–342.
IVANOV, B. A. 2001. Mars/Moon cratering rate ratio estimates. *Space Science Reviews*, **95**, 89–106.
IVANOV, M. A. & HEAD, J. W. 2001. Chryse Planitia, Mars: Topographic configuration from MOLA data and tests for hypothesized lakes and shorelines. *Journal of Geophysical Research*, **106**, 3275–3295.
JAKOSKY, B. & HABERLE, R. 1992. The seasonal behavior of water on Mars. *In*: KIEFFER, H. H., JAKOSKY, B. M., SNYDER, C. W. & MATTHEWS, M. S. (eds) *Mars*. University of Arizona Press, Tucson, 969–1016.
JAKOSKY, B. M. & PHILLIPS, R. J. 2001. Mars' volatile and climate history. *Nature*, **412**, 237–244.
JONES, J. G. 1969. Intraglacial volcanoes of the Laugarvatn region, southwest Iceland. *Quarterly Journal of the Geological Society of London*, **124**, 197–211.
JONES, J. G. 1970. Intraglacial volcanoes of the Laugarvatn region, southwest Iceland, II. *Journal of Geology*, **78**, 127–140.
KARGEL, J. S. & STROM, R. G. 1992. Ancient glaciation on Mars. *Geology*, **20**, 3–7.
KARGEL, J. S. & STROM, R. G. 1998. Clathrate hydrates on Earth and in the Solar System. *In*: SCHMITT, B., DE BERGH, C. & FESTOU, M. (eds) *Solar System Ices*. Kluwer, Dordrecht, 97–117.
KARGEL, J. S., BAKER, V. R., BEGÉT, J. E., LOCKWOOD, J. F., PÉWÉ, T. L., SHAW, J. S. & STROM, R. G. 1995. Evidence of ancient continental glaciation in the martian northern plains. *Journal of Geophysical Research*, **100**, 5351–5368.
KESZTHELYI, L. P. & MCEWEN, A. S. 2001. Recent flood volcanism on Mars: Implications for climate change, layered deposits, and lava-water interactions. *In*: *Eos, Transactions, AGU, 82(20), Spring Meeting Supplement*, Abstract V42A-03, S439.
KESZTHELYI, L. P., MCEWEN, A. S. & THORDARSON, T. 2000. Terrestrial analogs and thermal models for martian flood lavas. *Journal of Geophysical Research*, **105**, 15 027–15 049.
KRESLAVSKY, M. A. & HEAD, J. W. 1999. Kilometer-scale slopes on Mars and their correlation with geologic units: Initial results from Mars Orbiter Laser Altimeter (MOLA) data. *Journal of Geophysical Research*, **104**, 21911–21924.
KRESLAVSKY, M. A. & HEAD, J. W. 2000. Kilometer-scale roughness of Mars' surface: Results from MOLA data analysis. *Journal of Geophysical Research*, **105**, 26 695–26 712.

KRESLAVSKY, M. A. & HEAD, J. W. 2002. Fate of outflow channel effluents in the northern lowlands of Mars and their relation to the Hesperian-aged Vastitas Borealis Formation. *Journal of Geophysical Research*, in press.

KRESLAVSKY, M. A., MCEWEN, A. S., KESZTHELYI, L. P. & THORDARSON, T. 2001. Rootless cones on Mars indicating the presence of shallow equatorial ground ice in recent times. *Geophysical Research Letters*, **28**, 2365–2369.

LANAGAN, P. D., MCEWEN, A. S., KESZTHELYI, L. P. & THORDARSON, T. 2001. Rootless cones on Mars indicating the presence of shallow equatorial ground ice in recent times. *Geophysical Research Letters*, **28**, 2365–2368.

LAVIGNE, F. & THOURET, J. C. 2000. Les lahars: Dépôts, origines, et dynamique. *Bulletin de la Société Géologique de France*, **171**, 545–557.

LONGHI, J. 2000. Low-temperature phase relations in the CO_2–H_2O system with application to Mars. *Lunar and Planetary Science*, **31**, No. 1518.

LORENZ, V. 1973. On the formation of maars. *Bulletin Volcanologique*, **37**, 183–204.

LORENZ, V. 1986. On the growth of maars and diatremes and its relevance to the formation of tuff rings. *Bulletin of Volcanology*, **48**, 265–274.

LUCCHITTA, B. K. 1978. *Geologic map of the Ismenius Lacus quadrangle, Mars, scale 1:5 000 000*. USGS Miscellaneous Investigations Series Map **I-1065**.

LUCCHITTA, B. K. 1990. Young volcanic deposits in the Valles Marineris, Mars? *Icarus*, **86**, 476–509.

LUCCHITTA, B. K., FERGUSON, H. M. & SUMMERS, C. 1986. Sedimentary deposits in the northern lowland plains, Mars. *Journal of Geophysical Research*, **91**, 166–174.

LUCCHITTA, B. K., ISBELL, N. K. & HOWINGTON-KRAUS, A. 1994. Topography of Valles Marineris: Implications for erosional and structural history. *Journal of Geophysical Research*, **99**, 3783–3798.

LUCCHITTA, B. K., MCEWEN, A. S., CLOW, G. D., GEISSLER, P. E., SINGER, R. B., SCHULTZ, R. A. & SQUYRES, S. W. 1992. The global canyon system on Mars. *In*: KIEFFER, H. H., JAKOSKY, B. M., SNYDER, C. W. & MATTHEWS, M. S. (eds.) *Mars*. University of Arizona Press, Tucson, 453–492.

MAJOR, J. J. & NEWHALL, C. G. 1989. Snow and ice perturbation during historical volcanic eruptions and the formation of lahars and floods: A global review. *Bulletin of Volcanology*, **52**, 1–27.

MALIN, M. C. & EDGETT, K. S. 1999. Oceans and seas in the martian northern lowlands: High-resolution imaging tests of proposed coastlines. *Geophysical Research Letters*, **26**, 3049–3052.

MALIN, M. C. & EDGETT, K. S. 2000*a*. Sedimentary rocks on Mars. *Science*, **290**, 1927–1937.

MALIN, M. C. & EDGETT, K. S. 2000*b*. Evidence for recent groundwater seepage and surface runoff on Mars, *Science*, **288**, 2330–2335.

MASURSKY, H. 1973. An overview of geological results from Mariner 9. *Journal of Geophysical Research*, **78**, 4009–4030.

MASURSKY, H., BOYCE, J. V., DIAL, A. L., SCHABER, G. G. & STROBELL, M. E. 1977. Classification and time of formation of martian channels based on Viking data. *Journal of Geophysical Research*, **82**, 4016–4037.

MASURSKY, H., CHAPMAN, M. G., DIAL, A. L., JR. & STROBELL, M. E. 1986. Ages of rocks and channels in prospective martian landing sites of Mangala Valles region. *Lunar and Planetary Science*, **17**, 520–521.

MATHEWS, W. H. 1947. 'Tuyas', flat-topped volcanoes in Northern British Columbia, *American Journal of Science*, **245**, 560–570.

MAX, M. D. & CLIFFORD, S. M. 2001. Initiation of martian outflow channels: Related to the dissociation of gas hydrate? *Geophysical Research Letters*, **28**, 1787–1790.

MCCAULEY, J. F. 1978. *Geologic map of the Coprates quadrangle of Mars, scale 1:5 000 000*. USGS Miscellaneous Investigations Series Map **I-897**.

MCEWEN, A. S., MALIN, M. C., CARR, M. H. & HARTMANN, W. K. 1999. Voluminous volcanism on early Mars revealed in Valles Marineris, *Nature*, **397**, 584–586.

MEGE, D. & MASSON, P. 1996. A plume tectonics model for the Tharsis province, Mars. *Planetary and Space Science*, **44**, 1499–1546.

MELLON, M. T. & JAKOSKY, B. M. 1995. The distribution and behavior of martian ground ice during past and present epochs, *Journal of Geophysical Research*, **100**, 11 781–11 799.

MELLON, M. T., JAKOSKY, B. M. & POSTAWKO, S. E. 1997. The persistence of equatorial ground ice on Mars. *Journal of Geophysical Research*, **102**, 19 357–19 369.

MELOSH, H. J. & VICKERY, A. M. 1989. Impact erosion of the primordial atmosphere of Mars. *Nature*, **338**, 487–489.

MILLER, S. L. 1985. Clathrate hydrates in the Solar System. *In*: KLINGER, J., BENEST, D. & DOLLFUS, A. (eds) *Ices in the Solar System*. D. Reidel, Boston, 59–79.

MILTON, D. J. 1973. Water and processes of degradation in the martian landscape. *Journal of Geophysical Research*, **78**, 4037–4047.

MOUGINIS-MARK, P. J. 1985. Volcano/ground ice interactions in Elysium Planitia, Mars. *Icarus*, **64**, 265–284.

MOUGINIS-MARK, P. J. 1990. Recent water release in the Tharsis region of Mars. *Icarus*, **84**, 362–373.

MOUGINIS-MARK, P. J., WILSON, L. & HEAD, J. W. 1982. Explosive volcanism at Hecates Tholus, Mars: Investigation of eruption conditions. *Journal of Geophysical Research*, **87**, 9890–9904.

MOUGINIS-MARK, P. J., WILSON, L., HEAD, J. W., BROWN, S. H., HALL, J. L., & SULLIVAN, K. D. 1984. Elysium Planitia, Mars: Regional geology, volcanology and evidence for volcano-ground ice interactions. *Earth, Moon, and Planets*, **30**, 149–173.

MOUGINIS-MARK, P. J., WILSON, L. & ZIMBELMAN, J. R. 1988. Polygenic eruptions on Alba Patera, Mars, *Bulletin of Volcanology*, **50**, 361–379.

NEDELL, S. S., SQUYRES, S. W. & ANDERSEN, D. W. 1987. Origin and evolution of the layered deposits in the Valles Marineris, Mars. *Icarus*, **70**, 409–441.

PARKER, T. J., GORSLINE, D. S., SAUNDERS, R. S., PIERI, D. C. & SCHNEEBERGER, D. M. 1993. Coastal geomorphology of the martian northern plains. *Journal of Geophysical Research*, **98**, 11 061–11 078.

PARKER, T. J., SAUNDERS, R. S. & SCHNEEBERGER, D. M. 1989. Transitional morphology in the west Deuteronilus Mensae region of Mars: Implications for modification of the lowland/upland boundary. *Icarus*, **82**, 111–145.

PHILLIPS, R. J., ZUBER, M. T., SOLOMON, S. C. *ET AL.* 2001. Ancient geodynamics and global-scale hydrology on Mars. *Science*, **291**, 2587–2591.

PIERI, D. C. 1980. Martian valleys: Morphology, distribution, age, and origin. *Science*, **210**, 895–897.

PLAUT, J. J., KAHN, R., GUINNESS, E. A & ARVIDSON, R. E. 1988. Accumulation of sedimentary debris in the south polar region of Mars and implications for climate history. *Icarus*, **75**, 357–377.

RUSSELL, P. & HEAD, J. W. 2001. The Elysium/Utopia flows: Characteristics from topography and a model of emplacement, *Lunar and Planetary Science*, **32**, No. 1040.

SAKIMOTO, S. E. H. & GREGG, T. K P. 2001. Channeled flow: Analytic solutions, laboratory experiments, and applications to lava flows. *Journal of Geophysical Research*, **106**, 8629–8648.

SAKIMOTO, S. E. H., GARVIN, J. B. & WRIGHT, H. 2000. Topography of small volcanic edifices in the Mars northern polar region from Mars Orbiter Laser Altimeter Observations. *Lunar and Planetary Science*, **31**, No. 1971.

SAKIMOTO, S. E. H., GARVIN, J. B., BRADLEY, B. A., WONG, M. & FRAWLEY, J. J. 2001. Small martian north polar volcanoes: Topographic implications for eruption styles. *Lunar and Planetary Science*, **32**, No. 1808.

SAKIMOTO, S. E. H., FREY, H. V., GARVIN, J. B. & ROARK, J. H. 1999. Topography, roughness, layering, and slope properties of the Medusae Fossae Formation from Mars Orbiter Laser Altimeter (MOLA) and Mars Orbiter Camera (MOC) data. *Journal of Geophysical Research*, **104**, 24 141–24 154.

SCHULTZ, P. H. & GLICKEN, H. 1979. Impact crater and basin control of igneous processes on Mars. *Journal of Geophysical Research*, **84**, 8033–8047.

SCOTT, D. H. & TANAKA, K. L. 1986. *Geologic map of the western equatorial region of Mars, scale 1:15 000 000.* USGS Miscellaneous Investigations Series Map **I-1802-A**.

SCOTT, D. H., CHAPMAN, M. G., RICE, J. W. & DOHM, J. M. 1992. New evidence of lacustrine basins on Mars: Amazonis and Utopia Planitia, *Proceedings of Lunar and Planetary Sciences*, **22**, 53–62.

SHARP, R. P. & MALIN, M. C. 1975. Channels on Mars. *Geological Society of America Bulletin*, **86**, 593–609.

SKILLING, I. P. 1994. Evolution of an englacial volcano: Brown Bluff, Antarctica. *Bulletin of Volcanology*, **56**, 573–591.

SKILLING, I. P., CHAPMAN, M. G. & SMELLIE, J. L. 2001. Terrestrial subice volcanism and pre-flood basalt hydrovolcanism as models for magma-volatile interaction on Mars. *In*: *Eos, Transactions, AGU, 82(20), Spring Meeting Supplement*, Abstract V42A-09, **S440**.

SMELLIE, J. L. 2000. Subglacial eruptions. *In*: SIGURDSSON, H. (ed.) *Encyclopedia of Volcanoes*. Academic Press, San Diego, 403–418.

SMELLIE, J. L. & SKILLING, I. P. 1994. Products of subglacial volcanic eruptions under different ice thickness: Two examples from Antarctica. *Sedimentary Geology*, **91**, 115–129.

SOLOMON, S. C. 1978. On volcanism and thermal tectonics on one-plate planets. *Geophysical Research Letters*, **5**, 461–464.

SQUYRES, S. W., WILHELMS, D. E. & MOOSMAN, A. C. 1987. Large-scale volcano-ground ice interaction on Mars. *Icarus*, **70**, 385–408.

SWANSON, D. A., DUFFIELD, W. A. & FISKE, R. S. 1976. Displacement of the south flank of Kilauea Volcano: The result of forceful intrusion of magma into rift zones. *U. S. Geological Survey Professional Paper*, **963**(39), 1976.

TANAKA, K. L. 1986. The stratigraphy of Mars. *Journal of Geophysical Research Supplement*, **91**, 139–158.

TANAKA, K. L. & CHAPMAN, M. G. 1990. The relation of catastrophic flooding of Mangala Valles, Mars, to faulting of Memnonia Fossae and Tharsis volcanism. *Journal of Geophysical Research*, **95**, 14 315–14 323.

TANAKA, K. L. & DAVIS, P. A. 1988. Tectonic history of the Syria Planum province of Mars. *Journal of Geophysical Research*, **93**, 14 893–14 917.

TANAKA, K. L. & SCOTT, D. H. 1987. *Geologic map of the polar regions of Mars, scale 1:15 000 000.* USGS Miscellaneous Investigations Series Map **I-1802-C**.

TANAKA, K. L., BANERDT, W. B., KARGEL, J. S. & HOFFMAN, N. 2001*a*. Huge, CO_2-charged debris flow deposit and tectonic sagging in the northern plains of Mars. *Geology*, **29**, 427–430.

TANAKA, K. L., CHAPMAN, M. G. & SCOTT, D. H. 1992*b*. *Geologic map of the Elysium region of Mars, scale 1: 5 000 000*. USGS Miscellaneous Investigations Series Map **I-2147**.

TANAKA, K. L., ISBELL, N. K., SCOTT, D. H., GREELEY, R. & GUEST, J. E. 1988. The resurfacing history of Mars: A synthesis of digitized, Viking-based geology. *Proceedings of Lunar and Planetary Sciences*, **18**, 665–678.

TANAKA, K. L., KARGEL, J. S. & HOFFMAN, N. 2001*b*. Evidence for magmatically driven catastrophic erosion on Mars. *Lunar and Planetary Science*, **32**, No. 1898.

TANAKA, K. L., SCOTT, D. H. & GREELEY, R. 1992*a*. Global stratigraphy. *In*: KIEFFER, H. H., JAKOSKY, B. M., SNYDER, C. W. & MATTHEWS, M. S. (eds) *Mars*. University of Arizona Press, Tucson, 345–382.

THOMAS, P., SQUYRES, S., HERKENHOFF, K., HOWARD, A. & MURRAY, B. 1992. Polar deposits of Mars. *In*: KIEFFER, H. H., JAKOSKY, B. M., SNYDER, C. W. & MATTHEWS, M. S. (eds) *Mars*. University of Arizona Press, Tucson, 767–795.

THOMSON, B. & HEAD, J. W. 2001. Utopia Basin, Mars: Characterization of topography and morphology and assessment of the origin and evolution of basin internal structure. *Journal of Geophysical Research*, **106**, 23 209–23 230.

THORARINSSON, S. 1953. The crater groups in Iceland, *Bulletin Volcanologique*, **14**, 3–44.

THORARINSSON, S. 1967. *Surtsey: The new island in the North Atlantic*. New York, Viking Press.

WALKER, G. P. L., SELF, S. & WILSON, L. 1984. Tarawera, 1886, New Zealand: A basaltic plinian fissure eruption. *Journal of Volcanology and Geothermal Research*, **21**, 61–78.

WATTERS, T. R. 1988. Wrinkle ridge assemblages on the terrestrial planets. 1988. *Journal of Geophysical Research*, **93**, 15 599–15 616.

WATTERS, T. R. 1993. Compressional tectonism on Mars. *Journal of Geophysical Research*, **98**, 17 049–17 060.

WEBB, B. M., HEAD, J. W., KORTZ, B. E. & PRATT, S. 2001. Syria Planum, Mars: A major volcanic construct in the early history of Tharsis. *Lunar and Planetary Science*, **32**, No. 1145.

WEITZ, C. M. 1999. A volcanic origin for the interior layered deposits in Hebes Chasma, Mars. *Lunar and Planetary Science*, **30**, No. 1277.

WILLIAMS, R. M. & PHILLIPS, R. J. 2001. Morphometric measurements of martian valley networks from Mars Orbiter Laser Altimeter (MOLA) data. *Journal of Geophysical Research*, **106**, 23 737–23 751.

WILSON, L. & HEAD, J. W. 1988. The influence of gravity on planetary volcanic eruption rates. *Lunar and Planetary Science*, **29**, 1283–1284.

WILSON, L. & HEAD, J. W. 1994. Mars: Review and analysis of volcanic eruption theory and relationships to observed landforms. *Reviews of Geophysics*, **32**, 221–263.

WILSON, L. & HEAD, J. W. 1998. Evolution of magma reservoirs within shield volcanoes on Mars. *Lunar and Planetary Science*, **29**, No. 1128.

WILSON, L. & HEAD, J. W. 2000. Tharsis-radial graben systems as the surface manifestation of plume-related dyke intrusion complexes: Models and implications. *Lunar and Planetary Science*, **31**, No. 1371.

WILSON, L. & HEAD, J. W. 2001*a*. Heat transfer and melting in subglacial volcanic eruptions: implications for volcanic deposit morphology and meltwater volumes. *In*: SMELLIE, J. L. & CHAPMAN, M. G. (eds) *Volcano–Ice Interaction on Earth and Mars*. Geological Society, London, Special Publications, **202**, 5–26.

WILSON, L. & HEAD, J. W. 2001*b*. Giant dyke swarms and related graben systems in the Tharsis province of Mars, *Lunar and Planetary Science*, **32**, No. 1153.

WILSON, L. & MOUGINIS-MARK, P. 2001. Styles of phreatomagmatic activity adjacent to volcanic constructs on Mars. *In*: *Eos, Transactions, AGU, 82(20), Spring Meeting Supplement*, Abstract V42A-11, **S440**.

WYATT, M. B., HAMILTON, V. E., MCSWEEN, H. J., CHRISTENSEN, P. R. & TAYLOR, L. A. 2001. Analysis of terrestrial and martian volcanic compositions using thermal emission spectroscopy, 1, Determination of mineralogy, chemistry, and classification strategies. *Journal of Geophysical Research*, **106**, 14 711–14 732.

ZIMBELMAN, J. & EDGETT, K. S. 1992. The Tharsis Montes, Mars: Comparison of volcanic and modified landforms. *Proceedings of Lunar and Planetary Sciences*, **22**, 31–44.

ZIMBELMAN, J., HOOPER, D., CROWN, D. A., GRANT, J., SAKIMOTO, S. A. H. & FREY, H. 1999. Medusae Fossae Formation, Mars: An assessment of possible origins utilizing early results from Mars Global Surveyor. *Lunar and Planetary Science*, **30**, No. 1652.

ZUBER, M. T., SMITH, D. E., SOLOMON, S. C. *ET AL*. 1998. Observations of the north polar region of Mars from the Mars Orbiter Laser Altimeter. *Science*, **282**, 2053–2060.

ZUBER, M. T., SOLOMON, S. C., PHILLIPS, R. J. *ET AL*. 2000. Internal structure and early thermal evolution of Mars from Mars Global Surveyor topography and gravity. *Science*, **287**, 1788–1793.

The 1969 subglacial eruption on Deception Island (Antarctica): events and processes during an eruption beneath a thin glacier and implications for volcanic hazards

J. L. SMELLIE

British Antarctic Survey, High Cross, Madingley Road, Cambridge CB3 0ET, UK (e-mail: JLSM@bas.ac.uk)

Abstract: A short-lived eruption of basaltic andesite to andesite on Deception Island in 1969 occurred from a series of fissures underneath a glacier. The glacier was thin (*c.* 100 m) and the eruption created a large and sudden discharge of meltwater that overflowed the glacier, severely damaging buildings on the island. The eruption was unusually well documented and it illustrates several features of subglacial eruptions that are only poorly known and not well understood. In particular, overflowing meltwater is contrary to predictions based on existing simple hydrological models for eruptions beneath thin glaciers. The eruption is analysed in this paper and used as a model for the fluid dynamics and thermodynamics of eruptions beneath a thin glacier mainly composed of impermeable ice. It is suggested that, in eruptions of relatively fluid magmas with a low magma rise rate, volatiles and magma are able to decouple and subglacial melting is strongly influenced by the superheated magmatic and hydrothermal gases (mainly steam). Thus, melting is much faster than that due solely to coupled conductive (magma) and convective (meltwater) heat transfer. The influence of gas-driven melting also has an important effect on the shape of the meltwater cavity and may be at least partly responsible for the cylindrical ice chimneys developed above vents on Deception Island. The results of the study are important for reconstructing the shapes of englacial cavities melted above a vent. They also highlight the importance of glacier structure and densification, rather than simply glacier thickness, in determining the hydraulic evolution of an eruption. Even eruptions beneath thin glaciers can generate significant meltwater floods.

Deception Island is an active volcano situated just off-axis in Bransfield Strait, a very young (<2 Ma) ensialic marginal basin undergoing slow spreading at the northern tip of the Antarctic Peninsula (Fig. 1; Smellie 2001*a*). The volcano has a shield morphology, with a submerged basal diameter of 30 km and a total height of 1.5 km (Smellie 1990). It rises to 540 m above sea level at Mount Pond and 460 m at Mount Kirkwood. Both summits have extensive glaciers, and these and other smaller glaciers cover approximately half of the island. The centre of the island, known as Port Foster, is a flooded restless caldera about 10 km in diameter (Cooper *et al.* 1998). The caldera-forming eruption may have ejected >25–30 km^3 of magma, but all post-caldera eruptions appear to have been small volume (typically <0.05 km^3). They formed a variety of tuff cones and maars, together with cinder cones and associated lavas. Based on an extensive tephra record preserved in the glacier ice, post-caldera eruptions were frequent during the nineteenth century and occurred during several decades (Orheim 1972*a*, *b*). The volcano has been comparatively quiet during the twentieth century, with eruptions in two periods: 1906–1912, 1967–1970 and another, poorly dated, between 1931 and 1956 (Pallàs *et al.* 2001). Only the eruptions between 1967 and 1970 are well described (Baker *et al.* 1975).

Because of the large area of glacier ice on Deception Island, the potential for subglacial eruptions is high. In 1967 and 1970, eruptions occurred from multiple centres in shallow seawater and at onshore ice-free locations; only a single vent was situated beneath glacier ice. By contrast, the 1969 eruption occurred along a series of short en echelon fissures, which were situated beneath the Mount Pond glacier (Fig. 1). The 1969 eruption lasted less than two days. It was associated with a widespread flood of meltwater that severely damaged a British scientific station and caused its permanent abandonment. The eruption also completed the destruction (begun during 1967) of a Chilean station.

From: SMELLIE, J. L. & CHAPMAN, M. G. (eds) 2002. *Volcano–Ice Interactions on Earth and Mars*. Geological Society, London, Special Publications, **202**, 59–79. 0305-8719/02/$15.00

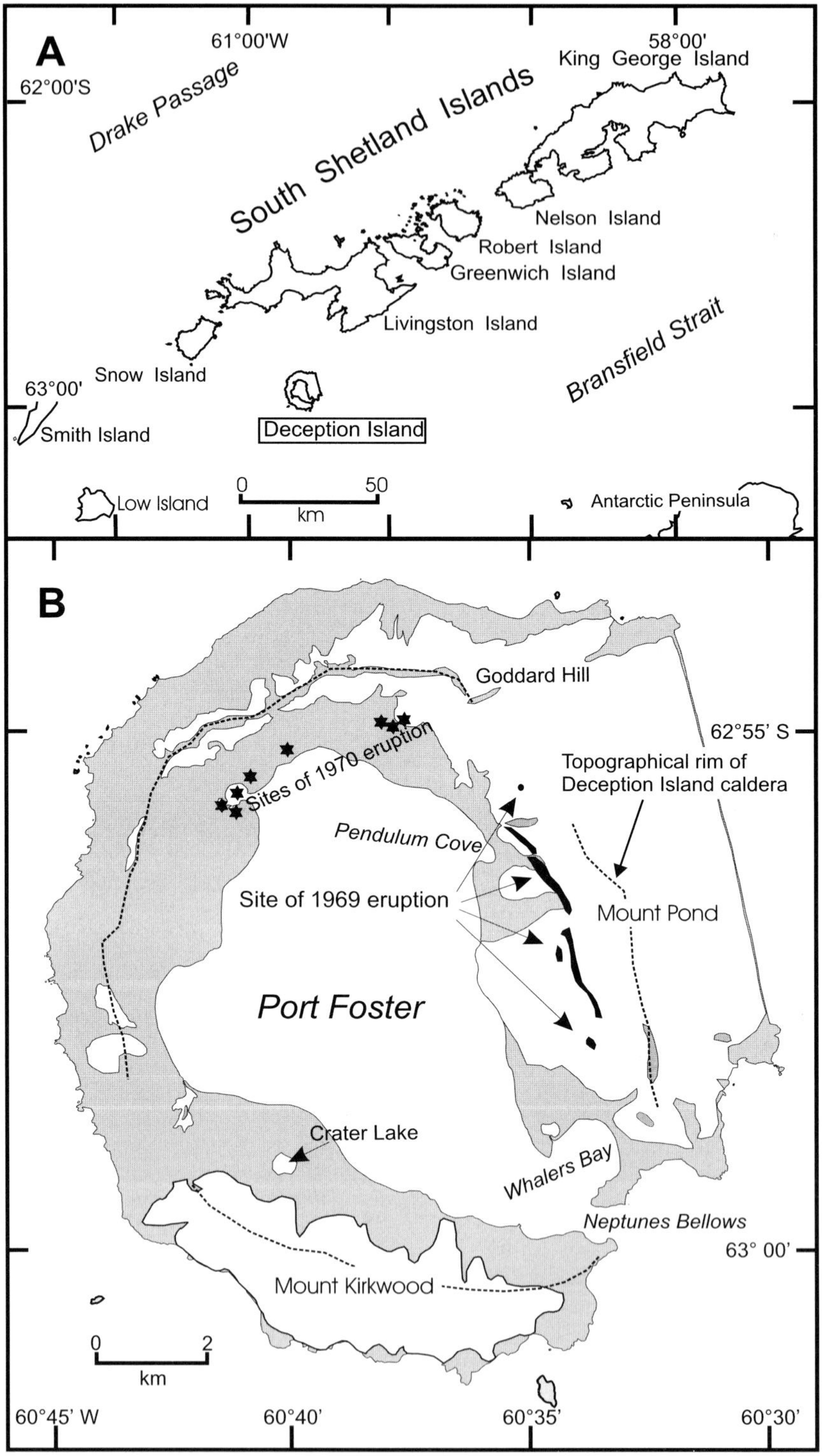

Fig. 1. Maps showing (**a**) the location of Deception Island, and (**b**) localities mentioned in the text, and the distribution of snow-free ground (grey) and glaciers (white) on the island.

Outside of Iceland, subglacial eruptions are not commonly observed or monitored, and the timings and sequences of events, particularly during the initiation and earliest eruption stages, are very poorly known. Five people were present at the British station on Deception Island when the 1969 eruption commenced, and they provided eye-witness accounts of events and a seismographic record. The island was also visited by geologists of a joint British Antarctic Survey–Royal Society expedition only two weeks after the eruption. As a result, an important record of the eruption exists (mainly Baker *et al.* 1975, and British Antarctic Survey archives). The eruption demonstrates several key aspects of subglacial eruptions that are only poorly documented and not well understood. Moreover, it shows features apparently at variance with hydrological predictions based on current models for subglacial eruptions. In this paper, the 1969 eruption on Deception Island is re-examined and re-interpreted, with emphasis on the implications for our broader understanding of hydrological conditions and melting processes. Implications for assessing natural volcanic hazards associated with eruptions beneath thin glaciers are also considered.

Description of the 1969 eruption on Deception Island

The following account is based on descriptions mainly by Baker & Roobol (in Baker *et al.* 1975), unpublished archival information of the British Antarctic Survey and observations of the author between 1987 and 1994. All times quoted are local.

The eruption occurred on 21 February 1969, from a series of short fissures extending about 5 km across the Mount Pond ice cap (Figs 2 & 3). Precursor and accompanying seismic activity was recorded at the British station at Whalers Bay (described later). At 0950, a white steam column was observed in the vicinity of the abandoned Chilean station at Pendulum Cove. After a few minutes, the column darkened and debris was observed falling from it, followed by a deep rumbling and the appearance of another column (otherwise undescribed) about 1 km E of the first one. The two columns merged and were associated with an intense electrical storm. The columns were comparatively low, calculated at only 4.5 km by Ortiz *et al.* (in Mazzuoli *et al.* 1989).

Snow began to fall, accompanied by showers of ash and ice-coated lapilli up to 4 cm in diameter (at Whalers Bay). A widespread flood occurred some time after 1000 but stream discharge around the British base was almost back to normal by 1205. The flood severely damaged several wooden buildings at the British station and helped to destroy the Chilean station (Fig. 4). The flood reached a peak depth exceeding 2.5 m facing Whalers Bay and moved two large tractors several hundred metres from the station to the edge of the bay. The flood also transported much gravelly sediment, numerous large blocks of ice and frozen masses of rock and ash. Parts of the old whaling station were swept into the bay and an elongate fibreglass accommodation building was half filled by debris, comprising small sub-angular blocks of ice which rapidly fused together into a solid mass. The electrical storm ended with the cessation of falling lapilli, but deep rumblings and fine black ashfall continued. The ash particles combined with snow to form a dense coating on walls and windows of the buildings. Evacuation of the five persons took place at 1630, with ash and snow still falling. When the site was revisited briefly on 23 February, under low cloud, ashfall had ceased and no tremors were felt. The eruption was over in less than 48 hours, but its precise duration is unknown.

Extensive examination of the area two weeks later showed that a widespread flood had washed the surface of the glacier below the newly-formed fissures. Abundant blocks of clear glacier ice were scattered over the washed surfaces, and depressions were covered by black and red ash and lapilli (Baker *et al.* 1975, plate IV*d*). Some sediment was deposited by the floodwaters but it was partially buried by later ash and lapilli fall. The initial flood appeared to have run off as a sheet. Later floodwater was concentrated in large steep-sided gullies up to 10 m deep cut in the ice and trending normal to the crevasse direction (Fig. 3c). The northernmost fissure showed no signs of eruptive activity. It contained a substantial volume of meltwater, some of which apparently entered from a portal at its up-slope (southern) end and exited through a prominent deep channel at its northern termination (Figs 3b & 5). Several eruptive sites consisted only of oval ice holes usually surrounded by scattered bombs and lapilli; all showed signs of floodwater that overflowed onto the ice surface (Fig. 6). Several overlapping pyroclastic cones were observed in the two largest open fissures (Figs 2, 3c & 7).

Successive 'strand lines' of debris were preserved on the low-lying ground forming a plain to the west of Ronald Hill. The highest strand line on the northern side of the plain occurred at 4.5 m above the level of the ground to the south (Fig. 8). Shallow kettle holes, up to 7 m in diameter and 2 m deep, mounds of melted-out

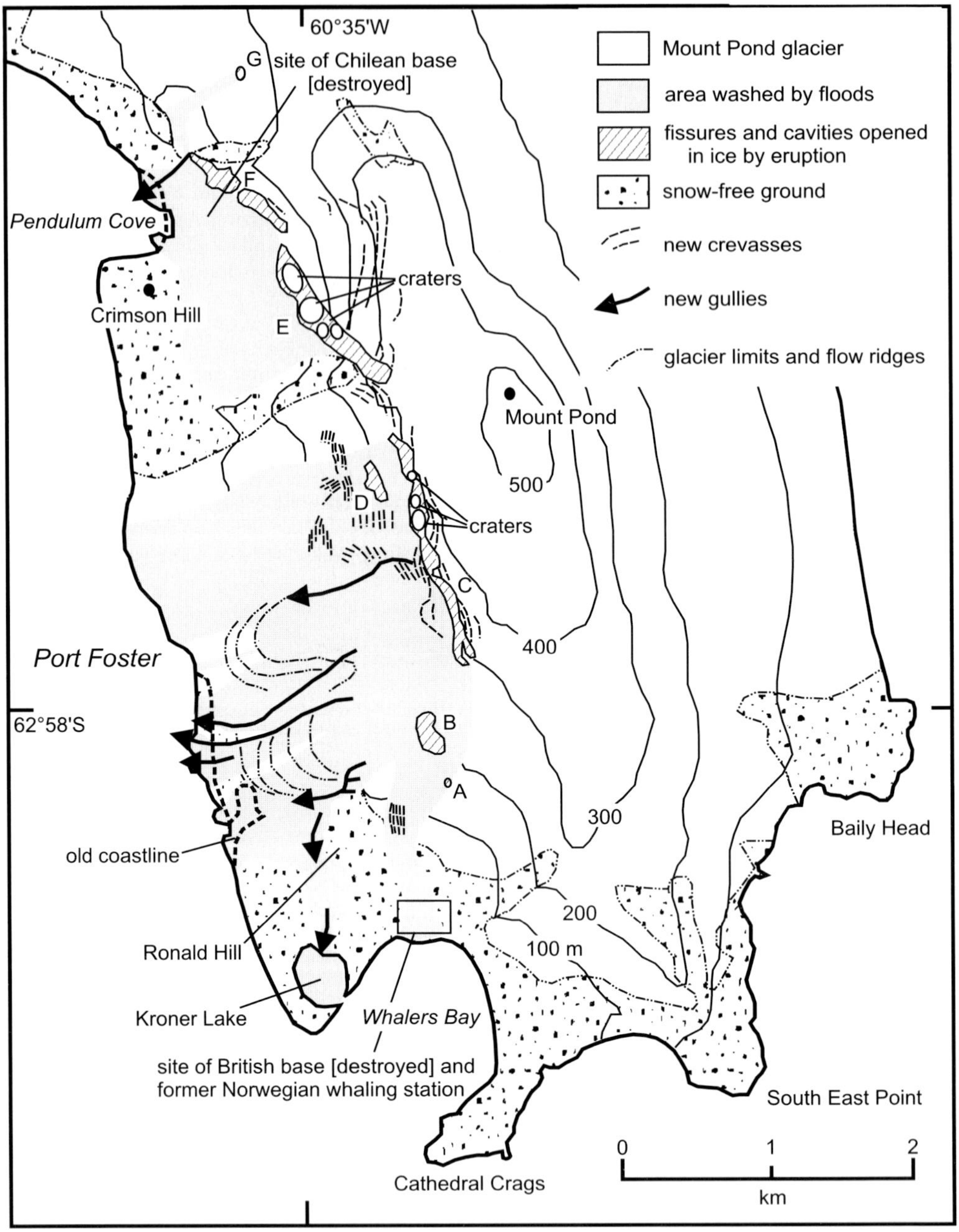

Fig. 2. Sketch map of the Mount Pond glacier showing the positions of fissures and ice chimneys formed during the 1969 eruption. The area washed by floods and associated new crevasses and gullies are also shown. The craters indicated are also the sites of pyroclastic cones. Modified after Baker *et al.* (1975).

morainic debris up to 5 m across, and blocks of lava up to *c.* 1.5 m in diameter, the latter with current-scoured hollows on their up-stream sides and downstream sediment 'tails', were also observed in 1994 strewn, across the southern side of the plain (Baker *et al.* 1975, plate IV*c*). The mounds extended up the slopes of the small hill SW of Ronald Hill, to a height *c.* 15 m above sea level (i.e. about 5 m above the level of the plain). The flood waters temporarily ponded in

(a)

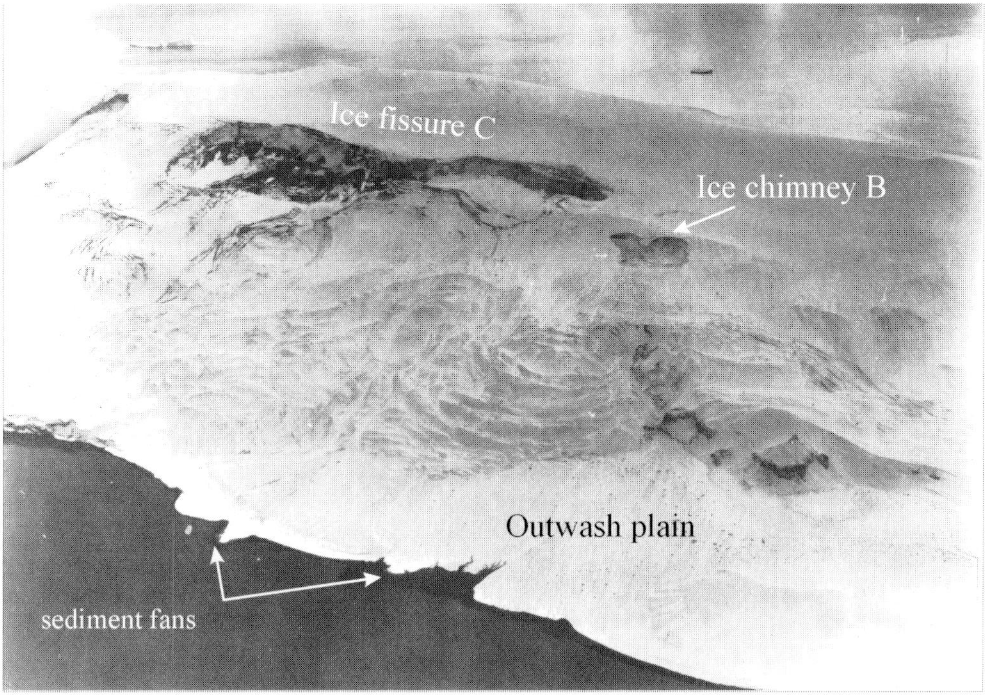

Fig. 3. Aerial photographs of the Mount Pond glacier, taken in March 1969 after a fresh snowfall. (**a**) View of the southern part of the system showing ice chimney B and fissure C. The series of coalesced coastal sediment fans is also prominent, as are numerous large dark permafrost blocks deposited from the meltwater flood on the large outwash plain. (**b**) Close view of fissure C showing dark tephra surrounding three cinder cones. The cones are constructed on the glacier surface and abut vertical ice walls on their eastern side. Several supraglacial channels are also prominent, carved in the glacier surface by meltwater overflowing the fissure, and orientated approximately orthogonal to pre-existing slope-contouring crevasses. (**c**) View of the northern part of the system, showing fissures E and F. There were no eruptions from fissure F whereas fissure E is largely filled by three cinder cones (still steaming), which draped the glacier with extensive dark-coloured tephra. Fissure C is seen in the background. Note the prominent outflow channel carved in glacier ice at the northern (left) end of fissure F and its prominent sediment fan. (**d**) View, looking NNE, over fissure F to ice chimney G. The northerly outflow channel in fissure F is a prominent feature, as is a narrow ice bridge near the centre of the fissure.

a former crater depression SW of Ronald Hill and carved a 200 m-long channel through to Kroner Lake, itself an old tuff ring. The channel measured *c.* 36 m wide and 2 m deep in 1994. The flood waters burst through the narrow tephra promontory that separated Kroner Lake from Whalers Bay, changing the lake to a shallow tidal inlet. Elsewhere, the broad sediment plain west of Ronald Hill was strewn with numerous large stranded ice and permafrost blocks up to 4 m in diameter (pictured in Baker *et al.* 1975, plate IV), and the coastline was extended up to 200 m by a series of coalesced alluvial fans (Fig. 9). Subsequent erosion resulted in almost all of the coastline resuming its pre-eruption position by 1974. The total volume of erupted products was small, estimated at *c.* 0.03 km^3, all pyroclasts. Lava effusion apparently did not occur. The erupted magma was mainly basaltic andesite but ranged up to silica-poor andesite (SiO_2 54–59 wt%; Roobol 1980). Its composition varied systematically along the fissure system, with the least-evolved compositions erupted centrally.

Description of fissures and vents produced in the Mount Pond glacier

The eruptions through the Mount Pond glacier produced two main morphological features (Figs 2 & 3). The most prominent were three gaping ice *fissures*. They were individually 500–1000 m in length, 100–150 m in width, and 70–100 m deep, with vertical to slightly overhanging ice walls. Steam-venting fumaroles were observed in all three, but were particularly vigorous in fissure E. Two groups of three overlapping pyroclastic

(b)

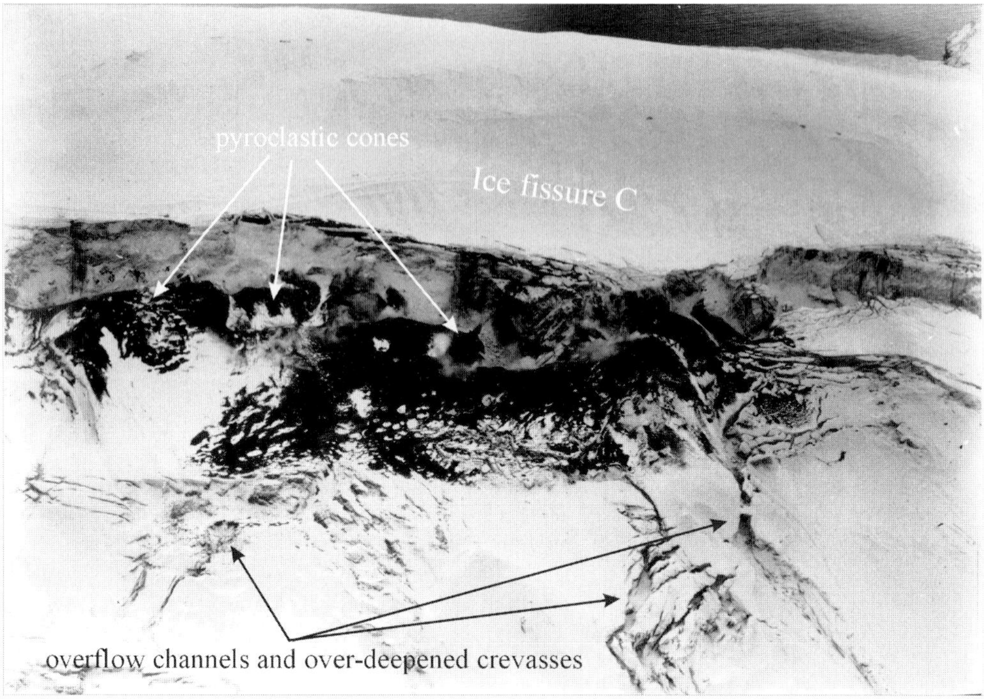

(c)

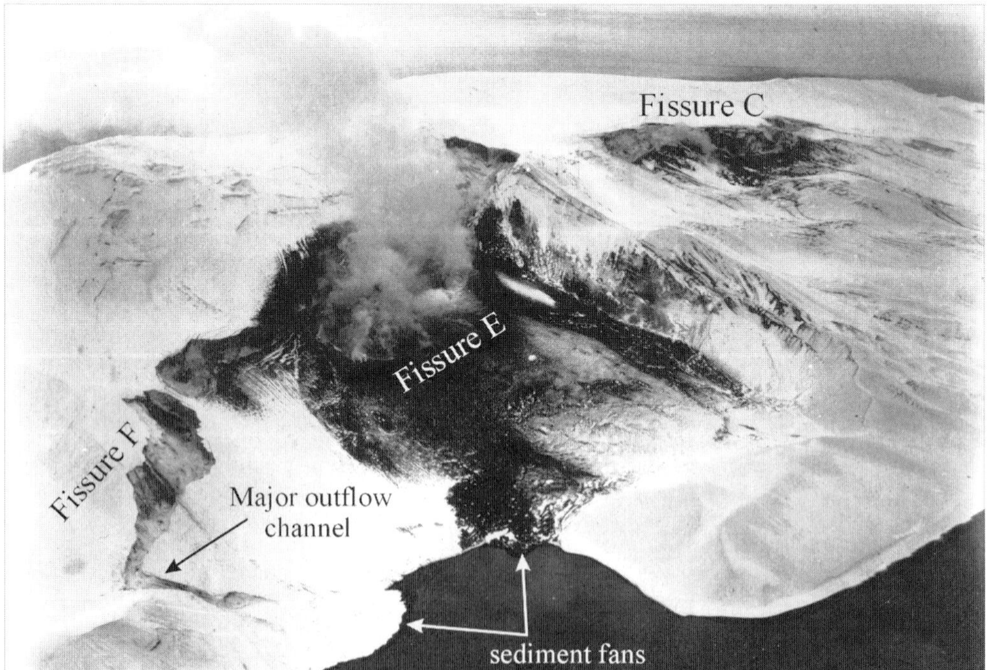

Fig. 3. (*continued*)

(d)

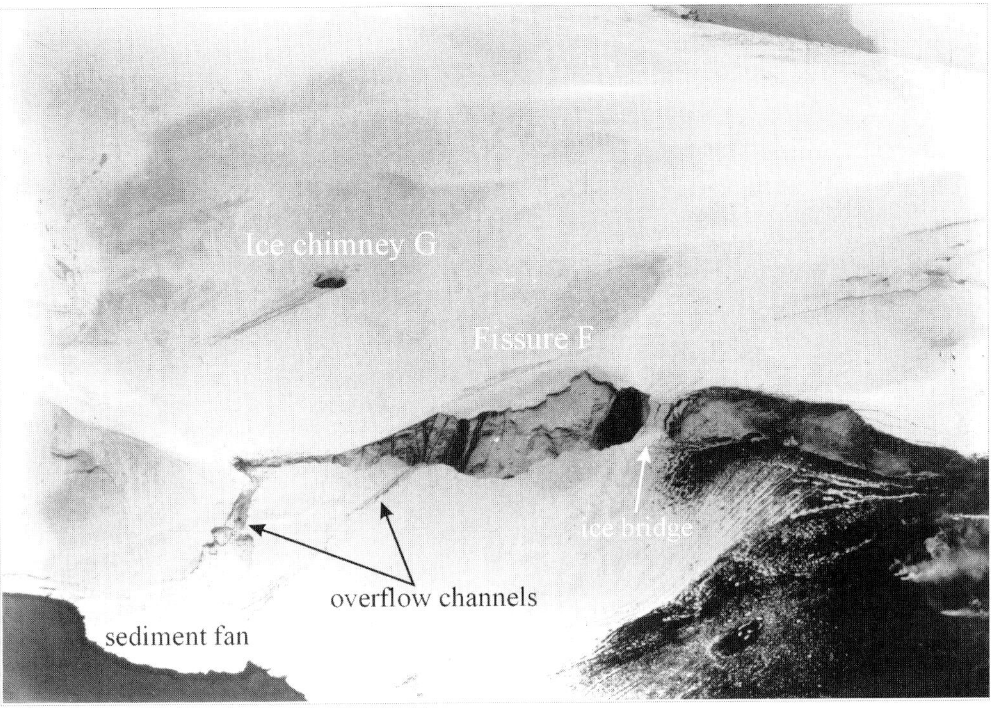

Fig. 3. (*continued*)

Fig. 4. View of the abandoned Chilean station at Pendulum Cove. The station, initially damaged by ash fall in 1967, was inundated by a lahar in 1969 and burnt down by hot lapilli and ash fall. The prominent sediment fan in the background is also a 1969 feature, originating in the outflow channel from fissure F (Figure 3c).

cones were constructed in fissures C and E and glowing lava was observed beneath the pyroclasts in both groups of cones. Crater depths varied from 10 to >100 m. There was no evidence of eruptive activity in fissure F. Four circular to ovoid ice *chimneys* were also present (features A, B, D and G in Fig. 2; Fig. 6), and a small chimney also occurred on the floor of fissure F. They were called vents by Baker *et al.* (1975) but chimney is used here to conform with the descriptive

Fig 5. View looking SE up fissure F. The fissure floor is covered in blocks of ice and a portal is present in the ice cliff at the far end. Steam still issues from the base of the ice cliffs on the right side. The ice cliff at right is about 50 m high.

terminology of Gudmundsson *et al.* (1997). The ice chimneys varied from 1 m (in fissure F) to 230 m in maximum diameter and were bounded by vertical to slightly overhanging ice walls typically 30–40 m high. Ice chimney B (and possibly D) appears to have been a composite of at least two overlapping structures.

The floors of the fissures and chimneys were covered by a jumble of ice blocks (e.g. Fig. 5). A narrow ice bridge spanned the central part of fissure F and the remains of a collapsed ice bridge were present near the middle of fissure C. A cavity described as a cave (probably a tunnel portal) was observed at the base of the ice cliff at the up-slope (southerly) end of fissure F and photographs suggest a similar feature existed at the northern end of compound chimney B. The presence of a small depression in the ice surface about 100 m south of fissure C is consistent with the presence of subglacial melting there, probably in a tunnel connected to fissure C. In addition, a shallow but prominent depression about 10 m across is present about half way up the tephra-strewn glacier east of the destroyed Chilean base, close to a prominent 1969 meltwater overflow channel from fissure E. The depression probably represents a pit formed from subglacial melting during the eruption. Its

Fig. 6. View of ice chimney G showing the distribution of pyroclasts and other debris washed out of the underlying ice cavity. The chimney is a semi-circular hole 50 m in diameter and 40 m deep, its floor covered by ice blocks. The largest bomb, situated 30 m NW of the hole (i.e. slightly down-slope), measured 120 × 90 × 70 cm.

preservation, with little modification, after more than 30 years suggests that the thick blanket of scoria has insulated the underlying ice from significant decay.

Description of the 1969 tephra

Tephra blanket

Baker *et al.* (1975) estimated that *c.* 0.02 km^3 of tephra (excluding near-vent deposits) was ejected during the 1969 eruption. The dominant pyroclasts are dark grey vesicular lapilli, with a small proportion of red vesicular lapilli and scarce accessory fragments of yellow lapilli tuff. Individual lapilli are highly vesicular, angular and mainly blocky-looking, with fracture-bounded surfaces. The deposits are distinguished by very good sorting and comparatively coarse grain size, with modal values of 4–8 mm at every site sampled. Maps of the tephra distribution, thickness variations, maximum clast size and sorting coefficient show a strong N–S geometry indicating control by a northerly wind. However, a thin layer of black ash was deposited during the final stages of the eruption and was distributed more to the SW, marking a late change to north-easterly winds. Most of the northern and parts of the eastern and western sides of the island escaped any tephra fall. Between Pendulum Cove and Whalers Bay, all of the snow-free ground is still draped in a uniform blanket of grey lapilli, which is slowly being reworked by wind and ephemeral streams. However, comparatively fragile contemporaneous features formed in the tephra are still well preserved. They include shallow kettle holes, and fluvial scour hollows and sediment tails associated with outsize blocks carried in the meltwater flood (described above), and low fluvial terraces

Fig. 7. Panorama looking east toward fissure C showing the sub-vertical ice walls on its east side and steaming cones formed of dark scoria. The cones are constructed on glacier ice. Their smooth surfaces indicate that they were unmodified by meltwater activity and that the cavity had drained and permitted essentially dry eruptions. Note the virtual absence of tephra on top of the backing ice cliff, consistent with Hawaiian–Strombolian activity and a low eruption fountain. The ice cliffs are about 30 m high.

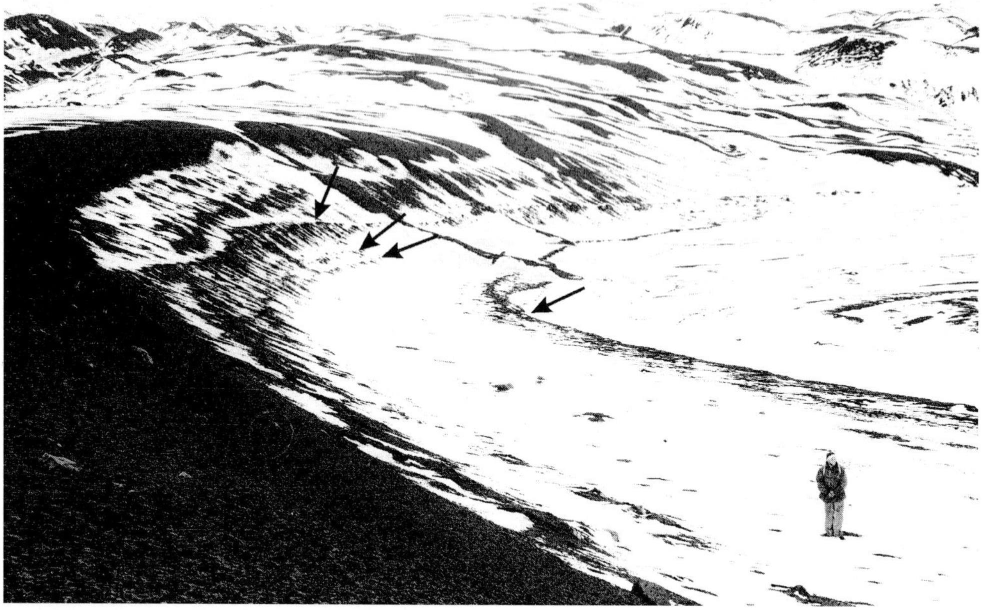

Fig. 8. Prominent strand lines left by the receding meltwater flood on an old degraded tuff cone on the north side of the outwash plain NW of Ronald Hill.

distinguished by conspicuous red lapilli at Whalers Bay, close to the local ice front.

Near-vent deposits

Near-vent deposits from eruptions beneath thin glaciers have not been described previously. Pyroclastic cones were constructed in fissure sections C and E and probably exist beneath all of the ice chimneys (e.g. the summit of a steaming pyroclastic mound had appeared at the site of chimney A by March 1974). Baker *et al.* (1975) estimated that 0.01 km^3 of magma were preserved in the near-vent deposits. The tephra accumulated on the glacier surface, to at least 20–50 m in thickness, down-slope of fissures C and E. Today, the cone-forming deposits are only exposed in fissure E. Although only the northernmost cone is well preserved, parts of

Fig. 9. Panorama showing coalesced sediment fans formed mainly during the period of peak meltwater discharge early in the eruption. The fans were fed by meltwater discharging through prominent channels, of which two prominent examples are seen in the foreground and middle distance carved into pre-eruption tephra deposits. The position of the pre-eruption coastline is also evident (arrow). The 1969 fans extended the coastline by up to 200 m but they had disappeared by erosion by 1974.

the central (largest) cone are still present, although much distorted by ice deformation. Elsewhere on Mount Pond, the ice fissures and chimneys have closed by ice flow since the eruption, with the exception of fissure F and the southernmost end of fissure E. From the contemporary accounts, craters within the southern fissure (C) were well formed on their west sides (where they overlapped the glacier ice), but abutted high vertical walls of ice to the east (Figs 3c & 7). Scoria deposits are generally thin to absent on topographically much higher glacier surfaces on the east (up-slope) sides of fissures.

In fissure E the cone-forming scoria deposits mantle old, slowly decaying glacier ice. They mainly comprise dark grey lapilli, with rare accidental fragments of yellow lapilli-tuff. Scoria size increases upward on the western slopes, from *c.* 1–2 cm east of the former Chilean station to 5–15 cm at the crater rim. Bombs 1–2 m across are also scattered over the surface and increase in size and abundance upslope. They vary from flattened cowpat shapes to spindle and ribbon bombs. Many have highly inflated cores and breadcrust textures are common. Baker *et al.* (1975) also described blocks of dense brownish lava around cones in fissure E. The topographically highest exposures, on the east side of the inner wall of the northernmost crater, are formed of unstratified bomb and scoria deposits, dark grey to base and bright red centrally, becoming brown to top. The deposits are dominated by weakly-welded, twisted and cowpat-shaped bombs, 5–30 cm in diameter, that are moderate to highly micro-vesicular. Many have chilled breadcrusted surfaces. Fine (ash-size) matrix is absent but a small proportion of poorly vesicular lapilli (0.5–2 cm) is present between bombs. Similar massive deposits are exposed nearby to the south and probably formed on the inner crater slopes of the central (largest) cone. They are mainly composed of strongly reddened rag-like and spinose scoria, and sparse bombs up to 1.6 m, some breadcrusted. Agglutinate-like masses of abundant weakly welded cowpat bombs are also conspicuous. Topographically lower exposures, probably related to the outer western flank of the central cone, comprise scoria, grey in the upper 0.5–1 m, bright red below. Those deposits are friable, mainly formed of bombs up to 40 cm across and finely vesicular coarse scoria (4–5 cm, ranging up to 15 cm). The scoria are equant and usually have relatively smooth (fluidal), chilled breadcrust-textured surfaces. Beds are very crudely stratified, with lenses of red fine lapilli 1–2 m in length and 5–15 cm thick. The lapilli in the lenses are blocky and micro-vesicular, with fracture-bounded surfaces. Upper parts of the deposit contain large ribbon bombs (to 2 m) dispersed in equant to oblate, spinose vesicular lapilli.

Interpretation of the 1969 tephra

The absence of fines, grading characteristics (sorting, grain-size variations), predominant clast shapes and vesicularity, and the abundance of bombs are all typical of weakly explosive magmatic eruptions, with the red coloration signifying oxidation under essentially dry conditions during rapid near-vent accumulation. There is no evidence for a significant involvement of external water during the eruption, although the earliest-formed tephra layers have never been described. The variable and sometimes comparatively low vesicularity, which was probably insufficient for frothing and disruption by magmatic volatiles, is typical of Strombolian eruptions of magmas with low viscosities and eruption rates (Houghton & Wilson 1989; Parfitt & Wilson 1995). The presence of blocks of dense (degassed) lava and bomb-rich agglutinate suggest the possibility of entrained recycled lava, possibly from ephemeral lava pond(s) (cf. Wilson *et al.* 1995). These interpretations are consistent with the low plume elevations, even though it is likely to have been enhanced by ingestion and heating of water-saturated air (cf. Sparks *et al.* 1997, p. 275), and the near-absence of scoria on the up-slope sides of fissures. The latter surfaces are about 70–100 m above the fissure floors, suggesting low tephra fountains characteristic of Hawaiian/Strombolian activity (Sparks *et al.* 1997, p. 266). Conversely, a limited influence of external water is indicated in the cone-flank deposits by the common breadcrust textures of many bombs and lapilli, and by layers of equant lapilli with broken surfaces. Similar lapilli, with broken surfaces, are also predominant in the tephra blanket. There, it may be a primary feature preserved during deposition or, alternatively, it may have been caused by impact with the substrate and/or later periglacial freeze–thaw processes.

Seismicity

Seismic signals associated with volcanic activity have a wide range of signatures. Empirically, the seismicity can be classified into two main families of signals based on the physics of the source process, i.e. differentiating between solid-state processes and those originating in fluids (Kumagai & Chouet 2000): (1) volcano-tectonic earthquakes, which represent the brittle response in the volcanic edifice associated with fluid movement (e.g. tensile or shear failure during faulting), and (2) tremor and long-period events, which are an oscillatory response to a steady non-linear process (e.g. variable liquid or gas pressures; e.g. Kieffer 1984; Julian 1994; Chouet 1996; Correig *et al.* 1997). Liquid water and steam are both important in the source processes that generate tremor and long-period events, and fluid effects (for example, flow through a channel with a series of constrictions and deformable walls) can cause pressure fluctuations (resonations) due to unsteady mass transport (Julian 1994; Chouet 1996). The seismicity can thus be viewed as a direct window into the dynamics of volcanic fluids (Kumagai & Chouet 2000).

The Deception Island volcano has a high heat flow within the caldera, corresponding to a temperature gradient of 360°C km^{-1} away from fumarolic areas (Orheim 1972*a*, *b*). It is also an island partly covered by glaciers and with a large caldera flooded with seawater. It is likely that both seawater and freshwater infiltrate the volcanic edifice and continuously replenish aquifers on the island (Vila *et al.* 1992). For example, the incidence of tremor is roughly correlated with days on which there was rapid snow melt (Almendros *et al.* 1997; Villegas *et al.* 1997). Corrieg *et al.* (1997) related tremor to vaporization of groundwater at depths of 2–3 km. This is close to likely magma chamber depths, estimated as *c.* 3–4 km by Orheim (1972*a*) by extrapolation of geothermal gradients. During 'normal' (quiescent) periods, the rate of energy release, averaged over a few days, is roughly constant and the stress drop is also very low, characteristics inconsistent with a solely tectonic origin (Vila *et al.* 1992). Thus, the volcano has been interpreted as a self-regulating system characterized by seawater and freshwater influxes along pores, fractures and faults, and thermal interaction with magmatic heat at depth. The interaction is envisaged as causing rapid phase changes (water flashing to steam) and generating pressure waves as a major source of tremor (Correig *et al.* 1997).

Geophysical investigations since 1987 have indicated that seismic activity on Deception Island is characterized by volcano-tectonic earthquakes, long-period and hybrid events, and tremor (Vila *et al.* 1992; Almendros *et al.* 1997). The 1969 seismogram and contemporary observations suggest that tremor and volcano-tectonic earthquakes were both involved in the activity (below), although the harmonic signature of very small volcano-tectonic earthquakes is not always separable in the comparatively crude record. Moreover, long-period and hybrid events (collectively termed LP/H events here) and tremor cannot be separately identified on the 1969 seismogram. The events identified here as tremor probably include LP/H events too; they are grouped together for descriptive purposes. Their distinction is probably unimportant as they all have similar spectra and probably share closely related origins (Julian 1994; Chouet 1996; Almendros *et al.* 1997; Kumagai & Chouet 2000).

Sequence of seismic events recorded at the British station in 1969

Uncommon and weak earthquake activity was felt across the island during January and early February. A particularly vigorous earthquake was experienced on 21 February at 0334 hours local time at the British scientific station. It was followed by tremor activity and a few much smaller earthquakes but, after 0750, the 'background' of tremor became essentially constant (Baker *et al.* 1975, and British Antarctic Survey unpublished). At 0832, a noticeable earthquake signalled the start of a period of multiple small earthquakes and tremor. The earthquakes peaked in intensity and magnitude at about 0910. Between 0910 and 0915, earthquake amplitude reduced but the frequency of occurrence increased. The earthquakes ceased abruptly at 0915, but continuous tremor activity persisted. The appearance of the white eruption column at 0950 coincided with another small earthquake recorded on the seismogram and the initiation of the second column was possibly associated with a second noticeable earthquake, at 0955. Thereafter, the seismogram shows only continuous tremor that are particularly strong between about 1005 and 1032. The seismometer apparently malfunctioned after about that time.

Interpretation

Background seismicity on Deception Island is seldom felt by people on the island. Therefore, the earthquakes experienced during January, although infrequent, were unusual and were a good indicator of volcanic unrest corresponding to precursory activity. The earthquakes may have been caused by incremental adjustments of the volcanic edifice accommodating increased magmatic pressures along local shear-planes, but they were insufficient to cause major crustal fracturing. It is inferred that the earthquake sources were probably located in the brittle rock surrounding the shallow magma reservoir. However, the earthquake at 0340 was far larger than any other experienced during January and February. It may represent an important crustal failure corresponding to the initiation of the fissure system, which ultimately erupted. At that time, the magma chamber is envisaged having become essentially 'unlocked'. Volatiles expelled ahead of the magma (see later) established an intimate connection with the shallow low-pressure hydrothermal system, causing groundwater to flash to steam and enhancing the tremor activity. The evolution to sustained tremor after 0750 hours may indicate unsteady, partially choked flow of superheated magmatic and hydrothermal steam and/or liquids driven by the pressure gradient between the magma chamber and the surface.

The increase in frequency and amplitude of volcano-tectonic earthquakes after 0832 was possibly due to intermittent dyke-driven crack propagation as the crack reached shallower depths (cf. Einarsson & Bransdottir 1984). The sudden disappearance of earthquakes at 0915, and their replacement by tremors only, may represent the time when the propagating dyke fracture reached the rock surface (cf. Julian 1994). Melting at the glacier base by superheated steam would have created a mechanical boundary (rock overlain by cavity or meltwater) preventing the fissure propagating directly into the ice. Eruption beneath the glacier presumably commenced at about that time (0915), assuming that the lag between the fracture front and magma was small. However, the glacier surface was not penetrated until 0950, when the first eruption column was first observed. Its white coloration indicates that it was a steam column. Transformation to an explosive, tephra-forming eruption did not occur until a few minutes later. A second column (at a different location) joined the first at about 0955. The strong tremor between about 1005 and 1032 was interpreted by Baker *et al.* (1975) as probably caused by meltwater-generated lahars entering the British station, in which the seismometer was installed.

Rates of magma rise, ice melting and peak meltwater discharge

Although the sequence of events interpreted above is not unique, it provides a basis for calculating qualitatively the magma rise rates and ice melting times for the eruption. The results are probably only order-of-magnitude correct. Assuming that magma began to rise at 0340 from a depth of *c.* 3–4 km, and reached the base of the glacier at about 0915, mean rise rates of 0.15–0.2 m s^{-1} are indicated. These are slow compared to mean rise rates calculated for basalt magma injected in the Krafla fissure eruption in Iceland by Einarsson & Bransdottir (1984; 0.4–0.5 m s^{-1}) but may reflect small differences in melt viscosity (basaltic andesite–andesite magma on Deception Island). Höskuldsson & Sparks (1997) used 1 m s^{-1} as a 'typical basalt ascent velocity', whereas more evolved magmas rise more slowly (typically 0.2–0.6 m s^{-1} for explosive eruptions; Rutherford & Gardner 2000). There is an elevation difference of *c.* 300 m between the surface expressions of fissures E (first eruption column) and C (second column), and first sightings of the columns were separated by five minutes. If the magma rose as a uniform dyke with a coincident upper surface between the fissures, the different eruption times for the two fissures implies a magma rise rate for fissure C of *c.* 1 m s^{-1}. This much faster rise rate implies either that (1) the leading edge of the dyke was not horizontal between the fissures, or (2) magma rose at different rates at different places.

Correlating the cessation of volcanic earthquakes with the arrival of the magma-driven propagating fissure at the base of the glacier enables the ice melting-rate to be calculated. In the fissures, the glacier has a mean thickness of about 70 m (range *c.* 50–100 m). However, crevasses are present and probably extend to at least 20–30 m (pinching out by creep deformation; Glen 1954), thus reducing to *c.* 40 m the likely thickness of unfractured impermeable ice. If melting began at *c.* 0915 (an unrealistic assumption; see later) and the glacier was penetrated by 0950, at least 35 minutes elapsed before 40–50 m of ice were penetrated and steam was able to vent to the atmosphere, i.e. 0.02 m s^{-1}. This is more than three times faster than the 1996 eruption at Gjálp (Iceland), in which 500 m of glacier ice was penetrated in about 30 hours (Gudmundsson *et al.* 1997), yielding a melt rate of *c.* 0.005 m s^{-1}. The rates for melting the glacier at Gjálp are comparable with the rate of deepening of the surface depression by subglacial melting during the eruption (50 m in 4 hours (Einarsson *et al.* 1997), equivalent to a 'melt rate' (actually ice collapse time) of *c.* 0.004 m s^{-1}).

From eyewitness accounts, runoff at the British base was at normal levels at 0945 hours, when the base was evacuated. However, the seismic record suggests that floodwater and lahars entered the base at 1016. It had certainly passed by before 1100, when large oil tanks from the whaling station were observed swept down to the beach, and had almost completely abated when the base area was revisited at 1205. It is unlikely that overflow had already commenced prior to the observation of the first eruption column at 0950, since the column signified the time at which melting penetrated through the glacier. The maximum period of flooding is therefore constrained to between 0950 and 1205. The distance from the nearest eruptive site (chimney B) to the British base is about 1100 m implying that the flood travelled at a minimum speed of $2.6\,\mathrm{km\,h^{-1}}$. It is assumed that the timing of the flooding was similar along the length of the fissure, which is probably good to a first approximation. Calculations using the volume of ice melted (*c.* $76 \times 10^6\,\mathrm{m^3}$ of ice), and assuming only about 70% was discharged during the first two hours (allowing for blocks of unmelted ice observed, and slow subsequent dissipation of heat at the vents), yields a minimum mean discharge rate of $6.6 \times 10^3\,\mathrm{m^3\,s^{-1}}$. For comparison, the discharge is similar to that which occurred during the first two hours of the 1996 Gjálp eruption, but it is much smaller than occurred later in that eruption, which peaked at about $45\,000\,\mathrm{m^3\,s^{-1}}$ (Einarsson *et al.* 1997).

Fluid dynamics of the 1969 subglacial eruption

Hydrology of subglacial eruptions – published models

Glacier hydrology, thickness and structure exert a profound influence on the sequence of events and the volcanic units formed during subglacial basaltic eruptions beneath temperate glaciers (Björnsson 1988; Smellie 2000, 2001*b*). Their influence is illustrated by recent investigations of subglacial volcanic sequences (Smellie *et al.* 1993; Skilling 1994; Smellie & Skilling 1994; Smellie 2001*b*). The products of eruptions beneath temperate ice can be divided into three sequence 'types', depending on original thickness and structure of the overlying glaciers. They comprise sequences formed from eruptions beneath 'thin' glaciers (<*c.* 150–200 m) and two types characteristic of 'thick' glaciers (i.e. ≫200 m).

Empirically, thin glaciers are formed mainly of snow, firn and fractured ice. They are permeable and any meltwater will drain away continuously beneath the glacier during an eruption. By contrast, water formed by melting beneath much thicker glaciers, formed predominantly of impermeable ice, accumulates within an englacial vault or lake overlying the erupting vent. The evolution of the system depends then on the glacier structure. In glaciers with a thin permeable upper layer, the surrounding ice barrier will be floated before the water level reaches the permeable layer and will be discharged subglacially, typically in a catastrophic flood known as a jokulhlaup. Conversely, in glaciers where firn, snow and fractured ice exceed 10–25% of the thickness of underlying unfractured ice, water will accumulate up to the base level of the permeable layer, and then flow out englacially (through the permeable layer). If meltwater production exceeds loss by diffusion through the permeable layer, then supraglacial flow will develop.

Based on these models, the Mount Pond glacier, <100 m thick, is a thin glacier and, empirically, any eruption should be characterized by continuous subglacial drainage. However, the 1969 eruption was associated with sudden, widespread and voluminous supraglacial flooding. Supraglacial flooding is not a feature of the models of Björnsson (1988) but it is a predicted consequence of eruptions beneath much thicker glaciers with a permeable upper layer (Smellie 2000, 2001*b*). However, there is no published description of a historical eruption within thick ice involving such widespread supraglacial flooding. In the only other documented eruption displaying an overflowing englacial vault (Gjálp, Iceland, 1996; Einarsson *et al.* 1997; Gudmundsson *et al.* 1997), the floodwaters were confined, by glacial surface topography, as a focused overflow which cut a narrow supraglacial channel into a glacier 400–600 m thick. No eruptions involving such flooding have been observed associated with a 'thin' glacier. The features of the 1969 eruption on Deception Island are thus apparently contradictory.

Structure and thermal regime of the Mount Pond glacier

Deception Island experiences a polar maritime climate and the Mount Pond glacier has a temperate thermal regime (Orheim 1972*a*, *b*; Orheim pers. comm.). The glacier is about 100 m thick, diminishing to about 50 m at the glacier snouts, with a mean equilibrium line at *c.* 200 m above sea level. Most of the fissures and ice chimneys occur at elevations between 200 and 400 m. A maximum thickness of *c.* 17 m of firn and snow occurs on the glacier at 400 m elevation, whereas below 200 m it is probably formed only of ice. Open crevasses also provide a permeable layer in the ice, probably extending to depths of *c.* 20–30 m (estimated). These observations yield an overall layered structure for the glacier at the fissures, comprising *c.* 70 m of impermeable ice overlain by 30 m of fractured ice, firn and snow.

Thus, the Mount Pond glacier does not correspond closely to the thin permeable glaciers envisaged by Smellie (2000, 2001*b*), which are permeable throughout, and is more typical of the layered 'thick' glaciers described by that author.

Fluid dynamics and thermodynamics of the 1969 eruption: a major role for volatiles?

Widespread surface discharge was an important characteristic of the 1969 eruption. The glacier was not floated, and meltwater overflowed along the length of the fissure. This suggests that (1) the rates of meltwater accumulation were very rapid and exceeded meltwater discharge via subglacial leakage, and (2) despite the presence of a substantial upper permeable layer (including fractured ice), discharge via the saturated permeable layer was slow and meltwater was forced to flow over the glacier surface. The absence of eroded portals at the glacier termini also suggests that subglacial meltwater discharge was comparatively small.

Open fractures were also a major conduit for supraglacial drainage (Fig. 3c). Many fractures became deeper and wider, and the flood was violent enough to rip up chunks of ice and dump them at lower elevations. Flood depths reached 17.5 m in deeper channels on the glacier, and were *c.* 4–5 m on the low-lying sediment plains to the west. Initially, the flood was broadly sheet-like because the overflowing fissures followed contours of the surface slope of the glacier, and they rapidly filled with water. By contrast, the Gjálp eruptive fissure was aligned approximately down-slope, leading to confined flow in a single narrow supraglacial channel (cf. Einarsson *et al.* 1997; Gudmundsson *et al.* 1997). As the runoff diminished on Deception Island, the water became focused in the crevasses and other supraglacial channels, which had been over-deepened probably by a combination of thermal erosion by meltwater, and friction caused by collisions of entrained rock and ice fragments. Thereafter, meltwater discharge became substantially reduced because most of the glacier overlying the vents (the major source of the meltwater) had been removed. Subglacial drainage became dominant and the eruptions were essentially dry, forming cinder cones that built up through the cavity and overlapped onto the glacier surface.

It was suggested above that the rising basaltic andesite magma beneath Mount Pond was preceded by a front of magmatic and hydrothermal volatiles (probably mainly superheated steam). This is only likely for such less viscous magmas with slow rise speeds, where the likelihood of volatile–magma separation and two-phase flow in the conduit or upper magma chamber is enhanced (Sparks *et al.* 1997), and further steam would be created when the magmatic gases interacted with aquifers. Gas temperatures are likely to have rapidly reached several hundred degrees centigrade in fumarolic vents at the bedrock surface, thus initiating melting at the base of the glacier prior to the arrival of the magma. For example, gas temperatures in the fissures exceeded 250°C 10 months after the eruption (Orheim 1970). Gas temperatures of several hundred °C at the vents correspond to minimum gas pressures of several MPa, easily enough to lift briefly the thin Mount Pond glacier, probably as a broad very low dome because of the high elastic stiffness of ice (Fig. 10). The gases would rapidly expand and decompress to basal ice pressures along the narrow gap opened up by cantilevering along the wet glacier base (a process anticipated by Nye 1976). The early-formed meltwater would also drain away from the vent, down-slope as a thin sheet, and exploiting any pre-existing channels cut in the ice or bedrock surface. The rapid removal of early meltwater from over the vent site would be favoured by the comparatively steep bedrock gradient beneath the Mount Pond glacier and may help to keep the vault roof 'dry'. In eruptions through horizontal bedrock, early meltwater may accumulate over the vent but it will still be able to migrate laterally along the ice–bedrock gap and will initially only be a narrow film unable to cool the gases significantly (see below).

The duration of the gas-driven melting period is uncertain but the seismic record suggests that abundant gases could have reached the bedrock surface in the 1969 fissures after 0750. That is the time at which tremor became constant and is interpreted as an indication of unsteady gas streaming in fractures (see above); magma probably did not reach the bedrock surface until shortly after 0915. Thus, a maximum period for gas-driven melting of *c.* 90 minutes is possible. In that time, the combination of melting and pressure-driven lifting would create a vault in the overlying glacier.

Published thermodynamic calculations for subglacial eruptions have shown that there is sufficient heat in conductively cooling lava to melt space in the overlying glacier (Allen 1980; Höskuldsson & Sparks 1997). However, if meltwater is unable to escape from the overlying vault, there is only just sufficient heat available since, although space is created by melting ice, mass

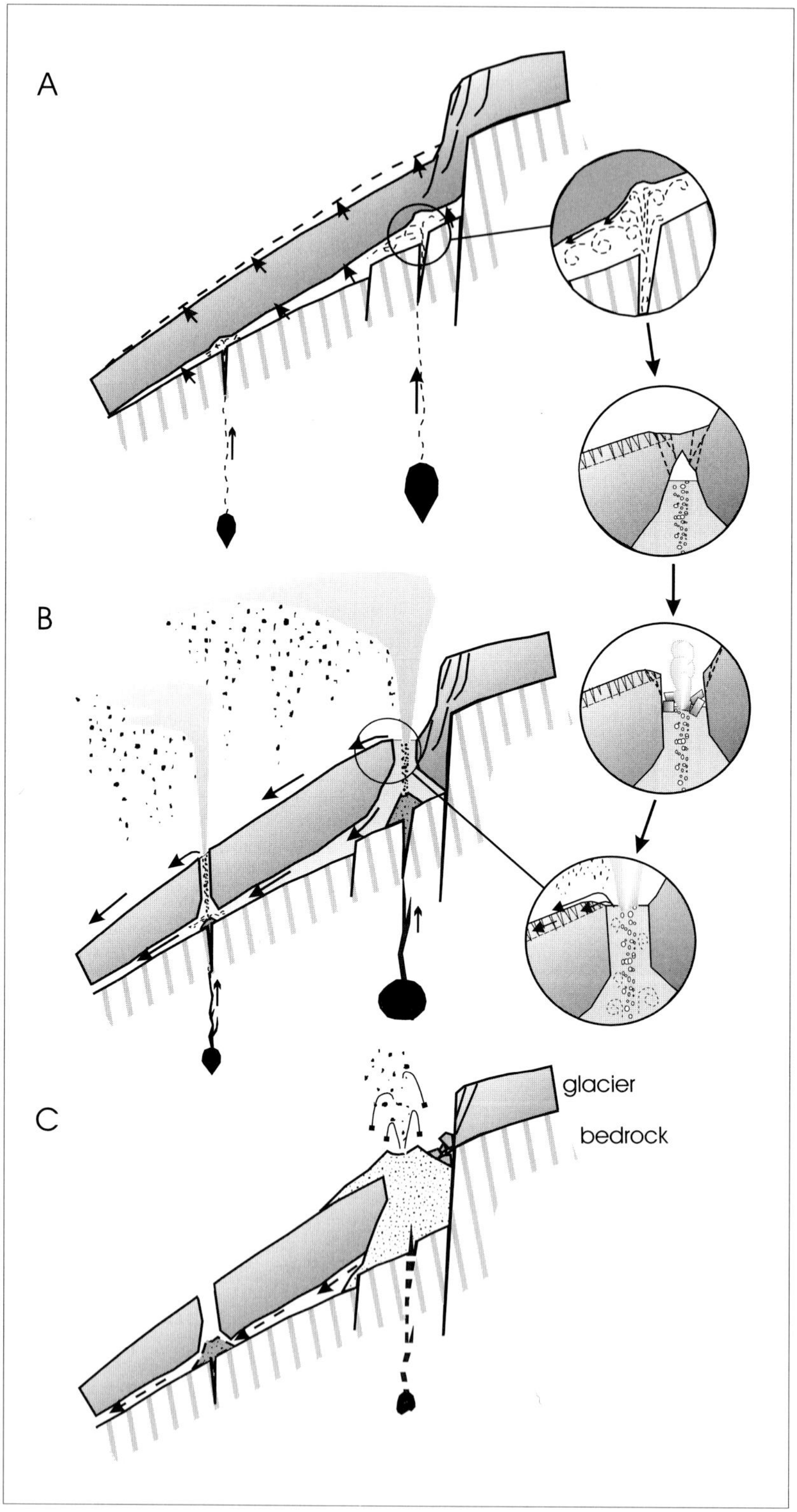
A
B
C
glacier
bedrock

(magma) is also being added to the vault. The published calculations also rely on the magma cooling to ambient temperatures (*c.* 0°C), which is unlikely (both problems acknowledged by those authors). The calculated ice-melting rates are also much slower than the very rapid melting rates which occur during observed eruptions, which are at least an order of magnitude faster (Gudmundsson *et al.* 1997, Smellie 2000). For example, Höskuldsson & Sparks (1997) suggested that, at the fastest rates of melting (for basaltic eruptions), it should take about 17 days to penetrate 500 m of ice. In the Gjálp eruption, it took just 30 hours. To increase the melting efficiency, Höskuldsson & Sparks (1997) considered laminar-flow convection in a water-filled vault, whereas Gudmundsson *et al.* (1997) appealed to highly turbulent convection of a mixture of quenched ash and meltwater. However, the latter implies vent explosivity during the earliest stages, which is unlikely to be a general case since many subglacial eruptions are initially effusive (pillow lava). They only become explosive later, once the cavity has penetrated the glacier surface, by which time the period of rapid melting is essentially complete. Observations of Surtseyan eruptions (analogous to subglacial eruptions in a water-filled cavity; Smellie & Hole 1997) suggest that many do not become highly explosive until the vent reaches very shallow depths (<10 m; cf. Kokelaar & Durant 1983; McInnes 2000).

The shape of the ice cavity is controlled by the difference in the melting rates of the roof and walls (Höskuldsson & Sparks 1997). In their analogue experiments, Höskuldsson & Sparks (1997) demonstrated that roof melting rates were approximately an order of magnitude slower than for the walls, partly as a consequence of cold air bubbles released by the melting ice becoming concentrated at the roof and thus forming an insulating layer. In experiments lacking an air-filled roof layer, a conical cavity resulted due to enhanced melting by hot fluid rising above the vent. The published thermodynamic models depend on heat fluxes based on conductive (magma) and convective (meltwater) heat transfer (see also Allen 1980). The importance of convective heat transfer by volatiles has not been considered, but there are clear indications that gas-driven processes are likely to result in a much faster ice-melting rate:

(1) Advection is much faster in gases than liquids (meltwater), hence heat transfer by gases is much faster than conduction from magma coupled with convection of surrounding meltwater in the cavity (i.e. the model of Höskuldsson & Sparks 1997).
(2) The heat capacity of steam-dominated gas is at least as high as an equivalent mass of basalt magma.
(3) Mass for mass, superheated steam can melt a similar volume of ice to that melted by conduction of heat from basalt magma chilled (unrealistically) to 0°C. For example, 1 kg of steam at 500°C can melt 9.5 kg of ice at 0°C; the mass of ice melted reduces if that ice is initially much colder (e.g. 8 kg of ice with a temperature of minus 30°C). However, there are major problems in extracting all of the heat from the basalt and the rate of heat exchange to melt ice in observed eruptions is apparently too high for simple conductive heat transfer from basalt (see above).
(4) Escaping gases will mechanically stir any meltwater overlying a vent, causing turbulent convection and efficient heat transfer from gas to water. The escaping gases will form rapidly expanding cooling bubbles. The internal pressure in many bubbles will probably drop below the hydrostatic pressure in the vault because of the inertia of the surrounding moving water carrying the bubbles through the equilibrium position. The bubbles will then partially collapse, and the process will repeat until the bubble either equilibrates, ponds against an obstruction or exits at the water surface.

Fig. 10. Series of cartoons illustrating the interpreted eruptive history at Mount Pond in 1969. The sections are orientated approximately east–west and traverse a typical fissure and ice chimney, corresponding approximately to features C and D in Figure 2. (**a**) Between *c.* 0750 and 0915 hours. The glacier is shown being lofted briefly by pressurized hot gases streaming ahead of advancing magma, and subglacial melting is initiated and focused particularly above each vent. (**b**) Circa 0950–1205 hours. Cavities above both vents and the subglacial space beneath the lofted glacier are filled with meltwater, which overflows supraglacially, through and (particularly) over the permeable upper layer of firn, snow and crevasses. Contemporaneous accounts suggest that explosive phreatomagmatic activity occurred during this early phase but no deposits were recovered. (**c**) Late afternoon on 21 February. The closing stages of the eruption are illustrated, with activity ceased at the ice chimney, but continuing to construct a supraglacial cinder cone at the fissure. Only minor meltwater is now present, draining subglacially, and eruptions are essentially dry (Hawaiian–Strombolian). The series of insets shows the inferred morphological evolution of the englacial cavity above the vents (see text for explanation). Not to scale.

At the excess bubble pressures envisaged and in shallow vaults like those in the Deception Island eruption, it is unlikely that most bubbles (a mixture of steam and juvenile gases) will condense completely and a layer of gases will accumulate at the vault roof. Oscillating bubbles will probably generate pressure waves similar to those formed by airguns used in seismic experiments (e.g. Ziolkowski *et al.* 1982). Those, in turn, may help to destabilize an unsupported vault roof (see below), although the magnitude of the pressure waves is unknown for the subglacial situation.

(5) Unlike heat being transferred from magma to ice via water, meltwater from ice melted by gases will form a transient film at the cavity roof. The film will run down-slope away from the melting site, continuously presenting an essentially 'dry' ice surface to the gases ready for further melting. Melting thus proceeds faster as an intervening water film does not need heating each time. Empirical calculations suggest that generally twice as much ice is melted in a 'dry' cavity compared with one that is water-filled.

(6) The gases will form a warm layer at the cavity roof. Its presence should lead to enhanced roof melting, with a conical or possibly cylindrical shape due to focusing of the hottest temperatures directly above the vent (Fig. 10). Like the Deception 1969 eruption, cylindrical ice-chimneys were observed during the 1996 Gjálp eruption in Iceland (Gudmundsson *et al.* 1997). They were envisaged forming by piston-like collapse along ring fractures into the underlying water-filled vault. Both sets of processes may have occurred in both of these eruptions. The rapid advance of a narrow conical or cylindrical vault roof by gas-driven melting, possibly associated with thermal softening, and bubble-generated seismic pressure waves would promote mechanical instability when the melting roof approaches to within a few tens of metres of the glacier surface. At that elevation, the overlying glacier is formed predominantly of fractured ice or firn/snow lacking mechanical strength and it will readily collapse along cylindrical fractures (Fig. 10).

It is therefore suggested that gas-driven ice-melting, even when the gases have to effervesce through a water-filled cavity, is likely to be a very rapid process, resulting in a conical cavity. Melting at the ice roof advances much more rapidly than the walls, ultimately leading to mechanical collapse of a cylindrical apex.

Evidence for other subglacial eruptions on Deception Island

Despite the abundance of glacier ice on Deception Island today, there is surprisingly little additional evidence for other subglacial eruptions on the island. Indeed, several localities indicate that, in the past, the area covered by ice was substantially smaller. For example, the Outer Coast Tuff Formation (previously Yellow Tuff Formation) is a relatively thick (*c.* 70 m) unit mainly formed of pyroclastic flow deposits (Hawkes 1961; Martí & Baraldo 1990; Smellie 2001*a*). It is the only voluminous island-wide stratigraphical unit known on the island. However, its pyroclastic origin, wide distribution and undisturbed attitude imply that it was erupted when the island was substantially ice-free. Similarly, thick post-caldera tuff cone deposits are exposed underneath parts of the Mount Kirkwood and Mount Pond glaciers, indicating that those glaciers have expanded since the tuff cone eruptions.

Conversely, a few signs indicate that some eruptions were subglacial. However, it is important to note that none of those eruptions, nor the 1969 eruption itself, have left a geological record that would be easily interpreted as evidence for subglacial eruptive activity (cf. Smellie *et al.* 1993; Smellie & Skilling 1994; Smellie 2000, 2001*b*).

A prominent deep and broad, flat-bottomed valley is situated *c.* 0.5 km W of Crater Lake. It is carved through thick deposits of lapilli tuff erupted from several post-caldera tuff cones and is presently occupied by a small meandering stream. The valley has its source in the caldera margin, which is a steep gullied headwall draped by thin post-caldera lavas and Strombolian tephras. The dimensions of the valley are such that it was more likely eroded mainly during a jökulhlaup released during an eruption beneath the Mount Kirkwood glacier, possibly that subsequently responsible for the lavas and Strombolian tephras observed in the caldera headwall.

In possibly the same or a related eruption, a prominent basaltic-andesite block lava was extruded during a fissure eruption from a series of cinder cones mapped on the flanks of Mount Kirkwood. The lava flowed through a channel cut across one of the Crater Lake tuff cones on the south side of Crater Lake. The water required to erode the path utilised by the lava was probably derived by eruption-related melting of the Mount Kirkwood glacier. The eruption has been dated to 1838 or early 1839 (Pallàs *et al.* 2001).

An eruption beneath the Mount Pond glacier also occurred sometime between 1931 and 1956 (Pallàs *et al.* 2001). Red and grey scoria from that eruption were mapped by Hawkes (1961) and the author, and are visible up to high elevations on the glacier in aerial photographs taken in 1956. From the Strombolian-like characteristics of the tephra, it seems likely that the eruption was similar to that in 1969 and it represents a precursor located in the same fissure(s). The destroyed Chilean station facing Pendulum Cove was constructed on an old sediment fan that partially infills the surrounding Crimson Hill crater. Crimson Hill is part of a large tuff cone dated by historical observations to the early nineteenth century (Roobol 1980). The fan is therefore younger than the early nineteenth century. It could also have formed during the 1931–1956 Mount Pond eruption (by analogy with fan-deltas formed in Pendulum Cove during 1969). In addition, the system of gullies at the north end of fissure F which were utilized by overflowing meltwater during the 1969 eruption, are also clearly shown in aerial photographs taken in 1956. This indicates that they are relict features inherited from a previous eruption (that between 1931 and 1956?) that also affected a precursor of fissure F.

In 1967, major flooding by meltwater was described adjacent to the 1967 tuff cone known as the 'land centre' (Baker *et al.* 1975). From the geographical distribution of new meltwater gullies and an absence of glacial ice on the eruptive site, the floodwater could not be attributed to the land centre eruption itself. There was no obvious source for the meltwater, but it was attributed to a site marked by a snow-filled hollow in the glacier ice about 1 km NNE of the land centre (cf. 1969 ice chimneys). That vent apparently did not eject any tephra.

During the very short-lived but violent eruption in 1970, the most north-easterly vent formed a crater about 300 m in diameter and 150 m deep in the thin margin of the Goddard Hill glacier (Baker *et al.* 1975). The explosion took place through at least 100 m of ice (Orheim 1972*b*; Baker *et al.* 1975). Baker *et al.* (1975) indicated that the floor of the 1967 land centre crater was covered by a mudflow at a late stage in the 1970 eruption. The source of the mudflow was not indicated but melting of the ice above the 1970 subglacial crater is the only likely source for the volume of meltwater required to mobilize and transport such a large mass flow.

Volcanic hazard implications for Deception Island and other areas affected by subglacial eruptions beneath thin glaciers

This study has demonstrated for the first time that there is a significant hazard associated with eruptions beneath some thin glaciers. Predicted meltwater hydrographs constructed from existing models for thin glacier eruptions would have suggested a comparatively low-volume meltwater discharge broadly continuous with the eruption and diminishing rapidly with time, a consequence of thin glaciers being permeable. However, if the glaciers are mainly formed of impermeable (unfractured) ice, meltwater discharge is likely to show a brief delay while the cavity is filling, followed by a sudden, relatively short-lived peak discharge when the cavity overflows supraglacially. The resulting jökulhlaup can significantly damage buildings and other proximal man-made structures.

Conclusions

The 1969 eruption on Deception Island comprised short-lived (<48 hours) explosive activity from several small pyroclastic vents along a series of *en echelon* fissures. The eruptions occurred beneath a thin glacier (<100 m thick). An important feature of the eruption was the very rapid generation of abundant meltwater, from melting of the glacier above the fissures. The meltwater overflowed onto the glacier surface and generated a flood that modified the glacier, extended the local coastline, and destroyed buildings in its path. Using an empirical interpretation of a contemporary seismogram of the eruption, it is suggested that magmatic and hydrothermal volatiles (mainly superheated steam) advanced ahead of the magma and melted a vault in the glacier base, over a maximum interval of *c.* 90 minutes. Heat transfer by decoupled volatiles is likely to be much faster than coupled conduction (from magma) and convection (from meltwater), thus explaining the previously enigmatic speed with which subglacial eruptions can melt through substantial ice thicknesses. The efficacy of volatile-driven melting is demonstrated on Deception Island by the presence of extensive melted-out fissures that contained vigorous fumaroles but lacked pyroclastic activity (e.g. fissure F and the southern half of fissure C).

Although meltwater percolation through a permeable upper layer will drain meltwater from the top of the vault or meltwater lake, the surface of the glacier on Deception Island showed extensive washing implying unconfined supraglacial overflow, at least in the initial stages of flooding. It seems that fluid-flow through a saturated permeable upper layer (snow, firn and/or crevassed ice) was too slow to prevent meltwater build up and, ultimately, permitted its sheet-like escape onto the glacier surface. Despite the eruption being subglacial, the preserved deposits are pyroclastic cones and tephra indistinguishable from those formed during Strombolian eruptions of basic–intermediate magmas (i.e. essentially dry and subaerial). Neither the products of the 1969 eruption, nor those from previous subglacial eruptions on the island, would be easily identified as subglacial. This is because, within *c.* 2–3 hours of the eruption commencing, the englacial vaults will have likely drained rapidly beneath the glacier as the local meltwater source (i.e. overlying glacier) was melted, leaving an essentially dry vault. The 1969 eruption also demonstrates that even eruptions beneath thin glaciers can generate locally devastating meltwater floods. Attention is drawn to the pre-volcanic glacier structure and densification (i.e. proportion of impermeable ice), rather than simply glacier thickness (i.e. 'thick' v. 'thin' glaciers in current models) during basalt–andesite eruptions. In the presence of superheated gases, those parameters will determine the hydraulic evolution of an eruption, and will have a major impact on volcanic hazard assessment and risk mitigation.

The author gratefully acknowledges the debt he has to the careful, detailed and remarkably prescient observations of the 1969 eruption published by Baker *et al.* (1975), without which this study would have been impossible. Particular thanks are also offered to P. Baker for access to his additional unpublished Deception Island records and for agreeing to their permanent transfer to the British Antarctic Survey archives; to O. Orheim for his invaluable comments on Mount Pond glacier structure; and to K. Mackinson, K. Nicholls, H. Roscoe, A. Smith and D. Vaughan at the British Antarctic Survey for their patience during conversations exploring the thermodynamics of gas-driven subglacial melting. Constructive reviews by C. Allen and B. Houghton (and his therapist) are also gratefully acknowledged.

References

ALLEN, C. C. 1980. Icelandic subglacial volcanism: thermal and physical studies. *Journal of Geology*, **88**, 108–117.

ALMENDROS, J., IBANEZ, J. M., ALGUACIL, G., DEL PEZZO, E. & ORTIZ, R. 1997. Array tracking of the volcanic tremor source at Deception Island, Antarctica. *Geophysical Research Letters*, **24**, 3069–3072.

BAKER, P. E., MCREATH, I., HARVEY, M. R., ROOBOL, M. J. & DAVIES, T. G. 1975. The geology of the South Shetland Islands: V. Volcanic evolution of Deception Island. *British Antarctic Survey Scientific Reports*, **78**.

BJÖRNSSON, H. 1988. Hydrology of ice caps in volcanic regions. *Vísindafélag Íslendinga, Societas Scientarium Islandica*, **45**.

CHOUET, B. A. 1996. Long-period volcano seismicity: its source and use in eruption forecasting. *Nature, London*, **380**, 309–316.

COOPER, A. P. R., SMELLIE, J. L. & MAYLIN, J. 1998. Evidence for shallowing and uplift from bathymetric records of Deception Island, Antarctica. *Antarctic Science*, **10**, 455–461.

CORREIG, A. M., URQUIZU, M., VILA, J. & MARTÍ, J. 1997. Analysis of the temporal occurrence of seismicity at Deception Island (Antarctica). *Pure and Applied Geophysics*, **149**, 553–574.

EINARSSON, P. & BRANSDOTTIR, B. 1984. Seismic activity preceding and during the 1983 volcanic eruption in Grímsvötn, Iceland. *Jökull*, **34**, 13–23.

EINARSSON, P., BRANSDOTTIR, B., GUDMUNDSSON, M. T., BJÖRNSSON, H. & GRONVOLD, K. 1997. Center of the Iceland hotspot experiences volcanic unrest. *EOS*, **35**, 369 and 374–375.

GLEN, J. W. 1954. The stability of ice-dammed lakes and other water-filled holes in glaciers. *Journal of Glaciology*, **2**, 316–318.

GUDMUNDSSON, M. T., SIGMUNDSSON, F. & BJÖRNSSON, H. 1997. Ice-volcano interaction of the 1996 Gjálp subglacial eruption, Vatnajökull, Iceland. *Nature, London*, **389**, 954–957.

HAWKES, D. D. 1961. The geology of the South Shetland Islands: II. The geology and petrology of Deception Island. *Falkland Islands Dependencies Survey Scientific Reports*, **27**.

HÖSKULDSSON, A. & SPARKS, R. S. J. 1997. Thermodynamics and fluid dynamics of effusive subglacial eruptions. *Bulletin of Volcanology*, **59**, 219–230.

HOUGHTON, B. F. & WILSON, C. J. N. 1989. A vesicularity index for pyroclastic deposits. *Bulletin of Volcanology*, **51**, 451–462.

JULIAN, B. R. 1994. Volcanic tremor: nonlinear excitation by fluid flow. *Journal of Geophysical Research*, **99**, 11 857–11 877.

KIEFFER, S. W. 1984. Seismicity at Old Faithful Geyser: an isolated source of geothermal noise and possible analogue of volcanic seismicity. *Journal of Volcanology and Geothermal Research*, **22**, 59–95.

KOKELAAR, B. P. & DURANT, G. P. 1983. The submarine eruption and erosion of Surtla (Surtsey), Iceland. *Journal of Volcanology and Geothermal Research*, **19**, 239–246.

KUMAGAI, H. & CHOUET, B. A. 2000. Acoustic properties of a crack containing magmatic or hydrothermal fluids. *Journal of Geophysical Research*, **105**, 25 493–25 512.

MARTÍ, J. & BARALDO, A. 1990. Pre-caldera pyroclastic deposits of Deception Island (South Shetland Islands). *Antarctic Science*, **2**, 345–352.

Mazzuoli, R., Omarini, R. H., Ortiz, R., Viramonte, J. G. & De Rosa, R. 1989. Mecanismos eruptivios de la actividad volcánica en la isla Decepción (Shetland del Sur, Antártida). *Revista Asociación Geológica Argentina*, **44**, 259–269.

McInnes, B. I. 2000. *Kavachi eruption – May 2000*. CSIRO Exploration and Mining, North Ryde, New South Wales, Australia (video).

Nye, J. 1976. Water flow in glaciers: jökulhlaups, tunnels and veins. *Journal of Glaciology*, **17**, 181–207.

Orheim, O. 1970. Glaciological investigations on Deception Island. *Antarctic Journal of the United States*, **5**, 95–97.

Orheim, O. 1972*a*. Volcanic activity on Deception Island, South Shetland Islands. *In*: Adie, R. J. (ed.) *Antarctic geology and geophysics*. Universitetsforlaget, Oslo, 117–120.

Orheim, O. 1972*b*. A 200-year record of glacier mass balance at Deception Island, southwest Atlantic Ocean, and its bearing on models of global climate change. *Ohio State University, Institute of Polar Studies Report*, **42**.

Pallàs, R., Smellie, J. L., Casas, J. M. & Calvet, J. 2001. Using tephrochronology to date temperate ice: correlation between ice-tephras on Livingston Island and eruptive units on Deception Island volcano (South Shetland Islands). *The Holocene*, **11**, 149–160.

Parfitt, E. A. & Wilson, L. 1995. Explosive volcanic eruptions – IX. The transition between Hawaiian-style lava fountaining and Strombolian explosive activity. *Geophysical Journal International*, **121**, 226–232.

Roobol, M. J. 1980. A model for the eruptive mechanism of Deception Island from 1820 to 1970. *British Antarctic Survey Bulletin*, **49**, 137–156.

Rutherford, M. J. & Gardner, J. E. 2000. Rates of magma ascent. *In*: Sigurdsson, H. (ed.) *Encyclopedia of volcanoes*. Academic Press, San Diego, 207–217.

Smellie, J. L. 1990. Graham Land and South Shetland Islands. Summary. *In*: LeMasurier, W. E. & Thomson, J. W. (eds.) *Volcanoes of the Antarctic plate and southern oceans. American Geophysical Union, Antarctic Research Series*, **48**, 303–312.

Smellie, J. L. 2000. Subglacial eruptions. *In*: Sigurdsson, H. (ed.) *Encyclopedia of volcanoes*. Academic Press, San Diego, 403–418.

Smellie, J. L. 2001*a*. Lithostratigraphy and volcanic evolution of Deception Island, South Shetland Islands. *Antarctic Science*, **13**, 188–209.

Smellie, J. L. 2001*b*. Lithofacies architecture and construction of volcanoes erupted in englacial lakes: Icefall Nunatak, Mount Murphy, eastern Marie Byrd Land, Antarctica. *In*: White, J. D. L. & Riggs, N. (eds) *Lacustrine volcaniclastic sedimentation*. International Association of Sedimentologists, Special Publication, **30**, 73–98.

Smellie, J. L., Hole, M. J. & Nell, P. A. R. 1993. Late Miocene valley-confined subglacial volcanism in northern Alexander Island, Antarctic Peninsula. *Bulletin of Volcanology*, **55**, 273–288.

Smellie, J. L. & Hole, M. J. 1997. Products and processes in Pliocene-Recent, subaqueous to emergent volcanism in the Antarctic Peninsula: examples of englacial Surtseyan volcano construction. *Bulletin of Volcanology*, **58**, 628–646.

Smellie, J. L. & Skilling, I. P. 1994. Products of subglacial volcanic eruptions under different ice thicknesses: two examples from Antarctica. *Sedimentary Geology*, **91**, 115–129.

Sparks, R. S. J., Bursik, M. I., Carey, S. N., Gilbert, J. S., Glaze, L. S., Sigurdsson, H. & Woods, A. W. 1997. *Volcanic plumes*. John Wiley and Sons, Chichester.

Vila, J., Martí, J., Ortiz, R., Garcia, A. & Correig, A. M. 1992. Volcanic tremors at Deception Island (South Shetland Islands, Antarctica). *Journal of Volcanology and Geothermal Research*, **53**, 89–102.

Villegas, M. T., Ortiz, R., Caselli, A. & Cohen, M. 1997. Chemical variations of fumarolic gases in Deception Island, South Shetland Isands, Antarctica. *In*: Ricci, C. A. (ed.) *The Antarctic region: Geological evolution and processes*. Terra Antartica Publication, Siena, 1077–1082.

Wilson, L., Parfitt, E. A. & Head III, J. W. 1995. Explosive volcanic eruptions – VIII. The role of magma recycling in controlling the behaviour of Hawaiian-style lava fountains. *Geophysical Journal International*, **121**, 215–225.

Ziolkowski, A., Parkes, G., Hatton, L. & Hauglands, T. 1982. The signature of an air gun array: Computation from near-field measurements including interactions. *Geophysics*, **47**, 1413–1421.

A brief overview of eruptions from ice-covered and ice-capped volcanic systems in Iceland during the past 11 centuries: frequency, periodicity and implications

GUDRÚN LARSEN

Science Institute, University of Iceland, Dunhaga 3, IS-107 Reykjavík, Iceland (e-mail: glare@raunvis.hi.is)

Abstract: Eruptions from partly ice-covered and ice-capped volcanic systems constitute nearly 60% of all known historical (i.e. in the past 11 centuries) eruptions in Iceland. Since the fourteenth century such eruptions have been reported in contemporary or near-contemporary documents. At least 120 historical eruptions have broken through the ice on the glaciated parts of five volcanic systems and have left tephra layers in ice and soil, or been recorded at the time. An unknown number of eruptions did not breach the overlying ice and left no record at all. Beginning as subglacial eruptions, most eruptions break through the ice in minutes, hours or days and can last from a few days to several months. A single vent or the whole length of a fissure may then emerge to emit highly-fragmented tephra in hydromagmatic explosions of varying strength. The volume of airborne tephra varies by at least four orders of magnitude with dispersal range varying from near-vent to transatlantic. In most of the eruptions the magma was of basaltic composition. Eruption frequency is highest within the Grímsvötn system where up to seven eruptions every 40 years have occurred during peaks of activity and at least 70 eruptions over historical time. The observed pattern of temporally and spatially close eruptions, separated by periods of low or no activity, leaves open the question whether pre-Holocene deposits of several closely spaced subglacial eruptions can be securely distinguished from those formed in a single subglacial eruption.

Ice caps cover substantial parts of the active volcanic zones in Iceland (Fig. 1). These include the NW half of Vatnajökull, situated above the centre of the Iceland hot spot (Wolfe *et al.* 1997), Mýrdalsjökull in the southern part of the Eastern Volcanic Zone (EVZ), Hofsjökull and Langjökull in central Iceland, and several smaller ice caps. Volcanic activity occurs in the ice-covered, as well as the ice-free, parts of the zones and eruptions within the ice caps are thus an integral part of volcanism in Iceland. The most recent examples of such eruptions are the 1996 Gjálp eruption and the 1998 Grímsvötn eruption (Gudmundsson *et al.* 1997; Sigmarsson *et al.* 2000).

This paper summarizes the current knowledge about the origin and frequency of eruptions in the ice-covered parts of volcanic zones within Iceland during the past 11 centuries. These eruptions constitute the majority of all eruptions in Iceland during that period and some of them are among the most destructive natural events in Icelandic history. The sources of information are also briefly explained. Finally, some implications of the observed eruption frequency pattern for the interpretation of older (pre-Holocene) subglacial deposits are touched upon.

Geological setting

Holocene volcanism in Iceland is confined to fairly distinct volcanic systems within the volcanic zones (Sæmundsson 1978, 1980; Jakobsson 1979; Gudmundsson 2000). Their number is close to 30, excluding the offshore systems. A volcanic system in its simplest form consists of a fissure swarm where basaltic magma is erupted, often on long fissures, and a central volcano where evolved magmas are also erupted. Eruption frequency is usually highest at the central volcanoes, and they tend to be the highest areas topographically of the volcanic systems; one reason why many central volcanoes are capped by ice. At present several central volcanoes and proximal parts of their fissure swarms are covered by ice while the remainder of the volcanic system is ice-free.

Five volcanic systems are the subject of the present paper (Fig. 2). Four of them lie within the Eastern Volcanic Zone (EVZ). The Grímsvötn and Bárdarbunga–Veidivötn systems in the north lie close to the centre of the Iceland hot spot (Fig. 1). The configuration of the Grímsvötn system below the ice is not known in detail

From: SMELLIE, J. L. & CHAPMAN, M. G. (eds) 2002. *Volcano–Ice Interaction on Earth and Mars*. Geological Society, London, Special Publications, **202**, 81–90. 0305-8719/02/$15.00

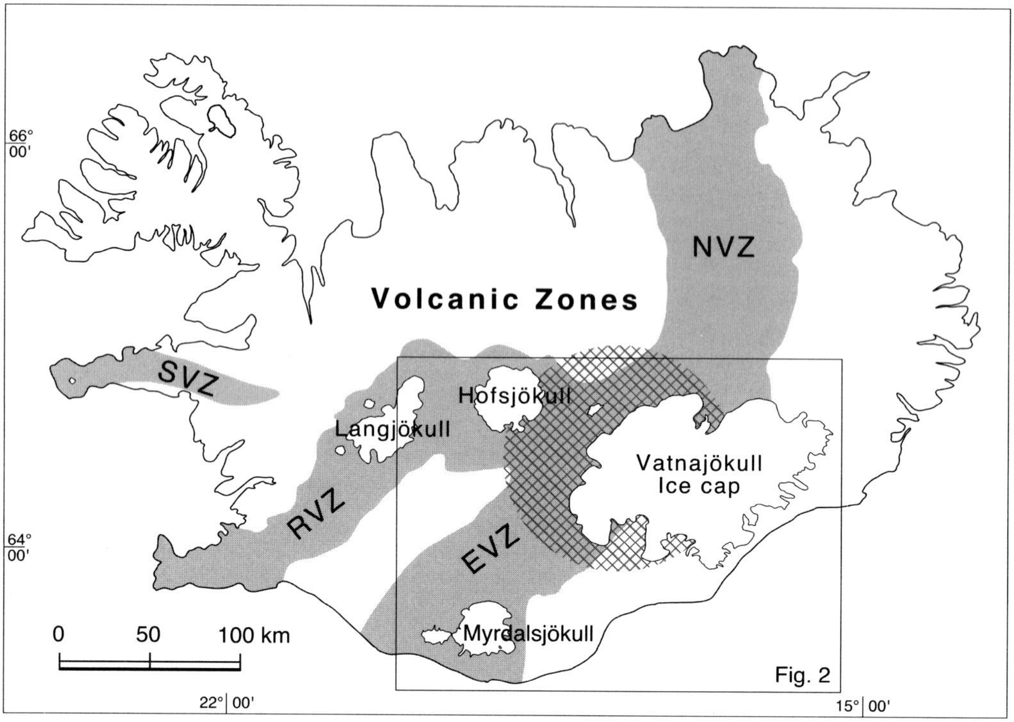

Fig. 1. Map of Iceland, showing the Quaternary volcanic zones (<0.7 Ma in grey after Sæmundsson 1979). Hatched area shows the location of the Icelandic mantle plume at 125 km depth (after Wolfe *et al.* 1997). EVZ, Eastern Volcanic Zone; NVZ, Northern Volcanic Zone; RVZ, Reykjanes Volcanic Zone; SNZ, Snæfellsnes Volcanic Zone. Also shown are the major ice caps and the position of Figure 2.

but at least one half of the system has an ice-cover, including the central volcano. The Katla and Eyjafjallajökull systems lie in the southern part of the EVZ which is defined as an off-rift or flank zone (Sæmundsson 1978, 1980). Öræfajökull is located on a flank zone in an intraplate setting to the east of the EVZ (Sæmundsson 1978, 1980). Each of the five volcanic systems has major element geochemical characteristics that allow their products to be recognized and correlated to source (Jakobsson 1979; Larsen *et al.* 1999). A system closely related to the Grímsvötn magma system, but producing more evolved magma, may exist within Vatnajökull (Steinthorsson *et al.* 2000), or alternatively the Grímsvötn system may occasionally erupt such magma (Sigmarsson *et al.* 2000), the last occurrence being the Gjálp eruption in 1996.

Ice thickness in the ice-covered parts of the five systems ranges from tens to hundreds of metres. Maximum thicknesses of 850 and 700 metres occur in the calderas of the Bárdarbunga and Katla central volcanoes, respectively, whereas a subglacial lake occupies the caldera of the Grímsvötn central volcano (Björnsson & Einarsson 1990; Björnsson *et al.* 2000).

Eruptions from ice-capped and ice-covered volcanic systems

Icelandic eruptions have been reported in contemporary or near-contemporary documents since the thirteenth century. A near-contemporary source, Gottskálksannáll, briefly mentioned an eruption in Knappafellsjökull in AD 1332 (Storm 1888). Knappafellsjökull, or Hnapparvallarjökull, are ancient names for the Öræfajökull volcano at the south margin of Vatnajökull (Thorarinsson 1974) or sometimes for the Vatnajökull ice cap in general. Thorarinsson (1974) convincingly argued that the 1332 eruption probably occurred in the Grímsvötn system. Several accounts mentioned an eruption in Sólheimajökull (an ancient name of Mýrdalsjökull) in AD 1245 and another eruption causing 'darkness' in AD 1262 (Storm 1888). In AD 1416, a contemporary source briefly described an

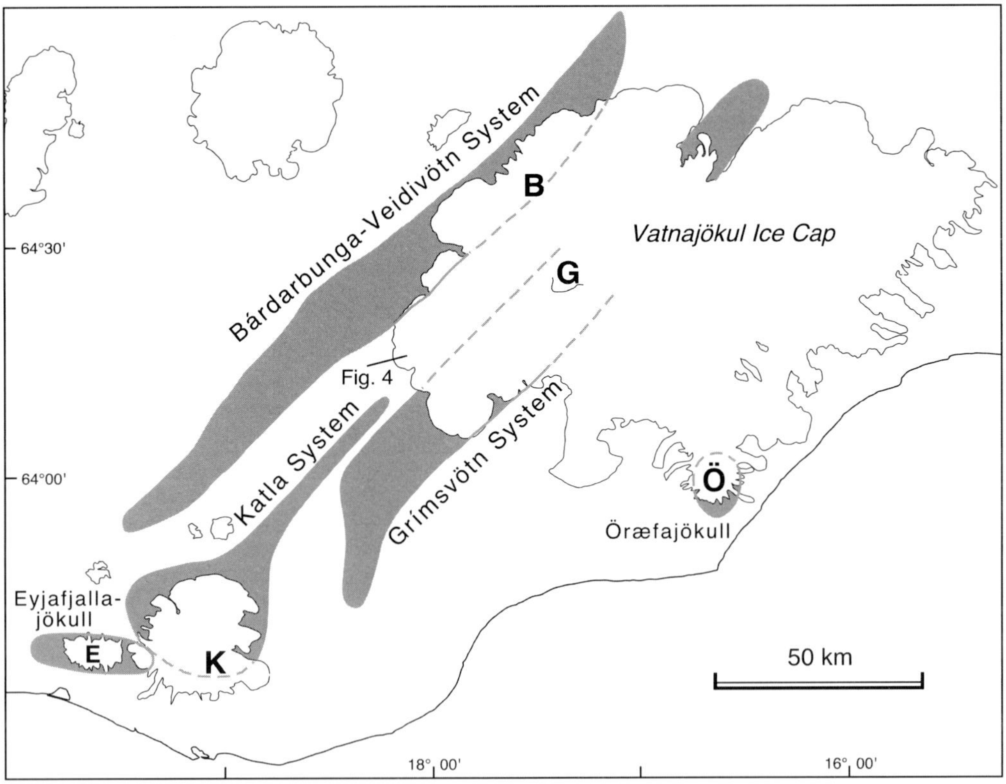

Fig. 2. Map of SE Iceland, showing the five partly ice-covered and ice-capped volcanic systems (shaded areas). Four systems lie within the EVZ, a rift zone. The Grímsvötn and Bárdarbunga–Veidivötn systems are present in the NE rift area, whereas the Katla and Eyjafjallajökull systems are situated in the off-rift or flank zones in the south-west. The Öræfajökull system is on a separate flank zone. The location of central volcanoes is indicated by capital letters: B, Bárdarbunga; E, Eyjafjallajökull; G, Grímsvötn; K, Katla; Ö, Öræfajökull. Also shown is the location of Figure 4.

eruption at Höfdárjökull (eastern part of Mýrdalsjökull) that 'burnt a great valley' into the ice and caused so much ashfall that 'damage almost occurred' (Storm 1888). Jökulhlaup is first mentioned in connection with an eruption in the disastrous AD 1362 Öræfajökull eruption (Storm 1888).

Detailed reports of eruptions occurring within the ice caps can be found in contemporary documents since AD 1600, describing precursors to and events during eruptions, as well as the consequences of tephra fall and jökulhlaups (e.g. Annales Islandici 1400–1800; Safn til sögu Íslands IV 1907–1915). These accounts are particularly good in the case of the Katla central volcano because of its location close to farming areas in southern Iceland. An attempt in 1756 to inspect the Katla 1755 eruption site failed but sites of the Eyjafjallajökull 1821–1823 and the Katla 1823 eruptions were inspected in the spring and summer of 1823 (Safn til sögu Íslands IV 1907–1915; Annáll nítjándu aldar 1912–1922).

Most eruptions in the glaciated parts of the volcanic systems begin subglacially, although volcanic fissures opening up near glacier margins may be partly subaerial. The majority break through the ice in minutes, hours or days, depending on magma extrusion rates, ice thickness and meltwater drainage. A single vent or part of a fissure may then emerge to erupt highly-fragmented tephra in hydromagmatic explosions of varying strength. The eruptions are predominantly basaltic and activity may last from days to months. Eruptions on the Grímsvötn system in the twentieth century have lasted from five days in 1983 to seven and a half months in 1903 (Thorarinsson 1974; Einarsson & Brandsdóttir 1984). Since AD 1625, Katla eruptions have lasted from 13 to *c.* 120 days (Larsen 2000). Eruption clouds reaching 14 km in height have been reported during the early stages of Katla eruptions (Eggertsson 1919)

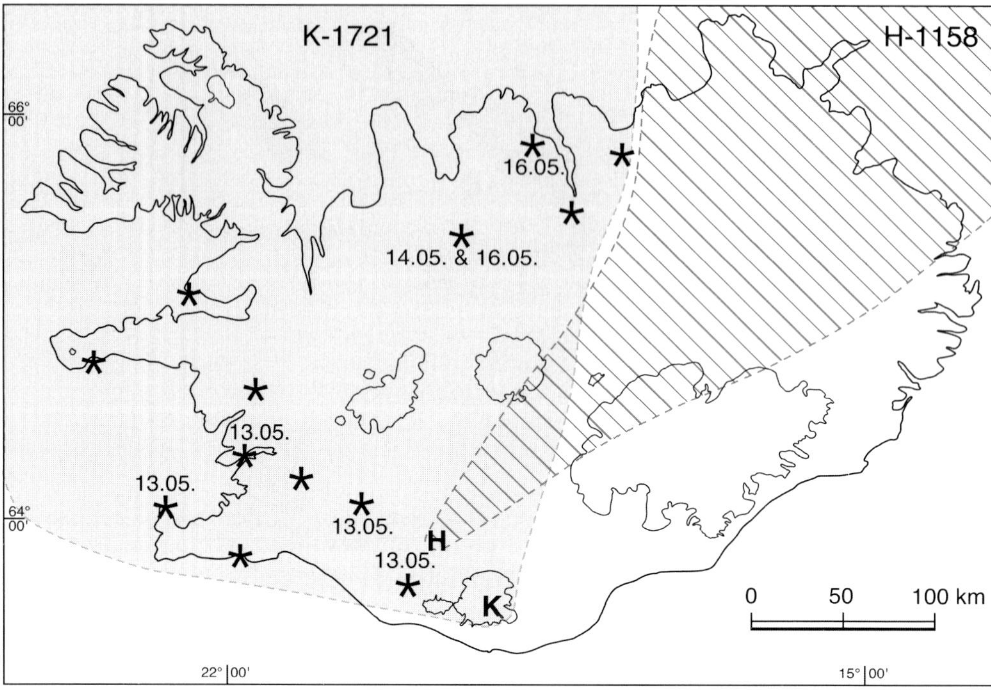

Fig. 3. Map showing the tephra sector (basaltic tephra, in grey) of the AD 1721 Katla eruption after the first six days (11–16 May). Stars indicate localities where tephra fall was recorded; dates are included where available (Thorarinsson 1955). Tephra fall is not mentioned in the NW peninsula and may have been negligible. Volume of the tephra (compacted) is close to 0.2 km^3. The tephra sector of the short-lived (probably <8 hours), subaerial plinian phase of the Hekla AD 1158 eruption (acid tephra, striped area) is shown for comparison. Compacted volume of the Hekla tephra is about 0.2 km^3. The dispersal of the Katla tephra is much wider because strong explosive activity, caused by interaction of basaltic magma and meltwater, continued for days.

and the resulting jökulhlaups are thought to have had peak discharge rates of as much as 300 000 $m^3\,sec^{-1}$ (Tómasson 1996).

If the volcanic fissure does not extend beyond the ice or lies close to the ice margin, the only observable products are airborne tephra and in some cases water-transported debris. Landforms at the vents that emerge during an eruption disappear below the ice within years or decades (Gudmundsson *et al.* 1997) because of the lateral movement of ice into the vent area. Volume and dispersal of the airborne tephra, and hence the potential for preservation, vary greatly. In the 1362 eruption of Öræfajökull, $\geq$10 km^3 of silicic tephra was emitted (calculated as freshly fallen, Thorarinsson 1958) but normally the volume of basaltic tephra is 1–3 orders of magnitude less (e.g. Katla tephra, Larsen 2000). Tephra production may occur at any time during a hydromagmatic basaltic eruption and depending on wind direction, several lobes (lobate deposit) and/or a wide sector may form (Fig. 3). In some cases tephra fall does not extend outside the ice caps while occasionally it has reached mainland Europe (Thorarinsson 1981).

The records of historical eruptions from ice-covered or ice-capped parts of the volcanic systems are thus of two kinds, tephra layers in ice and soil, and written descriptions of eruptive phenomena, such as tephra fall, eruption clouds, fire-lit sky, and possible associated phenomena, such as jökulhlaups, haze, lightning and thunder. These records, however, can only account for eruptions that broke through the ice and are either reported in documents or left a traceable tephra deposit. Many eruptions went unnoticed or barely managed to break the ice leaving very little evidence. An unknown number presumably did not break through at all.

An 1100 year eruption history – a short overview

Annals and other written sources have been extensively used to study the eruption history of volcanoes generally throughout Iceland (e.g. Thoroddsen 1925). The accounts report the location and date of specific eruptions and often describe the course of events, including the tephra fall and its effect in certain areas. Reports of eruptions in the Hekla and the ice-covered Katla central volcanoes, which lie close to farmlands and summer pastures in south Iceland, can generally be considered reliable (Thorarinsson 1967; Larsen 2000). Written sources also report eruptions of the ice-capped Öræfajökull and Eyjafjallajökull central volcanoes (Thorarinsson 1958; Larsen 1999).

There are also written sources describing eruptions within the Vatnajökull ice cap (Thorarinsson 1974). Vatnajökull is a large and remote area, poorly known until the twentieth century, and the written records are incomplete and often vague. The quality of the accounts varies and the older sources are generally less reliable and less detailed than more recent records. The timing of an event is considered more reliable than the location within the ice cap. Tephra layers in ice and soil are therefore often the best record of eruptions within Vatnajökull, although they are by no means complete (Larsen *et al.* 1996, 1998).

Detailed stratigraphical and areal mapping of historical tephra layers preserved in soil sections has been carried out with emphasis on major tephra layers from the Hekla and Katla central volcanoes (Thorarinsson 1967, 1980; Larsen & Thorarinsson 1977; Larsen 2000). Tephra layers originating within the EVZ can be correlated to their respective volcanic systems by using major element chemistry and mapping to suggest the source of individual tephra layers. Detailed descriptions of tephra fall can in many cases be safely correlated to specific tephra layers, providing accurately dated isochrons. Tephra layers from 14 out of 20 known Katla eruptions during the past 1100 years have been dated in this way (Thorarinsson 1975; Larsen 2000). The combination of the two kinds of records has thus provided a framework of accurately dated tephra layers that extends throughout most of Iceland and is particularly dense in southern Iceland. The key tephra layers can then be used to calculate or assess the likely eruption years of other tephra layers in soil and ice.

This approach has proved particularly useful in studies of tephra layers from the past nine centuries in the Vatnajökull ice. Over 80 horizons containing volcanic glass have been identified in the ablation areas of two outlet glaciers (Larsen *et al.* 1996, 1998) and in a 415 m ice core (Steinthorsson 1977), as well as from recent eruptions (Gudmundsson *et al.* 1997; Sigmundsson *et al.* 1999). The tephras form distinct horizons in the ice, with thickness ranging from millimeters to tens of centimeters. In the ablation areas these horizons crop out to form distinct dark bands on the ice that can be traced over large distances (Fig. 4). Those closest to the glacier snouts also contain moraine debris, but at least 79 horizons consist of tephra only (volcanic glass + crystals and occasionally small lithic fragments). The volcanic glass was analysed for its major element chemistry and is compositionally homogenous within individual layers with rare exceptions; the bands therefore are interpreted as true tephra layers (Larsen *et al.* 1996, 1998).

The great majority of the tephra layers in the Vatnajökull ice has the chemical characteristics of the Grímsvötn volcanic system, the only other major contributor is the Bárdarbunga–Veidivötn system. They are attributed to the ice-covered parts of the systems, because their grain characteristics indicate hydromagmatic fragmentation and the grain size and pattern of deposition indicate sources within Vatnajökull. Furthermore, tephra layers and lava flows of historical eruptions on the subaerial parts of these systems are already documented (Thorarinsson 1969; Thorarinsson & Sigvaldason 1972; Larsen 1984; Thordarson & Self 1993; Jóhannesson & Sæmundsson 1999) and one such tephra layer provides a dated horizon in the Vatnajökull ice (V 1477, Fig. 4). Four Katla tephra layers of known age are also present and together with a silicic tephra from Öræfajökull (Ö 1362, Fig. 4), they serve as key horizons, allowing the approximate ages of other tephra layers to be calculated and in many cases matched to eruptions reported in Annals and other written sources.

When the Vatnajökull ice–tephra record was integrated with the soil–tephra records it was found that tephra layers from 106 eruptions originating within the five volcanic systems during the past 11 centuries could be identified (Table 1). One hundred of the identified layers have already been chemically analysed and correlated to source, the remaining six tephra layers on the other hand have been provisionally correlated to source by field mapping. The total number of eruptions thus identified should be considered a minimum, as the Vatnajökull ice record does not extend beyond the twelfth century and many outlet glaciers still remain to be investigated.

Until the eighteenth century, reports directly referring to eruptions within the ice caps are far fewer than the tephra layers that have already been detected. Despite this, not all the reported eruptions have been linked with a tephra layer

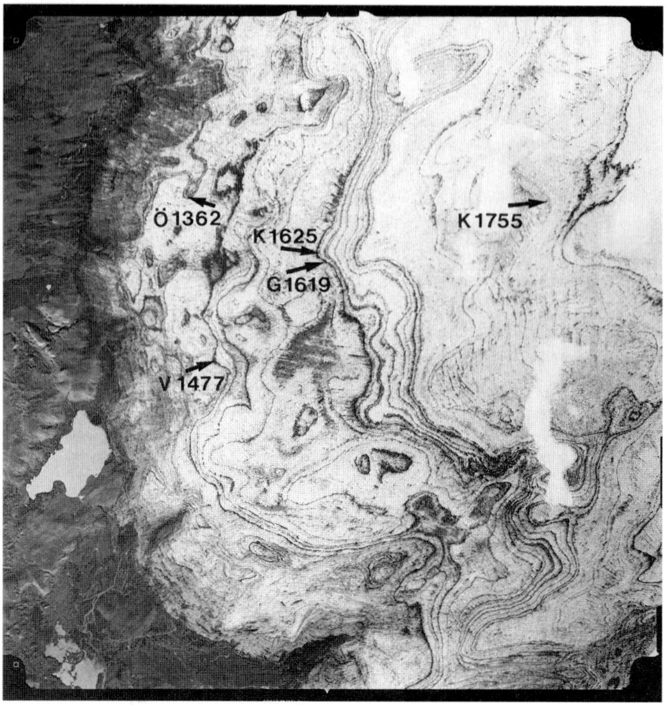

Fig. 4. A 1986 aerial photograph showing the ablation area and glacier margin in the southern part of Tungnaárjökull, western Vatnajökull, from 18000 ft. Ice flow is towards left. The glacier margin (far left) has a N–S orientation. The small lake at the glacier margin is about 0.5 km wide. Dark bands in the glacier are tephra horizons cropping out on the ice surface. Over 30 such bands can be discerned in Tungnaárjökull ice (although some are too faint to appear on this photograph) and most continue across the 20 km width of the glacier. Depending on the underlying topography, the tephra horizons may form sinuous near-parallel bands or more complex patterns, including concentric 'rings' in local ice domes. Sampling and stratigraphic measurements were done in selected, accessible sections in the northern part of Tungnaárjökull. Five key horizons are indicated by arrows, with the year of eruption and a letter indicating the system of origin. G, Grímsvötn; K, Katla; V, Veidivötn; Ö, Öræfajökull. The K 1625 layer appears as a thin band immediately above the much thicker G 1619 layer. The position of the poorly visible Ö 1362 layer is indicated. © (Landmælingar Íslands) National Land Survey of Iceland, licence L01110006.

whose age is calculated from its position in the ice stratigraphy. The written records greatly improved during the eighteenth century and, from 1800 onwards, descriptions of eruptions far outnumber the tephra layers detected in ice and soil sections. Published descriptions of 15 eruptions within the Vatnajökull ice cap, most of them AD post-1700, are considered reliable enough to be included although no tephra layers have yet been found. In this way, at least 121 historical eruptions from the ice-covered or ice-capped parts of the five volcanic systems are documented (Table 1). Another five eruptions took place on fissures that lie mostly or partly on the subaerial parts of these volcanic systems. These five include the two largest basaltic eruptions in Icelandic history, the tenth century Eldgjá eruption on the Katla system (Larsen 2000; Thordarson *et al.* 2001) and the eighteenth century Laki eruption on the Grímsvötn system (Thorarinsson 1969; Thordarson & Self 1993). In the Eldgjá eruption, some 18 km^3 of basaltic lava erupted over part of the fissure lying outside the Mýrdalsjökull ice cap, and at least 4 km^3 of basaltic tephra erupted over part of the fissure lying within the ice cap margins.

About 60% of the eruptions listed in Table 1 originated within the Grímsvötn system, which has by far the highest eruption frequency of all volcanic systems in Iceland. Some 20% of the eruptions originated within the Bárdarbunga–Veidivötn system and about 17% within the Katla system. Eyjafjallajökull and Öræfajökull volcanoes have only erupted twice each in historical time but the AD 1362 eruption of the latter represents the most disastrous volcano–ice interaction in Icelandic history when the surrounding district was almost completely laid

Table 1. *Historical eruptions from partly ice-covered and ice-capped volcanic systems*

Volcanic system	No. of eruptions from subglacial sections of volcanic systems	No. of eruptions from subaerial and subglacial sections of volcanic systems	Chemical composition of tephra†
Grímsvötn	70	1	Basaltic
Grímsvötn?	4		Intermediate
Bárdarb-Veidiv.	23	3‡	Basaltic
Katla	20	1	Basaltic
ÖræfaJökull,	2		Acid & intermed
EyjafjallaJökull,	2		Acid & intermed
Total	121*	5	

* Tephra from 106 eruptions from the subglacial parts of the volcanic systems have been identified, but written sources referring to another 15 eruptions are included although not yet supported by tephra layers.
† Tephra from 100 eruptions have been chemically analysed; the remaining 21 eruptions are assumed to be basaltic except in the case of Eyjafjallajökull.
?Source of tephra with intermediate composition is only known for the 1996 Gjálp eruption.
‡ Known eruption sites lie mainly outside Vatnajökull glacier, but probably include additional sources within the glacier.

Not included in the table are nearly 20 references in written documents that could be evidence of eruptions that broke through Vatnajökull in the period AD 1300–1910. Some descriptions undoubtedly refer to eruptions but they cannot be linked to specific sources. The subglacial eruption north of Grímsvötn in 1938, which did not break through the ice cover, is also not included in this compilation.

waste by jökulhlaups, lahars and tephra fall (Thorarinsson 1958).

No evidence, documented or evident in the ice or soil, of eruptions within the Hofsjökull or Langjökull ice-caps during the last 11 centuries has been found. A volcanic fissure at the western edge of Langjökull, active around AD 900, is now partly below-ice (Sæmundsson 1966; Jóhannesson 1989). The only products are lava flows, inferring that the fissure was not overlain by ice at the time of eruption (Jóhannesson 1989).

Discussion

Historical eruptions in Iceland, as known from their products (lava, tephra) and/or from written sources, exceed 200. About 60% of the known eruptions occurred in the ice-covered or ice-capped parts of five volcanic systems. Eruption frequency is highest on the ice-covered parts of the Grímsvötn and Bárdarbunga–Veidivötn systems, which lie above the Iceland hot spot, and correspond to $\geq$10 eruptions every 100 years. The eruption frequency in this region has, however, a period of 130–140 years with distinct highs and lows (Larsen *et al.* 1998). During peaks of activity as many as seven eruptions have been recorded over 40-year intervals on the Grímsvötn system and as many as 11 eruptions if both volcanic systems are involved (Larsen *et al.* 1998) while none may occur per 40 years during the lows (Fig. 5). Elsewhere on the rift zones, short episodes, characterized by numerous eruptions, occur from closely spaced fissures (e.g. nine effusive eruptions on the Krafla volcanic system between 1975 and 1984; Einarsson 1991; Sæmundsson 1991), separated by long intervals of no volcanic activity. Determining the number of individual effusive eruptions during earlier episodes can be difficult because their lavas may overlap, and they are underestimated in the existing records. However, the Vatnajökull ice record does not reach beyond the twelfth century and the number of tephra layers originating within Vatnajökull is therefore also underestimated. The opening statement of this paragraph is therefore considered valid for the time being.

Eruptions on the off-rift zones show little or no tendency towards periodicity. In historical time, Katla eruptions have occurred twice each century on average and, although intervals have varied from 13 to over 80 years, they are apparently not in phase with the hot spot and rift zone volcanism. Öræfajökull eruptions have, however, occurred during peaks of activity on the Grímsvötn and Bárdarbunga–Veidivötn systems.

The episodic or periodic nature of the plate-margin/hot-spot volcanism that has been observed in the historical eruptions (Björnsson *et al.* 1979; Sæmundsson 1991; Larsen *et al.* 1998) may have implications for interpretations of subglacially erupted deposits in Iceland. If the observed pattern of eruptions was also characteristic of past glacial periods, would it be possible to identify the products of individual eruptions, that

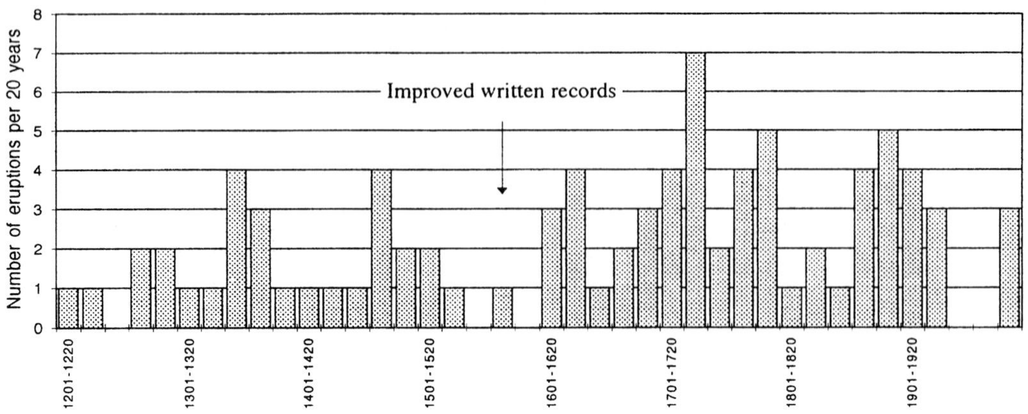

Fig. 5. Diagram showing the frequency of eruptions within Vatnajökull ice cap, 1200–2000 AD, which has a period of 130–140 years with distinct highs and lows (modified from Larsen *et al.* 1998). The double peak in the eighteenth century is the result of high activity on the Bárdarbunga–Veidivötn system in the first half of that century and high activity on the Grímsvötn system in the latter half. The remaining four peaks are the result of high activity on the Grímsvötn system. The Öræfajökull eruptions occurred during peaks of activity on the other two systems.

had occurred at 1–20 year intervals on closely spaced fissures, when the ice disappeared? For example, would it be possible to identify individual eruptions of the 1975–1984 Krafla episode if the eruptions had taken place during the last glacial? These nine eruptions occurred within an area 11 km long but less than 1 km wide and in many cases fissures from different eruptions partly overlapped (Sæmundsson 1991, figs 1–27 & 1–28). Evidence from recent eruptions within Vatnajökull shows that depressions formed in the ice surface persist for years (Gudmundsson & Björnsson, 1991; Gudmundsson *et al.* 2002). Thus, a second eruption occurring on the same fissure or on a fissure close by will take place under much reduced ice thickness. As a consequence, the confining pressure will be less than in the first eruption. This may affect the style of the eruption and the volcanic facies generated (e.g. the products of the early phase of the second eruption may be similar to the latter stage products of the first eruption). The possibility remains that subglacial deposits left by a number of small eruptions could be interpreted as sequences left by different phases of a single large eruption, resulting in an inaccurate interpretation of both the volcanic events and the eruptive environment.

Summary and conclusions

About 60% of all known eruptions over the past 11 centuries in Iceland occurred on ice-covered and ice-capped volcanic systems. Only five of the nearly 30 volcanic systems are involved: the Grímsvötn, Bárdarbunga–Veidivötn, Katla, Öræfajökull and Eyjafjallajökull systems.

Tephra deposits from some 106 eruptions on the glaciated parts of the five volcanic systems during the last 1100 years have been identified in ice and soil. They have been correlated to the respective systems by major element chemistry and/or mapping.

The amount of airborne tephra in these eruptions varies by at least four orders of magnitude, the largest being $\geq 10\,km^3$. Tephra fall has occasionally reached mainland Europe but does not extend outside the ice caps in some cases. The eruptions have lasted from a few days to several months.

Descriptions of 15 eruption-related phenomena within Vatnajökull are considered trustworthy although no tephra deposits have yet been found, bringing the total of verified eruptions to 121. Many more descriptions exist but are too vague to be included until supported by the discovery of tephra deposits in the ice or soil.

Eruption frequency is highest on the Grímvötn volcanic system, where over 70 verified eruptions have taken place during the past 1100 years; the second highest can be found on the Bárdarbunga–Veidivötn system. Both systems fall within the rift part of the EVZ.

Eruption frequency on volcanic systems within the rift zones is episodic. Several spatially close eruptions may occur within a time frame of decades followed by longer quiet intervals. Above the centre of the Iceland hot spot the eruption frequency has a period of 130–140 years.

This pattern of episodic volcanic activity is likely to have persisted within the rift zones during past glacial periods, leaving open the question, to what extent the products of small temporally and spatially close subglacial eruptions can be discerned from those of single large subglacial eruptions.

M. T. Gudmundsson critically read the manuscript and his thoughtful comments are gratefully acknowledged. K. Sæmundsson and D. McGarvie are thanked for their constructive reviews. The work on tephra layers in the Vatnajökull ice was supported by the Iceland Science Foundation (RANNÍS): Grant numbers 954710095, 954710096 and 954710097.

References

Annales Islandici 1400–1800, I–VI. Hid íslenzka bókmenntafélag, Reykjavík 1922–1987.

Annáll nítjándu aldar (Annals of the nineteenth century). Hallgrímur Pjetursson, Akureyri 1912–1922.

BJÖRNSSON, H. & EINARSSON, P. 1990. Volcanoes below Vatnajökull, Iceland: evidence from radio echo-sounding, earthquakes and jökulhlaups. *Jökull*, **40**, 147–174.

BJÖRNSSON, A., JOGNSEN, G., SIGURDSSON, S., THORBERGSSON, G. & TRYGGVASON, E. 1979. Rifting of the plate boundary in North Iceland 1975–1978. *Journal of Geophysical Research*, **84**, 3029–3038.

BJÖRNSSON, H., PÁLSSON, F. & GUDMUNDSSON, M. T. 2000. Surface and bedrock topography of Mýrdalsjökull, Iceland: the Katla caldera, recent eruption sites and routes of jökulhlaups. *Jökull*, **49**, 29–46

EGGERTSSON, S. 1919. Ýmislegt smávegis vidvíkjandi Kötlugosinu 1918. *Eimreidin*, **25**, 212–222.

EINARSSON, P. 1991. Umbrotin vid Kröflu 1975–89. *In:* GARDARSON, A. & EINARSSON, Á. (eds) *Náttúra Mývatns*. Hid íslenska náttúrufrædifélag, Reykjavík, 97–139.

EINARSSON, P. & BRANDSDÓTTIR, B. 1984. Seismic activity preceding and during the 1983 volcanic eruption in Grímsvötn, Iceland. *Jökull*, **34**, 13–23.

GUDMUNDSSON, Á. 2000. Dynamics of volcanic systems in Iceland: examples of tectonism and volcanism at juxtaposed hot spot and mid-ocean ridge systems. *Annual Review of Earth and Planetary Sciences*, **28**, 107–140.

GUDMUNDSSON, M. T. & BJÖRNSSON, H. 1991. Eruptions in Grímsvötn 1934–1991. *Jökull*, **41**, 21–46.

GUDMUNDSSON, M. T., SIGMUNDSSON, F. & BJÖRNSSON, H. 1997. Ice-volcano interaction of the 1996 Gjálp subglacial eruption, Vatnajökull, Iceland. *Nature*, **389**, 954–957.

GUDMUNDSSON, M. T., PÁLSSON, F. P., BJÖRNSSON, H. & HÖGNADÓTTIR, Þ. The hyaloclastite ridge formed in the subglacial 1996 eruption in Gjálp, Vatnajökull, Iceland: present day shape and future preservation. *In*: SMELLIE, J. L. & CHAPMAN, M. G. (eds) *Volcano–Ice Interaction on Earth and Mars*. Geological Society, London, Special Publications, **202**, 319–335.

JAKOBSSON, S. P. 1979. Petrology of recent basalts of the Eastern Volcanic Zone, Iceland. *Acta Naturalia Islandica*, **26**, 1–103.

JÓHANNESSON, H. 1989. Aldur Hallmundarhrauns í Borgarfirdi (Dating of Hallmundarhraun). *Icelandic Institute of Natural History Report*, **9**, 1–12.

JÓHANNESSON, H. & SÆMUNDSSON, K. 1999. *Ísland. Geological Map 1:1 000 000*. Icelandic Institute of Natural History, Reykjavík.

LARSEN, G. 1984. Recent volcanic history of the Veidivötn fissure swarm, southern Iceland – an approach to volcanic risk assessment. *Journal of Volcanology and Geothermal Research*, **22**, 33–58.

LARSEN, G. 1999. Gosid í Eyjafjallajökli 1821–1823 (The Eyjafjallajökull eruption of 1821–23). *Science Institute Research Report*, **RH-28-99**, 1–13.

LARSEN, G. 2000. Holocene eruptions within the Katla volcanic system, south Iceland: characteristics and environmental impact. *Jökull*, **49**, 1–28.

LARSEN, G. & THORARINSSON, S. 1977. H-4 and other acid Hekla tephra layers. *Jökull*, **27**, 28–46.

LARSEN, G., DUGMORE, A. J. & NEWTON, A. J. 1999. Geochemistry of historical-age silicic tephras in Iceland. *The Holocene*, **9**, 463–471.

LARSEN, G., GUDMUNDSSON, M. T. & BJÖRNSSON, H. 1996. Tephrastratigraphy of ablation areas of Vatnajökull Ice Cap, Iceland. *In:* COLBECK, S. C. (ed.) *Glaciers, Ice Sheets and Volcanoes: A Tribute to Mark Meier*. CRREL Special Report, **96-27**, 75–80.

LARSEN, G., GUDMUNDSSON, M. T. & BJÖRNSSON, H. 1998. Eight centuries of periodic volcanism at the center of the Icelandic hotspot revealed by glacier tephrostratigraphy. *Geology*, **26**, 943–946.

SÆMUNDSSON, K. 1966. Zwei neue C14-Datierungen Islandischer Vulkanausbruche. *Eiszeitalter und Gegenwart*, **17**, 85–86.

SÆMUNDSSON, K. 1978. Fissure swarms and central volcanoes of the neovolcanic zones of Iceland. *Geological Journal*, Special Issue, **10**, 415–432.

SÆMUNDSSON, K. 1980. Outline of the geology of Iceland. *Jökull*, **29**, 7–28.

SÆMUNDSSON, K. 1991. Jardfrædi Kröflukerfisins. *In:* GARDARSON, A. & EINARSSON, Á. (eds) *Náttúra Mývatns*. Hid íslenska náttúrufrædifélag, Reykjavík, 1–95.

Safn til sögu Íslands IV. 1907–1915. Hid íslenzka bókmenntafélag, Kaupmannahöfn og Reykjavík, 186–294.

SIGMARSSON, O., KARLSSON, H. & LARSEN, G. 2000. The 1996 and 1998 subglacial eruptions beneath the Vatnajökull ice sheet in Iceland: contrasting geochemical and geophysical inferences on magma migration. *Bulletin of Volcanology*, **61**, 468–476.

SIGMUNDSSON, F., GUDMUNDSSON, M. T., SVERRISDÓTTIR, G., ÓSKARSSON, N., EINARSSON, P., GRÖNVOLD, K. & HÖGNADÓTTIR, T. 1999. Style of basaltic eruptions at shallow ice/water depths: the 1998 Grímsvötn eruption, Vatnajökull ice cap, Iceland. *EOS*, **80**, F1084.

STEINTHORSSON, S. 1977. Tephra layers in a drill core from the Vatnajökull Ice Cap. *Jökull*, **27**, 2–27.

STEINTHORSSON, S., HARDARSON, B. S., ELLAM, R. M. & LARSEN, G. 2000. Petrochemistry of the Gjálp-1996 subglacial eruption, Vatnajökull, SE Iceland. *Journal of Volcanology and Geothermal Research*, **98**, 79–90.

STORM, G. 1888. *Islandske Annaler indtil 1578*. Det norske historiske Kildeskriftfond, Christiania.

THORARINSSON, S. 1955. Öskufall svo sporrækt vard og Kötlugosid 1721 (Ashfall so that footprints were traceable and the Katla eruption of 1721). *Náttúrufrædingurinn*, **25**, 87–98.

THORARINSSON, S. 1958. The Öræfajökull eruption of 1362. *Acta Naturalia Islandica*, **2**, 1–100.

THORARINSSON, S. 1967. The eruptions of Hekla in historical times. The eruption of Hekla 1947–1948. *Vísindafélag Íslendenga, Societas Scientarium Islandica*, **I**, 1–183.

THORARINSSON, S. 1969. The Lakagígar eruption of 1783. *Bulletin Volcanologique*, **33**, 910–929.

THORARINSSON, S. 1974. *Vötnin stríd* (*The swift flowing rivers*) Bókaútgáfa Menningarsjóds, Reykjavík.

THORARINSSON, S. 1975. Katla og annáll Kötlugosa (Katla and its historical eruptions). *Árbók Ferdafélags Íslands 1975*. Ferdafélags Íslands, Reykjavík, 125–149.

THORARINSSON, S. 1980. Langleidir gjósku úr thremur Kötlugosum. *Jökull*, **30**, 65–73.

THORARINSSON, S. 1981. Greetings from Iceland. *Geografiska Annaler*, **63A**, 109–118.

THORARINSSON, S. & SIGVALDASON, G. 1972. Tröllagígar og Tröllahraun (The Tröllagígar eruption). *Jökull*, **22**, 12–26.

THORDARSON, T. & SELF, S. 1993. The Laki (Skaftár Fires) and Grímsvötn eruptions in 1783–1785. *Bulletin of Volcanology*, **55**, 233–263.

THORDARSON, T., MILLER, D. J., LARSEN, G., SELF, S. & SIGURDSSON, H. 2001. New estimates of sulfur degassing and atmospheric mass-loading by the 934 AD Eldgjá eruption, Iceland. *Journal of Volcanology and Geothermal Research*, **108**, 33–54.

THORODDSEN, TH. 1925. *Die Geschichte der isländischen Vulkane. Det Kongelige danske Videnskabernes Selskabets Skrifter, 8. række. Naturvidenskabelig og mathematisk afdeling B IX*. Høst og Søn, København.

TÓMASSON, H. 1996. The Jökulhlaup from Katla in 1918. *Annals of Glaciology*, **22**, 249–254.

WOLFE, C. J., BJARNASON, I. T., VANDECAR, J. C. & SOLOMON, S. C. 1997. Seismic structure of the Iceland mantle plume. *Nature*, **385**, 245–247.

Basaltic pahoehoe lava-fed deltas: large-scale characteristics, clast generation, emplacement processes and environmental discrimination

I. P. SKILLING

Department of Geology and Planetary Sciences, 200 SRCC, University of Pittsburgh, Pittsburgh, PA 15260, USA (e-mail: Skilling@pitt.edu)

Abstract: Basaltic pahoehoe lava-fed deltas are important coastal constructions of many oceanic islands and continental flood basalt provinces. Whilst littoral processes associated with their formation have been described, little is known about subaqueous processes and products. This study is primarily focused on field studies of lava-fed deltas from the James Ross Island Volcanic Group (JRIVG), Antarctica, but also on other published studies of lava-fed deltas, and on information from studies of coarse-grained alluvial deltas. Seven coherent lava facies and eight subaqueously deposited clastic facies from the JRIVG are described and interpreted. Clastic facies are dominated by cobble-sized angular lithic and fluidal lithic–vitric breccias. The fluidal lithic–vitric breccias are derived from various slope failure processes acting on large-volume ponded lavas in the frontal crest area, or from gravity-driven ductile detachment of the margins of active pillow and sheet lavas on the steep subaqueous slope. Angular lithic breccias are generated mostly by similar brittle processes operating on cooled and jointed lavas ponded upslope. Subaqueous emplacement is mostly by density-modified grain flows with associated small buoyant plumes of finer sediment, and by high-density turbidity currents, many of which infill debris chutes. Basaltic lava-fed deltas have large-scale characteristics and processes that are similar to those of Gilbert-type and gravitationally-modified Gilbert-type alluvial deltas. Important contrasts that influence processes and facies in lava-fed deltas include the absence of any effluent force and the presence of hot clasts. Study of lava-fed delta deposits is important for palaeoenvironmental analysis because marine examples record relative sea level changes, and englacial examples provide evidence of minimum ice sheet thicknesses and meltwater levels. Characteristics that may be used to distinguish lava-fed deltas in marine and englacial lake environments are discussed, and littoral zone facies analysis is emphasized. Suggestions for future research include correlation of the physical parameters of subaerial and subaqueous lavas and the nature of cogenetic clastic products, submersible dives on active deltas, detailed facies analysis of individual tangential foreset beds, and comparative studies with steep coarse-grained alluvial deltas.

Basaltic lava-fed deltas are coarse-grained, poorly-sorted volcaniclastic wedges that are constructed when subaerial lava flows fragment on entering water that is sufficiently deep to allow the development of a steep subaqueous slope dominated by gravity-driven processes. The generation of lava-fed deltas is an important process in shoreline and flank growth of basaltic oceanic islands (Jones & Nelson 1970; Jones & McDougall 1973; Furnes & Sturt 1976; Peterson 1976; Lipman & Moore 1996; Schmincke *et al.* 1997; Smith *et al.* 1999) and of continental flood basalt provinces (Larsen & Pedersen 1990; Pedersen & Dueholm 1992; Pedersen *et al.* 1996; Planke & Alvestad 1999). Similar lava-fed deltas have also been recorded from environments interpreted as former englacial lakes (Mathews 1947; Sigvaldason 1968; Jones 1969, 1970; Wörner & Viereck 1987; LeMasurier *et al.* 1994; Skilling 1994; Smellie & Hole 1997, Werner & Schmincke 1999; Hickson 2000; Smellie 2001), and from ancient non-glacial lakes that were formed following the damming of rivers by subaerial lava flows (Schmincke 1967; Swanson 1967; Bishop 1985; Hamblin 1994; Pedersen *et al.* 1998). All deltas are important for palaeoenvironmental studies because they record evidence of former water levels. Marine lava-fed deltaic sequences record fluctuations in relative sea level, and englacial examples record meltwater levels, and indicate the presence and minimum thickness of the surrounding ice. Lava–fed deltas also provide information about magma fragmentation processes in shallow water. Despite their importance, an understanding of lava-fed deltas is far behind that of similar

From: SMELLIE, J. L. & CHAPMAN, M. G. (eds) 2002. *Volcano–Ice Interaction on Earth and Mars*. Geological Society, London, Special Publications, **202**, 91–113. 0305-8719/02/$15.00

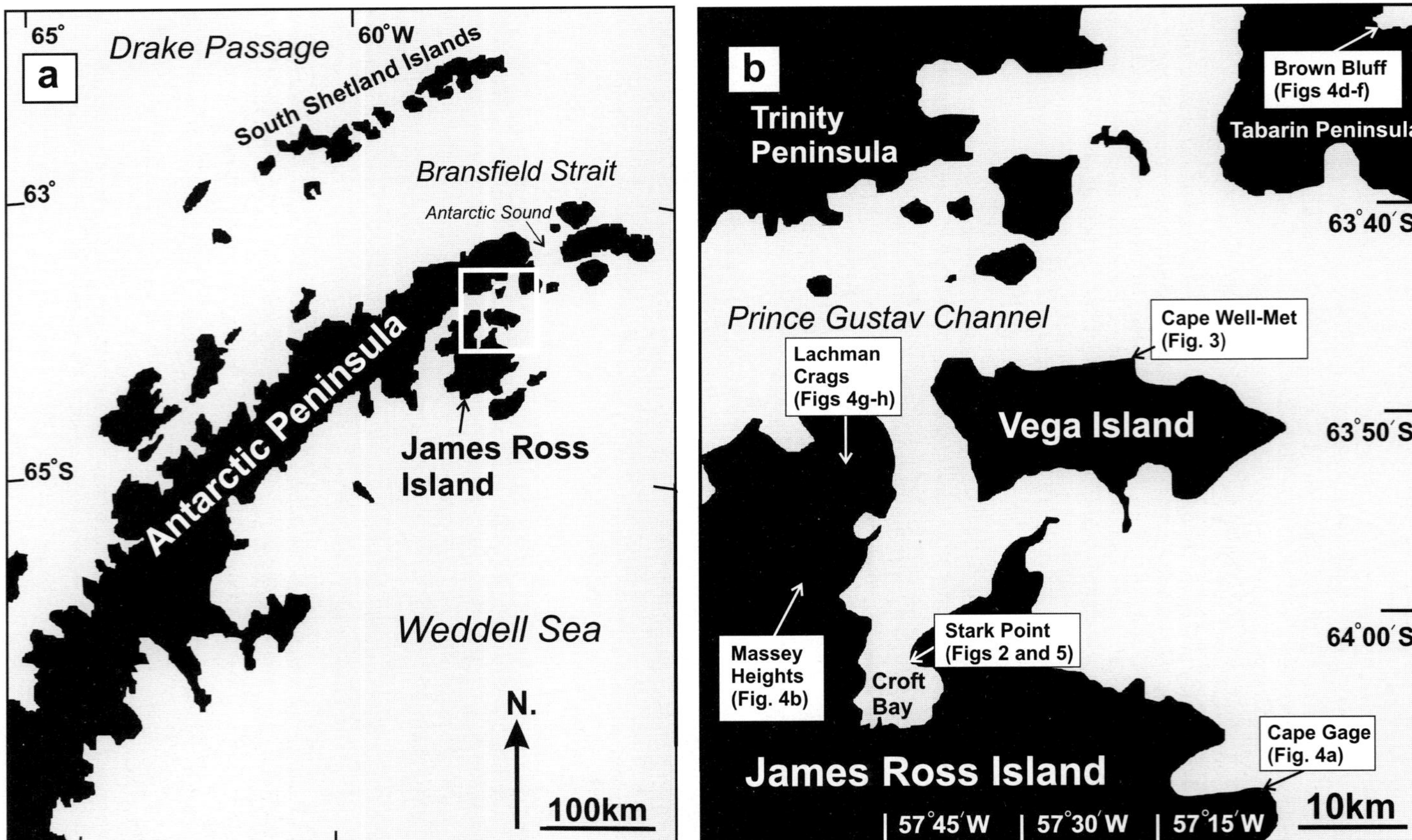

Fig. 1. Maps showing the location of the James Ross Island area, Antarctica, and the location of figures and localities of James Ross Island Volcanic Group (JRIVG) lava-fed deltas discussed in the text. Note that location of Figure 4c is uncertain.

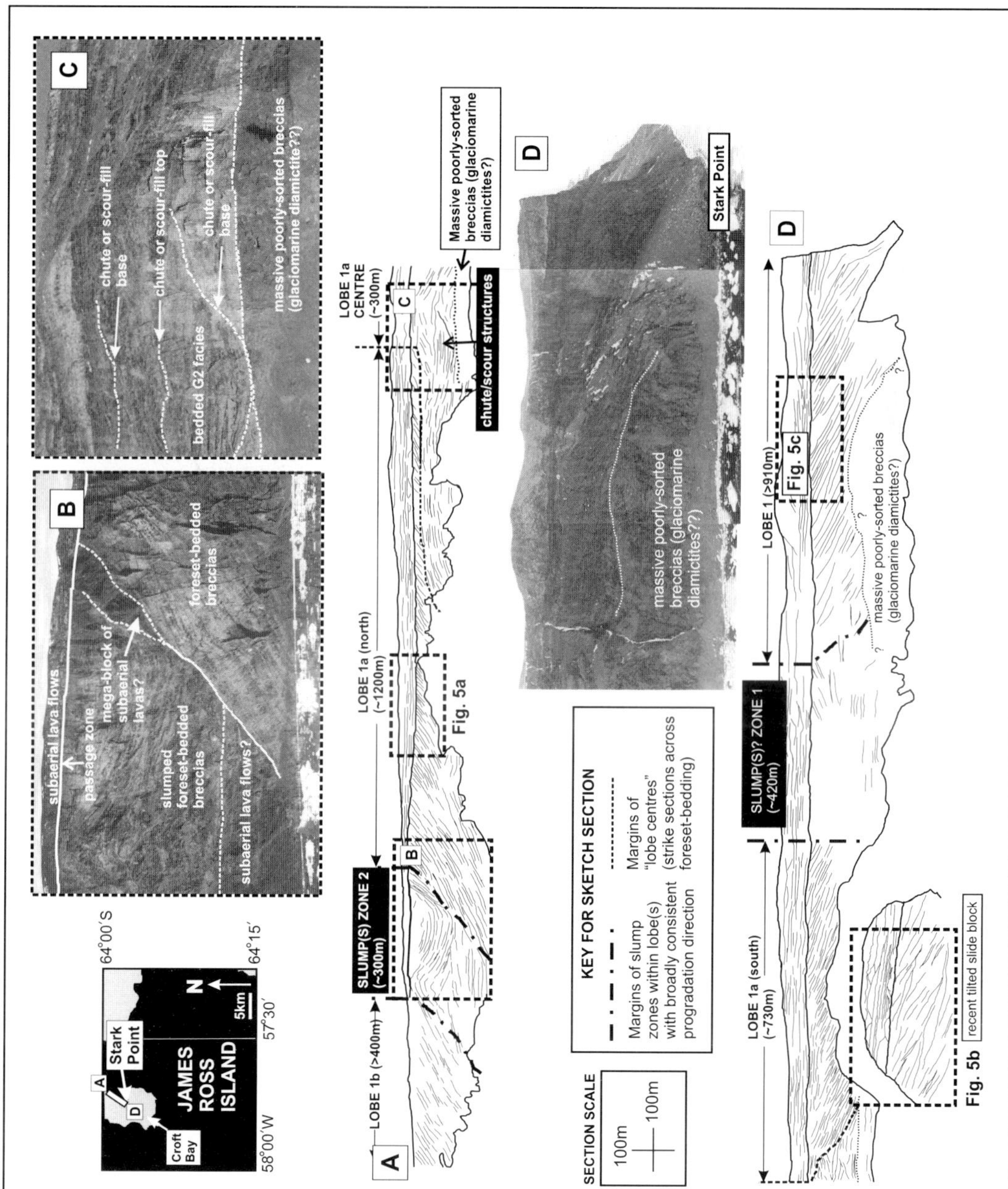

Fig. 2. Sketch section and photo-interpretation of lava-fed deltas in the Stark Point area, James Ross Island.

coarse-grained, steeply bedded alluvial deltas. Littoral processes and products of active, historical and ancient lava-fed deltas have been studied by several authors (Moore & Ault 1965; Fisher 1968; Furnes & Fridleifsson 1974; Furnes & Sturt 1976; Bluck 1982; Hon *et al.* 1993; Jurado-Cichay *et al.* 1996; Mattox & Mangan 1997), but the subaqueous processes and products are much less well known. Moore *et al.* (1973) and Tribble (1991) recorded the submarine processes of active lava-fed deltas, but their observations were restricted to depths of less than 50 m. Porebski & Gradzinski (1990) produced the only published study of an ancient submarine lava-fed delta. Non-marine lava-fed deltas are even less well known than their submarine equivalents. There are no studies of active non-marine examples, and studies of their ancient equivalents are also limited (Skilling 1994; Pedersen *et al.* 1998; Smellie 2001). This paper reviews and discusses the large-scale characteristics, clast generation and emplacement processes of basaltic pahoehoe lava-fed deltas, with particular reference to well-exposed examples from the James Ross Island Volcanic Group (JRIVG), Antarctica (Fig. 1), and published examples of lava-fed and comparable alluvial deltas. The characteristics that may be used to distinguish marine and englacial lava-fed deltas are also discussed, and some suggestions for future research given. Sedimentary grain sizes are used for the subaqueous facies of the JRIVG lava-fed deltas, to allow for comparison with alluvial deltas.

James Ross Island Volcanic Group (JRIVG)

The JRIVG is a suite of dominantly alkali basalts, with minor hawaiites, benmoreites and mugearites, which was erupted in a back-arc extensional setting, during and following cessation of subduction along the western margin of the Antarctic Peninsula (Nelson 1975; Smellie *et al.* 1988; Smellie 1999).

Published K–Ar dates suggest that the JRIVG was erupted from about 7 Ma BP up until about a few hundred thousand years ago (Sykes 1988). It crops out on James Ross Island, Vega Island, Tabarin Peninsula, Trinity Peninsula and surrounding islands in Prince Gustav Channel and Antarctic Sound (Fig. 1). Figure 1 illustrates JRIVG localities discussed in the text. Inland ice is present at most localities but coastal cliff and cirque wall exposures are excellent. The exposed part of the JRIVG is dominated by lava-fed deltas and overlying subaerial lava flows, but also includes several Surtseyan tuff cone sequences, and minor Strombolian cinder cones. There is evidence that the JRIVG deltas were formed in both an englacial lacustrine and submarine environment (Smellie 1999). The deltaic sequence at Brown Bluff (Fig. 1) has been interpreted as having been constructed within an englacial lake (Skilling 1994). The apparent occurrence of subaerial lava flows banked against foreset-bedded breccias in the Stark Point section (Fig. 2) suggests that catastrophic drainage occurred, and that this deltaic sequence may also be of englacial origin. A similar juxtaposition is present at Brown Bluff (fig. 7 in Smellie & Skilling 1994). However, such rapid changes of water level are not unequivocal evidence of an englacial origin (discussed later). Evidence of a submarine environment for at least some of the JRIVG deltas is suggested by the occurrence of apparent diamictites that immediately underlie the deltas on James Ross Island (Fig. 2). Similar diamictites elsewhere on James Ross Island have been interpreted as glaciomarine in origin (P. J Butterworth, pers. comm.). The broad contemporaneity of the diamictites and the deltas is demonstrated by lava-fed delta breccias at Whisky Bay, James Ross Island that ploughed into and partially mixed with unconsolidated diamictite (unpublished data of author). Marine fossils have also been discovered in JRIVG strata on James Ross Island (Bibby 1966).

Outcrops of lava-fed deltas in the JRIVG are most extensive on James Ross Island, Vega Island and Tabarin Peninsula. The largest exposed single-delta outcrop area is in SW James Ross Island, where it crops out in continuous cliff sections for over 15 km (Nelson 1975). Nelson (1975) described the occurrence of up to three stacked deltaic sequences from there and elsewhere on James Ross Island.

Large-scale characteristics

The large-scale characteristics of a delta include its dimensions, morphology and gross internal structure. These features and the rate of growth of lava-fed deltas are described and interpreted below with reference to the JRIVG and other published examples of lava-fed deltas.

Description

The subaerial dimensions of active lava-fed delta plains (new land) off Hawaii have been recorded by several authors (Moore *et al.* 1973; Peterson 1976; Tribble 1991; Mattox & Mangan 1997). For example, Mattox & Mangan (1997) noted that a delta 2.9 km long and 500 m wide was

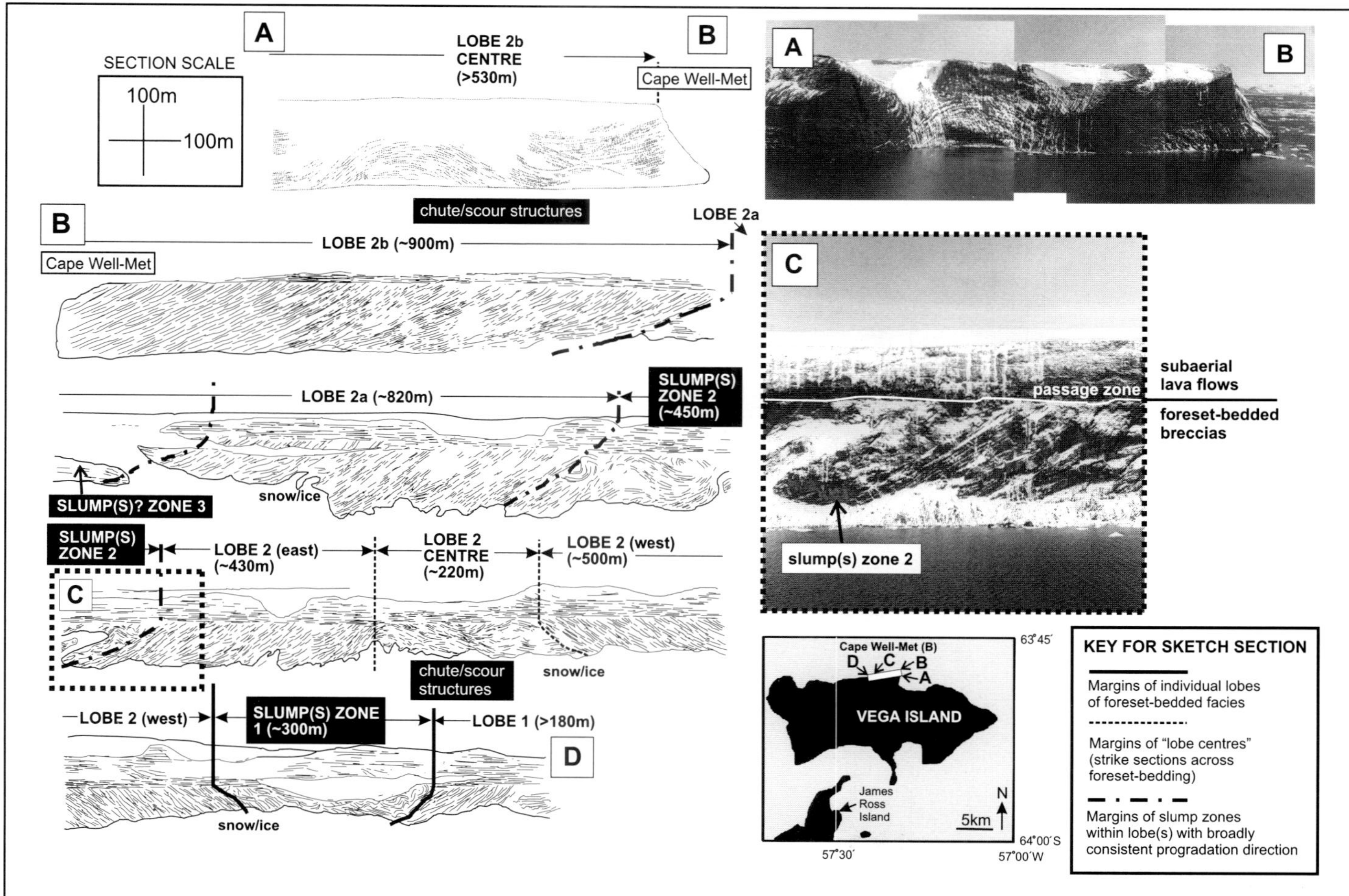

Fig. 3. Sketch section and photo-interpretation of lava-fed deltas on the north coast of Vega Island.

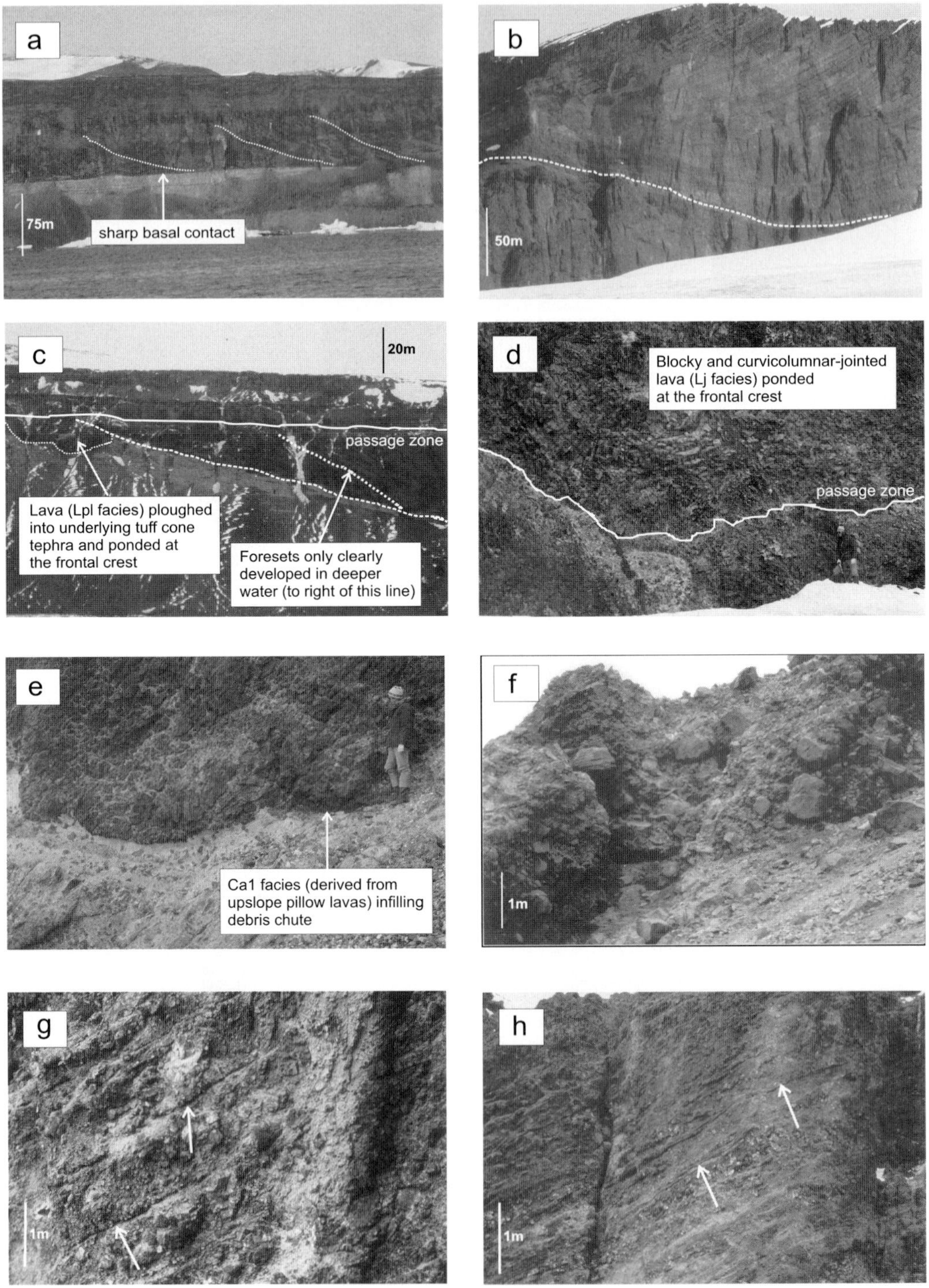
a
sharp basal contact
75m
b
50m
c
20m
passage zone
Lava (Lpl facies) ploughed into underlying tuff cone tephra and ponded at the frontal crest
Foresets only clearly developed in deeper water (to right of this line)
d
Blocky and curvicolumnar-jointed lava (Lj facies) ponded at the frontal crest
passage zone
e
Ca1 facies (derived from upslope pillow lavas) infilling debris chute
f
1m
g
1m
h
1m

constructed by lava flows from Kilauea between 1992 and 1994. However, subaerial dimensions are highly time-dependent, and land area may even be reduced at times, due to repeated collapse of the topset lavas. The gross subaqueous morphology and dimensions of lava-fed deltas from Kilauea is unknown. Gross morphology and dimensions may also be difficult to obtain for ancient deltas, because of lack of three-dimensional exposure. Lava-fed deltas, like most alluvial deltas or fans, are composed of coalesced lobes of sediment, where a lobe represents an individual depositional fan or cone generated, without a significant depositional break, from a group of closely spaced lava feeder streams, or a single stream. A new lobe is constructed when the feeder stream(s) shift significantly or an earlier lobe undergoes mass failure. The dimensions of lobes, if defined in this sense, can be more accurately estimated than the overall dimension of a particular delta. The north coast of Vega Island exposes a 4.3 km-long section through a lava-fed delta(s), which includes a section through such a lobe (Lobe 2 on Fig. 3). The lobe has a total exposed width of 1900 m, and a minimum height of 70 m, not including the upper subaerial lava flows. A similar section is exposed in cliff sections at Stark Point, James Ross Island (Lobe 1a on Fig. 2), where the lobe has a minimum width of 2650 m, and a height of *c.* 200 m.

All JRIVG lava-fed deltas studied by the author have, and all published papers on lava-fed deltas describe, a bipartite internal structure, with basal coarse-grained breccias, that are commonly steeply bedded (foreset-bedded), and have angular or asymptotic profiles (Figs 2 & 3). The breccias are overlain by broadly horizontal subaerial pahoehoe basaltic lava flows. In the JRIVG, and in many subaqueous to emergent sequences, the breccias commonly overlie Surtseyan eruption-fed subaqueous deposits (e.g. Jones 1969, 1970; Skilling 1994; Smellie & Hole 1997; Smellie 2001). The basal contact of the deltaic sequence is usually sharp and linear in dip sections (Fig 4a), except in the frontal crest area (Fig. 4c), and where the breccias overlie older deposits which have a significant surface topography (Figs. 2 & 4b), where it may be more irregular.

Large-scale slumping of the foresets is commonly displayed by JRIVG delta lobes (lobe 1a and lobe 2, Fig. 3). Slump zone 1 on Figure 3 is situated at the contact between two lobes with opposing foreset dip directions. All other slump zones in the Stark Point and Vega Island sections (Figs 2 & 3) are within foreset-bedded sequences, that display broadly the same dip directions (e.g. slump zone 2 between lobes 2 and 2a, Fig. 3). Slump zone 2 in the Stark Point section (Fig. 2) overlies a sequence with the appearance of subaerial lava flows, which are banked against older foreset-bedded breccias. Bedding in this slump appears to have been compressed against a rigid wedge-shaped megablock (subaerial lava flows?) that rests on the upper parts of the foresets of the adjacent lobe. Slump zone 1 (Fig. 2) lies directly on a poorly-sorted massive breccia deposit with a convex upper surface. It is unclear what this deposit is, but it may represent a glaciomarine diamictite, of the type that has been recorded from several localities in the JRIVG, where they commonly underlie foreset-bedded breccias (P. J. Butterworth, pers. comm.). Foreset-bedding is commonly truncated by surfaces which dip at similar angles to the foresets, and which often begin immediately below the passage zone (Fig. 5b, c), and extend for at least 150 m downslope. Strike sections across foreset beds, are exposed in the central parts of lobes 2 and 2b in the Vega Island section (Fig. 3), within lobe 1a in the Stark Point section (Fig. 2), and within a JRIVG delta lobe at Brown Bluff. All of these sections display scour or chute-fill structures.

The contact between the breccias and the lavas, which represents the former water level, was termed the 'passage zone' by Jones & Nelson

Fig. 4. Illustration of large-scale features (**a** & **b**) and facies of the JRIVG lava-fed deltas (**c–h**). (**a**) Sharp linear basal contact of foreset-bedded (dotted lines) breccias on underlying subaqueously-deposited Surtseyan eruption-fed tephra, Cape Gage, James Ross Island. Note the asymptotic nature of several beds. (**b**) Foreset bedded breccias onlapping more massive breccias (below dotted line) of unknown origin, Massey Heights, James Ross Island. (**c**) Transverse section across lava-fed delta illustrating lava (Lpl facies) that has ploughed into underlying Surtseyan tephra, and ponded at the frontal crest. An absence of clear foreset-bedding in the first *c.* 20 m of water is also illustrated, JRIVG, locality uncertain. (**d**) Lj facies ponded at the frontal crest, Brown Bluff, Tabarin Peninsula. (**e**) Ca1 facies within a debris chute, Brown Bluff, Tabarin Peninsula; (**f**) Ca3 facies interbedded with foreset-bedded Ca1 facies (not visible), Brown Bluff, Tabarin Peninsula. (**g**) Interbedded Ca1 and G1 facies (thinner units, arrowed), Lachman Crags, James Ross Island; (**h**) Possible examples of ductile compaction (arrows) of incompletely solidified or actively inflating pillow lava. Ca1 facies is interbedded mostly with G1 facies here. Lachman Crags, James Ross Island. Location of photographs is illustrated in Figure 1 and facies codes and their interpretation are given in Table 1.

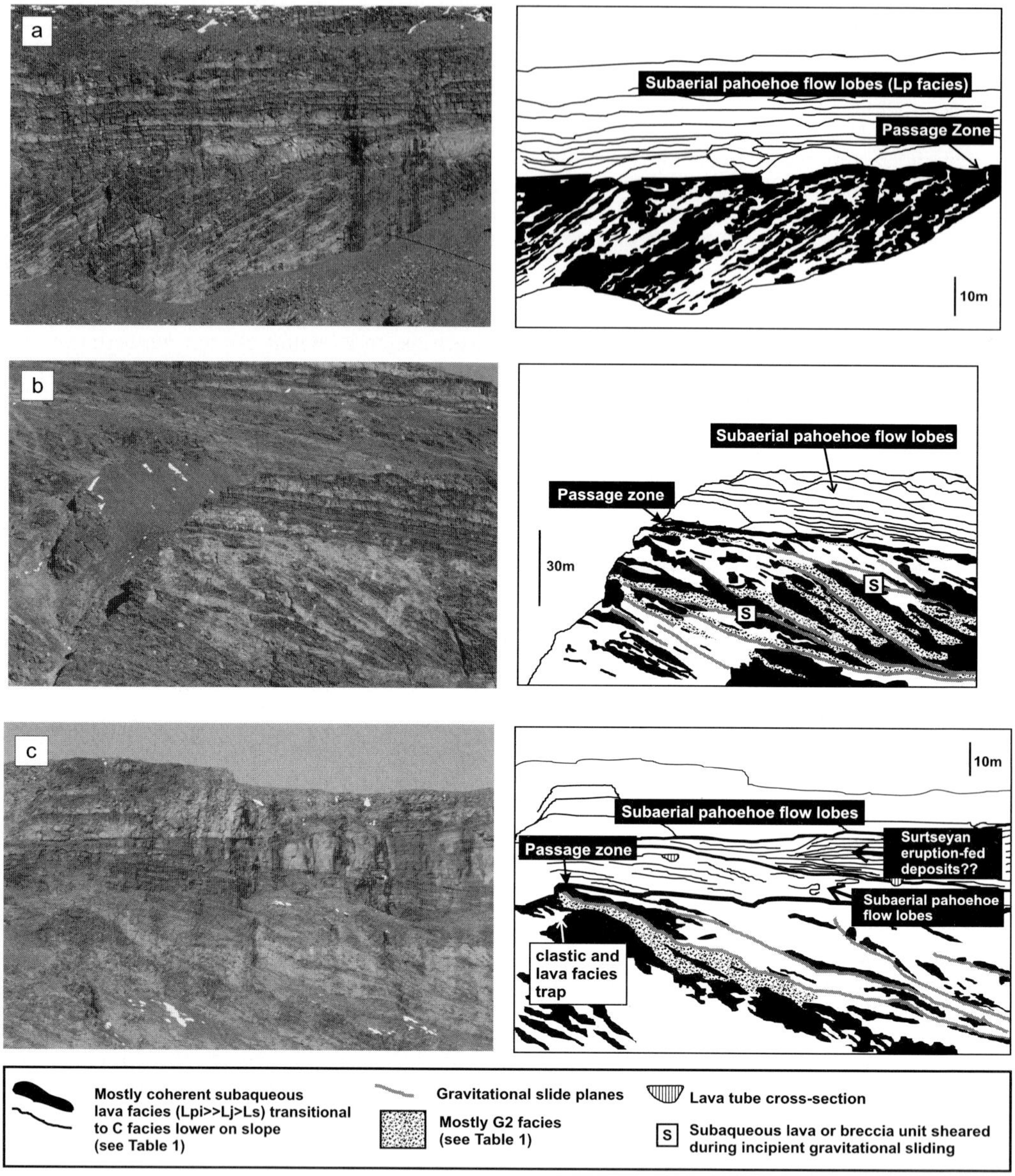

Fig. 5. Foreset-bedded sequences without prominent truncation surfaces (**a**) and with truncation surfaces (**b**, **c**). The truncation surfaces in these examples are interpreted as gravitational slide planes. Subaerial lava flows in (**a**) commonly pass directly into dipping subaqueous lava, but this is much less common in (**b**) and (**c**), suggesting that the connection between subaerial lavas and their subaqueous extensions was severed during sliding in the passage zone area. Sequences that display gravitational sliding (**b**, **c**) also contain a higher percentage of finer-grained facies (G2 facies), suggesting that large volumes of finer-grained sediment may be important in triggering some slides. Note the wedge of layered tephra in (**c**), which is photo-interpreted as Surtseyan eruption-fed deposits, that may have been generated by littoral explosions (process 5 on Figure 7a). Topographic barriers to downslope transport of sediment and lava that arise mostly from bodies of coherent lava (Lpi and Lj facies) ponded at the frontal crest or on the upper slope (see 'clastic and lava facies trap') are common in the JRIVG. Facies codes and their interpretation are given in Table 1.

(1970). The passage zone is horizontal or sub-horizontal in the JRIVG deltas studied by the author, but inclined, stepped or more complex passage zone structures have been recorded from elsewhere in the JRIVG and in other lava-fed deltas (Jones & Nelson 1970; Furnes & Fridleifsson 1974; Nelson 1975). Stacked sequences of lava-fed deltas are recorded from the JRIVG (Nelson 1975) and the Tertiary of West Greenland (Pedersen & Dueholm 1992).

Interpretation

The large-scale characteristics of an individual basaltic lava-fed delta are influenced by several, often interdependent factors, including the physical parameters of the lava (viscosity, volume, gas content and effusion rate), the location, number and spacing of feeding lava streams, coastal topography, immediate offshore bathymetry, tidal range, and coastal orientation with respect to predominant waves and/or fluvial dispersal axes. These factors cause spatial and temporal variations in the fragmentation mechanisms, volume, grain size, jointing and plasticity of clasts and their emplacement processes, and hence influence the large-scale characteristics of the delta. Other factors such as tectonics, subsidence, glacio-eustatic sea level change and climate change influence all deltaic sequences over a longer period of time.

The surface dimensions of the two-year old lava-fed delta (2×0.5 km) on Hawaii reported by Mattox & Mangan (1997) give some idea of the growth rate of basaltic pahoehoe lava-fed deltas. Though growth rates must be highly variable, these figures imply that the JRIVG deltaic sequences, illustrated in Figures 2 and 3, could have been constructed in only a few years. Deltas fed from continental flood basalts may grow more rapidly, especially if the vents were situated near the coast.

Relative water level changes cause variations in passage zone elevation. The effect of relative sea level changes on passage zone structure has been discussed by Jones & Nelson (1970), and Furnes & Fridleifsson (1974) discuss the effect of tides. Within any one submarine lobe or group of contemporaneous lobes, a maximum variation of *c.* 15 m may be due to tides, but larger variations may occur in marine and lacustrine deltas, as a consequence of rapid tectonic subsidence and drainage (discussed later). Passage zones that step downward in the progradation direction may be caused by downfaulting of the subaerial lavas (bench collapses) during delta progradation (discussed later), or by retrogressive slope failure during or after progradation. Stacked sequences of lava-fed delta deposits could arise from long-term processes, such as basin subsidence tectonics and glacio-eustatic sea level change, from rapid changes in englacial or non-glacial lacustrine levels, or even from large-scale sliding or slumping (Van der Straaten 1990).

Clast generation

Most clast generation processes described here, and their associated subaqueous emplacement processes, are illustrated conceptually in Figures 6a–d. Fragmentation of coherent subaerial lava at the shoreline can be non-explosive or explosive, brittle or ductile. Some fragmentation can take place prior to entry of a particular lava flow into the main body of water, including autoclastic flow-related fragmentation, various littoral fragmentation processes (discussed below) and gravity-driven ductile elongation, pinching and detachment (ductile or brittle) of low-volume lava streams that enter the water over a cliff or scarp (process 33, Fig. 6a). This process may generate low density scoria-like clasts, whose morphology can be modified by wave action (process 34, Fig. 6b) on entry into the water (Kalber 1995), but they are probably volumetrically insignificant. Crystalline clasts derived from several types of lava flow are commonly generated non-explosively (Figs 4d–f), along with vitric clasts of ash to block size. Moore *et al.* (1973) observed that small volume lava streams usually generated smaller clasts on entering the water than larger volume streams did. Clasts are generated mostly by brittle processes, including quench fragmentation, mechanical fragmentation due to movement or inflation of lava beneath a cooler crust and due to the downslope push of active lavas or mass flows higher on the slope. Other brittle processes include blocky peperite generation and fragmentation by wave and tidal activity. Non-explosive brittle fragmentation of coherent lavas commonly takes place along well-developed joints (Figs 4d & g), interpreted as cooling contraction joints resulting from chilling in water. Collapse of the lavas along these joint planes can be triggered by active lava movement or inflation, ploughing of the frontal crest deposits by subaerial lava flows (Fig. 4c), invasive flows (Pedersen *et al.* 1996), magma intrusion (Furnes & Fridleifsson 1979), marine erosional undercutting, slope failure of the underlying foresets or ice-related collapse (Fig. 7). Marine erosional undercutting is probably a particularly important process in undermining lava bodies in the littoral zone or frontal crests of lava-fed deltas in Hawaii. Subaerial lava clasts may also

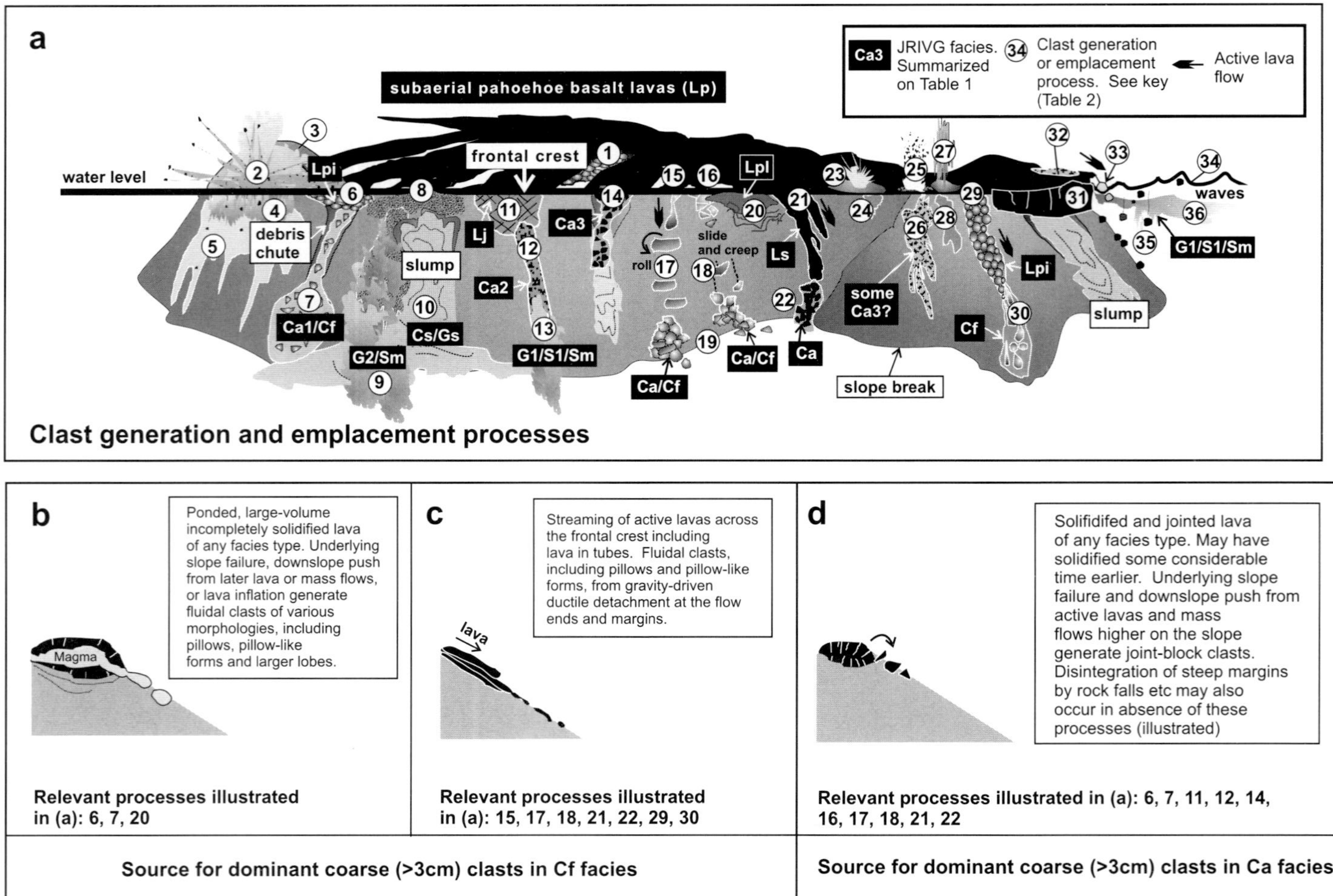

Fig. 6. (**a**) Conceptual depiction of the variety of clast generation and emplacement processes that may occur in a basaltic pahoehoe lava-fed delta. JRIVG facies described in this paper are also illustrated. Not all lava-fed delta processes are illustrated (see text and (**b**) and (**c**) for further details). Lt and Ll lava facies are also not illustrated. Note that large-scale gravitational slides and marine, fluvial and glacial processes that operate in alluvial deltaic environments are not illustrated. (**b**, **c**) interpretation of source for dominant coarse clasts in Cf facies. (**d**) Interpretation of source for dominant coarse clasts in Ca facies.

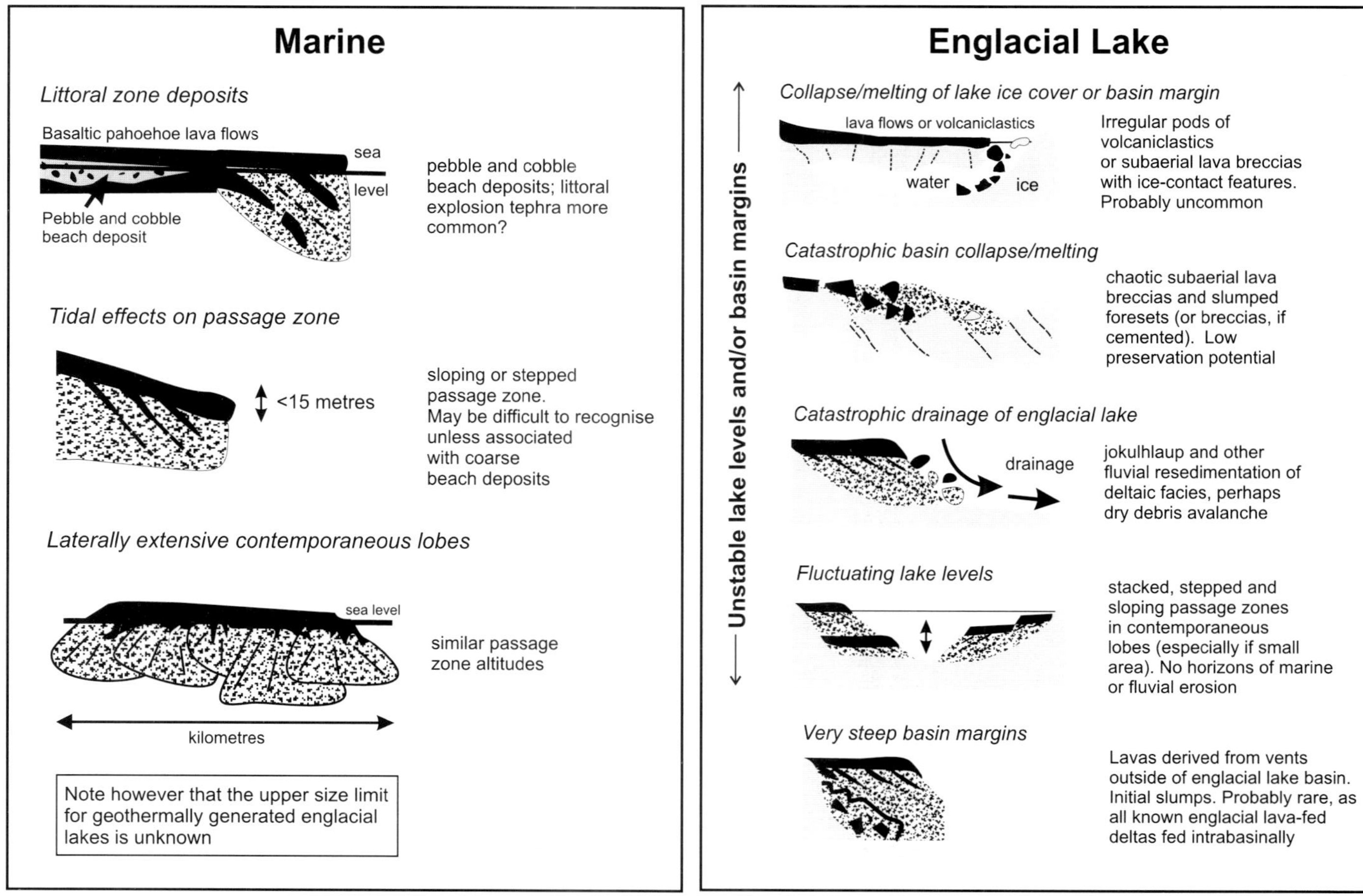

Fig. 7. Characteristics which may be used to distinguish marine and englacial basaltic lava-fed deltas. Note that non-glacial lacustrine deltas may also share many of the features of englacial examples, including catastrophic drainage and basin collapse (see text).

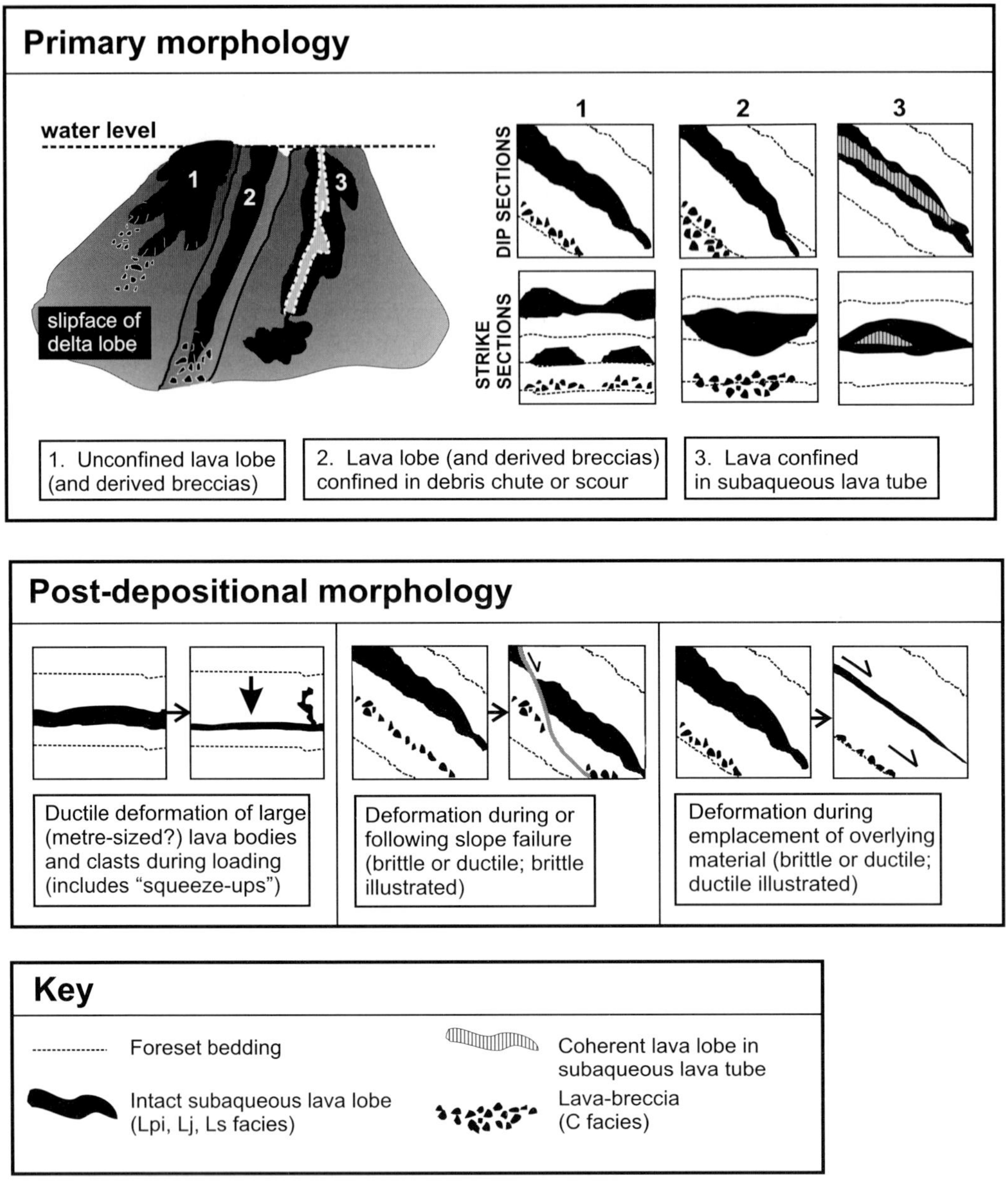

Fig. 8. Primary morphologies and post-depositional modification of subaqueous coherent lava lobes and their derived breccias. Note that post-depositional modification may deform the primary morphology of lava bodies and breccia units (Fig. 7b and c).

be generated by small tsunami-like events generated during bench collapse (Mattox & Mangan 1997). Non-explosive fragmentation may also be by ductile processes, the most important of which is gravity-driven elongation, pinching and ultimately detachment of the downslope ends of subaqueous lavas, especially actively-budding pillow lavas (fig. 7 in Moore *et al.* 1971). Gravity-driven elongation of subaqueous lavas on steep slopes may generate elongate relatively small pillows ('para-pillow lava' of Jones 1970; Walker 1992) from pillow lavas, and pillow-like or more irregular clasts or lobes with fluidal outlines from subaqueous sheet lavas.

The morphology of some of these ductile clasts may be modified during loading or downslope movement of the overlying or underlying beds (Fig. 8). Similar, but subaerial, ductile deformation of lava clasts during movement on steep slopes has been described by Sumner (1998). In the subaqueous environment such modification is only likely if the clasts or detached lobes are sufficiently large (>1 m?) to have retained a fluid interior during subsequent loading or slope failure.

Explosive fragmentation processes associated with the littoral zone of an active lava-fed delta are described by Mattox & Mangan (1997). They recorded four explosive processes which they termed tephra jets, lithic blasts, bubble bursts and littoral lava fountains. The first two of these processes they suggested were generated by wave-induced open water mixing of lava and seawater, and the second two by water flooding of lava in a confined area, such as a lava tube. Each of these processes produces a distinctive product, which should be preserved as interbeds between subaerial lava flows in ancient deltas, but none have been described to date. For this reason, it is not clear just how much of a typical basaltic lava-fed delta is composed of explosively generated clasts. During episodes of prolonged and localized explosive activity a littoral cone may be produced (Moore & Ault 1965, Fisher 1968; Jurado-Chichay *et al.* 1996). Littoral cone deposits commonly include a high percentage of coarse poorly vesicular agglutinated spatter. Spatter-rich beds or mega-blocks of spatter-rich material should also occur in the deposits of the littoral zone and subaqueous slopes of some ancient deltas. They have also not generally been recognized but an example is associated with englacial lava-fed deltas at Mount Murphy, Marie Byrd Land, Antarctica (J. L. Smellie, pers. comm.). Mattox & Mangan (1997) noted in their example, that Surtseyan-like tephra jets were the most common form of littoral explosion, and that they commonly followed bench collapse. This implies that slumped foreset beds and/or bench collapse structures in ancient deltas may be associated with explosion products, or their eruption-fed mass flow equivalents. Tephra jet eruptions produced angular ash and lapilli-sized clasts and spatter, that were associated with distinctive ash-sized Limu O' Pele and Pele's hair clasts (Hon *et al.* 1993). Figure 5c illustrates what may be Surtseyan eruption-fed deposits from littoral explosions associated with an englacial lava-fed delta in the JRIVG. Lenses of Surtseyan tephra have been described from foreset-bedded breccias (fig. 13 in Skilling 1994), and may have been derived from littoral explosion tephra, but could also in that case have been eroded from underlying subaqueous tuff cone deposits. Mattox & Mangan (1997) suggested that tephra jet explosions were restricted to conditions of relatively high-volume (*c.* 4 m^3/sec) focused lava entry at a single site, and also required contact between seawater and lava to be established over timescales of a few seconds or less. Tephra eruptions were commonly preceded by phreatic eruptions of lithic blocks ('lithic blasts') according to Mattox & Mangan (1997). This suggests that some lithic subaerial lava breccias interbedded with the onshore lavas or subaqueous breccias may have been fragmented explosively, especially if they are associated with slumped foreset beds. Lava bubble bursts and littoral lava fountains described by Mattox & Mangan (1997) may generate locally significant volumes of fluidal spatter, and associated Limu O' Pele and Pele's hair clasts.

Lithofacies descriptions (JRIVG)

Description and interpretation of the lithofacies of the JRIVG lava-fed deltas are summarized in Table 1. Fourteen lithofacies are recognized, divided into 7 coherent lava facies and 8 subaqueously-deposited clastic facies. Coherent lava facies include subaerial pahoehoe lava flows (Lp) and associated thin-layered lava flows (Ll facies). Other coherent lava facies include pod-like bodies of lava with zones of basal pillows and angular glassy lapilli or ash matrix (Lpl), pillow lavas (Lpi), curvicolumnar and blocky-jointed massive lavas (Lj, Fig. 4d), sheet-like flows with thicker glassy crusts (Ls) and lavas confined within a littoral or subaqueous lava tube (Lt). Continuity between subaerial lava flows (Lp) and dipping subaqueous lavas (Lpi $\gg$ Lj > Lt > Ls facies, Table 1) below the passage zone is commonly observed (Fig. 5a), although in the JRIVG examples studied, this is not the case for sequences that include surfaces that truncate the bedding of the underlying breccias (Figs 5b & c). The coherent subaqueous lavas, where present, vary in thickness from a few cm (Fig. 4h) to several metres, and may extend downslope for at least 200 m. Large bodies of Lpi, Lpl and Lj facies commonly occur on the passage zone or high on the delta slope (Figs 4c, d & 5c).

Steeply bedded (foreset-bedded) coarse-grained, clast-supported, poorly sorted breccias underlie the horizontal coherent lavas (Lp, Ll). These typically dip at angles of 20–40° in the upper 50 m of the foresets of the JRIVG deltas

Table 1. *Codes, descriptions and interpretation of the lithofacies of JRIVG lava-fed deltas. See text for further details of interpretation. Note that grain sizes S, G and C are mean values*

Facies group	Grain size	Code	Description	Interpretation
Coherent lava facies		Ll	Lensoid or pod-like lavas in passage zone area, with well-developed thin (<50cm) layering and thin internal glassy layers. No pillows or hydroclastic breccia	Subaerial lava flows (Lp) ponded (lava lake) in littoral area. Little or no interaction with water
Coherent lava facies		Lt	Massive lavas, often with radial jointing (in cross-section) infilling tube structure	Lava infilling lava tube (subaerial or subaqueous)
Coherent lava facies		Lp	Compound lava flows with pahoehoe surface crusts. Some reddened surfaces (<1-10m thickness)	Entirely subaerial lava flows (delta topsets)
Coherent lava facies		Lpl	Irregular or pod-like lava bodies in passage zone area whose bases may be mixed with underlying clastic facies (peperite). Tend to be large-volume and may be pillowed at their bases but resemble Lp	Subaerial lava flows (Lp) which ploughed into and mixed with sediment in the frontal crest area
Coherent lava facies		Lj	Irregular or pod-like bodies in the passage zone area or sheet-like units interbedded with Ca/Cf/G2 facies. Well-developed blocky or curvicolumnar joints. Tend to be large-volume (>2m thick)	Large-volume subaerial lavas (Lp) which ponded in the littoral zone/frontal crest and jointed due to cooling contraction in water
Coherent lava facies		Lpi	Pod-like pillowed bodies in the passage zone area (>3m thick) or more sheet-like pillowed flows (<1-3m thick) interbedded with Ca/Cf/G2 facies.	Pillow lavas. Ponded and lobes that travelled down the subaqueous slope
Coherent lava facies		Ls	Sheet-like lava flows interbedded with foreset-breccias (Ca/Cf/G2 facies). Not pillowed. Thickness varies from few cm to about 1.5m. Thick glassy crust (>5cm)	Subaqueous lava lobes that travelled down the subaqueous slope. Greater effusion rate than Lpi
Subaqueously-deposited clastic facies	cobble (C)	Ca	>50% cobble-sized clasts. Angular predominantly lithic lava clasts of sand to boulder size, of Lpi>>Lp>Lj types (no Ls-derived clasts recognised). Ca1: >50% Lpi-derived clast; Ca2: >50% Lj-derived clasts; Ca3: >50% Lp-derived clasts. Blocky-jointed clasts of indeterminate origin are common (Ca facies). Bedded with dips of >15-35 degrees. Matrix (10-50%) is dominantly sand-pebble sized poorly-vesicular glassy clasts. Beds are typically 15cm-2m thickness (thinner on average than Cf). Clast-supported, poorly sorted, no clast imbrication, massive but some rare indistinct planar stratification and rare basal reverse grading. Non-erosive broadly planar bases common. Irregular morphology to top of units is present sometimes, but tend to be convex topped. Wedge-shaped and tabular sheet-like units, but may also infill steep chute-like structures. Interbedded with G1, G2 and minor Sm	Gravity-driven joint-block fragmentation of lava facies. Density-modified grain flows (cohesionless debris flows), with some transitional to high-density turbidites (especially chute-fills). Matrix is quench and mechanically-fragmented vitriclasts. Bedding reflects collapse periodicity due for example to episodic slope failure. Thinner bedding than Cf facies due to derivation from smaller volume or lower aspect ratio cooled/jointed lava (Fig. 7d)
Subaqueously-deposited clastic facies	cobble (C)	Cf	As above, but lava clasts have dominantly fluidal outlines, including pillow-like forms. Some larger (>50cm?) Clasts may undergone post-depositional ductile modification. More variable clast sizes than Ca, usually coarser. Bed thicknesses are typically greater than Ca facies, and may be up to 5m	Similar origin to Ca, but source of coarse clasts is incompletely solidified lava bodies of all types (Fig. 7b, c).
Subaqueously-deposited clastic facies	cobble (C)	Cs	Lithologically similar to Ca and Cf facies, but often contains localised pods of Ca3 facies in particular. Folded and complexly deformed beds	Slumped Ca and Cf units and associated undermining of subaerial lavas (in particular) giving pods and lobes of Ca3
Subaqueously-deposited clastic facies	granule (G)	G1	Indistinct to distinct parallel lamination. Thin beds (<30cm, usually few cm). Always associated with tops of Ca or Cf beds, and Sm facies. S1 is sand-sized equivalent	Small-volume buoyant suspension plumes generated above Ca and Cf flows during downslope transport or from hot water
Subaqueously-deposited clastic facies	granule (G)	G2	>50% sand-pebble size, typically poorly vesicular, angular vitric clasts. Outsized lithic clasts are common. Massive, but some indistinct parallel stratification. Poorly sorted, no imbrication. Some normal graded tops. Beds of <1m to 12m thickness commonly tabular, but also infill steep chute-like structures. Pods of Sm facies	High-density turbidite deposits or density-modified grain flow deposits. Dominated by large volumes of quench and mechanically fragmented granule-sized vitriclasts. Episodic failure of frontal crest
Subaqueously-deposited clastic facies	granule (G)	Gs	Lithologically similar to G2 facies. Folded and complexly deformed beds	Slumped deposits of G2 facies
Subaqueously-deposited clastic facies	sand (S)	Sm	Silt-coarse sand sized. <1-75cm thick, irregular or pod-like morphology. Associated with G1 facies and within Ca and Cf faces units. Well-sorted. Massive, but may have some laminae draping clasts in surrounding facies	Sieve deposits derived from bouyant suspension plumes (G1, G2 facies)
Subaqueously-deposited clastic facies	sand (S)	S1	Sand-sized equivalent of G1 facies	As G1

(Fig. 5a). They are subdivided into several lithofacies, but consist mostly of several types of dominantly clast-supported, cobble-sized lithic lava breccias (C facies; equivalent to CH facies of Skilling 1994), which are described and interpreted in Table 1. They are equivalent to the 'flow-foot breccias' of Jones and Nelson (1970). Beds within any one area are commonly of similar thickness (Fig. 4g), convex-up in strike section, often wedge-shaped in longitudinal section, but their primary morphology may be modified by later processes (Fig. 6). In some areas the breccias are truncated by surfaces, with angles that are usually close to that of the foreset bedding (Figs 5b & c). C facies are subdivided into Ca (angular lithic breccias) Cf (fluidal lithic/vitric breccias) and Cs facies (slumped Ca and Cf). Breccias comprising mixtures of angular and fluidal clasts also occur in the JRIVG. All the C facies are commonly interbedded. Ca facies are further divided into three subfacies, termed Ca1, Ca2 and Ca3 facies, which are described in Table 1, and equivalent to the SC facies described by Skilling (1994). Ca facies often occurs as one or more beds within scour or steep chute structures, which are visible in strike sections across the foreset bedding (Ca1 facies: Fig. 4e; figs 5 and 6 in Skilling 1994). Cf facies are characterized by the presence of pillow, pillow-like and other irregular fluidal forms, which are dominantly vitric to finely crystalline, and of highly variable size (fig. 13 in Skilling 1994).

C facies breccias commonly alternate with finer units, of dominantly granule-size material, which are termed vitric granule tuffs (G facies; equivalent to GHb facies of Skilling 1994). G facies are subdivided into G1, G2 and Gs types. All types are composed dominantly of angular, poorly vesicular sideromelane glass fragments, which are identical to the matrix grains in the C facies breccias. G facies display a more tabular morphology than the C facies. G1 facies is commonly only a few cm in thickness (Fig. 4g), whereas G2 facies may be several metres in thickness. Two facies with a sand-sized mean grain size were recognized (Sm and S1).

Emplacement processes

Possible emplacement processes associated with basaltic pahoehoe lava-fed deltas, and those interpreted to be associated with known lithofacies in the JRIVG are illustrated in Figure 6. Interpretation of the emplacement processes of the JRIVG lithofacies described above is given in Table 1.

The development of pillow lavas rather than non-pillowed and sheet-like lavas in the littoral zone or on the subaqueous slope is probably due to a lower extrusion rate (Griffiths & Fink 1992; Gregg & Fink 1995). Coherent subaerial lavas with dipping subaqueous extensions (Lpi, Ls and Lj facies) represent flows that travelled over the frontal crest and maintained coherence subaqueously. Tribble (1991) noted that an active lava-fed delta slip-face was sometimes covered with 60–70% pillow lavas of elongate 'para-pillow' form (Jones 1970; Walker 1992), but suggested that 20% was typical. Invasive flows have also been reported from the littoral zone of lava-fed deltas (Pedersen *et al.* 1996), suggesting an additional mechanism by which dipping subaqueous extensions of subaerial lavas may be produced.

Gravity-driven processes dominate all steep-face deltas so consideration of the factors affecting slope stability is important. Many of these processes are common to steep coarse-grained alluvial deltas, such as oversteepening due to relatively greater amounts of deposition of sediment at the frontal crest compared to the delta slope, undercutting by marine erosion, progradation over irregular bottom topography and over-pressured pore water. However, some factors are specific to lava-fed deltas, including high-temperature clasts that may initiate fluidization and/or liquefaction of sand- or even granule-grade layers. Such layers are generally much less common than coarser facies, but may nevertheless be an important factor in some slope failures (Figs 5b & c). The ductile nature of some inflating subaqueous lava flows, feeder tubes and even detached large lava clasts may also trigger slope failure by lowering the coefficient of friction. Large-scale collapse may also be triggered by lava bodies ploughing into or invading sediment, or ponding behind earlier lavas, especially at the frontal crest (Figs 4c & 5c). Kauahikaua *et al.* (1993) suggested that inflation of subaqueous lava tubes within the breccia pile could be responsible for some slope failures.

Evidence of gravity-driven sliding and large-scale slumping is apparent in the JRIVG delta deposits. Surfaces which truncate foreset-bedding and sever the connections between subaqueous coherent lava lobes (Figs 5b & c) and their parent overlying subaerial lavas are interpreted as detachment surfaces associated with gravitational slides, with variable amounts of downslope displacement. Sheared bodies of pillow lava and derived breccia can be matched to nearby units below the slide plane in some cases (Fig. 5b), suggesting only incipient slide development. Possible gravitational slide planes have been reported from other lava-fed deltas (Porebski & Gradzinski 1990; Pedersen & Dueholm 1992; Pedersen *et al.* 1993; LeMasurier *et al.* 1994; Moore *et al.* 1995). Large-scale slumps have not been previously recorded from lava-fed deltas but are commonly described from alluvial deltas (Postma 1984; Nemec *et al.* 1988). The occurrence in the JRIVG of a slump zone located at the contact between two adjacent lobes (slump zone 1, Fig. 3) suggests that elevated depositional rates at the contact of simultaneously active lobes may be a factor in generating slumps in such areas. Other slump zones that apparently occur within the same lobe may have formed for several reasons. Slump zone 1 in the Stark Point section (Fig. 2) lies above an apparent diamictite unit(s) with a prominent surface topography, suggesting that progradation across areas with marked (>10 m?) bottom topography may also be important in slump generation. Some collapses of englacial deltas may also be related to catastrophic lake drainage, or collapse or melting of the confining ice walls or floor. Large-scale delta-front collapses of any origin could initiate the failure of overlying subaerial lavas, due to withdrawal of support, such that slumped foresets will commonly be associated with megablocks of subaerial lavas or breccias derived from subaerial lavas. Collapse of the foresets in Hawaii commonly downfault a block of subaerial lavas and underlying breccias only a short distance, a structure which is termed a 'bench collapse'

Table 2. *Key to clast-generation and emplacement processes illustrated in Figure 6a*

(1) pillow lava or hydroclastic breccia generated in the littoral zone in ponded water

(2) littoral tephra-jet explosion

(3) littoral cone, constructed from variety of littoral explosions types. Here associated with tephra jet explosions

(4) subaqueous fallout from tephra jet explosions

(5) subaqueous mass flows derived from 4

(6) pillow lavas ponded or ploughed into sediment at the frontal crest/upper slope

(7) high-density turbidite or density-modified grain flow fed from mass failure of (6). Here shown in a debris chute. Fluidal clasts generated if (6) is incompletely solidified

(8) ponded granule-grade quench and mechanically-fragmented lava

(9) high-density turbidite fed here from mass failure of (8)

(10) slump (derived from any facies, but especially Ca, Cf and G2 facies)

(11) Ponded, hydrofractured, blocky and curvicolumnar-jointed lava

(12) density-modified grain-flow fed here from mass failure of (11)

(13) small-volume suspension plume derived from 12 during downslope movement

(14) subaerial lava breccia derived here during slumping of underlying foresets

(15) subaqueous gravity-driven elongation, pinching and detachment of ductile lava clasts from small lava stream (process 33 is similar but subaerial).

(16) subaerial gravity-driven brittle detachment of small numbers of lava clasts

(17) rolling of individual clasts derived from process 15, 16 etc. Rolling of clasts may be associated with many other downslope processes

(18) sliding and creep of individual clasts. Sliding and creep of clasts may be associated with many other downslope processes

(19) ponding of coarse clasts at the slope break. This could happen with coarse clasts of any origin

(20) subaerial lava flow which has ploughed into and mixed with underlying sediment at its base

(21) subaqueous sheet lava

(22) mass flow derived from 21

(23) littoral "bubble burst" and associated small subaerial cone

(24) subaqueous deposits derived from 23

(25) littoral lithic-rich phreatic eruption. "Lithic blast" of Mattox and Mangan (1997)

(26) subaqueous deposits derived from 25

(27) littoral lava fountain

(28) subaqueous deposits derived from 27

(29) pillow lava

(30) density-modified grain-flow derived from gravity and inflation-driven downslope budding/pinching and ductile detachment. Here shown derived from the front of an active pillow lava.

(31) displaced/downfaulted block of subaerial lavas (bench collapse). Here shown as due to slumping of underlying foresets

(32) onshore subaerial lava breccia, generated by small tsunami-like events associated with failure of the subaqueous slope

(33) subaerial gravity-driven elongation, pinching and ductile detachment of small lava stream (process 15 is the subaqueous equivalent)

(34) moulding by wave activity of low density (scoria-like) ductile clasts derived by process 33

(35) subaqueous fallout of wave-entrained and perhaps wave-moulded clasts below wave base

(36) fine-grained sediment entrained in buoyant hot water plume (process 31, Fig. 6a). Such collapses are repeatedly generated during delta progradation (Mattox & Mangan 1997). Large blocks of subaerial lava flows within foreset-bedded breccias of a lava-fed delta have been described from West Greenland (Henderson 1977).

Several transport and depositional processes might operate on a steep, coarse-grained, cohesionless face (fig. 6 in Nemec 1990) but the steep bedding (>20°), occasional basal reverse-grading, non-erosional bases, lack of clast imbrication and traction structures, poor sorting and convex tops of the dominant C facies beds, imply that most of the JRIVG foreset beds are deposited by high-concentration density-modified grain flows (Lowe 1976, 1979). Their dominant clast content implies that they are mostly derived by gravity-driven brittle failure of the thoroughly jointed bodies of completely solidified coherent lavas higher on the slope (Fig. 6c). Thinner beds compared to Cf facies suggests derivation from smaller lava bodies or lobes (Fig. 6c), which is consistent with their brittle fragmentation. Tangential foreset bedding of C facies breccias (Fig. 3) implies that some fluid turbulence was involved in transport of sediment past the slope break, so some flows at least were probably transitional to high-density turbidity currents (Lowe 1982). Downslope wedging probably reflects the limited mobility of such high bulk density flows. Porebski & Gradzinski (1990) recorded a change from planar-angular foresets to tangential foresets with time in a marine lava-fed delta, and suggested on the basis of facies analysis that this reflected a shift in dominant processes from grain avalanching to later slump-generated density-modified grain flows and high-density turbidity currents. Such a change was not apparent in any of the JRIVG deltas studied. C facies breccias in the JRIVG that occur within eroded debris chutes, and which can display well-developed normally graded tops, were probably emplaced by high-density turbidity currents (Lowe 1982). Bedding visible in the lensoid structure (chute?) in Figure 2 may represent several high-density turbidites, which accumulated in the same chute.

Most of the large lava clasts in the JRIVG deltas were probably transported downslope within flows, but some clasts or groups of clasts probably also rolled, crept or slid down the foreset separately from the flows (processes 17 and 18, Fig. 6a). The irregular tops to C facies beds probably reflect such movements of a small number of the uppermost clasts after the main body of the flow had come to rest.

Cf facies clasts are derived from ductile fragmentation of upslope lava bodies. Many of these clasts resemble pillows or plastically deformed pillows (fig. 13 in Skilling 1994), which suggests their derivation from upslope pillow lavas. Many fluidal clasts in Cf facies were probably derived from actively budding pillows that broke off by gravity-driven elongation, pinching and detachment (process 30, Fig. 6a). Similar subaerial or subaqueous gravitational pinching of non-pillowed lava flows can, however, also generate clasts with pillow-like or more irregular outlines (processes 15 and 33, Fig. 6a). Large (metre-sized) incompletely solidified lava lobes and clasts may be plastically deformed by shearing, rotation and compaction during loading and movement, especially sliding of the overlying and underlying beds (Figs 4h & 8; fig 13 in Skilling 1994). The presence of thicker bedding and higher temperature clasts in Cf facies compared to Ca facies, suggests that they were generated from larger volume, perhaps inflated or tube-fed, lavas that were ponded upslope and had undergone less cooling prior to mass failure (Fig. 6b), or from continuous streaming of large volumes of lava over the frontal crest (Fig. 6c), a mechanism observed by Tribble (1991).

Interbedded G1 facies are interpreted as suspension deposits from low-volume buoyant plumes of finer sediment, that were winnowed from the underlying density-modified grain flows during downslope movement (Sohn *et al.* 1997), from hot water plumes (Sansone *et al.* 1991), and/or are deposits derived from post-emplacement sediment liquefaction (Postma 1984). Thicker-bedded G2 facies are interpreted as density-modified grain flow deposits or perhaps Ta layers of high-density turbidites (Lowe 1982). Depositional processes were similar to those that formed the C facies, but reflect a higher proportion of finer-grained glassy clasts generated at the frontal crest, perhaps as a consequence of many small-volume lava feeder streams (Moore *et al.* 1973). Some G facies may also represent mass flows of littoral eruption-fed tephra. Gravitational slide planes are most common in beds where the proportion of granule and sand-sized sediment is relatively high (Figs 5b, c), suggesting that these beds act as slide planes, perhaps due to their susceptibility to fluidization, sediment liquefaction or shear liquification. Other processes that could operate on the subaqueous slope of a basaltic pahoehoe lava-fed delta are illustrated in Figure 6a and keyed on Table 2.

Comparison with alluvial deltas

Lava-fed deltas are similar to coarse-grained, steeply bedded alluvial deltas, with Gilbert-type

(Gilbert 1885) or gravitationally-modified Gilbert-type profiles (Postma & Roep 1985; McCabe & Eyles 1988; Postma 1990; Postma 1995). Lava-fed deltas can be compared with Gilbert-type deltas with 'type-A' feeder streams (Postma 1990), which are characterized by a high percentage of bedload sediment, and the development of a steep subaqueous slope, dominated by gravity-driven processes. All Gilbert-type deltas also require water immediately offshore that is sufficiently deep to allow the development of a gravity-driven slip-face (Wright 1977). If water depths are shallow (<10 m) then bottom friction becomes dominant, which normally results in a gently inclined delta front (Hjulström-type profile) with dips of a few degrees, and the absence of a clear delta slope (Wright 1977; Postma 1990). No lava-fed deltas with Hjulström-type profiles have been described. Subaerial lava-fed hydroclastic breccias must occur in such shallow water settings but they would not resemble Gilbert-type deltas, and are more likely to be massive or chaotic due to en masse deposition and little or no resedimentation (Postma 1990). This suggestion is supported by the absence of foreset-bedding in the frontal crest area of one of the JRIVG deltas (Fig. 4c). Most coastlines of active volcanic oceanic islands have a steep bathymetry, which would favour slip-face-dominated deltas.

One of the most important differences between alluvial and lava-fed deltas is that the clasts in the latter are not transported into the basin by an effluent force, so that most of the coarser components fall immediately through the water at the point of entry, generating a steep subaqueous slope at the angle of repose. Lava-fed deltas are, in this sense, analogous to subaqueous debris or talus cones, where the sediment is mostly first emplaced onto the delta face by subaqueous debris fall. Wave and tidal energy and lava-heated water can suspend and transport some of the finer fraction in lava-fed deltas (Sansone *et al.* 1991). Suspension of a small volume of fine-grained clasts also occurs due to sorting during downslope transport in lava-fed deltas (see above). Products of such buoyant sediment plumes, however, are volumetrically insignificant because relatively little suspendable sediment is generated at the frontal crest. This implies that significant prodelta sedimentation may be absent in lava-fed deltas, and that if present, prodelta sediment is likely to have a sharp contact with the delta slope (Massari & Parea 1988; Prior & Bornhold 1990). However, the occurrence of finer-grained asymptotic foresets in some lava-fed deltas (Porebski & Gradzinski 1990; Skilling 1994) implies that sometimes there are sufficient finer-grained sediments to be entrained by fluid turbulence and deposited past the slope break. The lack of an effluent fluid and the high percentage of coarse- to fine-grained clasts also implies that delta front deposition and progradation will be very rapid compared to most alluvial deltas. This situation gives rise to a strongly concave-upwards profile and increases the potential for voluminous mass failures and gravity-driven mass flows (Postma 1984; Nemec *et al.* 1988).

Failure of the delta front may also be more common in basaltic lava-fed deltas than in most steep alluvial deltas, due to ponding of sediment and particularly lava, behind cooled subaqueous lava bodies in the frontal crest area and upper slope (Fig. 5c). The high density, often very coarse (metre-sized) lava bodies and clasts, their typical low vesicularity, and the possibility of thermally-induced fluidization and sediment liquefaction also favour high rates of slope failure. The fact that the majority of clasts are generated locally in the frontal crest region in lava-fed deltas also means that their deposits are more poorly sorted than those of most alluvial Gilbert-type deltas. Poor sorting also favours emplacement by mass flows and the development of debris chutes (Prior *et al.* 1981; Postma & Cruickshank 1988; Corner *et al.* 1990; Prior & Bornhold 1989, 1990). Orton and Reading (1993) noted that in steep alluvial deltas, chutes may arise from slumping or erosion by turbidity currents. Gravitational slides are also commonly reported from deposits of coarse-grained alluvial deltas (Postma 1984; Postma *et al.* 1988).

Environmental discrimination

In order to use data from lava-fed deltas for palaeoenvironmental studies, it is necessary to distinguish deltas that form in marine environments, englacial lakes and non-glacial lakes. Unequivocal marine or glacial sedimentary facies, fossils and post-depositional characteristics such as marine erosional platforms or evidence of deformation by ice are useful criteria, but may be absent. Environmental discrimination is particularly difficult with coastal lava-fed deltas, such as those of the JRIVG and some Icelandic examples. Figure 8 illustrates characteristics that may be used to recognize englacial lake and marine basaltic lava-fed deltas.

Detailed analysis of the littoral facies is the best method to identify marine and englacial lava-fed deltas. Deposits of key environmental significance occur in this area, and are also likely to be preserved between relatively erosion resistant subaerial lava flows. Well-sorted beach deposits comprising mostly sub-rounded to well-rounded lava pebbles and cobbles are described by several authors from marine lava-fed deltas (Furnes & Sturt 1976; Bluck 1982; Larsen & Pedersen 1990; Pedersen *et al.* 1996) and marine alluvial deltas (Postma & Cruickshank 1988). They may not be laterally extensive for several reasons, including their coarse grain size, and the typical steep offshore bathymetry of active volcanic coasts which means that wave energy is commonly dissipated back offshore, leading to little longshore displacement of sediment (Orton & Reading 1993). Coarse-grained sheet-like deposits with rounded clasts may also represent fluvial deposits, but these should be easy to distinguish from marine beach deposits. Pedersen *et al.* (1996) described an example of fluvial deposits on lava-fed delta tops and foresets, emplaced after the re-establishment of drainage from a former lava-dammed lake. The absence of pebble- and cobble-grade beach deposits would provide some support for an englacial origin, but their absence does not exclude a marine origin (Furnes & Fridleifsson 1974). Wave action is particularly strong along the coastlines of most volcanically active oceanic islands, which tend to have steep offshore bathymetry. Tidal effects on the passage zone of marine lava-fed deltas are described by Furnes & Fridleifsson (1974), but in the absence of associated pebbly beach deposits, the variations in passage zone altitude may not be easily recognized, even in areas with a large tidal range.

Products of littoral explosions may be less common in englacial or other lacustrine lava-fed deltas than in marine examples, because wave activity seems to be an important factor in their generation, and is also important in the generation of littoral lava tubes, which would provide confined conditions for explosive magma-water interaction (Papale & Dobran 1993; Mattox & Mangan 1997).

Marine and non-glacial lacustrine lava-fed deltaic sequences of similar age may extend laterally for kilometres, whereas known equivalent englacial lake sequences are much more restricted in volume, and typically extend for <2 km (Jones 1969, 1970; Skilling 1994; Smellie 2001). The lake generated during the small-volume englacial eruption at Gjálp in Iceland in 1996 was only 200–300 m wide (Gudmundsson *et al.* 1997). However, the upper size limit for geothermally generated englacial lakes is unclear. Passage zone altitudes in marine lava-fed sequences will also be consistent laterally across the width of a group of broadly contemporaneous lobes. Rapid changes in basin volume, bathymetry and water level typify small englacial lakes due to ice collapse, melting, drainage and rapid infilling with volcaniclastic sediments or tephra. These changes will be recorded in terms of the development of large differences in the passage zone elevation of associated delta lobes, including the emplacement of subaerial lavas in a dry moat (Fig. 2; Skilling 1994; fig. 7 in Smellie & Skilling 1994). Rapid drainage of the lake may also be associated with catastrophic flood (jökulhlaup) deposits and possibly dry debris avalanche deposits (Skilling 1994). However, rapid changes in water level are not unequivocal evidence of an englacial origin, as they also occur in non-glacial lacustrine and marine environments. Pedersen *et al.* (1998) described similar structures in a non-glacial lake, which they suggested was due to catastrophic rupture of a lava dam, and Pedersen & Dueholm (1992) also described subaerial lavas stacked against foresets in a marine lava-fed delta, which they interpreted as due to syn-volcanic basin uplift.

Variations in shoreline topography and bathymetry immediately offshore affect delta characteristics (see above). Marine and non-glacial lacustrine basins can have laterally extensive shallow margins, implying that shallow-water lava-fed deltas or breccias dumped *en masse* might develop in these environments. Under most conditions, however, englacial lakes will have relatively deep water and steep margins, due to calving of the basin walls, but shallow-water deltas may develop in small englacial lakes when subaerial lava flows prograde across a lake basin that is nearly filled with earlier volcaniclastic sediment. In this case lava could overflow the englacial lake basin, travel across the ice surface, and generate a lava-fed delta in an adjacent englacial lake. This would probably generate an initial chaotic breccia (Fig. 7), with lava clasts displaying ice-contact features (Lescinscky & Fink 2000), and a later Gilbert-type delta. However, all known examples of englacial lava-fed deltas are within subaqueous to emergent sequences, where they commonly overlie subaqueously emplaced Surtseyan tephra, which was derived from a vent(s) situated within the lake (Jones 1969, 1970; Skilling 1994; Smellie 2001). Steep foreset-like bedding is commonly developed in this tephra, which may be conformable with the overlying lava-fed delta foresets (fig. 8 in Skilling 1994), and will favour lava-fed deltas with Gilbert-type profiles. Similar subaqueous to emergent sequences can also occur in marine and non-glacial lacustrine settings.

Further research

Study of the volcanology and sedimentology of lava-fed deltas is in its infancy. Correlation between the physical parameters of subaerial lava flows and the nature of coherent or clastic products generated at the waters edge in active deltas is needed. We also have no observations of the processes or deposits on active deltas at depths below about 50 m, so submersible studies would be valuable. In ancient deltas, a detailed facies analysis of individual asymptotic foreset beds and prodelta deposits would provide information on the type of material delivered across the slope break and its emplacement process. Detailed studies of the influence of basin morphology, bathymetry and sediment supply on lava-fed delta architecture and facies are required. Study of lava-fed deltas would also benefit greatly from detailed comparative studies with coarse-grained alluvial deltas. We know very little about erosion and resedimentation of lava-fed deltaic sequences by marine, fluvial or glacial processes, including rapid lake drainage, which could serve as useful environmental discriminants. It is hoped that this paper will be a useful framework for future detailed sedimentological and volcanological studies of lava-fed deltas.

Conclusions

Basaltic lava-fed deltas are important for palaeoenvironmental studies but their processes and products, with the exception of the littoral explosive processes, remain poorly described and understood. The large-scale characteristics, and many of the subaqueous processes of lava-fed deltas are comparable to those of Gilbert-type and gravitationally-modified Gilbert-type alluvial deltas. Important contrasts with alluvial deltas that influence the processes on lava-fed examples are the lack of any effluent force and the high temperature of the clasts. There are several possible clast generation and emplacement mechanisms that could occur in lava-fed deltas, but study of lava-fed delta deposits of the JRIVG demonstrates that gravity-driven non-explosive fragmentation of coherent, often well-jointed, lavas was the dominant clast generation mechanism for these deltas. In the JRIVG examples, lavas were commonly ponded in the frontal crest area, prior to collapse, which was probably triggered by several mechanisms. Cobble-sized and larger clasts can be subdivided into angular (brittle) and fluidal (ductile) types. Fluidal clasts of this size were derived mostly by collapse of large volume, incompletely cooled or actively inflating ponded lavas, or by gravity-driven elongation, pinching and ductile detachment of the downslope ends of pillow and subaqueous sheet lavas. Emplacement in the JRIVG deltas was dominantly by density-modified grain flows, probably transitional to high-density turbidity currents. High-density turbidites are often associated with debris chutes, which are often described from similar alluvial deltas. Gravitational slides and large-scale slumps are commonly observed in the JRIVG deltas.

Fieldwork on the JRIVG was undertaken whilst the author was employed by the British Antarctic Survey. I wish to thank the reviewers J. White and J. Gilbert for their detailed comments, and J. Smellie and R. Batiza for useful discussion.

References

BIBBY, J. S. 1966. The stratigraphy of part of North-east Graham Land and the James Ross Island Group. *British Antarctic Survey Scientific Reports*, **53**.

BISHOP, P. 1985. Early Miocene flow-foot breccia from the Upper Lachlan Valley, New South Wales: characteristics and significance. *Australian Journal of Earth Sciences*, **32**, 107–113.

BLUCK, B. J. 1982. Hyalotuff deltaic deposits in the Ballantrae ophiolite of SW Scotland: evidence for crustal position of the lava sequence. *Transactions Royal Society Edinburgh: Earth Sciences*, **72**, 217–228.

CORNER, G. D., NORDAHL, E., MUNCH-ELLINGSEN, K. & ROBERTSON, K. R. 1990. Morphology and sedimentology of an emergent fjord-head Gilbert-type delta: Alta delta, Norway. *In*: COLELLA, A. & PRIOR, D. B. (eds) *Coarse-grained deltas*. International Association of Sedimentologists, Special Publications, **10**, 155–168.

FISHER, R. V. 1968. Puu Hou littoral cones, Hawaii. *Geologische Rundschau*, **57**, 837–864.

FURNES, H. & FRIDLEIFSSON, I. B. 1974. Tidal effects on the formation of pillow lava/hyaloclastite deltas. *Geology*, **2**, 381–384.

FURNES, H. & FRIDLEIFSSON, I. B. 1979. Pillow block breccia – occurrences and mode of formation. *Neues Jahrbuch fur Mineralogie und Palaeontologie*, **3**, 147–154.

FURNES, H. & STURT, B. A. 1976. Beach/shallow marine hyaloclastite deposits and their geological significance – an example from Gran Canaria. *Journal of Geology*, **84**, 439–453

GILBERT, G. K. 1885. The topographic features of lake shores. *Annual Report of United States Geological Survey*, **5**, 75–123.

GREGG, T. K. P. & FINK, J. H. 1995. Quantification of submarine lava-flow morphology through analog experiments. *Geology*, **23**, 73–76.

GRIFFITHS, R. W. & FINK, J. H. 1992. Solidification and morphology of submarine lavas: a dependence on extrusion rate. *Journal of Geophysical Research*, **97** 19 729–19 737.

GUDMUNDSSON, M. T., SIGMUNDSSON, F. & BJÖRNSSON, H. 1997. Ice-volcano interaction of the 1996 Gjálp subglacial eruption, Vatnajökull, Iceland. *Nature*, **389**, 954–957.

HAMBLIN, W. K. 1994. *Late Cenozoic lava dams in the western Grand Canyon*. Geological Society of America, Boulder, Memoirs, **183**.

HENDERSON, G. 1977. Features of the Tertiary volcanic rocks of the Niaqornat area, Nûgssuaq. Rapp. *Grønlands Geologiske Undersøgelse*, **79**, 17–25.

HICKSON, C. J. 2000. Physical controls and resulting geomorphological forms of Quaternary ice-contact volcanoes in western Canada. *Geomorphology*, **32**, 239–261.

HON, K., HELIKER, C. C. & KJARGAARD, J. I. 1993. The construction of pahoehoe lava deltas on Kilauea Volcano, Hawaii. *Eos*, **74**(3), 616.

JONES, J. G. 1969. Intraglacial volcanoes of the Laugarvatn region, south-west Iceland, I. *Quarterly Journal of the Geological Society of London*, **124**, 197–211.

JONES, J. G. 1970. Intraglacial volcanoes of the Laugarvatn region, southwest Iceland, II. *Journal of Geology*, **78**, 127–140.

JONES, J. G. & MCDOUGALL, I. 1973. Geological history of Norfolk and Philip Islands, southwest Pacific Ocean. *Journal of Geological Society of Australia*, **20**, 239–257.

JONES, J. G. & NELSON, P. H. H. 1970. The flow of basalt lava from air into water – its structural expression and stratigraphic significance. *Geological Magazine*, **107**, 13–19.

JURADO-CICHAY, Z., ROWLAND, S. K. & WALKER, G. P. L. 1996. The formation of circular littoral cones from tube-fed pahoehoe: Manoa Loa, Hawaii. *Bulletin of Volcanology*, **57**, 471–482.

KALBER, M. 1995. *Volcanoscapes Four: Kilaueas's Flow to Kamoamoa!* VHS video, Tropical Visions Video Incorporated, Hilo, Hawaii, USA.

KAUAHIKAUA, J., DENLIGER, R., FOSTER, J. & KESTHELYI, L. 1993. Lava delta instability: is it mass-wasting or is it triggered by lava flowing through tubes? *Eos*, **74**, 616.

LARSEN, L. M. & PEDERSEN, A. K. 1990. Volcanic marker horizons in the Maligât Formation on Disko and Nûgssuaq, and implications for the development of the southern part of the West Greenland basin in the early Tertiary. *Rapport Grønlands Geologiske Undersøgelse*, **148**, 65–73.

LEMASURIER, W. E., HARWOOD, D. M. & REX, D. C. 1994. Geology of Mount Murphy Volcano: an 8-m.y. history of interaction between a rift volcano and the West Antarctic ice sheet. *Bulletin of the Geological Society of America*, **106**, 265–280.

LESCINSKY, D. T. & FINK, J. H. 2000. Lava and ice interaction at stratovolcanoes: use of characteristic features to determine past glacial events and future volcanic hazards. *Journal of Geophysical Research*, **105**, 23 711–23 726.

LIPMAN, P. W. & MOORE, J. G. 1996. Mauna Loa lava accumulation rates at the Hilo drill site; formation of lava deltas during a period of declining overall volcanic growth. *Journal of Geophysical Research*, **101**, 11 631–11 641.

LOWE, D. R. 1976. Grain flow and grain flow deposits. *Journal of Sedimentary Petrology*, **46**, 188–199.

LOWE, D. R. 1979. Sediment gravity flows: their classification and some problems of application to natural flows and deposits. *Society of Economic Palaeontologists and Mineralogists Special Publication*, **27**, 75–82.

LOWE, D. R. 1982. Sediment gravity flows: II. Depositional models with special reference to the deposits of high-density turbidity currents. *Journal of Sedimentary Petrology*, **52**, 279–297.

MASSARI, F. & PAREA, G. C. 1988. Progradational gravel beach sequences in a moderate to high energy microtidal marine environment. *Sedimentology*, **35**, 881–913.

MATHEWS, W. H. 1947. 'Tuyas', flat-topped volcanoes in northern British Columbia. *American Journal of Science*, **245**, 560–570.

MATTOX, T. N. & MANGAN, M. T. 1997. Littoral hydrovolcanic explosions: a case study of lava-seawater interaction at Kilauea Volcano. *Journal of Volcanology and Geothermal Research*, **75**, 1–17.

MCCABE, A. M. & EYLES, N. 1988. Sedimentology of an ice-contact glaciomarine delta, Carey Valley, Northern Ireland. *Sedimentary Geology*, **59**, 1–14.

MOORE, J. G. & AULT, W. V. 1965. Historic littoral cones in Hawaii. *Pacific Science*, **19**, 3–11.

MOORE, J. G., CRISTOFOLINI, R. & GIUDICE, A. L. 1971. Development of pillows on the submarine extension of recent lava flows, Mount Etna, Sicily. *United States Geological Survey Professional Paper*, **750-C**, 89–97.

MOORE, J. G., HICKSON, C. J. & CALK, L. C. 1995. Tholeiitic-alkali transition at subglacial volcanoes, Tuya region, British Columbia, Canada. *Journal of Geophysical Research*, **100**, 24 577–24 592.

MOORE, J. G., PHILLIPS, R. L., GRIGG, R. W., PETERSON, D. W. & SWANSON, D. A. 1973. Flow of lava into the sea 1969–1971, Kilauea volcano, Hawaii. *Bulletin of the Geological Society of America*, **84**, 537–546.

NELSON, P. H. H. 1975. The James Ross Island Volcanic Group of north-east Graham Land. *British Antarctic Survey Scientific Reports*, **54**.

NEMEC, W. 1990. Aspects of sediment movement on steep delta slopes. *In*: COLELLA, A. & PRIOR, D. B. (eds) *Coarse-grained deltas*. International Association of Sedimentologists, Special Publications, **10**, 29–74.

NEMEC, W., STEEL, R. J., GJELBERG, J., COLLINSON, J. D., PRESTHOLM, E. & ØXNEVAD, I. E. 1988. Anatomy of collapsed and re-established delta front in lower Cretaceous of eastern Spitsbergen: gravitational sliding and sedimentation processes. *American Association of Petroleum Geologists Bulletin*, **72**, 454–476.

ORTON, G. J. & READING, H. G. 1993. Variability of deltaic processes in terms of sediment supply, with particular emphasis on grain size. *Sedimentology*, **40**, 475–512.

PAPALE, P. & DOBRAN, F. 1993. Magma-water interaction in closed systems and application to lava tunnels and volcanic conduits. *Journal of Geophysical Research*, **98**, 14 041–14 058.

PEDERSEN, A. K. & DUEHOLM, K. S. 1992. New methods for the geological analysis of Tertiary volcanic formations on Nuussuaq and Disko, central west Greenland, using multi-model photogrammetry. *Rapport Grønlands Geologiske Undersøgelse*, **156** 19–34.

PEDERSEN, A. K., LARSEN, L. M. & DUEHOLM, A. K. 1993. *Geological section along the south coast of Nuusuaq, central west Greenland.* 1 : 20000 coloured geological sheet, Geological Survey of Greenland, Copenhagen, Denmark.

PEDERSEN, A. K., LARSEN, L. M., PEDERSEN, G. K. & DUEHOLM, K. S. 1996. Filling and plugging of a marine basin by volcanic rocks: the Tunoqqu Member of the Lower Tertiary Vaigat Formation on Nuusuaq, central West Greenland. *Bulletin Grønlands Geologiske Undersøgelse*, **171**, 5–28.

PEDERSEN, G. K., LARSEN, L. M., PEDERSEN, A. K. & HJORTKJÆR, B. F. 1998. The syn-volcanic Naajaat lake, Paleocene of West Greenland. *Palaeogeography, Palaeoclimatology and Palaeoecology*, **140**, 271–287.

PETERSON, D. W. 1976. Processes of volcanic island growth, Kilauea Volcano, Hawaii. *Bulletin of the Geological Society of America*, **84**, 537–546.

PLANKE, S. & ALVESTAD, E. 1999. Seismic volcanostratigraphy of the extrusive breakup complexes in the Northeast Atlantic; implications from ODP/DSDP drilling. *In*: LARSEN *ET AL.* (eds) *Proceedings of the Ocean Drilling Program, Scientific Results.* Ocean Drilling Program, Texas, **163**, 3–16.

POREBSKI, S. J. & GRADZINSKI, R. 1990. Lava-fed Gilbert-type delta in the Polonez Cove Formation (Lower Oligocene), King George Island, West Antarctica. *In*: COLELLA, A. & PRIOR, D. B. (eds) *Coarse-grained deltas.* International Association of Sedimentologists, Special Publications, **10**, 335–354.

POSTMA, G. 1984. Slumps and their deposits in fan delta front and slope. *Geology*, **12**, 27–30.

POSTMA, G. 1990. Depositional architecture and facies of river and fan deltas: a synthesis. *In*: COLELLA, A. & PRIOR, D. B. (eds) *Coarse-grained deltas.* International Association of Sedimentologists, Special Publications, **10**, 13–27.

POSTMA, G. 1995. Causes of architectural variation in deltas. *In*: OTI, M. N. & POSTMA, G. (eds) *Geology of deltas.* Balkema, Rotterdam, 3–16.

POSTMA, G. & ROEP, T. B. 1985. Resedimented conglomerates in the bottomsets of Gilbert-type gravel deltas. *Journal of Sedimentary Petrology*, **55**, 874–885.

POSTMA, G., BABIČ, L., ZUPANIČ, J. & RØE, S. L. 1988. Delta-front failure and associated bottomset deformation in a marine, gravelly, Gilbert-type fan delta. *In*: NEMEC, W. & STEEL, R. J. (eds) *Fan Deltas: Sedimentology and Tectonic Settings.* Blackie and Son, Glasgow, 91–102.

POSTMA, G. & CRUICKSHANK, C. 1988. Sedimentology of a late Weichselian to Holocene terraced fan delta, Varangerfjord, northern Norway. *In*: NEMEC, W. & STEEL, R. J. (eds) *Fan Deltas: Sedimentology and Tectonic Settings.* Blackie and Son, Glasgow, 145–157.

PRIOR, D. B., WISEMAN, W. J. & BRYANT, W. R. 1981. Submarine chutes on the slopes of fjord deltas. *Nature*, **290**, 326–328.

PRIOR, D. B. & BORNHOLD, B. D. 1989. Submarine sedimentation on a developing Holocene fan delta. *Sedimentology*, **36**, 1053–1076.

PRIOR, D. B. & BORNHOLD, B. D. 1990. The underwater development of Holocene fan deltas. *In*: COLELLA, A. & PRIOR, D. B. (eds) *Coarse-grained deltas.* International Association of Sedimentologists, Special Publications, **10**, 75–90.

SANSONE, F. J., RESING, J. A., TRIBBLE, G. W., SEDWICK, P. N., KELLY, K. M. & HON, K. 1991. Lava-sea water interactions at shallow-water submarine lava flows. *Geophysical Research Letters*, **18**, 1731–1734.

SCHMINCKE, H. U. 1967. Flow direction in Columbia River Basalt flows and palaeocurrents of interbedded sedimentary rocks, South-Central Washington. *Geologische Rundschau*, **56**, 992–1020.

SCHMINCKE, H. U., BEHNCKE, B., GRASSI, M. & RAFFI, S. 1997. Evolution of the northwestern Iblean mountains, Sicily: uplift, Pliocene/Pleistocene sea level changes, palaeoenvironment and volcanism. *Geologische Rundschau*, **86**, 637–669.

SIGVALDASON, G. E. 1968. Structure and products of subaquatic volcanoes in Iceland. *Contributions to Mineralogy and Petrology*, **18**, 1–16,

SKILLING, I. P. 1994. Evolution of an englacial volcano: Brown Bluff, Antarctica. *Bulletin of Volcanology*, **56**, 573–591.

SMELLIE, J. L. 1999. Lithostratigraphy of Miocene-Recent, alkaline volcanic fields in the Antarctic Peninsula and eastern Ellsworth Land. *Antarctic Science*, **11**, 362–378.

SMELLIE, J. L. 2001. Lithofacies architecture and construction of volcanoes erupted in englacial lakes: Icefall Nunatak, Mount Murphy, eastern Marie Byrd Land, Antarctica. *In*: RIGGS, N. & WHITE, J. D. L. (eds), *Volcaniclastic Sedimentation in Lacustrine Settings.* International Association of Sedimentologists, Special Publications, **30**, 73–98.

SMELLIE, J. L., PANKHURST, R. J., HOLE, M. J. & THOMSON, J. W. 1988. Age, distribution and eruptive conditions of late Cenozoic alkaline volcanism in the Antarctic Peninsula and eastern Ellsworth Land: a review. *Bulletin of the British Antarctic Survey*, **80**, 21–49.

SMELLIE, J. L. & SKILLING, I. P. 1994. Products of subglacial eruptions under different ice thicknesses: two examples from Antarctica. *Sedimentary Geology*, **91**, 115–129.

SMELLIE, J. L. & HOLE, M. J. 1997. Products and processes in Pliocene-Recent, subaqueous to emergent volcanism in the Antarctic Peninsula: examples of englacial Surtseyan volcano construction. *Bulletin of Volcanology*, **58**, 628–646.

SMITH, J. R., MALAHOFF, A. & SHOR, A. N. 1999. Submarine geology of the Hilina Slump and morpho-structural evolution of Kilauea Volcano, Hawaii. *Journal of Volcanology and Geothermal Research*, **94**, 59–88.

SOHN, Y. K., KIM, S. B., HWANG, I. G., BAHK, J. J., CHOE, M. Y. & CHOUGH, S. K. 1997. Characteristics and depositional processes of large-scale gravelly Gilbert-type foresets in the Miocene Doumsan Fan Delta, Pohang Basin, SE Korea. *Journal of Sedimentary Petrology*, **67**, 130–141.

SUMNER, J. M. 1998. Formation of clastogenic lava flows during fissure eruption and scoria cone collapse: the 1986 eruption of Izu-Oshima Volcano, Japan. *Bulletin of Volcanology*, **60**, 195–212.

SWANSON, D. 1967. Yakima basalt of the Tieton River area, south-central Washington. *Bulletin of the Geological Society of America*, **78**, 1077–1110.

SYKES, M. A. 1988. New K-Ar age determinations on the James Ross Island Volcanic Group, north-east Graham Land, Antarctica. *British Antarctic Survey Bulletin*, **80**, 51–56.

TRIBBLE, G. W. 1991. Underwater observations of active lava flows from Kilauea volcano, Hawaii. *Geology*, **19**, 633–636.

VAN DER STRAATEN, H. C. 1990. Stacked Gilbert-type deltas in the marine pull-apart basin of Abarán, late Serravallian-early Tortonian, south-eastern Spain. *In*: COLELLA, A. & PRIOR, D. B. (eds) *Coarse-grained deltas*. International Association of Sedimentologists, Special Publications, **10**, 199–222.

WALKER, G. P. L. 1992. Morphometric study of pillow-size spectrum among pillow lavas. *Bulletin of Volcanology*, **54**, 459–474.

WERNER, R. & SCHMINCKE, H. U. 1999. Englacial versus lacustrine origin of volcanic table mountains: evidence from Iceland. *Bulletin of Volcanology*, **60**, 335–354.

WÖRNER, G. & VIERECK, L. 1987. Subglacial to emergent volcanism at Shield Nunatak, Mt. Melbourne Volcanic Field, Antarctica. *Polarforschung*, **57**, 27–41.

WRIGHT, L. D. 1977. Sediment transport and deposition at river mouths: a synthesis. *Bulletin of Geological Society of America*, **88**, 857–868.

Architecture and evolution of hydrovolcanic deltas in Marie Byrd Land, Antarctica

W. E. LE MASURIER

University of Colorado at Denver, CB 172, P.O. Box 173364, Denver, Colorado 80217–3364, USA (e-mail: wlemasur@carbon.cudenver.edu)

Abstract: The Marie Byrd Land volcanic province is a late Cenozoic alkaline basalt-trachyte volcanic field on the Pacific coast of West Antarctica. Most of these volcanoes are partially buried beneath the West Antarctic ice sheet, but in some, a combination of tectonic uplift and lowering of ice level has exposed basal hydrovolcanic sections produced by eruptions in an englacial environment. Some of the largest and best preserved hydrovolcanic structures are delta-like in form, with gentle distal slopes, and foreset bedded deposits composed of hyaloclastites, pillow breccias, pillow lavas, subaerial flows and air fall tephras. Three broad categories of processes related to delta evolution are described here; (1) flow of lava from a subaerial to an englacial environment; (2) intrusion of dykes and sills; and (3) edifice settling, which includes a variety of down-slope movement phenomena. This paper focuses on documenting post-depositional structures that are superbly exposed in these deltas. It describes the apparently province-wide lack of pillow lava cores in Marie Byrd Land englacial volcanoes, and factors that may be related to this anomaly, and it describes characteristics of hyaloclastites that are relevant to future glaciological, sedimentological and geophysical studies of the West Antarctic ice sheet.

The interpretation of volcanic edifices formed in glacial environments has been guided for over 50 years by the classic studies of table mountains (also called tuyas) in Iceland and British Columbia (Noe-Nygaard 1940; Mathews 1947; Kjartansson 1959; Sigvaldason 1968; Jones 1969, 1970). The model based on these studies consists of a basal section of pillow lavas, overlain by *c.* 200 m of hyaloclastite and pillow breccia deposits, overlain in turn by subaerial lava flows. The pillow lava–hyaloclastite transition has been interpreted to represent the onset of shallow water explosive activity: the 'volatile fragmentation depth' of Fisher & Schmincke (1984), which generally occurs at depths of 200 m or less for tholeiitic magmas (Kokelaar 1986; Wörner & Viereck 1987). The base of the subaerial flow sequence represents the position at which the volcanic pile emerged above ice or water level. Hydrovolcanic phenomena in Antarctica have been compared to this model since the earliest published studies (Rutford *et al.* 1968; Hamilton 1972; LeMasurier 1972).

The form and sub-sea level structure of oceanic islands, seamounts and guyots has been inferred by many workers to conform to the same model (Jones 1966; Cotton 1969; Moore & Fiske 1969; Bonatti & Tazieff 1970), and comparisons of submarine with subglacial volcanoes provide additional perspectives. For example, a study of the exposed cross-section of a seamount in the basement complex of La Palma (Canary Islands), by Staudigel & Schmincke (1984), has yielded evidence that intrusion of sills and dykes played a major role in the growth of La Palma seamount, and further, that the transition to shallow explosive activity took place at *c.* 800 m for the alkaline basalts at La Palma, in contrast to the much shallower depth usually cited for the Icelandic model, which is based largely on tholeiitic basalts. Because Marie Byrd Land (MBL) volcanic rocks are all alkaline, it is helpful to compare MBL hydrovolcanic structures with both Icelandic and marine counterparts, as presented below.

The purpose of this paper is to describe the main structural and lithologic characteristics of four lava-fed hydrovolcanic deltas in eastern Marie Byrd Land (Fig. 1), one of late Miocene age at Mount Murphy (Fig. 2), and three of late Pleistocene age at Mount Takahe (Fig. 3). Each is composed mainly of foreset-bedded hydrovolcanic deposits (e.g. pillow lavas, pillow breccias, hyaloclastites), and in three, there are well exposed down-dip transitions from subaerial lava to pillow lavas. Although there are substantial differences in the lithofacies assemblages

From: SMELLIE, J. L. & CHAPMAN, M. G. (eds) 2002. *Volcano–Ice Interaction on Earth and Mars*. Geological Society, London, Special Publications, **202**, 115–148. 0305-8719/02/$15.00

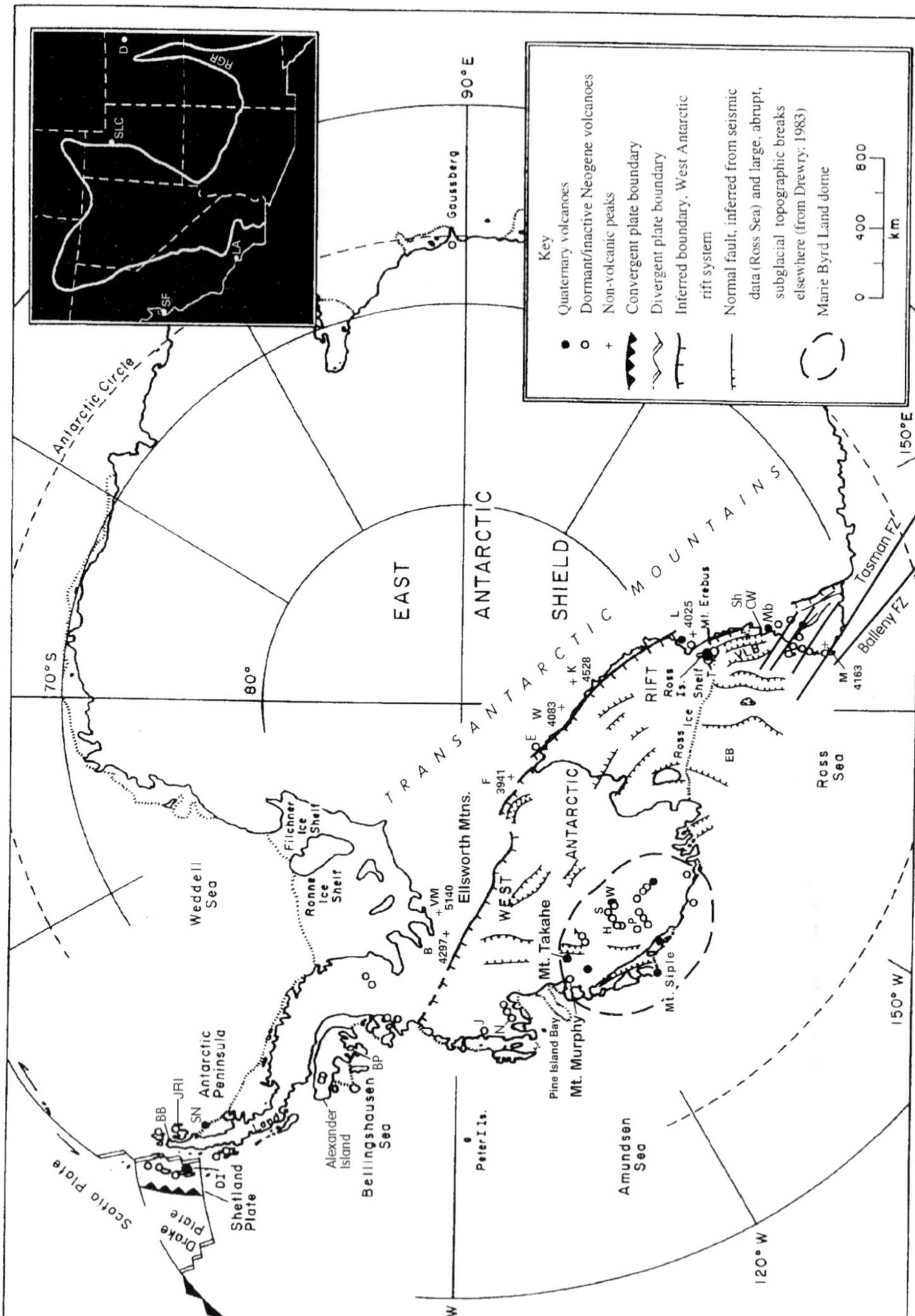
Key
Quaternary volcanoes
Dormant/inactive Neogene volcanoes
Non-volcanic peaks
Convergent plate boundary
Divergent plate boundary
Inferred boundary, West Antarctic rift system
Normal fault, inferred from seismic data (Ross Sea) and large, abrupt, subglacial topographic breaks elsewhere (from Drewry; 1983)
Marie Byrd Land dome
0 400 800
km
EAST ANTARCTIC SHIELD
TRANSANTARCTIC MOUNTAINS
WEST ANTARCTIC RIFT
Antarctic Circle
70° S
80°
90° E
90° W
120° W
150° W
150° E
Gaussberg
Weddell Sea
Filchner Ice Shelf
Ronne Ice Shelf
Ellsworth Mtns.
Antarctic Peninsula
Bellingshausen Sea
Alexander Island
Peter I Is.
Amundsen Sea
Pine Island Bay
Mt. Murphy
Mt. Takahe
Mt. Siple
Ross Ice Shelf
Ross Is.
Mt. Erebus
Ross Sea
Tasman FZ
Balleny FZ
Scotia Plate
Drake Plate
Shetland Plate
SLC
RGR
SF
LA
D

Fig. 2. Aerial view looking east to the late Miocene (8.3 Ma) Mount Murphy volcano from a distance of 20 km. Base diameter is 30 km, vertical relief is just over 2000 m. In the foreground from left to right are the volcanic nunataks Hedin Nunatak, Turtle Peak and Icefall Nunatak. The Mount Murphy hydrovolcanic delta is well exposed at the distal end of the SW ridge in the middle ground (arrow). Ice level is 400 m a.s.l. at the foot of the south-west ridge and descends to 200 m around the northernmost (left) outcrops. Pre-volcanic basement rock is exposed at ice level at the foot of the NW (left)-trending ridges. The Crosson ice shelf adjoins Mount Murphy to the north and east. (US Navy photograph, TMA 1719 F31 133).

found at each delta, as described below, they all appear to have been produced by the flow of subaerial lava into a subaqueous or subglacial environment, and can be compared with lava-fed deltas in Hawaii, Iceland, and elsewhere in Antarctica (Jones & Nelson 1970; Moore *et al.* 1973; Porebski & Gradzinski 1990; Skilling 1994). However the MBL deltas provide an added dimension. At Mt. Takahe, two deltas are composed of trachyte and benmoreite (Möll Spur and Gill Bluff), one is basaltic (Stauffer Bluff), and the Mt. Murphy delta is basaltic, providing an unusual opportunity to compare the influence of composition on textures and structures. Finally, an important part of this paper is concerned with documenting post-depositional structures and the processes they represent. At most localities these are so varied and pervasive that many primary features are obscured. Much of the present structure of MBL hydrovolcanic deltas appears to have been strongly influenced by these penecontemporaneous and secondary processes.

Geological setting

The Marie Byrd Land volcanic province lies within the West Antarctic rift system (Fig. 1), an asymmetric, intracontinental rift on the scale of

Fig. 1. Map of West Antarctica showing the locations of Mount Murphy and Mount Takahe, on the eastern flank of the Marie Byrd Land dome, West Antarctic rift system. The Transantarctic Mountains and Ellsworth Mountains, with elevations exceeding 4000 m form the southern boundary of the rift. Basement elevations within the rift are below sea level almost everywhere but on the Marie Byrd Land dome (see text), the crest of which is defined by basement rock exposures at +2700 m at Mount Petras (P). The rift system is defined by topographic expression, distribution of late Cenozoic alkaline volcanic fields, attenuated continental crust, and block faulting (LeMasurier & Rex 1990; Behrendt *et al.* 1991). Abbreviations: B, Mount Bentley; BB, Brown Bluff; BP, Beethoven Peninsula; CT, Central Trough; CW, Cape Washington; DI, Deception Island; E, Mount Early; EB, Eastern Basin; F, Mount Faure; FZ, fracture zones related to the Pacific–Antarctic ridge; J, Jones Mtns; JRI, James Ross Island; K, Mount Kirkpatrick; L, Mount Lister; M, Mount Minto; Mb, Mount Melbourne; N, Mount Nickens; Sh, Shield Nunatak; SN, Seal Nunataks; H, S, and W the three major volcanoes of the Executive Committee Range, Mount Hampton, Mount Sidley, Mount Waesche; VLB, Victoria Land Basin; VM, Vinson Massif; W, Mount Wade. Inset shows the Basin and Range Province and Rio Grande rift (RGR), in the western United States, at the same scale. Modified from LeMasurier (1990*a*).

Fig. 3. Mount Takahe, a late Quaternary (<300 ka) trachytic shield volcano, viewed looking east from a distance of about 25 km. Base diameter is 30 km; vertical relief from ice level is 2100 m. Ice level is 1400 m a.s.l., and the ice sheet is *c.* 1500 m thick at this locality. The Gill Bluff, Stauffer Bluff and Möll Spur deltas are the main occurrences of hydrovolcanic rock on this volcano. Elsewhere, trachytic flows are the predominant rock type, with a relatively few tuff cones and cinder cones. (US Navy photograph, TMA 1718 F33 022).

the Basin and Range and East African rift systems (LeMasurier 1978, 1990*a*; Behrendt *et al.* 1991*a*, *b*; LeMasurier & Landis 1996). Most of the West Antarctic rift is below sea level, even with the ice sheet removed and bedrock topography isostatically restored (Drewry 1983). The major exceptions are the block faulted Marie Byrd Land dome, centred under the volcanic province, and the 2000 km-long Transantarctic Mountains, which form the abrupt southern boundary of the rift. Rift structures apparently continue out to sea at the continental margin with no evidence for a northern boundary of un-extended continental crust (LeMasurier & Landis 1996).

The extent and nature of tectonic activity in the rift system during Cenozoic time is uncertain. The rift has been volcanically active throughout Neogene time, but it is also largely aseismic (Behrendt *et al.* 1991*a*). In the western Ross Sea region, fission track data indicate a history of episodic uplift for the Transantarctic Mountains (TAM), with the latest episode beginning at *c.* 50–45 Ma (Fitzgerald & Stump 1997). Recent sedimentological studies of drill cores from the Cape Roberts Project provide evidence that this uplift continued during the Oligocene (Smellie 2001). In the MBL volcanic province (Fig. 1), evidence from age v. elevation patterns of volcanic rocks that rest on the basement suggest that uplift of the MBL dome, accompanied by block faulting and volcanism, began at *c.* 25–30 Ma and has continued throughout the Neogene at a rate of *c.* 105–122 m Ma^{-1} (LeMasurier & Landis 1996). Reconstructions of Gondwanaland suggest that little extension has accompanied this uplift anywhere in the rift system in Cenozoic time (Lawver & Gahagan 1994); but recent magnetic anomaly data from the sea floor between New Zealand and Antarctica suggest 180 km of extension in the western Ross Sea basin between *c.* 43–26 Ma (Cande *et al.* 2000).

Volcanic and glacial deposits at Mount Murphy and Mount Takahe provide evidence for large scale changes in the level of the West Antarctic ice sheet during and after volcanic activity. At Mount Murphy these range from almost complete deglaciation, to a highstand *c.* 1300 m above the present level, between 24 Ma and 3.5 Ma (LeMasurier *et al.* 1994; see also Smellie 2000, 2001). At Mount Takahe, the elevations of subaerial topset beds on hydrovolcanic deltas suggest 300–500 m changes in ice level within the past *c.* 100 Ka (LeMasurier and Rex 1990; Wilch *et al.* 2002). Thus, the MBL hydrovolcanic deltas have apparently formed in an environment of slow uplift of the MBL dome, coupled with rapid fluctuations in ice level.

The age range of basaltic volcanic activity in MBL is from late Oligocene to recent. Felsic activity began around 19 Ma and continues to this day (LeMasurier 1990*c*). The delta deposits at Mt. Murphy are *c*. 8.3 Ma (LeMasurier *et al.* 1994); all of the dated rocks at Mt. Takahe are <0.3 Ma (LeMasurier & Rex 1990), and the three deltas around the base of the volcano have recently yielded $^{40}Ar/^{39}Ar$ ages of 21.9 ± 1.8 Ka for Gill Bluff, 15.2 ± 2.4 Ka for Möll Spur, and 66.4 ± 4.7 Ka for Stauffer Bluff (Wilch *et al.* 2000).

MBL volcanic rocks are highly alkaline, and range from ne-normative basalts to peralkaline varieties of trachyte, phonolite and rhyolite (LeMasurier 1990*a*). They occur most commonly as lava flows. The most common variety of pyroclastic rock is hyaloclastite, followed by Strombolian tephra. Thus, the larger volcanoes are either basaltic or felsic shield volcanoes, and the best exposed often have a hydrovolcanic base. The hydrovolcanic rocks described in this paper are either basaltic (basanite, basalt, hawaiite) or trachytic. Most contain phenocrysts (e.g. 0–33%), and virtually all vitric clasts in hyaloclastites contain abundant microlites, indicating sub-liquidus temperatures immediately prior to quenching. In spite of this, it is clear that these were relatively low viscosity liquids at the time of eruption. For example, phonolite flows at Mt. Sidley are about 2 m thick (LeMasurier 1990*b*), and the trachyte and benmoreite flows at Mount Takahe are *c*. 2–10 m thick. Low magmatic water contents are suggested by the scarcity of magmatic pyroclastic rocks throughout MBL and by the scarcity of hydrous phases in phenocryst assemblages.

Most MBL volcanoes are partially buried beneath the ice sheet. This distorts impressions of their size and structure and biases observations toward their upper parts. Near the coast, the surface elevation of the ice sheet is *c*. 400 m and it rises inland to over 2000 m around the southernmost volcanoes. Hydrovolcanic rocks are most abundant at the bases of volcanic sections. They are exposed down to the basement contact at Mount Murphy (Fig. 1), which rises immediately above the adjacent ice shelf, but hydrovolcanic rocks are found much less commonly in volcanoes further inland, where basal sections are buried beneath thick ice. This exposure problem is further complicated by the fact that most MBL volcanoes, even those as old as *c*. 14 Ma, are virtually undissected: an effect, presumably, of the lack of running water in this environment and the inadequacy of cold-based glaciers as agents of erosion. Among the 18 large (base diameter >10 km) volcanoes in the province (LeMasurier 1990*c*) only the coastal volcano Mount Murphy has been dissected deeply enough to expose internal structures. However, there are numerous smaller (base diameter <1 km) volcanic features scattered throughout the province, 25 Ma and younger, that are exposed in all stages of dissection (LeMasurier 1990*d*).

Terminology

Fisher & Schmincke (1984) has been the guide for most terminology used in this paper. The term hydrovolcanic is used here when the deposit being described includes pillow lavas, as well as hydroclastic materials, all produced by the interaction of lava and water or ice. In the descriptions of pillow lava, an attempt is made to distinguish those formed during the flow of subaerial lava into water, from those formed during subaquatic eruptions, by describing the latter as completely or wholly subaquatic. Nested pillow lavas are those whose shapes are moulded around the shapes of adjacent pillows, also described as closely packed pillows by Fisher & Schmincke (1984). This is in contrast to deposits containing isolated, matrix-supported, but still completely intact pillows. Pillow breccia is used in reference to deposits that contain fragments of pillows, whether fragmentation took place by internal expansion of magmatic gas (Sigvaldason 1968), by implosion of the brittle pillow crust due to water pressure (Moore 1975), or by disaggregation during downslope movement, i.e. no genetic implication is intended. Lava clasts described here as pillow fragments are recognized by the presence of curved, glass-crusted boundaries and remnants of radial joints.

Hyaloclastite is used in the broad sense of 'vitroclastic tephra produced by the interaction of water and hot magma or lava ...' (Fisher & Schmincke 1984). This is also consistent with the definition used by Batiza & White (2000). Blocky clast shapes that conform to the definition of Honnorez & Kirst (1975) are typical of nearly all MBL hyaloclastites, of both basaltic and trachytic compositions. Vesicularity is variable, as described below, but seldom influences clast shape. No genetic implication (i.e. fragmentation mechanism) is intended in the way the term is used in this paper, other than the universally held assumption of lava–water interaction. The term tephra is used synonymously with pyroclastic materials (Fisher & Schmincke 1984).

Form and structure of hydrovolcanic deltas

Mount Takahe is a trachyte shield volcano *c*. 30 km in base diameter that rises 2100 m above ice level (Fig. 3). The three deltas described below are roughly 25 km from each other, around the basal perimeter of the volcano, and the error bars on their ages do not overlap. They appear to have formed separately and independent of each other, and thus, it is possible that each delta formed in a somewhat different glacial environment. The morphology of each is well preserved, in contrast to the much older Mount Murphy delta, which is deeply dissected (Fig. 2). The original form of the latter can only be inferred.

Primary features

Form and structure

The deltas at Mt. Takahe occur as prominent exposures around the base of the volcano (Fig. 3). They are delta-like in plan view, roughly 2.5 km long by 1.5 km wide, with 7–8° constructional (or depositional) slopes, and 400–500 m of relief at their distal ends (Fig. 4). Each one is somewhat different from the others in gross internal structure. Möll Spur consists of foreset bedded trachytic hyaloclastite, pillow breccias, isolated pillows in hyaloclastite, and nested pillows (Fig. 5), all dipping distally at *c.* 30°, and overlain by much more gently dipping topset subaerial flows (Fig. 6). There is no obvious stratigraphic order to these rock units. However the nested pillows are at the top of the hydrovolcanic section, down-dip from the subaerial flow shown in Figure 6 evidently representing a transition from subaerial to subaquatic facies as the flow crossed the shoreline (Fig. 4).

Gill Bluff consists of gently dipping trachyte and benmoreite flows that pass abruptly across a near-vertical passage zone (Fig. 7) into foreset-bedded hyaloclastite breccias with interbedded, isolated pillows (Fig. 8), all dipping 20–25° distally. One locality of nested pillows was observed on the west side of the delta. The trachyte flows are moderately vesicular (i.e. vesicles are small but moderately abundant), but the cores and rims of pillows, and the vitric clasts, are non-vesicular. As at Möll Spur, there appears to be no systematic order to these rock units.

Stauffer Bluff consists of foreset bedded basaltic hyaloclastite tuffs and tuff breccias, pillow breccias and pillow lavas surmounted by laminated tuff cone deposits containing accretionary lapilli. Here too, pillow lavas are found

Fig. 4. View looking east to the Möll Spur hydrovolcanic delta, showing gentle southward slope of topset subaerial flows (Fig. 6). Jagged downslope (right) outcrops are foreset bedded pillow lavas (Fig. 5) and hyaloclastite. Relief from topsets to ice level is *c.* 400–500 m.

Fig. 5. Foreset bedded trachyte pillow lavas at Möll Spur, directly down-dip from the topset flow rock (see Fig. 4). Pillow lengths are *c.* 1–1.5 m (note ice axe for scale). Dark, brecciated glassy rinds are 2–3 cm thick and the pillows are non-vesicular. Bedding dips at *c.* 30° to the left.

near the top of the delta, close to the subaquatic/subaerial transition. At many localities on this delta downslope movement phenomena, described below, are more conspicuous than primary structures.

The hydrovolcanic delta at Mount Murphy is well exposed on three sides, but is not morphologically well defined (Fig. 2). Thus, the defining feature at this locality is the well exposed transition from subaerial flow rock to nested pillow lavas shown in Figure 9a. Basaltic hyaloclastite breccias, pillow lavas, and tillite predominate in the lower 150 m of section (Fig. 9a–c). They are overlain by 250 m of alternating hyaloclastites, palagonitized Strombolian tephras, heterolithic tuffs, and basaltic flow rock, overlain in turn by

Fig. 6. Single thick topset subaerial trachyte flow at Möll Spur. Pillow lavas shown in Figure 5 are upslope and to the left of the people in the foreground.

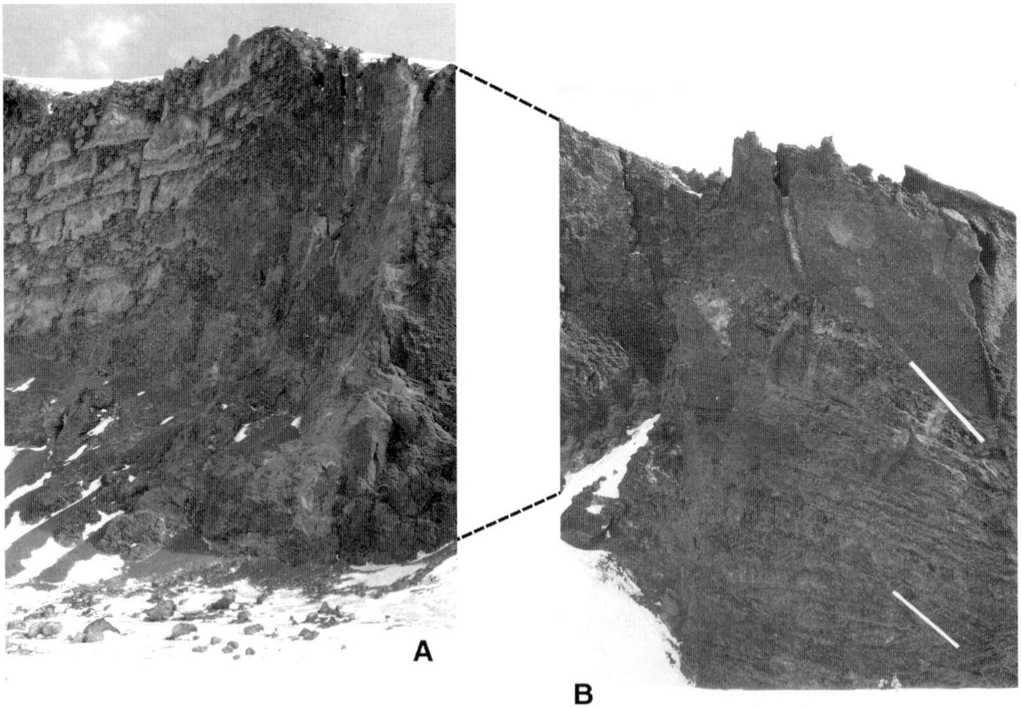

Fig. 7. Subaerial to hydrovolcanic facies transition at Gill Bluff. (**a**) Subhorizontal trachyte flows on the left pass laterally into distally dipping flow foot breccias on the right (**b**). The flow foot breccias include isolated unfragmented pillows roughly 2–4 m in longest dimension, some with thick (3–6 cm) glassy margins, embedded in trachytic hyaloclastite (Fig. 8). Note that the passage zone here (**a**) is roughly vertical. Downslope movement of coherent glide blocks is suggested by high angle truncations of bedding seen (white underline) at the base and top of this section (**b**). Vertical relief shown here is 400–500 m. Note figures in the lower right of (**b**) for scale.

Fig. 8. Isolated pillow in trachytic hyaloclastite at the base of the Gill Bluff section (Fig. 7b). The pillow lava is completely non-vesicular, and its thick (3–6 cm) glassy rim has been stripped off in some places, probably during transport. Post-depositional mass flow is represented by crude folds that envelop and trail the pillow, and by concentrations of coarser clasts in the cores of folds.

(a)

Fig. 9. (**a**) Basaltic hydrovolcanic basal unit (LeMasurier *et al.* 1994) at Mount Murphy SW ridge (see Fig. 2), in which subaerial to subaquatic facies changes are well displayed. At the lower left, two columnar jointed lava flows grade to the right into large globular lava bodies (kubbaberg), and then into nested pillow lavas. These are overlain by hyaloclastite (Fig. 10). The contorted dyke above the kubbaberg (mid-way up the cliff face on the left) appears to have been pulled apart, perhaps by soft sediment sliding when the volcanic pile was wet and unstable. To the right of the dyke, high angle beds of hyaloclastite and pillow lava truncate the bedding of underlying deposits (white underline), suggesting downslope movement of a glide block, later buried by beds with more normal attitudes. The dyke may follow the underside of the relatively well indurated glide block. Cliff height at the centre of the photo is *c.* 100 m

c. 1000 m of both massive and columnar jointed basalt flows (LeMasurier *et al.* 1994). These deposits dip *c.* 5–15° to the SSW with no clear topset-foreset relationships. Thus the term delta is not as clearly demonstrable here as it is at Mt. Takahe.

The structural and stratigraphic variability presented by these four deltas suggests rapidly changing lava-ice contact relations and variable eruption dynamics. At this large scale, there appear to be no differences in delta architecture related to lava composition.

Rock types

The common rock types in the Mount Murphy and Mount Takahe deltas are hyaloclastite, palagonitized Strombolian tephra, heterolithic tuff, pillow lava, and sheet lava. Hyaloclastite tuff and tuff breccia are the most common. Nested pillow lavas make up a small proportion (<10%?) of the outcrops at all four deltas, which is consistent with observations throughout this region. Among published descriptions of hydrovolcanic localities elsewhere in MBL and Ellsworth Land, nested pillow lavas have been reported only at Mount Nickens, in the Hudson Mountains (Rowley *et al.* 1990).

Hyaloclastites. Individual clasts in these deposits are usually blocky and angular in shape, and composed of sideromelane when the lava is basaltic. Most sideromelane clasts have been marginally altered to the yellow mineraloid palagonite. Figure 10 illustrates some common characteristics of MBL hyaloclastites. Non-vesicular, blocky, coarse-grained (1 mm–1 cm), microlite-rich vitric clasts, like those shown in Figure 10a, b and d are common in all the deltas, irrespective of composition. Those in Figure 10d represent the most commonly observed type of trachytic hyaloclastite at Gill Bluff and Möll

(b)

Fig. 9. (**b**) View to the left of (**a**), showing a closer view of the columnar jointed lava flows resting on pale brown hyaloclastite. The latter overlies tillite. Note figure at the base of the cliff for scale.

Spur. The palagonite rims (Fig. 10d) are similar to, but much thinner than, those surrounding sideromelane clasts.

A less common variety of basaltic hyaloclastite is represented by the sample in Figure 10c, from the tuff cone at the top of Stauffer Bluff. The deposit is thinly laminated and contains a significant component of lithic clasts of mixed lithology, visible in hand specimen and thin section (Table 1, column 4). The scoriaceous sideromelane clasts occur as cores of accretionary lapilli in a vesiculated tuff matrix. These characteristics and the field occurrence indicate that this variety of hyaloclastite is of phreatomagmatic origin, perhaps a base surge deposit (Lorenz 1970; White & Houghton 2000), produced when the summit of Stauffer Bluff was at, or just above ice level.

(c)

Fig. 9. (**c**) Nested pillow lavas at the centre of (**a**).

A trachytic deposit from Gill Bluff (Table 1, column 2), similar to the Stauffer Bluff deposit just described, consists of accretionary lapilli cored by pumice, together with nearly equal amounts of lithic clasts in a matrix of fine ash. This deposit too is likely to be of phreatomagmatic origin, but it is not associated in the field with a tuff cone. The trachytic pumice clasts are characterized by vesicles that are smaller and more densely spaced than those in basaltic scoriaceous clasts (e.g. Fig. 10c).

The photographs also illustrate the problems inherent in acquiring meaningful grain size distribution data. The palagonitized rims of vitric clasts are extremely fragile, and even the rare unaltered sideromelane clasts are unlikely to retain their original size during disaggregation and sieving. Petrographic methods of determining grain size distribution (Kellerhals *et al.* 1975) have not yielded results that are reproducible, or that could be verified against prepared standards. The use of image analysis software on backscattered images of these samples has not yet been attempted.

The modal analyses in Table 1 provide quantitative data on the petrographic characteristics of the hyaloclastites. Among these samples, vitric clasts (including palagonite after sideromelane) make up 52–81% of the rock, and the sum of lithic and crystal clasts is 0–10.2% excluding the phreatomagmatic tuff (column 2) described above. In basaltic hyaloclastites, vitric clasts range from predominantly sideromelane to predominantly tachylite. Tachylite is opaque, incipiently crystallized glass that is seldom altered to palagonite. It occurs beneath sideromelane in the rims of pillow lavas, and as an important component of Strombolian tephra, representing somewhat slower quenching than sideromelane. Tachylite clasts are common in hyaloclastites that contain pillows and lava globules (Fig. 11), presumably spalled from the rims of these bodies. It is also common in mixed hyaloclastite/Strombolian deposits (see below, palagonitized Strombolian/Surtseyan tephras), in which sideromelane is palagonitized and tachylite is not (e.g. Table 1, column 6). Columns 1–3 present data on representative trachytic hyaloclastites, illustrating the relatively small proportion of palagonite in deposits of this composition, and also the relatively high vesicularity of pumiceous deposits noted above. The proportion of microlites in glass was not counted in these modal analyses, but can be observed in the photos (Fig. 10a–d).

X-ray, SEM, and microprobe studies (Ellerman 1989) have shown that the major alteration products of vitric clasts, after palagonite, are clay minerals (principally smectite), zeolite (especially phillipsite), and carbonate. These alteration products are the same in MBL hyaloclastites of all ages (<0.3–25 Ma), and occur irrespective of rock composition. Dissolution of glass is the main source of the constituents. The percentages of these minerals shown in Table 1 are based on petrographic identification. The unambiguous determination that smectite is an important alteration product of hyaloclastites has significant implications regarding the mechanical behaviour of these deposits at the base of an ice sheet, as discussed below.

Table 2 presents density and porosity data for seven hyaloclastite samples, selected to represent compositional, textural, and age variation, and to provide data for geophysical modelling. These data show that the porosity of hyaloclastites is significantly higher than the values given for samples in Table 1 (pores + vesicles), because only the larger voids can be recognized petrographically. In addition, the data show that all these samples are low density deposits. Felsic hyaloclastites are lower density than basaltic hyaloclastites, as expected, and the older deposits are generally denser and less porous than the very young Mount Takahe deposits, probably due to diagenetic void filling. In Antarctica

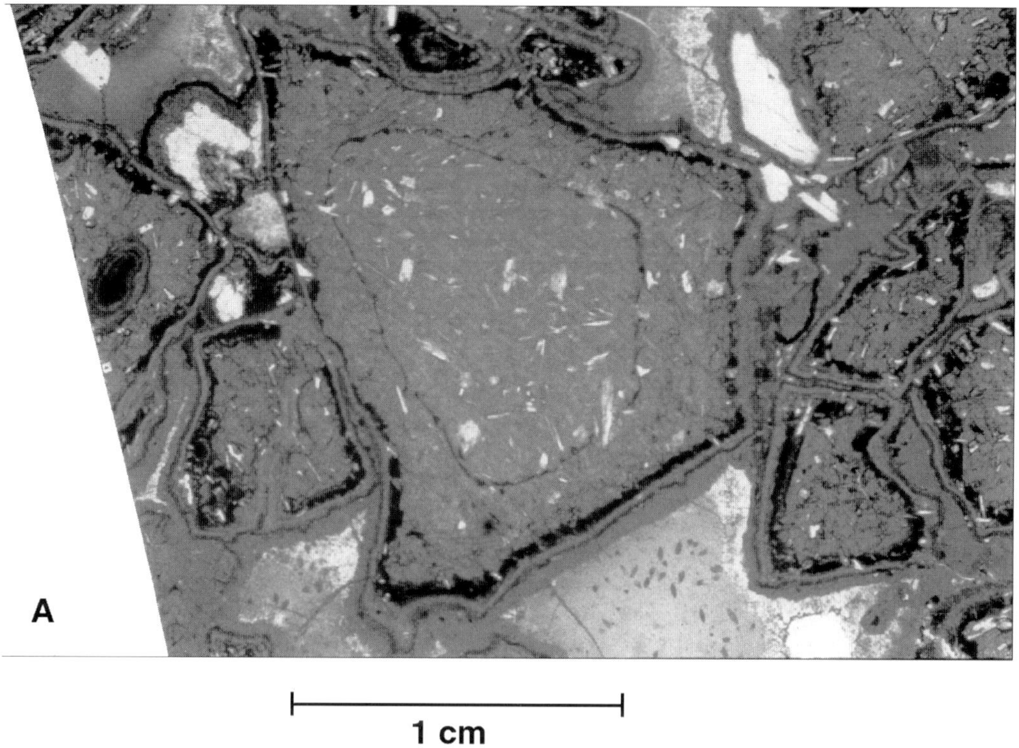

Fig. 10. Photomicrographs of hyaloclastites from MBL hydrovolcanic deltas. The scale of each view is 2 × 3 mm. (**a**) Relatively well-sorted basaltic hyaloclastite from the base of the Mount Murphy section (Fig. 9a) showing blocky, microlite-rich clasts of sideromelane altered to palagonite. The core of each clast is fresh sideromelane, and is surrounded concentrically by isotropic palagonite, and then by a thin rim of brown and yellow birefringent fibro-palagonite. Pore spaces are rimmed by zeolites. Most clasts in this slide are 1 mm–1 cm. Sample 85–32I. (**b**) Basaltic hyaloclastite from the base of Mount Murphy showing two relationships of tachylite. Clast in the centre is composed of microlite-rich sideromelane rimmed by tachylite, perhaps representing incomplete quenching, or incipient crystallization of the clast margin in a hot volcanic pile. To the side of the frame are clasts composed entirely of microlite-rich tachylite, perhaps spalled from the rims of pillows or lava globules (e.g. Fig. 11). Partly filled pore spaces are rimmed by highly birefringent smectite (?) and then colourless zeolite. Sample 85-33D. (**c**) Poorly-sorted, ash-rich basaltic hyaloclastite from Stauffer Bluff, Mount Takahe. A highly vesicular sideromelane clast partly rimmed by accreted ash is shown on the right, accompanied elsewhere in the slide by black, microlite-rich tachylite clasts. Smaller sideromelane and tachylite clasts are shown in the centre embedded in an ash matrix. The ash matrix is vesiculated, presumably by escaping gases (see text for discussion). Sample 65A. (**d**) Trachytic hyaloclastite from Gill Bluff, Mount Takahe. Blocky, non-vesicular, microlite-rich trachytic clasts with very thin palagonite rims are shown here. Grain sizes range from *c.* 1 cm down to fine ash. Two open pores are in the lower left. The low degree of palagonitization is common in trachytic hyaloclastites. Sample 85-2.

especially, density data are an important part of the description of hyaloclastites, because densities are needed to model accurately the extent, nature and distribution of volcanic deposits beneath the West Antarctic ice sheet. This in turn, is relevant to studies of the potential long-term behaviour of the ice sheet and tectonics of the rift system (e.g. Behrendt *et al.* 1994; Blankenship *et al.* 1993).

Pillow lavas. Trachytic pillow lavas observed in three-dimensional exposures at Gill Bluff appear to be somewhat tubular in shape, elongate down the dip of foreset beds, 2–4 m long by 1–1.5 m in diameter, and usually isolated in a hyaloclastite matrix, but occasionally nested. Many have glassy margins 3–6 cm thick, others are only partly enclosed by a glassy rim, and still others have no glassy rim, but are enclosed by vitric fragments that decrease in size away from the pillow. These features suggest spalling of glassy rims during downslope movement. The pillows, rims, and enclosing vitric fragments are virtually non-vesicular. Jointing within pillows is irregular

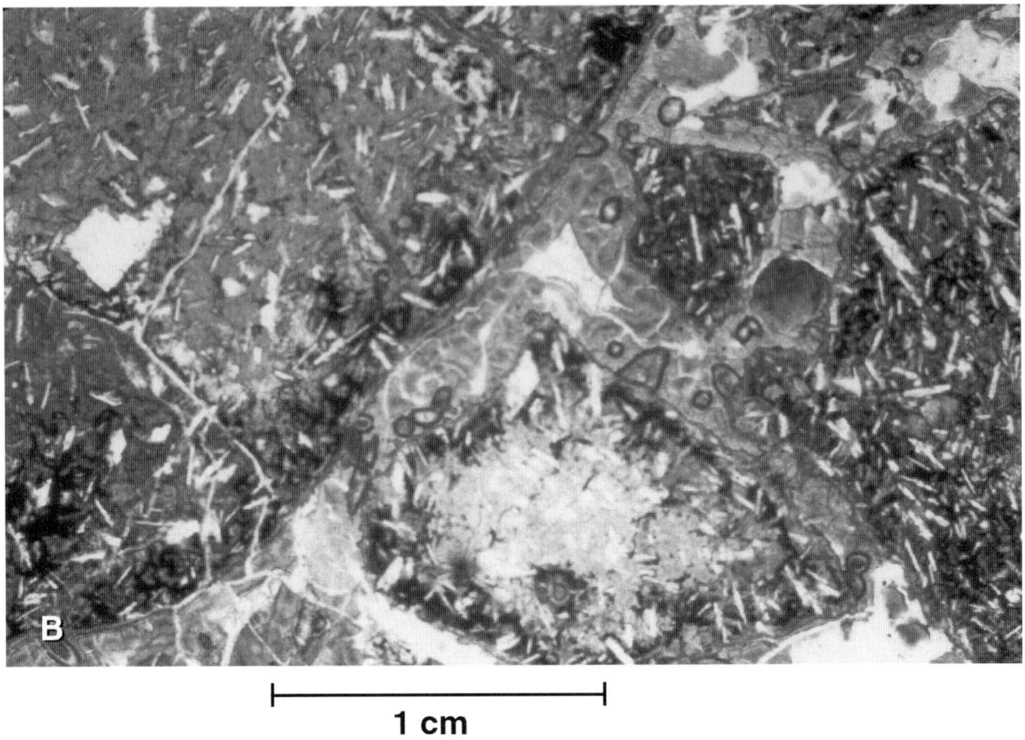

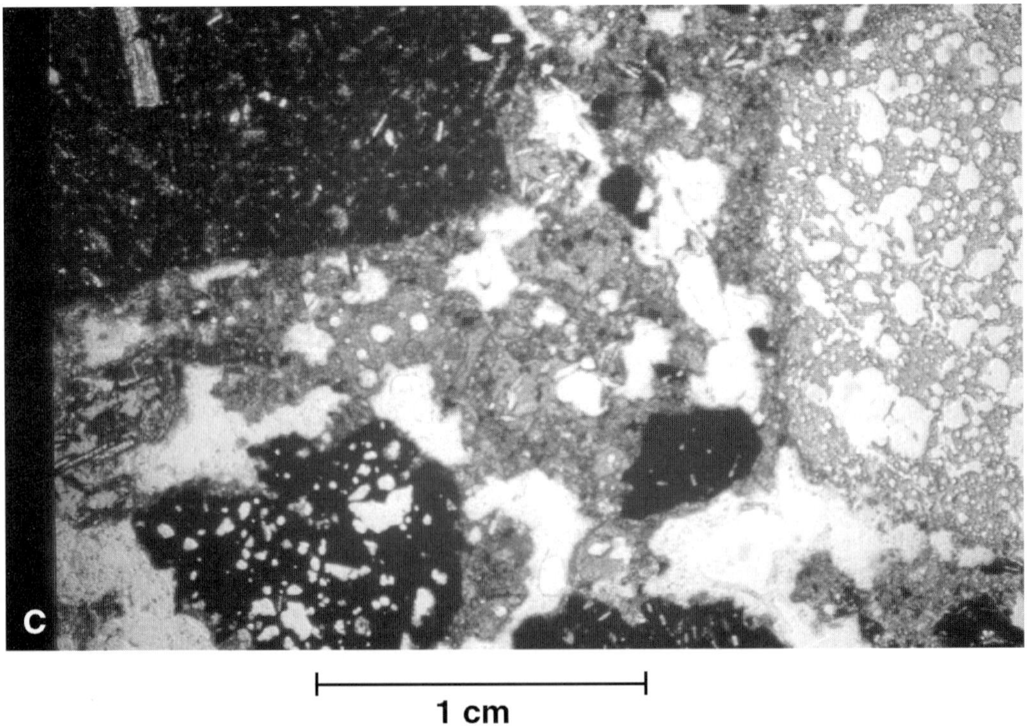

Fig. 10. (*continued*)

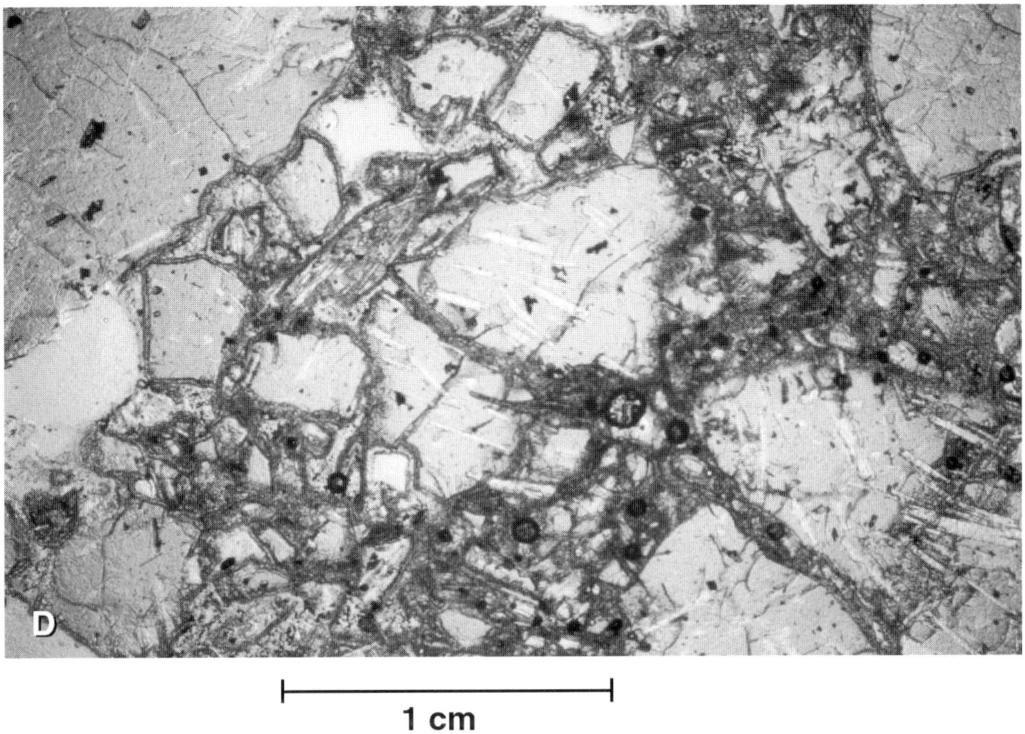

Fig. 10. (*continued*)

and blocky, and occasionally concentric (Figs 5 & 8). The trachytic pillows at Möll Spur appear to be similar in size and structure to those at Gill Bluff. However, smaller (0.5–1 m), irregularly shaped lobes are also found in this deposit. The pillows at the top of Möll Spur (Fig. 5) appear to be a distal extension of the topset subaerial flows (Fig. 6).

The basaltic pillows at Mount Murphy (Fig. 9c) are smaller than trachytic pillows (0.5–1 m in maximum dimension), but like the trachytic pillows they are virtually non-vesicular and surrounded by a 3–6 cm-thick glassy rim, which is markedly thicker than the rims on basaltic pahoehoe toes in this region. Some basaltic pillows show rudimentary radial jointing. At Mount Murphy the lateral transition from columnar jointed subaerial flows to large, bulbous, spheroidal bodies, and then to nested pillow lavas is well displayed (Fig. 9a).

Palagonitized Strombolian–Surtseyan tephra. These occur throughout a *c.* 300 m transition zone at Mount Murphy between the basal hydrovolcanic unit and the upper subaerial basalt section (LeMasurier *et al.* 1994). The distinguishing characteristics of this lithofacies are vesicular, fusiform bombs, up to 30 cm long, embedded in a yellow, palagonitized matrix of tachylite and subordinate sideromelane clasts (Fig. 10b; Table 1, column 6) that range from highly vesicular to nonvesicular. Palagonitization is restricted mainly to the sideromelane. A wet Strombolian to Surtseyan environment of eruption is inferred.

Heterolithic tuffs. These display a variety of textures, but all include a large proportion of lithic clasts of variable volcanic lithologies. Basaltic tuffs at Mount Murphy typically include nonvesicular basalt and tachylite clasts together with minor sideromelane, in matrix supported, massive bouldery deposits with grain sizes ranging from clay to *c.* 30 cm. They have been interpreted as lahar deposits (LeMasurier *et al.* 1994). The phreatomagmatic deposits at Gill Bluff and Stauffer Bluff are not included here. They contain a significant lithic component also, but they are predominantly vitric deposits (i.e. hyaloclastites), when fine ash matrix is included with larger clasts (e.g. Table 1, columns 2 and 4).

Differences in lava composition have not produced many distinctive structural or textural characteristics. Most hyaloclastites, whether basaltic or trachytic, are composed of blocky,

Table 1. *Modal analyses of hyaloclastites, in percent, based on 500 counts. Sample numbers in parentheses*

Sample no.	1 (85-2)	2 (64A)	3 (85-18B)	4 (65a)	5 (32 I)	6 (33D)	7 (61)	8 (58a)	9 (13a)	10 (6A)	11 (31c)
Vitric clasts											
sideromelane				14.8	17.8	16.2	7.0	25.0	11.2	15.0	17.0
tachylite				35.6	0	30.2	70.8	2.6	39.6	35.2	33.0
other	63	24	62.8								
Lithic clasts	0	29	0	5.4	1.8	0	0	0	0	0.4	0
Crystal clasts	<1	0	1.8	1.8	1	5.8	0	0	4.6	9.8	4.4
Fine ash matrix*	14.2	20.2	33.6	14.0	3.2	6.2	9.0	3.4	4.0	36.4	4.0
Open pores	4.4	16.4	0.6	15.2	1.8	4.0	3.0	10.0	3.2	0.2	18.2
Open vesicles	1.6	10.4	0	5.4	<1	0	0	0.8	2.2	1.0	6.6
Palagonite (after sideromelane)	10.8	0	1.2	7.6	53.0	11.8	3.0	47.8	26.4	1.8	14.8
Pore-fill cement		0	0								
zeolite	5.6			0	5.0	5.0	0.4	6.0	<0.1	0	1.0
carbonate	<1			0.2	1.2	6.0	2.0	2.0	8.6	0.2	1.0
clay				0	3.8	3.4	4.6	5.0	<0.2	0	0
opal					11.4	11.4				0	0
Pores + vesicles	6.0	26.8	0.6	20.6	1.8	4.0	3.0	10.8	5.4	1.2	24.8
Vitric clasts + palagonite excluding ash	73.8	24.0	64.0	58.0	70.8	58.2	80.8	75.4	77.8	52.0	64.8

1, Trachytic hyaloclastite between pillows, Gill Bluff, Mount Takahe.
2, Trachytic phreatic/phreatomagmatic tephra, Gill Bluff, Mount Takahe.
3, Trachytic hyaloclastite, Möll Spur, Mount Takahe.
4, Basaltic hyaloclastite from tuff cone deposit, top of Stauffer Bluff, Mount Takahe.
5, Basaltic hyaloclastite, base of Mount Murphy section.
6, Palagonitised Strombolian tephra, base of Mount Murphy section.
7, Globular basaltic tuff breccia, Turtle Peak, *c.* 8 Ma.
8, Basaltic hyaloclastite, late Oligocene, USAS Escarpment, 28.2 Ma.
9, Matrix of basaltic hyaloclastite tuff-breccia, base of summit section, Mount Petras, 22.2 Ma.
10, Tillite-bearing basaltic hyaloclastite tuff breccia, Shibuya Peak, 4.66 Ma.
11, Basaltic hyaloclastite from tuff cone deposit, Mount Hampton, <10 Ma.

* <1/16 mm (Fisher 1961).

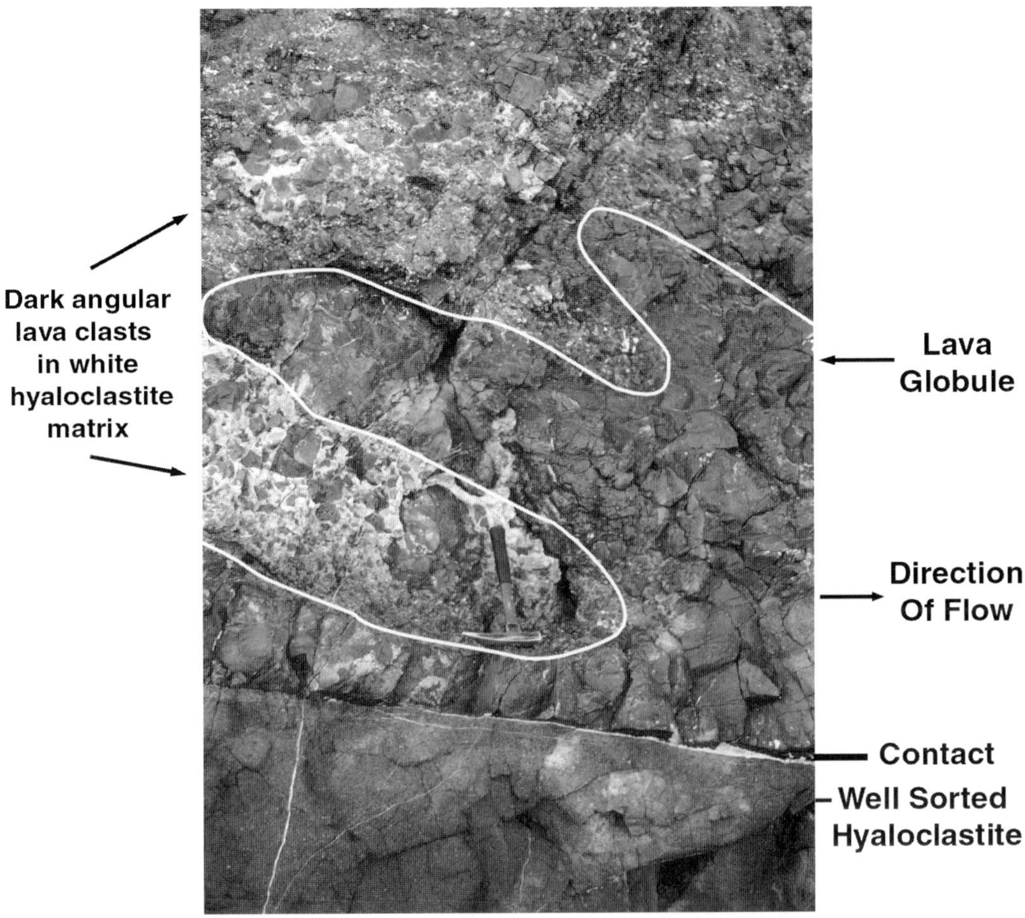

Fig. 11. Contorted and partially disaggregated lava globule in hyaloclastite breccia (outlined by white line) from the base of the Mount Murphy section (Fig. 9b). Angular lava fragments stand out against white hyaloclastite matrix. This relationship is interpreted to represent a transition from subaerial lava flow to hyaloclastite, by quenching and mechanical fragmentation, as lava flows off the volcano flank into an englacial lake environment. Well sorted hyaloclastite underlies the globule.

Table 2. *Densities and porosities of selected hyaloclastite samples from Marie Byrd Land**

Sample #	Comp.	Locality	Grain density	Sat'd blk density	Dry bulk density	Porosity (%)	Age (Ma)
6A	basalt	Shibuya Peak	2.67	2.48	2.37	11	4.66 ± 0.5
58A	basalt	USAS Escarpment	2.53	2.21	2.00	21	28.2 ± 1.2
61	basalt	Turtle Peak	2.65	2.40	2.25	15	6.4–11.3
64A	trachyte	Mount Takahe	2.37	1.64	1.11	53	<0.3
65A	basalt	Mount Takahe	2.74	2.26	1.99	27	<0.3
67B	trachyte	Mount Takahe	2.64	1.98	1.58	40	<0.3
67C	trachyte	Mount Takahe	2.42	2.15	1.96	19	<0.3

*All weight and volume values were determined three times and mean averages were used for calculations. Weights were determined using an analytical balance, and volumes were determined using a pychnometer. Saturation of samples with water was accomplished using a bell jar type vacuum. The data represent samples well enough indurated to retain cohesion during saturation and measurement. Analyst: J. C. Harrison, University of Colorado.

nonvesicular, microlite-rich vitric clasts. Pumiceous trachytic clasts are more vesicular than scoriaceous basaltic clasts, reflecting the higher water content of trachytic magma. The pillows in pillow lavas are larger in trachytic than in basaltic deposits, and palagonitization is more extensive in the basaltic deposits.

Post-depositional features

The most ubiquitous post-depositional phenomenon is alteration of sideromelane to palagonite, followed by the authigenic precipitation of zeolites, smectite, carbonate, and opal, as described above (Fig. 10; Table 1; Ellerman 1989). These processes have indurated and reduced the porosity of hyaloclastites to varying degrees. The dykes and sills, and downslope movement phenomena described below, show a variety of structural characteristics that seem to reflect the degree of induration, compaction and de-watering that characterized the deposit at the time of intrusion and/or deformation. Thus, diagenetic alteration, intrusion of dykes and sills, and downslope movement, or edifice settling, appear in some instances to have coincided with progradation of deltas, and in others, to have taken place long after progradation ceased, but with no obvious break in this continuum of processes. The relatively low porosity of older hyaloclastites (Table 2) suggests that palagonitization continued long after volcanism ceased.

Intrusion of dykes and sills

Pillowed intrusions. Figures 12 to 14 illustrate intrusive phenomena at Mount Takahe and Mount Murphy, in a sequence that is believed to represent increasing water content and decreasing competency of the volcanic pile into

Fig. 12. View of a pillowed basaltic sill or dyke intruded into hyaloclastite at Stauffer Bluff, Mount Takahe. (**a**) Large-scale field relations showing a more massive core flanked by pillow-like structures *c.* 1 m in diameter. Lighter coloured rock above and below the pillowed mass is a baked zone that envelopes the entire mass, and is more resistant than the adjacent unbaked rock. (**b**) Pillow-like bodies with thin (1–2 cm) glassy rims, little internal structure and few vesicles. The three pillows at the bottom of the photo are interconnected. (**c**) Pillow-like body showing a glassy peperitic margin, a gradation in the intensity of baking represented by darkening of enclosing rock away from the pillow (lower left), and a discordant relationship to multiple beds of the enclosing hyaloclastite. Baking is more conspicuous in very fine-grained rock (lower left) than in coarser grained rock (upper left). Note ice axe head in lower right for scale.

Fig. 12. (*continued*)

which lava was intruded. Figure 12 is interpreted to be a pillowed sill, and to represent a less water-rich environment of intrusion than the examples in Figures 13 and 14. The baked zone seen there has not been found adjacent to pillows in other MBL localities, and it appears to completely envelop the pillowed bodies (Fig. 12a). The ragged margins of pillows along some contacts (Fig. 12c) suggest incipient fragmentation, similar to, but less well developed than is seen in Figure 13. The glassy margins (Fig. 12b) are substantially thinner than those in true pillow lavas, and more like the margins in pahoehoe toes, but the intrusive relations (Fig. 12c) would seem to rule out this possibility. Pillowed dykes and sills have been described from several localities in Scotland and Wales (Kokelaar 1982) and from spectacular sea cliff exposures west of Auckland, New Zealand (Hayward 1979).

Contemporaneous intrusion and fragmentation. Figure 13 shows four views of dykes and sills that have undergone marginal quenching and fragmentation during emplacement. These features, and the repeated lateral protrusions of the intruding lava along bedding planes, suggest intrusion of lava into water-rich, poorly consolidated hyaloclastite beds, in which levels of neutral buoyancy were repeatedly reached during emplacement. The term peperite has been applied to tuffs and breccias formed by the mixing of hot

Fig. 13. (**a**) Irregularly shaped dykes and sills on the left and right intruding basaltic hyaloclastite at the base of the Mount Murphy section. Close-up views of the dyke on the right are shown in the next three frames. Note figures in the centre for scale. (**b**) Thick (>1 m) zones of marginal fragmentation are visible along the upper contact of the pod-like body in the centre (dashed arrow). Unlike the example in Figure 12, no wallrock alteration is apparent. Ice axe and hammer (solid arrows) on the lower left contact mark the location of the view in frame C. (**c**) Close-up view of a chilled and fragmented contact, that shows no wall-rock alteration. Large fragments of dense basaltic dyke rock (light colour) are partly enveloped by zones of darker fragmented glass. The intruded, unbaked hyaloclastite wall-rock is on the far left. (**d**) Angular, palagonitized glass-rich clasts in the fragmented contact zone above the pod-like body in the centre of 13B.

lava or intruding magma with wet sediment (Fisher & Schmincke 1984; Cas & Wright 1987; White *et al.* 2000), and the accompanying fluidisation process and its products have been thoroughly discussed by Kokelaar (1982). The term peperite can legitimately be applied to the examples shown here, but the boundary between intruded and intruding material is often difficult to draw, compared with examples where the intruded material is non-volcanic sediment. Furthermore, the fragmentation of some dykes and sills has been so complete that it becomes, in some cases, difficult to distinguish primary hyaloclastite beds from magmatically intruded sills, and diapiric from magmatically intruded dykes. The latter distinction requires finding a root for the dyke, either below the volcanic sequence (magmatic dyke), or within a subjacent layer (diapiric intrusion). Unusually good exposures are obviously essential.

Figure 14 shows two dykes at Stauffer Bluff that are composed entirely of massive hyaloclastite. The dyke in Figure 14a is coarser grained than the enclosing hyaloclastite, and seems likely to represent quenching and complete disintegration of a basalt lava intrusion, i.e. a more advanced stage of the process represented in Figure 13, or perhaps the edge or termination of a dyke concealed behind the exposure surface. The dyke in Figure 14b is more fine-grained than the enclosing hyaloclastite breccia, and its interpretation is ambiguous. It may be a finely fragmented intrusion that was originally lava, or alternatively, a diapiric intrusion of fine-grained, wet, very mobile hyaloclastite.

Edifice settling

Within the past decade, downslope movements of huge landslide blocks, with dimensions measurable in kilometres, have been documented on the flanks of large volcanoes such as Mauna Loa (Moore *et al.* 1995) and Mount Etna (Borgia *et al.* 1992). This relatively newly discovered

Fig. 13. (*continued*)

phenomenon has been described as sector collapse. Similar, but smaller scale, gravity-driven, downslope movement phenomena, from coherent blocks to cohesionless debris flow and grain flow deposits, are pervasive features in MBL hydrovolcanic deltas. They clearly belong in the same class of processes as sector collapse, and the driving mechanism is the same, but the MBL phenomena are described here as edifice settling processes, to distinguish them from the much larger scale sector collapse features. The following description presents five types of downslope movement phenomena in order of increasing cohesiveness, and presumably decreasing water content, of the deforming materials.

Grain flow and subaqueous mass flow deposits. Downslope flow of completely incoherent debris

Fig. 13. (*continued*)

is illustrated by the deposits in Figure 15. Three normally graded intervals are shown, each consisting of large rounded tuff breccia clasts and more angular basalt clasts at the base, grading upward to fine-grained laminated tuff at the top of the interval. The fine-grained top-most beds of each interval sag beneath the overlying coarse basal deposits of the next interval. The deposit as a whole is clast-supported, for the most part, with many large open pores; but spaces between many of the larger clasts are filled by laminated fine-grained tuff that appears to have been squeezed up by differential compaction beneath the weight of larger clasts. In discussing the differences between avalanche, debris flow, and grain flow deposits, Cas & Wright (1987) note that debris flow deposits frequently are matrix-supported and show grading of clasts, within a basal zone. Grain flow and avalanche deposits lack a muddy matrix: the former tend to be well-stratified and often show reverse grading, while the latter can sometimes be distinguished by the presence of megablocks (tens of metres or more in diameter). The lack of matrix in the Stauffer Bluff deposit (Fig. 15) suggests that it represents a subaqueous talus, or grain flow deposit, whereas the multiple, normally graded intervals suggest that the accumulation of talus was interrupted periodically by subaqueous granular mass flow (Cas & Wright 1987), or turbidity currents.

Soft sediment deformation. These weakly coherent, water saturated, deposits are represented by overturned isoclinal folds with downslope vergence (Fig. 16), and by concentrations of coarse clasts, and isolated pillows, that appear to be related to fold elements (Fig. 8).

Diapirism. This is suggested by the presence of cross-cutting bodies of fine-grained hyaloclastite. Figure 17 illustrates such a structure, in a deposit that was at one time highly mobile and undergoing deformation by soft sediment flow. In contrast to downslope flow phenomena, the upward flowage of diapiric structures is likely to result from differential compaction of low density deposits during edifice settling. The cross-cutting structure in Figure 17 is interpreted here to be diapiric because, compared to the structure in Figure 14b, this is more obviously a flowage phenomenon, represented especially well by lateral spreading (Fig. 17). These interpretations

Fig. 14. (**a**) A tabular dyke of coarse-grained, basaltic hyaloclastite breccia (upper right), probably produced by early dyke intrusion and fragmentation within a water-saturated, unconsolidated volcanic pile, Stauffer Bluff, Mount Takahe. (**b**) A dyke of fine-grained basaltic hyaloclastite (centre), also from Stauffer Bluff.

Fig. 15. Multiple, normally graded, basaltic mass flow deposits, Stauffer Bluff, Mount Takahe. The topmost bed of each interval is thinly laminated, very fine-grained tuff, that is highly deformed and squeezed up amongst the coarse basal clasts of the next overlying interval. Large basal clasts include lava joint blocks, pillow rinds, and *c.* 25–100 cm diameter blobs of rounded tuff breccia.

are weak, however, because no root for either structure could be found, either within or beneath the tuff sequence, which would clearly distinguish a diapir from a magmatic dyke.

Glide blocks. Downslope movement of large coherent blocks is suggested by the occurrence of well bedded sequences that truncate underlying beds at a high angle, as illustrated in Figures 7b and 9a. This relationship is the reverse of that seen in cross bedding, and is not easily explained by normal depositional processes, e.g. cut and fill. At Gill Bluff (Fig. 7b), the entire sequence appears to have experienced downslope flowage along with sliding of more coherent blocks.

Fault displacement. This is recognizable in well indurated hyaloclastite by the presence of thin lines and planes of light coloured, slickensided gouge (Fig. 18a–b). At Möll Spur, the fault sets shown in Figure 18 are appropriately oriented to reflect downslope movement and extension along the delta front, during distributive displacements along these fractures. The high-angle left-dipping (Fig. 18a) and right-dipping sets (Fig. 18b) form a conjugate normal fault pair consistent with a vertical maximum principle stress. The low angle sets are parallel to bedding. Wilch *et al.* (2002) have recently obtained a date of 15.2 ± 2.4 ka for this deposit, which is too young to reasonably attribute these fractures to tectonic stress. A more reasonable source of stress is removal of ice wall support during the lowering of ice level, which would accelerate gravity-driven collapse.

Discussion

The pillow lava problem

A comprehensive model for the architecture of MBL deltas remains elusive, because the cores of these structures are not exposed, and there is no convincing proxy for a core among any of the deeply eroded deposits elsewhere in MBL, with the possible exceptions of Icefall and Hedin nunataks (Smellie 2000, 2001), just west of Mount Murphy (Fig. 2). A comparison with hydrovolcanic centres elsewhere in Antarctica helps to put this in perspective. On the Antarctic Peninsula (Fig. 1), Skilling (1994) has studied the Pleistocene–Recent (<1 Ma) hydrovolcanic deposits at Brown Bluff volcanic centre, and Smellie & Hole (1997) have studied numerous Pliocene–Recent (<7 Ma) centres at Seal Nunataks and Beethoven Peninsula (Fig. 1). Composite cross-sections for centres at each of these localities consist of a pillow lava core, overlain by roughly 200 m of hydroclastic deposits, overlain in turn by subaerial flows, and surrounded by a distal mantle of hydroclastic delta deposits. This is, as they note, the classical Icelandic table mountain structure. Mount Nickens, in western Ellsworth Land (Fig. 1), apparently has the same gross structure, consisting of a core of vesicular pillow lavas overlain by *c.* 20 m of hyaloclastite; but has only been visited in reconnaissance (Lopatin & Poliakov 1974; Rowley *et al.* 1990). It is striking that, upon entering the MBL volcanic province from the east, the volume of volcanic products increases by an estimated three orders of magnitude (Hole & LeMasurier 1994), but there is no pillow lava core to any of the hydrovolcanic centres there, whether small or large, and no pillow lava resting on the basement

Fig. 16. Close-up of isoclinal folds at the base of the Gill Bluff exposure (lower right in Figure 7b). To the left of the ice axe base, and the right of the ice axe head, are two overturned isoclinal folds with axial planes dipping downhill, roughly parallel to the surface slope. Fold hinges point downhill. The two folds appear to be uncoupled along a horizontal slip surface that separates them from each other. Similar features can be seen, in a larger context, in Figure 7b.

Fig. 17. Hyaloclastite with deformed bedding cross-cutting massive-looking hyaloclastite. Stratification is also present within the latter, parallel to the skyline at the top of the photo. The upper surface of the deformed hyaloclastite appears to have spread laterally to the left and become involved in a small box fold to the left of the hammer. The deformed hyaloclastite is interpreted as a small diapiric intrusion.

Fig. 18. Pervasively faulted pillow lava/hyaloclastite complex at Möll Spur. (**a**) Two fault sets are marked by light coloured slickensided gouge. Bedding dips to the left, toward the delta front. The low angle fault set is parallel to bedding and seems to represent bedding plane slip; the high angle set is believed to represent normal extensional faulting. Scale is the same as for 18b. (**b**) Two fault sets, again marked by slickensided gouge, form a pattern that bounds pillow lavas rather than cutting through them, suggesting that the deposit was not thoroughly lithified at the time of faulting. The high angle fault set in this frame seems to be conjugate to the high angle set in 18a. Location is *c.* 10–20 m upslope from 18a; both exposures face east.

beneath any of the older (up to *c.* 28 Ma) hyaloclastite sections. In the equally voluminous volcanic provinces along the western coast of the Ross Sea, Wörner & Viereck (1987) described the table mountain-like structure of Shield Nunatak (Fig. 1), where the section is *c.* 200 m thick, but comment that the lack of pillow lavas is notable. However, these authors described a pillow lava core from the volcanic centre at nearby Cape Washington (Wörner & Viereck, 1989), Stump *et al.* (1980, 1990) described a pillow lava core to the table mountain volcano Mount Early (Fig. 1), and an Italian expedition has recently discovered pillow lavas at Shield Nunatak (P. Armienti, pers. comm., 2001). Still other pillow lava cores occur around the base of Mount Erebus (Fig. 1), most notably at Turks Head, on the coast SW of Mount Erebus summit (Luckman 1974).

While acknowledging that the apparent absence of pillow lava cores in MBL could be an effect of either inadequate exposure, or inadequate depth of eruption, it is striking that there is no lack of pillow core exposures among volcanoes in the western Ross Sea or Antarctic Peninsula regions, where chemical compositions and field conditions are similar to those in MBL. While many of the large MBL volcanoes are undissected, many of the smaller centres have been deeply eroded and still no pillow core has been found. Some are exposed down to the basement contact, and some have been thoroughly investigated (Smellie 2000, 2001). Among those, the detailed study of Icefall Nunatak by Smellie (2001) may serve to illustrate the point. He described a small area at the base of the nunatak composed of basaltic lithic breccias that include dispersed, non-vesicular basalt pillows, some of which are intact and some fragmented. He interpreted this lithofacies as corresponding 'structurally, but not in lithofacies, to the predominantly effusive pillow volcano stage observed in many subaqueous volcanoes' (Smellie 2001).

In MBL, nested pillow lavas have been documented only at Mount Takahe and Mount Murphy. As described above, these are non-vesicular pillows that occur at the top, and in the middle portions of foreset bedded delta fronts, and in two cases (Mount Murphy and Möll Spur) their relationship to up-dip subaerial flows is obvious. It seems clear that MBL pillow lavas formed when subaerial lava, after degassing, flowed down into a subglacial environment. Their origin is analogous to the examples described by Moore *et al.* (1973), of pillow lavas that form when lava flows into the sea, and is therefore not controlled by a volatile fragmentation depth (VFD). The apparent lack of a core of nested vesicular pillow lavas in MBL englacial volcanoes suggests that they represent somewhat different conditions of eruption than the examples in Iceland, the Antarctic Peninsula, or the western Ross embayment. Some possible explanations related to conditions of eruption are discussed below.

La Palma seamount provides an alternative model of subaquatic volcanism that addresses the question of whether variation in VFD, related to magma composition, could explain the absence of pillow lava cores in MBL. At La Palma, Staudigel & Schmincke (1984) estimated that the VFD was *c.* 800 m and was related to the increased water content of alkaline compared to tholeiitic basalt. The La Palma sequence is composed predominantly of ne-normative alkaline basalts and hawaiites, and subordinate mugearites and trachytes, all similar to the compositions of MBL volcanic rocks. The thickest sections of predominantly hydrovolcanic rock in MBL are found in the deltas described above. Therefore, it is possible that none of the exposed MBL deposits formed under >800 m of ice, and if the water content of MBL and La Palma magmas was roughly the same, this factor alone could explain the absence of pillow cores in MBL. However, the possibility of a greater VFD, related to alkaline composition, fails to explain why pillow cores are found in all the late Cenozoic volcanic provinces of Antarctica except MBL. These rocks are all remarkably alike in chemical composition. Only the proportions of a few large ion lithophile elements can be used to distinguish Antarctic Peninsula rocks from those throughout the West Antarctic rift (Hole & LeMasurier 1994).

It seems equally unlikely that the ice sheet was thinner in MBL than in the volcanic provinces where pillow lava cores are found. Ice thicknesses have been estimated to be >400 m at Beethoven Peninsula, 500–600 m at Seal Nunataks (Smellie & Hole 1997), and 200 m greater than today at Shield Nunatak (Wörner & Viereck 1987). In MBL, ice level is interpreted to have fluctuated rapidly during the growth of the Mount Murphy delta, from present ice level to *c.* 300 m above the present level (LeMasurier *et al.* 1994). The Mount Takahe deltas have recently been estimated to represent ice levels between 400 m and 575 m above present level (Wilch *et al.* 2002), and in late Miocene time, Smellie (2000) estimated that ice level was perhaps 100 m above its present level upstream (south) of Mount Murphy, completely covering Icefall and Hedin Nunataks, and Turtle Rock (Fig. 2), while they were forming.

Hyaloclastites have been recovered from Lo'ihi Seamount and the submarine rift zone of Kilauea at depths of 1200 to 2200 m, far below the normal VFD (Clague *et al.* 2000). The CO_2 and H_2O contents of the vitric clasts are consistent with eruption depths of 1200 m and more, rather than with downslope transport from shallower depths. Clast shapes are dominated by angular, sand and silt sized grains that include thin, bubble wall fragments. They bear a close resemblance to limu o Pele, which form when instantaneous mixing of seawater and melt results in steam explosions, usually as lava streams enter the sea. Clague *et al.* (2000) show that this mechanism can operate down to depths of 1200–1400 m if some mechanical perturbation stimulates mixing; but it should be noted that this mechanism need not operate to the exclusion of pillow lava formation: the two are found together at Lo'ihi. In MBL hyaloclastites, clast shapes do not resemble limu o Pele (Fig. 10a–d), and vesicles and bubble wall fragments are generally scarce. Thus, although pervasive turbulent mixing might be postulated to have promoted fragmentation in MBL, and prevented pillow formation, there is no evidence to support this mechanism.

A preliminary study of hyaloclastites associated with pillow lavas suggests that the presence of abundant microlites in glass rims might have inhibited pillow formation. Figure 19 shows the petrographic characteristics of hyaloclastites from three localities in Iceland and Antarctica where pillow lavas form the cores of table mountain-like structures. These may be compared with MBL samples in Figure 10. A consistent feature is the absence of microlites in hyaloclastites that occur in table mountains with pillow lava cores. This suggests the possibility that microlite-rich glass has a lower tensile strength than crystal-free glass, causing pillow crusts to crumble when subjected to stress. Direct observations of pillow growth show that the pillow crust expands as fresh lava is fed into the pillow (Moore 1975). Small crystal inclusions in a glassy crust would not accommodate such expansion as easily as hot glass, thereby concentrating tensile stresses in the surrounding

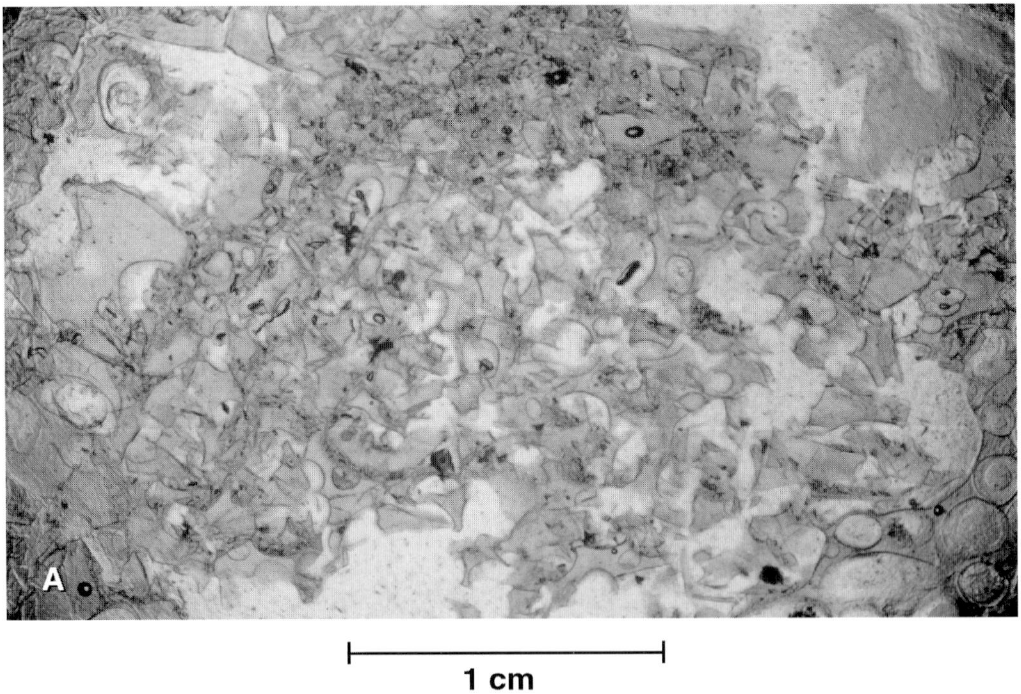

Fig. 19. Photomicrographs of hyaloclastites associated with wholly subaqueous pillow lavas. The scale of each frame is 2 × 3 mm. (**a**) Basaltic hyaloclastite from the Icelandic table mountain Herdubreid (Sigvaldason 1968), approximately 200 m below the subaerial summit section (underlying pillow lava contact is concealed). (**b**) Basaltic hyaloclastite overlying the pillow lava section, Mount Nickens, Hudson Mountains, Ellsworth Land, Antarctica. Sample provided courtesy of F. A. Wade. (**c**) Blocky clasts of sideromelane from hyaloclastite immediately above the pillow lava section, Turks Head, Ross Island, Antarctica. Courtesy of P. Luckman.

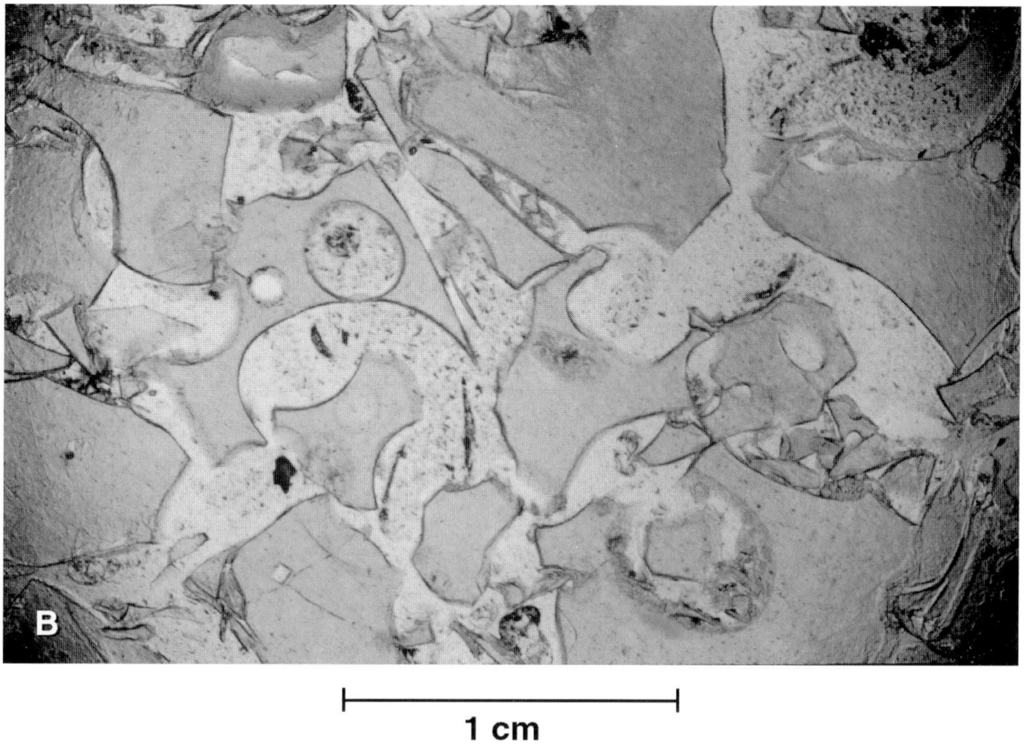

Fig. 19. (*continued*)

glass. Studies of Mauna Loa lava have also shown that early formation of crystals during an eruption causes an increase in lava viscosity (Lipman *et al.* 1985), which in turn, might facilitate the disintegration of pillow crusts during expansion. Loss of volatiles has been shown to control the timing of crystal precipitation (Lipman *et al.* 1985; Simakin *et al.* 1999) This implies that MBL lava lost volatiles earlier, or was poorer in volatiles originally, than the lava in volcanoes where pillow cores formed. In summary, it seems that pillow lavas formed in MBL englacial environments only when lava degassed subaerially, completely relieving internal gas pressure, but even then, they formed only rarely. The proposed possibility that microlites in glass inhibited pillow formation can easily be tested in other hydrovolcanic environments, and should be.

Delta evolution

Formation of hyaloclastite. The major process involved in the formation of MBL deltas is the flow of subaerial lava into an englacial lake, followed by the formation of hyaloclastite tuffs and tuff breccias, and occasionally, pillow lavas, much like the flow of lava into the sea at Kilauea (Moore *et al.* 1973). As with the pillow lavas, the best evidence that hyaloclastites formed in this manner is the down-dip transition from subaerial flows to hyaloclastite (Fig. 7). Figure 11 shows a lava globule that was undergoing quench fragmentation, and producing hyaloclastite in the process. Spalling of glassy rinds from pillows is essentially the same process, in lava where the glassy crust is able to maintain better cohesion. Non-vesicular, blocky, vitric clasts (Fig. 10a, b & d) appear to be the main products of this process, and the resulting deposit is the most common variety of hyaloclastite observed in MBL deltas. The less common pumiceous and scoriaceous clasts are most likely the products of phreatic and phreatomagmatic tuff cone eruptions, that took place when the volcanic pile was partly or wholly emergent.

Hawaiian studies are helpful in interpreting the factors that might have controlled the formation of hyaloclastite v. pillow lavas in this shallow water setting. Moore *et al.* (1973)

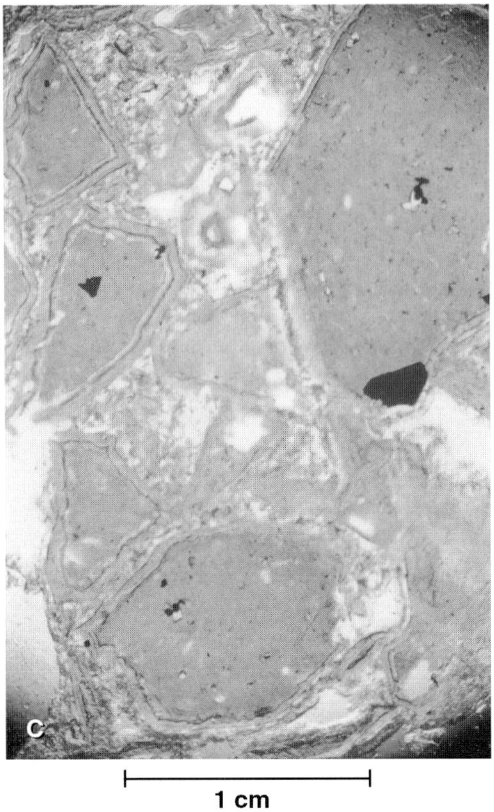

Fig. 19. (*continued*)

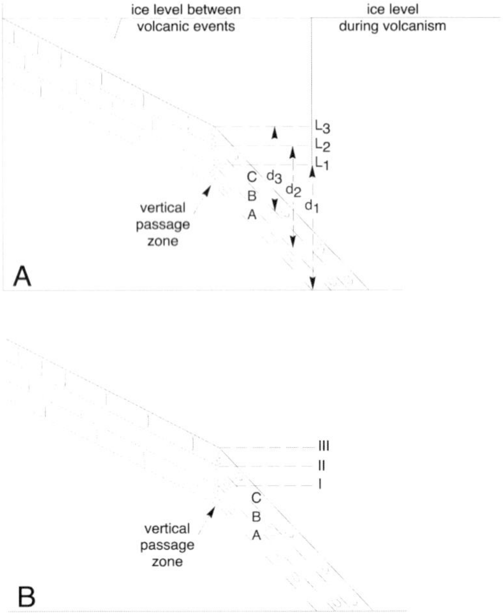

Fig. 20. Diagrammatic representation of two alternative interpretations of conditions that might have produced the vertical passage zone, shown in Figure 7 at Gill Bluff, Mount Takahe. Both panels depict the transition from subaerial trachyte flows, on the left, to more steeply dipping hyaloclastites that include isolated pillows, on the right. Panel A represents delta growth with a relatively stable ice level and good subglacial drainage of meltwater. L_1 and d_1 represent englacial lake level and depth, during deposition of layer A. L_2 and d_2 represent lake level and depth during deposition of layer B. L_3 and d_3 represent lake level during deposition of C. Panel B represents delta growth during a time of rising ice level. Levels I II, and III, represent ice levels during the deposition of layers A B, and C respectively. See text for discussion.

have observed that only the largest-volume flows retain cohesion, and yield pillow lavas when they enter the sea. Most of these flows are quenched and shattered to sand and rubble, and steam explosions (Clague *et al.* 2000) together with wave action aid in the process. These observations suggest that pillow lavas in MBL were produced from the largest-volume flows, and hyaloclastites from the rest.

The englacial environment. Unlike Hawaii, the glacial environment is subject to periodic draining and refilling of meltwater lakes, a process that has been directly observed in Iceland (Gudmundsson *et al.* 1997). This phenomenon provides a reasonable explanation for a variety of features found in MBL deltas. The vertical passage zone at Gill Bluff (Fig. 7) has previously been interpreted to represent a transition to a subglacial environment that was bounded by steep ice walls, with good subglacial drainage of meltwater (LeMasurier & Rex 1990). This contrasts with examples at James Ross Island, Antarctic Peninsula (Fig. 1) and Iceland (Jones & Nelson 1970), where a relatively constant water level helped maintain a horizontal passage zone. Figure 20 illustrates two alternative interpretations of conditions that might have produced the relations at Gill Bluff. Sketch A elaborates on the interpretation of LeMasurier & Rex (1990). It relies on emptying and refilling of an englacial lake, with a fairly constant ice level at or above the top of Gill Bluff. In this model, an englacial lake would be created during each eruptive episode, into which the hydrovolcanic materials would be deposited. Between each episode the lake would drain (a jokulhaup) and the ice would flow back to its original position, as has been observed at Grimsvötn volcano (M. T. Gudmundsson, pers. comm.). Lake level for the next episode would presumably lie somewhat above that of the previous episode, as the floor of the lake rose by delta progradation. Sketch b shows the vertical passage zone and a series of rising ice levels, following an interpretation presented by Jones & Nelson (1970) to explain vertical interdigitations of lava and flow foot breccia, somewhat like the field relations at Gill Bluff. This interpretation would require a 400 m rise of ice level while the delta was forming, or at least a rise through the upper *c.* 200 m seen in the exposed passage zone. The recently acquired date of 21.9 ka for Gill Bluff (Wilch *et al.* 2002) implies that such a rise would have been very rapid, followed by a drop of *c.* 400m to the present ice level.

In all four of the deltas described above, the lack of stratigraphic order among deposits that represent different environments of eruption, can be interpreted as an effect of ice level fluctuation, or a product of periodic draining and refilling of englacial lakes, or a combination of both. The best examples are found at Gill Bluff and Mount Murphy. At Mount Murphy, repeated intercalations of columnar jointed lava, palagonitized Strombolian/ Surtseyan deposits, and hyaloclastite, suggest that conditions at the delta front fluctuated between completely emergent and completely submerged (see also LeMasurier *et al.* 1994). At Gill Bluff, hyaloclastites composed of blocky, non-vesicular, vitric clasts (Fig. 10d) and isolated pillows, derived from updip subaerial flows, are intercalated with hyaloclastites containing lithic clasts and pumiceous accretionary lapilli, of probable phreatic and phreatomagmatic origin. This heterogeneous assemblage is consistent with the rapidly changing environment that seems to be represented by the vertical passage zone (Fig. 20a), but at this delta, downslope movements have so pervasively disrupted primary depositional relationships that the original relationships between these varied deposits cannot be determined. At Stauffer Bluff, by contrast, the vesiculated tuffs with armoured lapilli of scoriaceous sideromelane (Fig. 10c), are clearly the products of phreatomagmatic eruptions that built the tuff cone at the top of the bluff. These deposits have experienced some downslope slumping, but they are located where one would expect to find deposits that represent emergence above an englacial lake, and the environment they represent can be interpreted with more confidence

Role of intrusive phenomena. Intrusion of dykes and sills apparently took place at all stages in the consolidation and dewatering of hydrovolcanic deposits. Intrusions into water-rich, unconsolidated tuffs resulted in fluidisation and the development of broad zones of marginal fragmentation that sometimes merge indistinguishably with the intruded hyaloclastite. In more consolidated and dewatered deposits, dykes and sills developed pillowed structures, and baked, peperitic contact zones. This process undoubtedly added new material to the MBL deltas, but the amount cannot be estimated as it was at La Palma, by measuring sill thicknesses (Staudigel & Schmincke 1984). Sills and dykes in the MBL deltas were too fragmented during intrusion to permit even rough estimates of total thickness, and furthermore, the La Palma model shows that intrusive thickening was confined almost exclusively to the volcano core, which is not exposed in the deltas described here.

Applications to studies of glacial history and glacier dynamics

Volcanic rocks provide datable horizons that are especially useful for studying glacial history in the Antarctic environment. Moraines are much less common in Antarctica than in glaciated temperate regions, and datable materials are rare due to the near absence of terrestrial life. For more than 30 years, geologists have attempted to reconstruct the chronology of Antarctic glaciation by determining the ages of hydrovolcanic sections (e.g. Rutford *et al.* 1968; Hamilton 1972; LeMasurier 1972; Stump *et al.* 1980). Ages, elevations, and facies relationships at subaerial/ subglacial transition zones have been used to infer ice level fluctuations, as well as glacial chronology (Wörner & Viereck 1987; LeMasurier *et al.* 1994; Wilch *et al.* 2002; Fig. 20a), and some studies have used hydrovolcanic facies to infer the size and thermal characteristics of the glaciers that produced them (Smellie *et al.* 1993; Smellie 2000, 2001). Some controversial topics covered by these studies are the age of the West Antarctic ice sheet, and the extent and timing of deglaciations, especially during the Pliocene. These are still topics of ongoing research.

It is widely believed that the mechanical behavior of bedrock has an influence on ice sheet dynamics. It has been proposed, for example, that the locations of ice streams in the West Antarctic ice sheet are controlled by easily deformable deposits such as weakly consolidated till, or marine sediment (MacAyeal 1992; Iverson 1999). It is clear from the characteristics described above that hydrovolcanic deposits are weak and easily deformable, by virtue of their abundant glass fragments as well as the pervasive alteration of the glass to smectites, zeolites and carbonate, described above. Aeromagnetic surveys over large areas of the West Antarctic rift system suggest the presence of large volumes of volcanic rock beneath the ice sheet (Behrendt *et al.* 1994), and sedimentological studies of samples recovered from McMurdo Sound sediments include, in some cases, significant amounts of smectite (e.g. Ehrmann 1998). These diverse studies converge around the possibility that volcano–ice interactions beneath large ice sheets may not only produce distinctive deposits, but may also influence the behavior and stability of the ice sheet (e.g. Blankenship *et al.* 1993). Sedimentologists studying basal sediments from drill cores in the West Antarctic ice sheet should keep in mind that the provenance of smectites found in these samples may be hyaloclastites, rather than weathered basement rock, and evaluate the mechanical properties of the source rock accordingly.

Field work in Antarctica was supported by NSF grants DPP 80-20836 and DPP 77-27546. A comparative study of Icelandic tablemountains and their deposits was supported by NSF Grant OPP9720411, administered by the Office of Polar Programs, and greatly facilitated by the hospitality of M. T. Gudmundsson during a visit to Grimsvötn volcano, Iceland. I am happy to acknowledge the help of J. C. Harrison, who provided the specific gravity data, C. Lee, for hours of assistance with the lab study of hyaloclastites, and K. Kellogg, for discussions about fault mechanisms and their role in downslope movement. The thorough and constructive comments of P. Armienti, I. Skilling and J. Smellie greatly improved the manuscript, and are gratefully acknowledged.

References

BATIZA, R. & WHITE, J. D. L. 2000. Submarine lavas and hyaloclastite. *In:* SIGURDSSON, H. (ed.) *Encyclopedia of Volcanoes*. Academic Press, San Diego, 361–381.

BEHRENDT, J. C., LEMASURIER, W. E., COOPER, A. K., TESSENSOHN, F., TRÉHU, A. & DAMASKE, D. 1991*a*. Geophysical studies of the West Antarctic rift system. *Tectonics*, **10**, 1257–1273.

BEHRENDT, J. C., LEMASURIER, W. E., COOPER, A. K., TESSENSOHN, F., TRÉHU, A. & DAMASKE, D. 1991*b*. The West Antarctic rift system: A review of geophysical investigations. *American Geophysical Union, Antarctic Research Series*, **53**, 67–112.

BEHRENDT, J. C., BLANKENSHIP, D. D., FINN, C. A., BELL, R. E., SWEENEY, R. E., HODGE, S. M. & BROZENA, J. M. 1994. CASERTZ aeromagnetic data reveal late Cenozoic flood basalts (?) in the West Antarctic rift system. *Geology*, **22**, 527–530.

BLANKENSHIP, D. D., BELL, R. E., HODGE, S. M., BROZENA, J. M., BEHRENDT, J. C. & FINN, C. A. 1993. Active volcanism beneath the West Antarctic ice sheet and implications for ice sheet stability. *Nature*, **361**, 526–529.

BONATTI, E. & TAZIEFF, H. 1970. Exposed guyot from the Afar rift, Ethiopia. *Science*, **168**, 1087–1089.

BORGIA, A., FERRARI, L. & PASQUARÈ, G. 1992. Importance of gravitational spreading in the tectonic and volcanic evolution of Mount Etna. *Nature*, **357**, 231–235.

CANDE, S. C., STOCK, J. M., MÜLLER, R. D. & ISHIHARA, T. 2000. Cenozoic motion between East and West Antarctica. *Nature*, **404**, 145–150.

CAS, R. A. F. & WRIGHT, J. V. 1987. *Volcanic successions, modern and ancient*. Allen & Unwin (Publishers), London.

CLAGUE, D. A., DAVIS, A. S., BISCHOFF, J. L., DIXON, J. E. & GEYER, R. 2000. Lava bubble-wall fragments formed by submarine hydrovolcanic explosions on Lo'ihi Seamount and Kilauea Volcano. *Bulletin of Volcanology*, **61**, 437–449.

COTTON, C. A. 1969. The pedestals of oceanic volcanic islands. *Geological Society of America Bulletin*, **80**, 749–760.

DREWRY, D. J. 1983. *Antarctica: Glaciological and geophysical folio*. Scott Polar Research Institute, Cambridge.

EHRMANN, W. 1998. Implications of late Eocene to early Miocene clay mineral assemblages in McMurdo Sound (Ross Sea, Antarctica) on paleoclimate and ice dynamics. *Paleogeography, Paleoclimatology, Paleoecology*, **139**, 213–231.

ELLERMAN, P. J. 1989. *Depositional environments and post-depositional alteration of Cenozoic hyaloclastites in Antarctica*. PhD thesis, University of Colorado, USA.

FISHER, R. V. 1961. Proposed classification of volcanoclastic sediments and rocks. *Geological Society of America Bulletin*, **72**, 1409–1414.

FISHER, R. V. & Schmincke, H-U. 1984. *Pyroclastic rocks*. Springer-Verlag, New York.

FITZGERALD, P. G. & STUMP, E. 1997. Cretaceous and Cenozoic episodic denudation of the Transantarctic Mountains, Antarctica: New constraints from apatite fission track thermochronology in the Scott Glacier region. *Journal of Geophysical Research*, **102**, 7747–7765.

GUDMUNDSSON, M. T., SIGMUNDSSON, F. & BJÖRNSSON, H. 1997. Ice–volcano interaction of the 1996 Gjálp subglacial eruption, Vatnajökull, Iceland. *Nature*, **389**, 954–957.

HAMILTON, W. 1972. The Hallett Volcanic Prov-ince, Antarctica. *US Geological Survey Professional Paper*, **456C**, 1–62.

HAYWARD, B. W. 1979. Ancient undersea volcanoes: a guide to the geological formations at Muriwai, West Auckland. *Geological Society of New Zealand Guidebook*.

HOLE, M. J. & LEMASURIER, W. E. 1994. Tectonic controls on the geochemical composition of Cenozoic mafic alkaline volcanic rocks from West Antarctica. *Contributions to Mineralogy and Petrology*, **117**, 187–202.

Honnorez, J. & Kirst, P. 1975. Submarine basaltic volcanism: morphometric parameters for discriminating hyaloclastites from hyalotuffs. *Bulletin of Volcanology*, **39**, 1–25.

Iverson, N. R. 1999. Till rheology and the stability of ice sheets on soft beds. *Geological Society of America, Abstracts with Programs*, **31**, 203.

Jones, J. G. 1966. Intraglacial volcanoes of southwest Iceland and their significance in the interpretation of the form of the marine basaltic volcanoes. *Nature*, **212**, 586–588.

Jones, J. G. 1969. Intraglacial volcanoes of the Laugar-vatn region, south-west Iceland, I. *Quarterly Journal Geological Society of London*, **124**, 197–211.

Jones, J. G. 1970. Intraglacial volcanoes of the Laugarvatn region, southwest Iceland, II. *Journal of Geology*, **78**, 127–140.

Jones, J. G. & Nelson, P. H. H. 1970. The flow of basalt lava from air into water – its structural expression and stratigraphic significance. *Geological Magazine*, **107**, 13–19.

Kellerhals, R., Shaw, J. & Arora, V. K. 1975. On grain size from thin sections. *Journal of Geology*, **83**, 79–96.

Kjartansson, G. 1959. The Moberg Formation. *Geografisker Annaller*, **41**, 135–169.

Kokelaar, B. P. 1982. Fluidization of wet sediments during the emplacement and cooling of various igneous bodies. *Journal Geological Society, London*, **139**, 21–33.

Kokelaar, P. 1986. Magma–water interactions in subaqueous and emergent basaltic volcanism. *Bulletin of Volcanology*, **48**, 275–289.

Lawver, L. A. & Gahagan, L. M. 1994. Constraints on timing of extension in the Ross Sea region. *Terra Antarctica*, **1**, 545–552.

LeMasurier, W. E. 1972. Volcanic record of Cenozoic glacial history of Marie Byrd Land. *In:* Adie, R. J. (ed.) *Antarctic Geology and Geophysics*. Universitetsforlaget, Oslo, 251–260.

LeMasurier, W. E. 1978. The Cenozoic West Antarctic rift system and its associated volcanic and structural features. *Geological Society of America Abstracts with Programs*, **10**, 443.

LeMasurier, W. E. 1990*a*. Late Cenozoic volcanism on the Antarctic plate: an overview. *American Geophysical Union, Antarctic Research Series*, **48**, 1–17.

LeMasurier, W. E. 1990*b*. Mount Sidley. *American Geophysical Union, Antarctic Research Series*, **48**, 203–207.

LeMasurier, W. E. 1990*c*. Marie Byrd Land Summary. *American Geophysical Union, Antarctic Research Series*, **48**, 146–163.

LeMasurier, W. E. 1990*d*. Satellitic Volcanic Centers. *American Geophysical Union, Antarctic Research series*, **48**, 234–252.

LeMasurier, W. E. & Landis, C. A. 1996. Mantle plume activity recorded by low-relief erosion surfaces in West Antarctica and New Zealand. *Geological Society of America Bulletin*, **108**, 1450–1466.

LeMasurier, W. E. & Rex, D. C. 1990. Mount Takahe. *American Geophysical Union, Antarctic Research Series*, **48**, 169–174.

LeMasurier, W. E., Harwood, D. M. & Rex, D. C. 1994. Geology of Mount Murphy volcano: An 8-m.y. history of interaction between a rift volcano and the West Antarctic ice sheet. *Geological Society of America Bulletin*, **106**, 265–280.

Lipman, P. W., Banks, N. G. & Rhodes, J. M. 1985. Degassing-induced crystallization of basaltic magma and effects on lava rheology. *Nature*, **317**, 604–607.

Lorenz, V. 1970. Some Aspects of the Eruption Mechanism of the Big Hole Maar, Central Oregon. *Geological Society of America Bulletin*, **81**, 1823–1830.

Lopatin, B. G. & Poliakov, M. M. 1974. Geology of the volcanic Hudson Mountains, Wallgreen Coast, West Antarctica (translated from the Russian). *Antarktika*. **13**, 36–51. Separate printing, USSR Academy of Science, Joint Antarctica Studies Commission, Izdatelstvo 'Nauka', Moskva.

Luckman, P. G. 1974. *Products of submarine and subglacial volcanism in the McMurdo Sound region, Ross Island, Antarctica*. BSc Hons thesis, Victoria University of Wellington, New Zealand.

MacAyeal, D. R. 1992. Irregular oscillations of the West Antarctic ice sheet. *Nature*, **359**, 29–32.

Mathews, W. H. 1947. 'Tuyas', flat-topped volcanoes in northern British Columbia. *American Journal of Science*, **245**, 560–570.

Moore, J. G. 1975. Mechanism of formation of pillow lava. *American Scientist*, **63**, 269–277.

Moore, J. G. & Fiske, R. S. 1969. Volcanic substructure inferred from dredge samples and ocean bottom photographs, Hawaii. *Geological Society of America Bulletin*, **80**, 1191–1202.

Moore, J. G., Bryan, W. B., Beeson, M. H. & Normark, W. R. 1995. Giant blocks in the South Kona landslide, Hawaii. *Geology*, **23**, 125–128.

Moore, J. G., Phillips, R. L., Grigg, R. W., Peterson, D. W. & Swanson, D. A. 1973. Flow of lava into the sea 1969–1971, Kilauea Volcano, Hawaii. *Geological Society of America Bulletin*, **84**, 537–546.

Noe-Nygaard, A. 1940. Subglacial volcanic activity in ancient and recent times – studies in the palagonite system of Iceland. *Folia Geografica Danica*. **1**, 1–67.

Porebski, S. J. & Gradzinski, R. 1990. Lava-fed Gilbert-type delta in the Polonez Cove Formation (Lower Oligocene), King George Island, West Antarctica. *In:* Colella, A. & Prior, D. (eds) *Coarse-grained deltas*. International Association of Sedimentologists, Blackwell Science, Oxford, Special Publications, **10**, 335–351.

Rowley, P. D., Laudon, T. S., LaPrade, K. E. & LeMasurier, W. E. 1990. Hudson Mountains. *American Geophysical Union, Antarctic Research Series*, **48**, 289–293.

Rutford, R. H., Craddock, C. & Bastien, T. W. 1968. Late Tertiary glaciation and sea-level changes in Antarctica. *Paleogeography, Paleoclimatology, Paleoecology*, **5**, 15–39.

Sigvaldason, G. E. 1968. Structure and products of subaquatic volcanoes in Iceland. *Contributions to Mineralogy and Petrology*, **18**, 1–16.

SIMAKIN, A. G., ARMIENTI, P. & EPEL'BAUM, M. B. 1999. Coupled degassing and crystallization: experimental study at continuous pressure drop, with application to volcanic bombs. *Bulletin of Volcanology*, **61**, 275–287.

SKILLING, I. P. 1994. Evolution of an englacial volcano: Brown Bluff, Antarctica. *Bulletin of Volcanology*, **56**, 573–591.

SMELLIE, J. L. 2000. Subglacial eruptions. *In:* SIGURDSSON, H. (ed.) *Encyclopedia of Volcanoes*. Academic Press, San Diego, 403–418.

SMELLIE, J. L. 2001. Lithofacies architecture and construction of volcanoes erupted in englacial lakes: Icefall Nunatak, Mount Murphy, eastern Marie Byrd Land, Antarctica. International Association of Sedimentologists, Blackwell Science, Oxford, Special Publications, **30**, 9–34.

SMELLIE, J. L. 2001. History of oligocene erosion, uplift and unroofing of the Transantarctic Mountains deduced from sandstone detrital modes in CRP-3 drillcore, Victoria Land basin, Antarctica. *Terra Antarctica*, **8**, 481–489.

SMELLIE, J. L. & HOLE, M. J. 1997. Products and processes in Pliocene-Recent, subaqueous to emergent volcanism in the Antarctic Peninsula: examples of englacial Surtseyan volcano construction. *Bulletin of Volcanology*, **58**, 628–646.

SMELLIE, J. L., HOLE, M. J. & NELL, P. A. R. 1993. Late Miocene valley-confined subglacial volcanism in northern Alexander Island, Antarctic Peninsula. *Bulletin of Volcanology*, **55**, 273–288.

STAUDIGEL, H. & SCHMINCKE, H-U. 1984. The Pliocene seamount series of La Palma/Canary Islands. *Journal of Geophysical Research*, **89**, 11 195–11 215.

STUMP, E., BORG, S. G. & SHERIDAN, M. F. 1990. Mount Early. *American Geophysical Union, Antarctic Research Series*, **48**, 138–139.

STUMP, E., SHERIDAN, M. F., BORG, S. G. & SUTTER, J. F. 1980. Early Miocene subglacial basalts, the East Antarctic ice sheet, and uplift of the Transantarctic Mountains. *Science*, **207**, 757–759.

WHITE, J. D. L. & HOUGHTON, B. 2000. Surtseyan and related phreatomagmatic eruptions. *In:* SIGURDSSON, H. (ed.) *Encyclopedia of Volcanoes*. Academic Press, San Diego, 495–511.

WHITE, J. D. L., MCPHIE, J. & SKILLING, I. 2000. Peperite: a useful genetic term. *Bulletin of Volcanology*, **62**, 65–66.

WILCH, T. I., MCCUDDY, S. M. & MCINTOSH, W. C. 2000. Middle and late Wisconsinan expansions of the West Antarctic ice sheet at Mt. Takahe volcano (abs). *In*: *Volcano/Ice Interaction on Earth and Mars Abstract Volume*, August 13–15 2000, University of Iceland, Reykjavik.

WÖRNER, G. & VIERECK, L. 1987. Subglacial to emergent volcanism at Shield Nunatak, Mt. Melbourne Volcanic Field, Antarctica. *Polarforschung*, **57**, 27–41.

WÖRNER, G. & VIERECK, L. 1989. The Mt. Melbourne Volcanic Field (Victoria Land, Antarctica) I. Field Observations. *Geologische Jahrbuch*, **E38**, 369–393.

Facies analysis of proximal subglacial and proglacial volcaniclastic successions at the Eyjafjallajökull central volcano, southern Iceland

S. C. LOUGHLIN

British Geological Survey, West Mains Road, Edinburgh EH9 3LE, UK
(e-mail: s.loughlin@bgs.ac.uk)

Abstract: The long-lived (at least 0.78 Ma) Eyjafjallajökull volcano has been constructed during a period of dramatic climatic shifts, and has produced subaerial lavas and cinder cones, pillow lavas, hyaloclastite, monogenetic volcaniclastic sediments and polygenetic glacio-fluvial sediments. Subaqueous lithofacies are dominant and nine distinct cogenetic lithofacies associations are identified within the successions that are interpreted as the proximal deposits of subglacial eruptions. These associations are typically bound by unconformities and lie directly on thin diamictites or glaciated surfaces; epiclastic sedimentary rocks are absent. The lithofacies include subaqueous sheet lava flows, lobate flows, pillow lavas, breccias, hyaloclastite and hyalotuff generated by eruption of mainly basaltic lavas. Lavas associated with massive hyaloclastite breccias commonly lie on or intrude cogenetic redeposited hyalotuff, indicating rapid changes in style of activity from explosive to effusive. This suggests that the vent was initially subaqueous but became subaerial as meltwater drained away down-slope beneath temperate ice sheets. In other instances, probably under thicker ice, the eruptions were effusive throughout but lava was subjected to intense steam explosivity. There is abundant evidence that volcaniclastic material was transported downslope by mass flow, grain flow and traction currents (i.e. by running water). Lithofacies associations that ponded in ice-dammed water commonly comprise thick redeposited tephra overlain by a 'passage zone' of hyaloclastite and then subaerial lavas. These deposits are voluminous but conditions suitable for ponding of large volumes of water were not common in the evolution of this large volcano.

During the early stages of volcano growth, eruptive deposits were emplaced on the gentle slopes of the growing cone during both glacial and interglacial periods. Glacial erosion modified the growing edifice and, as a result, younger deposits (<600 ka) tend to be valley-confined. The unconformities produced during glacial advance represent significant time gaps within the succession. Observations support the hypothesis that there were higher volcanic production rates during periods of deglaciation, probably as a result of the rapid decrease in lithostatic pressure as ice melted.

The Eyjafjallajökull volcanic centre is a large, E–W-orientated complex in southern Iceland, which reaches a height of 1651 m and covers an area of about 400 km^2 (Fig. 1). It has been active for over 0.78 Ma (Kristjansson *et al.* 1988) during both glacial and interglacial periods. Mapping has shown that complex associations of lava, hyaloclastite and volcaniclastic lithofacies, resulting from subaqueous volcanic activity, form a large part of the edifice (e.g. Jonsson 1988). Evidence for a subaqueous eruption environment includes pillow lavas, abundant glassy (sideromelane) clasts, lavas with thick entablatures indicative of water saturation, and associated water-lain sedimentary rocks (Sæmundsson 1970; Long & Wood 1986). Sideromelane is extensively altered to palagonite giving deposits a distinctive yellowish appearance; this occurs as a result of warm water-saturated conditions (Jakobsson 1978). Owing to the elevation and topography of the volcano, most of these deposits are assumed to be the result of subglacial volcanic activity (Loughlin 1995). Following a subglacial eruption, deposition of eruption products may occur englacially, subglacially, supraglacially, or in a proglacial subaerial or submarine environment. In general, the finer grained debris is transported by meltwater to distal environments, which in the case of this coastal volcano is the marine environment. However, most of the assemblages described in this paper are dominated by lavas and coarse breccias that were deposited in a proximal environment. The common association

From: SMELLIE, J. L. & CHAPMAN, M. G. (eds) 2002. *Volcano–Ice Interaction on Earth and Mars*. Geological Society, London, Special Publications, **202**, 149–178. 0305-8719/02/$15.00 © The Geological Society of London 2002.

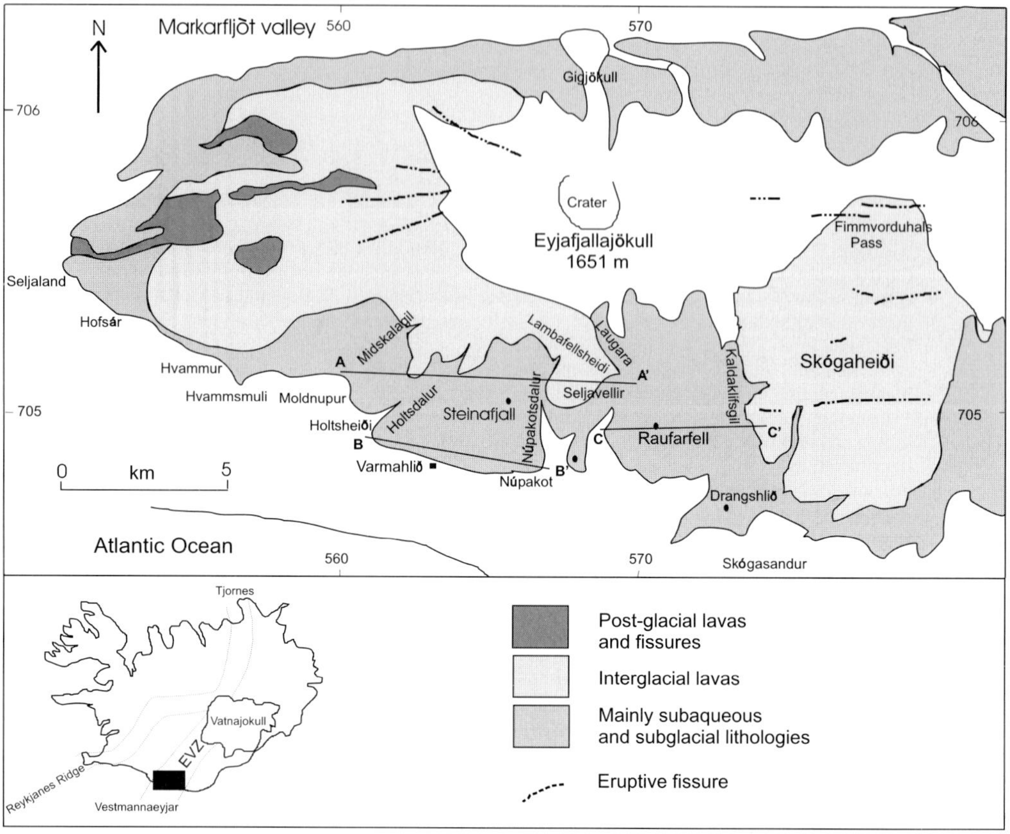

Fig. 1. Simplified map of the Eyjafjallajökull volcano (after Jonsson 1988) showing the distribution of postglacial, interglacial and intraglacial deposits. Grid references refer to 1:50 000 map of Eyjafjallajökull, published by the Defense Mapping Agency Hydrographic/ Topographic Center, Washington, DC in cooperation with the Iceland Geodetic Survey, 1990. EVZ, Eastern Volcanic Zone. Locations of schematic profile sections (Fig. 5) also shown.

of striated bedrock, diamictites and/or subaerial lavas with assemblages of hyaloclastite and hyalotuff is considered to be evidence against submarine deposition of these proximal deposits (e.g. Walker & Blake 1966; Carswell 1983). The occurrence of steep angular unconformities and the absence of moraines and fluvial sediments suggests that the majority of preserved deposits were emplaced in a subglacial rather than a proglacial environment (e.g. Walker & Blake 1966).

In this paper, distinctive common lithofacies associations on Eyjafjallajökull central volcano will be described and interpreted in terms of eruptive processes, eruptive environment, mode of emplacement, depositional environment and ice extent and thickness. When combined with isotopic dating, this kind of analysis of the temporal and spatial variation in ice cover on large volcanic centres is of value in assessing climatic variations.

Geological background

The onland extension of the Mid-Atlantic Ridge in Iceland, known as the 'axial rift zone', is characterized by relatively thin crust (8–10 km, Pálmason 1971; Gebrande *et al.* 1980), high heat flow (up to 160° km^{-1}, Pálmason 1973; Pálmason *et al.* 1979), well developed tensional features (Sæmundsson 1979) and the production of tholeiitic basalts. In contrast, off-axis volcanic zones, known as 'flank zones', are situated on older, thicker crust (20–30 km, Pálmason 1971), heat flow is lower (Pálmason 1973), tensional features are poorly developed and transitional alkali to alkali basalts are produced. The Eyjafjallajökull volcano is situated in the southern part of the Eastern Volcanic Zone (EVZ, Fig. 1) which extends SW from Vatnajökull to the Vestmannaeyjar off the south coast (Jakobsson 1979). The northern part of the EVZ has the characteristics of an axial rift zone, whereas the southern part has the characteristics of a flank zone. It has been interpreted as a southwards propagating rift. The volcano has erupted mainly transitional alkali basalts with minor ankaramite to quartz trachyte compositions (e.g. Jakobsson 1979; Loughlin 1995) and has a low magma productivity compared with volcanoes in the axial rift zone and the large nearby Katla volcano. The last eruption (of mainly acid tephra) occurred between 1821 and 1823 from the central crater.

At present, an ice cap covers the summit of the Eyjafjallajökull volcano down to an elevation of about 1000 m. Valley glaciers extend to lower elevations both north and south of the crater. The volcano is bounded to the south by coastal sandur deposits and to the north and west by the braided river system of Markarfljòt (Fig. 1). The southern margin of the edifice is sharply defined by steep cliffs (<500 m high), which formed by a combination of wave action (when sea level was high) and glaciation. The southern flanks of the volcano are highly dissected by glacial valleys and deep river gorges some of which expose old deposits and an active hydrothermal system in the core of the volcano. The western flanks of the volcano slope gently down to cliffs up to 500 m high above the alluvial plain. The cliffs were probably carved by a glacier that advanced down the Markarfljòt valley during the Younger Dryas (Kjartansson 1958). To the east, an E–W-orientated ridge (the Fimmvörduháls Pass) joins Eyjafjallajökull to Myrdalsjökull, the ice cap which covers the large, active volcanic centre of Katla.

The Plio-Pleistocene and the late Pleistocene in Iceland was characterized by glacial and interglacial periods (Sæmundsson 1979). Thick ice sheets covered much of Iceland at the maximum of the last glaciation (*c.* 20 000 BP), reaching about 1000 m thick in central Iceland (Walker 1965) and thinning towards the coast. During the Weichselian glaciation (11–10 000 BP) Iceland was again covered by an ice cap. The ice retreated rapidly and sea level rose to 100 m above its present level (Einarsson & Albertsson 1988). Marine deposits are found more than 100 m above present sea level in southern Iceland (Sigmundsson 1991). As a result of rapid isostatic rebound (over 1000 years) the sea regressed to 30 m below present sea level by 9000 BP (Sigmundsson 1991). It seems likely that sea level has not risen by much more than 100 m in the last 1 Ma.

Subglacial volcanism: previous work

An eruption may be defined as subglacial if the vent is situated beneath a glacier and yet the resulting volcanic edifice is constructed englaci ally (Smellie 2000*a*). During subglacial eruptions the confining ice causes complex interactions between meltwater, ice, lava, pre-existing volcanic and clastic deposits and the topography (Nielson 1936; Smellie & Hole 1997; Smellie 2000*a*). Subglacial eruptions generate distinctive landforms (e.g. Noe-Nygaard 1940; Mathews 1947; Sigvaldasson 1968; Jones 1969*a*); shield eruptions tend to form flat-topped table mountains and fissure eruptions form steep-sided ridges. Both of these structures typically comprise assemblages of basaltic pillow lavas, hyaloclastite and hyalotuff (in Iceland, these assemblages are known collectively as 'moberg' or the 'hyaloclastite formation'). This association is now widely accepted to have a subglacial origin (e.g. Mathews 1947; Saemundsson 1967; Jones 1969*a*, *b*) although, in many cases, the complex assemblages of hyaloclastite and lava may actually have formed in englacial lakes (e.g. Skilling 1994; Smellie & Skilling 1994; Werner & Schmincke 1999; Smellie 2000*b*).

At the onset of an eruption, meltwater and volcaniclastic debris will either collect in a subglacial meltwater vault or englacial lake, flush through subglacial passages, or flow along the margins of valley glaciers and over the glacier surface. These processes are influenced by factors such as the thickness of the ice, the sub-glacier topography and the eruption rate (Smellie 2000*a*, *b*). Thin ice (<150 m thick) is permeable so ice melts rapidly at the onset of an eruption, and meltwater and eruption products tend to be transported away from the vent (Smellie & Skilling 1994). During transport, the lava, water, ice and clastic material interact in a variety of ways, eventually forming distinctive deposits that are commonly topographically confined (Smellie *et al.* 1993; Smellie 2000*a*). Some associations are stacked vertically suggesting successive phases of a single volcanic event; in other cases, very thick successions are built up during a single eruptive event (Smellie *et al.* 1993; Smellie & Skilling 1994; Smellie 2000*a*). The fracturing and disintegration of lava to form breccias and hyaloclastite, and explosive eruptions which generate hyalotuff ensure highly effective heat transfer from lava to water. This

Table 1. *Summary descriptions and interpretations of lithofacies examined on Eyjafjallajökull volcano.*

Lithofacies	Code	Description	Interpretation
DIAMICTITE Heterolithic	Dmm	Matrix-supported, massive, poorly-sorted, rounded to sub-angular polymict. Clasts <1 m diameter, <70 vol%; sharp bed boundaries, beds <6 m thick; Matrix of silt to coarse sand includes palagonitised glass shards and lapilli, lithic fragments and crystals.	Meltout tills, flow tills or debris flows derived from glacial and volcanic material. If a large proportion of clasts are monogenetic, debris flow may have been triggered by a volcanic eruption.
	Dcm	Clast-supported, massive, poorly-sorted, rounded to subangular polymict clasts; beds <2 m thick; matrix of sand to gravel sized lithic fragments.	Meltout till or flow till.
	Dmm(s)	Matrix-supported, massive, poorly to moderately sorted; as Dmm but outsize clasts are rare. Shearing may be intense.	Lodgement till, shearing due to movement of overlying ice.
GRAVEL 8–256 mm Monomict	Gms	Massive, matrix-supported beds <10 m thick; poorly sorted, angular to sub-angular, vesicular and poorly vesicular lithic clasts, some scoriaceous material. Matrix of palagonitized glass shards and fine-grained scoriaceous fragments.	Subaqueous deposition of subaerial pyroclasts and hyaloclastite in volcanically-induced debris flow deposits, slumping of near-vent tephra or other gravity mass flow.
	Gm	Massive or crudely bedded, poorly sorted, clast-supported. Clasts sub-rounded to angular lithics, pillow fragments, crystals. Matrix similar. Beds <1.5 m thick.	Rapid deposition from suspension or coarse hyperconcentrated flood flow deposits.
GRIT 2–8 mm Monomict	GRm	Massive, well-sorted, beds 5 cm–1 m thick, vesicular to non-vesicular palagonitized glass shards, may be some scoria, sharp contacts, fills hollows etc.	Primary lapilli tuff (?), examples with weak bedding may be reworked
	GRh	Horizontal beds <5 cm thick of vesicular to non-vesicular palagonitized glass shards, lithics, crystals. Bed contacts may be gradational, units may be laterally extensive <150 m.	Reworked hyaloclastite and/or tephra, deposition by traction currents; laterally extensive units may be sheetflood deposits.
	GRt	Large-scale trough cross-bedded vesicular to poorly vesicular palagonitized glass shards, lithics, crystals, outsize clasts common (pillow fragmants), gravelly interbeds are also common	Channel-confined deposition from high velocity currents, cross set size indicates high sediments suspension load and a substantial water depth.
	GRp	Large or small scale planar cross-bedded vesicular to poorly vesicular palagonitized glass shards, lithics, crystals, gravel clasts are common; beds <2 m thick. Foresets dip <20°.	Deposition from traction currents (small scale bedforms) or currents of high suspension load and high velocity (large scale bedforms).
	GRl	Very low angle cross-bedding (<15°), pebbly beds mostly phreatomagmatic tephra.	Transitional dune to upper flow regime deposits.

SAND 0.063–2 mm	Sm	Massive, well-sorted, beds 5–60 cm thick, diffuse planar lamination, grains dominated by vesicular clasts, or bubble wall shards.	Airfall tephra, homogenized planar-bedded sands.
LAVA	Ls	Simple flows, margins are blocky, lobate or pillowy; 0.5–50 m thick. May be blocky, columnar, massive, may be gradational with Lk.	Coherent lavas, may be topographically confined, lie on wet sediments, majority are subaerial.
	Lc	Compound flows, massive or columnar, commonly with reddened surfaces intercalated with cinders or thin lenses of reworked palagonitized glass shards and lapilli.	Subaerial compound lavas, lenses of reworked palagonitized volcaniclastics occur in transitional subaerial/subaqueous environments.
	Lk	Cube-jointed and hackly-jointed lava (kubbaberg), may form isolated lobes or thick entablature.	Water-saturated, fragmentary lava. May be subaqueous, flooded in a river valley or subglacial.
	Lp	Lobate and pillow lavas, irregularly-shaped radially-jointed pseudo-pillows, ragged para-pillows; <30% interstitial hyaloclastite.	Subaqueous lava flows, feeder zones etc.
	Lb	Clast-supported coarse breccia, poorly sorted angular lithic blocks in coarse sand-gravel matrix. May contain radially-jointed 'globes'. Commonly gradational upslope with Lp and/or Lk.	Breccias, formed by gravitational collapse of Lp or Lk.
Hyaloclastite and hyaloclastite breccia	Bhm	Massive, matrix-supported, poorly sorted, angular to sub-angular clasts of basalt and basalt pillows, some jigsaw fit clasts and lava lobes. Closely associated with Ls or Lp. Matrix of palagonitised glass shards and crystals.	Forms during interaction of lava with water and ice. *In-situ* envelope of hyaloclastite around lava bodies, or proximal mass flow/slump deposits.
	Bpm	Same as Bhm but clasts dominated by round, elongate or ragged pillows.	As Bhm, but occurs in close proximity to feeder lava bodies. May be generated by slumping or collapse .
	Bhs	Matrix-supported, bedded hyaloclastite breccia. May be large or smale scale bedforms. Matrix material is more abundant than in Bhm.	Resedimented hyaloclastite breccia, debris avalanches, debris flows, and grain flows; commonly found downslope from Bhm.

Lithofacies codes used are based on Eyles et al. *(1983), Miall (1977) and Maizels (1993).*

facilitates the thermal erosion of the subglacial ice tunnels by meltwater (Björnsson 1975, 1988; Höskuldsson & Sparks 1997).

Thick ice (>150 m) is mostly impermeable and at the onset of an eruption meltwater may be sealed in subglacial vaults (e.g. Skilling 1994). Rapid melting of the roof may ultimately form an englacial lake (for full discussion, see Smellie 2000). Werner and Schmincke (1999) proposed that several table mountains, previously described as 'monogenetic' subglacial table mountains, were in fact constructed over thousands or tens of thousands of years in changing eruptive environments which may have included large stable lakes. However, making the distinction between lacustrine and englacial deposits is extremely difficult. The former presence of ice may only be clearly appreciated when it has laterally confined lavas or associated volcaniclastic sediments (e.g. Walker & Blake 1966; Skilling 1994; Smellie & Hole 1997; Smellie 2000). Stable ice-marginal lakes formed by climatic changes rather than by geothermal heating were present in Iceland during deglaciation (late Younger Dryas and early Preboreal; Geirsdottir *et al.* 2000) and their probable existence during any deglaciation must be taken into account when interpreting lava and hyaloclastite successions.

Subglacial eruptions have rarely been observed. The best-documented example is a subglacial fissure eruption at Gjálp, Iceland that began on 30 September 1996 under a 500 to 750 m-thick ice sheet (Gudmundsson *et al.* 1997). Two days later the eruption broke through the ice at one location, thus implying melting rates of several hundred metres in a few hours. Nevertheless, most of the 7 km-long eruptive fissure remained under the ice and by the end of the eruption on 13 October 1996 it had built a 300 m-high ridge of hyaloclastite (Gudmundsson *et al.* 1997).

Until recently, relationships between the products of subglacial eruptions and pre-existing topography have only been studied by a few authors. Walker & Blake (1966) described a thick 'palagonite breccia mass' at Dalsheiði, southern Iceland, which lies on a glacially scoured palaeovalley floor. They interpreted this as a valley-confined subglacial eruption in which lava flowed through an englacial tunnel at the base of the valley. Carswell (1983) described thick hyaloclastite and hyalotuff deposits confined in a broad palaeovalley immediately west of Solheimajökull and interpreted them as products of subglacial fissure eruptions. Recent authors have described and interpreted deposits in more detail. Smellie *et al.* (1993) used lithofacies and succession analysis to interpret proximal, valley-confined volcanogenic sequences at Mount Pinafore in Antarctica. These sequences reflect deposition from flowing water and include heterogeneous masses of lava and hyaloclastite and complex associations of volcaniclastic material. Smellie *et al.* (1993) suggested that these deposits formed under thin, temperate ice sheets (100–150 m thick). Vertically stacked lava and hyaloclastite sequences at Mount Pinafore were interpreted to have formed in subglacial tunnels in a manner similar to esker formation (Smellie & Skilling 1994). Lescinsky & Fink (2000) described lava and ice interaction below thin valley glaciers at stratovolcanoes.

This study of Eyjafjallajökull volcano is one of the first to describe proximal lithofacies associations at a topographically complex, long-lived, large volcanic centre that has been active during several glacial and interglacial periods.

Methodology

The volcano is composed of mainly gently dipping to horizontal stratified volcanic deposits, which generally show limited lateral continuity. Mapping was initiated in order to determine a reliable stratigraphy for geochemical studies. During the mapping and logging of the complex lava–hyaloclastite assemblages it became clear that certain distinctive proximal assemblages were common (all deposits are preserved in obvious relation to their likely crater/fissure source). Nine distinct lithofacies associations that occur repeatedly both spatially and temporally are described. This is not a complete description of the variation in lithofacies associations at the volcano but represents the commonest proximal cogenetic lithofacies associations most likely to be related to the presence of ice. Each association is described in terms of its constituent lithofacies; all of these are believed to be genetically related on the basis of similarities in petrography and geochemistry. Despite having different ages and locations the processes that formed each lithofacies association type must have been similar. Understanding these processes is important in order to determine palaeoenvironments, ice extent and thickness and to study the potential impact of future activity at the volcano.

Lithofacies descriptions

Details of lithofacies observed at Eyjafjöll are summarized in Table 1. Four major lithofacies groups are recognized:

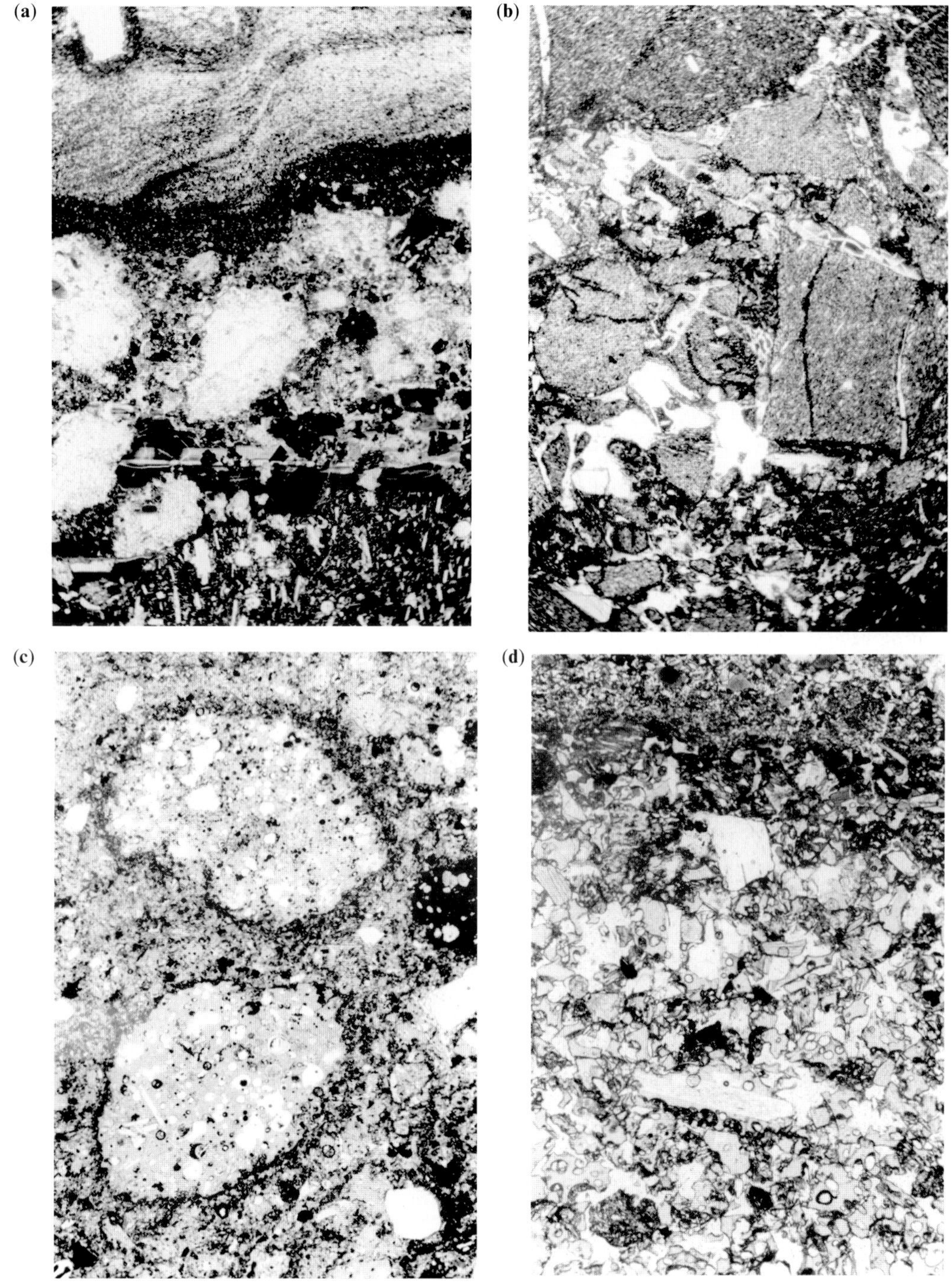

Fig. 2. Hydroclast morphologies represented in redeposited volcaniclastic deposits at the Eyjafjallajökull volcano. (**a**) Hyaloclastite breccia from Skógaheiði [755 480]. Lower lithic clast is angular with planar edges, in centre are vesicular hyalotuff lapilli formed by steam explosivity; above is silt-sized glass. Rock is cemented by goethite. Field of view 12 × 7.2 mm. (**b**) *In situ* hyaloclastite at the base of a ponded lava flow [685 492] (Fig. 10). Clasts are non-vesicular with planar edges. Such clasts are often redeposited by meltwater. Field of view 12 × 7.2 mm. (**c**) Armoured lapilli in a massive deposit of hyalotuff on Fagrafell [704 479]. Clast shapes suggest that vesicles did not play a major role in fragmentation. Steam explosivity was probably more important. Field of view 6 × 3.7 mm. (**d**) Redeposited blocky and poorly vesicular glass shards and crystals from the summit of Raufarfell [706 494]. Field of view 6 × 3.7 mm.

Diamictite

As a result of the ubiquitous basaltic source rock in Iceland, poorly sorted polymict conglomerates with very different depositional environments (e.g. glacial, glacio-fluvial, mass flow) may look very similar (Geirsdottir 1991). It is crucial that the correct genetic interpretation of volcaniclastic sediments associated with probable subglacial deposits is made before palaeoenvironmental or climatic models are constructed (e.g. Dreimanis & Schluchter 1985). In this paper, diamictites are described using a simplified version of the lithofacies code used by Eyles *et al.* (1983).

Volcaniclastic sediments

Vitric clasts and lithic volcanic clasts that are redeposited by meltwater (Fig. 2a) are collectively called volcaniclastic sediments here and are divided up by grain size (gravels, grits and sands, see Table 1). They are described using a system based on that of Miall (1977) and also that of Maizels (1993) for glaciofluvial sediments. Deposits are typically stratified and bedforms suggest deposition by hyperconcentrated flows and traction currents.

The vitriclasts may be composed of blocky and angular, non-vesicular, fragments of brown glass (sideromelane) or black glass (tachylite; Macdonald 1972; Fig. 2b). The lithic and vitric clasts have predominantly planar edges (Honnorez & Kirst 1975) and are typically monogenetic. These clasts represent redeposited hyaloclastite and hyaloclastite breccia. Hyaloclastite forms as a result of *in-situ* non-explosive quenching and granulation of magma when it is exposed to water (e.g. Honnorez 1961; Rittman 1962; Honnorez & Kirst 1975; Kokelaar 1986).

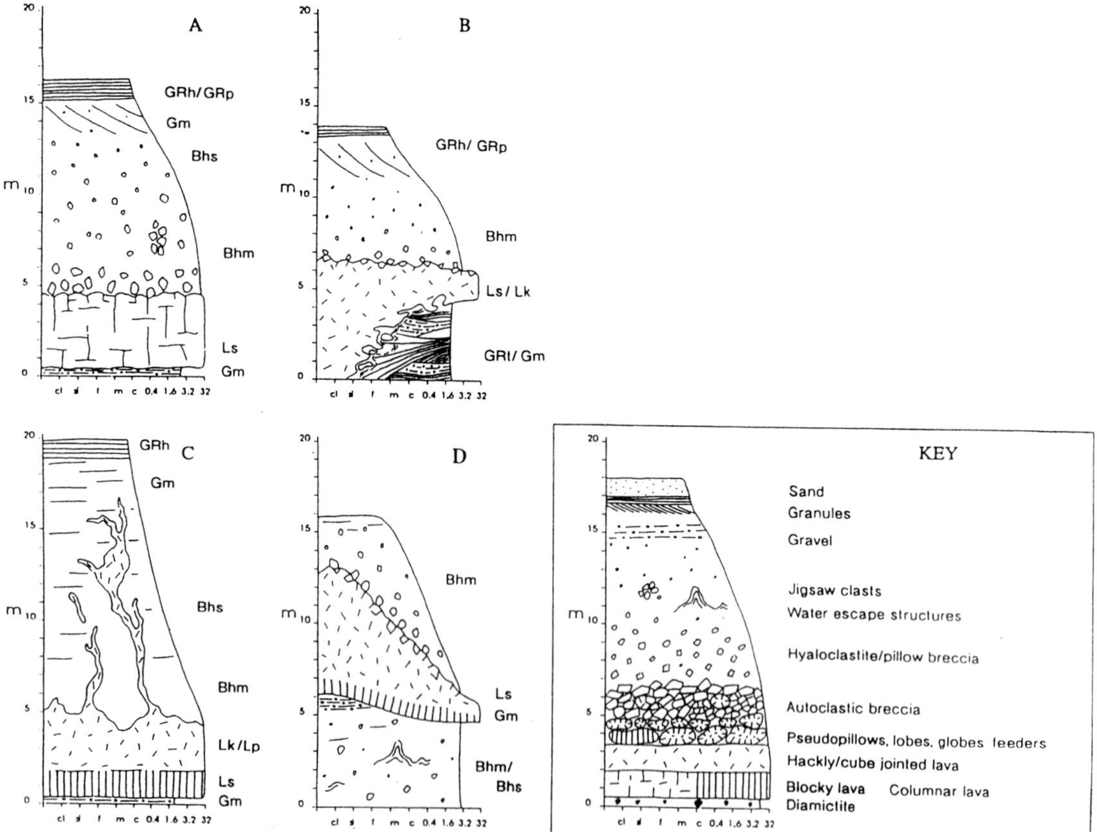

Fig. 3. Schematic vertical profile diagrams illustrating the nine common proximal lithofacies associations identified in successions at the Eyjafjallajökull volcano and described in the text.

Redeposited 'hyalotuff' is composed of angular glass lapilli with few planar edges and numerous corners and inflections formed by exploding vesicles and/or steam explosions (Fig. 2a, c, d; Honnorez & Kirst 1975). Deposits are dominated by sand size vitric clasts but may be associated with scoriaceous lithic clasts. Hyalotuffs form as a result of the explosive interaction between magma and water.

Lava and hyaloclastite breccia

Lavas show a variety of features including rectilinear and fanned columns, blocky jointing, radial jointing, pseudopillows and pillow lavas (Fig. 3). 'Hyaloclastite breccia' is defined here as a generally massive lithofacies comprising abundant poorly sorted lithic clasts (typically >40mm) in a sandy matrix of vitric and lithic clasts. These deposits are derived from granulation, brecciation and spalling of associated blocky-jointed or pillow lava, and by steam explosivity as lava interacted with shallow water or water-saturated sediments. Clastic debris may build up around a vent or lava then collapse to form composite mass flows associated with effusive and intrusive lava facies (see Bergh & Sigvaldason 1991; Smellie *et al.* 1993; Smellie & Skilling 1994).

Lithofacies associations

The lithofacies associations are described and interpreted below. Diagrammatic representations of each association are given in Figure 3 and their distribution is shown in Figure 4. Cross sections through the volcano showing their relationships are shown in Figure 5 and sketches of the western flanks of the volcano are shown in Figure 6. Photographs of specific examples are given in Figures 7–12.

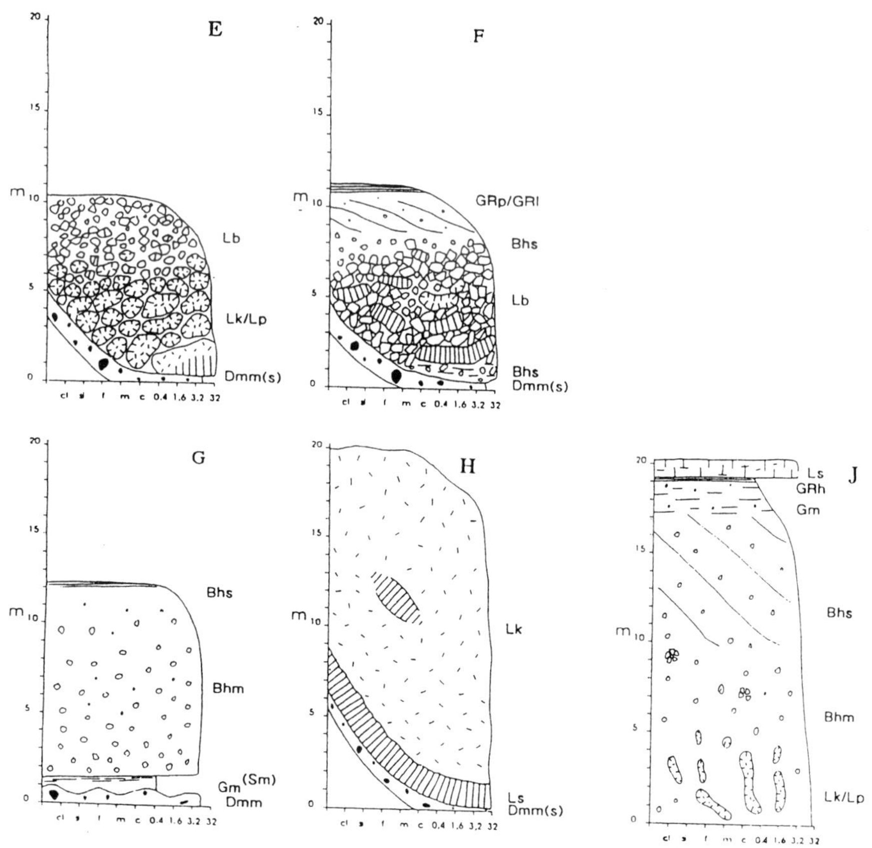

Fig. 3. (*continued*).

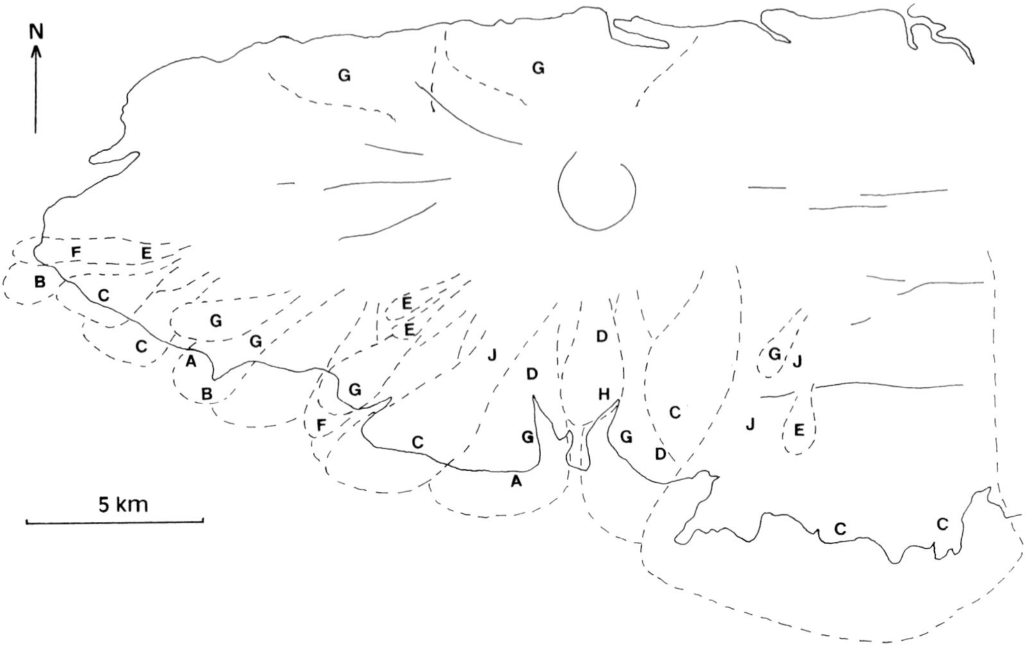

Fig. 4. Outline map of the Eyjafjallajökull volcano showing the general distribution of lithofacies associations described in this paper.

Lithofacies association 'A'

Hyaloclastite breccias overlying discontinuous lavas that lie, in turn, on resedimented, syneruptive volcaniclastics or diamictite. Commonly repetitive, forming thick, vertically stacked successions of lithofacies GRh/GRp, Ls, Bhm/Bhs, GRp, Gm (e.g Núpakot [658 475]).

Description. This association varies from only 4–5 to over 20 m thick and commonly occurs as repeated units (Figures 3 & 7; Table 2). It has a generally low aspect ratio (thickness ('height')/lateral extent ('length')) and was emplaced on low angle (<20°) slopes. At Núpakot, and elsewhere, the association commonly lies directly on thin grey discontinuous diamictite (Dmm(s)) which locally (usually at the base of a succession) lies on an irregular striated surface. At Núpakot, there is a thick succession of subaerial lavas below this surface. A thin diamictite layer commonly overlies the lithofacies association if it is a single deposit, or overlies the uppermost succession if there is a vertical stack.

The basal unit may or may not be present; it comprises discontinuous massive to horizontally bedded, locally open framework monomict gravels and grits (Gm, GRh) less than 4 m thick.

The overlying lava (Ls) has an irregular, locally brecciated base. Lavas are laterally discontinuous with individual lobes rarely broader than about 40 m but they may be several kilometres in length. They tend to be slightly convex in profile, that is, thickest along the axis of the flow (up to 20 m). The thickest lavas tend to be columnar jointed, others are blocky jointed or comprise lobate and pillowed lavas. The upper surface of the lava is typically brecciated, blocky or lobate and grades upward into massive hyaloclastite breccia.

The hyaloclastite breccia (Bhm) comprises clasts of different sizes (from millimetres to decimetres) derived partly from the blocky upper surface of the lava. Most clasts are angular and non-vesicular to weakly vesicular. Radially jointed lava occurs as large isolated bodies within the breccia. Where the lava is absent, the breccia lies with an erosive or non-erosive basal contact on underlying volcaniclastic deposits or diamictite.

Interpretation. The lowest gravelly units (Gm/GRh) represent deposition from hyperconcentrated grain flows. They are confined in shallow channels tens of metres across.

The margins of the lavas were in continuous contact with water and/or ice as they advanced.

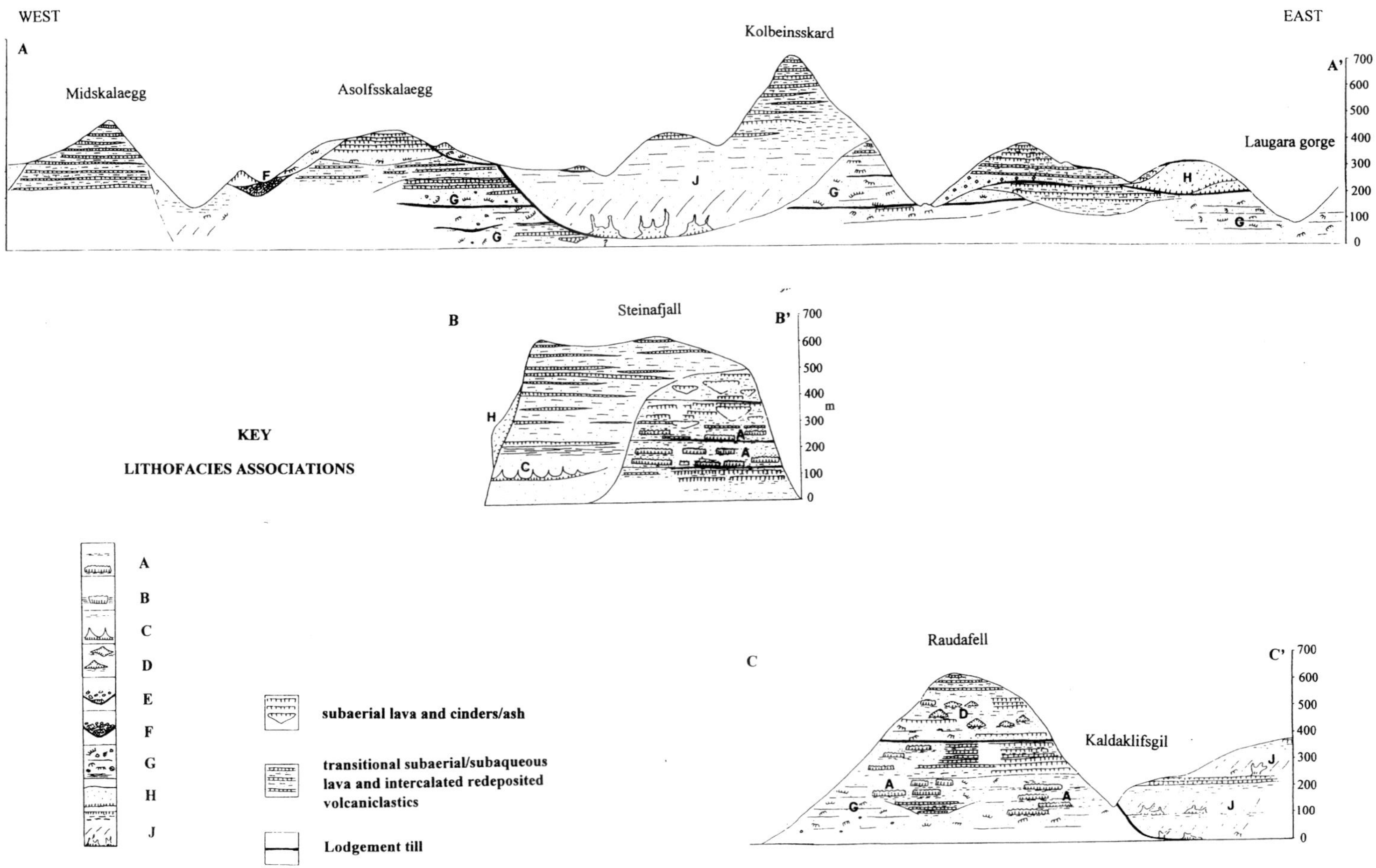

Fig. 5. East–west sections through proximal deposits showing an abundance of angular unconformities, U-shaped valleys and till development. The locations of some lithofacies associations are marked and the character of adjacent deposits shown. Elevation is in metres above sea level. The location of these cross-sections is shown on Figure 1.

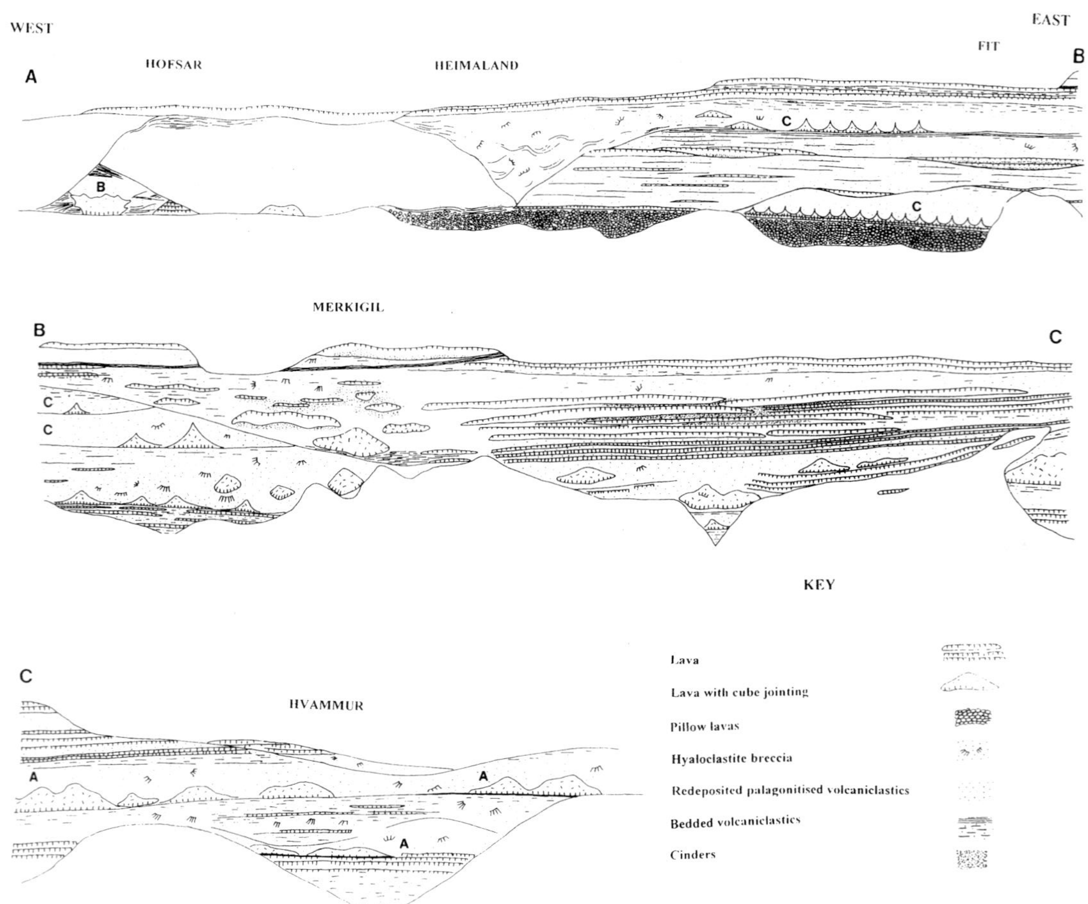

Fig. 6. Simplified sketch (not to scale) of the relationships between units on the western flanks of the volcano between Hofsár and Hvammur (for location see Fig. 1). The base of the crags is *c.* 60 m above sea level and crags are, on average, 200 m high. These successions include shallow submarine, proglacial and subaerial deposits as well as some potential subglacial deposits. There are numerous channels containing relatively fine-grained redeposited hyaloclastite and hyalotuff, this is described as a medial environment 12–15 km from the summit crater. Basal tills are not well-developed in this area.

Fig. 7. Looking east at a vertical stack of lithofacies association A near Núpakot [658 475].

This caused them to fracture, brecciate and develop a carapace of hyaloclastite breccia. Lava lobes spalled off the lavas into the water saturated hyaloclastite breccia. In places, the advancing lava incorporated water-saturated breccia and cogenetic volcaniclastic material, suggesting that the flow front was at least partially intrusive. The coherence of some jointed lava lobes and stringers extending from the lava into the breccia suggests that, locally, little relative movement of the breccia took place after the lava flows stopped advancing. Weak stratification, sorting and a reduction in clast size in the upper part of the breccia (Gm, GRh, GRp) reflects waning meltwater flow at the end of a flooding event. The uppermost cross-bedded and horizontally bedded gravels and granules (Bhs/Gm) represent reworking of the finer grained grits and sand by traction currents in channels. This late-stage reworking is due to shallow running water, in either a subglacial or subaerial environment.

The hyaloclastite breccias extend laterally for tens or hundreds of metres but they are not laterally homogenous. Nevertheless, the hyaloclastite breccia facies is typically a sheet-like deposit would require extensive melting or lofting of the ice for its emplacement. The stratified hyaloclastite breccia facies (Bhs) and stratified basal gravels are generally laterally discontinuous suggesting emplacement along shallow channels.

It is proposed that, due to the lack of any epiclastic material in these deposits and the absence of moraine, that they were deposited in a subglacial environment. The lava must have been generated by a mainly effusive eruption, either under thick ice that inhibited explosive activity (in cases where the basal phreatomagmatic facies is absent) or from a vent beneath relatively thin ice (<150 m) that melted during the early stages of the eruption. In the latter case, the initial activity was probably pheatomagmatic and generated tephra-loaded

Fig 8. Lithofacies association B at Hofsár [507 535]. The exposed face is about 8 m high.

Fig. 9. Looking north at lithofacies association C at Varmahlið [623 478], just west of Núpakot. Note extensive sheet lava (Ls) that back-injects overlying volcaniclastic sediments. Rock face is *c.* 100 m high.

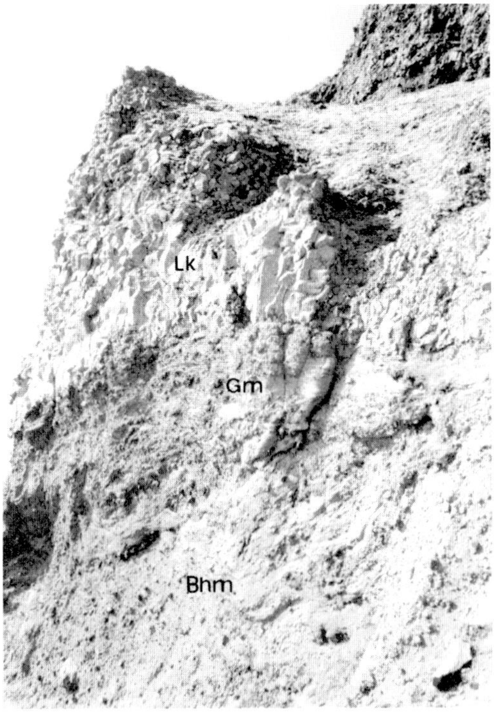

Fig. 10. Lithofacies association D on the south side of Raufarfell [703 478]. The lava unit (Lk) is 5 m thick at its thickest point.

meltwater. The meltwater may have ponded briefly in the vent region if conditions were favourable, or flowed continuously down-slope beneath the glacier. The meltwater enlarged channels through the ice and partially eroded the substrate, and most of the tephra was deposited offshore. Lava erupted into a slurry of water-saturated tephra and resulting steam explosions generated abundant angular glassy and lithic clasts. The lava flowed down-slope and followed the meltwater path along channels. The ice would be thinner further down-slope, thus facilitating the emplacement of extensive lateral deposits.

This association is commonly repeated with an apparent lack of unconformities and no change in petrographic characteristics, implying that large volumes of material were deposited in pulses and in rapid succession. The pulses of activity may be due to separate volcanic events, variations in eruption rate, blocking of subglacial channels or to progressive sealing of ice causing a renewed build-up of meltwater in the vent region. The thorough fracturing and brecciation of the lava suggests highly effective heat transfer between the meltwater and the lava. The re-forming of ice after an eruption explains the local presence at the top of the deposits of thin (<0.5 m) glass-rich diamictites.

Fig. 11. Looking west at lithofacies association F near Seljaland [509 546]. The deposit is *c*. 10 m thick. The stream seen flows along its northern margin at a contact between it and a postglacial subaerial lava.

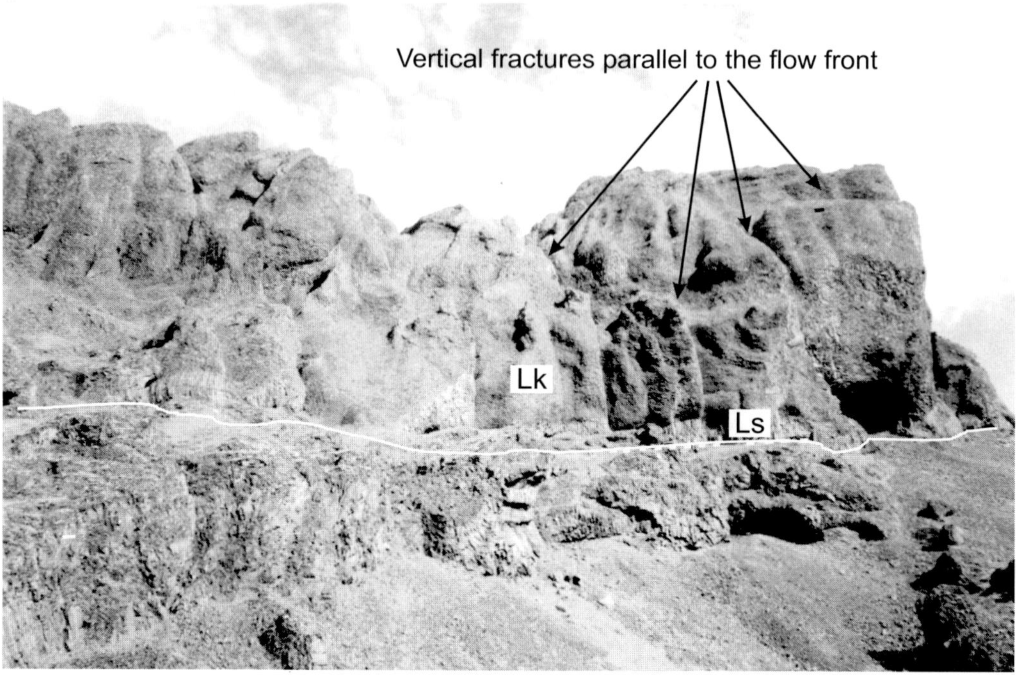

Fig. 12. Looking north, at lithofacies association H at Seljavellir [685 494] on the lower slopes of Lambafellsheiði. The cliffs are at least 80 m in height and lava (Lk) flowed downslope towards the observer.

Distribution. Lithofacies Association 'A' is very common in horizontally stratified older successions in the steep cliffs at the southern margins of the volcano, where it was deposited on gentle slopes up to 10°. Similar associations occur in the gently sloping younger successions to the west, on slopes of less than 20° (Figs 4 & 5).

Lithofacies association 'B'

Lava and hyaloclastite breccia cutting trough cross-bedded volcaniclastic sedimentary rocks. Comprises lithofacies GRt, GRm, Ls, Lk, Bhm, GRh, GRp (e.g Hofsár [509 535] and Hvammsmuli [560 505]).

Description. The association at Hofsár (Figs 3 & 8) consists of large-scale trough cross-bedded gravels and grits (cross-bed sets <2 m high) and clast-supported gravels (GRt, Gm) containing angular basalt clasts (<5 cm). The association is cut by an intrusive lava (Ls, Lk) at least 15 m in diameter, which grades upward through hyaloclastite breccia (Bhm) into massive volcaniclastic sediments of sand to grit size (Table 1).

Aligned clasts in the bedded sedimentary rocks indicate a palaeoflow direction downslope towards the SSW. Lava lobes and fingers project laterally from the lava into the bedded sedimentary rocks (GRt, Gm) over a 2 m-thick contact zone in which the sedimentary rocks are massive or chaotically bedded. This disrupted zone is in sharp contact with the trough cross-bedded sedimentary rocks. The lava is surrounded by a thin halo of massive sandstone a few centimetres thick. The entire association is at least 50 m in width and 10 m in height. Isolated basalt lobes in the hyaloclastite breccia appear to have spalled off, or be protrusions from, the main lava body below and they lack glassy margins. The margins of the hyaloclastite breccia also truncate the bedded sedimentary rocks (GRt, Gm). Clast size within the hyaloclastite breccia decreases upwards. The association is overlain by channel-confined reworked volcaniclastic sedimentary rocks.

A broadly similar deposit at Hvammsmuli is more laterally extensive, exceeding 100 m in width. In it, a laterally extensive cube-jointed and brecciated lava body grades upwards into hyaloclastite breccia (Bhm) and this assemblage of lava and hyaloclastite sharply truncates early bedded sedimentary rocks (GRt, Gm) which are now preserved as lenses at the base and margins of the deposit. Some of the basal and marginal

Table 2. *Summary descriptions and interpretations of the nine proximal lithofacies associations identified in this paper*

Lithofacies association	Description	Interpretation
A	Thin basal monomict gravels and granules containing varied vitriclasts, planar bedded. Lava (jointed, brecciated) overlain by resedimented hyaloclastite and pillow breccias grading upwards into stratified monomict gravels.	Vitriclasts in basal gravels suggest phreatomagmatic and/or magmatic explosivity. Hyaloclastite breccias interacted with coherent lava bodies during flow. Upwards fining suggests waning flow conditions. Vitriclasts in upper deposits dominated by platy shards.
B	Lava and resedimented hyaloclastite (& pillow) breccias intruding trough-cross-bedded monomict gravels. <2 m thick contact zone contains lava lobes, massive or chaotic bedded sands and gravels.	Stratified gravels may represent channelled flood flow deposits, lava bodies intruded the wet sediments in the channel, possibly by fluidization. Lack of glassy rinds suggests sediments were hot. Vitriclasts in gravels suggest phreatomagmatic and/or magmatic explosivity.
C	Sheet-like monomict resedimented hyaloclastite breccias and gravels with basal intrusive lava sheet showing very irregular upper surface (vertical lobes and fingers extending upwards) and *in-situ* hyaloclastite.	Similar to B but sheet-like. Lava confined laterally by topography (not ice). Initially subglacial/subaqueous explosive eruption (varied vitriclasts) becoming effusive due to changes in vent area or properties of magma. Lava intruded base of wet sediments.
D	Cube-jointed lava 'tubes' lying on thick, massive resedimented hyaloclastite breccias and gravels. Commonly occurs on slopes <20°.	Formed in subglacial tunnels. Sediment accumulation occurs at base of tunnel and meltwater cuts up into ice enlarging tunnel again. Lava is emplaced develops a sheath of hyaloclastite.
E	Topographically-confined coherent pillows/lobes with interstitial hyaloclastite. Lies on diamictite or striated surface.	Topographically confined, subglacial near-vent feeder-zone.
F	Topographically-confined resedimented hyaloclastite/pillow breccia, pillow fragments/blocky lava clasts, minimal interstitial hyaloclastite. Lies on diamictite.	Flood water carried finer sediments away down slope. Clasts formed by autobrecciation and collapse, convex upper surface suggests confinement, by ice?
G	Massive, chaotic deposits of resedimented hyaloclastite breccia, tephra, brecciated lava, lava lobes, incorporating mega clasts of diamictite and lying on steep slopes (<40°).	Mass flow, caused by emplacement in subaqueous environment, probably a valley glacier.
H	Lavas with very thick entablature and thin basal colonnade which occur on steep valley sides. Marginal hyaloclastite and mega-clasts of palagonitized volcaniclastics.	Lava ponded between steep valley side and valley-confined glacier.
J	Thick, massive and cross-bedded resedimented hyaloclastite and tephra. May have a cap of hyaloclastite and lava.	Fissure eruption into a subglacial meltwater vault. Redeposition by grainflow on foresets, mass flow. Vitriclasts suggest phreatomagmatic and/or magmatic explosivity changing, throughout eruption.

sedimentary rocks are deformed into tight folds. The contact zone between the sedimentary rocks and lava is sharp. This deposit is well-exposed on three sides and extends down-slope at less than 5° in a SSW direction, forming a promontory. It lies on a thin diamictite which in turn overlies a succession of subaerial lavas. The association is overlain by subaerial lavas.

Interpretation. Trough cross-bedded sediments (GRt) suggest deposition in channels, although the channel base is not exposed at Hofsár. The size of the cross-beds at Hofsár implies large scale bedforms (dunes) which formed in deep water during a period of high flow strength.

The massive sands surrounding the subaqueous lava body and the disruption and truncation of bedding suggest that the lava intruded and locally fluidized wet bedded sediments (Kokelaar 1982). Following Kokelaar's model, the intrusive lava body may have moved forward and generated an envelope of steam-saturated massive sediment that flowed away from the flow front, over the top and along the sides and even below parts of the lava, allowing it to progress forward.

The lava body at Hvammsmuli is sheet-like in comparison and was emplaced in a channel originally occupied by cross-bedded and horizontally bedded volcaniclastic sediments. Deformation of sedimentary lenses preserved beneath and at the margins of the lava suggests that the thick, extensively fractured lava body 'bulldozed' the sediments aside. A large raft (<10 m long) composed almost entirely of altered yellowish vesicular shards, is preserved in the hyaloclastite breccia suggesting that it was partially lithified before being ripped up and incorporated in the breccia.

There is no evidence of lateral confinement by ice, although at Hvammsmuli the sediments lie directly on a grey diamictite. At Hofsár, the base of the succession is not exposed but subaerial lavas occur adjacent to the outcrop and may have formed the channel margin. The preservation of the sedimentary structures and close association with subaerial lavas suggests that it was deposited above sea level. The gravels and glass-rich sands include angular non-vesicular clasts indicative of granulation and vesicular clasts and shards indicative of phreatomagmatic activity. Meltwater carried the sands and gravels down-slope from the vent. The effusive lava flowed from the vent following subglacial channels and forming lava tubes. Lavas intruded recently emplaced wet sediments at breaks in slope in a proximal to medial environment. An increase in eruption rate would have facilitated such a process.

This association is interpreted as the medial deposit of a subglacial eruption. The depositional slopes were gentle to flat and encouraged deposition from hyperconcentrated flood flows in channels. These deposits are relatively fine-grained sands and gravels and are intruded or eroded by later lava and hyaloclastite breccias. The absence of moraines in the successions, and the common presence of thin grey, sheared diamictite (Dmm(s)) directly below the basal sands and gravels implies subglacial emplacement. However, the base of the Hofsár example (at an elevation of 20 m above sea level) is not seen and it contains abundant fine-grained sediments, and a proglacial depositional environment is also possible. Deposits higher up the succession at Hofsár are dominated by broad massive channel-fill deposits of yellow, altered sandy hyaloclastite separated by steep angular unconformities (Fig. 5).

It is proposed that this lithofacies association was deposited in subglacial to proglacial channels. The successions are commonly overlain by subsequent (unrelated) subaerial lavas implying that the ice receded following emplacement. The subaerial lavas protected the deposits from erosion.

Distribution. This lithofacies association occurs in successions located several kilometres from the central vent (Figs 4 & 5). It commonly occurs at breaks in slope from greater than to less than 10°, at low elevations and is channel-confined. Flow directions suggest that this association was derived mainly from flank or fissure eruptions.

Lithofacies association 'C'

Topographically confined sheet lava and hyaloclastite with feeders that intrude overlying re-sedimented syn-eruptive volcaniclastics. Lithofacies Ls, Lk, Lp, Bhm, Bhs, Gm, GRh. (e.g Varmahlið [610 483], nr. Hvammur [557 511], Fit [520 525]).

Description. This association may vary in thickness from 5 to over 100 m although deposits in all areas have a very similar structure (Figs 3 & 9). It typically contains a thin, discontinuous basal unit of massive or planar-bedded gravels (Gm) and/or diamictite up to 0.5 m thick or else lava rests directly on a glacially scoured substrate.

The lava (Ls) base is commonly sharp, though it may be blocky with poorly developed pillows. It typically has well developed vertical columns at the basal contact with a cube-jointed entablature above (Lk) and weak columnar jointing at the upper margin. The upper surface is irregular, with flame structures, fingers and lobes of hackly lava protruding into a mantle of massive hyaloclastite and stratified volcaniclastic sedimentary rocks. Budding and spalling from these apophyses creates an abundance of isolated lava bodies in the lower part of the hyaloclastite. The basal lava forms a sheet that may be several tens of metres or hundreds of metres in breadth, thinning towards its margins. In thick deposits, the thinned lateral margins of the lava sheet commonly curve up into the overlying hyaloclastite breccia and massive volcaniclastic sediments at the palaeovalley margins.

Massive and horizontally bedded volcaniclastic sedimentary rocks (Gm, GRh) overlie the lava with localized brecciation at the lava margins (Bhm). Basalt lobes and feeders connected to the basal lava intrude the uppermost stratified units.

The minimum volume of the deposit at Varmahlið is 0.6 km^3. The deposit at Fit is much thinner in comparison and, although not exposed in three dimensions, is probably much smaller in volume.

Interpretation. The occurrence of very large volumes of volcaniclastic sediments composed mainly of hyalotuff suggests a subaqueous and explosive origin for this material. Phreatomagmatic tephra might normally be expected to be carried by meltwater to the distal flood plain environment, but in this case the sediments accumulated in very gently sloping to flat-based palaeo valleys or channels.

The volcaniclastic sediments are more strongly bedded towards the top and essentially massive in lower parts. This may be caused by the waning flood and increasing dilution of the flows, perhaps due to additional meltwater. The contact with the lava body, though irregular, is relatively sharp. Assuming that the stratified sediments mark the waning of a flood, field relationships imply that lava intruded after the main flood flow had subsided. Intrusions may have been emplaced concurrently with the basal sheet lava or they could be later breakouts from the surface; the relative timing is difficult to ascertain. These lavas are laterally confined only by palaeotopography and are essentially sheet flows.

A phreatomagmatic onset to the eruption would have generated large volumes of tephra around the vent and contributed to heat transfer and hence rapid and effective melting of the ice. Large volumes of water-satuated tephra and hyaloclastite were transported down-slope in sheet-like mass flows. The very thick and extensive associations are valley-confined with basal diamictite defining the palaeovalley floor and sides. The whole valley glacier would have to rise by more than 150 m to accommodate the deposits at Varmahlið or it may have thinned and even melted during progressive deposition of the stratified volcaniclastic material. The association could have been deposited in a submarine environment (e.g. Bergh & Sigvaldason 1991) but that would require sea level at least 300 m higher than present which seems unlikely. These successions show substantial similarities to geographically less extensive associations described from Alexander Island (Smellie *et al.* 1993), which were interpreted as subglacial deposits. The basal lava was intruded below the wet sediment pile forming thick, ponded sheets that resisted rapid cooling. Rapid extrusion rates would aid this process.

The thinner examples of this association were emplaced on the western flanks of the volcano in channels beneath thin receding ice sheets. Thicker examples were confined to deep U-shaped valleys in which glaciers may have been thick enough to impede transport of the voluminous tephra. These topographically confined deposits can be traced from the outcrops on the southern margin of the volcano up-slope to more proximal areas where they are exposed on hillsides and on the sides of river gorges. Thick valley glaciers probably extended well beyond the present southern margins of the volcano.

Distribution. This lithofacies association is common in successions younger than 600 ka that are controlled by topography. Large-volume examples may lie in flat-based U-shaped palaeovalleys that extended south and SW from the summit area (Figs 4 & 5). Smaller-volume examples occur on the west flanks of the volcano in broad channels formerly beneath ice sheets that may have been extensions of the summit ice cap.

Lithofacies association 'D'

Cube-jointed lava 'tubes' lying on thick resedimented volcaniclastic sediments and hyaloclastite breccias comprising lithofacies Bhm, Gm, Ls, Lk (e.g. Raufarfell [703 478], Lambafellsheiði [688 504], Núpakotsdalur [658 510]).

Description. This association is typically up to 15 m thick (Figs 3 & 10). It is characterized by elongate lava tubes (Ls, Lk) lying upon laterally discontinuous volcaniclastic lithofacies (Gm). Massive hyaloclastite breccia occurs above and below the lava tubes. Several such associations are typically aligned and interleaved. Together, they may extend laterally for several hundred metres. This association commonly occurs within or in close proximity to thick successions of thinly planar-bedded hyaloclastite and thin lavas.

At Raufarfell (Fig. 5), the association is very gently sloping to almost flat-lying. The lower unit is a thick, massive or locally stratified hyaloclastite breccia (Bhm). There may be dewatering structures and localized chaotic bedding. The top metre or so of this unit locally consists of planar cross-bedded or horizontally bedded volcaniclastic sediments (Gm), which are capped by sharp-based lava (Ls). The lava tubes are triangular in cross-section; their surfaces are blocky rather than pillowy. They have well-developed columns arranged perpendicular to the basal contact (and locally the upper and lateral contacts) and an internal entablature (Lk). Massive hyaloclastite breccia (Bhm) mantles the lava tube.

At Lambafellsheiði, linear ridges up to 200 m long and 20 m wide are formed by lava tubes that splay out downslope like fingers. The tubes dip at up to 20° and feed large ponded lava bodies further down-slope. The lava tubes are lens-shaped in cross-section. Some associations are stacked vertically. Much of the surrounding hyaloclastite has been removed by subsequent erosion. The lavas are intermediate in composition and cube-jointed, although patches of columnar jointing occur.

The Núpakotsdalur deposits are inaccessible, but the distinctive cross-section of the lava tubes is clearly visible there and elsewhere.

Interpretation. Chaotic bedding and dewatering structures suggest water-saturated deposition en masse. Slump structures may also have formed after melting of ice walls of a subglacial tunnel or ice blocks incorporated in the breccias. The better-stratified upper parts of the hyaloclastite breccia suggest diminished and more pulsed sediment supply by meltwater, following deposition of the initial mass flow. The later deposition was from traction currents, possibly with dunes migrating downstream.

The distinctive cross-sectional shape of the lava tubes may reflect the shape of a subglacial Röthlisberger-type tunnel (Röthlisberger 1972). Röthlisberger tunnels form when turbulent meltwater erodes the overlying ice. Reduction in flow rate or turbulence in the meltwater results in ice deformation and closure. Esker formation is proposed to occur in R-tunnels (Bannerjee & McDonald 1975). Sediment accumulation occurs at the floor of the tunnel, reducing its cross-section and causing increased pressure and turbulence of the water, and erosion upward into the ice. In that way, thick sediments can be deposited in tunnels that only ever contained relatively shallow water. Smellie & Skilling (1994) suggested that volcaniclastic assemblages at Mt Pinafore, Antarctica with similar characteristics were deposited in subglacial tunnels in a similar way to esker formation. Evidence suggests that this lithofacies association was also emplaced in subglacial tunnels

The lava tubes were probably insulated from the ice by the surrounding breccia, allowing the lava to cool slowly and develop columnar or cube jointing (kubbaberg, Lk). These lava bodies may form blockages in the subglacial tunnels forcing subsequent meltwater to take different pathways generating sheet-like deposits of hyaloclastite. A system of anastamosing tunnels is formed, similar to those produced during normal esker deposition (Banerjee & McDonald 1975). Such blockages may cause different 'surge' events during meltwater discharge at the more distal glacial margin. Such a process may explain the interleaving of such deposits (e.g. at Lambafellsheidi), where tunnels may have been continuously shifting. Eskers develop best where ice is active but receding (Hambrey 1994) and they are commonly destroyed by meltwater. The lava protects underlying sediments from erosion; palagonitisation and secondary mineralisation also increase the preservation potential.

Distribution. Lithofacies Association 'D' tends to be localized and may change laterally and downslope into other lithofacies associations (e.g. C, H). It occurs throughout the stratigraphy and typically in close association with striated glaciated surfaces and diamictites and thinly planar bedded hyaloclastites and occasionally thin brecciated lavas (Figs 4 & 5).

Lithofacies association 'E'

Topographically confined lava/basalt lobe/pillow/feeder zone with interstitial hyaloclastite comprising lithofacies Dmm (s), Lk/Lp, Lb (e.g. Midskalagil [617 519]).

Description. This association is valley-confined and formed of massive to pillow lava that is elongate downslope (Fig. 3). The basal unit is typically a diamictite (e.g. Dmm(s)) that defines the outline of a U-shaped palaeovalley. The slopes of these narrow, proximal, steep-sided palaeovalleys may reach 40°.

Lava directly overlies the diamictite with a sharp, non-erosive contact. The lava may be massive with crude columnar jointing (Lk) particularly along valley axes. However, it is mainly composed of radially jointed basalt lobes/feeders and pseudo-pillows (Lp) in a sparse matrix of hyaloclastite. The palaeovalley surface preserved at Miðskalagil is up to 200 m across and 30 m deep. Other similar deposits lie in palaeovalleys of similar proportions high on the flanks of the volcano.

Interpretation. The sharp contact between the basal diamictite (likely lodgement till) and the lava suggests that any loose material was flushed down the steep valley by meltwater before the main lava body was deposited. The basal part of the lava in the centre of the valley is crudely columnar suggesting ponding and slow conductive cooling. Above the basal zone the lava has intense and blocky fracturing associated with hyaloclastite that suggests water-enhanced chilling. The lava unit comprises feeders for an advancing flow front and pseudo-pillows formed by inundation of flood water. Down-slope, the lava unit is intensely fractured and brecciated. The fine hyaloclastite matrix is absent, presumably removed by meltwater, and remobilization of the breccia blocks in debris flows may result in the deposits seen in lithofacies association F (Fig. 11). The presence of basal tillite, the lack of epiclastic sediments and evidence that large amounts of water were involved all suggest a subglacial origin and subglacial emplacement for this proximal association. If the eruption rate is high, the lava may eventually intrude wet sediments deposited in medial environments. Depending on the topography and depositional environment this would generate lithofacies association B.

Distribution. This lithofacies association was observed on the upper western and south-western flanks of the volcano and probably formed during subglacial fissure eruptions (Figs 4 & 5).

Lithofacies association 'F'

Topographically confined breccia, pillows/blocky lava clasts with minimal interstitial hyaloclastite comprising lithofacies Dmm(s), Lb, GRp, GRl (e.g. Seljaland [509 546]).

Description. A good example of this lithofacies association occurs on the south bank of the Seljaland stream (Figs 3 & 11). The flow has a lensoid cross-section and lies on approximately 1 m of massive, poorly sorted, clast-supported diamictite that defines a gently curving palaeo-valley/subglacial channel base. The dip of the deposits is at least 15. The diamictite contains subangular to rounded polymict clasts up to 40 cm in diameter and there are clasts/lenses of reworked silt beds in the sandy matrix. The volcaniclastic deposit is a chaotic mixture of irregularly columnar lava, broken columns and radial-jointed pillows fragments in a sparse matrix of vitric sands, silts, and crystals (plagioclase, olivine and pyroxene reflecting phenocryst-rich nature of source lava). The clasts and blocks show no evidence of quenching. The massive blocky deposit is locally capped by low angle cross-bedded sands and gravels (GRp/GRl).

Interpretation. The basal diamictite is probably a melt-out till as it lacks the shear texture of lodgement till and is clast-supported. Collapse of lava, disintegration of pseudo-pillows and lava blocks due to the slope and interaction with meltwater generated a water-saturated debris flow. The lack of sandy sediments in this unit implies that flood water carried most finer sediments away. The uppermost cross-bedded sediments were deposited from turbulent meltwater with a low sediment concentration during the late stages of the event. The width to height ratio and lensoid cross-section of the deposit, suggest that it was confined on deposition, probably in subglacial channels near the margin of the ice cap.

Distribution. This lithofacies association occurs on the lower slopes of the volcano (Figs 4 & 6). It is a proximal/medial deposit, emplaced on the volcano's flanks but reflecting the disintegration of lavas farther up-slope (e.g. lithofacies association E).

Lithofacies association 'G'

Massive redeposited chaotic hyaloclastite breccia and hyalotuff commonly overlying thin layers of redeposited volcaniclastics, or with a thin basal sand layer comprising lithofacies Dmm, Gm, Sm, Bhm (e.g. Fit [524 526], Hvammsmuli [561 503], Moldnupur [602 495]).

Description. This lithofacies association usually occurs on gentle to moderate slopes. The basal unit is typically diamictite (Dmm) (Fig. 3). At Fit, the diamictite is 0.5 to 1 m thick. This is overlain by erosive-based thin gravel (Gm; <50 cm) dominated by palagonitised vitric lapilli, lithic fragments and crystals. The gravels contain faint laminations. The contact with overlying breccia (Bhm) is sharp. At Hvammsmuli, the gravels are overlain by a thin layer (<40 cm) of volcanic ash and scoria (Sm) which has a diffuse contact with overlying hyaloclastite breccia. At other localities, the hyaloclastite breccia may rest directly on the substrate. The basalt tends to be vesicular and there are a variety of hyaloclast types.

Interpretation. The presence of thin layers of reddish sandy tephra suggests a subaerial depositional environment or tephra and scoria resedimented by traction currents. They may represent the subglacial deposition of material derived from the glacier surface. The massive, blocky hyaloclastite breccia unit may or may not be genetically related to the underlying scoria sands. The breccia comprises large volumes of lava clasts probably derived by steam fragmentation, brecciation and granulation. There is no evidence of a basal lava such as occurs in lithofacies associations A, B and C. The hyaloclastite breccias may be derived from the collapse of piles of intensely fragmented lava generated by steam explosivity in the vent area, or perhaps the collapse of a brecciated subaqueous lava dome (e.g. Smellie *et al.* 1998). The presence of vesicular clasts and vesiculated glass shards suggests the former may be more likely. Based on the structure of the deposits, transport was by mass flow.

The hyaloclastite breccia (Bhm) has protected and helped to preserve the finer grained underlying sediments. At Hvammsmuli, the overlying hyaloclastite breccia appears to have loaded the thin sand layer suggesting that they were deposited in rapid succession.

This proximal lithofacies association may represent deposits formed in subglacial or ice marginal areas on moderate to level slopes.

Distribution. Massive redeposited hyaloclastite breccia is very abundant throughout the succession but particularly in younger topographically controlled successions (Figs 4, 5 & 6). The occurrence of a thin bed of redeposited scoria (e.g Gm, Sm) between diamictite and a thick hyaloclastite breccia implies that subaerial vents were active and coexisted with valley glaciers.

Lithofacies association 'H'

Lavas with very thick entablature and thin basal colonnade that crop out on steep valley sides comprising lithofacies Dmm(s), Ls, Lk (e.g. 'Svarthamrar' at Seljavellir [685 494] and Holtsheii [618 496]).

Description. This association is dominated by lava lithofacies and is relatively uncommon (Figs 3 & 12).

The association at Lambafellsheidi lies on steep valley sides of up to 40°. The lavas are fed by lava tubes on higher slopes (i.e. lithofacies asssociation D). The lava is intermediate in composition. At the base of the association, a thin (<0.5 m) veneer of hyaloclastite (Fig. 3b) intervenes and lies on up to 2 m of diamictite. Small-scale pillow buds from the overlying lava spall into the hyaloclastite. The lower 0.5–2 m of the lava lithofacies Ls is composed of well-developed columns up to 0.5 m wide which are perpendicular to the basal topography (colonnade). Locally there are voids at the base of the lava, up to 3 m in diameter and 2 m high, with glassy internal surfaces and intense jointing. Joints radiate out perpendicular to this surface and increase in separation with distance from the void.

Above the colonnade is an unusual 'entablature' of up to 60 m of columnar-jointed lava. The columns are <20 cm in width and are all consistently orientated perpendicular to the vertical 'flow front' and large planar sub-vertical fractures that cut the lava mass. Large clasts of lithified palagonitized volcaniclastic material (up to 5 m) are enclosed within the upper part of the entablature.

At Holtsheiði, the lavas are very similar in structure and lie on a steep valley side (dipping about 30°) but the source for this unit has not been identified.

Interpretation. To form such a thick entablature a large quantity of water would have been required to cause rapid cooling; this may have been supplied by flooding (Saemundsson 1970; Long & Wood 1986). The lava 'froze' against a steep valley side and not on the valley floor. It has a sub-vertical flow front up to 50 m high suggesting that it was supported laterally. Ice is the only plausible support mechanism so it appears that the preserved lava was intruded between the valley side and an ice mass. The basal hyaloclastite layer may have formed by brecciation and granulation of the flow front as it advanced down slope.

Lava is not an effective heat conductor once its surface is chilled so only just enough ice would melt to create space for the lava (Bjornsson 1988). The lava may originally have been erupted in an elevated subaerial environment and then flowed down into the sides of a valley-confined alpine-style glacier. Alternatively, the lava could have flowed through subglacial tunnels and sheets on gentle slopes before ponding on a steep subglacial slope. Smellie *et al.* (1993) observed lavas in Antarctica very similar to those at Seljavellir and considered them to be ponded subglacial lavas.

The volcaniclastic blocks caught up in the lava were probably incorporated as it flowed over a steep substrate of volcaniclastic material. The voids at the base of the lava may have been clastic blocks that have weathered out, or more probably they are cavities once filled by blocks of ice.

The planar sub-vertical lava fractures are parallel to the contact at the flow front that caused the ponding (i.e. an almost vertical wall of ice). The fractures appear to have acted as cooling surfaces for the entablature. It is proposed that the ponded magma initially cooled very slowly inwards from the contacts with the ice. The main part of the flow maintained its heat for a considerable period but cooled *in situ* to the point of brittle failure. When ice support was gradually removed from the flow front, the loss of support caused extensional fractures to open in the partially solidified lava. Rapid inundation of these fractures with meltwater caused the formation of further sets of columns perpendicular to the new vertical fractures. The fractures are up to 1 m wide but may originally have been filled with fragmented lava or hyaloclastite.

Distribution. There are only two good examples of this lithofacies association, each occuring on the steep sides of a glacial valley (Figs 4 & 5). Valley-confined ice at the base of a slope is essential for this lithofacies association to form.

Lithofacies association 'J'

Thick, massive and large-scale cross-bedded hyaloclastite breccias and syn-eruptive volcaniclastic deposits comprising lithofacies Lk/Lp, Bhm, Gm, Grh, Ls (e.g. Skógaheiði [725 480], Holtsdalur [625 504]). Each unit is at least 50 m thick, thinning gradually down-slope to about 20 m near the truncated southern margin of the volcanic edifice.

Description. Skógaheiði is a broad area that slopes south from the E–W ridge between Eyjafjallajökull and Myrdalsjökull (1050 m) to sea level. Carswell (1983) described in detail thick deposits in this area and surmised that they were subglacial, topographically confined deposits probably fed by E–W fissures. The Kaldaklifsgil gorge on the western margin of Skógaheiði (Fig. 1) exposes thick deposits of hyaloclastite breccia and hyalotuff which form lithofacies association J (Figs 3 & 13). The gorge has formed along an unconformity between the steep mountainside of Raufarfell and the younger subglacial valley-fill deposits of Skógaheii. Traces of a basal diamictite are exposed on the eastern slopes of Raufarfell which dip steeply eastwards beneath the thick deposits of Skógaheiði and mark the western margin of the glaciated palaeovalley. The topographic high which forms the easternmost margin of the Skógaheiði deposits is Jökulgil, a prominent ridge near Solheimajökull (Carswell 1983).

The Skógaheiði lithofacies association is composed of two thick eruptive units, each composed of massive hyalotuff, hyaloclastite and

Fig. 13. Looking SE down Kaldaklifsgil gorge, west of Skógaheiði. Lithofacies association J is exposed in the gorge sides. The gorge is over 100 m deep.

locally hyaloclastite breccia, which grade upwards into cross-bedded and horizontally-bedded volcaniclastic sandstones and gravels. Each unit is about 50 m thick and similar in appearance (Fig. 13). At the base of the lower unit there are isolated pillows, ragged blocky lobes, feeders of lava, irregular patches of hyaloclastite breccia (Bhm) and abundant angular lithic blocks. The lava is mostly vesicular. These are the lateral equivalents of the deposits described by Carswell (1983) near Solheimajökull, 6 km to the east.

In places, this association is capped by a thin 'passage zone' of hyaloclastite breccias and subaerial lavas. These lavas may represent the subaerial emergence of the volcanic pile; there is no obvious unconformity between the subaqueous and subaerial successions. The lavas are then overlain by a basal diamictite and a further succession of palagonitized volcaniclastic sandstones and gravels similar to those below but with distinctive lensoid cross sections. The deposits frequently contain lenses of laminated sediments and are cemented by palagonite and goethite, suggesting that late stage warm fluids were abundant during and after deposition.

The Holtsdalur deposits are dominated by crudely bedded volcaniclastic sandstones, gravels and hyaloclastite breccias. Deposits are at least 150 m thick and the proportion of basalt lobes and feeders appears to be less than at Skógaheiði. These deposits were confined to steep-sided flat-based valley glaciers and lie on diamictite horizons. The total volume of the Holtsá deposit is at least 0.9 km^3.

Interpretation. These thick deposits were the result of subglacial fissure eruptions beneath broad, gently sloping to flat-based valley-confined ice sheets. There is no evidence of beach deposits or tidal effects on the eruptive products (e.g. Furnes & Sturt 1976).They comprise large quantities of fragmental, palagonitized volcaniclastic debris. In order to accumulate such thicknesses of fragmentary material, efficient fragmentation processes are required and large quantities of water (i.e. phreatomagmatic activity as a result of eruption into shallow water or thin ice). Hyaloclastite, pillow lavas, hyalotuff and associated facies presumably built up around a fissure vent and were redeposited by grain flows, slumping and as mass flows.

Thick deposits of mainly phreatomagmatic tephra with some basal lava feeders were emplaced across a large area. The two units may represent collapses of a large tephra pile around the vent. The final lavas at the top of the succession imply either that the deposits were above a ponded water level (*c.* 200 m above sea level) or that the ice cover had melted. The extent of the subaerial lavas suggests the latter. Thin diamictites on the lavas suggest that ice reformed after their emplacement.

The thick massive deposits (<150 m) at Holtsá may have formed beneath a thick valley-confined temperate ice (greater than 150 m) judging by their thickness. However, exposure is poor. The thinner deposits (about 100 m) at Skogar, which show horizontal bedding and localized large-scale low angle cross-bedding, were probably formed below mainly permeable temperate ice (i.e. about 150 m thick). Mass flow deposits near the base progressed upwards to low angle cross beds and planar bedding. The latter were presumably deposited during the waning of the meltwater floods.

The vents for this activity were almost certainly along E–W fissures which are common in this area and run at right angles to the slope. Significant volcanic activity along these fissures may actually have detached the valley glacier from its source on the Fimmvorðuhals plateau.

Distribution. This lithofacies association is found at localities where extensive valley-confined glaciers were present. The deposits are proximal, extensive and formed as a result of large-scale, fissure eruptions. Lithofacies J is exposed mainly in two localities where ice can be demonstrated to have formed in large flat-based palaeovalleys (Figs 4 & 5). These valleys dropped steeply from high on the volcano's flanks and ran out for several kilometres at low elevations.

Discussion

Large volcanic centres in Iceland are constructed over long time periods (typically up to 1 Ma, Sæmundsson 1979) during glacial and interglacial periods. The extent and thickness of ice cover on a large volcano is variable, but is typically dominated by a summit ice cap and flanking alpine-style valley-confined glaciers at lower elevations which wax and wane with climatic changes. At present, the climate in Iceland results in 'temperate' or 'wet-based' glaciers (Björnsson 1975, 1988). Katla is presently covered by the Myrdalsjökull ice cap that reaches thicknesses of 400–500 m (Höskuldsson & Sparks 1997). The Vatnajökull ice sheet is over 500 m thick (Gudmundsson *et al.* 1997). In contrast, the Eyjafjallajökull ice cap is currently 250 m thick above the crater and about 100 m thick on the flanks (Strachan 2001). This shows that even during interglacial periods, eruption products may show evidence of interaction with ice and meltwater. For this reason, any palaeoenvironmental reconstructions must be made carefully.

Glacier hydrology

Smellie *et al.* (1993) suggested that a glacier less than approximately 150 m thick may be composed of mostly snow and firn, through which meltwater and volcanic debris may escape continuously from the vent area during an eruption. Meltwater may flow along the bedrock–glacier interface, widening and adding to the pre-existing system of anastamosing tunnels or sheets (Nye 1976; Björnsson 1988) or it may escape over the glacier surface as observed at Katla in 1918 (Jonsson 1982). Under thicker ice, overburden pressure around a vent may trap initial geothermal meltwater in a subglacial vault (Björnsson 1975, 1988). However, later modelling by Smellie (2000*a*, *b*) highlighted basal meltwater leakage from the vault coincident with overflowing and creation of an englacial lake. Thorarinsson (1965) noted that geothermal heat from Grimsvötn was sufficient to produce a surface depression many years before or after an eruption. Katla produced catastrophic floods that reached a peak discharge in 1918 of 200 000 $m^3 s^{-1}$ (Hoskuldsson & Sparks 1997) and yet no subglacial lakes have been detected there (Björnsson *et al.* 1993). This implies that once an eruption has begun, ice melting may increase very rapidly and this is consistent with the experimental work and theoretical considerations of Höskuldsson & Sparks (1997) on effusive eruptions. They calculated rates of heat transfer for basaltic lavas beneath ice and calculated that the walls of an ice cavity could melt rapidly at rates of 1–40 m per day. Nevertheless, this is still too slow to explain real observed eruptions (e.g. Gjálp in 1996 or Deception Island in 1969; Smellie 2000*b*, 2002). These observations suggest that at the beginning of a volcanic eruption under thick ice (>200 m for an effusive eruption (Allen 1980)) a water-filled cavity is likely to exist, if only for a few hours. The products of such an eruption will pile up around the vent within a water-filled vault. Under thick ice, eruptions may not be explosive and are likely to build a pile of pillow lava and hyaloclastite breccia. The water pressure in the vault will eventually exceed that of the enclosing ice, thus briefly 'floating' the glacier (Björnsson 1988) and giving rise to a jökulhlaup. Höskuldsson & Sparks (1997) suggested that the stability of such a cavity depends on the effusion rate of lava and the total heat exchange efficiency, and that pressure changes caused by rapid melting of the cavity can be much larger than those related to ice topography and thickness.

Phreatomagmatic activity

Phreatomagmatic activity facilitates efficient heat transfer and could cause rapid ice melting (Höskuldsson & Sparks 1997). An eruption beneath relatively thin ice sheets (<150 m; Smellie & Skilling 1994) would cause phreatomagmatic activity. Tephra deposited in the vent area may be redeposited by a variety of mass flow processes (Smellie & Skilling 1994). Tephra deposited on the glacier surface around the vent will increase supraglacial melting and will ultimately be reworked into the subglacial meltwater system. This process may explain why hyalotuff is relatively abundant in the uppermost, stratified beds of some associations. Much tephra is transported away from the eruption site by meltwater to be redeposited on coastal sandur or in the marine environment (Maizels 1993).

The processes and products of magma–water interactions depend largely upon the volatile content of a magma, its viscosity (and hence its composition) and hydrostatic pressure (e.g. Jones 1969*a*; Allen 1980; Kokelaar 1986). At Eyjafjallajökull, vesicular hyaloclasts are abundant, especially in thin hyperconcentrated flood flow deposits preserved at the base of some lithofacies associations (e.g. A, B) or in the massive fragmented hyaloclastite units of others (e.g. lithofacies associations C, G). This is typical of shallow subaqueous environments where clasts are formed by a variety of processes such as contact-surface steam explosivity, bulk interaction steam explosivity and magmatic explosivity (Kokelaar 1986). Phreatomagmatic activity was common in the initial stages of many (but not all) subglacial/subaqueous eruptions at Eyjafjallajökull (e.g. lithofacies associations C, J) generating hyalotuff and crystal fragments. Meltwater and tephra flowed away from the vent, eroding the ice and enlarging pre-existing meltwater channels into which lava flowed.

Lava and hyaloclastite breccia

Many of the lithofacies associations described here include complex associations of lava and hyaloclastite. Very similar associations have been described in detail by Smellie *et al.* (1993) and Smellie & Skilling (1994) who interpreted them as the products of subglacial eruptions. The work presented here suggests that similar cogenetic lithofacies associations may be common on all large, long-lived volcanoes constructed during several glacial and interglacial periods. Bergh & Sigvaldason (1991) described unconfined sheet-like deposits and interpreted them as shallow marine. The effusive nature of the lava implies either that the vent area was subaerial as a result of initial melting of the ice cap, or that it erupted into a deep englacial lake. Because of the lack of evidence for ponded lithologies and the abundant evidence of redeposition by mass flows and traction currents it is assumed that most eruptions took place beneath 'thin' ice (<150 m; Smellie & Skilling 1994).

Lava flowing down slope in subglacial channels would undergo quenching and brecciation (spall fragmentation), and would generate clasts from sand size to large lithic blocks that are rapidly removed from the site of fragmentation. Polygonal and irregular fractures caused by thermal stresses in the rapidly cooling lava would form, facilitating the brecciation process (thermal shock granulation). The continued movement of lava in the ductile interior of a tube or sheet flow would impose stress on the cooling outer surface and increase fracturing. Steam explosivity (bulk interaction steam explosivity; Kokelaar 1982, 1986) in these fractures would enhance fracturing. The heat transfer to meltwater is efficient, potentially increasing the meltwater temperature well above 0°C (Höskuldsson & Sparks 1997) although temperature of the meltwater decreases with distance from the vent as a result of further melting of ice. The advancing lava may form protrusions into the adjacent hyaloclastite which may become detached as a result of shear stresses. These lobate bodies develop prismatic jointing and break up in the upper parts of the unit but may remain partially intact in lower parts, near the lava. Most of the hyaloclastite breccia deposits are sheet-like, implying lofting or total melting of the ice. Some are mass flow deposits and others developed *in situ* adjacent to subaqueous lava flows. Down-slope from lava bodies, breccias may develop as pillow lavas and blocky lavas disintegrate and the debris moves as a mass flow.

There have been several papers discussing the stability of subglacial sheet flows of meltwater relative to flows confined in tunnels (e.g Shreve 1972; Björnsson 1975; Nye 1976; Weertman & Birchfield 1983; Shoemaker 1992). Both processes appear to have been active at this volcano. Nye's model (1976), which is supported by most authors, suggests that water flowing in a sheet is unstable and tends to become channelized. Any variation in thickness of the water sheet will result in heterogeneous heat production (due to turbulence and friction) and melting, and ultimately flow localization. Larger passages will then tend to increase in size at the expense of smaller ones because they contain comparatively more heat relative to the wall area (Röthlisberger 1972; Shreve 1972). To sustain these subglacial passages, the rate of melting of the ice must equal or exceed the rate of closing by plastic deformation. Therefore, if the flow rate and/or turbulence of the meltwater decreases, the tunnels will begin to close (Nye 1976). This suggests that ice will reform after subglacial deposition of volcanic products. Thin diamictites will develop on the surface of the deposits but without evidence of major glacial erosion at their base.

Effusion rates

At Eyjafjallajökull, channelled lavas and associated hyaloclastite breccias typically extend several kilometres from the vent. The lava surface may be blocky jointed or irregular, with flame structures and lobes advancing upwards into the overlying hyaloclastite. The blocky lavas commonly have the appearance of subaerial aa lava flows and a morphology similar to the levéed flows generated by analogue experiments (Gregg & Fink 1995). The experimental flows were generated by high effusion rates and low cooling rates on slopes in which the high flow velocities prevented crust formation. In the Eyjafjallajökull successions, a thick layer of hyaloclastite breccia may have insulated the basal lava thus preventing formation of a distinctive lava crust.

Proximal deposits on the upper flanks suggest that lava was also supplied to the flow front via topographically confined lavas, lavas and pillow lavas which generally lack glassy selvedges (lithofacies association E). Massive and pillowed feeder lavas tend to transform down-slope into broken pillows and ultimately breccia deposits (e.g. lithofacies association F). Pillowed and lobate lavas represent eruptions of low effusion rate (Ballard *et al.* 1979) although in proximal areas a massive basal lava may represent a higher effusion rate. Subaqueous sheet flows form during eruptions with high effusion rates (Ballard *et al.* 1979) and are usually massive with vertical or partly radial columnar jointing (e.g. lithofacies association C). Where lobes of lava extend upwards from the lava (e.g. lithofacies association C), shear stresses between the lava and overlying hyaloclastite/volcaniclastic deposits would have been low probably owing to the flat or gentle slopes (<10°).

Observations at Eyjafjallajökull fit well with analogue experiments on lava flow morphology carried out by Gregg & Fink (1995). They found that sheet flows developed best on steeper slopes, at high effusion rates and with low cooling rate. In contrast, they found that pillow lavas were produced at low effusion rates, on low slopes and at high cooling rates. Lobate flows formed under intermediate conditions. The down-slope transformation from lobate or sheet flows on steeper slopes, to pillowed flows and ultimately brecciation on gentler slopes can therefore be related to a combination of local effusion rates (these are highest in the vent region and lowest at the flow margins) and angle of slope.

Temporal and spatial development of the lithofacies associations

The lithofacies associations are indicative of subglacial eruptions, and, for the main part, subglacial deposition and they occur throughout the succession. The earliest activity at the volcano was subaerial, based on the oldest observed deposits but lithofacies association B is present above them and implies that subglacial activity occurred shortly afterwards. The contact is marked by a thick and well-developed till. A large edifice composed of crudely horizontally stratified hyaloclastite assemblages and thin subaerial lavas built up gradually. This probably occurred during the advance and retreat of a summit ice cap. Major glaciations carved deep U-shaped valleys into the volcano's southern flanks. The largest glacial valley is between Steinafjall and Raufarfell (Fig. 1). This is an area of high geothermal activity and may have experienced a sector collapse. This would have created a large scar that then encouraged developing ice sheets to flow from the summit area in that direction. Repeated glacial advances and retreats carved more valleys into the edifice, deepened existing ones and deep gorges were carved by meltwater. Most of these palaeovalleys were filled with subglacial assemblages that lie directly on a thick basal till. They are commonly overlain by further thin diamictites and thin stratified hyaloclastites. The thin diamictites are fine-grained grey lodgement tills but they are poorly developed implying that they simply formed over subglacial deposits after an eruption stopped. The presence of the thick basal till under these successions and evidence for only relatively thin ice during emplacement of the subglacial deposits suggests that the volcanism occurred mainly during deglaciation. At some localities, thin successions of subaerial lavas lie directly on hyaloclastite assemblages. In some cases they appear to be cogenetic (based on petrography and chemistry) implying deposition in a proglacial or emergent environment. Their preservation implies that deposition was during a deglaciated episode. Where the lava is not cogenetic, the common lack of intercalated glaciofluvial sediments implies that they were emplaced shortly after glacial retreat, possibly from a different eruptive vent to the underlying hyaloclastite deposits. They protected the underlying clastic units from erosion. Thick successions of subaerial lavas are rare suggesting that more volcanic activity occurred during deglaciation as proposed elsewhere (e.g. Sigvaldason *et al.* 1992).

Volcanic production rates in Iceland appear to be related to glaciations (Sigvaldason *et al.* 1992). During and after glacier melting, at the beginning of the Holocene, volcanic productivity within the axial rift zones increased by a factor of 20–30 and the propagating rift below Eyajafjallajökull was also affected (Sigvaldason *et al.* 1992). Under equilibrium conditions, magmatic pressure at the roof of a magma chamber and lithostatic pressure are equal but an increase in magmatic pressure or a decrease in lithostatic pressure may cause an eruption. Evidence suggests that as a glacier melts, lithostatic pressure is reduced and crustal movements caused by isostatic rebound may cause intense volcanism until new pressure equilibrium is reached (Sigvaldason *et al.* 1992). Thorarisson (1953) suggested that the release of a 100 m-deep meltwater lake from the ice-dammed Grimsvötn caldera triggered volcanic eruptions. This was confirmed by Wallman *et al.* (1988) who modelled the effects of a 100 m sea level draw down on a magma body in equilibrium conditions and showed that this could lead to dyke propagation and an eruption.

Some major unconformities beneath thick till horizons may correspond to major time gaps coeval with glacial periods (Wiese 1992). Precise dating of rocks above and below these unconformities could constrain the timing of glaciations. Major unconformities of this nature are most common immediately south of the volcano's summit. The topography is now complex in this area and volcanic deposits are correspondingly more complex too.

The E–W 'spine' of the volcano defined by multiple E–W fissures is clearly long-lived and most volcanic activity may have been generated from there. The western flanks are generally smooth implying that they have only been affected by advance and retreat of the summit ice cap and no major glacial valleys have formed.

Successions there are dominated by channelled flows and sheet flows. The lower parts of the succession were probably submarine (see pillow lavas on Fig. 5) but above an elevation of *c.* 100 m subaerial lavas, cinder cones and hyaloclastite assemblages predominate. Numerous unconformities and channels suggest rapidly changing meltwater paths. Lavas, showing irregular fanning columns and voids may represent interglacial lavas that have interacted with meltwater from a thin summit ice cap.

Subglacial volcanic debris is typically deposited in the centre of channels or valleys. This disrupts and blocks the meltwater drainage causing it to find new paths. Typically new drainage begins at the margins of the deposits opening new channels that are then filled by the products of the next eruption. This process can be observed on a relatively minor scale on the SW slopes. On a larger scale, meltwater followed the margins of palaeovalley fill assemblages and created deep gorges which are gradually widened and then utilized by advancing ice sheets. These ultimately form a new steep-sided flat-based glacial valley (e.g. Figs 12 & 13; see also Walker & Blake 1966). This rapid erosion of volcanic edifices subjected to multiple glacial and interglacial periods causes volcanic successions to be complex and lateral correlations difficult.

Conclusions

The Eyjafjallajökull volcano has been active for almost 1 Ma, through several glacial and interglacial periods. Nine proximal lithofacies associations are interpreted to indicate subglacial eruptions and subglacial deposition of volcanic products. These distinctive assemblages comprise resedimented tephra, massive hyaloclastite, sheet lava and blocky lava. The exact form depends on the slope, the effusion rate of lava and the thickness of ice. The assemblages show little evidence of ponding and abundant evidence for emplacement by running water implying emplacement under relatively thin ice (<150 m). Some deposits were emplaced beneath valley-confined glaciers, others beneath the extensive ice cap. Emplacement was in sheets, channels and tunnels cut into the ice by meltwater. This was facilitated by highly efficient heat transfer between lava and meltwater. The deposits commonly lie directly on tills with no intervening glacio-fluvial sediments. Thin poorly developed tills may develop above them as ice reforms after an eruption but there is no evidence of glacial advance and erosion on those surfaces. Unrelated subaerial lavas commonly form on top of the assemblages, again with no intervening epiclastic sediments suggesting emplacement rapidly after ice withdrawal. It seems that much of the volcanic succession at Eyjafjallajokull was emplaced during deglaciation as implied from other volcanic centres in Iceland.

Thanks to G. Sigvaldason and W. McIntosh for constructive review comments that greatly improved the manuscript. Thanks to K. Sæmundsson and S. Jakobs son for good advice and to H. Emeleus, R. Weller and N. Hindmarch for assistance during fieldwork.

References

ALLEN, C. C. 1980. Icelandic subglacial volcanism; thermal and physical studies. *Journal of Geology*, **88**, 108–117.

BALLARD, R. D., HOLCOMB, R. D. & VAN ANDEL, TJ. H. 1979. The Galapagos at 86°W: Sheet flows, collapse pits and lava lakes of the rift valley. *Journal of Geophysical Research*, **84**, 5407–5422.

BANNERJEE, I. & MCDONALD, B. C. 1975. Nature of esker sedimentation. *In*: JOPLING, A. V. & MCDONALD, B. C. (eds) *Glaciofluvial and Glaciolacustrine Sedimentation*. Society of Economic Palaeontology and Mineralogy, Special Publication, **23**, 132–154.

BERGH, S. G. & SIGVALDASON, G. E. 1991. Pleistocene mass-flow deposits of basaltic hyaloclastite on a shallow submarine shelf, South Iceland. *Bulletin of Volcanology*, **53**, 597–611.

BJÖRNSSON, H. 1975. Subglacial water reservoirs, jökulhlaups and volcanic eruptions. *Jökull*, **25**, 1–26.

BJÖRNSSON, H. 1988. Hydrology of ice caps in volcanic regions. *Vísindafélag Íslendinga, Societas Scientarium Islandica*, **45**.

BJÖRNSSON, H., PALSSON, F. & GUDMUNDSSON, M. T. 1993. Myrdalsjökull: yfirborh, botn og rennslisleihir jökulhlaupa. *In*: LARSEN, G. (ed.) *Kotlustefna, rannsoknir a eldvirkni undir Myrdalsjökli*. University of Iceland, Science Institute, Reykjavik, **RH-3-93**, 11–13 (in Icelandic).

CARSWELL, D A. 1983. The volcanic rocks of the Solheimajokull area, southern Iceland. *Jokull*, **33**, 61–71.

DREIMANIS, A. & SCHLUCHTER, C. 1985. Field criteria for the recognition of till or tillite. *Palaeogeography Palaeoclimatology Palaeoecology*, **51**, 7–14.

EINARSSON, T. & ALBERTSSON, K. J. 1988. The glacial history of Iceland during the past three million years. *Philosophical Transactions of the Royal Society of London*, **B318**, 637–644.

EYLES, C. H., EYLES, N. & MIALL, A. D. 1983. Lithofacies types and vertical profile models; an alternative approach to the description and environmental interpretation of glacial diamict and diamictite sequences. *Sedimentology*, **30**, 393–410.

FURNES, H. & STURT, B. A. 1976. Beach/shallow marine hyaloclastite deposits and their geological significance – An example from Gran Canaria. *Journal of Geology*, **84**, 439–453.

GEBRANDE, H., MILLER, H. & EINARSSON, P. 1980. Seismic structure of Iceland along RRISP. *Journal of Geophysics*, **47**, 239–249.

GEIRSÓTTIR, A. 1991. Diamictites of late Pliocene age in western Iceland. *Jökull*, **40**, 3–24.

GEIRSDÓTTIR, A., HARDARDÓTTIR, J. & SVEINBJÖRNSDÓTTIR, A. E. 2000. Glacial extent and catastrophic meltwater events during the deglaciation of Southern Iceland, *Quaternary Science Reviews*, **19**, 1749–1761.

GREGG, T. K. P. & FINK, J. H. 1995. Quantification of submarine lava-flow morphology through analog experiments, *Geology*, **23**, 73–76.

GUDMUNDSSON, M. T., SIGMUNDSSON, F. & BJÖRNSSON, H. 1997. Ice-volcano interaction of the 1996 Gjálp subglacial eruption, Vatnajökull, Iceland. *Nature*, **389**, 954–957.

HAMBREY, M. 1994. *Glacial Environments*. University College London Press, London.

HONNOREZ, J. 1961. Sur l'origine des hyaloclastites. *Bulletin of the Belgian Geological Society*, **70**, 407–412.

HONNOREZ, J. & KIRST, P. 1975. Submarine basaltic volcanism: morphometric parameters for discriminating hyaloclastites from hyalo-tuffs. *Bulletin of Volcanology*, **39**, 1–25.

HÖSKULDSSON, A. & SPARKS, R. S. J. 1997. Thermodynamics and fluid dynamics of effusive subglacial eruptions. *Bulletin of Volcanology*, **59**, 219–230.

JAKOBSSON, S. P. 1978. Environmental factors controlling the palagonitization of the Surtsey tephra, Iceland. *Bulletin of the Geological Society of Denmark*, **27** (Special Issue), 91–105.

JAKOBSSON, S. P. 1979. Petrology of Recent basalts of the Eastern Volcanic Zone, Iceland. *Acta Naturalia Islandica*, **26**, 103.

JONES, J. G. 1969*b*. Pillow lavas as depth indicators. *American Journal of Science*, **267**, 181–195.

JONSSON, J. 1982. Notes on the Katla volcanoglacial debris flows. *Jökull*, **32**, 61–68.

JONSSON, J. 1988. *Geological Map of Eyjafjöll*. Research Institute Neðri Ás Hveragerði, Hveragerði.

KJARTANSSON, G. 1958. Endaslepp hraun undir Eyjafjollum. (Lava flows deprived of their distal ends in Eyjafjoll, South Iceland). *Natt. Fraed*, **28**, 127–140.

KOKELAAR, B. P. 1982. Fluidization of wet sediments during emplacement and cooling of various igneous bodies. *Journal of the Geological Society of London*, **139**, 21–33.

KOKELAAR, P. 1986. Magma–water interactions in subaqueous and emergent basaltic volcanism. *Bulletin of Volcanology*, **48**, 275–289.

KRISTJANSSON, L., JOHANNESSON, H., EIRIKSSON, J. & GUDMUNDSSON, A. I. 1988. Brunhes-Matuyama paleomagnetism in three lava sections in Iceland. *Canadian Journal of Earth Sciences*, **25**, 215–225.

LESCINSKY, D. T. & FINK, J. 2000. Lava and ice interaction at stratovolcanoes: Use of characteristic features to determine past glacial extents and future volcanic hazards. *Journal of Geophysical Research*, **105**, 23 711–23 726.

LONG, P. E. & WOOD, B. J. 1986. Structures, textures and cooling histories of Columbia River basalt flows. *Bulletin of the Geological Society of America*, **97**, 1144–1155.

LOUGHLIN, S. C. 1995. *The evolution of the Eyjafjöll volcanic system, southern Iceland*. PhD thesis, University of Durham, UK.

MACDONALD, G. A. 1972. *Volcanoes*. Prentice-Hall, Englewood Cliffs, New Jersey.

MAIZELS, J. 1993. Lithofacies variations within sandur deposits; the role of runoff regime, flow dynamics and sediment supply characteristics. *Sedimentary Geology*, **85**, 299–325.

MATHEWS, W. H. 1947. 'Tuyas', flat topped volcanoes in northern British Columbia. *American Journal of Science*, **245**, 560–570.

MIALL, A. D. 1977. A review of the braided-river depositional environment. *Earth Science Reviews*, **13**, 1–62.

NIELSON, N. 1936. A volcano under an ice cap, Vatnajökull, Iceland. *Geographical Journal*, **40**, 6–23.

NOE-NYGAARD, A. 1940. Sub-glacial volcanic activity in ancient and recent times. *Folia Geographica Danica Tom*, **1**, 5–67.

NYE, J. F. 1976. Water flow in glaciers; jokulhlaups, tunnels and veins. *Journal of Glaciology*, **76**, 181–207.

PÁLMASON, G. 1971. Crustal Structure of Iceland from explosion seismology. *Society Science Islandica Rit*, **40**, 187.

PÁLMASON, G. 1973. Kinematics and heat flow in a volcanic rift zone, with application to Iceland. *Geophysical Journal of the Royal Astronomical Society*, **33**, 451–481.

PÁLMASON, G., ÁRNORSSON, S., FEIðLEIFSSON, I. B., KRISTMANSDÓTTIR, H., SÆMUNDSSON, K., STEFANSSON, V., STEINGRIMSSON, B., TOMASSON, J. & KRISTJANSSON, L. 1979. The Iceland crust; evidence from drill hole data on structure and processes. *In*: TALWANI, M., HAY, W. & RYAN, W. B. F. (eds) *Deep Sea Drilling Results in the Atlantic Ocean: Oceanic Crust*. American Geophysical Union, Washington, Maurice Ewing Series, **2**, 43–65.

RITTMAN, A. 1962. *Volcanoes and their activity*. John Wiley and Sons, New York.

RÖTHLISBERGER, H. 1972. Water pressure in intra- and subglacial channels. *Journal of Glaciology*, **11**, 177–203.

SÆMUNDSSON, K. 1967. Vulcanismus und Tektonik des Hengill-Gebietes in Sudwest-Island. *Acta Naturalia Island*, **11**, 1–105.

SÆMUNDSSON, K. 1970. Interglacial lava flows in the lowlands of southern Iceland and the problem of two-tiered columnar jointing. *Jökull*, **20**, 62–76.

SÆMUNDSSON, K. 1979. Outline of the geology of Iceland. *Jokull*, **29**, 7–28.

SHOEMAKER, E. M. 1992. Subglacial floods and the origin of low relief ice sheet lobes. *Journal of Glaciology*, **38**(128), 105–112.

SHREVE, R. L. 1972. Movement of water in glaciers. *Journal of Glaciology*, **11**, 205–214.

SIGMUNDSSON, F. 1991. Post-glacial rebound and asthenosphere viscosity in Iceland. *Geophysical Research Letters*, **18**, 1131–1134.

SIGVALDASON, G. E. 1968. Structure and products of subaquatic volcanoes in Iceland. *Contributions to Mineralogy and Petrology*, **18**, 1–16.

SIGVALDASON, G. E., ANNERTZ, K. & NILSSON, M. 1992. Effect of glacier loading/deloading on volcanism: postglacial volcanic production rate of the Dyngjufjöll area, central Iceland. *Bulletin of Volcanology*, **54**, 385–392.

SKILLING, I. P. 1994. Evolution of an englacial volcano: Brown Bluff Antarctica. *Bulletin of Volcanology*, **56**, 573–591.

SMELLIE, J. L. 2000*a*. Subglacial eruptions. *In*: SIGURDSSON, H. (ed.) *Encyclopedia of Volcanoes*. Academic Press, San Diego, 403–418.

SMELLIE, J. L. 2000*b*. Lithofacies architecture and construction of volcanoes erupted in englacial lakes: Icefall Nunatak, Mount Murphy, eastern Marie Byrd Land, Antarctica. *In*: WHITE, J. D. L. & RIGGS, N. (eds) '*Lacustrine Volcaniclastic Sedimentation*'. International Association of Sedimentologists, Special Publications, **30**, 73–98.

SMELLIE, J. L. 2002. The 1969 subglacial eruption on Deception Island (Antarctica): events and processes during an eruption beneath a thin glacier and implications for volcanic hazards. *In*: SMELLIE, J. L. & CHAPMAN, M. G. (eds) *Volcano–Ice Interaction on Earth and Mars*. Geological Society, London, Special Publications, **202**, 59–79.

SMELLIE, J. L. & SKILLING, I. P. 1994. Products of subglacial volcanic eruptions under different ice thicknesses: two examples from Antarctica. *Sedimentary Geology*, **91**, 115–129.

SMELLIE, J. L. & HOLE, M. J. 1997. Products and processes in Pliocene–Recent, subaqueous to emergent volcanism in the Antarctic Peninsula: examples of englacial Surtseyan volcano construction. *Bulletin of Volcanology*, **58**, 628–646.

SMELLIE, J. L., HOLE, M. J. & NELL, P. A. R. 1993. Late Miocene valley-confined subglacial volcanism in northern Alexander Island, Antarctic Peninsula. *Bulletin of Volcanology*, **55**, 273–288.

SMELLIE, J. L., MILLAR, I. L., REX, D. C. & BUTTERWORTH, P. J. 1998. Subaqueous, basaltic lava dome and carapace breccia on King George Island, South Shetland Islands, Antarctica. *Bulletin of Volcanology*, **59**, 245–261.

STRACHAN, S. A. 2001. *Geophysical investigation of the Eyjafjallajökull glacio-volcanic system using radio echo sounding*. PhD thesis, Edinburgh University, UK.

THORARINSSON, S. 1953. Some new aspects of the Grimsvotn problem. *Journal of Glaciology*, **214**, 267–276.

THORARINSSON, S. 1965. Changes of the water-firn level in the Grimsvötn caldera 1954–1965. *Jökull*, **15**, 109–119.

WALKER, G. P. L. 1965. Some aspects of Quaternary volcanism in Iceland. *Transactions of the Leicester Literary Philosophical Society*, **59**, 25–39.

WALKER, G. P. L. & BLAKE, D. H. 1966. The formation of a palagonite breccia mass beneath a valley glacier in Iceland. *Quarterly Journal of the Geological Society, London*, **122**, 45–61.

WALLMANN, P. C., MAHOOD, G. A. & POLLARD, D. D. 1988. Mechanical models for correlation of ring-fracture eruptions at Pantelleria, Straits of Sicily, with glacial sea-level drawdown. *Bulletin of Volcanology*, **50**, 327–339.

WEERTMAN, J. & BIRCHFIELD, G. E. 1983. Stability of sheet water flow under a glacier. *Journal of Glaciology*, **29**, 374–384.

WERNER, R. & SCHMINCKE, H.-U. 1999. Englacial vs lacustrine origin of volcanic table mountains: evidence from Iceland. *Bulletin of Volcanology*, **60**, 335–354.

WIESE, P. K. 1992. *Geochemistry and geochronology of the Eyjafjöll Volcanic System*. MSc thesis, University of Oregon, USA.

Glacial influences on morphology and eruptive products of Hoodoo Mountain volcano, Canada

B. R. EDWARDS[1] & J. K. RUSSELL[2]

[1] *Department of Geology, Dickinson College, Caroline, PA 17013-2896, USA (e-mail: edwardsb@dickinson.edu)*

[2] *Igneous Petrology Laboratory, Department of Earth & Ocean Sciences, University of British Columbia, Vancouver, British Columbia V6T 1Z4, Canada*

Abstract: Hoodoo Mountain volcano (HMV), a Quaternary composite volcano in northwestern British Columbia, is a well-exposed example of peralkaline, phonolitic ice-contact and subglacial volcanism. Its distinctive morphology and unique volcanic deposits are indicative of subglacial, within-ice, and/or ice-contact volcanic eruptions. Distinct ice-contact deposits result from three different types of lava–ice interaction: (1) vertical cliffs of lava, featuring finely jointed flow fronts up to 200 m in height, resulted from lava flows being dammed and ponded against thick masses of ice; (2) pervasively-jointed, dense lava flows, lobate intrusions, and domes associated with mantling deposits of poorly-vesiculated breccia are derived from volcanic eruptions contained beneath relatively thick ice; and (3) an association of pervasively-jointed, highly-vesicular lava flows or dykes encased by vesicular hyaloclastite of identical composition formed by eruption under and/or through relatively thin ice. The distribution of these three deposit types largely explains the distinctive morphology of Hoodoo Mountain and can be used to reconstruct variations in ice thickness surrounding the volcano since *c.* 85 ka. Our analysis suggests that at *c.* 85 ka Hoodoo Mountain erupted underneath ice cover of at least several hundred metres. At *c.* 80 ka eruptions were no longer subglacial, but the edifice was surrounded by ice at least 800 m high that dammed lava flows around the perimeters of the volcano. After a period of eruptions showing no apparent evidence for ice interaction, from <80 to >40 ka, subglacial eruptions began again, signalling the build-up of regional ice levels. Local ice thickness during these eruptions may well have been over 2 km thick.

The morphology of volcanoes and the character of volcanic deposits ascribed to subglacial volcanism are functions of the type (i.e. wet-based v. dry-based; Smellie 2000) and thickness of ice (i.e. thin v. thick; Smellie & Skilling 1994), the length of time over which the edifice developed and the history of ice fluctuations during that development, and the chemical composition of the erupted lavas. Our current understanding of subglacial eruption processes and our models for the formation of subglacial volcanic deposits are heavily influenced by observations and data collected on relatively small, basaltic volcanoes erupted in glacial environments (e.g. Mathews 1947; Jones 1969*a*, 1970; Allen *et al.* 1982; Smellie 2000). Basaltic subglacial deposits are the most widespread and have been extensively described from Iceland (Jones 1969*a*, 1970; Allen *et al.* 1982; Bergh & Sigvaldason 1991; and references therein), British Columbia (Mathews 1947; Allen *et al.* 1982; Hickson *et al.* 1995; Moore *et al.* 1995; Souther 1992; Hickson 2000; and references therein), South America, (Larsson 1940), Antarctica (LeMasurier 1976; Smellie *et al.* 1993; Skilling 1994; Smellie & Hole 1997), and Hawaii (Porter 1987). Smellie (2000) has recently provided a comprehensive review of the unique features of mafic subglacial deposits.

Many subglacial volcanoes, however, are not basaltic in composition. These include andesitic stratovolcanoes of the Cascade volcanic arc (Mathews 1951, 1952*a*, *b*; Lescinsky & Fink 2000) and in Argentina (Larsson 1940), dacitic and rhyolitic centres in Iceland (Gronvold 1972; Furnes *et al.* 1980; Tuffen *et al.* 2001) and peralkaline phonolitic to rhyolitic edifices in north-western British Columbia (Souther 1992; Edwards 1997; Edwards *et al.* 2000) and in Antarctica (LeMasurier 1976, 1990). Since many of these studies have documented features not commonly seen at basaltic subglacial centres (e.g. Furnes *et al.* 1980), the models and principles developed for subglacial basaltic volcanism

From: SMELLIE, J. L. & CHAPMAN, M. G. (eds) 2002. *Volcano–Ice Interaction on Earth and Mars*. Geological Society, London, Special Publications, **202**, 179–194. 0305-8719/02/$15.00

cannot necessarily be used as a paradigm for interpreting the processes attending subglacial eruption of more siliceous lavas.

Hoodoo Mountain volcano (HMV) is a Quaternary phonolitic volcano situated in northwestern British Columbia (Fig. 1; Souther 1992; Edwards 1997; Edwards & Russell 1997; Edwards *et al.* 2000). The volcano comprises fragmental and non-fragmental phonolitic volcanic rocks and was formed during the last 100 ka. Much of its eruptive history involved volcano–ice interaction, which accounts for its distinctive shape and many of its volcanic deposits. Detailed studies of HMV offer unique advantages over other peralkaline subglacial volcanic edifices, especially those found in Antarctica, because it is relatively free of ice at most elevations. Hoodoo Mountain volcano exposes a nearly complete stratigraphic sequence from the base of the volcano to the top (except for the present-day ice cap), allowing for detailed study of its overall morphology, as well as the origin of volcanic deposits resulting from subglacial volcanism (Edwards 1997; Edwards & Russell 1997).

The purpose of this paper is threefold. First, described in detail are three associations of volcanic deposits that result directly from interactions between phonolitic volcanism and ice at Hoodoo Mountain. Second, the physical characteristics of these deposits are used to constrain conceptual models for their formation. Third, the morphology of the volcano and the distribution of these ice-contact volcanic deposits are used to establish the syn-eruptive presence and thickness of bounding ice masses through time at HMV. In this regard, the volcanology of

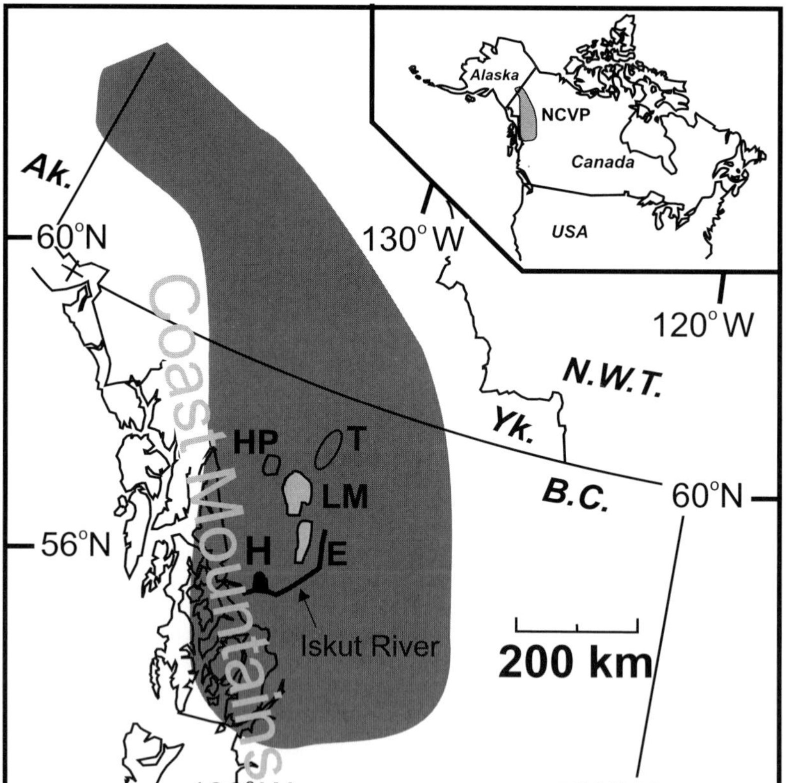

Fig. 1. Map showing the location of Hoodoo Mountain volcano (H). It is situated in the southern part of the northern Cordilleran volcanic province (NCVP in inset; Edwards & Russell 2000) and is considered to be part of the Stikine subprovince, which includes Mount Edziza (E) and Level Mountain (LM). The Tuya–Teslin volcanic district (T) comprises several of the tuya volcanoes that were initially described by Mathews (1947). Other abbreviations: Ak, Alaska; N.W.T, North West Territories; B.C., British Columbia; HP, Heart Peaks.

Hoodoo Mountain traces variations in the thicknesses of regional ice sheets within the Iskut region since *c.* 85 ka.

Hoodoo Mountain volcano (HMV)

Hoodoo Mountain is located immediately north of the Iskut River, in the Coast Mountains of north-western British Columbia, Canada (Fig. 1) and comprises a maximum volume of 17.3 km^3 of Quaternary volcanic rocks. HMV lavas have maintained a relatively uniform chemical and mineralogical composition over 85 ka. All lava and glass samples from HMV are classified as phonolite or trachyte, are nepheline and acmite normative, and have a (micro-) phenocryst assemblage of alkali-feldspar, clinopyroxene, and magnetite. The phonolite at HMV was probably derived from alkali olivine basalt via assimilation–fractional crystallization processes (unpublished information of the authors). The stratigraphy comprises fragmental and non-fragmental volcanic rocks including fine- to medium-grained subglacial and subaerial lava flows, domes, spines, dykes, and pyroclastic deposits (Fig. 2). The formation of HMV overlaps, in space and time, with the formation of the Iskut volcanic field, which encompasses nine small-volume, basaltic centres (Russell & Hauksdottir 2000). HMV is also part of the peralkaline Stikine subprovince of the northern Cordilleran volcanic province, which includes Edziza and Level Mountain volcanic complexes (Fig. 1; Edwards & Russell 2000).

Hoodoo Mountain volcano has had at least six eruptive cycles since *c.* 85 ka (Edwards & Russell 1997, 2000). Individual eruptive cycles have been distinguished on the basis of changes in the physical character of eruptive products (e.g. lava flows v. lava flows and domes with associated breccias v. pyroclastic deposits), observed stratigraphic relationships, and results from $^{40}Ar/^{39}Ar$ geochronology (cf. Villeneuve *et al.* 1998; Edwards *et al.* 2000). The six eruptive episodes are summarized as follows (Fig. 2a):

(a) Volcanism began at Hoodoo at least as early as *c.* 85 ka with eruptions that produced aphanitic lava flows, domes, and breccia deposits.
(b) By *c.* 80 ka volcanism was manifest as a series of aphanitic lava flows that formed massive cliffs at low elevations near the base of the volcano.
(c) Pyroclastic deposits, up to 100 m thick on the north and west sides of the edifice, formed between 80 and 54 ka.
(d) Aphanitic lava flows erupted at *c.* 54 ka directly on top of the pyroclastic deposits.
(e) Between 54 and 30 ka, volcanism produced two distinct types of lava–breccia associations, one between 54 and 40 ka, and the other probably slightly younger, between 40 and 30 ka.
(f) Eruptions of highly porphyritic lava flows as young as 9 ka mark the last eruptive cycle at HMV.

Three types of glacial volcanic deposits at HMV

The focus of this research is on associated volcanic deposits produced during the (a), (b) and (e) eruptive cycles at HMV. These three eruptive intervals comprise deposits that have characteristics indicative of lava–ice interactions. Specifically, they represent the products of three distinct ice-contact relationships: (1) the damming and ponding of lava flows against ice; (2) subglacial volcanic eruptions producing an association of lavas and breccias; and (3) sub- to supraglacial eruptions driven, in part, by vesiculation and expansion of a magmatic gas phase. Each of the three deposit associations below are described below.

Ice-dammed lava (IDL)

One of the most prominent features of Hoodoo Mountain is the set of cliffs that bound and define the base of the volcano (Kerr 1948; Souther 1992; Edwards *et al.* 2000). These cliffs comprise massive, aphanitic to sparsely porphyritic trachyte and phonolite lavas that show prominent flow banding and pervasive, well-developed jointing. One of the cliff-forming lavas is dated at *c.* 80 ka (Villeneuve *et al.* 1998). This stratigraphic unit is directly overlain by pyroclastic deposits and lava flows that appear to result from younger subaerial eruptions at *c.* 54 ka.

The lava cliffs vary in height from 50 m to greater than 200 m (Fig. 3a) and they almost completely circumscribe the volcano (Fig. 2). On the north side of HMV, the lower cliffs are partly buried by recent, unconsolidated deposits of glacially-derived sediments and Twin Glacier. On the eastern and south-eastern sides the lower cliffs are obscured by the youngest (post-glacial, *c.* 9 ka) lava flows (Fig. 2a). The lava cliffs create the lowest 'step' in the overall terraced topographic profile of HMV, which is particularly clear on the eastern side of the

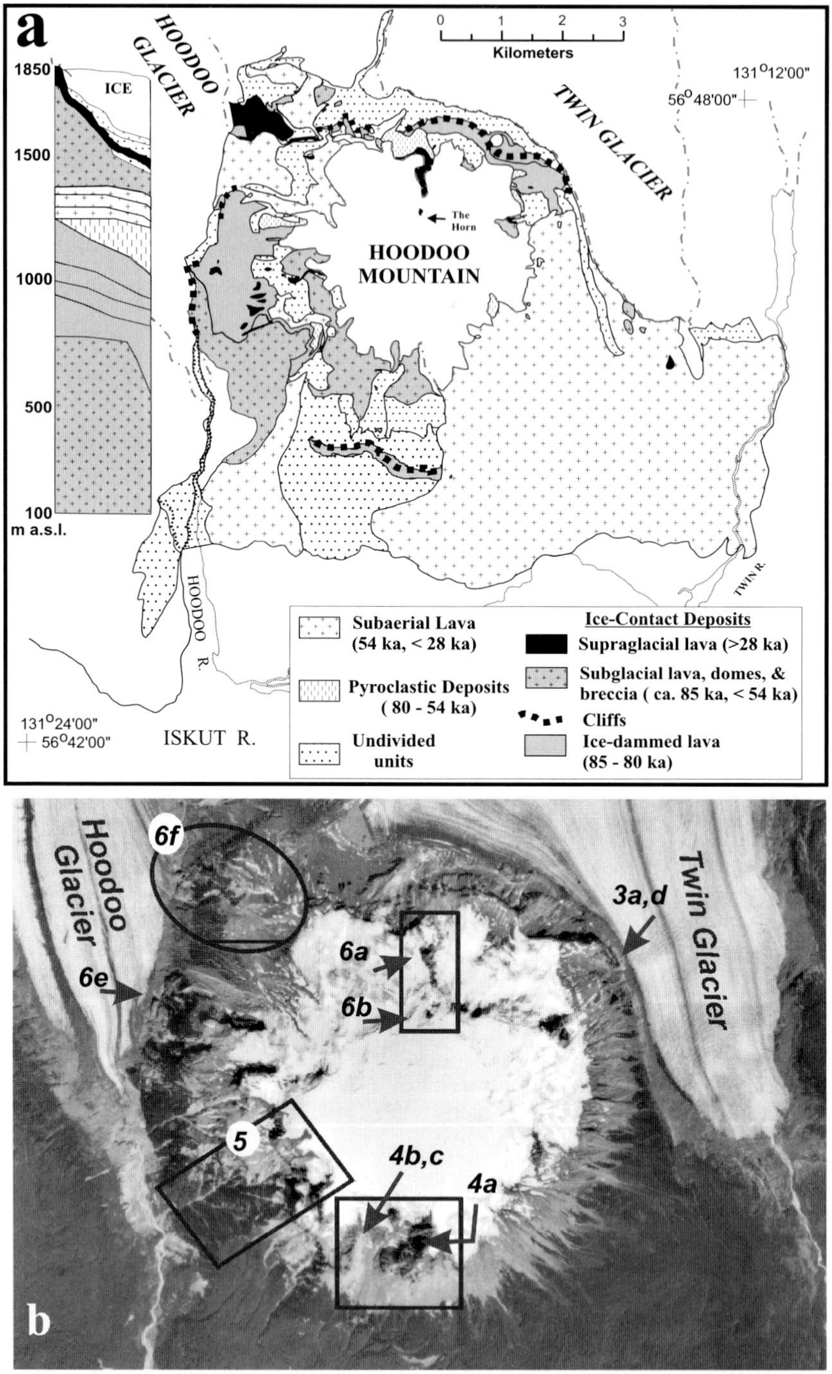

Fig. 2. Geology and physiography of Hoodoo Mountain volcano. (**a**) Geological map and simplified stratigraphic column, modified from Edwards *et al.* (2000) and highlighting ice-contact deposits described in the text. Locations of prominent cliffs defined by ice-dammed lava flows also indicated (coarse dashed line; see Fig. 3). (**b**) Aerial photograph of Hoodoo Mountain (derived from British Columbia Airphoto BC82022) showing the physiography of the volcano in plan view. Boxes delineate areas containing the outcrops with features shown in figures in this paper or discussed in text. Numbers correspond to figure numbers.

Fig. 3. Field photographs showing the character of the lower and upper cliffs produced by damming and ponding of lava flows by ice at Hoodoo Mountain (see Fig. 2b for location). (**a** & **d**) Views of prominent cliffs, 100–200 m in height, forming the base of HMV, interpreted as original quenched surfaces of lava flows that ponded against adjacent wall of former ice. (**b**) Cliffs composed of dense, aphanitic phonolitic lava featuring pervasive, fine-scale, horizontal columnar jointing. Note ice axe in bottom right for scale. (**c**) Intense radiating cooling joints on cliff face; field of view of photograph is 40–50 m.

mountain (Fig. 3a). The elevation of the upper surfaces to these cliffs of lava varies between 700–800 m asl on the west and south side, and about 1200 m asl on the north side. Taken together, these prominent cliffs of massive lava flows constitute the 'lower cliffs' and they are a discrete stratigraphic interval relative to the younger deposits of lava and breccia (see below) that form a second 'upper' set of cliffs near the top of HMV.

The cliffs of massive lava show a series of important genetic features. Specifically, samples from some cliff faces show the lava to be cryptocrystalline (nearly glassy) and pervasively jointed; both features are indicative of accelerated cooling or quenching. The cooling joint patterns and the variable nature of the orientations of columnar jointing are, in particular, diagnostic. Typically the cliff faces comprise a myriad of polygonal cooling joints with diameters generally less than 30 cm (Fig. 3b). The joints are oriented perpendicular to the vertical cliff face (cf. 'cordwood' jointing of Lescinsky & Fink 2000); in several faces the cliffs are characterized by radiating patterns of small-diameter columnar jointing (Fig. 3c). The cliffs commonly appear to be formed by a single lava flow, although several exposures (south side of HMV; Fig. 2) show at least two separate lavas. Massive slope failures on the SW flanks of HMV have exposed cross-sections through the length of the lavas that form these 'lower cliffs'. In cross section, it appears that the lava flows began as relatively thin (<30 m) flows but thickened substantially as they moved downslope from the vent. The marked thickening occurs despite the overall slope of the substrate remaining virtually constant.

The observations presented above are consistent with the inferences of previous workers (Kerr 1948; Souther 1992) that these lower cliffs represent ice-dammed lava flows. The lava flows, which appear to have originated from vents now buried beneath the main mass of the volcano, flowed down a relatively gentle incline over previously erupted subglacial deposits of lava and breccia (see next section), and then were dammed by thick ice along the flanks of the volcano. Assuming that all of the lower cliffs formed during one eruptive episode, Hoodoo Mountain must have been completely surrounded by ice at *c*. 80 ka.

Subglacial lava flows, domes and breccia (SGL)

Another distinctive manifestation of volcano–ice interaction at HMV is an association of highly jointed lava flows, domes and spines that are commonly encased in deposits of low-vesicularity, monomict breccia (Fig. 4). Rocks of this association are found at the very base and top of the edifice. At lower stratigraphic levels (and elevations) the association is buried except

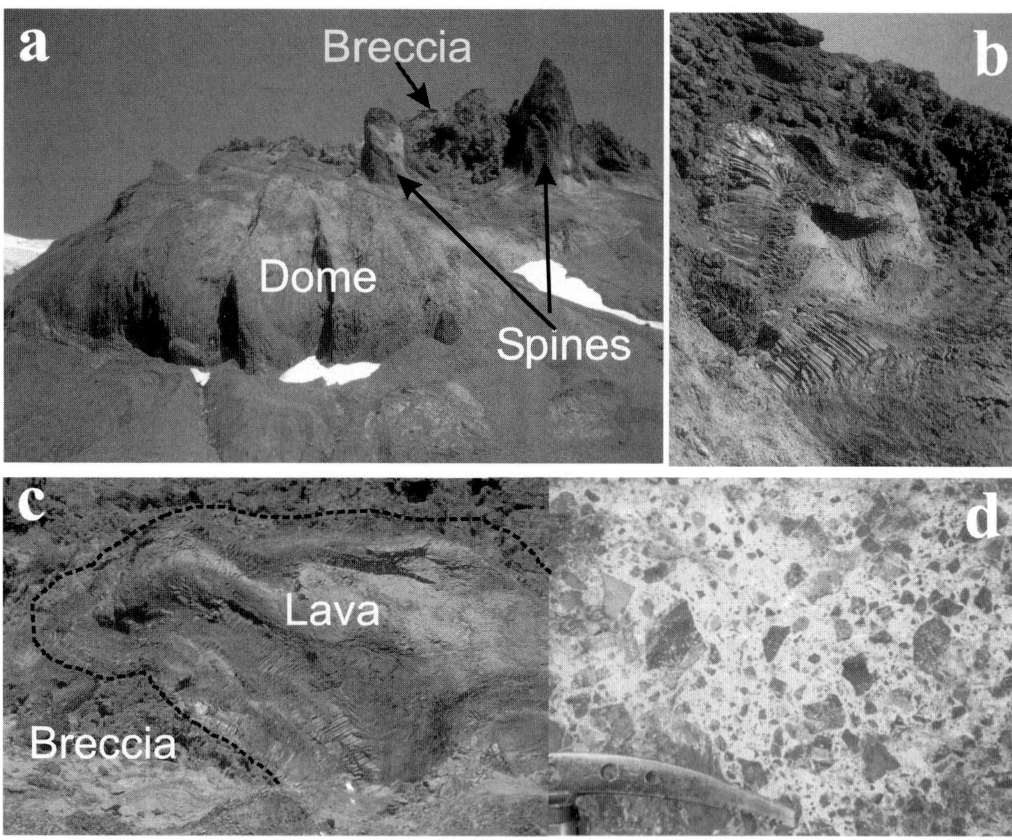

Fig. 4. Photographs of non-vesiculated lava flows, domes and associated breccias (see Fig. 2b for locations). These deposits probably formed during eruption beneath ice that was thick enough to inhibit vesiculation. (**a**) Bulbous, pervasively jointed lava dome and spines overlain by breccia of similar composition. The lava is dome-shaped and about 60 m in height. (**b**) Intense radially-oriented jointing in lava lobe encased by breccia. Lobe is approximately 10–15 m in height. (**c**) View of pervasive columnar jointing paralleling the contact between the lava lobe and breccia. Columnar joints are about 1–2 m in length. (**d**) Close view of angular, non-vesiculated clasts that comprise the monomict breccias that encase massive bulbous lava lobes and domes. Head of ice axe is about 30 cm long.

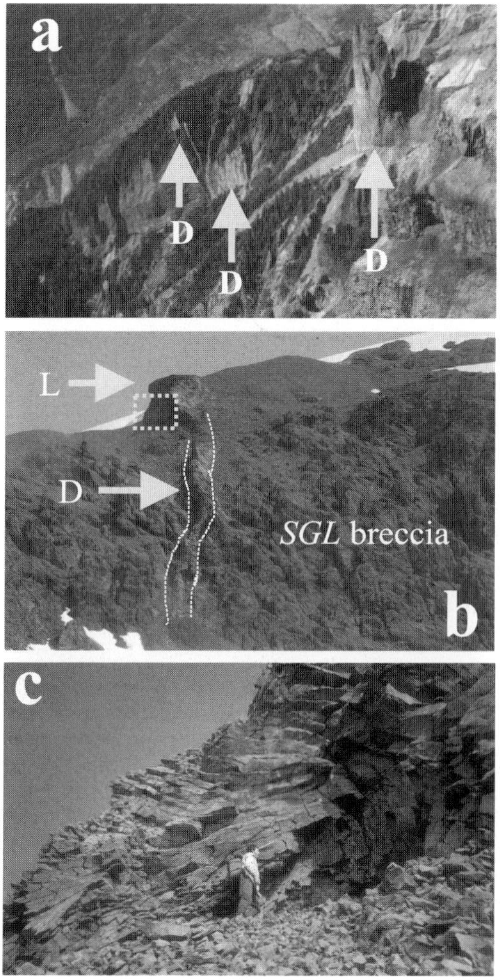

Fig. 5. Field photographs showing features of dykes that feed highly vesiculated lavas and breccias (cf. Fig. 2b). (**a**) Westward view of a large (20 m thick), discontinuously exposed, dyke (D; arrows) of phonolite that cuts through older HMV lava flows. (**b**) View of 50 m-high rock face that exposes a near-vertical feeder-dyke (D) that ultimately becomes a lava flow (L). Lava overlies and dyke intrudes breccias associated with the poorly-vesiculated subglacial lavas. HMV ice cap is visible in top right corner of photograph. (**c**) Detailed view of ice-chilled lava flow showing well-developed radially-oriented columnar jointing. The location of this exposure is shown in (**b**) by outlined box.

where exposed in the walls of a massive canyon on the SW flank of HMV (Fig. 2b). The deposits at the top of the volcano form a second set of cliffs that partially circumscribe the edifice. These 'upper' cliffs crop out between about 1300 and 1700 m asl, but are below the Hoodoo ice cap (Fig. 2). They provide the best exposures of SGL deposits.

The lava flows and domes are massive, non-vesicular and composed of aphanitic to slightly (1–5%) porphyritic black trachyte and phonolite. The associated breccia deposits are monomict, poorly sorted, and made up of lapilli- to block-sized, angular to subrounded, aphanitic, low-vesicularity clasts of trachyte/phonolite in a matrix of ash-size fragments (Fig. 4d). The breccia varies locally from matrix- to clast-supported. Crude layering is observed locally and is defined by accumulations of block-sized clasts. The yellow to reddish-orange color of the matrix, in addition to discoloured rims on freshly broken clasts, is interpreted as indicating varying degrees of incipient clay(?) alteration of the matrix material.

Several different lava forms are exposed in the upper cliffs. Locally the lava appears to have been extruded as bulbous domes and vertical spines (Fig. 4a). The monomict breccia enclosing the lava bodies appears to have formed synchronously with extrusion. The lava bodies are always pervasively jointed with cooling joints that are consistent with highly irregular cooling surfaces (Fig. 4b, c). A particularly well-exposed lava lobe is pervasively jointed, comprising small diameter, columnar joints (< 50 cm in diameter) whose orientation is perpendicular to the irregular contact with the breccia (Fig. 4c).

This paper interprets the association of intensely jointed lava and monomict breccia to have formed contemporaneously during subglacial eruptions. Spatial relationships between the enclosing breccia and the lava are consistent with the breccia forming contemporaneously with the lava. The angularity of the breccia clasts and the pervasive nature, type and orientation of the lava jointing are all consistent with formation in an environment characterized by rapid cooling by water and/or ice. The limited vesicularity of the lavas and breccia clasts, as well as the topographically restricted nature of the deposit, suggest that the eruptions took place under relatively thick ice (sufficient to suppress extensive vesiculation and exceeding the height of the cliffs, i.e. ice >500 m thick).

Sub- to supraglacial lava (SSL)

The upper stratigraphy of HMV hosts a second distinctive association between lava and pyroclastic deposits. This association comprises dykes, lava flows, domes, and well-indurated

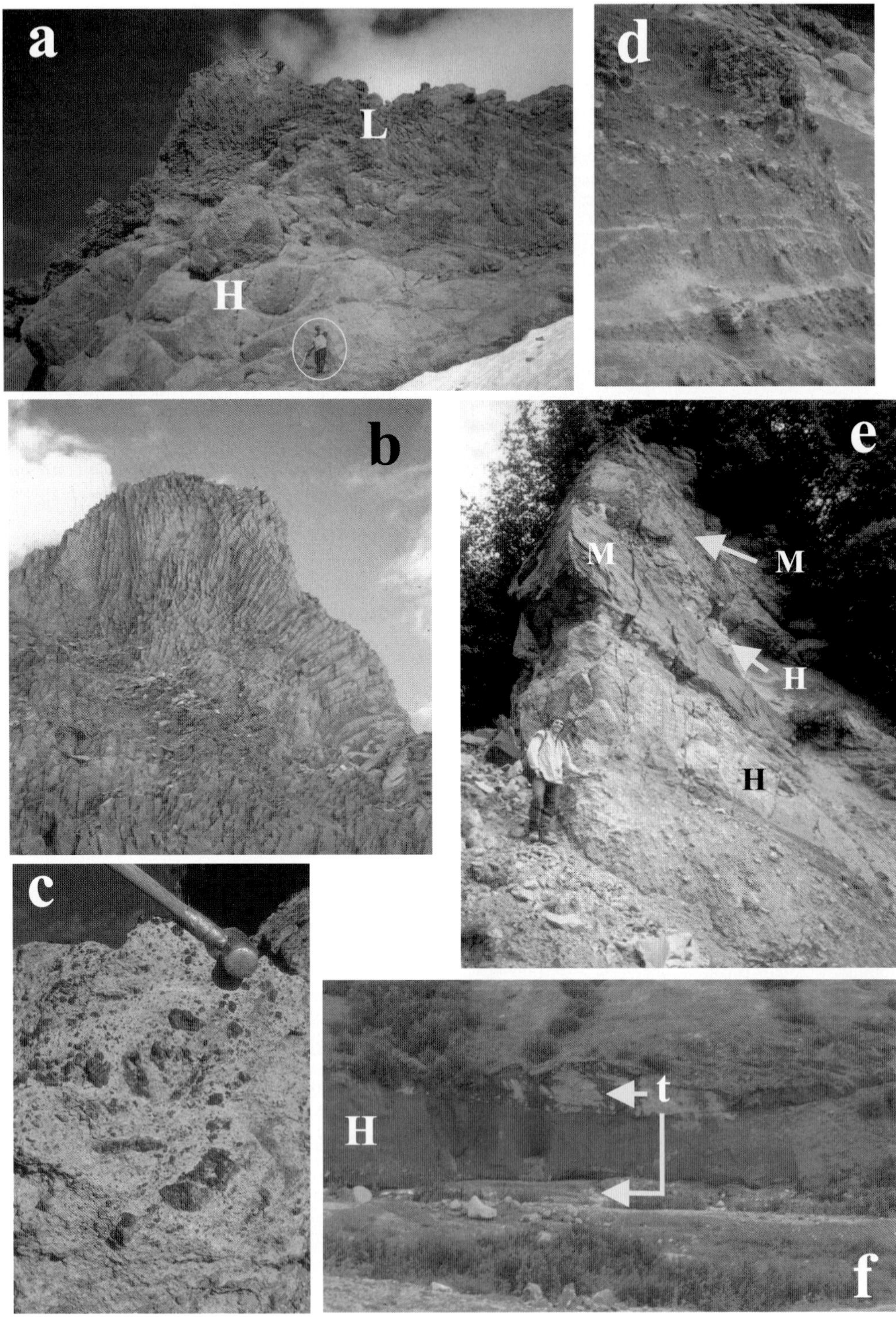
a
L
H
d
b
e
M
M
H
H
c
t
H
f

and unconsolidated pyroclastic deposits (Figs 5 & 6). The deposits are dated by $^{40}Ar/^{39}Ar$ geochronology at between 40 and 30 ka (Fig. 2; Villeneuve *et al.* 1998). The association is volumetrically small but occurs over a much broader range of elevations than the previously described SGL association (Fig. 2). For example, the SGL deposits from the upper cliff sequence are restricted to a narrow range of elevations (e.g. 1100 to 1700 m) that reflect their stratigraphic position. In contrast, deposits of the SSL association from the same stratigraphic level are found in exposures near the top of the edifice at 1720 m asl (e.g. the Horn) as well as at 700 m asl, near the base and distal edge of the volcano and near Hoodoo Glacier (Fig. 2a, b).

These deposits share a number of similarities with the SGL association. They both comprise aphanitic trachyte to phonolite lava, and massive lava flows and domes show radiating, pervasive columnar jointing and are narrow and domical in cross-section (Fig. 5b, c & 6b). However, the dykes, lava flows and associated fragmental deposits of the SSL association are different in one critical way: most of these units are moderately to highly vesicular, whereas the SGL deposits have comparatively lower vesicularities. Additionally, locally dykes can be traced directly into lava flows (Fig. 5b).

The pyroclastic deposits in the SSL association comprise two varieties. The first is well indurated, massive lapilli tuff, and contains vitric, vesicular lapilli and bombs (Fig. 6c). Typically, it forms a carapace around massive lava units and it has a characteristic yellowish white colour, which is interpreted as indicating hydrothermal (likely clay) alteration of ash-size glass particles in the matrix analogous to palagonitization of basaltic hyaloclastite. Locally pumiceous bombs are elongate and contain stretched vesicles that can be several centimetres long. The Horn (Fig. 2a) is a nunatak made of this pyroclastic unit with a core of lava. The pyroclastic deposit forms a carapace to the intensely jointed mass of lava (Fig. 6a, b) and is also found interlayered with brown to grey polymictic diamictite near the base of the west side of HMV (Fig. 6e). The diamictites comprise 10–25% subangular to subrounded clasts (from <1 cm to >50 cm) of HMV lava as well as clasts obviously not derived from HMV lavas. The contact between the diamictite and the pyroclastic deposits is irregularly shaped but marked by a sharp contrast in colour between the orange-yellow pyroclastic unit and the grey diamictite. Where a diamictite lens overlies the pyroclastic units, it locally contains subrounded pumiceous clasts.

The second type of pyroclastic deposit comprises green to black, non-indurated, ash- to bomb-size clasts, which locally appear to be moderately graded (Fig. 6d) and weakly stratified. The non-indurated pyroclastic deposits are only found in two locations on the lower flanks of the edifice. The deposits occur directly beneath lava flows of the association, and do not exhibit signs of extensive glass alteration. At the NW base of HMV, the deposits form a 2–3 m thick unit between two glacial tills (Fig. 6f).

The SSL association is inferred to have formed contemporaneously during subglacial to supraglacial eruptions beneath a thin ice cover. The vesicularity of the lava and pyroclastic deposits is consistent with low confining pressures and/or high volatile contents during eruption, in contrast to the poorly-vesiculated SGL deposits. Although the pyroclastic deposits in the SSL association are inferred to result from the same eruptive event, they are diverse in character and were deposited over a wide range of elevations. This aspect of the deposits is a strong indication that the eruptions breached the surface of the ice and were able to flow down the side of the edifice. The well-indurated pyroclastic deposits probably represent the proximal facies of fragmental deposits resulting from explosive lava–ice interaction. High heat content and abundant water facilitated alteration and concomitant cementation. In contrast, unconsolidated, graded and crudely stratified deposits of pyroclasts, found only near the base of the

Fig. 6. Photographs of vesiculated lava and palagonitized hyaloclastite deposits on the northern flank of HMV (cf. Fig. 2b). Deposits result from eruptions that were initiated beneath relatively 'thin' ice and may ultimately have breached the ice surface. (**a**) Field photograph of deposits near the Horn (Fig. 2a) showing the close association between vesicular, strongly-jointed lava (L) and veneer of yellow, altered hyaloclastite (H). (**b**) Pronounced, coarse–medium scale (*c.* 0.5 m), radially-oriented cooling joints in lava forming the Horn nunatak. (**c**) Close view of altered hyaloclastite showing angular, dark vitric clasts. Hammer is about 35 cm long. (**d**) Moderately stratified deposits of hyaloclastite showing variations in clast size and size sorting attributed to the effects of water transport. Large block is *c.* 0.5 m in diameter (**e**) Two lenses of yellow, altered hyaloclastite (H) interfingering with lenses of green mudstone (M) located immediately north of the Wall (see Fig. 2a for location). (**f**) View of lens, *c.* 3 m thick, of water-transported hyaloclastite (H) between deposits of glacial till (t).

edifice and overlain by lava flows from the same eruptive episode, represent the more distal, water-transported facies.

Comparison to other glacial volcanic deposits

The three associations described above, although unique in many respects, share common traits with glacial volcanic deposits. For example, Mathews (1952*a*) reported lava cliffs at Clinker Mountain up to 1500 feet high, formed by unusually thick flows at relatively low elevations. He interpreted the cliffs as having formed by flows being dammed by valley-filling glacial ice. Thus, the interpretation of the IDL association, which is very similar to deposits formed by lava flows at Clinker Mountain in southwestern British Columbia (Mathews 1952*a*), as lava flows dammed by ice is consistent with previous interpretations of similar features.

The SGL association shares some characteristics with subaqueous basaltic and subglacial rhyolitic lithofacies. Bergh & Sigvaldason (1991) described basaltic lithofacies (isolated and broken pillow hyaloclastite breccia, and lobate basalt hyaloclastite breccia) that comprise a mixture of monomict breccia and pervasively columnar jointed lava lobes very similar in appearance to outcrops of the SGL association. Bergh & Sigvaldason (1991) inferred a subaqueous marine eruption environment followed rapidly (at high temperatures) by high concentration mass flows for the genesis of both lithofacies. Tuffen *et al.* (2001) described a rhyolitic lithofacies of monomict breccia and lava lobes (Breccia B), interpreted as forming by hot mass flows during subglacial eruption. Given the similarities between the two lithofacies mentioned above and the SGL association at Hoodoo Mountain, it seems likely that hot mass flows resulting from gravitational collapse could also have been important in forming the SGL association. This paper favours a subglacial environment of formation for the following reasons: (a) the thickness of the deposits (up to >500 m); (b) the low vesicularity of the lava and breccia clasts; (c) the pervasive nature and small diameter of columnar jointing; and (d) the lack of any obvious marine fossils. However, it is not possible to rule out subaqueous eruptions into an englacial lake.

Previous workers have inferred that in water depths greater than about 500 m (Jones 1969*b*; Moore 1970), vesiculation in undegassed(?) basaltic lavas is largely suppressed (Moore *et al.* 1995). Although this depth is not well constrained for phonolite lavas, and is dependent the composition of the magma, especially its water content, it gives a crude approximation for the water–ice depth that may have existed above the SGL deposits and which may have largely suppressed their vesiculation. The morphology of the lava and the mantling relationship between the lava and breccia also seem most consistent with eruption under a substantial thickness of ice. Lava domes and highly irregularly shaped lava bodies (Fig. 5) all totally surrounded by breccia, indicate that these deposits formed by confined flow. Finally, the pervasive nature, size, and orientation of jointing in the lava flows and domes are most consistent with cooling in a water-rich environment (Long & Wood 1986; Bergh & Sigvaldason 1991; DeGraf & Aydin 1993).

The SSL association is the most unusual of the three subglacial associations at HMV. It shares some broad characteristics with previously described subaqueous to emergent volcanism reported for basaltic rocks from Antarctica (Smellie & Hole 1997) in that the association contains glassy breccia, massive lava flows, vesiculated pyroclastic material, and locally important dykes. However, as opposed to formation in an englacial lake (e.g. Smellie & Hole 1997), the SSL association appears to have largely erupted on the flanks of HMV, where lava and pyroclastic debris was able to move downslope, as opposed to accumulating in one location. The moderate to high degree of vesiculation in the pyroclastic units of the SSL association indicates that the thickness of overlying ice was not enough to suppress bubble nucleation and rapid growth. Based on the previously mentioned studies (Jones 1969*b*; Moore 1970) the ice depths were possibly much less than 500 m. Many of the large bombs have rinds of glass that are >2 cm thick, implying formation under conditions where the glass transition temperature was readily achieved. More importantly, the morphology of associated lava and breccia deposits is most consistent with eruption under thin ice in two ways. First, the half-domed shape of relatively small lava flows, pervasively fractured with joints that are horizontal on lower surfaces and radiate to vertical at top (Fig. 5b), are most consistent with the lava flow flowing down beneath the ice or in a channel melted through the ice (e.g. Lescinsky & Fink 2000). The pervasive, small diameter columnar joints indicate rapid cooling as with the SGL association. Second, the flow and its pyroclastic carapace at the Horn and flows along the northwest side of HMV appear to have formed while moving down a slope similar to the current topography of the mountain, along a confined path. The non-palagonitized, non-indurated pyroclastic deposits show evidence of some size sorting and are cautiously interpreted as deposits of pyroclasts transported in meltwater channels cut through or below thin ice formed during the initial stages of eruption. Further evidence of this derives from isolated deposits of diamictite interlayered with altered pyroclastic deposits. The apparent ease of down-slope movement, coupled with the high vesicularity of the pyroclasts, is most consistent with eruption into and/or through ice thinner than that present during formation of the SGL association.

Implications for edifice morphology

The combination of subaerial and subglacial eruptive products from the six recognized eruptive cycles (Table 1) has produced the most remarkable aspect of HMV, its morphological form. In plan view, HMV is symmetrical, circular, and approximately 6 km in diameter at its base (Fig. 2b). Views of all sides of the volcano (Fig. 7) illustrate the unique, step like shape of

Table 1. *Summary descriptions and interpretations of associations at HMV*

Association[1]	Description	Age[2] (in ka)	Interpretation
SSL	Lava flows; locally overlie all other deposits[3]	28–9	Subaerial
	Lava lobes, flows and dikes and monomict breccia; moderately to highly vesicular and/or amygdaloidal, pervasively jointed with irregularly oriented, fine (<0.3 m) columnar joints; breccia is varies from matrix- to clast-supported, with highly vesicular, elongated clasts up to 0.5 m long; matrix is whitish yellow; total unit thicknes *c.* 30–50 m	30–40	Sunglacial: relatively thin ice
Upper SGL	Lava lobes and domes and monomict breccia; Low vesicularity, and pervasively jointed with fine (<0.3 m) columnar joints of highly variable orientation; breccia is dominantly matrix-supported, with angular to subangular, low vesicularity clasts identical in appearance to associated massive lava; clasts range in size from 0.5 to >20 cm; matrix is orange to yellow; locally lava lobes are totally encapsulated by breccia; total unit thickness *c.* 400 m	<54 >40	Subglacial: relatively thin ice
	Lava flows; locally overlies lower SGL deposits[3]	54	Subaerial
	Lapilli-tuff; locally overlies IDL deposits[3]	<80 >54	Subaerial (?)
IDL	Lava flows; flow thicknesses vary from *c.* 30 m to >200 m forming large cliffs; flow faces show extensive entabulature development, often with horizontal oriented columnar joints; total unit thickness *c.* 200 m	80	Ice-dammed: subaerial lava flows erupted near summit (*c.* 1300–1400 m asl), dammed by ice at *c.* 700 m asl
Lower SGL	Lava flows and domes interlayered with monomict breccia, low vesicularity with pervasive small diameter (*c.* 0.3 m) columnar joints; breccia clasts are angular, matrix- to clast-supported, similar in appearance to associated lava; total unit thickness *c.* 500 to 1000 m	85	Subglacial: relatively thick ice

[1] See Figure 2.
[2] Villeneuve *et al.* (1998); Edwards *et al.* (2000).
[3] Not described in this work; cf. Edwards (1997), Edwards *et al.* (2000).

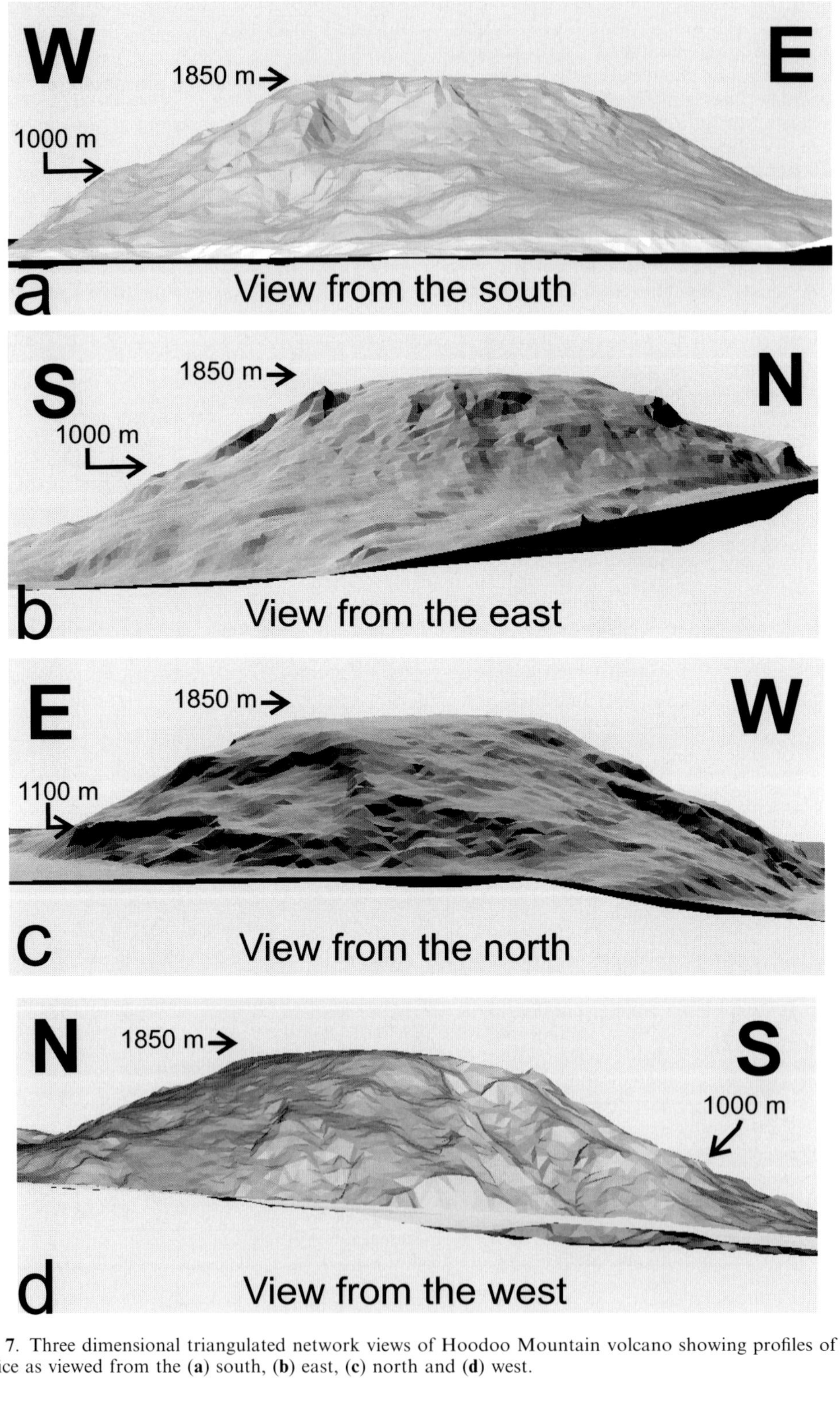

Fig. 7. Three dimensional triangulated network views of Hoodoo Mountain volcano showing profiles of the edifice as viewed from the (**a**) south, (**b**) east, (**c**) north and (**d**) west.

the edifice. It features a broad, rounded summit at 1850 m asl and an ice cap which is 3 km in diameter. Radar surveys have confirmed that the mountain has a flat top beneath the 120–150 m thick ice cap (Russell *et al.* 1998). At least two sets of prominent cliffs partly circumscribe the volcano, producing a discontinuous, step-like topographic profile (Figs 3a & 7). For example, the base of the volcano is largely delimited by a series of 100–200 m-high cliffs, except on the south-eastern side of the edifice. There, lava flows from the youngest eruptions form a veneer over the pre-existing topography; the slope of the SE flank is smooth from near the summit (*c.* 1800 m) to the Iskut River at *c.* 100 m asl (Fig. 7a). The top of the lower set of cliffs defines a broad bench at approximately 1000 m elevation that terminates against an upper set of cliffs. The second set of vertical cliffs is between 50 and 100 m high and surrounds the summit.

The overall flat-topped morphology of HMV led Souther (1991) to refer to Hoodoo Mountain as a 'tuya'. Although Hoodoo Mountain volcano does not fit the classic tuya model (i.e. Mathews 1947; Hickson 2000) because of its complex stratigraphy, its unique morphology apparently results from repeated interactions between volcanism and proximal ice-sheets over the past 100 ka. Thus, we consider it to be a tuya *sensu lato*.

Implications from HMV for glacier fluctuations in the Iskut region

The three distinct styles of lava–ice interaction described above are a reflection of local environmental conditions, mainly the presence and thickness of ice deposits during eruption. As a result, it is possible to use the physical character of these ice-contact volcanic deposits, their distribution, and their relative and numerical ($^{40}Ar/^{39}Ar$) ages to put preliminary constraints on the thickness of ice masses in the Iskut region since *c.* 85 ka (summarized in Table 1; Fig. 8).

The oldest known volcanic deposits, formed at *c.* 85 ka, are part of the SGL lithofacies association. These deposits exist up to elevations of

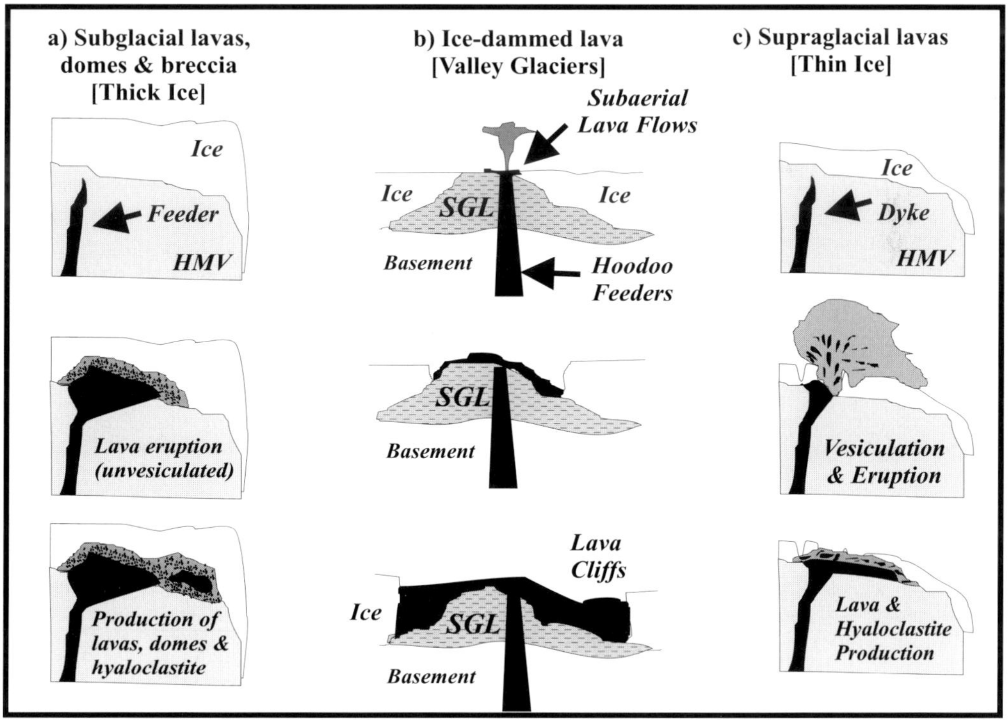

Fig. 8. Cartoon illustrating styles of volcanism that may have produced the three different types of ice-contact deposits described in the text. (**a**) Ice-contained massive, poorly vesiculated lavas, domes and associated breccias. (**b**) Ice dammed subaerial lavas. (**c**) Highly vesicular lavas and hyaloclastite. Thinner ice and possible ice breaching could have triggered explosive eruptions.

about 800–900 m asl and their presence and distribution establishes that, at this time, the local ice sheet had a minimum elevation of about 1300–1400 m asl (700–800 m of SGL deposits plus 500 m of ice). This translates into a minimum ice thickness of about 500 m covering HMV and 1200 m in the adjacent valleys (Fig. 2b). A schematic model has been developed, in this paper, for the origin of this style of relatively 'thick ice' eruption (Fig. 8a).

Situated above the basal *SGL* deposits are cliffs of the ice-dammed lavas that formed at about 80 ka (IDL; Fig. 2a). The presence and distribution of these deposits indicate that valley glaciers filled all of the river valleys surrounding HMV at *c.* 80 ka, assuming that all of the IDL were formed during the same eruptive cycle. Since valley glaciers currently bound the north, west and east sides of the edifice, this is not a surprising conclusion. However, the cliffs on the south side of the edifice, if correlative with corresponding cliffs on the north and west sides, indicate that the Iskut River valley also was filled with ice at *c.* 80 ka. Eruptions between 80 and 54 ka, which show no signs of ice-interaction (Edwards & Russell 2000), indicate that any glaciers present were confined to elevations below about 1300 m (the minimum elevation of the subaerial deposits).

The second sequence of SGL deposits, formed between 54 and 40 ka, indicate a substantial thickening of ice, with a minimum thickness of about 500 m overlying the summit of HMV and possibly up to 2000 m of ice in valleys adjacent to the edifice. The latter minimum estimate is based on the assumption that HMV was not acting as a highpoint of snow/ice accumulation, but that the ice on top of HMV was being fed by the large ice fields such as those that are currently located about 10 km N of HMV.

The distribution and characteristics of SSL deposits, formed at *c.* 40–30 ka, are indicative of eruption beneath relatively thin ice and perhaps are even a direct indication of eruptions breaching the ice surface. Figure 8c is a schematic model illustrating the sequence of events during these 'thin' ice eruptions.

Although the constraints on the absolute thickness of ice surrounding HMV since *c.* 85 ka are qualitative, the observations within this paper help to constrain the formation processes for these deposits. Understanding the variations in volcanic ice-contact and ice-proximal deposits at HMV will permit a much more detailed reconstruction of regional glacier fluctuations in the Iskut region over the last 100 ka. Future field, experimental, and geochronometric studies will allow several questions to be addressed further, including more accurately determining the relationship between ice thickness and inhibition of vesiculation in phonolitic magmas, determining if all of the lower cliffs at HMV formed during a single eruptive episode (as inferred above), and correlation between regional glacial stratigraphy in the Canadian Cordillera and local ice fluctuations in the Iskut region.

Conclusions

The morphology of Hoodoo Mountain volcano and the character of its volcanic deposits can largely be ascribed to three distinct phases of interaction between volcanism and glacial ice. Eruptions thought to be initially subaerial encountered ice probably from bounding valley glaciers and were substantially thickened to form impressive cliffs that partly circumscribe HMV. Thick ice cover produced a topographically-confined association of low vesicularity lava and breccia at least twice in the eruptive history of HMV. Eruption beneath thinner ice produced the association of lava with highly vesiculated deposits of pyroclastic breccia. The characteristics of volcanic units deposited throughout the entire eruptive history of HMV indicate an extended period of glacial-volcano interactions over the past 85 ka. Inferred fluctuations of ice thickness vary from relatively thick (1400 m?) between 85 and 80 ka near HMV and in the Iskut valley, to somewhat thinner (<1300 m?) between 80 and 54 ka, to a final build-up of >2000 m (?) between 54 and 40 ka.

This research was supported by the Geological Survey of Canada through the Iskut Field Mapping Program (R. G. Anderson) and the NSERC Research Grants program (89820 to JKR). The senior author was supported by the University of British Columbia through a University Graduate Fellowship (1995–1997). Our research benefited from discussions with C. Hickson and M. Stasiuk, and from constructive reviews of this paper by Thom Wilch and John Smellie.

References

ALLEN, C. C., JERCINOVIC, M. J. & ALLEN, J. S. B. 1982. Subglacial volcanism in north-central British Columbia and Iceland. *Journal of Geology*, **90**, 699–715.

BERGH, S. G. & SIGVALDASON, G. E. 1991. Pleistocene mass-flow deposits of basaltic hyaloclastite on a shallow submarine shelf, South Iceland. *Bulletin of Volcanology*, **53**, 597–611.

DEGRAFF, J. M. & AYDIN, A. 1993. Effects of thermal regime on growth increment and spacing of contraction joints in basaltic lava. *Journal of Geophysical Research*, **98**, 6411–6430.

EDWARDS, B. R. 1997. *Field, kinetic and thermodynamic studies of magmatic assimilation in the northern cordilleran volcanic province, northwestern British Columbia.* PhD thesis, University of British Columbia.

EDWARDS, B. R. & RUSSELL, J. K. 1997. Terrestrial subglacial magmatism: glacial influences on volcanic morphology and eruption products at the Hoodoo Mountain volcanic complex, northwestern British Columbia. *Geological Society of America, Abstracts with Programs*, **29**, 137.

EDWARDS, B. R., ANDERSON, R. G., RUSSELL, J. K., HASTINGS, N. L. & GUO, Y. T. 2000. *The Quaternary Hoodoo Mountain Volcanic Complex and Paleozoic and Mesozoic basement rocks, parts of Hoodoo Mountain (NTS 104B/14) and Craig River (NTS 104B/11) map areas, northwestern British Columbia.* Geological Survey of Canada, Open File Map **3721**, 1:20 000 scale.

EDWARDS, B. R. & RUSSELL, J. K. 2000. Distribution, nature and origin of Neogene-Quaternary magma tism in the northern Cordilleran volcanic province, Canada. *Geological Society of America. Bulletin*, **112**, 1280–1295

FURNES, H., FRIDLEIFSSON, I. B. & ATKINS, F. B. 1980. Subglacial hyaloclastites – on the formation of acid hyaloclastites. *Journal of Volcanology and Geothermal Research*, **8**, 95–110.

GRONVOLD, K. 1972. *Structural and petrochemical studies in the Kerlingarfjoll region, central Iceland.* PhD thesis, Oxford.

HICKSON, C. J. 2000. Physical controls and resulting morphological forms of Quaternary ice-contact volcanoes in western Canada. *Geomorphology*, **32**, 239–261.

HICKSON, C. J., MOORE, J. G., CALK, L. & METCALFE, P. 1995. Intraglacial volcanism in the Wells Gray – Clearwater volcanic field, east-central British Columbia, Canada. *Canadian Journal of Earth Science*, **32**, 838–851.

JONES, J. G. 1969*a*. Intraglacial volcanoes of the Laugarvatn region, south-west ICELAND, I. *Journal of the Geological Society of London*, **124**, 197–211.

JONES, J. G. 1969*b*. Pillow lavas as depth indicators. *American Journal of Science*, **267**(2), 181–195.

JONES, J. G. 1970. Intraglacial volcanoes of the Laugarvatn region, southwest Iceland, II. *Journal of Geology*, **78**, 127–140.

KERR, F. A. 1948. *Lower Stikine and western Iskut River areas.* Geological Survey of Canada, Memoirs, **248**.

LARSSON, W. 1940. Petrology of interglacial volcanics from the Andes of Northern Patagonia. *Bulletin of the Geological Institute of Upsala*, **28** 194–405.

LEMASURIER, W. E. 1976. Intraglacial volcanoes in Marie Byrd Land. *Antarctic Journal of the United States*, **11**, 269–270.

LEMASURIER, W. E. 1990. Marie Byrd Land, Summary. *In*: LEMASURIER, W. E. & THOMSON, J . W. (eds) *Volcanoes of the Antarctic Plate and Southern Oceans.* American Geophysical Union, Antarctic Research Series, **48**, 147–163.

LESCINSKY, D. T. & FINK, J. H. 2000. Lava and ice interaction at stratovolcanoes: Use of characteristic features to determine past glacial events and future volcanic hazards. *Journal of Geophysical Research: Solid Earth and Planets*, **105**(B), 23 711–23 726.

LONG, P. E. & WOOD, B. J. 1986. Structures, textures, and cooling histories of Columbia River basalt flows. *Geological Society of America Bulletin*, **97**, 1144–1155.

MATHEWS, W. H. 1947. Tuyas, flat-topped volcanoes in northern British Columbia. *American Journal of Science*, **245**, 560–570.

MATHEWS, W. H. 1951. The Table, a flat-topped volcano in southern British Columbia. *American Journal of Science*, **249**, 830–841.

MATHEWS, W. H. 1952*a*. Ice-dammed lavas from Clinker Mountain, southwestern British Columbia. *American Journal of Science*, **250**, 553–565.

MATHEWS, W. H. 1952*b*. Mount Garibaldi, a supraglacial Pleistocene volcano in southwestern British Columbia. *American Journal of Science*, **250**, 81–103.

MOORE, J. G. 1970. Water content of basalt erupted on the ocean floor. *Contributions to Mineralogy and Petrology*, **28**, 272–279.

MOORE, J. G., HICKSON, C. J. & CALK, L. C. 1995. Tholeiitic-alkalic transition at subglacial volcanoes, Tuya region, British Columbia, Canada. *Journal of Geophysical Research*, **100**, 24 577–24 592.

PORTER, S. C. 1987. Pleistocene subglacial eruptions on Mauna Kea. *In*: DECKER, R. W., WRIGHT, T. L. & STAUFFER, P. H. (eds) *Volcanism in Hawaii.* United States Geological Survey Professional Paper, **1350**, 587–598.

RUSSELL, J. K. & HAUKSDOTTIR, S. 2000. Estimates of Crustal Assimilation in Quaternary lavas from the northern Cordillera, British Columbia. *Canadian Mineralogist*, **39**, 275–297.

RUSSELL, J. K., STASIUK, M. V., SCHMOK, J., NICHOLLS, J., PAGER, T., RUST, A., CROSS, G., EDWARDS, B. R., HICKSON, C. J. & MAXWELL, M. 1998. *The ice cap of Hoodoo Mountain volcano, northwestern British Columbia: estimates of shape and thickness form surface radar surveys.* Geological Survey of Canada, Current Research, **1998-A**, 55–63.

SKILLING, I. P. 1994. Evolution of an englacial volcano: Brown Bluff, Antarctica. *Bulletin of Volcanology*, **56**, 573–591.

SMELLIE, J. L. 2000. Subglacial eruptions. *In*: SIGURDSSON, H. (ed.) *Encyclopedia of volcanoes.* Academic Press, San Diego, 403–418

SMELLIE, J. L., HOLE, M. J. & NELL, P. A. R. 1993. Late Miocene valley-confined subglacial volcanism in northern Alexander Island, Antarctic Peninsula. *Bulletin of Volcanology*, **55**, 273–288.

SMELLIE, J. L. & SKILLING, I. P. 1994. Products of subglacial volcanic eruptions under different ice thicknesses: two examples from Antarctica. *Sedimentary Geology*, **91**, 115–129.

SMELLIE, J. L. & Hole, M. J. 1997. Products and processes in Pliocene–Recent, subaqueous to emergent volcanism in the Antarctic Peninsula: examples of

englacial Surtseyan volcano construction. *Bulletin of Volcanology*, **58**, 628–646.
SOUTHER, J. G. 1991. Hoodoo Mountain. *In*: WOOD, C. A. & KIENLE, J. (eds) *Volcanoes of North America, United States and Canada*. Cambridge University Press, Cambridge 127–128.
SOUTHER, J. G. 1992. *The late Cenozoic Mount Edziza Volcanic Complex, British Columbia*. Geological Survey of Canada Memoir, **420**.
TUFFEN, H., GILBERT, J. & MCGARVIE, D. 2001. Products of an effusive subglacial rhyolite eruption: Blahnukur, Torfajokull, Iceland. *Bulletin of Volcanology*, **63**, 179–190.
VILLENEUVE, M. E., WHALEN, J. B., EDWARDS, B. R. & ANDERSON, R. G. 1998. Time-scales of episodic magmatism in the Canadian Cordillera as determined by ^{40}Ar–^{39}Ar geochronology. *Geological Association of Canada, Abstracts with Programs*, **23**, 192.

Effusive intermediate glaciovolcanism in the Garibaldi Volcanic Belt, southwestern British Columbia, Canada

M. C. KELMAN[1], J. K. RUSSELL[1] & C. J. HICKSON[2]

[1] *Igneous Petrology Laboratory, Department of Earth and Ocean Sciences, University of British Columbia, 6339 Stores Road, Vancouver, British Columbia V6T 1Z4, Canada (e-mail: mkelman@eos.ubc.ca)*

[2] *Geological Survey of Canada, 101-605 Robson Street, Vancouver, British Columbia V6B 5J3, Canada*

Abstract: The Garibaldi Volcanic Belt (GVB) in southwestern British Columbia is dominated by intermediate composition volcanoes in a setting that has been intermittently subjected to widespread glaciation. The glaciovolcanic features produced are distinctive, and include flow-dominated tuyas, subglacial domes, and ice-marginal flows. Flow-dominated tuyas, which are intermediate in composition, are unlike conventional basaltic tuyas; they consist of stacks of flat-lying lava flows, and lack pillows and hyaloclastite. They are inferred to represent subglacial eruptions that ultimately breached the ice surface. Subglacial domes occur as steep-sided masses of heavily-jointed, glassy lava, and represent eruptions that were entirely subglacial. Ice-marginal flows derive from subaerial flows that were impounded against ice.

Two unique aspects of GVB glaciovolcanic products are the presence of flow-dominated tuyas and the apparent scarcity of primary fragmental deposits. These unique features result from lava composition, the minimization of direct lava–water contact during eruptions, and topography. Composition influences morphology because eruption temperature decreases, and viscosity and glass transition temperature both increase with silica content. The result of this is that silicic subglacial volcanoes melt less water and are less likely to trap it near the vent, leading to the formation of structures whose shapes are strongly influenced by the surrounding ice. Topography also enhances meltwater drainage, favours lava flow impoundment in ice-filled valleys, and may, through erosion, influence the observed distribution of fragmental glaciovolcanic deposits.

Southwestern British Columbia is a region of extensive Neogene–Quaternary volcanism, and the waxing and waning of continental-scale ice sheets, as well as the presence of alpine glaciers, during Quaternary time has led to many interactions between volcanoes and ice. These interactions explain numerous geomorphological features of southwestern British Columbia's Garibaldi Volcanic Belt (GVB).

In this paper, we aim to (1) introduce the GVB and its glaciovolcanic landforms, (2) highlight the unique aspects of GVB glaciovolcanic deposits, and (3) discuss the origins of some of the more distinctive glaciovolcanic landforms in the GVB. Glaciovolcanic features of the GVB are substantially different from those described in Iceland (Jones 1966, 1969, 1970; Furnes *et al.* 1980; Tuffen *et al.* 2001), Antarctica (Smellie *et al.* 1993; Smellie & Skilling 1994; Smellie 2000), and other parts of British Columbia (Mathews 1947; Souther 1992; Hickson 1987; Hickson *et al.* 1995; Edwards & Russell 2002; Edwards *et al.* 2002). The three main landforms resulting from intermediate composition glaciovolcanism are: flow-dominated tuyas, subglacial domes and ice-marginal lava flows. Pillows and fragmental material are rare or absent at all three types of landform. The unique aspects of GVB glaciovolcanic landforms reflect lava compositions and terrain.

The landforms and lithologic features formed by glaciovolcanism are distinctive. Where these deposits are preserved, they can serve as an important palaeoclimatological tool, because they provide an indicator of the past presence of ice. Additionally, an understanding of how volcanoes and ice interact is important for hazard assessments, since subglacial eruptions have the potential to release catastrophically large volumes of water (as jökulhlaups) due to

From: SMELLIE, J. L. & CHAPMAN, M. G. (eds) 2002. *Volcano–Ice Interaction on Earth and Mars*. Geological Society, London, Special Publications, **202**, 195–211. 0305-8719/02/$15.00

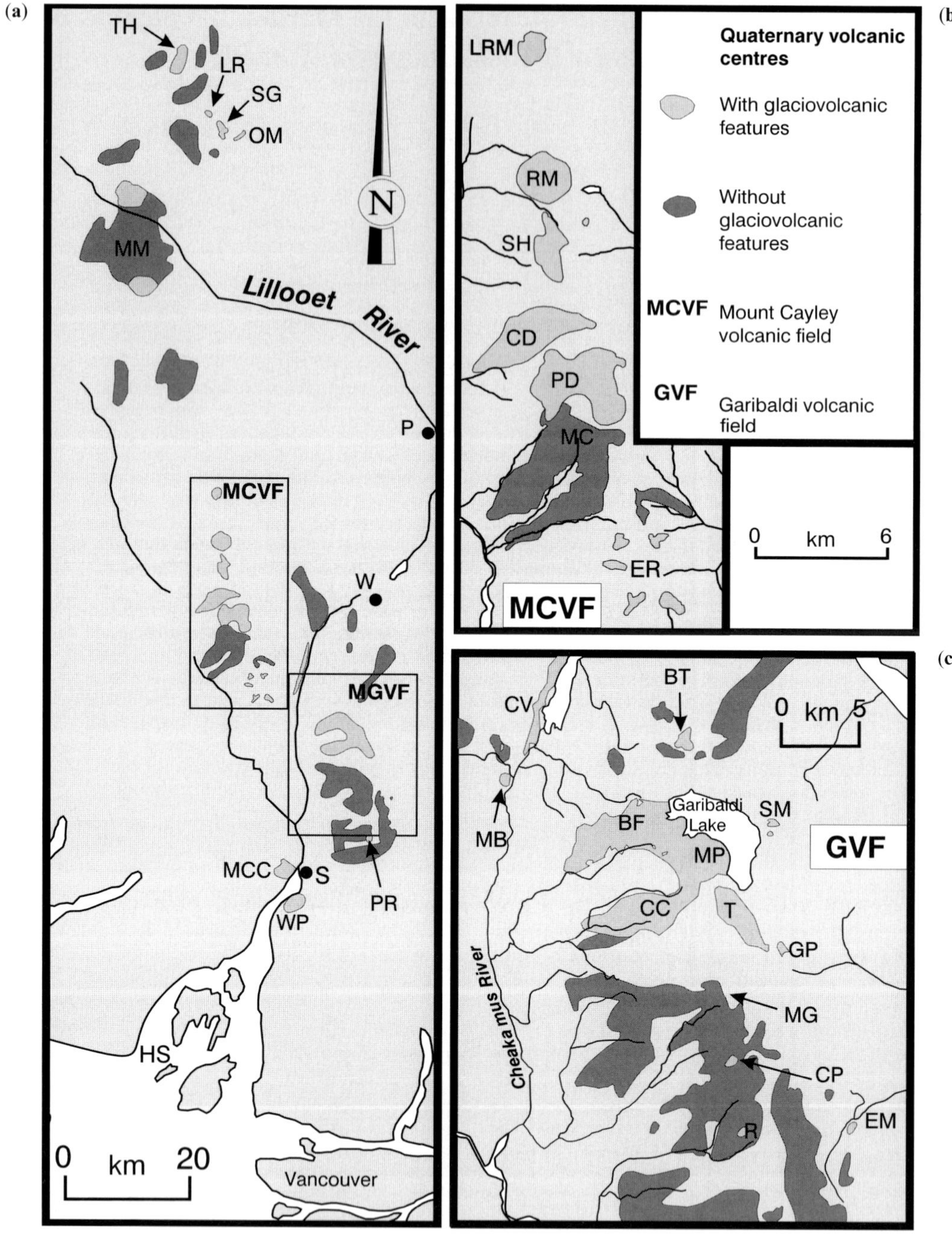

Fig. 1. (**a**) Map of the Garibaldi Volcanic Belt, showing the distribution of volcanic centres (after Hickson 1994). (**b**) Map of the Mount Cayley volcanic field (after Souther 1980). (**c**) Map of the Garibaldi volcanic field. Individual glaciovolcanic centres are abbreviated as follows: BF, Barrier flow; BT, Black Tusk; CC, Culliton Creek flow; CD, Cauldron Dome; CP, Columnar Peak; CV, Cheakamus Valley flows; EM, Eenostuck mass; ER, Ember Ridge; GP, Glacier Pikes; HS, Howe Sound; LR, Logan Ridge; LRM, Little Ring Mountain; MB, Mount Brew; MCC, Monmouth Creek complex; MC, Mount Cayley; MG, Mount Garibaldi; MM, Mount Meager; MP, Mount Price; OM, Ochre Mountain; P, Pemberton; PD, Pali Dome; PR, Paul Ridge; R, Round Mountain; RM, Ring Mountain; S, Squamish; SG, Salal Glacier volcanic complex; SH, Slag Hill; SM, Sphinx Moraine; T, Table; TH, Tuber Hill; W, Whistler; WP, Watts Point. Silverthrone Mountain, at the northern extent of the Garibaldi Volcanic Belt, lies to the north of the area covered by this map.

rapid, large-scale melting of ice. Finally, distributions of products of glaciovolcanism can be used to constrain temporal and spatial linkages between volcanism and glaciation.

Garibaldi Volcanic Belt

The Garibaldi Volcanic Belt (Fig. 1) of southwestern British Columbia is the northern extension of the Cascade magmatic arc of the western United States (Green *et al.* 1988; Guffanti & Weaver 1988; Read 1990; Sherrod & Smith 1990; Hickson 1994). It extends from Mount Garibaldi, which is located near the head of Howe Sound, northward through the Salal Glacier volcanic complex to Silverthrone Mountain. Volcanic deposits in the GVB range from Miocene to Holocene in age, and are principally a result of the subduction of the Juan de Fuca plate beneath the North American plate (Green *et al.* 1988; Rohr *et al.* 1996). There are at least eight separate volcanic complexes in the GVB (Table 1), which include stratovolcanoes, isolated flows, domes, spines, cones and tuyas ranging in composition from high-alumina basalt to rhyolite (Fig. 2). The most recent eruption occurred at Mount Meager at 2360 BP (Clague *et al.* 1995; Leonard 1995).

Glaciovolcanic features have been recognized at Watts Point (Bye *et al.* 2000), the Monmouth Creek complex, Garibaldi volcanic field, (Mathews 1951, 1952*a*, *b*, 1958; Green 1977), the Mount Cayley volcanic field (Souther 1980; Kelman *et al.* 2001), Mount Meager and the Salal Glacier region (Lawrence *et al.* 1984; Roddick & Souther 1987; Green *et al.* 1988; Fig. 1). The sizes, ages and lithologic attributes of these landforms are summarized in Table 2.

Isotopic studies on deep-sea sediments show that there have been eight major climatic cycles in southern British Columbia over the past 800 000 years (Shackleton & Opdyke 1973). Many of these cycles were accompanied by widespread, though not necessarily continental-scale, glaciations (Clague *et al.* 1982); ice may have been confined to regions between mountain ranges (Clague 1986). There have been at least two major glaciations in the Canadian Cordillera during the past 100 000 years (Fulton 1984). The Fraser Glaciation, in southwestern British Columbia, began at about 25 000–30 000 BP, peaked at about 14 000–14 500 BP, and was in retreat by about 10 000 BP (Fulton 1971; Clague 1980, 1981). The timing of earlier glaciations is poorly known, but stratigraphic evidence for pre-Wisconsinan ice sheets in the western Canadian Cordillera does exist (e.g. Denton & Stuiver 1967; Klassen 1978; Armstrong 1981). Past glacial episodes and their timing with respect to glaciovolcanic events in the GVB are shown in Figure 3. Further dating of glaciovolcanic deposits within southwestern British Columbia would elucidate Quaternary glaciochronologies.

Glaciovolcanic landforms

Tuyas

Within the GVB, there are a number of tuya-shaped edifices which lack the internal stratigraphy described for basaltic tuyas (Mathews 1947; Jones 1966, 1969). These volcanic landforms, herein termed flow-dominated tuyas, comprise masses of intensely jointed, intermediate composition lava flows that have flat tops and steep sides, and have few or no pillow lavas or

Table 1. *Quaternary volcanic centres of the Garibaldi Volcanic Belt*

Major centres and fields	Composition*	Age	Source†
Silverthrone volcanic field	BA to R	1 to 0.4 Ma	1
Franklin Glacier complex	D, A	3.9 to 2.2 Ma	1
Salal Glacier volcanic field	AOB, H	1 to 0.6 Ma	2
Mount Meager–Elaho Valley	B to D	2.2 Ma to 2.4 ka	3, 4, 5
Mount Cayley volcanic field	BA to RD	0.5 to 0.3 Ma	3
Garibaldi volcanic field	B to D	1.3 Ma to 10 ka	3, 6, 7
Monmouth Creek complex	BA to D	nd‡	8
Watts Point volcanic field	D	130 to 90 ka	3

*AOB, alkali olivine basalt; H, hawaiite; B, basalt; BA, basaltic andesite; A, andesite; D = dacite; RD, rhyodacite; R, rhyolite.
† 1, Souther & Yorath 1991; 2, Lawrence *et al.* 1984; 3, Green *et al.* 1988; 4, Leonard 1995; 5, Clague *et al.* 1995; 6, Mathews 1958; 7, Brooks & Friele 1992; 8, Green 1994.
‡ nd, not determined.

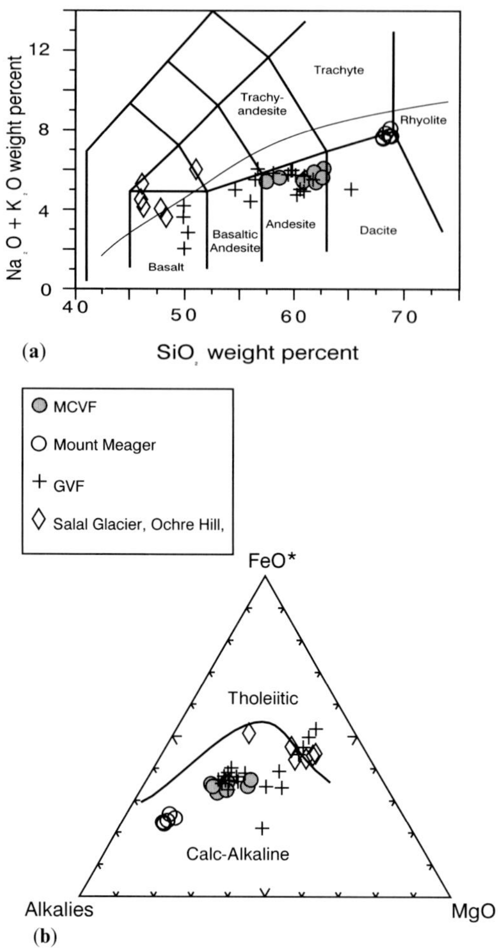

Fig. 2. Chemical classification of volcanic rocks from the Garibaldi Volcanic Belt. Data are for the Mount Cayley volcanic field (Kelman *et al.* 2001), Mount Meager (Stasiuk & Russell 1990; Hickson *et al.* 1999), Garibaldi volcanic field (Mathews 1957; Green 1977), and Salal Glacier volcanic field, Ochre Hill and Logan Ridge (Lawrence *et al.* 1984). Abbreviations used are the same as in Figure 1. (**a**) Total alkalis v. silica. Fields are from LeBas *et al.* (1986). The curved line is the alkaline–subalkaline division of Irvine & Baragar (1971). (**b**) AFM diagram (Irvine & Baragar 1971), showing the dominance of calc-alkaline compositions in the Garibaldi Volcanic Belt.

fragmental deposits. The Table (Fig. 4a), located to the north of Mount Garibaldi (Fig. 1), exemplifies this type of landform. It consists of a nearly vertical-sided stack of flat-lying hornblende-phyric andesite lava sheets, partially coated by thin, nearly vertical, lava sheets. Horizontal or nearly horizontal columns occur at numerous locations along the periphery of the mass.

Two other flow-dominated tuyas occur in the northern Mount Cayley volcanic field (Fig. 1). Ring Mountain is a nearly circular feature of unknown age, composed of a sequence of flows of glassy, plagioclase–orthopyroxene–phyric andesite. The uppermost, horizontal flows are coarsely jointed and separated by layers of oxidized scoria. On Ring Mountain's western side, outcrops contain highly variable, fine-scale jointing and are locally broken down into many small spires and knobs. Little Ring Mountain (Fig. 4b, c) is smaller but otherwise similar in morphology and jointing characteristics to Ring Mountain, and may be of similar age. Its eruptive products are fine-grained to glassy black plagioclase-augite-phyric andesite flows (Kelman *et al.* 2001).

Cauldron Dome, in the central Mount Cayley volcanic field (Fig. 1), has a complex structure. It has an elliptical shape in plan view, and comprises a flat-topped stack of hypersthene-phyric andesite flows. Lower flows are coarsely jointed and commonly oxidized; their margins form vertical cliffs. The upper Cauldron Dome flows have more intense, variable, and fine-scale jointing than the lower flows, are commonly glassy, and form near-vertical cliffs featuring many small spires and knobs.

The shapes and joint characteristics of flow-dominated tuyas cannot be explained by the apparent palaeotopography. For example, these steep sided masses were clearly not the result of lava filling palaeovalleys. Thus, they are inferred to have originated from subglacial eruptions.

The top surfaces of the Table, Ring Mountain, and Little Ring Mountain represent subaerial emplacement, and these edifices are likely monogenetic. Cauldron Dome, however, probably results from multiple eruptions. Its lower flows appear to be subaerial; these flows form precipitous cliffs whose steepness is a result of erosion, and there is no evidence for quenching by ice. The upper Cauldron Dome flows appear to have formed subglacially, although it is unclear whether the uppermost surface of Cauldron Dome represents subglacial or subaerial eruption.

Flow-dominated tuyas, thus, are characterized by (a) stacked, relatively flat-lying lava flows; (b) upper surfaces which are subaerial in origin; (c) horizontal columns at the lateral margins, indicating vertical cooling surfaces; (d) irregular, fine-scale, columnar jointing at the lateral margins, indicating rapid cooling; and (e) edifice shapes and joint patterns which cannot be explained by known palaeotopography. These features are consistent with subglacial eruptions that ultimately breached the ice surface.

Table 2. *Summary and description of glaciovolcanic features in the Garibaldi Volcanic Belt*

Name	Feature*	Rock type†	Area (m)	Height (m)	Age/Glacial period	Source‡
Tuber Hill	T	B	2000 × 2000	<450	0.598 ± 0.015; 0.731 ± 0.037; 0.760 ± 0.033	1
Little Ring Mountain	T	A	120 × 120	270	?	2
Ring Mountain	T	A	200 × 200	590	?	2
Slag Hill	SGD, T	A, BA	1540 × 3000	<700	?	3
Cauldron Dome	T	A	4100 × 2100	<400	0.49 ± 0.08	3, 4
Ember Ridge (6 domes)	SGD	A	180 × 260 to 1060 × 1080	<300	?	3
Cheakamus Valley flows	SGF	B	<1600 × 50–100	10–20	Fraser Glaciation	5, 6
Mount Brew	SGD	A	?	?	?	4
Sphinx Moraine complex	SGD	BA	?	>30	pre-Fraser Glaciation	6
Table	T	A	300 × 180	250	Fraser Glaciation	7
Columnar Peak	SGD	D	?	?	?	4, 8
Eenostuck Mass	SGD	BA	800 × 180	90	Fraser Glaciation?	8
Monmouth Creek complex	?	D, BA	1000? × 600?	~150	?	8
Watts Point	SGD	D	700 × 600	< 100	0.09 ± 0.03; 0.13 ± 0.03	4, 9

* T, tuya; SGD, subglacial dome; SGF, subglacial flow.
† B, basalt; BA, basaltic andesite; A, andesite; D, dacite.
‡ 1, Roddick & Souther 1987; 2, Kelman *et al.* 2001; 3, Souther 1980; 4, Green *et al.* 1988; 5, Mathews 1952*b*; 6, Green 1977; 7, Mathews 1951; 8, Mathews 1958; 9, Bye *et al.* 2000.

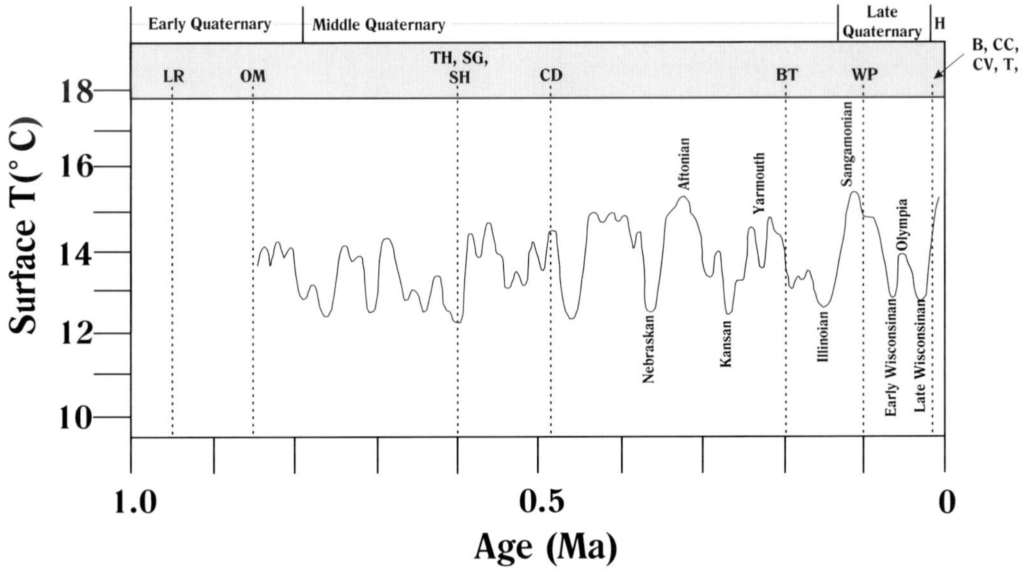

Fig. 3. Timing of Garibaldi Volcanic Belt glaciations and glaciovolcanic events during the last 1 million years. Curve shows surface temperature variations during the last 850 000 years (Gribbin 1990). Peaks represent interglacial intervals whereas troughs represent glacial intervals. Dashed vertical lines show glaciovolcanic events in the Garibaldi Volcanic Belt whose ages are known (Table 2), using the abbreviations in Figure 1. Uncertainties in K–Ar dates are variable and are as great as 0.1 Ma for some dates, but are not shown for simplicity. Some events for which there are no radiometric dates available are placed in the Late Wisconsinan because they contain glaciovolcanic features with no evidence of subsequent glacial overriding. H, Holocene.

Fig. 4. Glaciovolcanic features in the Garibaldi Volcanic Belt. (**a**) The Table, a flow-dominated tuya in the Garibaldi volcanic field. Its height is approximately 260 m. (**b**) Columnar joints on the east side of Little Ring Mountain. Columns at the top are subvertical, whereas columns near the bottom are nearly horizontal. Columns near the bottom are approximately are 30–50 cm across. Scale is also indicated by the climber at the lower left of the image. (**c**) Little Ring Mountain, a flow-dominated tuya in the northern Mount Cayley volcanic field. Its height is approximately 250 m. (**d**) One of the six subglacial domes of the Ember Ridge deposit in the Mount Cayley volcanic field. Height of the visible outcrop is approximately 30 m. (**e**) The Barrier, an ice-marginal flow in the Garibaldi volcanic field, which truncates abruptly in a 200 m cliff overlooking the valley floor. The photograph is taken standing down-valley, on deposits from a major landslide from the Barrier during the mid-nineteenth century. (**f**) Ice-contact cliffs at the eastern margin of Pali Dome, approximately 100 m high. Material at the top of the cliffs, and the debris below, is glassy. Jointing along the cliff tops is fine-scale, and column orientations are locally variable.

Subglacial domes

Subglacial domes are another prominent morphological feature common to the GVB, consisting of steep-sided, dome-shaped lava flows with fine-scale jointing. Pillows and fragmental material, such as hyaloclastite, are scarce or absent.

The Mount Cayley volcanic field hosts several good examples of subglacial domes. Ember Ridge (Fig. 4d), consists of six irregular masses of aphanitic to glassy hornblende–phyric andesite. Each mass has a colloform surface, with margins consisting of steeply inclined or vertical cooling units; individual flows are up to 60 m thick, and the summit regions of most outcrops are eroded into irregular knobs and spires. Jointing is fine-scale and complex; columns average 15–25 cm in diameter, although they are commonly significantly smaller (i.e. 5–10 cm) near the top surfaces of some flows.

A more complex series of flows occurs at Slag Hill, near the northern end of the Mount Cayley volcanic field (Fig. 1), the easternmost lobe of which has the characteristics of a subglacial dome. Slag Hill consists of piles of black, glassy, bulbous augite-phyric basaltic andesite flows. Columns have small diameters, variable orientations, and commonly occur as locally radiating masses. The northwestern lobe of Slag Hill is nearly flat-topped, with near-vertical, finely-jointed margins, while the eastern lobe has a more rounded morphology. A small, isolated bluff at the northern end of the Slag Hill pile is flat-topped and steep-sided, and has fine-scale irregular jointing around its margins.

The region around Mount Garibaldi (Fig. 1) also contains numerous dome-like masses of andesite or dacite lava at Mount Brew, the Eenostuck complex, the Sphinx Moraine complex, Round Mountain, Columnar Peak and on the ridge south-east of the Table (Mathews 1958; Green 1977).

The Watts Point volcanic field is a pile of flows at the southernmost end of the GVB (Fig. 1). It comprises 0.02 km^3 of heavily-jointed, sparsely hornblende or pyroxene-phyric dacite lava and breccia (Bye *et al.* 2000). Evidence of quenching is common; columns are 5–40 cm in diameter, with locally radiating patterns, and flows are glassy to fine-grained.

The Monmouth Creek complex (Fig. 1) is a prominent, complex, and enigmatic feature of unknown age west of Squamish. It consists of at least four en echelon dikes which form the ribs of spires rising 60–180 m above a main lava mass (Mathews 1958). The tallest of these spires, the Castle, features continuous horizontal, radiating columnar joints. The spires are mantled by agglutinated breccia near their bases, and jointing extends into the agglutinate. The uppermost flows and spires are dacite (Mathews 1958; Green 1994).

All of the features described above, based on their morphologies, and irregular, commonly vertical cooling surfaces, likely formed by eruption into a confined setting. They cannot be explained by the apparent palaeotopography, and are inferred to have been confined by ice. The Monmouth Creek complex probably represents a series of subglacial domes and dykes, while the other features are subglacial domes.

Vent locations appear to be beneath or near most subglacial domes, based on jointing patterns and the glassiness of some lavas (e.g. Ember Ridge, Fig. 1), and many subglacial domes represent a single extrusive episode. The Ember Ridge deposits likely represent discrete extrusive events which were approximately coeval, based on their similar morphologies, jointing styles, and degrees of erosion (Souther 1980). The Slag Hill deposits are more complex and may represent multiple eruptions in the same location; the western lobe and isolated bluff of Slag Hill have peripheries consistent with eruption against ice, although their top surfaces suggest that eruptions breached the ice surface, and are thus similar to flow-dominated tuyas. The north-eastern lobe of Slag Hill, however, has the characteristics of a subglacial dome.

Characteristics of subglacial domes are (a) fine-scale (<25 cm) columnar jointing; (b) horizontal columns, indicating vertical cooling surfaces; (c) other fine-scale jointing (e.g. flaggy jointing); (d) irregular joint patterns, or joints forming radiating masses, on the top surfaces or sides of flows, indicating locally irregular cooling surfaces; (e) scarcity or lack of primary fragmental material; and (f) edifice shapes and joint patterns that cannot be explained by the apparent palaeotopography. Subglacial domes are interpreted as lava masses that did not breach the ice surface during eruption.

Subglacial flows

Subglacial flows are laterally extensive lava flows with steep, horizontally-jointed margins. The Cheakamus River Valley, on the western margin of the Garibaldi volcanic field (Fig. 1), contains the only known example of this feature. The Cheakamus River Valley has been subjected to episodic eruptions of flat-lying basalt flows. The uppermost flows consist of olivine-plagioclase-phyric basalts with sinuous, anastomosing outcrop patterns, and are commonly described as 'esker-like' (Mathews 1958). An underlying till is radiocarbon dated at 34 200 ± 800 BP, which correlates with the Olympia Interstade, a non-glacial interval which immediately preceded the Fraser Glaciation (Fulton *et al.* 1976). Columnar jointing is ubiquitous, and is horizontal along the steep sides of the flows and vertical beneath the blocky flow tops. At several locations, pillows or pillow-like features are present in the basal parts of the flows, and some portions of flows are underlain by hyaloclastite breccia.

The Cheakamus Valley lavas appear to have been highly fluid, based on their flow termini, which are commonly less than a metre in thickness. They appear to have been emplaced subglacially, based on the age of the underlying till, the presence of pillows near the bases of some flows, which indicates subaqueous eruption, the horizontal jointing at flow margins, which indicates a vertical cooling surface, and the scale of jointing, which indicates rapid cooling. The source of the Cheakamus Valley flows is unknown (Green *et al.* 1988).

Subglacial flows are characterized by (a) fine-scale jointing; (b) horizontal columns at lateral margins; (c) locally irregular or radial jointing; and (d) outcrops with low aspect ratios, which cover wide areas, and cannot be explained by the apparent palaeotopography. These flows are inferred to have been extruded subglacially, and to have flowed some distance from the vent in tunnels or trenches within the ice sheet (Mathews 1958).

Ice-marginal flows

Ice-marginal flows have subaerial origins, but have steep or vertical, intensely jointed, over-thickened margins, resulting from interaction with ice.

The best-described example of an ice-marginal flow in the GVB is the Barrier (Fig. 4e), located to the NE of Mount Garibaldi, near the Mount Price complex (Fig. 1). An oxidized, rubble-topped, andesite flow stretches from a cone on the western side of Mount Price, Clinker Peak, into the upper reaches of the Rubble Creek Valley. There it terminates in a 200 m cliff, surrounded by debris. A similar cliff occurs in the upper reaches of the Culliton Creek Valley. The current face of the Barrier is a scar from a major landslide during the mid-nineteenth century (Mathews 1952*a*). Other nearby examples of ice-marginal flows of andesite or dacite occur at Paul Ridge, Glacier Pikes and Black Tusk (Green 1977; Green *et al.* 1988).

Pali Dome, in the Mount Cayley volcanic field (Fig. 1) consists of coarsely plagioclase–hypersthene ± hornblende–phyric andesite flows whose apparent source is currently ice-covered (Souther 1980). Proximal parts of flows have large diameter columns (up to 90 cm) overlying scoriaceous, oxidized, flow breccia. Distal portions of flows, however, are finely-jointed; column diameters are commonly less than 20 cm and may be horizontal or in complex radiating masses. Flow termini comprise nearly vertical cliffs which are 100–200 m high, and are locally broken into small spires and knobs (Fig. 4f). The terminal cliffs of the south-eastern flows are flanked by more than 70 m of glassy debris.

Several ice-marginal glaciovolcanic features occur in the Salal Glacier area north of Mount Meager (Fig. 1). This small volcanic field is compositionally distinctive because its lavas are alkaline (Fig. 2). At Logan Ridge, two basanitoid pillowed flows, separated by an 8 m-thick tuff layer, form a steep cliff face (R. B. Lawrence pers. comm., 1979; Lawrence *et al.* 1984). At Ochre Mountain, south-east of Logan Ridge (Fig. 1), there is a thin (<8 m thick) alkali olivine basalt flow with a brecciated (and partially palagonitized) base. The flow shows pervasive but poorly-developed jointing.

Ice-marginal lava flows are inferred to have formed when subaerially-erupted lavas flowed downhill and pooled against ice which occupied lower elevations, based on their joint sizes and orientations, and the lack of palaeotopographic features against which they could have cooled; the Barrier was the first such recognized feature in the GVB (Mathews 1952*a*). Flows such as those at the Barrier and in the Culliton Creek Valley clearly followed the retreat of ice from higher elevations, based on the lack of ice-contact or glacial erosion features at these altitudes. They must, however, have predated the total disappearance of the ice sheet and thus are inferred to have formed during the waning stages of the Fraser Glaciation (Mathews 1952*a*). The tuff layer at Logan Ridge was interpreted as airfall by Lawrence (pers. comm., 1979). However, its thickness, partial palagonitization, relationship to the pillow lavas above and below, and the nearby presence of hyaloclastite suggest that it could be waterlain, suggesting that the Logan Ridge deposit is also an ice-marginal flow.

Ice-marginal flows are useful in delineating the margins of ice sheets. Evidence for ice impoundment of flow fronts includes (a) abundant small-diameter, chaotically-oriented, columnar joints; (b) fine-scale horizontal jointing; and (c) unusually thick and vertically-faced flow termini which cannot be explained by the apparent palaeotopography. These features are diagnostic, but can easily be lost through erosion because of the inherently unstable nature of ice-impounded flows.

Comparisons with other settings

The best analogues for the glaciovolcanic features found in the GVB derive from the United States Cascades, which has similarities in tectonic setting, lava compositions, topography, and glacial history. Lescinsky & Fink (2000) described many glaciovolcanic features in the Cascades which appear to be similar to those of the GVB.

Tuyas similar to the flow-dominated tuyas of the GVB have not yet been described from other glaciovolcanic regions. This is probably because most descriptions are from regions that are dominated by basaltic edifices (e.g. Jones 1969; Skilling 1994; Smellie & Hole 1997). The only conventional tuyas within the GVB are of basaltic composition (e.g. Tuber Hill; Roddick & Souther 1987). This suggests a correlation between morphology and composition. Most of the non-basaltic glaciovolcanic systems studied have been rhyolites (e.g. Furnes *et al.* 1980, Tuffen *et al.* 2001), and potential analogues to flow-dominated tuyas have not been identified in these systems. Phonolitic glaciovolcanism has been investigated at Hoodoo Mountain volcano in northwestern British Columbia, a steep-sided edifice that, beneath its ice cap, is flat-topped (Edwards & Russell 1994, 1995, 2002; Edwards *et al.* 2002). However, this is a long-lived stratovolcano, which is a poor analogue for the smaller, shorter-lived, monogenetic intermediate centres of the GVB.

The best analogues for the subglacial domes of the GVB are andesitic to rhyolitic domes formed by lava–ice interaction in the United States Cascades, but few studies have focused on their glaciovolcanic attributes (Guffanti & Weaver 1988; Lescinsky & Fink 2000). GVB subglacial domes are unlike the more extensively studied rhyolitic subglacial edifices of Iceland (Furnes *et al.* 1980; Tuffen *et al.* 2001). Icelandic subglacial rhyolitic volcanoes, such as Bláhnúkur, in south-central Iceland, commonly contain lava lobes 5–10 m long set in breccia or hyaloclastite (Furnes *et al.* 1980; Tuffen *et al.* 2001), whereas subglacial domes in the GVB consist of coherent lava masses several hundred metres across, with little or no associated pyroclastic material. At Bláhnúkur, however, there are also several larger (up to 400 m long and 20 m thick) columnar jointed rhyolite flows whose margins indicate cooling against steeply inclined ice walls, and these flows appear to be grossly similar to lava domes and masses found in the GVB (e.g. Ember Ridge).

Features analogous to the Cheakamus River Valley esker-like flows have not yet been identified outside the GVB. Their morphology and outcrop patterns may result from relatively rare events where special conditions for ice thickness, effusion rate, lava properties, and slope, are met. Conversely, the absence of these flow features in the literature may be a reflection of their fragility; the Cheakamus Valley flows are relatively thin, and much of the lava is pervasively jointed, making it easily eroded.

Features indicative of ice-marginal volcanism have been identified in many settings but are notably abundant in the United States Cascades (e.g. Conrey 1991; Lescinsky & Sisson 1998; Lescinsky & Fink 2000) where the relief and topography are similar to that found in the GVB.

Discussion

The differences in morphology and internal characteristics between GVB glaciovolcanic landforms and those described in other regions appear to be primarily a function of composition. Composition controls many important lava properties, including liquidus temperature, viscosity, volatile content, heat capacity, and glass transition temperature. Ultimately, these properties control effusion rates and eruption style and, thus, the forms of many volcanic deposits. In subaqueous settings, for example, viscous siliceous lava flows tend to be thicker and have larger pillows than basaltic flows (Pichler 1965; Furnes *et al.* 1980).

Topography plays a lesser but still significant role in the development of glaciovolcanic deposits in the GVB. Firstly, steep topography enhances erosion rates. Secondly, in an intermittently-glaciated setting such as the GVB, where there are many high-altitude vents, instances of subaerially erupted lavas moving downslope and ponding against ice should be common. This process is recorded, both in the GVB and the Cascades, by numerous exposures of ice-marginal lava flows. Thirdly, topography must influence the ice depth at which eruptions occur. Glaciers should be thickest in the valleys and thinner at high elevations. Eruptions occurring at high elevations, therefore, have less ice available to be melted and are more likely to breach the ice surface. Finally, steep topography will enhance meltwater drainage. The ability of a subglacial volcano to melt and retain water is an important control on its final lithological characteristics.

Heat transfer considerations

Allen (1980) investigated the heat balance for subglacial basaltic eruptions. He demonstrated that, for average magmatic temperatures and effusion rates, the heat derived from basalt is more than sufficient to melt an ice cavern large enough to accommodate the growing lava pile. However, the heat budgets attending the eruption of mafic and felsic magmas beneath ice can be substantially different (e.g. Hoskuldsson & Sparks 1997).

Heat transfer differences between mafic and felsic magmas arise because of their differences in eruption temperature (*Te*), and the magnitude of the interval between *Te* and the calorimetrically defined glass transition temperature (*Tg*). The value of *Tg* marks the transition from the melt to the glassy state and is defined by marked changes in molar heat capacity, thermal expansivity, and other second order thermodynamic properties; it also corresponds to an increase in viscosity (typically $>10^{13}$ Pa s^{-1}). The calorimetric *Tg* is an important limiting value for magmatic processes because, above *Tg*, rates of nucleation, crystallization, and vesiculation are fast enough to compete with most magmatic and eruptive time scales. Conversely, at temperatures below *Tg*, glass forms, and crystallization and vesiculation may be suppressed. *Tg* is important to heat budgets because crystallization is suppressed below this temperature. This eliminates potential latent heats of crystallization (*L*). Latent heats of vitrification are essentially zero and, thus, the transition from melt to glass does not contribute to the heat budget.

We have calculated the liquidus temperatures and glass transition temperatures for a variety of mafic to intermediate volcanic rock types (Table 3). The basalt compositions are based on glass analyses from Icelandic basalts erupted at the Laki fissure (Thordarson *et al.* 1996) and Herdubreid (Moore & Calk 1991). The intermediate composition analyses are of andesite and dacite samples from the GVB (Fiesinger 1975; Kelman *et al.* 2001). Liquidus temperatures are calculated using MELTS: a model used for the computation of multi-component phase

Table 3. *Compositions of lavas from Garibaldi volcanic belt and Iceland used to compare the differences in liquidus temperatures and glass transition temperatures*

Sample† Rock type†	KR-1[1] A	KR-2[1] A	KR-4[1] A	KR-10[1] D	KR-11[1] D	GV31[2] D	GV22[2] D	gl-1[3] B	gl-4[3] B	Hr[3] B
SiO_2	62.18	58.63	62.60	65.99	67.57	64.41	64.81	49.68	49.13	48.70
TiO_2	0.63	0.64	0.62	0.45	0.41	0.52	0.37	2.96	2.00	1.38
Al_2O_3	17.97	18.09	17.93	16.72	16.26	16.38	16.32	13.05	12.03	15.40
Fe_2O_3	4.90	5.83	4.94	4.10	3.68	1.32	1.59	–	–	–
FeO	–	–	–	–	–	2.64	2.20	13.78	13.41	10.70
MnO	0.08	0.10	0.08	0.08	0.08	0.04	0.01	0.22	0.23	0.18
MgO	2.27	4.28	2.25	1.88	1.60	2.46	1.41	5.78	8.47	8.23
CaO	5.73	6.29	5.74	4.27	3.88	5.03	4.37	10.45	11.37	12.80
Na_2O	4.38	4.68	4.37	4.55	4.46	4.92	4.82	2.84	2.47	2.08
K_2O	1.55	1.01	1.56	1.79	1.89	1.63	1.73	0.42	0.39	0.13
P_2O_5	0.27	0.25	0.27	0.18	0.16	0.17	0.20	0.28	0.17	0.16
H_2O^-	0.15	0.26	0.01	0.08	0.22	0.27	1.18	–	–	–
H_2O^+	0.09	0.56	0.11	0.01	0.00	0.07	0.60	–	–	–
Total	100.20	100.63	100.47	100.10	100.21	99.86	99.61	99.46	99.67	99.76
Liquidus (1 atm) and Glass Transition Temperatures (°C) Computed at QFM										
Fe_2O_3‡	0.78	0.92	0.78	0.66	0.59	0.71	0.66	2.33	2.29	1.83
FeO‡	3.71	4.42	3.74	3.1	2.78	3.19	3.04	11.68	11.35	9.05
T_L (anhydrous)	1179	1188	1178	1146	1133	1160	1155	1166	1195	1216
T_L (hydrous)	1162	1135	1170	1140	1118	1136	1047	–	–	–
Tg	1013	1010	1013	1010	1009	1009	1013	900	797	825
T_L–*Tg*	149	125	157	130	109	127	34	266	398	391

* Sample locations: 1, Mount Cayley volcanic field (Kelman *et al.* 2001); 2, Southern Garibaldi Volcanic Belt (Fiesinger 1975); 3, Iceland–Laki (Thordarson *et al.* 1996) and Herdubreid (Moore & Calk 1991).

† A, andesite; D, dacite; B, basalt glass.

‡ Ferric-ferrous contents recalculated at liquidus temperatures for the QFM oxygen buffer.

Table 4. *Physical constants used in calculation of heat budgets attending cooling of basalt and dacite magmas*

	Basalt		Dacite		Ratios dacite/basalt
	Melt	Glass	Melt	Glass	
Te (°C)[1]	1200	–	1125	–	–
Tg (°C)[2]	–	800	–	1000	–
Cp (J mol^{-1} K^{-1})[3]	50.52	44.00	51.80	44.00	–
L (J mol^{-1})[4]	26 200	–	29 139	–	–
% crystallization	30	–	0	–	–
ΔH–Tg (kJ mol^{-1})[5]	28.94	–	6.48	–	0.22
ΔH – total (kJ mol^{-1})[6]	64.14	–	50.475	–	0.79

1, eruption temperature; 2, glass transition temperature; 3, heat capacity; 4, latent heat of crystallization; 5, heat released from Te to Tg; 6, heat released from Te to 0°C.

equilibria in silicate melts (Ghiorso & Sack 1995). All liquidus calculations were performed at 1 bar pressure, and ferric : ferrous ratios of the melts were fixed by imposing QFM buffered oxygen fugacities (Table 4). Values of T_L were calculated for both anhydrous and hydrous (based on the measured H_2O contents) melt compositions.

We have also calculated the values for the glass transition temperatures of these same melts (Table 4) using the empirical model developed by Russell & Nicholls (1992, 1996). This model predicts the thermodynamic glass transition temperature (Tg) as a function of melt composition and is based on a database of over 750 experimental calorimetric heat content measurements on silicate melts and glasses of diverse composition. The model reproduces the experimentally measured values of Tg to within 30°C, although it does not account for the effects of H_2O.

Values of Tg for basaltic liquids are calculated to vary between 800 and 900°C, implying a T_L–T_g interval of 260 to 400°C. Conversely, the calculated values of Tg for the intermediate rocks vary between 1005 and 1015°C, and suggest a much smaller interval between T_L and T_g (105–160°C). Results of the calculations are summarized in Figure 5. Firstly, the calculated values of T_L and Tg are compared directly in Figure 5a. Basaltic melts have higher liquidus temperatures and lower values of Tg than intermediate compositions. The same data are plotted against the melt composition (SiO_2 content) in Figure 5b. In general, T_L decreases with increasing SiO_2 content, while Tg increases with increasing SiO_2. The result is that T_L–T_g decreases markedly as SiO_2 content increases. Melts that have small values of T_L–T_g require only slight cooling to produce glass whilst bypassing crystallization.

Below, we compare semi-quantitatively the energetics of extruding basalt and dacite magmas beneath ice. The heat that could be released by basalt intruding and cooling beneath ice (Fig. 5a) is summarized as:

$$\Delta H_B = \int_{Te}^{Tg} Cp_B \, dT + \int_{Te}^{Tg} L_B \, dT + \int_{Tg}^{Ti} Cp_g \, dT \qquad (1)$$

where Cp_B and Cp_g are the average molar heat capacities (J mol^{-1} K^{-1}) of basalt melt and basalt glass, and L_B is the average molar latent heat of crystallization (J mol^{-1} K^{-1}) (cf. Russell *et al.* 1995). The three integral terms in Equation 1 represent the sensible heat (S, Fig. 6a) released to the ice by cooling of the basalt from Te to Tg, the latent heat (L, Fig. 6a) liberated by crystallization between Te and Tg, and the heat released by conductive cooling of the basalt pile from Tg to the ambient ice temperature (Ti), respectively. Figure 6a shows model results for a basalt erupted at 1200°C and with a Tg of 800°C (Table 3). We have allowed for 30% crystallization, assuming that the accelerated cooling rates will inhibit crystallization even in a basaltic system. These calculations suggest a total heat release of just over 60 kJ mol^{-1} for a basalt magma.

The heat budget for cooling of dacite beneath ice features several differences. Firstly, dacite will have a lower eruption temperature and could have a different heat capacity (Cp_D). More importantly, the interval separating Te and Tg for dacite melts is likely to be much smaller than that found in basalts (Fig. 5, Table 4). For example, a typical dacite may have values of $Te = 1125$°C and $Tg = 1000$°C; thus, the dacite Te–T_L interval is 125°C v. 400°C for basalts (Fig. 5, Table 4). The implication is that there is substantially less opportunity for crystallization of the dacite and this reduces the total heat

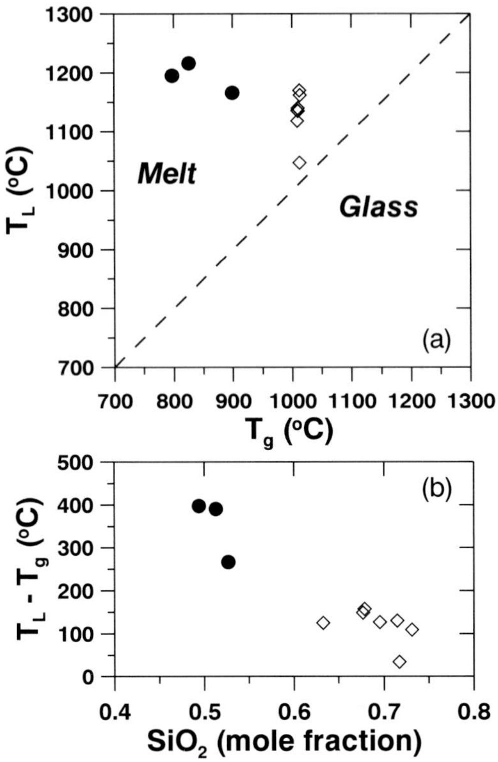

Fig. 5. Comparison of calculated 1-atmosphere liquidus (T_L) and glass transition temperatures (Tg) for melt compositions listed in Table 3. Values of T_L are calculated at 1 bar pressure and at the QFM oxygen buffer using the thermodynamic model MELTS (Ghiorso & Sack 1995). Table 1 reports T_L values for both anhydrous and hydrous compositions; the latter uses the measured H_2O contents of the samples. (**a**) Calculated values of T_L v. Tg for basaltic lavas from Iceland (solid circles) and intermediate lavas from the Garibaldi Volcanic Belt (open diamonds; see Table 4). Dashed line corresponds to equivalence between T_L and Tg where melt would pass through the glass transition temperature. (**b**) Plot showing the difference between T_L and Tg mapped against melt composition (mole fraction SiO_2). In general T_L decreases with increasing SiO_2 content whereas Tg increases with increasing SiO_2 content. The result is that T_L–Tg decreases markedly with increasing SiO_2 content. Melts that have small values of T_L–Tg require only slight cooling to produce glass whilst bypassing crystallization.

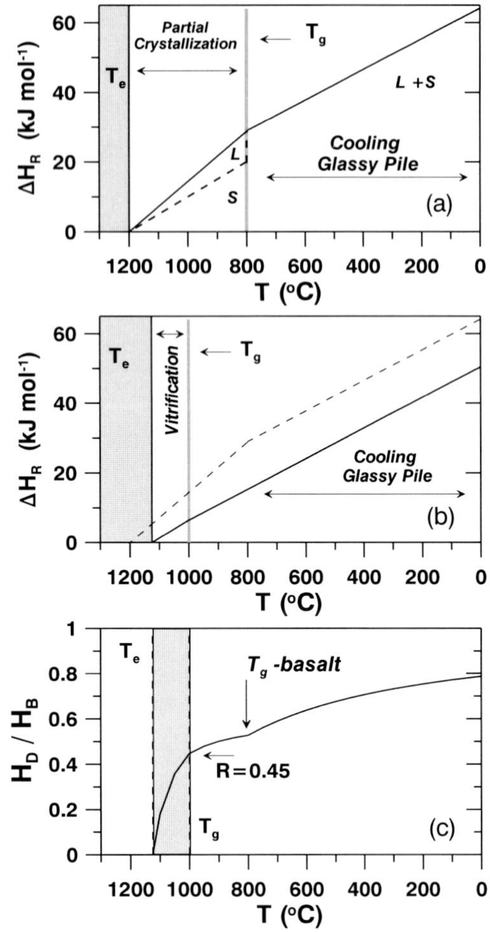

Fig. 6. Model values of heat that would be released by eruption and cooling of lava beneath ice and that could be used to melt ice (see text and Table 4). Thermodynamic calculations are summarized as: (**a**) Heat released (ΔH_R) by a basalt erupted at 1200°C undergoing 30% crystallization before reaching the glass transition temperature ($Tg = 800$°C) and then cooling to ambient ice temperature. Total heat released is a combination of sensible (S) and latent heat of crystallization (L). (**b**) ΔH_R for dacite erupted at 1125°C that reaches its Tg (1000°C) without crystallization and then cools within the ice. Basalt curve (dashed) is shown for reference. (**c**) The ratio of heat released by dacite v. that of basalt plotted against temperature. At the calorimetric Tg, the dacite will have released 45% of the total heat released by the basalt at an equivalent temperature. Over the entire path, dacite releases 70–80% of the total heat released by basalt.

available for melting of the ice. Below, we have modelled the end member system where the accelerated cooling rates prevent crystallization of the dacite by:

$$\Delta H_D = \int_{Te}^{Tg} Cp_D \, \mathrm{d}T + \int_{Tg}^{Ti} Cp_g \, \mathrm{d}T \qquad (2)$$

The results of this computation are shown in Figure 6b (solid line) and are compared to the heat released by cooling of basalt (dashed line). The dacite releases considerably less heat over its cooling history because the total sensible heat loss is less (*Te* is lower) and because there is essentially no latent heat of crystallization.

The ratio of ΔH_R for dacite to basalt is shown in Figure 6c. Over the entire cooling interval, dacite magmas release less total heat (80%) than would corresponding volumes of basalt. However, over the interval *Te*–*Tg*, the heat released by dacite is less than half (45%) of the heat released by basalt over the same temperature interval. Furthermore, at the *Tg* of dacite, the basalt magma is still well above its *Tg*. This is significant because it is the temperature interval above *Tg* that is important to the eruptive process; no crystallization can occur below *Tg*. These calculations point to an important conclusion, that dacite is less able to produce large subglacial caverns; the ice sheet may play a substantially greater role in shaping the volcanic pile than it does in the case of basaltic volcanism.

The scarcity of fragmental deposits

There are apparently few pyroclastic glaciovolcanic deposits (i.e. hyaloclastite) preserved in the GVB, except at basaltic centres. Either such deposits are rare in the GVB, or their scarcity is a result of removal. As such deposits occur in lavas of mafic to felsic compositions elsewhere (e.g. Pichler 1965; Furnes *et al.* 1980; Tuffen *et al.* 2001; Edwards & Russell 2002; Edwards *et al.* 2002), composition cannot be the sole explanation. Below, we discuss the possible factors that may contribute to this apparent scarcity.

Primary fragmentation mechanisms. The formation of primary fragmental volcanic deposits requires the presence of volatile components in the magma or magma–water contact. Kokelaar (1986) divided subaqueous clast-forming processes into four types: (1) explosive release of magmatic volatiles; (2) explosive expansion and collapse of steam formed at magma–water contact surfaces; (3) explosive expansion of steam following enclosure in magma; and (4) cooling-contraction. The first process should depend on the primary volatile content of a magma and its ascent history. The two clast-forming processes related to steam may be potentially controlled by the pressure exerted by overlying ice or water during eruption, or by the amount of direct water–lava contact. The final clast-forming process, cooling-contraction, requires only the presence of water, and may occur at any water depth in a magma of any volatile content.

Primary volatile content and magma ascent history are poorly known for most GVB glaciovolcanic edifices. Green (1977) estimated the water content of Garibaldi Lake region hornblende andesites to be 3.8–4.5 wt% and volatile contents at other GVB volcanoes, although not known, may well be similar. Efficient pre-eruptive degassing would explain a lack of fragmental deposits formed due to primary volatile exsolution. However, there are numerous occurrences of non-glaciovolcanic pyroclastic deposits within the GVB (Mathews 1952*b*, 1958; Read 1977; Souther 1980; Green *et al.* 1988; Stasiuk & Russell 1990; Stasiuk *et al.* 1996; Hickson *et al.* 1999). This suggests that GVB magmas are capable of supporting explosive volcanism. Thus, lack of magmatic volatiles is unlikely to be the reason for the scarcity of intermediate composition fragmental glaciovolcanic deposits in the GVB.

Kokelaar (1986) observed that primary magmatic explosivity and the two steam-related clast-forming processes (magma–water contact surface explosivity and steam entrapment explosivity) are diminished as water depth increases. Thus, great ice thicknesses might inhibit clast formation by these two methods. However, eruptions throughout the GVB have occurred during all phases of glaciation, and eruption must have occurred under ice sheets of a wide range of thicknesses. If ice thickness was the only control on hyaloclastite production, fragmental deposits would form when eruption occurred under thin ice of shrinking glaciers.

If not suppressed by pressure, however, fragmentation related to steam-expansion and cooling-contraction should occur during subglacial eruptions, regardless of a lava's volatile content, unless there is insufficient direct water–lava contact. This would occur if cooling units grew mostly by endogenous intrusion, or if water drained continuously and efficiently from subglacial caverns. Hoskuldsson & Sparks (1997) calculated rates of ice melting in basalt and rhyolite eruptions. They proposed that, if heat exchange efficiency in a basaltic subglacial eruption were greater than approximately 80%, a negative pressure would result for cavities melted into the ice, allowing water to accumulate in the cavity. For less efficient heat exchange, which would be more likely for rhyolitic compositions, cavity pressure would be positive, and water would drain from the cavity. These results suggest that, in, silicic and intermediate subglacial eruptions, less water may be present at vents. This could explain the scarcity of fragmental material in the resultant deposits. Additionally, as discussed above, less heat will be released during eruption of intermediate lavas than in eruption of basaltic lavas, so less water will be produced through melting, regardless of whether it accumulates or drains. Finally, vents that are perched at high elevations, situated under thin ice and surrounded by steep slopes, are unlikely to support the accumulation of meltwater; the permeable ice and steep topography will combine to keep the ice mass over the vent well-drained (Smellie *et al.* 1993; Smellie & Skilling 1994; Smellie 2000).

GVB subglacial domes lack hyaloclastite, do not form pillows, have steep flow margins featuring fine-scale, chaotically-oriented cooling joints, and commonly have bulbous or colloform surfaces. These features suggest that the lavas have been efficiently impounded by ice but have not necessarily been erupted into standing bodies of water. Outside the GVB, a possible example of eruptive products shaped more by ice than by water occurs at Bláhnúkur, in south-central Iceland, where columnar jointed rhyolite flows show limited evidence for direct lava–water interaction. Calculated potential flow velocities at Bláhnúkur are greater than melting rates would have been by 3–4 orders of magnitude, indicating that ice would have constrained the lava, an hypothesis which is supported by field observations such as joint orientations (Tuffen *et al.* 2001).

Preservation potential. An alternative explanation for the scarcity of fragmental deposits at glaciovolcanic edifices within the GVB may be erosion. Due to high uplift rates and steep, irregular, topography, rates of mass wasting in the GVB are extremely high, and most of the large edifices, such as Mount Cayley, have undergone multiple landslide events (Clague & Souther 1982; Jordan 1987; Evans & Brooks 1991; Lu 1992). The poor consolidation of unwelded fragmental deposits makes them extremely vulnerable to erosion. Additionally, the proportion of supraglacial eruptions, some of which may have been explosive, may be much greater than is suggested by the geologic record, because such deposits are very unlikely to be preserved. Thus, the record of glaciovolcanism in the GVB may not be representative of the original size and number of deposits. However, since fragmental glaciovolcanic deposits do occur at basaltic edifices in the GVB, erosion cannot be the only explanation for their scarcity at intermediate composition edifices.

In summary, the scarcity of fragmental glaciovolcanic deposits in the GVB could potentially be attributed to one or more of the following: suppression of primary magmatic explosivity through insufficient primary volatile content of magmas, pre-eruptive degassing, or high ambient pressure due to thick ice or deep water; suppression of steam-expansion related explosive processes through eruption under thick ice or deep water, or insufficient direct contact between erupting magma and water (due to endogenous growth of cooling units or to efficient draining of water during eruption); suppression of cooling-contraction fragmentation due to insufficient direct contact between magma and water; or, lack of deposit preservation. The factor most likely to be responsible for the scarcity of fragmental glaciovolcanic deposits in the GVB is lack of direct lava–water contact; lack of deposit preservation is probably also influential. Further data on magma volatile contents and effusion rates for the GVB should help to clarify this issue.

Conclusions

Glaciovolcanic processes in the GVB are distinctive, and the products are dominated by landforms different from those of more thoroughly studied regions (e.g. Iceland, the Tuya–Teslin region of northern British Columbia, etc.). Tuyas in the GVB, except those of basaltic composition, are flow-dominated, featuring atypical internal stratigraphy which consists of stacks of finely jointed flows without pillows and with little or no hyaloclastite, and are products of subglacial eruptions which ultimately breached the ice surface. Subglacial domes (rounded, steep-sided, finely jointed lava piles which likely accumulated above or near vents during entirely subglacial eruptions) and ice-marginal features, formed when lava erupted from subaerial vents and pooled against ice downslope, are also common.

The most important control on glaciovolcanic edifice morphology is probably composition, because intermediate lavas, being more viscous and closer to their glass transition temperatures at the time of eruption, release less sensible heat in cooling to the ambient ice temperature, and release less latent heat from crystallization. The efficiency of heat transfer will also influence the accumulation or drainage of meltwater around the vent. Thus, intermediate composition lavas are more likely to pile up as domes directly over vents (as they do subaerially), and edifice morphologies are likely to be strongly influenced by the overriding ice during eruption. The scarcity of fragmental glaciovolcanic deposits at intermediate composition centres in the GVB is probably due to minimization of lava–water interaction during eruption, and to high erosion rates.

Further analysis of glaciovolcanism in the GVB is ongoing and includes: (1) more detailed mapping and sampling of specific glaciovolcanic features for which there are few data; (2) detailed comparative study of the many flow-dominated tuyas and subglacial domes throughout the GVB; (3) quantitative thermodynamic modelling of the interactions between intermediate lava and ice. This ongoing analysis of glaciovolcanism in southwestern British Columbia is also expected to contribute substantially to climate studies by providing additional radiometric dates for glaciovolcanic deposits. These results, when combined with existing databases, will refine our current knowledge of ice distributions in the Canadian Cordillera throughout space and time and will constrain the timing of major glacial events.

This research was supported by the NSERC Research Grants program (R89820 to J. K. Russell) and the Geological Survey of Canada through Project No. 303071 (C. J. Hickson). Funding for the senior author's participation in the Conference on Volcano–Ice Interaction derived, in part, from the Department of Earth & Ocean Sciences at the University of British Columbia. The senior author thanks D. Lui for field assistance, L. Fox for logistical support, and the reviewers J. Moore, J. Stix, and especially J. Smellie. Sampling of volcanic centres in the northern portion of the GVB was made possible through collaboration with N. Green.

References

ALLEN, C. C. 1980. Icelandic subglacial volcanism: thermal and physical studies. *Journal of Geology*, **88**, 108–117.

ARMSTRONG, J. E. 1981. Post-Vashon Wisconsinan glaciation, Fraser Lowland, British Columbia. *Geological Survey of Canada Bulletin*, **322**.

BROOKS, G. R. & FRIELE, P. A. 1992. Bracketing ages for the formation of the Ring Creek lava flow, Mount Garibaldi volcanic field, southwestern British Columbia. *Canadian Journal of Earth Sciences*, **29**, 2425–2428.

BYE, A., EDWARDS, B. R. & HICKSON, C. J. 2000. Preliminary field, petrographic, and geochemical analysis of possible subglacial, dacitic volcanism at the Watts Point volcanic centre, southwestern British Columbia. *Geological Survey of Canada, Current Research*, **2000-A20**.

CLAGUE, J. J. 1980. Late Quaternary geology and geochronology of British Columbia. Part 1: radiocarbon dates. *Geological Survey of Canada Paper*, **80-13**.

CLAGUE, J. J. 1981. Late Quaternary geology and geochronology of British Columbia. Part 2: summary and discussion of radiocarbon-dated Quaternary history. *Geological Survey of Canada Paper*, **80-35**.

CLAGUE, J. J. 1986. The Quaternary stratigraphic record of British Columbia – evidence for episodic sedimentation and erosion controlled by glaciation. *Canadian Journal of Earth Sciences*, **23**, 885–894.

CLAGUE, J. J., HARPER, J. R., HEBDA, R. J. & HOWES, D. E. 1982. Late Quaternary sea levels and crustal movements, coastal British Columbia. *Canadian Journal of Earth Sciences*, **19**, 597–618.

CLAGUE, J. J. & SOUTHER, J. G. 1982. The Dusty Creek landslide on Mount Cayley, British Columbia. *Canadian Journal of Earth Sciences*, **19**, 524–539.

CLAGUE, J. J., EVANS, S. G., RAMPTON, V. N. & WOODSWORTH, G. J. 1995. Improved age estimates for the White River and Bridge River tephras, western Canada. *Canadian Journal of Earth Sciences*, **32**, 1172–1179.

CONREY, R. M. 1991. *Geology and petrology of the Mt. Jefferson area, High Cascade Range, Oregon*. PhD thesis, Washington State University.

DENTON, G. H. & STUIVER, M. 1967. Late Pleistocene glacial stratigraphy and chronology, northeastern St. Elias Mountains, Yukon Territory, Canada. *Geological Society of America Bulletin*, **78**, 485–510.

EDWARDS, B. R. & RUSSELL, J. K. 1994. Preliminary stratigraphy of Hoodoo Mountain volcanic centre, northwestern British Columbia. *Cordilleran and Pacific margin*. Geological Survey of Canada, Current Research, **1994-A**, 69–76.

EDWARDS, B. R. & RUSSELL, J. K. 1995. Revised stratigraphy for the Hoodoo Mountain volcanic centre, northwestern British Columbia. *Cordilleran and Pacific margin*. Geological Survey of Canada, Current Research, **1995-A**, 105–115.

EDWARDS, B. R. & RUSSELL, J. K. 2002. Glacial influences on morphology and eruptive products of Hoodoo Mountain volcano, Canada. *In*: SMELLIE, J. L. & CHAPMAN, M. G. (eds) *Volcano–Ice Interaction on Earth and Mars*. Geological Society, London, Special Publications, **202**, 179–194.

EDWARDS, B. R., RUSSELL, J. K. & ANDERSON, R. G. 2002. Subglacial, phonolitic volcanism at Hoodoo Mountain volcano, northern Canadian Cordillera. *Bulletin of Volcanology*, **64**, 254–272.

EVANS, S. G. & BROOKS, G. R. 1991. Prehistoric debris avalanches from Mount Cayley volcano, British Columbia. *Canadian Journal of Earth Sciences*, **28**, 1365–1374.

FIESINGER, D. W. 1975. *Petrology of the Quaternary volcanic centres in the Quesnel Highlands and Garibaldi Provincial Park areas, British Columbia*. PhD thesis, University of Calgary.

FULTON, R. J. 1971. Radiocarbon geochronology of southern British Columbia. *Geological Survey of Canada Paper*, **71-37**.

FULTON, R. J. 1984. Quaternary glaciation, Canadian Cordillera. *In*: FULTON, R. J. (ed.) *Quaternary Stratigraphy of Canada: a Canadian Contribution to IGCP Project 24*. Geological Survey of Canada Paper, **84-10**, 39–48.

FULTON, R. J., ARMSTRONG, J. E., FYLES, J. G. & EASTERBROOK, D. J. 1976. Stratigraphy and palynology of late Quaternary sediments in the Puget Lowland, Washington. *Geological Society of America Bulletin*, **87**, 153–156.

FURNES, H., FRIDLEIFSSON, I. B. & ATKINS, F. B. 1980. Subglacial volcanics – on the formation of acid hyaloclastites. *Journal of Volcanology and Geothermal Research*, **8**, 95–110.

GHIORSO, M. S. & SACK, R. O. 1995. Chemical mass transfer in magmatic processes; IV, A revised and internally consistent thermodynamic model for the interpolation and extrapolation of liquid-solid equilibria in magmatic systems at elevated temperatures and pressures. *Contributions to Mineralogy and Petrology*, **119**, 197–212.

GREEN, N. L. 1977. *Multistage Andesite Genesis in the Garibaldi Lake Area, Southwestern British Columbia.* PhD thesis, University of British Columbia.

GREEN, N. L. 1994. Mechanism for middle to upper crustal contamination: evidence from continental-margin magmas. *Geology*, **22**, 231–234.

GREEN, N. L., ARMSTRONG, R. L., HARAKAL, J. E., SOUTHER, J. G. & READ, P. B. 1988. Eruptive history and K–Ar geochronology of the late Cenozoic Garibaldi volcanic belt, southwestern British Columbia. *Geological Society of America Bulletin*, **100**, 563–579.

GRIBBIN, J. R. 1990. *Hothouse Earth.* Grove Weidenfeld, New York.

GUFFANTI, M. & WEAVER, C. S. 1988. Distribution of late Cenozoic volcanic events in the Cascade Range: Volcanic arc segmentation and regional tectonic considerations. *Journal of Geophysical Research*, **93**, 6513–6529.

HICKSON, C. J. 1987. *Late Cenozoic rocks of the Wells Gray – Clearwater area, British Columbia.* PhD thesis, University of British Columbia.

HICKSON, C. J. 1994. Character of volcanism, volcanic hazards, and risk, northern end of the Cascade magmatic arc, British Columbia and Washington state. *In*: MONGER, J. W. H. (ed.) *Geology and Geologic Hazards of the Vancouver Region, Southwestern British Columbia.* Geological Survey of Canada Bulletin, **481**, 231–250.

HICKSON, C. J., MOORE, J. G., CALK, L. & METCALFE, P. 1995. Intraglacial volcanism in the Wells Gray – Clearwater volcanic field, east-central British Columbia, Canada. *Canadian Journal of Earth Sciences*, **32**, 838–851.

HICKSON, C. J., RUSSELL, J. K. & STASIUK, M. V. 1999. Volcanology of the 2350 B.P. eruption of Mount Meager Volcanic Complex, British Columbia, Canada: implications for hazards from eruptions in topographically complex terrain. *Bulletin of Volcanology*, **60**, 489–507.

HOSKULDSSON, A. & SPARKS, R. S. J. 1997. Thermodynamics and fluid dynamics of effusive subglacial eruptions. *Bulletin of Volcanology*, **59**, 219–230.

IRVINE, T. N. & BARAGAR, W. R. A. 1971. A guide to the chemical classification of the common volcanic rocks. *Canadian Journal of Earth Sciences*, **8**, 523–548.

JONES, J. G. 1966. Intraglacial volcanoes of southwest Iceland and their significance in the interpretation of the form of the marine basaltic volcanoes. *Nature*, **212**, 586–588.

JONES, J. G. 1969. Intraglacial volcanoes of the Laugarvatn region, southwest Iceland, I. *Quarterly Journal of the Geological Society, London*, **124**, 197–211.

JONES, J. G. 1970. Intraglacial volcanoes of the Laugarvatn region, southwest Iceland, II. *Journal of Geology*, **78**, 127–140.

JORDAN, P. 1987. Impacts of mass movement events on rivers in the southern Coast Mountains, British Columbia: a summary report. *Environment Canada Water Resources Branch*, **IWD-HQ-WRB-55-87-3**.

KELMAN, M. C., RUSSELL, J. K. & HICKSON, C. J. 2001. Preliminary petrography and chemistry of Quaternary volcanoes of the Mount Cayley volcanic field, British Columbia. *Geological Survey of Canada, Current Research*, **2001-A.**

KLASSEN, R. W. 1978. A unique stratigraphic record of late Tertiary–Quaternary events in southeastern Yukon. *Canadian Journal of Earth Sciences*, **15**, 1884–1886.

KOKELAAR, P. 1986. Magma–water interactions in subaqueous and emergent basaltic volcanism. *Bulletin of Volcanology*, **48**, 275–289.

LAWRENCE, R. B., ARMSTRONG, R. L. & BERMAN, R. G. 1984. Garibaldi Group volcanic rocks of the Salal Creek area, southwestern British Columbia: alkaline lavas on the fringe of the predominantly calc-alkaline Garibaldi (Cascade) volcanic arc. *Journal of Volcanology and Geothermal Research*, **21**, 255–276.

LEBAS, M. J., LE MAITRE, R. W., STRECKEISEN, A. & ZANETTI, B. 1986. Chemical classification of volcanic rocks. *Journal of Petrology*, **68**, 277–279.

LEONARD, E. M. 1995. A varve-based calibration of the Bridge River tephra fall. *Canadian Journal of Earth Sciences*, **32**, 2098–2102.

LESCINSKY, D. T. & FINK, J. H. 2000. Lava and ice interaction at stratovolcanoes: Use of characteristic features to determine past glacial extents and future volcanic hazards. *Journal of Geophysical Research*, **105**, 23 711–23 726.

LESCINSKY, D. T. & SISSON, T. W. 1998. Ridge-forming, ice-bounded lava flows at Mount Rainier, Washington. *Geology*, **26**, 351–354.

LU, Z. Y. 1992. Prehistoric debris avalanches from Mount Cayley volcano, British Columbia: Discussion. *Canadian Journal of Earth Sciences*, **29**, 1342–1343.

MATHEWS, W. H. 1947. 'Tuyas', flat-topped volcanoes in northern British Columbia. *American Journal of Science*, **245**, 560–570.

MATHEWS, W. H. 1951. The Table, a flat-topped volcano in southern British Columbia. *American Journal of Science*, **249**, 830–841.

MATHEWS, W. H. 1952*a*. Ice-dammed lavas from Clinker Mountain, southwestern British Columbia. *American Journal of Science*, **250**, 553–565.

MATHEWS, W. H. 1952*b*. Mount Garibaldi, a supraglacial Pleistocene volcano in southwestern British Columbia. *American Journal of Science*, **250**, 81–103.

MATHEWS, W. H. 1957. Petrology of Quaternary volcanics of the Mount Garibaldi map-area, southwestern British Columbia. *American Journal of Science*, **255**, 400–415.

MATHEWS, W. H. 1958. Geology of the Mount Garibaldi map-area, southwestern British Columbia, Canada. Part II. Geomorphology and Quaternary volcanic rocks. *Bulletin of the Geological Society of America*, **69**, 161–178.

MOORE, J. G. & CALK, L. C. 1991. Degassing and differentiation in subglacial volcanoes, Iceland. *Journal of Volcanology and Geothermal Research*, **46**, 157–180.

PICHLER, H. 1965. Acid hyaloclastites. *Bulletin of Volcanology*, **28**, 293–310.

READ, P. B. 1977. Meager Creek volcanic complex, southwestern British Columbia. *Geological Survey of Canada Paper*, **77-1A**, 277–281.

READ, P. B. 1990. Mount Meager Complex, Garibaldi Belt, southwestern British Columbia. *Geoscience Canada*, **17**, 167–174.

RODDICK, J. C. & SOUTHER, J. G. 1987. Geochronology of Neogene volcanic rocks in the northern Garibaldi Belt, British Columbia. Radiogenic Age and Isotope Studies, Report 1. *Geological Survey of Canada Paper*, **87-2**, 21–24.

ROHR, K. M. M., GOVERS, R. & FURLONG, K. P. 1996. A new plate boundary model for the Pacific-North American-Juan de Fuca triple junction, Slave-Northern Cordillera Lithospheric Evolution (SNORCLE) and Cordilleran Tectonics Workshop. *In*: *Lithoprobe Report 50: British Columbia, Canada*. Lithoprobe Secretariat for the Canadian Lithoprobe Program, 213–214.

RUSSELL, J. K., EDWARDS, B. R. & SNYDER, L. D. 1995. Volatile production possibilities during magmatic assimilation: Heat and mass-balance constraints. *In*: J. F. H. Thompson (ed.) *Short Course on Magmas, Fluids, and Ore Deposits*. Mineralogical Association of Canada, 1–24.

RUSSELL, J. K. & NICHOLLS, J. 1992. The glass transition temperature in natural systems: Empirical calibration and application. *EOS*, **73**, 600.

RUSSELL, J. K. & NICHOLLS, J. 1996. Petrological perspectives on the calorimetrically-defined glass transition temperature (Tg). *The Physics of Explosive Volcanic Eruptions, Arthur Holmes European Research Conference, Santorini, Program with Abstracts*, **2**.

SHACKLETON, N. J. & OPDYKE, N. D. 1973. Oxygen-isotope and palaeomagnetic stratigraphy of equatorial Pacific core V28–238: oxygen isotope temperatures and ice volumes on a 10^5 year and 10^6 year scale. *Quaternary Research*, **3**, 39–55.

SHERROD, D. R. & SMITH, J. G. 1990. Quaternary extrusion rates of the Cascade Range, northwestern United States and southern British Columbia. *Journal of Geophysical Research*, **95**, 19474–19645.

SKILLING, I. P. 1994. Evolution of an englacial volcano: Brown Bluff, Antarctica. *Bulletin of Volcanology*, **56**, 573–591.

SMELLIE, J. L. 2000. Subglacial eruptions. *In*: SIGURDSSON, H. (ed.) *Encyclopedia of Volcanoes*. Academic Press, San Diego, 403–418.

SMELLIE, J. L. & Hole, M. J. 1997. Products and processes in Pliocene–Recent, subaqueous to emergent volcanism in the Antarctic Peninsula: examples of englacial Surtseyan volcano construction. *Bulletin of Volcanology*, **58**, 628–646.

SMELLIE, J. L., HOLE, M. J. & NELL, P. A. R. 1993. Late Miocene valley-confined subglacial volcanism in northern Alexander Island, Anarctic Peninsula. *Bulletin of Volcanology*, **55**, 273–288.

SMELLIE, J. L. & SKILLING, I. P. 1994. Products of subglacial volcanic eruptions under different ice thicknesses: two examples from Antarctica. *Sedimentary Geology*, **91**, 115–129.

SOUTHER, J. G. 1980. Geothermal reconnaissance in the central Garibaldi Belt, British Columbia. *Geological Survey of Canada, Current Research, Part A*, **80-1A**, 1–11.

SOUTHER, J. G. 1992. Volcanic regimes. *In*: GABRIELSE, H. & YORATH, C. J. (eds) *Geology of the Cordilleran Orogen in Canada*. Geological Survey of Canada, Ottawa, 459–490.

SOUTHER, J. G. & YORATH, C. J. 1991. Neogene Assemblages *In:* GABRIELSE, H. & YORATH, C. J. (eds) *Geology of the Cordilleran Orogen in Canada*. Geological Survey of Canada, Ottawa, 373–401.

STASIUK, M. V. & RUSSELL, J. K. 1990. The Bridge River assemblage in the Meager Mountain volcanic complex, southwestern British Columbia. *Geological Survey of Canada, Current Research, Part E*, **90-1E**, 227–233.

STASIUK, M. V., RUSSELL, J. K. & HICKSON, C. J. 1996. Distribution, nature, and origins of the 2400 BP eruption products of Mount Meager, British Columbia: Linkages between magma chemistry and eruption behaviour. *Geological Survey of Canada Bulletin*, **486**, 1–27.

THORDARSON, T., SELF, S., ÓSKARSSON, N. & HULSEBOSCH, T. 1996. Sulfur, chlorine, and fluorine degassing and atmospheric loading by the 1783–1784 AD Laki (Skaftár Fires) eruption in Iceland. *Bulletin of Volcanology*, **58**, 205–225.

TUFFEN, H., GILBERT, J. & MCGARVIE, D. 2001. Products of an effusive subglacial rhyolite eruption: Bláhnúkur, Torfajökull, Iceland. *Bulletin of Volcanology*, **63**, 179–190.

Physical volcanology of a subglacial-to-emergent rhyolitic tuya at Rauðufossafjöll, Torfajökull, Iceland

H. TUFFEN[1,2], D. W. McGARVIE[1], J. S. GILBERT[2] & H. PINKERTON[2]

[1] *Department of Earth Sciences, The Open University, Milton Keynes, MK7 6AA, UK (e-mail: h.tuffen@lancaster.ac.uk)*

[2] *Department of Environmental Science, Lancaster University, Lancaster LA1 4YQ, UK*

Abstract: This paper presents the first modern volcanological study of a subglacial-to-emergent rhyolite tuya, at SE Rauðufossafjöll, Torfajökull, Iceland. A flat-topped edifice with a volume of *c.* 1 km^3 was emplaced in Upper Pleistocene time beneath a glacier >350 m thick. Although it shares morphological characteristics with basaltic tuyas, the lithofacies indicate a very different eruption mechanism.

Field observations suggest that the eruption began with vigorous phreatomagmatic explosions within a well-drained ice vault, building a pile of unbedded ash up to 300 m thick. This was followed by a subaerial effusive phase, in which compound lava flows were emplaced within ice cauldrons. Small-volume effusive eruptions on the volcano flanks created several lava bodies, with a variety of features (columnar-jointed sides, subaerial tops, peperitic bases) that are used to reconstruct spatially-heterogeneous patterns of volcano-ice interaction. Volcaniclastic sediments exposed in a stream section provide evidence for channelised meltwater drainage and fluctuating depositional processes during the eruption.

Models are developed for the evolution of SE Rauðufossafjöll, and the differences between subglacial rhyolitic and basaltic eruption mechanisms, which are principally caused by contrasting hydrological patterns, are discussed.

Subglacial rhyolitic tuyas are widespread in Iceland, occurring at central volcanoes (e.g. Torfajökull, Kerlingarfjöll) and fissure zones (e.g. Hágöngur, Prestahnúkur). Although rhyolitic tephra widely dispersed throughout northern Europe has been attributed to Quaternary subglacial rhyolite eruptions in Iceland (Dugmore *et al.* 1995; Lacasse *et al.* 1995; Zielinski *et al.* 1997; Larsen *et al.* 1998; Hafliðason *et al.* 2000), very few descriptions of ancient subglacial rhyolite sequences have been published (Gronvöld 1972; Sæmundsson 1972; Furnes *et al.* 1980). A recent study of Bláhnúkur, a small-volume (<0.1 km^3), entirely subglacial rhyolitic edifice at Torfajökull, has provided evidence for drainage of meltwater away from the vent area during the eruption (Tuffen *et al.* 2001). At Bláhnúkur, the eruption style appears to have switched from explosive magma–water interaction to the effusion of ice-constrained lava flows as melting enlarged cavities in the basal ice.

Subglacial basaltic tuyas are more widespread (Iceland, Antarctica, British Columbia) and better understood than their rhyolitic counter-parts. Fieldwork at numerous localities (e.g. Noe-Nygaard 1940; Jones 1969; Jones 1970; Smellie *et al.* 1993; Smellie & Skilling 1994; Werner *et al.* 1996, Smellie & Hole 1997) has led to well-constrained models of basaltic tuya evolution beneath thick (>150 m) temperate glaciers. At each locality, meltwater appears to have ponded during the eruption in a subglacial or ice-bound lake (Smellie 2000). This has led to the emplacement of pillow lithofacies at the base of the sequence, followed by glassy pyroclasts produced by increasingly explosive magma–water interactions and displaying sedimentary features characteristic of reworking in a body of static water. This suite of lithofacies is thought to reflect decreasing confining pressure as the edifice grew vertically within a meltwater lake (e.g. Jones 1969; Skilling 1994; Werner *et al.* 1996). Subaerial lava flows may unconformably overlie the fragmental deposits, signalling an abrupt change in the eruption environment, thought to be triggered by drainage of the meltwater lake (e.g. Smellie 2000).

Basaltic eruptions beneath thin ice (<150 m) produce a discrete set of lithofacies, due to the different mechanical properties of thin glaciers.

From: SMELLIE, J. L. & CHAPMAN, M. G. (eds) 2002. *Volcano–Ice Interaction on Earth and Mars*. Geological Society, London, Special Publications, **202**, 213–236. 0305-8719/02/$15.00

Meltwater tends to readily drain away from the eruption site, and pillow lavas are scarce (Smellie *et al.* 1993; Smellie & Skilling 1994; Smellie 2000).

Comparison between subglacial rhyolite and basalt eruption mechanisms

Hoskuldsson & Sparks (1997) presented a simplified heat-exchange model for subglacial effusive eruptions, which calculated the volume of ice melted per unit volume of magma erupted. Changes in the total volume of the system (ice + meltwater + magma) were calculated, and used to estimate the resultant variations in pressure. This model assumes that all the energy from the cooling magma is transferred to the ice via convecting meltwater, and does not account for ice deformation. It provides a useful prediction of the hydrological patterns that may develop during eruptions of rhyolite and basalt under temperate glaciers (Table 1a). The model suggests that meltwater accumulation is likely in a basaltic eruption. This is due to the high temperature of basaltic magma (*c.* 1200°C), which is capable of melting up to 14 times its own volume of ice, which is more than sufficient to accommodate the volume of magma added. This net volume decrease is expected to cause a pressure reduction, which will encourage meltwater accumulation. Rhyolitic magma has a lower temperature (*c.* 850°C) and can thus melt <10 times its own volume of ice. Positive pressure changes are predicted, which will favour drainage of meltwater.

Eruptions of rhyolitic and basaltic magmas under ice are thus likely to produce contrasting suites of lithofacies, due to differences in the physical environment. The contrasting physical properties of the magmas are also likely to influence subglacial eruption dynamics and the nature of the products (Table 1b).

This paper has tested these predictions by examining the lithofacies produced during construction of a subglacial rhyolitic tuya at Rauðufossafjöll, and comparing the inferred eruption mechanisms with those of basaltic tuyas.

Table 1a. *Effects of magma temperature on hydrology (after Hoskuldsson & Sparks 1997)*

Property of eruption	Rhyolite	Basalt
Magma temperature	800–900°C	1100–1200°C
Melting potential	≤10 times own volume of ice	≤14 times own volume of ice
Volume change	10 v ice → 10 × $\rho_i/\rho_w v$ water + 1 v magma ∴ change = +0.17 v	14 v ice → 14 × $\rho_i/\rho_w v$ water + 1 v magma ∴ change = −0.162 v
Pressure change	positive	negative
Hydrological patterns	meltwater tends to drain away	meltwater tends to collect at vent

ρ_w, density of water (1000 k m^{-3}); ρ_i, density of ice (917 kg m^{-3}); v, unit volume.

Table 1b. *Important differences between rhyolitic and basaltic subglacial eruptions (after Tuffen et al. 2001)*

Magma property	Rhyolite	Basalt	Implications
Magma viscosity*	10^6–10^7 Pa s	10^3–10^4 Pa s	Rhyolitic eruptions tend to be more explosive, larger aspect ratio lava flows
Effusion rate*	10^1–10^2 m^3 s^{-1}	10^1–10^4 m^3 s^{-1}	Inward ice creep more significant during rhyolitic eruptions because edifice growth is slower
Distribution‡	Iceland, mostly at central volcanoes	Antarctica, Iceland, British Columbia	Products of subglacial basaltic eruptions better studied than rhyolitic.
Recent eruptions	None observed	Gjálp 1996, Iceland†	Insight gained on basaltic eruptions, not on rhyolitic.

* Hoskuldsson & Sparks 1997; † Gumundsson *et al.* 1997; ‡ Smellie 2000.

Geological setting

Torfajökull central volcano is located at the southern terminus of the Eastern Rift Zone, in south-central Iceland (Fig. 1a). It is the largest silicic complex in Iceland, measuring 18 by 12 km and has erupted >250 km^3 of peralkaline rhyolite (Sæmundsson 1972, 1988; McGarvie *et al.* 1990). Since activity began *c.* 1 Ma ago, eruptions during glacial and interglacial periods have produced a variety of volcanic landforms (Sæmundsson 1972, 1988; McGarvie *et al.* 1990), which now comprise a highly dissected upland plateau, at an elevation of 600–1300 m. Torfajökull has erupted 11 times in the Holocene, most recently in AD 1477, producing rhyolitic lava

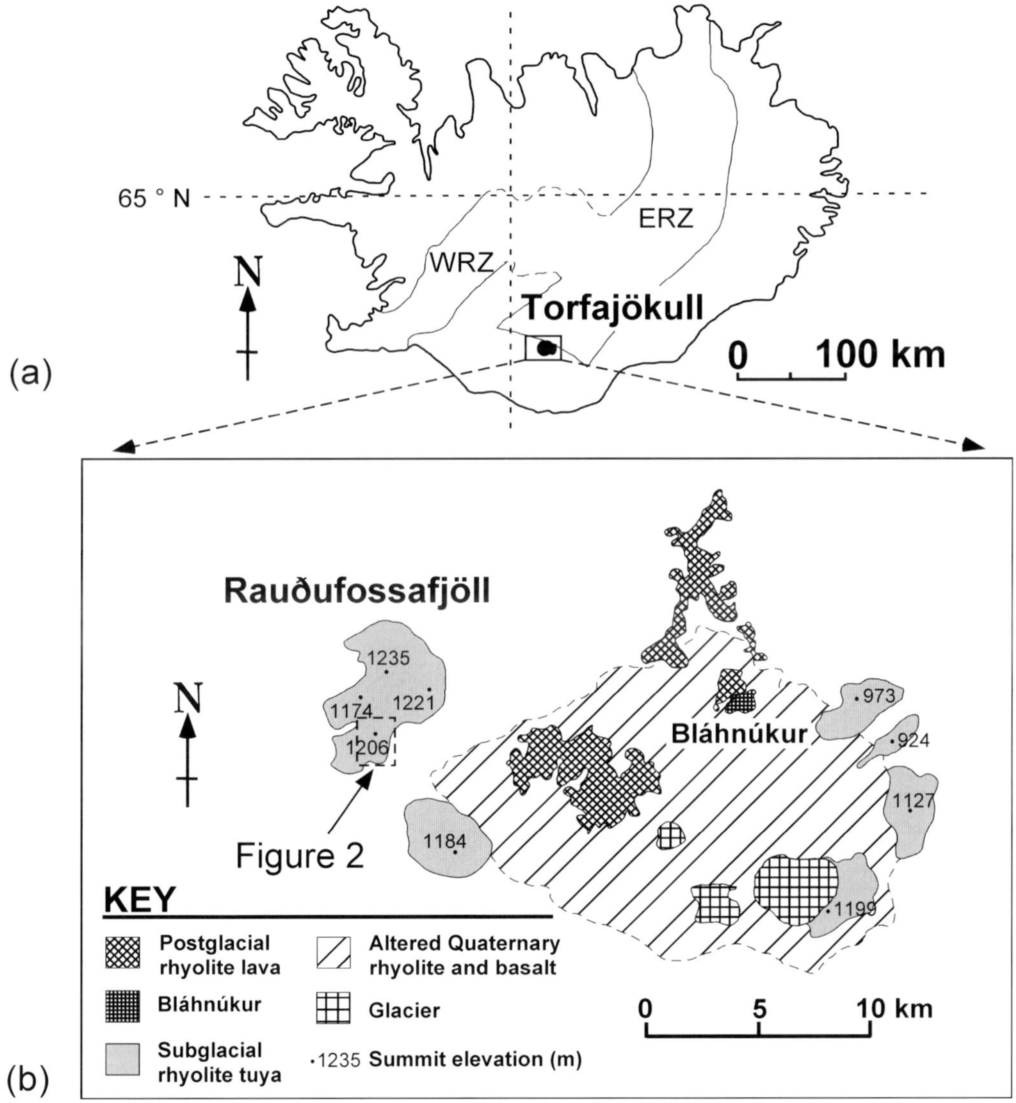

Fig. 1. (**a**) Map showing the location of Torfajökull central volcano in south-central Iceland, at the southern terminus of the Eastern Rift Zone (ERZ). WRZ, Western Rift Zone. (**b**) Simplified geological map of the Torfajökull central volcano, indicating the position of SE Rauðufossafjöll, which is shown in detail in Figure 2. The subglacial rhyolite tuyas indicated are all of similar composition, with a total volume of *c.* 17 km^3 (McGarvie 1984). The unshaded regions outside Torfajökull central volcano consist of subglacial and subaerial rhyolite formations from the Vatnafjöll and Veiðivötn fissure systems.

flows with a combined volume of <0.1 km^3 (Larsen 1984). An active geothermal field persists today, with numerous hot acidic springs, and seismic tomography indicates the likely presence of a magma chamber at 2–4 km depth (Soosalu & Einarsson 1997).

An incomplete ring of flat-topped rhyolite volcanoes surrounds the hydrothermally altered interior of the complex (Fig. 1b). These rise between 370 and 550 m above the surrounding land and have summit elevations of 924 to 1235 metres above sea level. All these volcanoes are of similar composition, suggesting that they may have been emplaced in a single eruptive episode during the last glacial period (McGarvie 1984). Rauðufossafjöll, situated at the western margin of the complex (Fig. 1b), is the most voluminous (*c.* 6 km^3). It consists of four separate flat-topped edifices that have developed on two parallel NE–SW fissures. The southeastern edifice, which is focused on in this paper, has the best exposure, due to multiple failures of its western flank and resulting cliff sections over 100 high. Elsewhere, flanks are mostly covered by scree derived from the flat tops.

Morphology of SE Rauðufossafjöll

Southeast Rauðufossafjöll consists of a NE–SW-trending flat-topped ridge, 1.5 km long and 35–250 m wide, surrounded by a broad apron of scree (Fig. 2, Fig. 3). The flat top rises 350–450 m above the surrounding topography. The total area of the edifice is *c.* 4 km^2, of which the flat top makes up only 0.5 km^2. A gently inclined plateau, at 900 metres elevation on the south eastern flank, is up to 500 m wide and dips at about 5 to the south. The volcano is bounded to the north and west by neighbouring flat-topped rhyolite volcanoes of Rauðufossafjöll, and to the south and east by subglacial and subaerial basaltic formations (Fig. 2).

Evidence for a subglacial environment

The following features suggest that SE Rauðufossafjöll was erupted under ice:

(1) The fragmental lithofacies at the base of the volcano show evidence for magma–water interaction, such as perlitized obsidian and blocky ash shards. No evidence exists for a palaeo-topography that could have confined a non-glacial lake (see also Jones 1969; Smellie & Skilling 1994; Smellie & Hole 1997; Tuffen *et al.* 2001). The current elevation of 800–1206 m in the absence of any tectonic structures consistent with uplift, is a convincing argument against a submarine setting. Furthermore, marine fossils are absent. Glacier melting is thus the most likely source of water. Eruption within an ice-dammed lake (Werner & Schmincke 1999) is rejected, since extensive lacustrine deposits are absent.

(2) Columnar-jointed rhyolite lava bodies occur at up to 1150 m elevation. Their morphologies and joint orientations are best explained by chilling against ice walls (Lescinsky & Sisson 1998; Tuffen *et al.* 2001).

Lithofacies descriptions and interpretations

Rhyolitic ash

Although fragmental rhyolite deposits appear to make up much of the lower portion of SE Rauðufossafjöll, they are only poorly exposed at a few localities on the eastern flank, between 780 and 1000 m and at 1150 m on the western flank (Fig. 2). All deposits are unwelded and poorly consolidated.

Over 40 m thickness of massive, well-sorted pale grey ash crops out at the SE base of the eastern plateau (Fig. 2). Shards are mostly 10–100 μm in diameter and generally contain less than 20% vesicles by volume. They are blocky in morphology, with sharp corners and elongate bubble walls (Fig. 4a). The deposit contains up to 5% volume of angular chips of dense black obsidian 1–10 mm across.

Ash exposed at 900 m elevation due south of the south top (Fig. 2) is similar in grain size and morphology, but contains *c.* 40% angular clasts of pale grey, pumiceous glassy rhyolite up to 10 cm in diameter. Also within the ash are elongate, highly sheared ribbons of dense black obsidian 0.5–1 m long and 0.1–0.2 m wide, with pale grey, pumiceous margins 1–5 cm in width. The pumiceous margins of some obsidian ribbons are highly fragmented and surrounded by a 'cloud' of angular pumiceous clasts.

The proportion of pumiceous clasts is much higher in ash exposed at 1000 m elevation on the north side of Blautakvísl gully (Fig. 2). This is a massive clast-supported breccia containing angular clasts of pumiceous obsidian 1–20 cm and, exceptionally, 1 m in length. The ash matrix contains blocky ash shards mostly 10–100 μm in diameter (Fig. 4b).

Interpretation. The blocky morphology, wide range of vesicularity but predominantly low vesicularity of the ash at SE Rauðufossafjöll

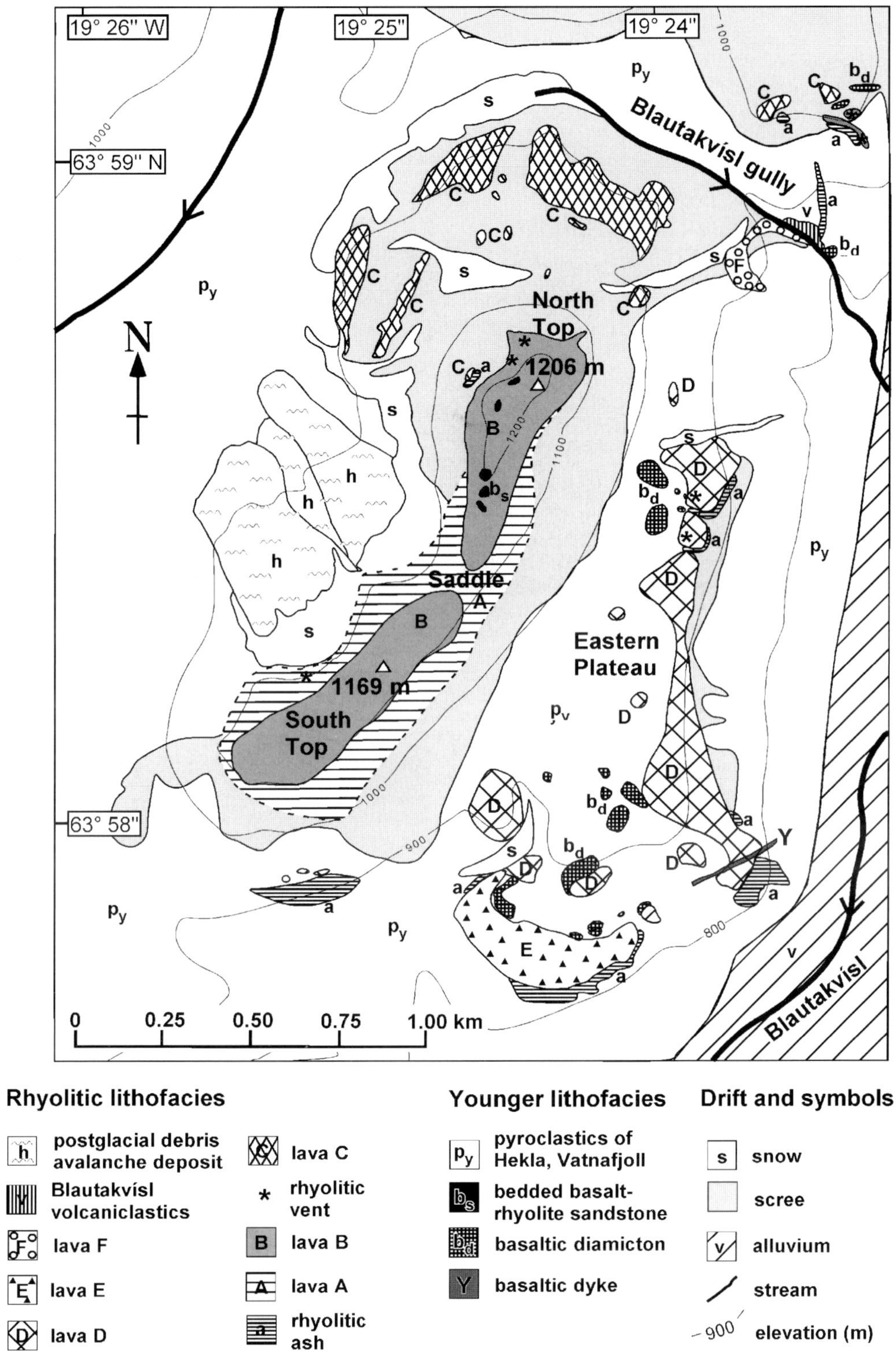

Fig. 2. Simplified solid and drift geological map of SE Rauðufossafjöll, indicating the major lithological units. Much of the map area is covered by Holocene pyroclastic fall deposits (white), principally from Hekla and Vatnafjöll volcanoes.

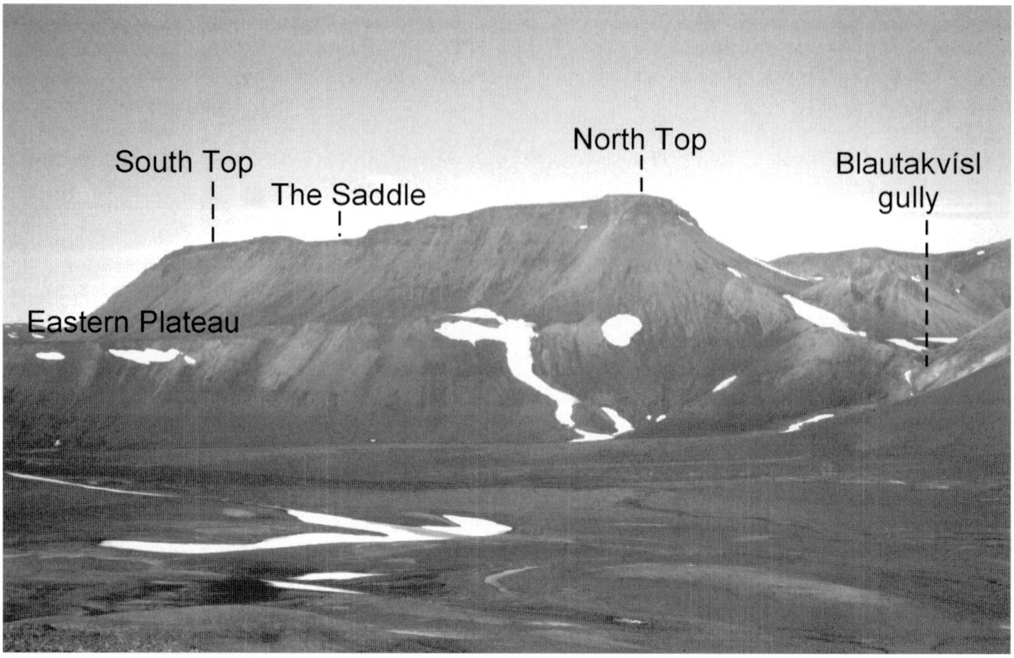

Fig. 3. View of SE Rauðufossafjöll from the ENE. The prominent flat top is formed by rhyolite lava flows, and rises 400 m above the low-lying alluvial plain in the foreground.

Fig. 4. (**a**) SEM image of rhyolite ash shards from the base of the eastern plateau. Shards are blocky, angular, generally contain <20% vesicles and are typically 50–100 μm across. (**b**) SEM image of rhyolitic ash shards from Blautakvísl gully. Note the angular, blocky shard morphologies.

suggests that fragmentation was driven principally by magma–water interaction, rather than by degassing of magmatic volatiles (Heiken & Wohletz 1985; Wohletz 1986). Glacial meltwater is the most likely source of water in the vicinity, hence the ash is interpreted as the product of explosive magma–water interaction within an ice vault. The evolution of the ice vault during the subglacial ash-producing phase of the eruption is discussed later. Confinement of phreatomagmatic explosions by ice walls may explain the presence of vent-proximal deposits dominated by fine ash (<100 μm), which was unable to escape the vent area.

The massive, poorly-sorted deposits are interpreted as the products of low-temperature pyroclastic surges, due to the lack of welding and internal structure (Cas & Wright 1987). Dense

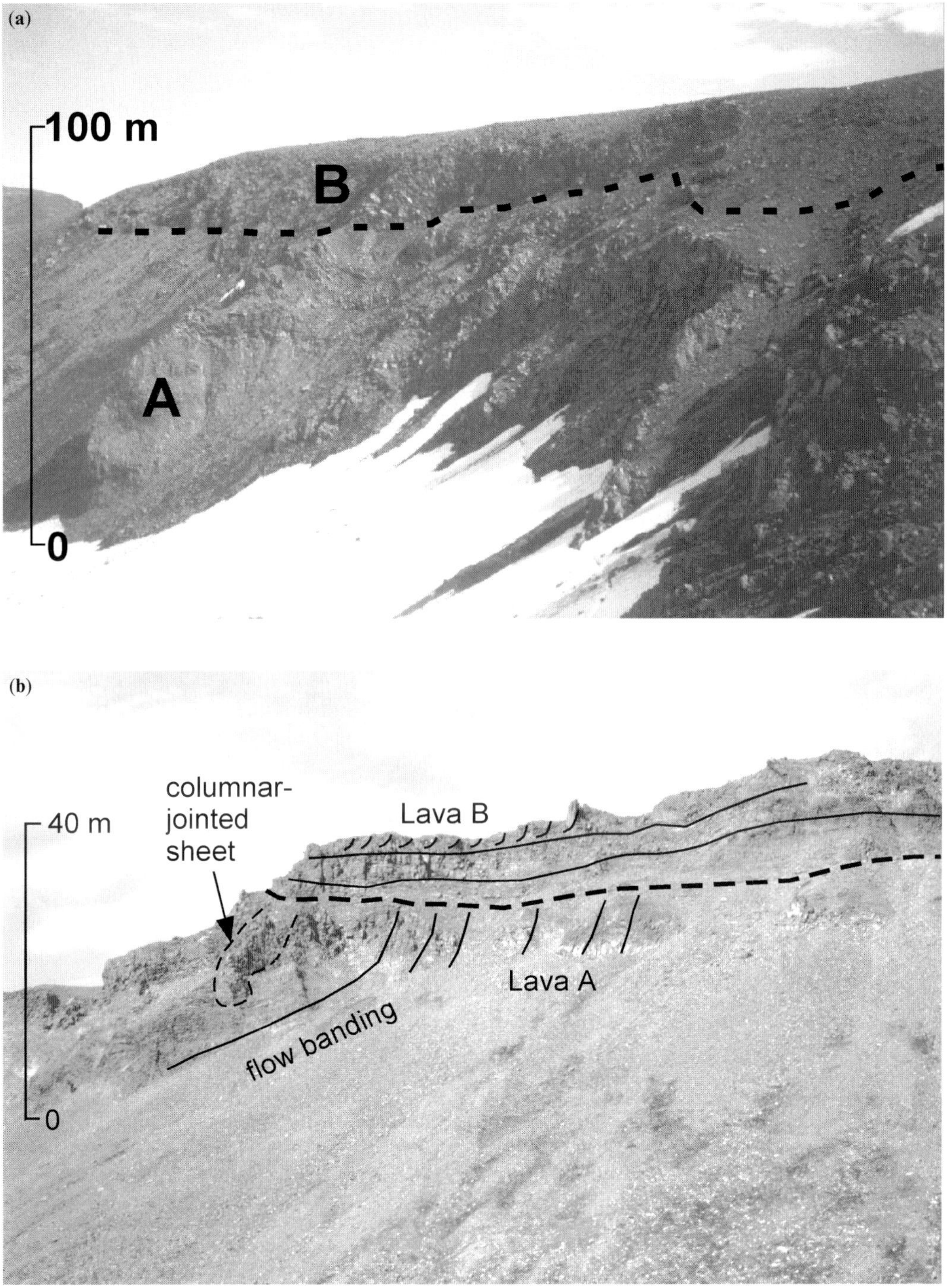

Fig. 5. (**a**) View of the southern part of the west flank of SE Rauðufossafjöll, affected by postglacial flank collapse. Cliffs of lava A (bottom left) up to 80 m thick are overlain by lava B which is subhorizontal and *c*. 10 m thick. (**b**) Close-up of lavas A and B on the eastern flank of SE Rauðufossafjöll, looking west to the Saddle from the eastern plateau (Fig. 2). Flow banding (indicated by solid black lines) is steeply ramped in lava A and near-horizontal at the base of lava B. The dashed black line indicates the contact between the lavas.

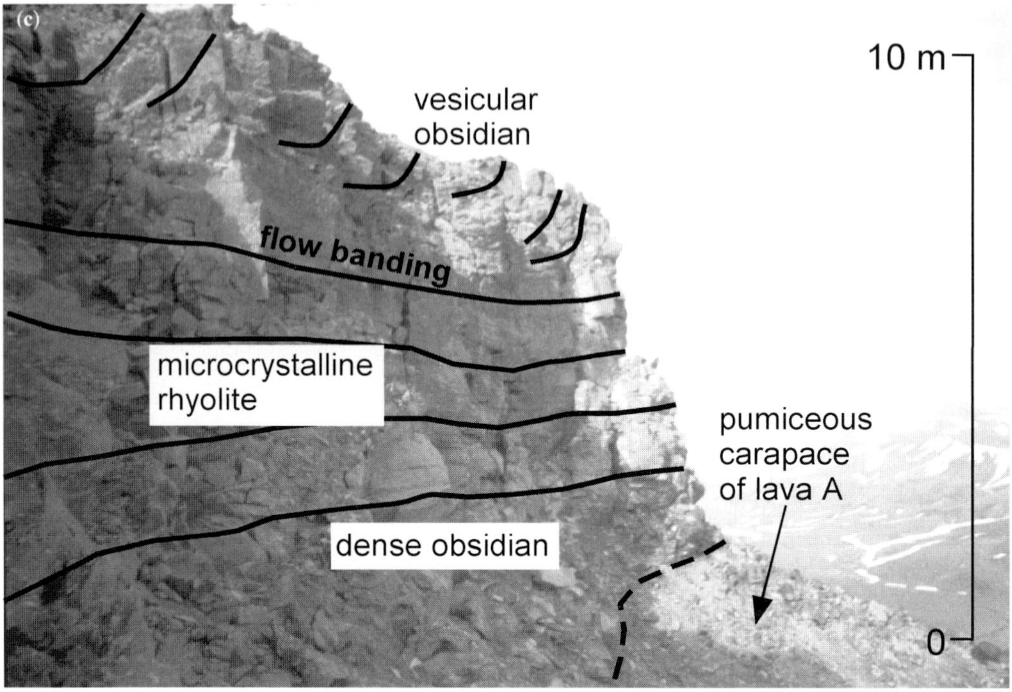
(c)
10 m
0
vesicular
obsidian
flow banding
microcrystalline
rhyolite
dense obsidian
pumiceous
carapace
of lava A

(d)
obsidian
wall
fb
obsidian
wall

clasts of obsidian observed in some outcrops may be spatter-fed material (Stevenson & Wilson 1997) that was entrained in ash-dominated surges. Well-sorted fine-grained ash on the eastern plateau may be either a fine-grained pyroclastic surge deposit or an epiclastic deposit (Cas & Wright 1987). The former interpretation is preferred, due to the presence of outsized clasts of dense obsidian, which are likely to be segregated from the fine ash fraction during epiclastic reworking.

Lava A

This is a 1.5 km-long rhyolite lava flow that crops out at between 1000 and 1120 m elevation on the main flat-topped ridge (Figs 2 & 3). It is well exposed on the west flank where it forms cliffs up to 100 m high (Fig. 5a), but is mostly concealed by scree on the south and east flanks. Lava A is overlain by subhorizontal rhyolite lava flows (lava B), and its base is not exposed. The majority of the lava flow consists of non-vesicular microcrystalline rhyolite. Flow banding is well developed and steepens from near-horizontal in the lowest exposures to near-vertical at the lava top (Fig. 5b). The uppermost 4–5 m consists of tightly flow-folded obsidian. Bands of pale pumiceous obsidian *c.* 10 cm wide are estimated to contain *c.* 40% elongate vesicles, and are interbanded with non-vesicular obsidian. The pumiceous obsidian is patchily oxidised, giving it a reddish hue.

A sheet-like lava body 5 m thick and 20 m long appears to be continuous with, and a part of, the upper part of lava A 50 m N of the Saddle (Fig. 5b). It has contorted flow banding and poorly developed columnar joints that are normal to a sub-planar surface that dips at around 40 down the eastern flank. It drapes the outer margin of lava A.

Interpretation. Lava A has a near-horizontal, glassy upper surface with tightly folded flow banding and heterogeneous vesicularity. Such features are typical of subaerial rhyolite lava flows (e.g. Fink 1983). The downslope-dipping, columnar-jointed lava body attached to the edge of the main lava flow is best interpreted as an ice-contact 'dribble' that has spilled down the surface of lava A possibly into a gap between lava A and a nearby ice wall (see also Mathews 1951). This is the only evidence within the lava carapace for an ice-contact setting, although the high aspect ratio (*c.* 15:1) and ramped flow banding are also consistent with a degree of topographic confinement. The lava is elongate in a NE–SW direction (Fig. 2), parallel to the subglacial rhyolite fissures at Rauðufossafjöll and the regional tectonic trend. This may indicate either effusion from a number of vents aligned NE–SW, or that a highly elongate ice cauldron had formed above the subglacial tephra pile. This paper favours the former explanation, since lava B lithofacies provides evidence for the effusion of lava from multiple vents on the ridge of SE Rauðufossafjöll.

Lava B

Lava B appears as two separate lava bodies that make up the upper part of the flat cap of south-east Rauðufossafjöll (Figs 2 & 3). Each is approximately 0.75 km long, up to 250 m wide and between 8 and 100 m thick, with an estimated combined volume of 10^7 m^3. A considerable portion of the lava flows has been removed by post-glacial debris avalanches on the western flank. The sides of the lava flows have been almost completely removed by erosion, creating a blanket of scree beneath. Based upon the volume of collapsed material, it is estimated that the lava flows were originally up to twice their current width. Flow interiors consist of non-vesicular microcrystalline rhyolite with well-defined platy flow banding.

The northern lava flow is *c.* 8 m thick at its most southerly exposure, with near-horizontal flow banding in the base and interior which steepens to near-vertical in the top 2 m (Fig. 5c). A 2 m-thick, highly sheared obsidian base is well exposed directly north of the Saddle (Fig. 5b, c). This lava flow thickens considerably to the north, where the maximum exposed thickness is *c.* 100 m and the base is concealed by scree. Flow banding is steeply inclined in

Fig. 5. (**c**) Detail at X on Fig. 5b. View shows the oxidized, pumiceous upper carapace of lava A overlain by lava B. Lava B comprises a 1.5 m-thick basal zone of dense obsidian, a microcrystalline rhyolite interior, and a vesicular obsidian upper carapace 2 m thick. Flow banding in lava B is near-horizontal in the base and interior, and steeply ramped in the upper portion. (**d**) Vent 150 m west of the north top, seen from the south. It crops out *c.* 40 m below the upper carapace of lava B which it appears to feed (although erosion has removed any portion of this lava that may have directly overlain the vent). Near-vertical obsidian walls (black) are *c.* 5 m thick. Platy fracture in the mid grey interior portion of the vent is near-vertical and parallel to flow banding (fb, dashed black line). Ice axe is 0.6 m long (arrow).

Fig. 6. (**a**) View of SE Rauðufossafjöll from the NE. Lava C forms a prominent ridge on the tuya flank. The lava crops out as cliffs up to 100 m high (cliff bases indicated by dashed black lines), above extensive scree slopes. (**b**) Columnar-jointed lava 'cliff' 90 m high at locality X (Figure 6a). Column orientations are indicated (black lines). Figure for scale (arrow).

the thick northern portion. The upper surface of the northern lava flow is best preserved in its southern portion, where the upper 2–3 m is glassy (Fig. 5c), with a heterogeneous population of elongate vesicles typically 10 mm long. Elsewhere the upper carapace is either missing or obscured by pyroclastic deposits from Hekla and Vatnafjöll.

A prominent near-vertical sheet-like vent is close to the north-western edge of the northern lava flow (Figs 2 & 5d). This has obsidian walls *c.* 5 m thick, cut by anastamosing pale grey tuffisite veins. Veins are 1–50 mm wide and filled with cross-bedded clastic material, comprising ash shards, crystal fragments and 1–10 mm ellipsoidal obsidian blebs. Elongate, crenulate pale patches in the intact obsidian contain crystal fragments and appear to represent strongly sheared earlier generations of tuffisite veins. Flow banding can be followed from the vent steeply downslope to the NW.

The southern lava flow is less well exposed, but shares many features with the northern flow, including a 2–3 m-thick, glassy upper carapace and steeply ramped flow banding in the thicker portions. The maximum exposed thickness is *c.* 75 m. The lava flow base dips SW at *c.* 10° where exposed at its northeastern margin (Fig. 5b).

Interpretation. Lava B is interpreted as the product of subaerial lava effusion, due to the absence of evidence for interaction with ice. The northern and southern lava flows are considered to be separate eruptive units; they appear not to have been joined. This implies that the lavas were erupted from at least two discrete vents, only one of which is exposed. These vents are likely to have been aligned parallel to the NE–SW axis of the ridge. We are unable to infer the position of the palaeo-ice surface during the effusion of the summit lavas. Lava A appears to have been thoroughly quenched when it was overlain by lava B.

Lava C

About ten microcrystalline lava bodies up to 250 m long and 80 m thick are distributed around the northern flank of the volcano at 1000 m elevation (Figs 2 & 6a). They typically consist of a near-horizontal section on the upslope side (upper section) and a steeply dipping section on the downslope side (lower section) which forms a crumbling cliff 50–90 m high (Fig. 6b). Upper sections consist of pale grey microcrystalline rhyolite with tightly folded flow banding. Flow banding in lower sections is sub-planar and dips downslope at 40–80°. The outer 4–5 metres of lower sections are cut by well-developed columnar joints spaced 15–20 cm apart. These dip gently into the local slope and are approximately normal to flow banding (Fig. 6b). The exposure of lava C 500 m WNW of the northern summit has an upper section that dips inwards towards the summit of SE Rauðufossafjöll and is cut by columnar joints similar to those described above. No obsidian, perlite or tuffisite were observed at any of the lava C exposures.

Interpretation. The orientation of columnar joints in the lower portions of lava C suggests emplacement against steeply inclined near-planar ice walls (Mathews 1951; Lescinsky & Sisson 1998; Tuffen *et al.* 2001, 2002). These ice walls probably exceeded 80 m in height in places. It is likely that chilled obsidian margins formed during emplacement of the lavas, and were subsequently removed by erosion. No evidence for magma–water interaction, such as perlitization or fragmentation is present. It is arguable as to whether lava bodies entered pre-existing cavities in the ice, or whether cavities formed in advance of the lavas as high heat flux preceded their emplacement (Tuffen *et al.* 2002). Orientations of columnar joints in the lava body on the west flank suggest that it was also emplaced beneath an ice roof. None of the other type C lava bodies show evidence for the presence of an ice roof. The relative timing of the emplacement of lava C and the main flat-topped edifice is unclear.

Lava D

Three separate lava bodies are exposed on the Eastern Plateau, forming a prominent break in slope trending approximately north–south (Fig. 2). The largest exceeds 1 km in length, 250 m in width and 80 m in thickness. Lava bases, where exposed, are peperitic (Fig. 7a). Matrix-supported breccia beneath the lavas consists of angular clasts of dense black obsidian and pale grey pumiceous obsidian 1–5 cm in diameter suspended in a pale grey, massive matrix of fine-grained rhyolitic ash (Fig. 4a). The breccia contains localized ‘rafts’ of well-sorted, planar bedded ash up to 2 m across and 0.3 m thick, which are highly sheared and disrupted. These bedded rafts were only observed within 1 m of the lava base.

The lavas have a basal zone of dark grey, non-vesicular perlitized obsidian, typically 0.5 m in thickness. The lower part consists of ‘clouds’

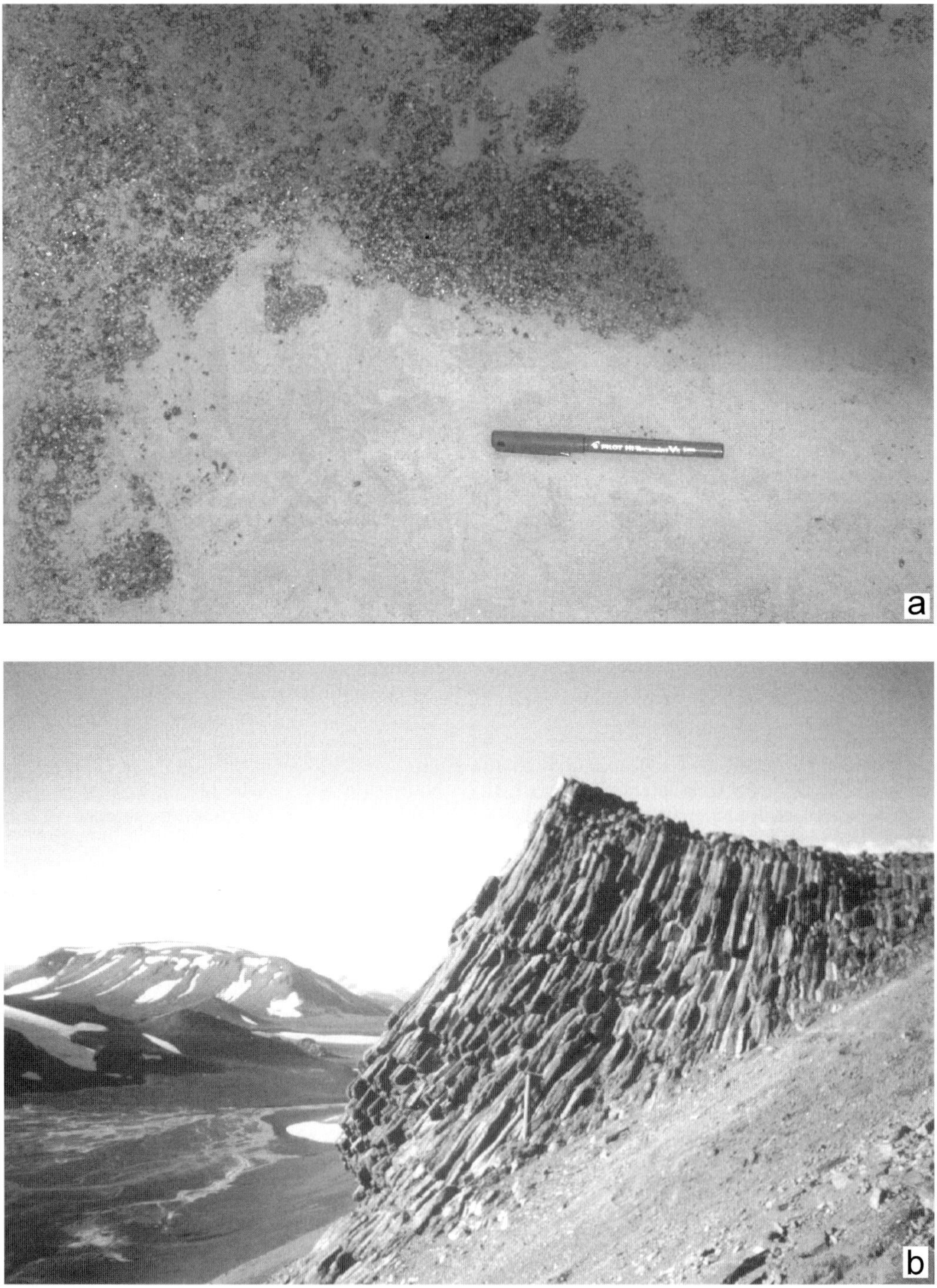

Fig. 7. (**a**) Close-up view of the peperitic base of lava D at the SE limit of the eastern plateau. Jigsaw-fit perlitized obsidian breccia (dark grey, top and left) is in contact with well-sorted fine-grained ash (pale grey). Pen is 15 cm in length. (**b**) View of columnar-jointed, perlitized obsidian on the east side of lava D. Banding normal to columns is caused by variations in the degree of perlitization. Ice pick is 60 cm long.

of angular obsidian clasts 1–10 cm across suspended in a poorly-sorted, unbedded ash-lapilli matrix. This grades upwards into a zone of jigsaw-fit obsidian breccia, cut by angular veins of unstratified pale grey ash. Above the lava base an inclined sheet of mid-grey obsidian makes up the bulk of the lava D exposures. The obsidian is variably perlitized, comprising 0–80% spherical bead-like relicts of black glass and 20–100% altered pale grey glass. It is cut by columnar joints, which are locally spectacular (Fig. 7b). Columns are polygonal, 8–12 cm across, and aligned normal to planar surfaces mostly dipping at 30–62° down the present-day slope. At one locality, the planar surface dips gently into the slope (Figs 7b & 8). The orientation of flow banding in this part of the lava is highly variable and is seldom normal to the columnar joints. The top 5–10 m of exposed lava below the plateau crest lacks columnar joints and is not perlitized. It is glassy, with near-vertical flow banding and interleaved zones of dense black obsidian and pale pumiceous obsidian (40–50% vesicles). A pale, clast-supported breccia composed of pumiceous and dense obsidian, up to 2 m thick, overlies the top of the northernmost lava flow (Fig. 8). The breccia grades downwards into the intact, tightly flow-folded lava flow top. Although the eastern plateau is mostly covered by Holocene pyroclastic fall deposits and basaltic diamicton, apparently unrelated to the eruption of the Rauðufossafjöll complex, intact pumic-

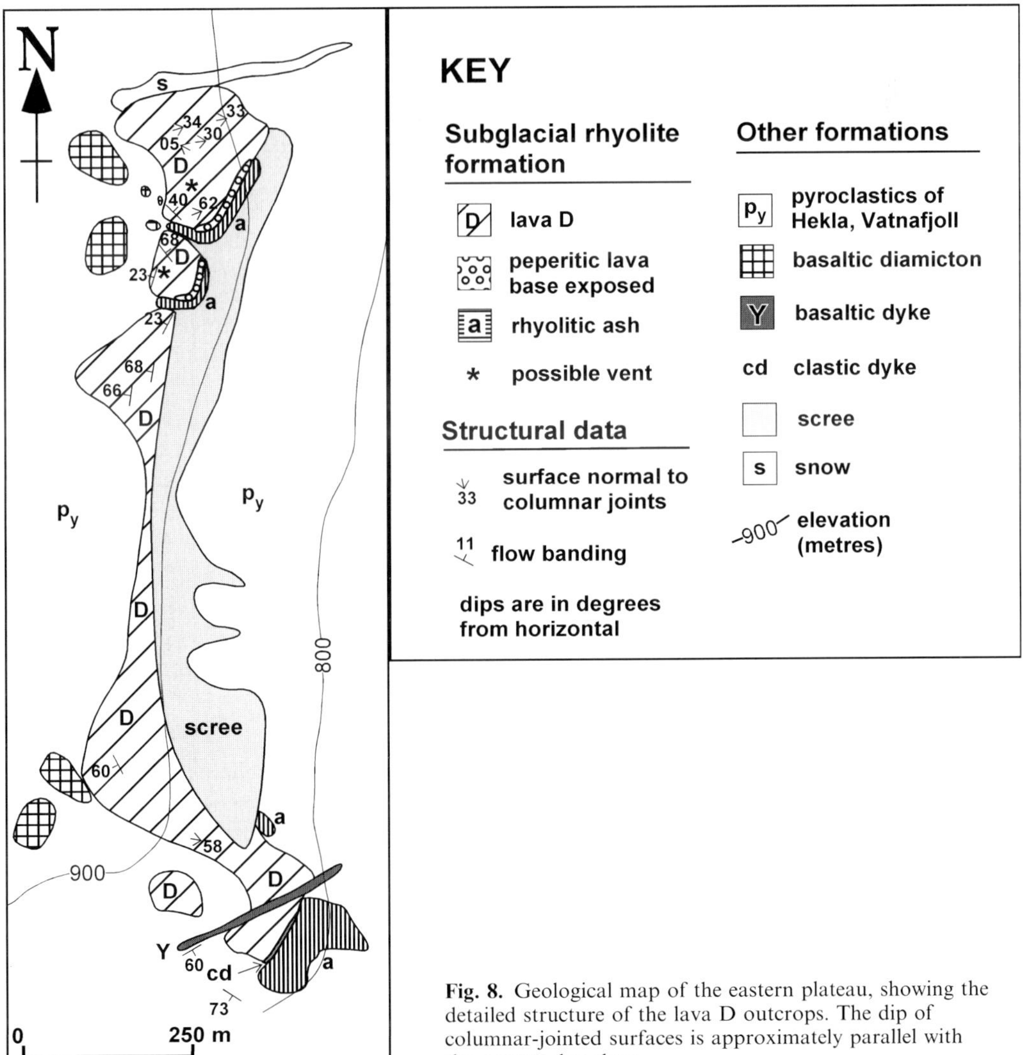

Fig. 8. Geological map of the eastern plateau, showing the detailed structure of the lava D outcrops. The dip of columnar-jointed surfaces is approximately parallel with the present-day slope.

eous obsidian crops out as a number of knolls 40–50 m across (Fig. 2).

Interpretation. The nature of the lava flows on the eastern plateau is thought to reflect an unusual emplacement environment. Peperitic flow bases may indicate that the lavas flowed over wet, poorly-consolidated breccias (Kokelaar 1982; McPhie *et al.* 1993). The columnar-jointed obsidian is interpreted as an ice-contact feature (Mathews 1951; Lescinsky & Sisson 1998; Tuffen *et al.* 2001), formed as the lavas chilled against ice walls that dipped at 30–62° down the local present-day slope. Pervasive perlitic alteration indicates that water interacted with the lava body (e.g. Davis & McPhie 1996); a process that is greatly accelerated when the lava is still hot (Friedman *et al.* 1966). Any fragmental material generated at the ice-lava contact has not been preserved. Patterns of vesicularity and flow banding in the flow tops are characteristic of subaerial rhyolite lava flows (Fink 1983). Stratified ash deposits directly underlying the lava indicate that epiclastic reworking occurred prior to emplacement of the lava, probably by flowing meltwater streams.

Lava E

Lava E crops out on the southern flank of the eastern plateau (Fig. 2), in low-lying outcrops that cover an area approximately 500 m by 250 m. The best exposure is in a gully section at the south-western margin of the eastern plateau. The lava is *c.* 30 m thick, with near-horizontal top and base (Fig. 9a). It overlies >10 m of massive, poorly-sorted matrix-supported pumice-obsidian breccia, which is pale brown in colour and has a fine-grained matrix of blocky ash. A unit of well-sorted, delicately cross-laminated fine rhyolitic sandstone *c.* 0.5 m thick directly underlies the rhyolite lava body. The lowest part of the lava consists of non-vesicular, deformed clasts of dense black obsidian surrounded by unbedded pale brown ash (Fig. 9b). Clasts are elongate, folded in a ductile manner and frequently bounded by flow-banding planes. They range from 0.5 cm to 20 cm in length and the deposit is of variable thickness, typically 0.5–1 m. It grades upwards through a zone of jigsaw-fit obsidian breccia < 1 m thick into a zone of intact, dense black obsidian. The obsidian is cut by anastomosing veins 1–30 mm wide, infilled by near-white cross-laminated ash. This is in turn overlain by flow-banded obsidian *c.* 25 m thick, of which the upper 15 m is columnar-jointed. Vesicularity increases from 0 in the lava base and interior to >40% in the uppermost 5 m. Columnar joints are 8–15 cm across and normal to a near-planar surface that dips downslope at *c.* 45°. Joint surfaces in the upper 10 m are covered by a veneer of white ash.

At one locality, the columnar-jointed upper surface is cut by a fracture 10 m deep and 2 m wide (Fig. 9a, c). The walls of the fracture are mostly parallel to joint planes (Fig. 9c). It is filled with broken columns and irregular blocks of obsidian 10 cm–1 m across, interspersed with bedded gravelly sandstones that drape over the larger obsidian blocks. The obsidian in the upper 1 m of the void walls contains *c.* 40% coalescing, near-spherical vesicles. Vesicles within 30 cm of the void wall are filled with white ash. Locally, vesicle walls have ruptured, creating jagged black bubble-wall shards 1–10 mm long suspended in white ash. The lava top is covered by pale grey, poorly-sorted pumice-obsidian breccias supported by an ash matrix.

The south-eastern outcrop consists of a complex association of columnar-jointed black obsidian cut by ash-filled veins and a variety of volcaniclastic sandstones and breccias. Two separate lava bodies *c.* 10 m thick were identified, each displaying textural patterns similar to the 30 m-thick lava body described above. Lenses of grey-brown sandstone, 1–2 m thick, underlie the base of the lower lava body. These comprise well-sorted, planar-bedded pale brown ash and units of 'mixed' sandstone. 'Mixed' sandstones contain 0–40% black, jagged obsidian shards 1–10 mm across within a pale brown sand-grade ash matrix (Fig. 9d). Obsidian shards are locally randomly orientated, although in places they display marked horizontal imbrication. Bedding in the sandstones is locally faulted and folded.

Interpretation. Lava bases are similar to the bases of lava D and are typical of peperitic lavas/sills that have flowed over/through a wet, poorly consolidated clastic substrate (Kokelaar 1982; White & Busby-Spera 1987). Vesicular lava tops are suggestive of low pressure conditions (Hunns & McPhie 1999) and show evidence of infiltration of permeable, vesiculating obsidian by mobile ash. The top and sides of the lava bodies display columnar joint patterns typical of lava–ice interaction (Lescinsky & Sisson 1998; Tuffen *et al.* 2001). It appears that the columnar-jointed upper carapace of the north-western lava body burst open prior to complete quenching, creating a fracture that was filled by material spalled from its walls and by ash-dominated sediments that were possibly washed in by flowing water. Sediments directly

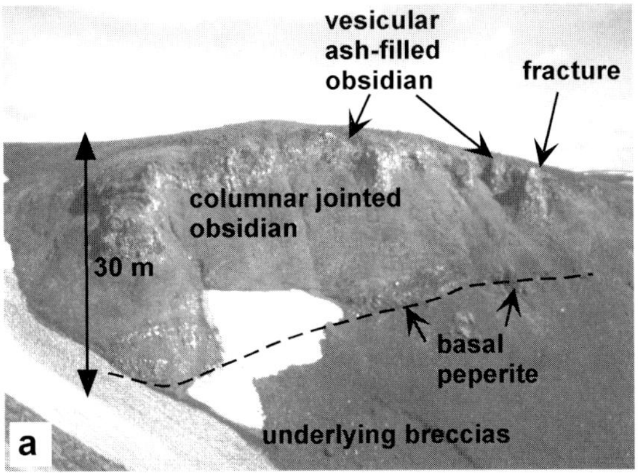

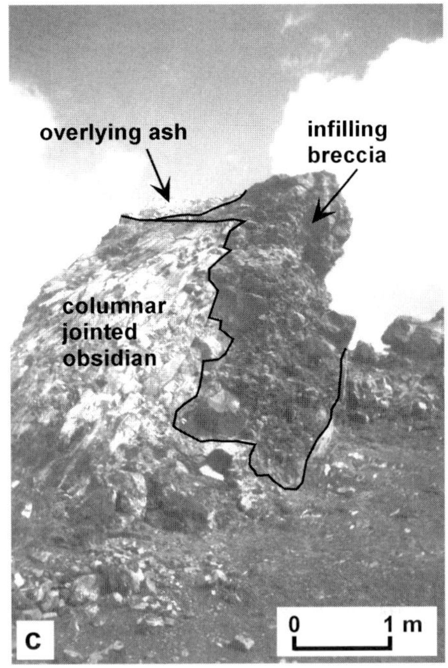

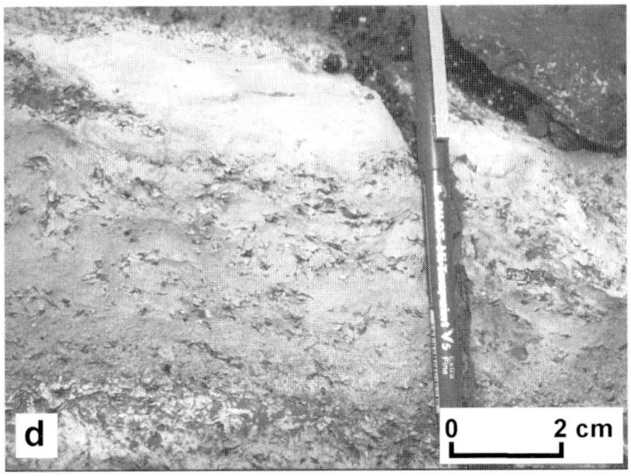

Fig. 9. (**a**) View of the northwestern outcrop of lava E on the eastern plateau, looking ESE from an altitude of 900 m. (**b**) Base of lava E showing dense obsidian clasts (dark grey) surrounded by ash (white). (**c**) Breccia-filled fracture in the columnar-jointed upper carapace of lava E. The fracture is bounded by surfaces that have followed columnar joints. The pale colour of the columnar-jointed obsidian (left of picture) is caused by a veneer of ash that coats joints planes. (**d**) Mixed fine-grained ash (white) and obsidian shards (black) forming sediments underlying lava E. The proportion of obsidian shards varies between 0 and *c.* 25% within the pictured section.

underlying the southwestern lava, which are interpreted as water-lain, contain obsidian clasts that may be derived from the vesicular lava top.

Combining the observations and inferences above, a similar scenario is envisaged for the emplacement of the peperitic lavas to that of lava D. Subaerial lava bodies flowed over water-saturated sediments towards an ice wall (Fig. 10). Meanwhile, an ash–water mixture washed over the surface of the advancing lava, possibly flowing downslope from the main edifice 500 m to the NW. This draped a thin deposit of wet ash over the top of the lava, which interacted with the hot, vesiculating upper carapace. It also picked

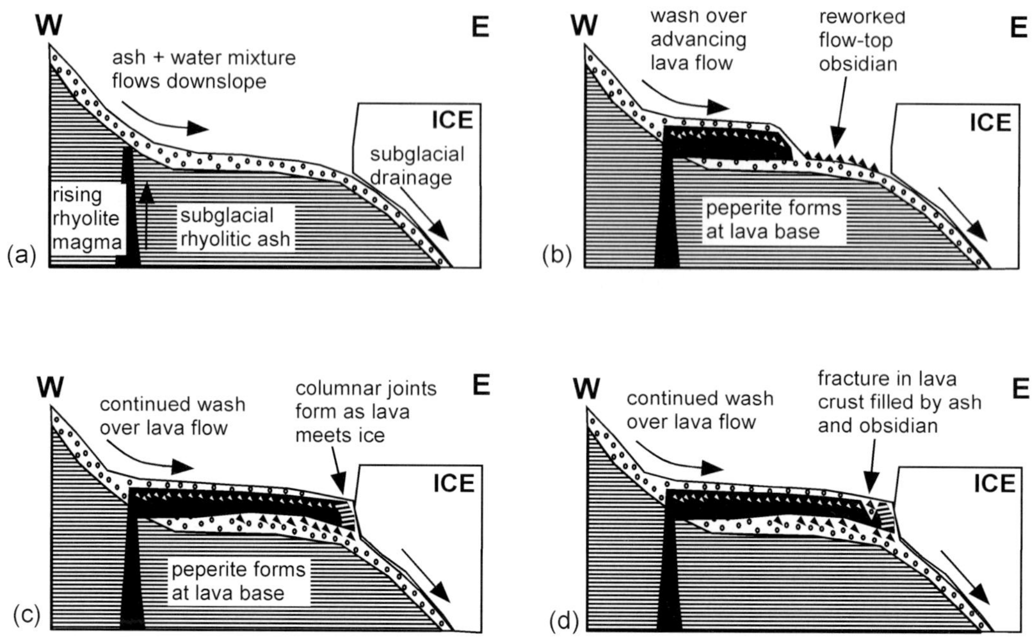

Fig. 10. Cartoons illustrating a possible model for the emplacement of lava E. See text for explanation.

up clasts of fragmented, vesicular obsidian from the lava surface and redeposited them at the foot of the lava (Fig. 10c). The lava then chilled against an ice wall, and columnar joints formed. Brittle failure of the columnar-jointed nose of the lava, possibly triggered by recession of the adjacent ice walls, formed a fracture into which sediments were washed by flowing meltwater (Fig. 10d).

Lava F

A dome-like exposure of pale grey obsidian crops out at 915 m elevation on the NE flank of the volcano (Fig. 2). The dome stands approximately 20 m above the surrounding slope and measures 100 m by 50 m. The obsidian contains *c.* 20% elongate, coalescing vesicles up to 10 mm in length, giving it a 'woody' texture. It consists of 95–100% grey, perlitized obsidian and 0–5% spherical, unaltered bead-like relicts of black glass 3–5 mm across. The central portion of the dome appears to consist of mostly intact obsidian and is surrounded by monomict breccia. The breccia drapes the sides of the dome and forms a sheet 1–4 m thick that extends up to 250 m downslope to the east (Fig. 2). It is clast-supported, crudely bedded and moderately well sorted, and contains angular clasts 5–25 cm across of woody, perlitized obsidian identical to the dome material. Lava F breccia overlies all other units in Blautakvísl gully.

Interpretation. Pervasive perlitization of lava F suggests that the lava encountered water, possibly whilst still hot (Friedman *et al.* 1966; Davis & McPhie 1996). The apron of monomict breccia is thought to be talus derived from gravitational collapse of chilled, perlitized lava from the dome (Cas & Wright 1987). Although no direct evidence for subglacial or subaqueous emplacement is present, perlitization was probably caused by interaction with glacial meltwater. It is proposed that lava F represents a small-volume lava dome that is positioned directly above its (unexposed) feeder vent. This vent is colinear with the vents from which lava D and lava E were erupted (Fig. 2), leading to speculation that lavas D E and F were formed late in the development of SE Rauðufossafjöll, as activity became restricted to lava effusion from a N–S-trending chain of vents.

Blautakvísl volcaniclastic sediments

Much information about eruptive and depositional processes can be gained from examining sequences of volcaniclastic sediments emplaced during a subglacial eruption (e.g. Smellie & Skilling 1994; Tuffen *et al.* 2001). Thus the complex succession of volcaniclastic sediments that crops out in the Blautakvísl stream gully, c. 900 m NE of the north summit (Fig. 2), is of fundamental importance for our understanding of the evolution of SE Rauðufossafjöll. The stream gully is the lowest topographic point on the 6 km-long NE–SW-trending ridge formed by SE Rauðufossafjöll and the adjacent subglacial rhyolite tuya to the NE. The sediments are 15 m thick and have dips of 20–30° to the SE. They lie in an erosive channel cut into the underlying ash. Only the northern side of the channel is exposed (Fig. 11a). The sediments comprise a succession of laterally discontinuous units, which display a wide variety of clast types, sizes, sorting and bedding characteristics, summarized in Table 2 and illustrated in Figure 11b.

Interpretation. The wide range in sedimentology and clast types (Table 2, Fig. 11b) suggests a variety of depositional mechanisms and clast sources. The youngest unit, lithology 7 was derived from gravitational collapse of a perlitic lava dome lithofacies (lava F). It is monomict, and grades into the intact portion of the lava dome, and is interpreted as a syn-eruptive unit. Thus it is proposed that the entire volcaniclastic succession at Blautakvísl was emplaced during the eruption of SE Rauðufossafjöll. The presence of water-lain sediments is strong evidence for meltwater drainage during the eruption, focussed at a topographic low on the subglacial rhyolite ridge. Variable clast types in different units of the sedimentary sequence point to variable eruption styles (e.g. forming pumice, microcrystalline rhyolite and perlitic obsidian), and a variable 'catchment area' from which clasts were derived. Lithologies 2, 4 and 6 show evidence for deposition by flowing water and are interbedded with lithologies 1, 3, 5 and 7 which were derived from dense gravity flows (Table 2, Fig. 11b). The debris flow deposits may have been generated during subglacial melting events in which recession of supporting ice walls destabilized growing piles of volcanic detritus (Tuffen *et al.* 2001). The tops of many of the debris flow deposits were reworked by meltwater.

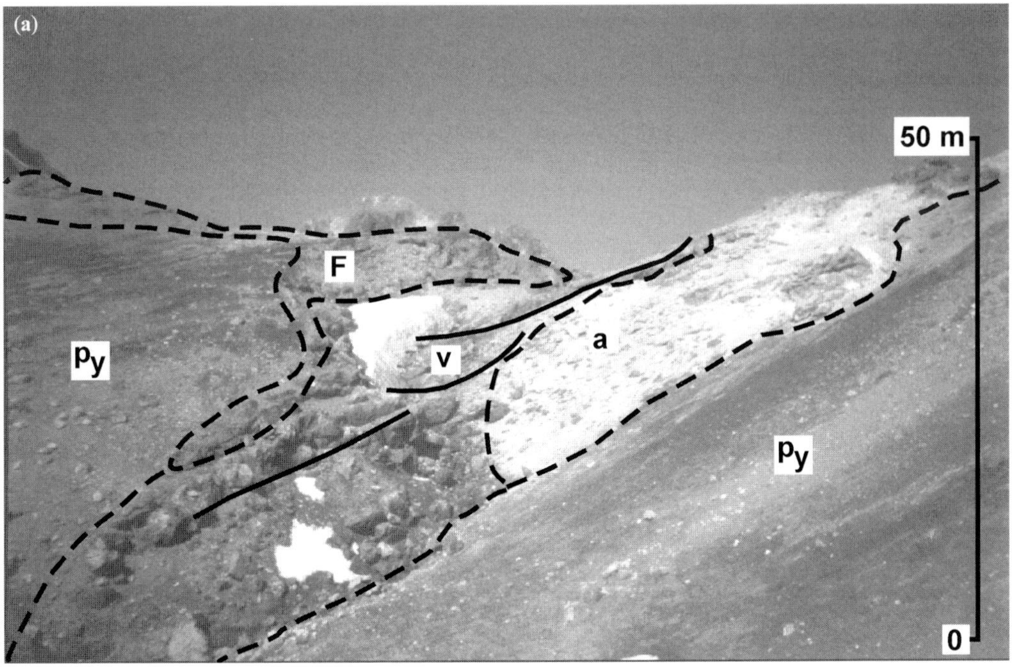

Fig. 11. (**a**) View of Blautakvísl gully, looking NW. Volcaniclastic sediments (v) extend to the base of the section shown and erosively overlie massive pale ash (a). Bedding in the sediments dips at approximately 20–30° to the left (solid black lines). A veneer of perlitic breccia overlying the sediments (F) is thought to be derived from collapse of lava F (top left corner). Pyroclastic deposits from Hekla and Vatnafjöll volcanoes are labelled p_y. Dashed black lines indicate geological contacts. (**b**) Simplified graphic log showing the major units in the Blautakvísl gully sedimentary section. See Table 2 for detailed descriptions and interpretations.

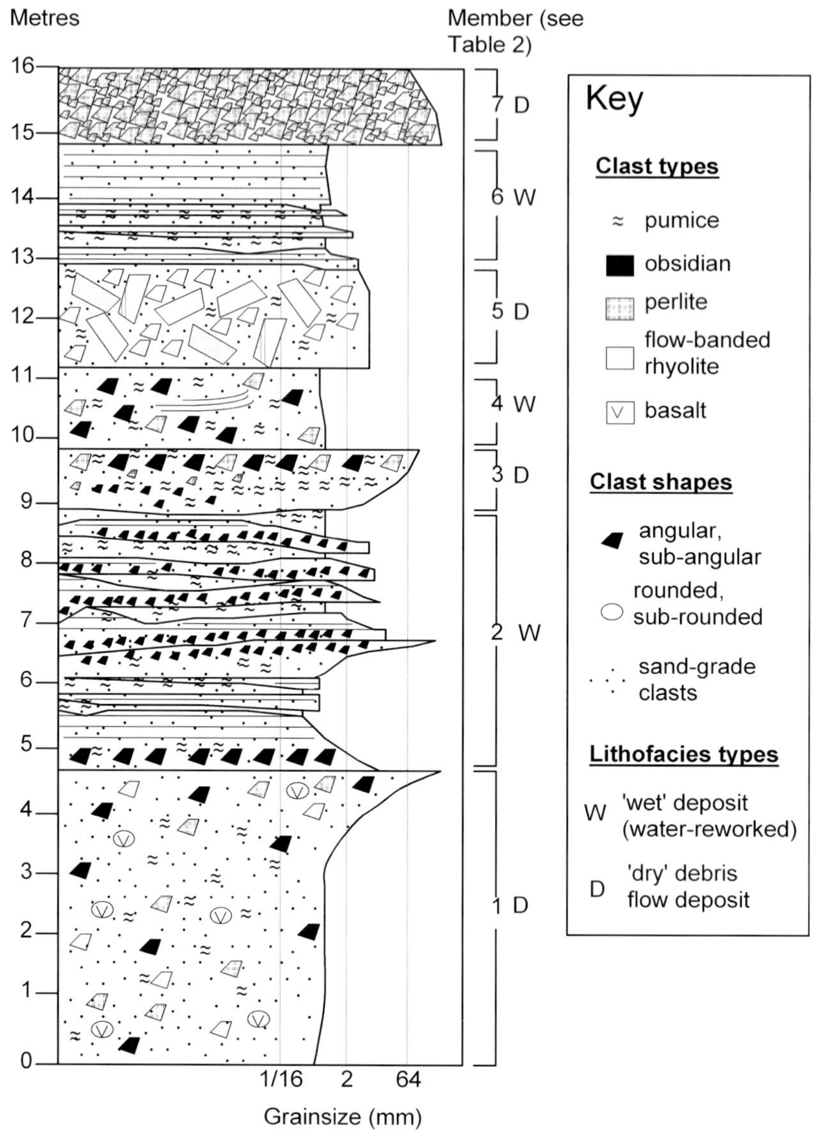

Fig. 11. (*continued*).

Discussion

In developing a model for the evolution of the SE Rauðufossfjöll rhyolite tuya this paper has assumed that it is the product of a single eruptive event that occurred during only one glacial period. No evidence of either glacial or interglacial deposits intercalated with the eruptive units or major erosional surfaces has been found. On the basis of geochemical coherence, McGarvie (1984) also argued that Rauðufossafjöll and the other rhyolite tuyas fringing Torfajökull (Fig. 1b) constituted a single eruptive episode.

Construction of the SE Rauðufossafjöll rhyolite tuya

Subglacial ash-producing phase.

A subglacial eruptive environment is proposed for the ash deposits on the lower flanks of SE Rauðufossafjöll, citing the following evidence:

Table 2. *Summary characteristics of volcaniclastic sediments in Blautakvísl gully*

Lithology	Description	Clast types	Other observations	Interpretation
7	1–4 m-thick crudely-bedded poorly-sorted clast-supported monomict perl breccia	Angular perl 5–25 cm	Perl woody, 40% vesicles, identical to Lava F perl.	Rock avalanche deposit from homogeneous body of perlitic obsidian (lava F)*
6	2 m-thick planar-bedded well-sorted obs + pum gravelly sandstone. Beds 1–5 cm thick	Sub-angular sand-grade obs + cry, sub-rounded pum 1–5 mm	Laterally continuous over >10 m	Traction current deposits, pulsing sediment supply*†
5	1.5 m-thick massive poorly-sorted matrix + clast-supported rhy breccia	Angular flow-banded rhy 0.1–80 cm, sub-rounded pum 1–5 cm. Matrix sub-angular coarse sand-grade obs + cry	Laterally continuous over 10 m	Deposit from a cohesive debris flow* derived primarily from collapse of a rhyolite lava body
4	1.6 m-thick massive poorly-sorted matrix-supported polymict breccia, lenses of planar-laminated well-sorted gravelly sandstone	Breccia: sub-angular perl + obs 0.5–20 cm, sub-rounded pum 1–5 cm. Gravelly sandstone: sub-angular perl 1–10 mm	Laterally continuous over 10 m	Episodic aggradation from debris flows, intermittent reduced particle flux and channelised water-reworking*
3	1 m-thick reverse-graded massive poorly-sorted polymict breccia. Matrix-supported sandstone at base, clast-supported pebble breccia at top	Base: sub-angular coarse sand-grade obs + cry. Top: sub-angular obs + perl + pum 0.5–50 cm, fine sand-grade obs + cry matrix	Contains lenses of rounded pum	Grain flows of obsidian and pumice followed by cohesive debris flows*‡
2	4 m-thick sequence comprising 1–10 cm beds of matrix-supported pum + obs conglomerate, poorly-sorted obsidian breccia and lenticular-bedded gravelly obsidian sandstone	Conglomerate: sub-rounded pum 0.5–4 cm, obs + cry < 2 mm Gravelly sandstone: sub-angular obs + cry 1–5 mm. Obs breccia: sub-angular obs + cry 2 mm–5 cm	Obs breccia grades upwards into gravelly sandstone	Dense gravity currents of obsidian, tops reworked by running water. Interspersed with pumice-bearing grain flows*‡ Fluctuating water and sediment flux
1	5 m-thick massive poorly-sorted matrix-supported polymict breccia.	Sub-angular perl + rhy + obs + pum 0.5–20 cm Sub-rounded bas 2–4 cm. Matrix fine sand-grade obs + cry	Clasts coated by sand matrix, sand-filled veins 1–3 mm wide	Fluidized debris flow deposit* wide clast source. Matrix remobilised post-emplacement. Basalt clasts possibly dervied from basement

Description and interpretation of the principal lithologies within the volcaniclastic sedimentary sequence in Blautakvísl gully.
Abbreviations: bas, basalt; cry, crystal; obs, obsidian; perl, perlitic obsidian; pum, pumiceous rhyolite; rhy, microcrystalline rhyolite. Lithology 1 is the oldest, lithology 7 the youngest.
* Cas & Wright 1987; † Smellie *et al.* 1993; ‡ McPhie *et al.* 1993.

(1) they are locally intruded by perlitic and/or peperitic lava bodies, suggesting that the environment was water-rich (most probably glacial meltwater), and (2) they are overlain by lava flows that show distinctive ice-contact features. Fragmentation was by explosive hydrovolcanic activity within an ice vault (Fig. 12a). Blocky shard morphologies suggest that they were formed mainly by magma–meltwater interaction, with less important degassing of magmatic volatiles (Heiken & Wohletz 1985). Although some water was present in the vault, the lack of bedding and distinctive sedimentary lithofacies (e.g. sediment gravity-flow deposits) suggests

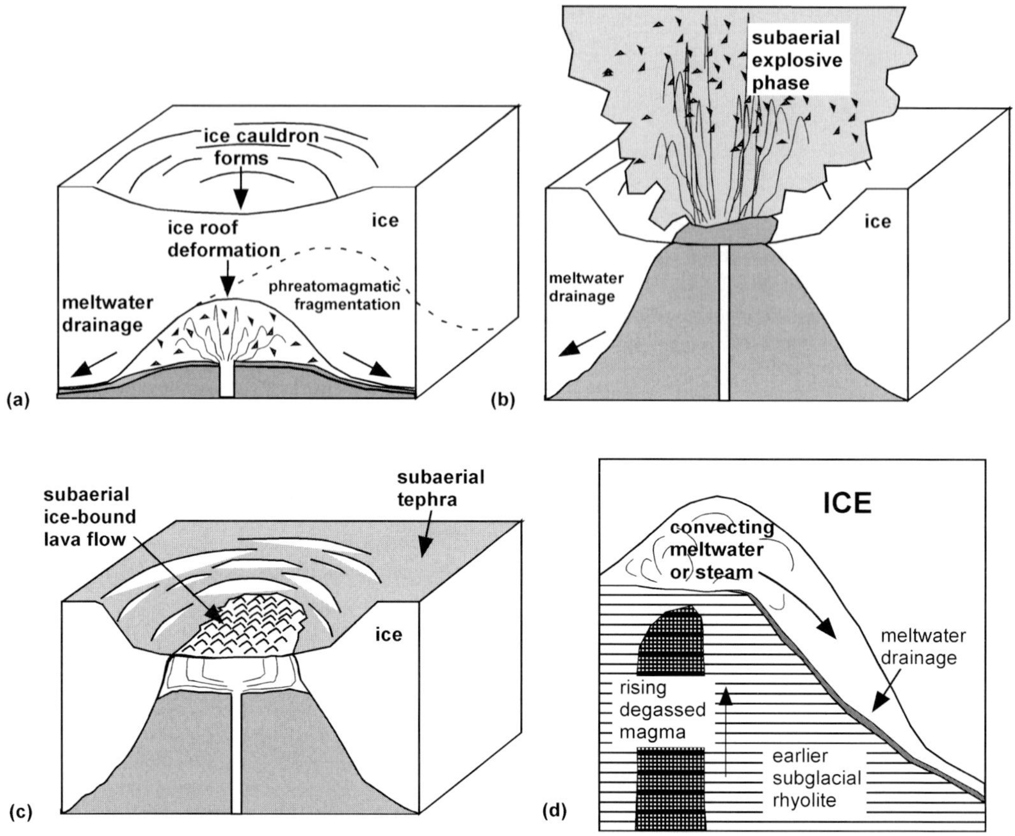

Fig. 12. Cartoons illustrating the possible evolution of SE Rauðufossafjöll. See text for explanation. The existence of a subaerial plinian eruptive episode (in (b)) is conjectural.

that a body of standing water did not develop (cf. Skilling 1994; Smellie & Hole 1997; Smellie 2000).

In the model, meltwater drained from the vent area (as in the 'leaky vault' scenario of Smellie 2000) during phreatomagmatic explosions confined within an ice vault (Fig. 12a). Pressure may have been near-atmospheric if exiting meltwater was connected hydrologically with the glacier snout (Hooke 1984; Björnsson 1988). The void space formed by melting and subsequent meltwater escape was filled by a pile of poorly-consolidated tephra. The bulk of the thermal energy of the magma may have been transferred to the ice by convecting steam or meltwater, with minor mechanical abrasion as tephra collided with the ice walls. Vault volume was controlled by the comparative rates of melting and ice deformation (Kiver & Steele 1975; Cutler 1998). The vault may have become completely filled if its volume was unable to increase sufficiently rapidly to accommodate the growing pile of tephra. If this occurred, the vault pressure may have increased to near-glaciostatic ($\leq$4 MPa for 400 m ice thickness), causing a switch to an intrusive phase of eruption.

This eruption mechanism contrasts with that inferred for a small-volume subglacial rhyolite eruption at Bláhnúkur, Torfajökull (Tuffen *et al.* 2001). The Bláhnúkur eruption was an order of magnitude less voluminous than SE Rauðufossafjöll. In it, lava bodies intruded fragmental debris generated by steam-quenching (hyaloclastite), with minor explosivity (Furnes *et al.* 1980). The ice roof appears to have always been <20 m above the growing edifice (Tuffen *et al.* 2001), and a large ice vault never developed. It is proposed that melting rates were considerably higher during the subglacial eruption at SE Rauðufossafjöll. This may be due to the greater role of fragmentation, which increases the rate of energy exchange from magma to ice (Guðmundsson *et al.* 1997). Furthermore, the magma discharge rate at SE Rauðufossafjöll was probably far higher, since the total erupted volume was an order of magnitude greater and, in general, the magma discharge rate is thought to be roughly proportional to the total erupted volume (Pyle 1999). This would have led to a higher heat flux at the glacier base, and more rapid melting.

Transition to subaerial eruption

The only evidence that the eruption pierced the roof of the ice vault and became subaerial comes from the characteristics of the lava flows on the summit ridge. Unfortunately, the base of these lava flows and contacts with the underlying ash are not exposed. Any unconsolidated tephra is susceptible to erosion, whereas a robust lava cap has a much higher preservation potential. Therefore, the suite of lithofacies preserved at ancient subglacial rhyolite volcanoes may be strongly influenced by the relative timing and magnitude of explosive and effusive phases of eruption (Eichelberger *et al.* 1986). In an entirely explosive eruption, a subglacial explosive phase may be followed by a subaerial explosive phase, but no lava cap is produced. Evidence for subaerial activity may not be preserved at the volcano, which is thus liable to be misinterpreted as entirely subglacial.

It is only possible to speculate as to whether SE Rauðufossafjöll produced a major subaerial explosive eruption (Fig. 12b). Analysis of the age and composition of silicic tephra layers in ice cores (Zielinski *et al.* 1997), deep-sea sediments (Lacasse *et al.* 1995) or peat bogs (Dugmore *et al.* 1995) and correlation with Icelandic source volcanoes may provide the answer. However, tephra from smaller-magnitude subaerial eruptions may only accumulate on the adjacent glacier surface and not be represented in the tephrochronological record.

The summit lava flows (lava A and B) were probably emplaced within ice cauldrons (Fig. 12c). The height of the surrounding ice walls (i.e. depth of the ice cauldron) cannot be accurately reconstructed from the information available.

Columnar-jointed lava phase (lava C)

The columnar-jointed lava bodies that crop out at 900–1000 m elevation on the north and west flanks appear to have chilled against steeply-inclined ice walls. Lack of vesicles suggests that they were largely degassed. In the model, these lavas were emplaced at a late stage of the eruption, as effusive activity began from vents situated on the flanks of the edifice. Lavas, rather than phreatomagmatic tephra were produced due to a low eruption rate. High heat flux preceding lava emplacement melted cavities in the ice above the new flank vents, creating 'moulds' into which the lavas flowed (Tuffen *et al.* 2002). Heat transfer via convecting steam or meltwater melted a steep-walled subglacial cavity (Kiver & Steele 1975). Drainage of meltwater would be most likely for vents located on steep slopes (Fig. 12d), possibly allowing low-pressure steam-filled cavities to develop. This lithofacies is similar to the columnar-jointed lava flow lithofacies of Bláhnúkur (Tuffen *et al.* 2001).

Peperitic lava phase (lava D lava E)

The upper carapace of lava D is characteristic of subaerial rhyolite lava flows, whereas its sides appear to have chilled against inclined ice walls, and its base has flowed over wet, poorly-consolidated sediments. To explain this unusual combination of features, it is proposed that lava D was formed from a series of effusive vents active on the edge of the eastern plateau, creating lavas that flowed over waterlogged breccias before freezing against an ice wall. The presence of 'rafts' of bedded ash directly underlying the lavas suggests that the massive breccias were locally redeposited by running meltwater prior to lava effusion.

One possibility is that lava D was emplaced after the construction of the main edifice. In this scenario, the glacier may have receded to *c.* 900 m surface elevation in the entire SE Rauðufossafjöll area, forming a 'moat' of ice around the edifice. Lava D then flowed subaerially before abutting the ice walls. However, wholesale retreat of ice from around the lavas of the summit ridge would probably trigger major instability and spalling of debris, forming a talus deposit on top of the fragmental material and underlying lava D. Such a talus deposit was not observed, leading us to suspect that the surface underlying lava D was protected from spalling debris, possibly by a roof of ice.

Comparison with basaltic tuya sequences

Rhyolitic tuyas such as SE Rauðufossafjöll are morphologically similar to basaltic tuyas. Both consist of a subglacial portion, dominated by fragmental deposits, that is overlain by near-horizontal subaerial lava flows (Table 3). However, the fragmental deposits have contrasting sedimentological features. These indicate that

Table 3. *Differences between basaltic and rhyolitic tuyas*

Basaltic tuya	Rhyolitic tuya
Subaerial lava flows 1.5–20 m thick*	Subaerial lava flows 10–150 m thick¶
Hyaloclastite – increasingly vesicular with height, commonly bedded in turbidite sequences (Surtseyan succession)† Pillow lava – high-pressure lithofacies (deep water/filled vault)†	Poorly-sorted, unbedded phreatomagmatic ash¶
Energy initially transferred from magma to ice via convecting meltwater, then by steam	Energy transferred primarily by steam
Fragmental subglacial deposits well exposed due to palagonitization‡	Fragmental subglacial deposits poorly exposed and mostly scree-covered¶
Meltwater accumulates at the vent area†	Meltwater drains away from the vent area.
High-magnitude jökulhlaups likely due to meltwater accumulation§	Low-magnitude jökulhlaups likely due to gradual meltwater release¶
Presence of ice above volcano increases likelihood of tephra-producing eruption¶	Likelihood of tephra-producing eruption not greatly increased by presence of ice¶

* Smellie & Hole 1997; † Smellie 2000; ‡ Jones 1969, § Björnsson 1988; ¶ This study.

the ice vault formed during a rhyolitic eruption is well-drained of meltwater, whereas meltwater tends to accumulate during a basaltic eruption. This has profound effects on the mechanisms of magma–ice interaction and edifice construction.

A possible explanation is that a rhyolitic eruption will commence with an ice-confined phreatomagmatic phase. Upon transition to a subaerial eruption, little meltwater is likely to be available for interaction with rising magma. Therefore, explosive subaerial activity may only occur if the magma is sufficiently volatile-rich to generate a 'dry' magmatic eruption. This may explain the similar morphology of distal rhyolitic ash shards formed during subaerial and subglacial-to-emergent eruptions in Iceland (Hafliðason *et al.* 2000).

Basaltic tuya-building eruptions typically begin with the effusion of pillow lavas. Magma–water interaction becomes increasingly explosive as the confining pressure decreases (e.g. Jones 1969; Smellie 2000). Once the ice surface is breached and a subaerial eruption ensues, meltwater flooding the vent is likely to interact explosively with rising magma, triggering an eruption of ash and steam (Guðmundsson *et al.* 1997). Distal tephra from subglacial-to-emergent eruptions of basalt has blocky shard morphologies consistent with phreatomagmatic fragmentation (Hafliðason *et al.* 2000). Since Icelandic basaltic tuyas are thought to be the subglacial equivalents of effusive shield volcanoes, the presence of an overlying glacier increases the likelihood of an explosive subaerial eruption.

Conclusions

Lithofacies associations within the subglacial rhyolite tuya at SE Rauðufossafjöll indicate different eruption characteristics from those at basaltic tuyas. In the initial subglacial phase, explosive magma–water interaction generated a pile of fine-grained ash within a well-drained ice vault. After the glacier surface was breached, subaerial lava flows were emplaced within ice cauldrons above the *c.* 2 km-long eruptive fissure. Effusive eruptions on the north and west flanks of the tuya created columnar-jointed lava flows, which were emplaced within steeply-inclined subglacial cavities. Lavas on the east flank are interpreted as subaerial flows that travelled over wet sediments before freezing against an ice wall. Volcaniclastic sediments at the northern margin of the tuya were deposited during channelized meltwater drainage from the vent area during the eruption. Differences between rhyolitic and basaltic tuya sequences are best explained by contrasting hydrological patterns caused by the temperature differences between the two magma types.

We are grateful to F. Eiríksson, B. Bjarnadóttir and Á. Kristinsdóttir of Ferðafelag Íslands for their assistance during fieldwork. Discussions in the field with B. F. Houghton and J. L. Smellie helped to focus our ideas. B. Edwards and A. Hoskuldsson are thanked for reviews which greatly improved the manuscript. We thank K. Mee for fieldwork assistance. H. T. was supported by an Open University research studentship.

J. S. G.'s fieldwork was supported by a Lancaster University travel grant. D. McG. was supported by an Open University research grant.

References

BJÖRNSSON, H. 1988. Hydrology of ice caps in volcanic regions. *Vísindafélag Íslendinga, Societas Scientarium Islandica*, **45**.

CAS, R. A. F. & WRIGHT, J. V. 1987. Volcanic Successions. Chapman & Hall, London.

CUTLER, P. M. 1998. Modelling the evolution of subglacial tunnels due to varying water input. *Journal of Glaciology*, **44**, 485–497.

DAVIS, B. K. & MCPHIE, J. 1996. Spherulites, quench fractures and relict perlite in a Late Devonian rhyolite dyke, Queensland, Australia. *Journal of Volcanology and Geothermal Research*, **71**, 1–11.

DUGMORE, A. J., LARSEN, G. & NEWTON, A. J. 1995. 7 Tephra isochrones in Scotland. *Holocene*, **5**, 257–266.

EICHELBERGER, J. C., CARRIGAN, C. R. & WESTRICH, H. R. 1986. Non-explosive silicic volcanism. *Nature*, **323,** 598–602.

FINK, J. H. 1983. Structure and emplacement of a rhyolitic obsidian flow – Little Glass Mountain, Medicine Lake highland, Northern California. *Geological Society of America, Bulletin*, **94**, 362–380.

FRIEDMAN, I., SMITH, R. L. & LONG, W. D. 1966. Hydration of natural glass and formation of perlite. *Geological Society of America, Bulletin*, **77**, 323–328.

FURNES, H., FRIDLEIFSSON, I. B. & ATKINS, F. B. 1980. Subglacial volcanics – on the formation of acid hyaloclastites. *Journal of Volcanology and Geothermal Research*, **8**, 95–110.

GRÖNVOLD, K. 1972. *Structural and petrochemical studies in the Kerlingarfjöll region, central Iceland.* PhD thesis, University of Oxford.

GUÐMUNDSSON, M. T., SIGMUNDSSON, F. & BJÖRNSSON, H. 1997. Ice–volcano interaction of the 1996 Gjálp subglacial eruption, Vatnajökull, Iceland. *Nature*, **389**, 954–957.

HAFLIÐASON, H., EIRíKSSON, J. & VANKREVELD, S. 2000. The tephrochronology of Iceland and the North Atlantic region during the Middle and Late Quaternary: a review. *Journal of Quaternary Science*, **15**, 3–22.

HEIKEN, G. & WOHLETZ, K. 1985. *Volcanic ash*. University of California Press, Berkeley.

HOOKE, R. L. 1984. On the role of mechanical energy in maintaining subglacial water conduits at atmospheric pressure. *Journal of Glaciology*, **30**, 180–187.

HOSKULDSSON, A. & SPARKS, R. S. J. 1997. Thermodynamics and fluid dynamics of effusive subglacial eruptions. *Bulletin of Volcanology*, **59**, 219–230.

HUNNS, S. R. & MCPHIE, J. 1999. Pumiceous peperite in a submarine volcanic succession at Mount Chalmers, Queensland, Australia. *Journal of Volcanology and Geothermal Research*, **88**, 239–254.

JONES, J. G. 1969. Intraglacial volcanoes of the Laugarvatn region, south-west ICELAND, I. *Quarterly Journal of the Geological Society, London*, **124** 197–211.

JONES, J. G. 1970. Intraglacial volcanoes of the Laugarvatn region, southwest Iceland, II. *Journal of Geology*, **78**, 127–140.

KIVER, E. P. & STEELE, W. K. 1975. Firn caves in the volcanic craters of Mount Rainier, Washington. *National Speleological Society, Bulletin*, **37**, 45–55.

KOKELAAR, B. P. 1982. Fluidization of wet sediments during the emplacement and cooling of various igneous bodies. *Journal of the Geological Society, London*, **139**, 21–33.

LACASSE, C., SIGURDSSON, H., JOHANNESSON, H., PATERNE, M. & CAREY, S. 1995. Source of Ash-Zone-1 in the North-Atlantic. *Bulletin of Volcanology*, **57**, 18–32.

LARSEN, G. 1984. Recent volcanic history of the Veiðivötn fissure swarm, southern Iceland – an approach to volcanic risk assessment. *Journal of Volcanology and Geothermal Research*, **22**, 33–58.

LARSEN, G., GUðMUNDSSON, M. T., BJÖRNSSON, H. 1998. Eight centuries of periodic volcanism at the center of the Iceland hotspot revealed by glacier tephrostratigraphy. *Geology*, **26,** 943–946.

LESCINSKY, D. T. & SISSON, T. W. 1998. Ridge-forming, ice-bounded lava flows at Mount Rainier, Washington. *Geology*, **26**, 351–354.

MCGARVIE, D. W. 1984. Torfajökull – a volcano dominated by magma mixing. *Geology*, **12**, 685–688.

MCGARVIE, D. W., MACDONALD, R., PINKERTON, H. & SMITH, R. L. 1990. Petrogenetic evolution of the Torfajökull volcanic complex, Iceland. 2. The role of magma mixing. *Journal of Petrology*, **31**, 461–481.

MCPHIE, J., DOYLE, M. & ALLEN, R. 1993. *Volcanic textures. A guide to the interpretation of textures in volcanic rocks.* Tasmanian Government Printing Office, Tasmania.

MATHEWS, W. H. 1951. The Table, a flat-topped volcano in southern British Columbia. *American Journal of Science*, **249**, 830–841.

NOE-NYGAARD, A. 1940. Sub-glacial volcanic activity in ancient and recent times (studies in the palagonite system of Iceland, no. 1). *Folia Geographica Danica*, **1**, 1–67.

PYLE, D. M. 1999. Sizes of volcanic eruptions. *In:* SIGURDSSON, H. (ed.) *Encyclopaedia of Volcanoes.* Academic Press, San Diego, 263–269.

SÆMUNDSSON, K. 1972. Jarðfræðiðattur um Torfajökulssvæðið. *Natturufræðingurinn*, **42**, 81–99 (in Icelandic).

SÆMUNDSSON, K. 1988. Jarðfræðiðattur um Torfajökulsöræfi. *Arbok Ferð*afelag Islands, 164–180 (in Icelandic).

SKILLING, I. P. 1994. Evolution of an englacial volcano: Brown Bluff, Antarctica. *Bulletin of Volcanology*, **56**, 573–591.

SMELLIE, J. L. 2000. Subglacial eruptions. *In*: Sigurdsson H. (ed.) *Encyclopaedia of volcanoes.* Academic Press, San Diego, 403–418.

SMELLIE, J. L. & Hole, M. J. 1997. Products and processes in Pliocene–Recent, subaqueous to emergent volcanism in the Antarctic Peninsula: examples of englacial Surtseyan volcano construction. *Bulletin of Volcanology*, **58**, 628–646.

SMELLIE, J. L. & SKILLING, I. P. 1994. Products of subglacial volcanic eruptions under different ice thicknesses – 2 examples from Antarctica. *Sedimentary Geology*, **91**, 115–129.

SMELLIE, J. L., HOLE, M. J. & NELL, P. A. R. 1993. Late Miocene valley-confined subglacial volcanism in northern Alexander Island, Antarctic Peninsula. *Bulletin of Volcanology*, **55**, 273–288.

SOOSALU, H. & EINARSSON, P. 1997. Seismicity around the Hekla and Torfajökull volcanoes, Iceland, during a volcanically quiet period 1991–1995. *Bulletin of Volcanology*, **59**, 36–48.

STEVENSON, R. J. & WILSON, L. 1997. Physical volcanology and eruption dynamics of peralkaline agglutinates from Pantelleria. *Journal of Volcanology and Geothermal Research*, **79**, 97–122.

TUFFEN, H., GILBERT, J. S. & MCGARVIE, D. W. 2001. Products of an effusive subglacial rhyolite eruption: Bláhnúkur, Torfajökull, Iceland. *Bulletin of Volcanology*, **63**, 179–190.

TUFFEN, H., PINKERTON, H., GILBERT, J. S. & MCGARVIE, D. W. 2002. Melting of the glacier base during a small-volume subglacial rhyolite eruption: evidence from Bláhnúkur, Iceland. *Sedimentary Geology*, **149**, 183–198.

WERNER, R., SCHMINCKE, H. U. & SIGVALDASON, G. 1996. A new model for the evolution of table mountains: volcanological and petrological evidence from Herðubreid and Herðubreidartogl volcanoes (Iceland). *Geologische Rundschau*, **85**, 390–397.

WERNER, R. & SCHMINCKE, H.-U. 1999. Englacial vs lacustrine origin of volcanic table mountains: evidence from Iceland. *Bulletin of Volcanology*, **60**, 335–354.

WHITE, J. D. L. & BUSBY-SPERA, C. 1987. Deep marine arc apron deposits and syndepositional magmatism in the Alisitos Group at Punta Cono, Baja California, Mexico. *Sedimentology*, **34**, 911–927.

WOHLETZ, K. H. 1986. Explosive magma–water interactions: thermodynamics, explosion mechanisms and field studies. *Bulletin of Volcanology*, **48**, 245–264.

ZIELINSKI, G. A., MAYEWSKI, P. A., MEEKER, L. D., GRÖNVOLD, K., GERMANI, M. S., WHITLOW, S., TWICKLER, M. S. & TAYLOR , K. 1997. Volcanic aerosol records and tephrochronology of the Summit, Greenland, ice cores. *Journal of Geophysical Research*, **102**, 26 625–26 640.

Lithofacies analysis and $^{40}Ar/^{39}Ar$ geochronology of ice–volcano interactions at Mt. Murphy and the Crary Mountains, Marie Byrd Land, Antarctica

THOMAS I. WILCH[1] & WILLIAM C. McINTOSH[2]

[1] *Department of Geological Sciences, Albion College, Albion MI 49224, USA (e-mail: twilch@albion.edu)*

[2] *Department of Earth and Environmental Science, New Mexico Institute of Mining and Technology, Socorro, NM 87801, USA*

Abstract: Palaeoenvironmental reconstructions and $^{40}Ar/^{39}Ar$ geochronology of volcanism at Mt. Murphy and the Crary Mountains in eastern Marie Byrd Land (MBL), West Antarctica, provide records of changing ice levels of the West Antarctic Ice Sheet (WAIS) since the late Miocene. Interpretations of eruptive and depositional environments are based on lithofacies studies and indicate whether the volcanoes erupted below, near or above the level of the ice sheet. Seventy-seven new $^{40}Ar/^{39}Ar$ dates offer a precise chronological framework for the ice volcanic history.

Late Miocene (9–8 Ma) basal volcanic sequences at Mt. Murphy and the Crary Mountains (Mt. Rees and Mt. Steere) exhibit fluctuations between 'wet' ice-contact lithofacies and 'dry' subaerial lithofacies. The 'wet' lithofacies include pillow lava and hyaloclastite breccia; the 'dry' lithofacies include massive and deuterically oxidized lava and associated welded breccia deposits. The sequences at Mt. Murphy include several erosion surfaces and tillites, which are inferred to represent fluctuations in the WAIS. At Mt. Rees and Mt. Steere, the alternating lithofacies form the constructional slopes of the volcano and are inferred to represent interactions with local slope ice that occurred above the level of the regional ice sheet.

The Miocene to Pleistocene volcanic history of the area provides a proxy record of ice-level changes in West Antarctica, with the following three major conclusions. First, the oldest evidence for a large-scale WAIS is from Late Miocene (*c.* 9 Ma) glaciovolcanic sequences at Mt. Murphy and several other sites in Marie Byrd Land. The combined Mt. Murphy and Crary Mountains records indicate that ice-level expansions of the WAIS were more extensive at coastal sites than at inland sites. Second, the present-day WAIS appears to be in a near maximum configuration that has existed at several times since 9 Ma but was rarely exceeded. Finally, a significant expansion of the WAIS above its present-day level occurred at 590 ± 15 ka, when ice levels were 550 m higher at the coastal volcano, Mt. Murphy.

Volcanism and glaciation have been active geological forces in Marie Byrd Land (MBL), West Antarctica, since middle Cenozoic time (LeMasurier 1990; Anderson & Shipp 2001). About 50 middle to late Cenozoic alkaline volcanic centres, including 19 large, polygenetic volcanoes (2364–4181 m above sea level), are exposed as nunataks in the West Antarctic Ice Sheet (WAIS; Fig. 1). The MBL volcanic nunataks are located on the north flank of the large-scale intracontinental rift system that extends from the Ross Sea basin to deep marine basins beneath the WAIS (Behrendt 1999). Wilch & McIntosh (2000) suggested that the oldest terrestrial indications for glacial ice in West Antarctica are middle Oligocene (29–27 Ma) tuff cone deposits at Mt. Petras. These Oligocene deposits were inferred to result from ice–volcanic interactions with a thin, local ice cap or ice and snow on slopes. Direct terrestrial evidence for the timing of the subsequent full-scale development of the WAIS is based on the volcanic record and is the subject of this study.

Many MBL volcanoes preserve records of syn-eruptive interactions with ice that can be used to infer palaeo-ice-levels of the WAIS (LeMasurier 1972*a*, *b*). The premise of this volcanological approach is when volcanoes erupt below, at or above the level of an ice sheet, the resulting rocks exhibit specific textural features and structures that are diagnostic of their eruptive and/or depositional environments (e.g. Smellie & Skilling

From: SMELLIE, J. L. & CHAPMAN, M. G. (eds) 2002. *Volcano–Ice Interaction on Earth and Mars*. Geological Society, London, Special Publications, **202**, 237–253. 0305-8719/02/$15.00

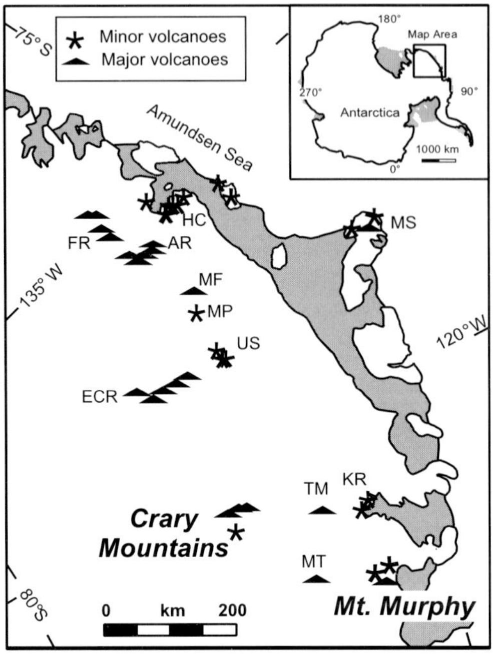

Fig. 1. Map of Marie Byrd Land, West Antarctica (base map from Drewry 1983), showing major polygenetic and minor monogenetic volcanoes of the Marie Byrd Land Volcanic Province. Mt. Murphy and the Crary Mountain volcanoes are located in eastern Marie Byrd Land. Volcano abbreviations: AR, Ames Range; ECR, Executive Committee Range; FR, Flood Range; HC, Hobbs Coast nunataks; KR, Kohler Range; MF, Mt. Flint; MP, Mt. Petras; MS, Mt. Siple; MT, Mt. Takahe; TM, Toney Mountain; and US, USAS. Escarpment.

1994; Wilch & McIntosh 2000; Smellie 2001). In particular, subglacially emplaced rocks imply higher former ice-levels relative to the present-day outcrop level. The horizontal transitions from subglacial to subaerial depositional environments, termed passage zones (after Jones 1969), mark the water level at the time of the eruption and provide a minimum estimate of the syn-eruptive palaeo-ice-level. Thus, records of the age, elevation and depositional environment of volcanic rocks provide snapshot views of the syn-eruptive level of the ice sheet. LeMasurier (1972*a*, *b*) and LeMasurier & Rex (1982, 1983) reconstructed a volcanic record of glaciation in West Antarctica, based on regional reconnaissance fieldwork and K/Ar geochronology. Recent studies by LeMasurier *et al.* (1994), Wilch & McIntosh (2000) and Smellie (2001) have presented more detailed accounts of glaciovolcanic interactions at Mt. Murphy and Mt. Petras. In this paper, we compare and contrast the records of glaciovolcanism of two eastern MBL volcanic centres: Mt. Murphy and the Crary Mountains (Fig. 1).

Lithofacies terminology and analysis

Definitions of volcanic and lithofacies terminology generally follow suggestions by Fisher & Schmincke (1984) and McPhie *et al.* (1993). The lithofacies are subdivided into four classes: coherent lavas, autoclastic deposits, pyroclastic deposits, and sedimentary deposits. Coherent lava and autoclastic deposits are the most common lithofacies exposed at the volcanoes. For this study, lithofacies were subdivided into 'dry' and 'wet' end-members, based on the absence or presence of features characteristic of interactions with external water either during the eruption or during deposition.

Lavas at the Crary Mountains and Mt. Murphy include a wide range of alkaline to peralkaline compositions from basanite to trachyte and phonolite, and rare rhyolites. The 'dry' lava lithofacies includes lavas with reddened, brecciated bases and pahoehoe tops that are interpreted to result from subaerial lava effusion, without recognizable water interaction. Many inferred subaerial lavas appear to be clastogenic, i.e. derived from agglutinated pyroclastic spatter. Two 'wet' lava lithofacies, pillow lava and hackly, jointed lava, are interpreted as water-cooled and generally associated with underwater depositional environments (see e.g. Fuller 1933). Two variations of pillow lava lithofacies are recognized: compound pillowed flows that form either nested sets with minor interpillow hyaloclastite breccia, or distinct lobes with abundant (>10%) interpillow hyaloclastite breccia. Pillow lavas are common at the Crary Mountains and Mt. Murphy, but are very rare elsewhere in Marie Byrd Land (Wilch 1997). The hackly, jointed lava lithofacies includes slightly glassy, compound and simple flows and intrusive bodies, with irregular to hackly jointing and rare crude pillow structures. Some of these jointed lavas resemble 'para-pillows' of Icelandic flow-foot delta sequences (Jones 1970), which Walker (1992) suggested form when thin, subaqueous lavas fail to form pillows because they are flowing too fast down steep underwater slopes or are cut off from their sources. We interpret these jointed lavas and intrusive rocks as water-cooled (Skilling 1994), although a subaqueous origin cannot be inferred in all cases. Lava apophyses locally deform bedding and exhibit hackly jointing. These apophyses are interpreted as dykes or incursive flows that intruded wet volcaniclastic sediments. In some cases, lavas, particularly those that are thick, massive or poorly exposed, lack diagnostic features of subaerial or subaqueous environments. Unless these lavas are associated with other subaqueous deposits, they have been tentatively interpreted as subaerial lavas.

Autoclastic deposits are formed by mechanical granulation during lava emplacement, and include 'dry' flow breccias associated with subaerial lava emplacement and 'wet' hyaloclastite breccias associated with subaqueous pillow lava emplacement. Both types of breccia deposits are easily identified by their close association with coherent lava facies. Subaerial lava flow breccias are recognized by welding textures and reddening caused by deuteric oxidation. The term 'hyaloclastite' requires discussion because, in previous reconnaissance studies of Marie Byrd Land volcanoes (LeMasurier 1972*a*, *b*; LeMasurier & Rex 1982, 1983), it was broadly defined as a fracture-bounded, glassy, fragmental rock and also included most palagonitized or altered fragmental rock. Hyaloclastites were interpreted as indications of subglacial environments and higher palaeo-ice-levels. This broad definition of hyaloclastite has two major weaknesses. First, the definition does not differentiate clasts produced by passive granulation in a subaqueous setting from those produced by hydromagmatic explosivity in an emergent to subaerial environment. Second, palagonite is a product of hydration and alteration of quenched sideromelane glass and is common in a variety of hydrovolcanic environments from deep subaqueous to emergent and subaerial (e.g. Fisher & Schmincke 1984). Palagonitized deposits have also been identified in dry, Strombolian subaerial volcanoes, where they are attributed to post-eruptive alteration by steam or surface water (Houghton & Schmincke 1986; McIntosh & Gamble 1991; LeMasurier *et al.* 1994). For this study, the term hyaloclastite is restricted to breccia dominated by angular, blocky, poorly vesiculated ($<25\%$ vesicles) vitriclasts that commonly exhibit a jigsaw-fit arrangement (following Honnorez & Kirst 1975; see e.g. Skilling 1994; Smellie & Skilling 1994). No particular fragmentation method is implied by the term, although empirical observation suggests that hyaloclastites are typically formed by cooling-contraction granulation (*sensu* Kokelaar 1986) in combination with mechanical granulation processes. For this study, hyaloclastite breccias are classified as autoclastic deposits despite the fact that they may be formed wholly or in part by cooling-contraction granulation when water comes in contact with lava (Carlisle 1963; Kokelaar 1986).

Pyroclastic deposits are also common in MBL sequences and are dominated by two types: pyroclastic-fall and pyroclastic-surge deposits. Pyroclastic fall deposits are products of magmatic and hydromagmatic explosions and typically form crudely to well stratified deposits that mantle topography. A magmatic eruptive mechanism is inferred where pyroclasts are well-sorted and exhibit uniform, moderate to high vesicularity and angular to fluidal shapes. Reddening by deuteric oxidation and welding are common features and are interpreted as indications of close proximity to a subaerial vent (Walker & Croasdale 1972). A hydromagmatic eruptive mechanism is inferred from relatively poor sorting, fine grain sizes, variable vesicularity, armoured lapilli, predominance of sideromelane glass (in basalts), and blocky forms (Fisher & Schmincke 1984). The explosively erupted hydromagmatic deposits are referred to as hyalotuffs (following Honnorez & Kirst 1975). In MBL, base surge deposits were previously unrecognized in late-stage flank deposits and in smaller monogenetic volcanoes. The MBL base surge deposits were identified by comparison to base surge deposits described elsewhere (e.g. Fisher & Schmincke 1984) and typically form moderate to poorly sorted, planar, cross-stratified, or massive beds that thicken and thin laterally. Individual clasts in base surge deposits resemble those in hydromagmatic fall deposits, except that they are often much more rounded due to turbulent lateral transport. Base surge and hydromagmatic fall deposits are commonly interbedded. Well-preserved armoured lapilli, bedding plane sags and sand wave bed forms suggest that many of the deposits have not been reworked since the time of eruption.

Sedimentary deposits are rare in MBL and include deposits formed by primary glacial, mass-flow and fluvial processes. Sedimentary deposits are characterized by their heterolithic clasts, subrounded clast shapes, sedimentary structures, and stratigraphic context. Despite the glaciated setting of the MBL volcanic province, interbedded glacial tills/tillites and glacial-erosional unconformities are rare, except at Mt. Murphy.

$^{40}Ar/^{39}Ar$ geochronology

Methods

The geochronology at Mt. Murphy and the Crary Mountains is based on $^{40}Ar/^{39}Ar$ dating analyses of seventy-seven mafic to felsic alkaline lava flow and bomb samples. Thin sections of lava and bomb samples were inspected for mineralogy and alteration under a cross-polarizing

Table 1. *$^{40}Ar/^{39}Ar$ furnace step-heating and laser-fusion results*

ID[1]	Site	Rock type[2]	sam (meth)[3]	Total fusion age ±2 s.d.	Isochron age ±2 s.d.	$^{40}Ar/^{36}Ar$ ±2 s.d.	D	%^{39}Ar	Mean (plateau) age ± 2 s.d.	%^{39}Ar	Geology notes
Mt. Murphy											
90-69	Sechrist Peak	basn	gms (F)	0.66 ± 0.15	0.599 ± 0.086	294 ± 15	0.97	66.0	**0.587 ± 0.043**	66.0	west flank, 1299 m
90-139	Sechrist Peak	basn	gms (F)	0.60 ± 0.18	0.601 ± 0.028	292.9 ± 3.6	1.27	96.4	**0.592 ± 0.040**	96.4	dense interior of flow, 1247 m
90-34	main shield section	trach	gms (F)						**8.22 ± 0.04**	90.0	trachyte lava, 725 m
90-37	main shield section	basn	gms (F)	8.86 ± 0.37	8.58 ± 0.09	301.6 ± 4.6	47.6	58.0	***8.80 ± 0.20***	79.9	lava, 684 m
90-37	main shield section	basn	gms (F)	9.65 ± 0.16	8.72 ± 0.07	304.5 ± 3	21.7	81.5	***8.86 ± 0.15***	81.5	lava, 684 m
90-39	main shield section	basn	gms (F)	9.48 ± 0.26	9.23 ± 0.06	300.1 ± 3.8	0.82	85.3	**9.26 ± 0.14**	85.3	lava & hyaloclastitie, 555 m
90-39	main shield section	basn	gms (F)	9.21 ± 0.21	9.18 ± 0.05	246.9 ± 7.8	63.90	89.3	***9.03 ± 0.12***	92.0	lava & hyaloclastitie, 555 m
90-33	main shield section	basn	gms (F)	9.09 ± 0.13	9.18 ± 0.12	278.1 ± 17	0.45	58.3	***9.07 ± 0.09***	58.3	striated lava, 527 m
90-33	main shield section	basn	gms (F)	9.57 ± 0.34	9.38 ± 0.09	296.8 ± 4	3.8	98.9	**9.39 ± 0.12**	98.9	striated lava, 527 m
90-50	main shield section	basn	gms (F)	9.31 ± 0.21	9.31 ± 0.05	299.2 ± 2.8	25.50	87.0	***9.35 ± 0.13***	87.0	pillow lava, *c.* 498 m
90-50	main shield section	basn	gms (F)	9.13 ± 0.18	9.27 ± 0.05	301.4 ± 3.4	12.0	78.0	***9.32 ± 0.13***	78.0	pillow lava, *c.* 498 m
90-99	Hedin Nunatak	basn	gms (F)	6.23	***6.20 ± 0.24***	305.2 ± 6.4	0.50	100.0	no plateau		subaerial lava, upper tuya
90-110	Hedin Nunatak	basn	gms (F)	12.0 ± 5.0	***6.50 ± 0.12***	310 ± 0.9	3.69	100.0	no plateau		lowest tindar
90-47	Icefall Nunatak	basn	gms (F)	6.46 ± 0.25	6.43 ± 0.11	300.6 ± 5.2	1.96	58.8	**6.52 ± 0.13**	58.8	high subaerial lava
90-48	Icefall Nunatak	basn	gms (F)	11.30 ± 4.60	***6.80 ± 0.20***	301.9 ± 2.4	2.58	95.7	7.10 ± 0.25	51.9	lobe in low hyaloclastite
90-94	Turtle Peak	basn	gms (F)	5.17 ± 0.58	4.59 ± 0.09	299.7 ± 2.2	0.93	96.6	**4.70 ± 0.15**	96.6	v. vesicular pahoehoe top
90-92	Turtle Peak	basn	gms (F)	5.69 ± 0.22	5.71 ± 0.10	289.5 ± 5.6	29.30	90.9	***5.65 ± 0.23***	90.9	upper pillow
90-87	Turtle Peak	basn	gms (F)	16.80 ± 2.80	***5.95 ± 0.60***	316.3 ± 3.8	1.87	99.5	9.36 ± 0.60	64.7	lower flow-foot breccia
Mt. Rees											
92-174	N Trabucco	trach	ano (L)	**8.94 ± 0.06**						100.0	lava flow top of section
92-175	N Trabucco	trach	ano (L)	**8.95 ± 0.06**						100.0	lava flow top of section
92-6	Trabucco Cliff	trach	ano (L)	**8.98 ± 0.07**						100.0	massive flow, >3
92-15	Trabucco Cliff	trach	ano (L)	**9.01 ± 0.04**						100.0	lava flow middle of section
92-1	Trabucco Cliff	basn	gms (F)	9.42 ± 0.17	7.96 ± 0.08	351.0 ± 4.2	253	100.0	***9.13 ± 0.53***	100.0	lobe hyaloclastite
92-36	Tasch Peak	phono	ano (L)	**7.52 ± 0.06**						100.0	phonolite dyke
92-38	Tasch Peak	basn	gms (F)	8.24 ± 0.05	8.12 ± 0.06	313.0 ± 9.0	154	96.3	***8.21 ± 0.13***	96.3	pillow hyaloclastite, 2230 m

92-34	Tasch Peak	mafic	gms (F)	8.42 ± 0.05	8.04 ± 0.06	347.3 ± 9.4	13.2	84.1	***8.23 ± 0.09***	84.1	intrusive hyaloclastite, 2265 m
92-41	Tasch Peak	mug	gms (F)	9.20 ± 0.14	8.42 ± 0.24	305.5 ± 28	5.05	49.9	***8.50 ± 0.16***	49.9	hyaloclastite lobes
92-31	Tasch Peak	ph-teph.	gms (F)	9.13 ± 0.06	8.90 ± 0.06	309.5 ± 2.8	554	100.0	***8.98 ± 0.28***	100.0	pillow lobe interior
92-28	Tasch Peak	ben	gms (F)	9.23 ± 0.08	9.06 ± 0.07	298 ± 6.2	1.0	54.1	**9.08 ± 0.06**	54.1	spatter-fed lava, 1844 m
92-23	Tasch Peak	ben	gms (F)	8.94 ± 0.06	8.85 ± 0.01	309.2 ± 1.8	39.1	88.6	***8.91 ± 0.12***	100.0	pillow lobe, 1817 m
92-59	Tasch Peak	basn	gms (F)	9.44 ± 0.38	9.03 ± 0.06	302.5 ± 1.8	20.0	100.0	***9.34 ± 0.24***	65.0	lava flow, 1739 m
92-117	SW Summit		gms (F)	6.78 ± 0.18	6.92 ± 0.05	286.3 ± 1.8	96.9	94.5	***6.82 ± 0.26***	94.5	lava flow, 1643 m
92-14	SW Summit		gms (F)	6.72 ± 0.14	7.10 ± 0.06	283.5 ± 2.4	35.6	99.3	***6.92 ± 0.21***	99.3	lava flow
92-112	north end	trach	gms (F)	12.52 ± 0.08	NO ISOCRHON				***9.02 ± 0.19***	51.5	lava flow, striae (240)
92-109	north end	trach	gms (F)	8.76 ± 0.10	8.76 ± 0.06	282.4 ± 5.4	5.52	94.1	***8.70 ± 0.08***	94.1	lava flow, striae (328)
Mt. Steere											
92-118	west side	phono	ano (L)	**7.55 ± 0.05**						100.0	flow banded lava
92–95	moraine	mug	gms (F)	6.83 ± 0.07	5.72 ± 0.11	303.3 ± 45.3	0.27	60.6	**5.74 ± 0.04**	60.6	flow banded lava
92-64	NE outcrop	phono	gms (F)	8.05 ± 0.04	8.03 ± 0.05	301.1 ± 2.6	18.4	100.0	***8.06 ± 0.08***	98.9	dyke
92-53	NE outcrop	trach	ano (L)	**8.24 ± 0.08**						100.0	lava flow dome
92-93	NE outcrop	trach	ano (L)	**8.25 ± 0.08**						100.0	dyke
92-63	NE outcrop	trach	ano (L)	**8.27 ± 0.08**						100.0	dyke, 145, 67E salt and pepper dyke
92-51	NE outcrop	trach	ano (L)	**8.33 ± 0.06**						100.0	lava flow, exposed plug
92-107	N. arete	trach	ano (L)	**8.34 ± 0.06**						100.0	dyke, 2412 m
92-104	N. arete	trach	ano (L)	**8.35 ± 0.08**						100.0	flow banded lava, 2278 m
92-108	N. arete	trach	ano (L)	**8.37 ± 0.06**						100.0	flow banded lava, 2412 m
92-91	East Side	trach	ano (L)	**8.46 ± 0.09**						100.0	flow banded lava
92-181	East Side	phono	ano (L)	**7.91 ± 0.20**						100.0	dyke, fine grained
92-182	East Side	rhyo	ano (L)	**8.52 ± 0.06**						100.0	flow banded lava
92-183	East Side	rhyo	ano (L)	**8.52 ± 0.06**						100.0	flow banded lava
92-178	East Side	trach	ano (L)	**8.55 ± 0.04**						100.0	flow banded lava
92-169	Lie Cliff, SE ridge	bas	gms (F)	6.26 ± 0.07	5.93 ± 0.01	341.4 ± 2.7	1132	80.5	***6.41 ± 0.43***	76.0	lava flow, 1814 m
92-165	Lie Cliff, SE ridge	basn	gms (F)	7.42 ± 0.14	7.38 ± 0.10	296.5 ± 18.2	4.39	65.3	***7.378 ± 0.07***	65.3	lava, near 92-162 1736 m
92-162	Lie Cliff, SE ridge	bas	gms (F)	6.65 ± 0.09	6.70 ± 0.03	266.8 ± 102.6	5.3	60.7	6.70 ± 0.05	60.7	lava, near 92-165, 1736 m
92-80	Lie Cliff	trach	ano (L)	**7.82 ± 0.06**						100.0	dyke, 1.5 m wide
92–85	Lie Cliff	haw	gms (F)	8.50 ± 0.14	8.65 ± 0.09	267.9 ± 5.8	84.6	90.5	***8.38 ± 0.33***	90.5	lava flow, 1631 m
92-82	Lie Cliff	haw	gms (F)	9.02 ± 0.11	8.06 ± 0.07	329.8 ± 9.4	213.9	82.1	***8.28 ± 0.21***	82.1	hyaloclastite lobe, 1600 m
92-79	Lie Cliff	haw	gms (F)	8.41 ± 0.06	8.43 ± 0.08	296.9 ± 2.0	2.34	93.1	**8.435 ± .061**	93.1	

Table 1. (*continued*)

ID[1]	Site	Rock type[2]	sam (meth)[3]	Total fusion age ±2 s.d.	Isochron age ±2 s.d.	$^{40}Ar/^{36}Ar$ ±2 s.d.	D	$\%^{39}Ar$	Mean (plateau) age ± 2 s.d.	$\%^{39}Ar$	Geology notes
92–86	Lie Cliff	haw	gms (F)	8.62 ± 0.27	8.16 ± 0.13	302.8 ± 2.8	5.6	93.4	***8.52 ± 0.23***	93.4	lava flow, 1558 m
92-89	Lie Cliff, NW ridge	trach	ano (L)	7.68 ± 0.06						100.0	dyke, intrudes entire section
92-193	Lie Cliff, NW ridge	haw	gms (F)	8.41 ± 0.10	8.02 ± 0.01	311.9 ± 1.0	28.2	64.7	**8.19 ± 0.18**	83.4	feeder dyke, lava, 1798 m
92-192	Lie Cliff, NW ridge	phono	ano (L)	8.22 ± 0.07						100.0	feeder dyke, lava
92-194	Lie Cliff, NW ridge	haw	gms (F)	8.47 ± 0.10	8.02 ± 0.05	313.4 ± 2.2	5.98	98.6	***8.19 ± 0.22***	98.6	glassy breccia, 1753 m
92-189	Lie Cliff, NW ridge	haw	gms (F)	8.93 ± 0.27	7.84 ± 0.08	309.5 ± 1.8	19.4	100.0	***8.45 ± 0.46***	100.0	glassy lava, 1747 m
92-190	Lie Cliff, NW ridge	haw	gms (F)	9.17 ± 0.35	7.06 ± 0.07	319.6 ± 1.8	12.6	100.0	***8.27 ± 0.64***	100.0	pillow lava, 1743 m
92-186	Lie Cliff, NW ridge	haw	gms (F)	8.93 ± 0.39	8.42 ± 0.13	294.8 ± 2.8	0.196	61.5	8.40 ± 0.11	61.5	subaerial lava, 1646 m
Mt. Frakes											
92-145	Morrison Rocks	haw	gms (F)	2.00 ± 0.07	1.98 ± 0.03	256.6 ± 5.2	145.0	98.5	***1.81 ± 0.10***	98.5	subaerial lava
92-142	Morrison Rocks	haw	gms (F)	2.02 ± 0.11	1.89 ± 0.08	287.6 ± 19.8	12.1	59.9	***1.815 ± 0.05***	59.9	subaerial lava
92-128	Morrison Rocks	haw	gms (F)	2.48 ± 0.11	2.43 ± 0.04	313.9 ± 8.4	7.34	67.4	***2.517 ± 0.06***	67.4	subaerial lava
92-125	Morrison Rocks	haw	gms (F)	2.68 ± 0.27	2.61 ± 0.07	293 ± 2.4	1.70	52.3	***2.535 ± 0.09***	52.3	subaerial lava
92-130	Morrison Rocks	haw	gms (F)	3.78 ± 0.09	3.89 ± 0.04	293.4 ± 6.8	1.37	51.5	**3.876 ± 0.03**	51.5	subaerial lava
92-122	Morrison Rocks	phono	ano (L)	**4.17 ± 0.05**						100.0	lava flow
92-121	Morrison Rocks	phono	ano (L)	**4.18 ± 0.04**						100.0	lava flow
92-127	Morrison Rocks	phono	ano (L)	**4.25 ± 0.03**						100.0	lava flow
92-151	English Rock	basn	gms (F)	0.158 ± 0.039	0.035 ± 0.019	293.7 ± 4.7	0.1	78.4	**0.032 ± 0.010**	78.4	parasitic cone
92-151	English Rock	basn	gms (F)	0.122 ± 0.049	0.024 ± 0.012	302.3 ± 9.2	0.5	86.1	**0.035 ± 0.010**	86.1	parasitic cone
92-157	English Rock	basn	gms (F)	0.78 ± 0.11	±	301 ± 58	53.5	93.4	***0.826 ± 0.079***	93.4	parasitic cone
92-157	English Rock	basn	gms (F)	0.858 ± 0.072	0.92 ± 0.02	268.6 ± 4.0	10.5	91.8	***0.851 ± 0.036***	91.8	parasitic cone
92-159	English Rock	basn	gms (F)	1.62 ± 0.04	1.63 ± 0.02	281 ± 12	7.36	71.9	***1.60 ± 0.02***	71.9	parasitic cone

The ages in bold are the best age estimates for each sample. The ages in plain text bold are of higher quality than those in italics bold. The higher quality is indicated by either highly concordant single crystal laser fusion ages or agreement of total fusion, isotope correlation and plateau ages for furnace step-heating analytical data. The lower quality ages are furnace step-heating ages that show some inconsistency among the total fusion, isotope correlation and plateau ages. The plateau ages of the the higher quality data meet the Fleck *et al.* (1977) criteria; the plateau ages of the lower quality data are 'selected mean ages' that do not meet Fleck *et al.* (1977) criteria.

[1] ID refers to sample identification numbers. Samples that are listed twice were analysed in duplicate.

[2] Rock type abbreviations: bas, basalt; basn, basanite, ben, benmoreite; haw, hawaiite; mug, mugearite; phono, phonolite; ph-tph, phono-tephrite; rhyo, rhyolite; trach, trachyte. Rock types are based on XRF analysis (Wilch, 1997) except those with asterisks, which are based on visual identification.

[3] sam (meth) refers to sample type and method of analysis. Ano, anorthoclase crystal concentrate; gms, groundmass concentrate; plg, plagioclase crystal concentrate; (F), furnace step-heating method; (L), laser-fusion analysis.

petrographic microscope. Groundmass aliquots were separated from the mafic samples and anorthoclase crystal aliquots were separated from the felsic samples. Groundmass and anorthoclase-crystal aliquots were prepared using standard crushing, sieving, acid-treatment, magnetic separation and heavy-liquid separation. Prior to irradiation, samples were hand-picked and packed in copper-foil ampules.

$^{40}Ar/^{39}Ar$ dating sample preparation and analyses were conducted at the New Mexico Geochronology Research Laboratory at the New Mexico Institute of Mining and Technology, Socorro, according to methods described in Wilch *et al.* (1999). Fish Canyon Tuff sanidine (FC-1, 27.84 Ma; Deino & Potts 1990) was used as a neutron-flux monitor. The sample ages were corrected for blank, background, mass discrimination, radioactive decay and interfering reactions. All analytical uncertainties are reported at the 2σ confidence level, and the decay constants and isotopic abundances used are those suggested by Steiger & Jaeger (1977). The furnace incremental heating method was used for dating mafic and intermediate groundmass concentrates, and the CO_2 laser-fusion method was used to date single-crystal anorthoclase phenocrysts separated from silicic samples.

Age data

Table 1 summarizes the $^{40}Ar/^{39}Ar$ geochronology results of 25 samples dated by laser-fusion and 52 samples dated by furnace step heating. Wilch (1997) provides complete $^{40}Ar/^{39}Ar$ and XRF data tables for these samples. The best age estimates for each sample are listed in bold type in Table 1. The age results were interpreted according to the following guidelines. Very few laser-fusion ages of anorthoclase crystals from individual samples differed by more than their 2σ analytical errors. These rare discordant ages are attributed to xenocrysts (older ages), contaminants such as low potassium plagioclase (low-precision older or younger ages), or alteration (typically low-precision younger ages). Weighted-mean single-crystal laser-fusion ages and errors were calculated according to the method of Samson & Alexander (1987). For most furnace step-heating results, plateau ages using the plateau criteria of Fleck *et al.* (1977), or 'selected mean ages' were interpreted as the best estimates of the eruption ages. Selected mean ages represent the weighted mean of near-plateau segments of age spectra, where typically one step lies slightly outside of 2σ uncertainty. In a few cases, isotope correlation ages provided the best eruption age estimates for samples yielding discordant age spectra and isochrons with high trapped $^{40}Ar/^{36}Ar$ content and low MSWD (mean standard weighted deviation) values.

Dating results from satellite nunataks just west of Mt. Murphy provide examples of significant differences between $^{40}Ar/^{39}Ar$ and K/Ar ages. Initial K/Ar dating of Turtle Peak resulted in apparent ages ranging from middle Eocene to middle Miocene (LeMasurier 1972*b*; LeMasurier & Rex 1982). More recent K/Ar ages of Turtle Peak are 6.70 ± 0.64 Ma and 7.9 ± 1.6 Ma for rocks from the base of the section and 9.8 ± 3.0 Ma for rocks situated 100 m above the base (LeMasurier *et al.* 1994). In contrast, the $^{40}Ar/^{39}Ar$ ages of Turtle Peak are significantly younger and more consistent with stratigraphy: 5.95 ± 0.60 Ma for the basal section, 5.65 ± 0.23 Ma for the middle section, and 4.70 ± 0.15 Ma for uppermost section. K/Ar ages of Hedin Nunatak and Icefall Nunatak are 17.7 ± 6.0 Ma and 11.3 ± 4.0 Ma, respectively, whereas $^{40}Ar/^{39}Ar$ ages are much younger at 6.50 ± 0.12 and 6.20 ± 0.24 Ma for Hedin Nunatak and 6.80 ± 0.10 Ma and 6.52 ± 0.08 for Icefall Nunatak. K/Ar dating of these nunataks has been difficult due to low radiogenic yields and secondary alteration (LeMasurier *et al.* 1994). Both of these problems are common in glassy basaltic rocks. In fact, some additional samples from the satellite nunataks of Mt. Murphy were analysed by the $^{40}Ar/^{39}Ar$ dating method but yielded highly discordant, uninterpretable age spectra with much older total fusion ages, so those results were rejected.

An additional advantage of the $^{40}Ar/^{39}Ar$ method comes in dating young samples (<1 Ma). Two groundmass samples from Sechrist Peak at Mt. Murphy yielded $^{40}Ar/^{39}Ar$ ages of 587 ± 43 ka and 592 ± 40 ka (Table 1). The K/Ar age of this site is 940 ± 280 ka (LeMasurier *et al.* 1994), which lies outside of $\pm 2\sigma$ uncertainty of the more precise $^{40}Ar/^{39}Ar$ ages. In general, the $^{40}Ar/^{39}Ar$ ages of recent volcanism differ significantly from previous K/Ar ages of the same sites and offer a more refined and reliable chronology of middle to late Pleistocene eruptions in MBL.

Finally, the $^{40}Ar/^{39}Ar$ method can be used to date very young K-rich rocks (<100 ka), thus permitting chronologies to be established through the critical Wisconsinan glacial cycle (Wilch *et al.* 1999). Fairly high levels of precision and reproducibility were obtained from several samples, including two separate analyses of a basanite groundmass sample from the base of Mt. Frakes, which yielded plateau ages of 32 ± 10 ka and 35 ± 10 ka.

Glaciovolcanic reconstructions

The polygenetic volcanoes of Mt. Murphy and the Crary Mountains contain abundant evidence for glaciovolcanic interactions and changing ice levels. Mt. Rees and Mt. Steere in the Crary Mountains and Mt. Murphy all exhibit evidence for fluctuating ice levels in the late Miocene. Mt. Murphy also shows signs of changing ice levels in the Pliocene and Pleistocene. Late Pleistocene basanitic cinder cone volcanic deposits near the level of the WAIS at Mt. Frakes, an undissected Plio-Pleistocene shield volcano in the Crary Mountains, afford upper limits on ice expansion during the last glacial maximum.

The wide diversity and complexity of lithofacies and stratigraphic relationships at Mt. Murphy, Mt. Rees and Mt. Steere volcanoes are attributed to their long eruptive histories during periods of fluctuating regional and local ice. In addition, thickening and thinning of the ice sheet at the coastal Mt. Murphy setting resulted in a more complete record of ice-sheet and volcano interactions than was produced by smaller ice sheet changes at the inland Crary Mountains. Detailed descriptions (below) of glaciovolcanic sequences at Mt. Murphy and the Crary Mountains (Mt. Rees, Mt. Steere and Mt. Frakes) illustrate important differences in the character of volcanic deposits and the nature of ice level changes at inland v. coastal sites on the ice sheet.

Mt. Murphy

Mt. Murphy (75°15′–75°30′S; 110°–111°W) is a complex polygenetic coastal volcano with the most extensive glaciovolcanic record of all volcanoes in MBL (Fig. 2). The heavily dissected volcano has a summit elevation of 2703 m asl.

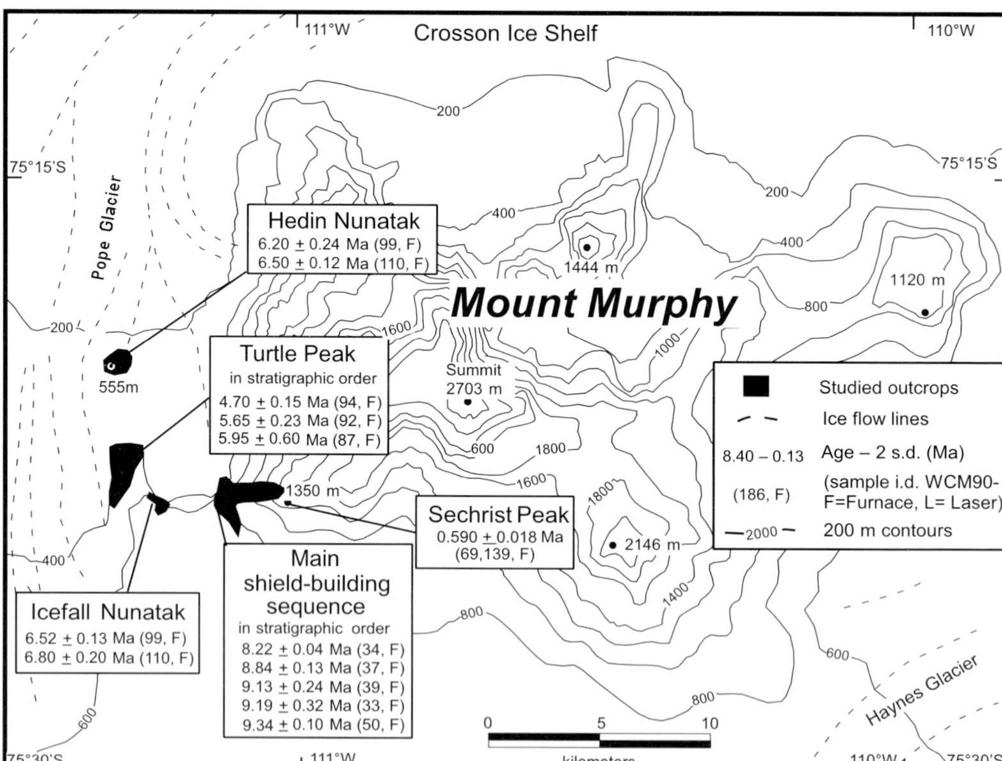

Fig. 2. Topographic map of Mt. Murphy volcano showing rock exposures described by LeMasurier *et al.* (1994) and new $^{40}Ar/^{39}Ar$ dates. Ice flow (dashed lines) around Mt. Murphy is funnelled through the Pope and Haynes glaciers to the Crosson Ice Shelf. Spot elevation (black dots) are in metres above sea level. Base map is the Mount Murphy quadrangle (1972), 1:250 000-scale USGS Reconnaissance Series, Antarctica, US Geological Survey.

Because of its location, Mt. Murphy produces a major ice-damming effect, with an inland ice level at *c.* 800 m asl and a coastal ice level at *c.* 200 m asl. The ice sheet drains around Mt. Murphy through steep icefalls of the Pope and Haynes glaciers. The most extensive rock exposures and evidence for glaciovolcanic interactions at Mt. Murphy are on the western flanks and in satellite nunataks that stand out as interfluves in the Pope Glacier (e.g. Smellie 2001). LeMasurier *et al.* (1994) used volcanic sequences at the lower SW flank of the Mt. Murphy shield, at Sechrist Peak and at three satellite nunataks as evidence for higher palaeo-ice-levels during Miocene to Pleistocene times. We dated 17 samples from these sequences by the $^{40}Ar/^{39}Ar$ method and offer a significantly revised chronology of inferred higher ice levels compared to the previous K/Ar chronology of LeMasurier *et al.* (1994).

A stratigraphic section at the base of the main shield of Mt. Murphy exemplifies the complex interactions between a growing polygenetic volcano and an active ice sheet (Fig. 3). The section consists of >1300 m of late Miocene pillow lava and hyaloclastite breccia with intercalated palagonitized hyalotuff and Strombolian tuff. The alternation between pillow lava and tuff was interpreted to represent rapid ice-level variations in the late Miocene, with ice high-stands up to 300 m above today's ice level (LeMasurier *et al.* 1994). A 5 m-thick striated-clast-bearing heterolithic tillite is sandwiched between overlying pillow and hyaloclastite deposits and underlying striated basalt lava near the base of the volcano (LeMasurier *et al.* 1994). Additional striated lava surfaces and tillites are exposed in the middle and upper sections of the sequence. The transitions between 'wet' and 'dry' volcanic lithofacies are abrupt and between units in this sequence, rather than gradational and within units, as they are in sequences at the Crary Mountains. Three K/Ar ages reported from this sequence are out of stratigraphic order but have large overlapping uncertainties (LeMasurier *et al.* 1994). New $^{40}Ar/^{39}Ar$ ages of this sequence are generally consistent with stratigraphy and suggest that the alternations between subaqueous and subaerial eruptions occurred mostly between 9.34 and 8.84 Ma (Table 1, Fig. 2).

Interpretations of WAIS palaeo-ice-levels from the Late Miocene main shield outcrops at Mt. Murphy are complicated by three factors. First, the main shield sequences are located on the west side of Mt. Murphy, where ice is currently descending from high upstream levels at *c.* 800 m asl to low downstream levels of *c.* 200 m asl. The elevations of these sequences (up to *c.* 800 m asl) are about the same as the elevation of the regional ice sheet on the upstream side. Therefore, these outcrops may record fluctuations of an ice sheet that was similar in size to today's ice sheet. Second, because Mt. Murphy is at the coast where glaciers are feeding into the Crosson Ice Shelf, the local ice configurations may be very responsive to changes in sea level. A drop in sea level might result in thickening of ice at Mt. Murphy. Third, the coastal position of Mt. Murphy may facilitate draining of englacial lakes formed by volcanism, and passage zone sequences may be much lower than the inland palaeo-ice-level. Given these three complicating factors, the main shield stratigraphic sequence indicates ice level fluctuations but the complex setting of Mt. Murphy precludes making quantitative interpretations about regional palaeo-ice-levels based solely on the main shield sequence.

Pleistocene-aged (940 ± 280 ka) hydrovolcanic deposits, situated at 1250 m above sea level at Sechrist Peak, are interpreted as products of emergent tuff cone eruptions by LeMasurier *et al.* (1994; Fig. 3). The emergent eruptions were attributed to interaction between magma and an expanded WAIS at 940 ka and implied that the ice sheet was 600 m above its present level at that time. Two $^{40}Ar/^{39}Ar$ ages are significantly younger and more precise than the K/Ar age and pinpoint the inferred ice-sheet high-stand at 590 ± 15 ka.

Three latest Miocene–Pliocene (6.80–4.70 Ma, Table 1) nunataks located just west of Mt. Murphy (Icefall Nunatak, Hedin Nunatak and Turtle Peak) form interfluves between Pope Glacier and an unnamed glacier, which descend from the polar plateau to the Crosson Ice Shelf. The nunataks are composed, in part, of pillow lava and hyaloclastite breccia, suggesting higher syn-eruptive local ice levels relative to today (LeMasurier *et al.* 1994; Smellie 2001). LeMasurier *et al.* (1994) interpreted these sequences as table-mountain remnants. They exhibit multiple passage zones at 250 m to 600 m above present local ice level (Smellie 2001). The elevations of the uppermost passage zones are near the present level of the regional ice sheet. However, uncertainties associated with the coastal interfluve setting preclude establishing absolute WAIS palaeo-ice-level elevations. These uncertainties include the timing of volcanism relative to erosion on interfluves, possible tectonic uplift or subsidence, and the effects of changes in sea level on local ice levels. The satellite nunataks are conservatively re-interpreted as indications of the presence of an active late Miocene ice-sheet, without specific implications for palaeo-ice-levels.

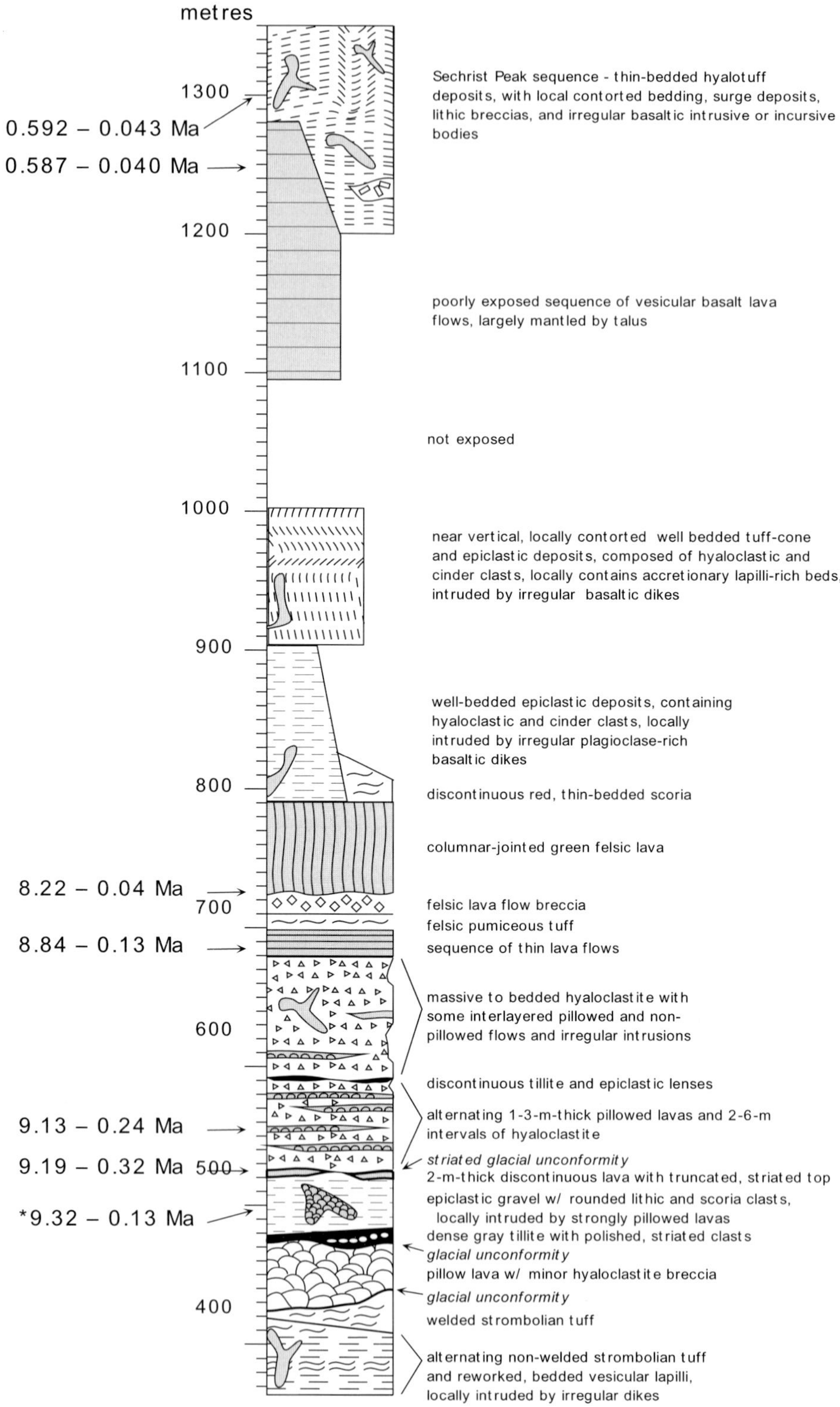
metres
1300
1200
1100
1000
900
800
700
600
500
400
0.592 – 0.043 Ma
0.587 – 0.040 Ma
8.22 – 0.04 Ma
8.84 – 0.13 Ma
9.13 – 0.24 Ma
9.19 – 0.32 Ma
*9.32 – 0.13 Ma
Sechrist Peak sequence - thin-bedded hyalotuff deposits, with local contorted bedding, surge deposits, lithic breccias, and irregular basaltic intrusive or incursive bodies
poorly exposed sequence of vesicular basalt lava flows, largely mantled by talus
not exposed
near vertical, locally contorted well bedded tuff-cone and epiclastic deposits, composed of hyaloclastic and cinder clasts, locally contains accretionary lapilli-rich beds, intruded by irregular basaltic dikes
well-bedded epiclastic deposits, containing hyaloclastic and cinder clasts, locally intruded by irregular plagioclase-rich basaltic dikes
discontinuous red, thin-bedded scoria
columnar-jointed green felsic lava
felsic lava flow breccia
felsic pumiceous tuff
sequence of thin lava flows
massive to bedded hyaloclastite with some interlayered pillowed and non-pillowed flows and irregular intrusions
discontinuous tillite and epiclastic lenses
alternating 1-3-m-thick pillowed lavas and 2-6-m intervals of hyaloclastite
striated glacial unconformity
2-m-thick discontinuous lava with truncated, striated top
epiclastic gravel w/ rounded lithic and scoria clasts, locally intruded by strongly pillowed lavas
dense gray tillite with polished, striated clasts
glacial unconformity
pillow lava w/ minor hyaloclastite breccia
glacial unconformity
welded strombolian tuff
alternating non-welded strombolian tuff and reworked, bedded vesicular lapilli, locally intruded by irregular dikes

Crary Mountains: Mt. Rees, Mt. Steere and Mt. Frakes

The Crary Mountains (76°37′–76°53′S; 117°20′–118°20′W), located about 250 km inland from the coast, consist primarily of three large coalesced volcanoes that are aligned from NNW to SSE: Mt. Rees, Mt. Steere, and Mt. Frakes (Fig. 4). The degree of dissection of the volcanoes increases to the north, with deep cirques cut into the north side of Mt. Steere and the east side of Mt. Rees. On the basis of reconnaissance fieldwork and four K/Ar ages, the Crary Mountains were characterized as late Miocene to early Pliocene shield volcanoes composed of bimodal alkaline lava and hyaloclastite (LeMasurier & Thomson 1990). Using samples collected for this study, Panter *et al.* (2000) conducted an isotopic analysis of basalts from Crary Mountains and attributed the volcanism to a HIMU-type mantle plume source. In this paper, we incorporate detailed stratigraphy and $^{40}Ar/^{39}Ar$ geochronology (60 analyses) into a more comprehensive record of volcanism in the Crary Mountains (Table 1, Fig. 4).

Mt. Rees is an elongate, eroded late Miocene (mostly 9–8 Ma) polygenetic volcano with a summit elevation of 2709 m (Fig. 4). Thick stratigraphic sequences at Trabucco Cliff (Fig. 5) and Tasch Peak (Fig. 6) consist mostly of mafic to intermediate volcanic rocks with subordinate interlayered felsic lavas. The mafic to intermediate rock outcrops are characterized by two alternating lithofacies: (1) 'dry' brecciated and unbrecciated lavas with red-oxidized bases and, (2) 'wet' palagonitized glassy hyaloclastite breccias and pillow lavas (Fig. 7). The $^{40}Ar/^{39}Ar$ ages at several levels in the Trabucco Cliff sequence agree with stratigraphic order (from 9.13 ± 0.53 to 8.94 ± 0.06 Ma), although the analytical uncertainties overlap (Figs 4 & 5). A short interval of eruptions is consistent with the lack of unconformities in the sections. No tillites or glacially striated surfaces (typically associated with wet-based glaciations) were observed within the subglacial volcanic sequences. A >400 m stratigraphic section is exposed along Tasch Peak ridge, east of the summit of Mt. Rees (Fig. 6). The $^{40}Ar/^{39}Ar$ ages from several levels are mostly in stratigraphic order and range from 9.34 ± 0.24 to 8.21 ± 0.13 Ma. These mostly hawaiitic to basanitic, alternating lavas and hyaloclastite breccias are cut by a trachytic dike, dated to 7.52 ± 0.06 Ma.

Mt. Steere is a late Miocene (9–6 Ma) polygenetic volcano with an intact summit caldera at 3558 m above sea level. The north and NE sides of the volcano are deeply dissected by cirques, which expose flow-banded rhyolite, trachyte and phonolite lavas and breccias, cut by numerous felsic to intermediate dykes. Basanite lavas and breccias with subordinate trachyte lava are exposed in the lowest 800 m of the volcano on the east side of Mt. Steere at Lie Cliff. Samples from Lie Cliff and adjacent ridge exposures to the north and south yielded $^{40}Ar/^{39}Ar$ ages from 8.55 ± 0.04 Ma to 5.74 ± 0.04 Ma. Basanite deposits near the bottom of the sections and trachytic lavas near the top dominate these basal Mt. Steere exposures. Basanite rocks at Lie Cliff and adjacent ridge outcrops resemble the glaciovolcanic sequences at Mt. Rees, with alternating 'wet' and 'dry' lithofacies and no syn-eruptive tillites or striated surfaces. The volcanic sequences in Lie Cliff and an adjacent ridge to the north were erupted over a short interval, with eruptions that cannot be differentiated by $^{40}Ar/^{39}Ar$ dating, and overlapping ages ranging from 8.52 ± 0.23 Ma to 8.19 ± 0.18 Ma. These mafic and intermediate eruptions occurred during the same interval as trachyte, phonolite and rhyolite eruptions elsewhere on Mt. Steere and Mt. Rees. A second eruption interval, characterized by alternating 'wet' and 'dry' lithofacies, occurred between 7.38 ± 0.07 Ma and 6.41 ± 0.43 Ma and is recorded in ridge outcrops located just southeast of Lie Cliff.

The alternating 'wet' and 'dry' lithofacies sequences at Mt. Rees and Mt. Steere are inferred to represent fluctuations between subaerial and subaqueous (into glacial meltwater) depositional environments. The alternations are difficult to interpret in terms of changing ice sheet levels, because the wet-to-dry transitions are parallel to the constructional slopes of the volcanoes (Figs 5, 6 & 7) and are tentatively interpreted as products of interactions between lavas and local slope ice or snow. The indistinguishable, high precision $^{40}Ar/^{39}Ar$ ages in many parts of the sequences indicate that they accumulated over short time intervals. The multi-story alternating lithofacies comprise slope-forming constructional sequences that lack interbedded glacial deposits and glacial unconformities. Skilling

Fig. 3. Composite lithostratigraphic section west of Bucher Peak, Mt. Murphy, including Miocene main shield-building sequence and overlying Pleistocene Sechrist Peak tuff cone sequence. $^{40}Ar/^{39}Ar$ dates shown to left of sequence (also see Fig. 2 and Table 1). * Indicates that the dating sample collected from 0.5 km SW of line of section.

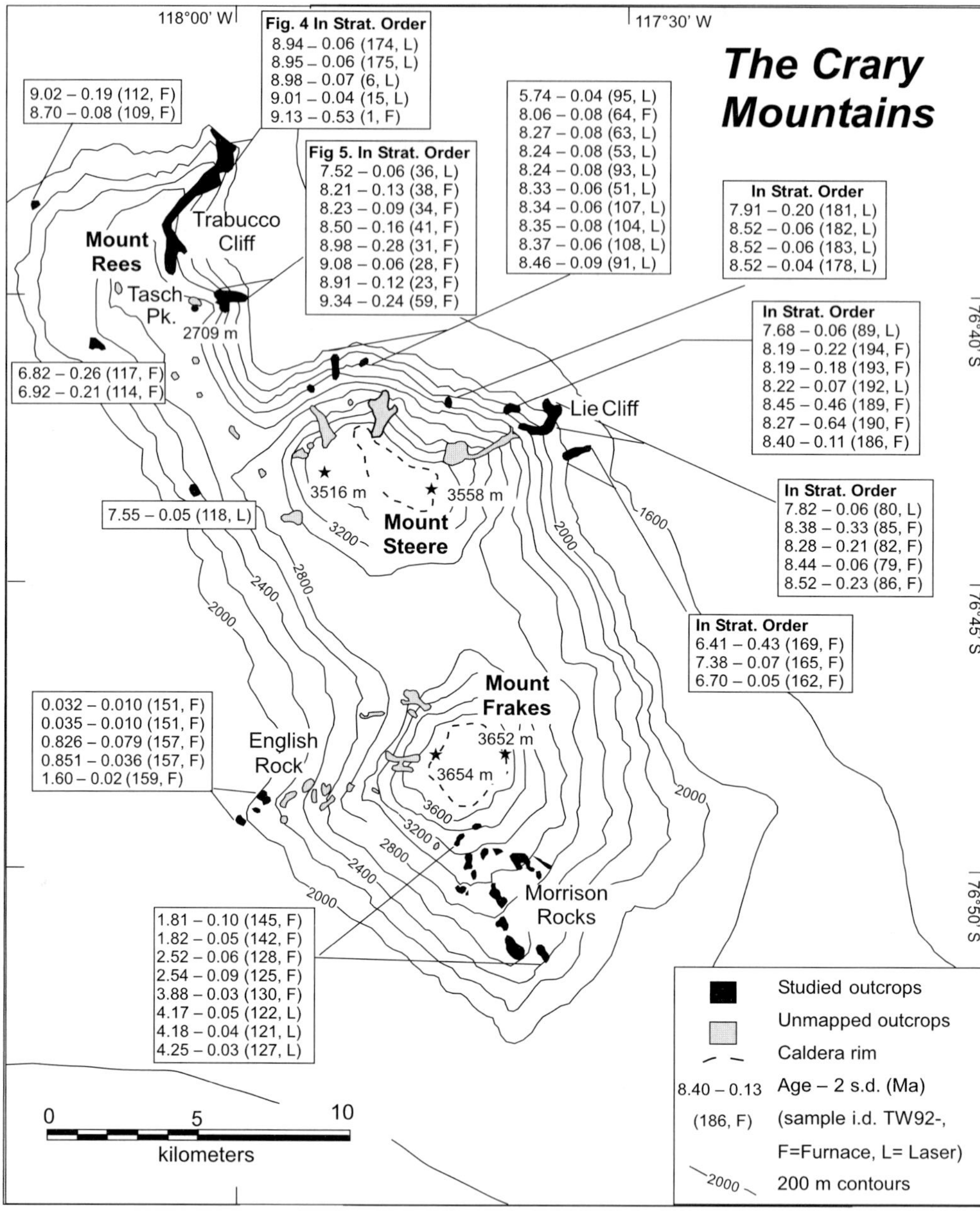

Fig. 4. Topographic and rock outcrop map and $^{40}Ar/^{39}Ar$ chronology of Crary Mountains. $^{40}Ar/^{39}Ar$ ages at Mt. Rees and at Lie Cliff area of Mt. Steere are from alternating sequences of 'wet' and 'dry' lithofacies and are in stratigraphic order. $^{40}Ar/^{39}Ar$ ages of other Mt. Steere sites and of Mt. Frakes are from isolated outcrops and are listed from youngest to oldest. Base map is the Crary Mountains quadrangle (1973), 1:250 000-scale USGS Reconnaissance Series, Antarctica, United States Geological Survey. Stars indicate elevation localities from USGS base map.

(1994) and Smellie & Hole (1997) suggested that sloping transitions from subaqueous to subaerial lithofacies could form after drainage of an englacial lake. However, at Mt. Rees and Mt. Steere, there is very little evidence for ponded or flowing water. The up-sequence transition from wet to dry lithofacies can be explained by lavas interacting with slope ice and snow to form pillows and hyaloclastites until they melt the slope ice or build above its level and form dry

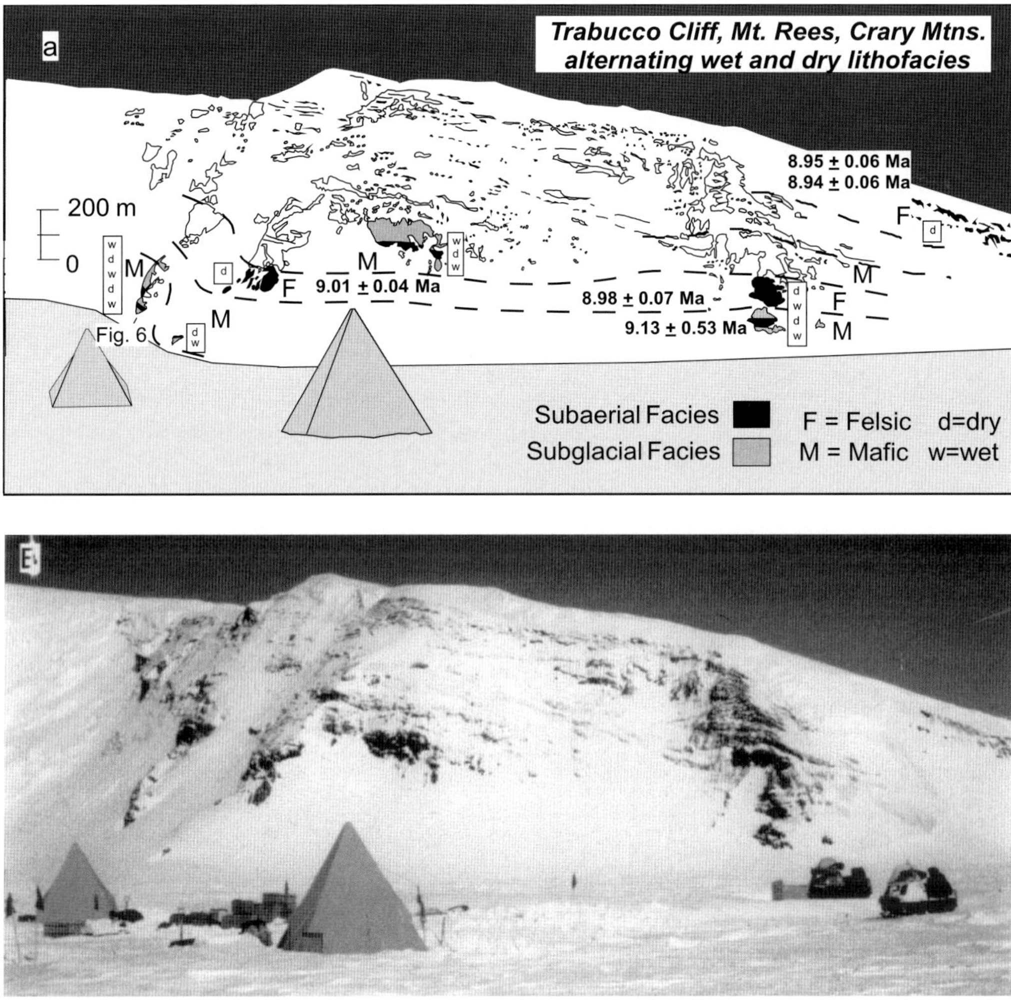

Fig. 5. Sketch (**a**) and photograph (**b**) of Trabucco Cliff outcrops at Mt. Rees, Crary Mountains, showing location of alternating 'wet' and 'dry' lithofacies, compositions and $^{40}Ar/^{39}Ar$ ages. The cliff section is about 600 m in height. In boxes, w and d denote wet and dry lithofacies, respectively. The dashed lines show major breaks in lithology. See Figure 4 for location of cliff section.

lavas. The up-sequence transition from dry to wet lithofacies is more difficult to interpret and may have resulted from lava flowing through open channels or tunnels in the ice, or from re-establishment of slope ice during eruptive intervals. The lack of glacial tills and unconformities suggests that the ice was thin or cold-based. We suggest that these interactions occurred on the slopes of the growing Mt. Steere and Mt. Rees volcanoes above the level of the ice sheet. The Alexander Island model of alternating lithofacies produced during valley-confined subglacial volcanism provides an analogue for these Crary Mountain sequences (Smellie *et al.* 1993).

In contrast to Mt. Steere, the Pliocene to latest Pleistocene (4.25 ± 0.03 Ma to 33.5 ± 7.4 ka) Mount Frakes shield volcano is completely undissected, with a circular 4 km-diameter summit caldera at 3654 m asl (Fig. 4; Table 1). The Mt. Frakes outcrops consist entirely of subaerially erupted phonolite, hawaiite and basanite. The absence of glaciovolcanic sequences at Mt. Frakes may simply reflect the lack of dissection. Mt. Frakes has lower flank, late-stage parasitic vents that provide reliable measures of regional palaeo-ice-levels. Late-stage basanite and hawaiite cinder cone deposits crop out on the south side of Mt. Frakes at Morrison Rocks

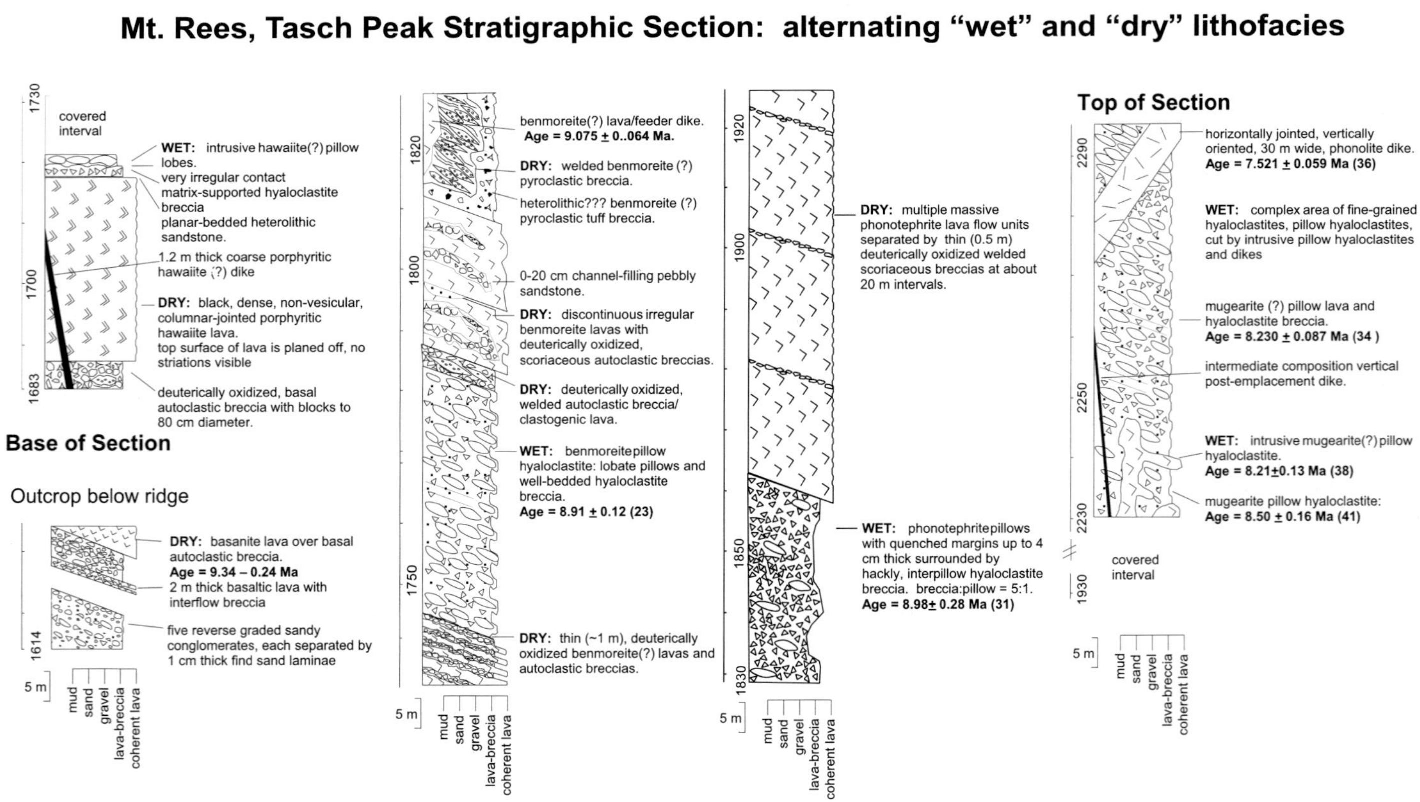

Fig. 6. Stratigraphic section of Tasch Ridge, Mt. Rees, Crary Mountains, showing alternating 'wet' and 'dry' lithofacies and $^{40}Ar/^{39}Ar$ ages. The ages are listed with 2σ uncertainties. The sample number is indicated in parentheses following the age (add TW92- for complete number). See Figure 4 for location of cliff section.

Fig. 7. Views of alternating 'wet' subaqueous and 'dry' subaerial volcanic lithofacies exposed in a Trabucco Cliff outcrop. Note that transitions between subaqueous hyaloclastite and subaerial lava and breccia are sloping. See Figure 4 for location of outcrop.

and on the western side of Mt. Frakes at English Rock. The youngest cinder cone deposits at English Rock, with a mean age of 33.5 ± 7.4 ka, are situated *c.* 150 m above the level of the ice sheet. These deposits limit syn-eruptive ice-sheet expansion to < 150 m above the present ice level.

Implications for late Miocene ice sheet history

The late Miocene interval from 9 to 8 Ma marks a pulse in polygenetic volcanism in MBL and provides the earliest substantial evidence for a widespread WAIS (LeMasurier 1990; Wilch 1997). Late Miocene (9–8 Ma) glaciovolcanic sequences are exposed at the inland Crary Mountains (Mt. Steere and Mt. Rees) and at the coastal Mt. Murphy volcano, as well as at monogenetic volcanoes of western MBL (LeMasurier & Thomson 1990; Wilch 1997). The Mt. Murphy main shield sequence records fluctuating syn-eruptive (mostly 9.34 ± 0.10 to 8.84 ± 0.13 Ma) local palaeo-ice-levels up to 300 m higher than today's local ice level. Relatively low-elevation striated glacial unconformities and interbedded tillites record fluctuating ice flow across the growing volcano during this interval. In the Hobbs Coast region of western Marie Byrd Land, Late Miocene glacial unconformities overlain by subglacial pillow lava and hyaloclastite sequences are exposed at Bowyer Butte (LeMasurier 1990) and Kennel Peak (Wilch 1997).

Coeval glaciovolcanic sequences at Mt. Rees and Mt. Steere in the Crary Mountains (9–8 Ma) apparently resulted from slope-ice interactions and imply that abundant local slope ice extended to near or below the level of the modern ice sheet. The Mt. Rees and Mt. Steere reconstructions together with the Mt. Murphy record are consistent with late Miocene WAIS levels similar to modern WAIS levels. In summary, the record of 9 to 8 Ma volcanism at Mt. Murphy, Mt. Rees and Mt. Steere provides substantial evidence for a widespread glaciation in West Antarctica by 9 Ma.

Marine geologic records corroborate the volcanic record of widespread glaciation in West Antarctica in the Late Miocene. In the Ross Sea, Anderson & Shipp (2001) summarized marine geological evidence for extensive Late Miocene glaciation of continental shelves in the Ross Sea region. Marine data from the Weddell Sea also imply that development of a large-scale WAIS occurred *c.* 10–8 Ma. Significant glacial erosion in West Antarctica during this interval is indicated by increased deposition of hemipelagic sediments and ice-rafted detritus and rapid deposition of turbidite sequences (Kennett & Barker 1990; Kennett 1995). The marine sediment record 'suggests considerable climatic/cryospheric instability in the source area' during that interval (Kennett & Barker 1990, p. 956).

The coincidence of increased volcanism and widespread glaciation in West Antarctica at 9–8 Ma may reflect a cause-and-effect relationship. Sigvaldason *et al.* (1992) suggested that 'vigorous crustal movements' in response to isostatic changes caused by 'glacier deloading' may have caused higher rates of late Quaternary volcanism in Iceland. McGuire *et al.* (1997) recognized increased late Quaternary volcanicity in coastal and island volcanoes in the Mediterranean during periods of rapid sea level change, and suggested that rapidly changing sea level can cause edifice slope failure and consequent decompressive expulsion of magma stored at shallow depths. Kennett & Barker (1990) described the early WAIS (10–6 Ma) as quite dynamic with frequent growth and decay cycles. Regional isostatic crustal readjustments and sea-level changes during the latest Miocene may have contributed to the apparent increase in volcanism after

10 Ma. Although such an association is speculative, the coincidence of inception of a dynamic WAIS and increased volcanism in Marie Byrd Land is worth consideration.

Conclusions

Major conclusions of the ice–volcanic records of the Crary Mountains and Mt. Murphy include:

(1) Palaeoenvironmental reconstructions and $^{40}Ar/^{39}Ar$ geochronology of Late Miocene volcanic sequences at Mt. Rees and Mt. Steere in the Crary Mountains suggest interaction between lava flows and slope ice over intervals of volcano construction. The slope-ice interactions are inferred on the basis of ubiquitous dipping passage zones in the stratigraphic sequences and an absence of glacial unconformities and till. The slope-ice interpretation of the volcanic sequences implies that the Late Miocene ice sheet was at or below today's ice level.

(2) New $^{40}Ar/^{39}Ar$ ages of Mt. Murphy ice-volcanic sequences suggest Late Miocene fluctuations in local ice level at this coastal volcano. The sub-horizontal character of passage zones combined with the presence of several glacial unconformities and till units in the stratigraphic sequences indicates the presence of a fluctuating wet-based ice sheet during volcano growth. The coastal setting of Mt. Murphy complicates interpretation of the ice volcanic environment and makes difficult the reconstruction of regional palaeo-ice-level elevations for the WAIS at that locality.

(3) Together the Crary Mountains and Mt. Murphy record provide evidence for a widespread West Antarctic Ice Sheet in the late Miocene (9.3–8.2 Ma). This evidence is corroborated by other Late Miocene glaciovolcanic sequences from across Marie Byrd Land. The onset of increased glaciovolcanism in Mare Byrd Land coincides with the first occurrences of abundant ice-rafted detritus at 10–8 Ma (Kennett 1995). An important implication of the record of syn-eruptive palaeo-ice-levels at the Crary Mountains and Mt. Murphy is that former WAIS expansions were more extensive at coastal sites than at inland sites.

(4) A middle Pleistocene ice-sheet high stand of +550 m is inferred at the coastal volcano, Mt. Murphy, on the basis of late-stage 590 ± 15 ka tuff cone deposits at Sechrist Peak. Nearly coeval (573 ± 5 ka) inland ice levels at Mt. Berlin (125 km from the coast) were probably no more than 200 m above present-day local ice levels (Wilch *et al.* 1999). Sea level lowering and resulting coastal ice sheet expansion may have caused the extreme thickening of the WAIS at Mt. Murphy. This ice-sheet high stand may correspond to global ice expansion (and eustatic lowering) at isotopic stage 16 in the marine record.

This work was supported by the National Science Foundation (NSF-DPP918806), with additional funding from the New Mexico Geochronological Research Laboratory. We thank U.S. Navy VXE-6 squadron, Antarctic Support Associates, and Ken Borek Air Ltd. for logistical support; N. Dunbar, K. Panter, and T. Teeling for field assistance in the Crary Mountains; R. Esser, M. Heizler, and L. Peters for assistance with $^{40}Ar/^{39}Ar$ geochronology; and I. Skilling, J. Smellie and L. Viereck for helpful reviews.

References

ANDERSON, J. B. & SHIPP, S. B. 2001. Evolution of the West Antarctic Ice Sheet. *In*: ALLEY, R. B. & BINDSCHADLER, R. A. (eds) *The West Antarctic Ice Sheet: behavior and environment*. American Geophysical Union, Antarctic Research Series, **77**, 45–57.

BEHRENDT, J. C. 1999. Crustal and lithospheric structure of the West Antarctic Rift System from geophysical investigations – a review. *Global and Planetary Change*, **23**, 25–44.

CARLISLE, D. 1963. Pillow breccias and their aquagene tuffs. Quadra Island, British Columbia. *Journal of Geology*, **71**, 48–71.

DEINO, A. & POTTS, R. 1990. Single-crystal $^{40}Ar/^{39}Ar$ dating of the Olorgesailie Formation, Southern Kenya Rift. *Journal of Geophysical Research*, **95**, 8453–8470.

DREWRY, J. 1983, *Antarctica: Glaciological and Geophysical Folio*. Scott Polar Research Institute, University of Cambridge, Cambridge.

FISHER, R. V. & SCHMINCKE, H. 1984. *Pyroclastic Rocks*. Springer-Verlag, Berlin.

FLECK, R. J., SUTTER, J. F. & ELLIOT, D. H. 1977. Interpretation of discordant $^{40}Ar/^{39}Ar$ age spectra of Mesozoic tholeiites from Antarctica. *Geochimica et Cosmochimica Acta*, **41**, 15–32.

FULLER, R. E. 1931. The aqueous chilling of basaltic lava on the Columbia River Plateau. *American Journal of Science*, **21**, 281–300.

HONNOREZ, J. & KIRST, P. 1975. Submarine basaltic volcanism: morphometric parameters for discriminating hyaloclastites from hyalotuffs. *Bulletin of Volcanology*, **39**, 1–25.

HOUGHTON, B. F. & SCHMINCKE, H.-U. 1986. Mixed depositions of simultaneous strombolian and phreatomagmatic volcanism: Rothenberg volcano, East Eifel volcanic field. *Journal of Volcanology and Geothermal Research*, **30**, 117–130.

JONES, J. G. 1969. Intraglacial volcanoes of the Laugarvatn Region, south-west ICELAND, I. *Quarterly Journal of the Geological Society, London*, **124**, 197–211.

JONES, J. G. 1970. Intraglacial volcanoes of the Laugarvatn Region, south-west Iceland, II. *Journal of Geology*, **78**, 127–140.

KENNETT, J. P. 1995. A review of polar climatic evolution during the Neogene, based on the marine sediment record. *In*: VRBA, E. S., DENTON, G. H., PARTRIDGE, T. C. & BURCKLE, L. H. (eds) *Paleoclimate and evolution, with emphasis on human origins*. Yale University Press, New Haven and London, 49–64.

KENNETT, J. P. & BARKER, P. F. 1990. Latest Cretaceous to Cenozoic climate and oceanographic developments in the Weddell Sea, Antarctica: an ocean-drilling perspective. *In*: BARKER, P. F. & KENNETT, J. P. (eds) *Proceedings of the Ocean Drilling Program, Scientific Results*. Ocean Drilling Program, Texas, **113**, 937–958.

KOKELAAR, B. P. 1986. Magma-water interactions in subaqueous and emergent basaltic volcanism. *Bulletin of Volcanology*, **48**, 275–289.

LEMASURIER, W. E. 1972*a*. Volcanic record of Cenozoic glacial history in Marie Byrd Land. *In*: ADIE, R. J. (ed.) *Antarctic Geology and Geophysics*. Universitetsforlaget, Oslo, 251–260.

LEMASURIER, W. E. 1972*b*. Volcanic record of Antarctic glacial history: Implications with regard to Cenozoic sea levels. *In*: PRICE, R. J. & SUGDEN, D. E. (eds) *Polar Geomorphology*. Special Publication, **4**, 59–74.

LEMASURIER, W. E. 1990. Marie Byrd Land. *In*: LEMASURIER, W. E. & THOMSON, J. W. (eds) *Volcanoes of the Antarctic plate and southern oceans*. American Geophysical Union, Antarctic Research Series, **48**, 146–163.

LEMASURIER, W. E. & REX, D. C. 1982. Volcanic record of Cenozoic glacial history in Marie Byrd Land and western Ellsworth Land: Revised chronology and evaluation of tectonic factors. *In*: CRADDOCK, C. (ed.) *Antarctic Geoscience*. University of Wisconsin Press, Madison, 725–734.

LEMASURIER, W. E. & REX, D. C. 1983. Rates of uplift and the scale of ice level instabilities recorded by volcanic rocks in Marie Byrd Land, West Antarctica. *In*: OLIVER, R. L., JAMES, P. R. & JAGO, J. B. (eds) *Antarctic Earth Sciences*. Australian Academy of Science, Canberra, 660–673.

LEMASURIER, W. E. & THOMSON, J. E. 1990. *Volcanoes of the Antarctic Plate and Southern Oceans*. American Geophysical Union, Antarctic Research Series, **48**.

LEMASURIER, W. E., HARWOOD, D. M. & REX, D. C. 1994. Geology of Mount Murphy Volcano: An 8-m.y. history of interaction between a rift volcano and the West Antarctic ice sheet. *Bulletin of the Geological Society of America*, **106**, 265–280.

MCGUIRE, W. J., HOWARTH, R. J., FIRTH, C., SOLOW, A. R., PULLEN, A. D., SAUNDERS, S. J., STEWART, I. S. & VITA-FINZI, C. 1997. Correlation between rate of sea-level change and frequency of explosive volcanism in the Mediterranean. *Nature*, **389**, 473–476.

MCINTOSH, W. C. & GAMBLE, J. A. 1991. A subaerial eruptive environment for the Hallett Coast volcanoes. *In*: THOMSON, M. R. A., CRANE, J. A. & THOMSON, J. W. (eds) *Geological evolution of Antarctica*. Cambridge University Press, Cambridge, 657–661.

MCPHIE, J., DOYLE, M. & ALLEN, R. 1993. *Volcanic Textures: A Guide to the Interpretation of Textures in Volcanic Rocks*. Centre for Ore Deposit and Exploration Studies, University of Tasmania, Hobart, Tasmania.

PANTER, K. S., HART, S. R., KYLE, P., BLUSZTANJN, J. & WILCH, T. I. 2000. Geochemistry of Late Cenozoic basalts from the Crary Mountains: characterization of mantle sources in Marie Byrd Land, Antarctica. *Chemical Geology*, **165**, 215–241.

SAMSON, S. D. & ALEXANDER, C. E. 1987. Calibration of the interlaboratory $^{40}Ar/^{39}Ar$ dating standard, Mmhb-1. *Isotope Geoscience*, **66**, 27–34.

SIGVALDASON, G. E., ANNERTZ, K. & NILSSON, M. 1992. Effect of glacier loading/deloading on volcanism: postglacial volcanic production rate of the Dyngjufjoll area, central Iceland. *Bulletin of Volcanology*, **54**, 385–392.

SKILLING, I. P. 1994. Evolution of an englacial volcano: Brown Bluff, Antarctica. *Bulletin of Volcanology*, **56**, 573–591.

SMELLIE, J. L. 2001. Lithofacies architecture and construction of volcanoes erupted in englacial lakes: Icefall Nunatak, Mount Murphy, eastern Marie Byrd Land, Antarctica. International Association of Sedimentologists, Special Publications, **30**, 9–34.

SMELLIE, J. L. & HOLE, M. J. 1997. Products and processes in Pliocene-Recent, subaqueous to emergent volcanism in the Antarctic Peninsula: examples of englacial Surtseyan volcano construction. *Bulletin of Volcanology*, **58**, 628–646.

SMELLIE, J. L. & SKILLING, I. P. 1994. Products of subglacial volcanic eruptions under different ice thicknesses, two examples from Antarctica. *Sedimentary Geology*, **91**, 115–129.

SMELLIE, J. L., HOLE, M. J. & NELL, P. A. R. 1993. Late Miocene valley-confined subglacial volcanism in northern Alexander Island, Antarctic Peninsula. *Bulletin of Volcanology*, **55**, 273–288.

STEIGER, R. H. & JAEGER, E. 1977. Subcommission on Geochronology: Convention of the use of decay constants in geo- and cosmochronology. *Earth and Planetary Science Letters*, **36**, 359–362.

WALKER, G. P. L. 1992. Morphometric study of pillow-size spectrum among pillow lavas. *Bulletin of Volcanology*, **54**, 459–474.

WALKER, G. P. L. & CROASDALE, R. 1972. Characteristics of some basaltic pyroclastics. *Bulletin of Volcanology*, **35**, 303–317.

WILCH, T. I. 1997. *Volcanic history of the West Antarctic Ice Sheet*. PhD thesis, New Mexico Institute of Mining and Technology.

WILCH, T. I. & MCINTOSH, W. C. 2000. Eocene and Oligocene volcanism at Mt. Petras, Marie Byrd Land: implications for middle Cenozoic ice sheet reconstructions in West Antarctica. *Antarctic Science*, **12**, 477–491.

WILCH, T. I., MCINTOSH, W. C. & DUNBAR, N. W. 1999. Late Quaternary Volcanic Activity in Marie Byrd Land: Potential $^{40}Ar/^{39}Ar$-dated time horizons in West Antarctic ice and marine cores. *Bulletin of the Geological Society of America*, **111**, 1563–1580.

Volatiles in basaltic glasses from a subglacial volcano in northern British Columbia (Canada): implications for ice sheet thickness and mantle volatiles

J. E. DIXON[1], J. R. FILIBERTO[1,4], J. G. MOORE[2] & C. J. HICKSON[3]

[1] *Division of Marine Geology and Geophysics, Rosenstiel School of Marine and Atmospheric Science, University of Miami, 4600 Rickenbacker Causeway, Miami FL 33149, USA (e-mail: jdixon@rsmas.miami.edu)*
[2] *US Geological Survey, MS-910, 345 Middlefield Road, Menlo Park, CA 94025, USA*
[3] *Geological Survey of Canada, 101–605 Robson Street, Vancouver, British Columbia V6B 5J3, Canada*
[4] *Present address: Department of Geosciences, SUNY, Stonybrook, NY 11794-2100, USA*

Abstract: Dissolved H_2O, CO_2, S and Cl concentrations were measured in glasses from Tanzilla Mountain, a 500 m-high, exposed subglacial volcano from the Tuya–Teslin region, north central British Columbia, Canada. The absence of a flat-topped subaerial lava cap and the dominance of pillows and pillow breccias imply that the Tanzilla Mountain volcanic edifice did not reach a subaerial eruptive phase. Lavas are dominantly tholeiitic basalt with minor amounts of alkalic basalt erupted at the summit and near the base. Tholeiites have roughly constant H_2O (*c.* 0.56 ± 0.07 wt%), CO_2 (<30 ppm), S (980 ± 30 ppm) and Cl (200 ± 20 ppm) concentrations. Alkalic basalts have higher and more variable volatile concentrations that decrease with increasing elevation (0.62–0.92 wt% H_2O, <30 ppm CO_2, 870–1110 ppm S and 280–410 ppm Cl) consistent with eruptive degassing. Calculated vapour saturation pressures for the alkalic basalts are 36 to 81 bars corresponding to ice thicknesses of 400 to 900 m. Maximum calculated ice thickness (*c.* 1 km) is at the lower end of the range of predicted maximum Fraser glaciation (*c.* 1–2 km), and may indicate initiation of volcanism during the waning stages of glaciation. Temporal evolution from tholeiitic to alkalic compositions may reflect compositional gradients within a melting column, instead of convective processes within a stratified magma chamber. The mantle source region for the subglacial volcanoes is enriched in incompatible elements similar to that for enriched mid-oceanic ridge basalt (e.g. Endeavour Ridge) and does not contain residual amphibole. Thus, metasomatic enrichment most likely reflects small degree partial melts rather than hydrous fluids.

Lavas erupted under glaciers form quenched glassy rinds on pillows that preserve information about eruptive, and in some cases, pre-eruptive volatile contents (Allen *et al.* 1982; Hickson *et al.* 1995; Moore *et al.* 1995). Quantification of pre-eruptive magmatic water and carbon dioxide concentrations in subglacial volcanic glasses may be used to answer important questions about ice thickness and timing of degassing-related magmatic volume expansion (e.g. convection in magma chamber or transition from passive to explosive eruptive modes).

This paper details measurements of dissolved water, carbon dioxide, sulphur and chlorine concentrations in glasses from Tanzilla Mountain, an exposed subglacial volcano from the Tuya–Teslin region, north central British Columbia, Canada. These data are used to constrain overlying ice thickness, to test models of magma chamber degassing during depressurization, and to infer mantle volatile contents.

Geological setting

More than 36 small basaltic subglacial volcanoes and eroded volcanic remnants occur in northern British Columbia in a region bounded by 58–60°N latitude and 131–123°W longitude (Fig. 1; Mathews 1947; Gabrielse 1970; Allen *et al.* 1982; Moore *et al.* 1995; Hickson 2000). This area of

From: SMELLIE, J. L. & CHAPMAN, M. G. (eds) 2002. *Volcano–Ice Interaction on Earth and Mars.* Geological Society, London, Special Publications, **202**, 255–271. 0305-8719/02/$15.00

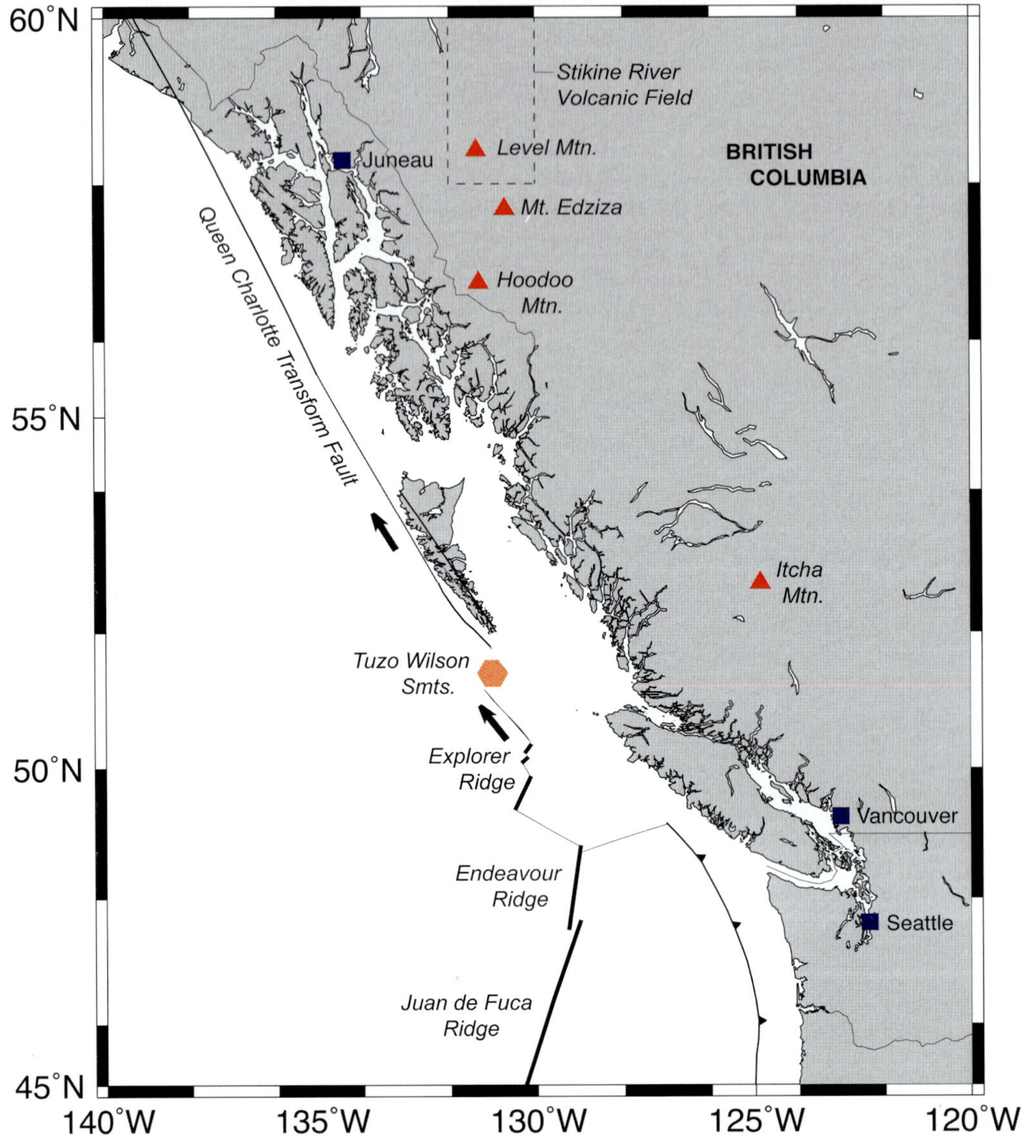

Fig. 1. Map of the NE Pacific region showing the location of some Quaternary volcanoes in British Columbia (solid triangles) (after Hickson 1990 and Allen 1990), NE Pacific spreading centres, the Queen Charlotte Transform Fault, and the Tuya–Teslin region in the Stikine Volcanic Belt. Box shows study area enlarged in

Quaternary (?) volcanism has been included in the Stikine volcanic belt (Souther 1977; Hickson 1990) or the northern Cordilleran volcanic province (Edwards & Russell 1999), which extends as far south as 55°N. Volcanism in this region is usually ascribed to regional tectonic extension resulting from unresolved plate motion on the major transcurrent plate boundary 400 km to the west, the dextral Queen Charlotte transform fault (Souther 1990; Edwards & Russell 1999, 2000). However, current estimates of Pacific–North America motion predict a velocity for the Pacific plate in this region that is more northerly than the NNW strike of the Queen Charlotte fault, implying regional compression. Using the geological (3 Ma average) model of DeMets & Dixon (1999), we calculate Pacific motion relative to North America at an azimuth of about

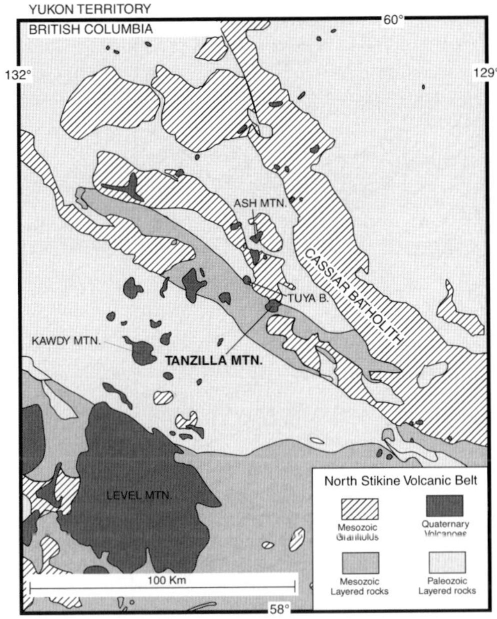

Fig. 2. Map showing the distribution of quaternary basaltic volcanic centres in the Tuya–Teslin region (modified after Souther *et al.* 1979).

345°, whereas the Queen Charlotte fault strikes at about 325–330°. Moreover, DeMets and Dixon (1999) showed that this motion has changed negligibly over the last 3 million years, and plate reconstructions suggest no significant changes in Pacific–North America motion for about the last 8 million years (Atwater & Stock 1998). This suggests that the cause of volcanism in the region may be more complicated.

Recently, 8 volcanoes from this region have been mapped and sampled in detail (Fig. 2; Hickson *et al.* 1995; Moore *et al.* 1995; this study). Individual volcanoes range from 200–700 m in height, 2–8 km in diameter, and from <0.5–10 km^3 in volume. Each is believed to be monogenetic, but some of the centres are fairly voluminous; however, no evidence of erosion was found within individual stratigraphic piles that might suggest episodic eruptions, or significant repose periods between eruptions. Had they not erupted under ice, a field of cinder cones or (given their size) small shield volcanoes would probably have formed. The centres may be as old as Pliocene, but observations based on glacial features and other geomorphic evidence suggest that most are Quaternary and at least some date from the Fraser glaciation, which lasted from 25 to 10 ka.

The volcanoes reflect complex interaction with the overlying ice sheet, and generally contain three dominant strato-volcanic units. These are (from the bottom up): (1) coherent, glassy pillow lava (erupted below ice/water level); (2) glassy phreatomagmatic tuffs and breccias; and (3) subaerial lava flows (erupted above ice/water level). Unit (1) is commonly associated with large (5–20 m in diameter), elongate feeding pipes and interlayered lenses of water-laid deposits; unit (2) commonly contains layers of pillow lava, pillow breccia and scattered cauliflower bombs, and is cut by dykes; and unit (3) may be associated with reddish cinders.

Many of these subglacial volcanoes share an interesting compositional stratigraphy, tholeiitic at the base and alkalic at the top (Hickson *et al.* 1995; Moore *et al.* 1995). The tholeiitic and alkalic magmas cannot be related by low-pressure crystal fractionation and apparently originated by differing degrees of deep melting of a mantle source (Moore *et al.* 1995). Moore *et al.* (1995) hypothesized that this stratigraphy may be related to the different pressure of volatile exsolution of tholeiitic and alkalic magmas within a stratified magma chamber. One of the goals of this study is to collect H_2O and CO_2 concentration data in glasses to evaluate the published model.

Tanzilla Mountain is located at 59°04′N and 130°24′W and rises *c.* 500 m above the surrounding terrain. Topography and sample locations are shown in Figure 3. Based on the absence of a flat-topped subaerial lava cap and the dominance of pillows and pillow breccias (Figs 4a, b), it is proposed that Tanzilla Mountain did not reach the phreatomagmatic or subaerial eruptive phases. Thus, Tanzilla Mountain provides the best candidate for preservation of pre-eruptive volatile contents within the glassy rinds of pillows.

Analytical techniques

Infrared spectroscopy

Concentrations of dissolved water and carbon dioxide were measured using transmission Fourier-transform infrared spectroscopy (FTIR). In general, precision of infrared analyses is about ±2% for total water and ±7–10% for molecular water and carbonate. For this study, however, concentrations of CO_2 are all near or below the detection limit of *c.* 30 ppm. The accuracy of the total water analyses is the same as reported in Dixon *et al.* (1991), about ±10%. The accuracy of the molecular water analyses is estimated to be about ±20%.

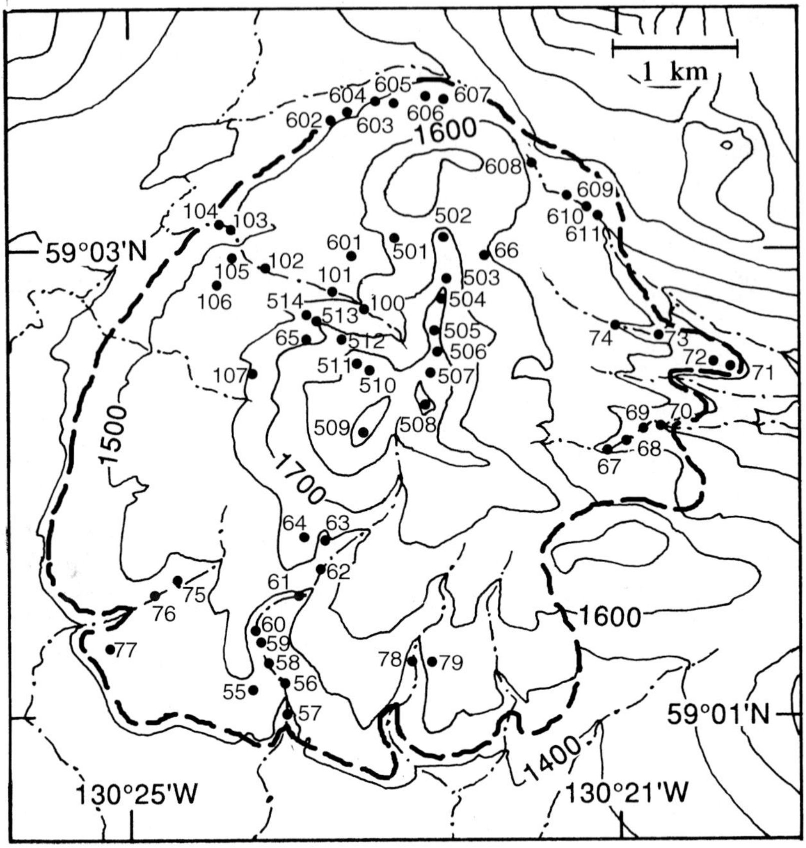

Fig. 3. Topographic map of Tanzilla Mountain showing locations of analysed samples.

Electron microprobe

Concentrations of sulphur and chlorine were determined on a 9-spectrometer ARL electron microprobe using natural and synthetic standards and instrumental parameters described in Moore *et al.* (1995). Mean atomic number calculations, based on the backgrounds measured on high and low mean atomic number standards, were used to obtain the background counts. Sulphur analyses determined by these procedures are consistent with those measured by electron probe in MORB and seamount glasses (Wallace & Carmichael 1992) based on interlaboratory comparison of glasses from Loihi seamount (D. Clague, pers. comm.). Error in the S analyses is estimated to be ±6% based on analysis of a standard with comparable S content (mean of 14 analyses on standard VG-2 is 0.127 ± 0.008 wt%, ±6% relative). Error in the Cl analyses is estimated to be ±8% based on analysis of standards A-99 (mean of 17 analyses is 0.024 ± 0.002 wt%, ±8% relative) and VG-2 (mean of 12 analyses is 0.031 ± 0.002 wt%, ±7% relative).

Results

Major and minor oxides

Tanzilla Mountain glass compositions are similar to glass compositions from Ash Mountain, South Tuya and Tuya Butte (Moore *et al.* 1995) and have two distinct compositional groups (Fig. 5a): a higher silica, lower alkali group of tholeiitic basalts, and a lower silica, higher alkali group of alkalic basalts. All glasses are fairly differentiated (Fig. 5b), with the alkalic basalts being more differentiated (4.6–5.0 wt% MgO) than the tholeiitic basalts (5.6–6.2 wt% MgO). Concentrations of incompatible minor oxides K_2O and P_2O_5 in Tanzilla Mountain tholeiitic

Fig. 4. Photographs of Tanzilla Mountain volcanic rocks. (**A**) Large coherent pillows overlying pillow breccia located near the base of Tanzilla Mountain. Note radial jointing defining individual large pillow. (**B**) Breccia formed of basalt pillow fragments in a yellowish matrix of finer-grained palagonitized glassy material. Hammer is 36 cm in length.

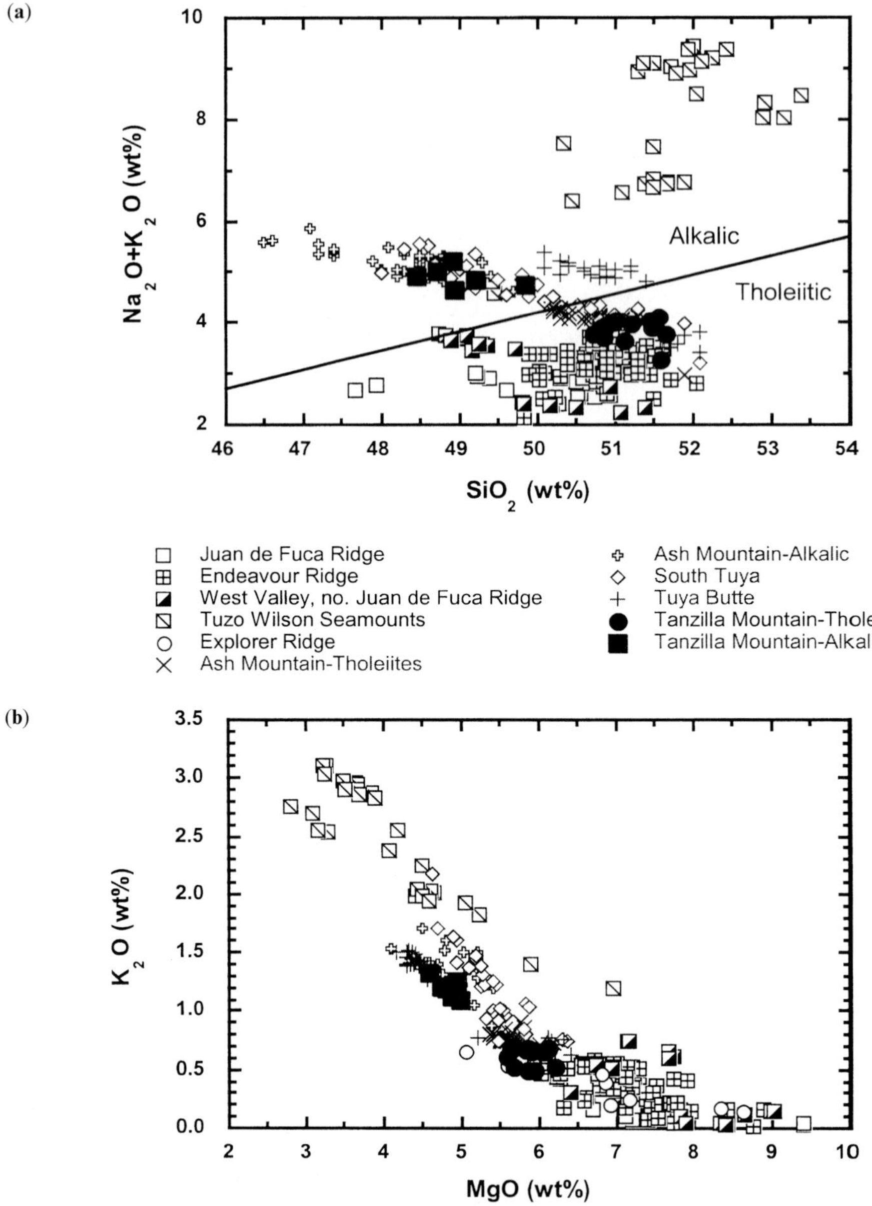

Fig. 5. (**a**) Total alkalis v. SiO_2 diagram for Tanzilla Mountain (this study) and other subglacial volcanoes (Moore *et al.* 1995). Tanzilla Mountain glasses have a bimodal distribution of tholeiitic and alkalic compositions. Line separating alkalic from tholeiitic compositions from Macdonald & Katsura (1964). (**b**) K_2O v. MgO diagram showing that Tanzilla Mountain alkalic basalts are more differentiated (lower MgO) than the tholeiitic basalts. Symbols for subglacial volcanic glasses are the same as in (**a**). Other data shown include alkalic lavas and intrusions from the northern Canadian Cordillera (filled and open triangles; Francis & Ludden 1990, 1995); the Tuzo Wilson Seamount Chain (open square with diagonal line; Allan *et al.* 1993); the West Valley segment of the northern Juan de Fuca Ridge (half-filled square; Cousens *et al.* 1995); Explorer Ridge (open circle; Michael *et al.* 1989); Endeavour Ridge (open square with cross; Karsten *et al.* 1990); and Juan de Fuca Ridge (open square; Dixon *et al.* 1988). Juan de Fuca Ridge basalts are derived from a transitional MORB source (Dixon *et al.* 1988). Endeavour and Explorer Ridge basalts are derived from a moderately enriched MORB source (Michael *et al.* 1989; Karsten *et al.* 1990). Tuzo Wilson Seamount Chain basalts are derived from a strongly enriched source (Allan *et al.* 1993). Subglacial lavas are more differentiated than basalts from the NE Pacific ridges. Note also that the alkalic lavas from the northern Canadian Cordilleran have much higher K_2O concentrations than the subglacial volcano and ridge-crest lavas.

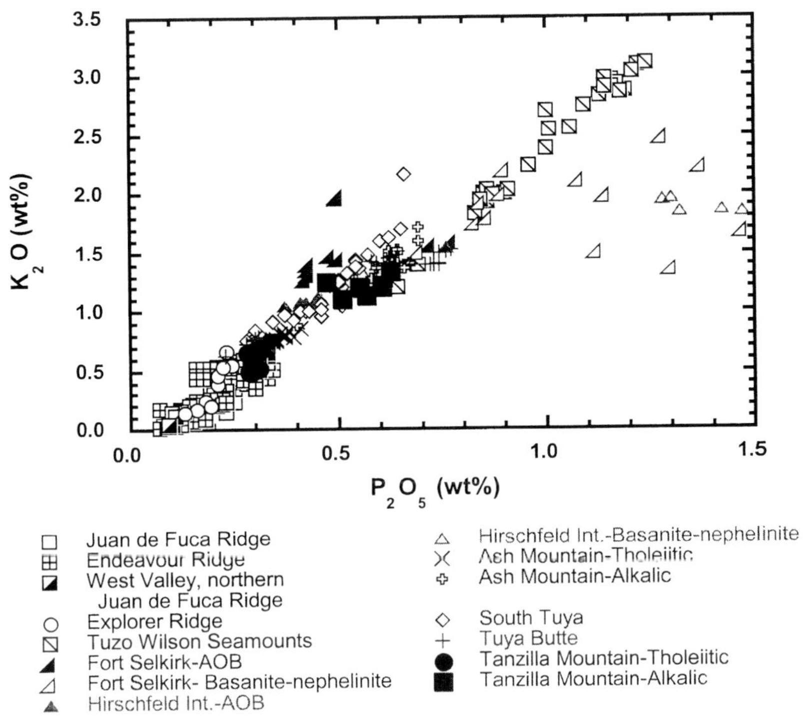

Fig. 6. Diagram showing P_2O_5 v. K_2O for subglacial volcanoes, Canadian Cordilleran alkalic lavas, and NE Pacific lavas. Symbols and references are the same as in Figures 4 and 5. The subglacial volcanoes lie on the linear trend defined by the NE Pacific ridge basalts and British Columbia alkali olivine basalts indicating incompatible element behaviour during differentiation and partial melting of a source region in which amphibole is not a residual phase. The lower slope defined by olivine nephelinites from British Columbia is consistent with partial melting of a source region having residual amphibole (Francis & Ludden 1995).

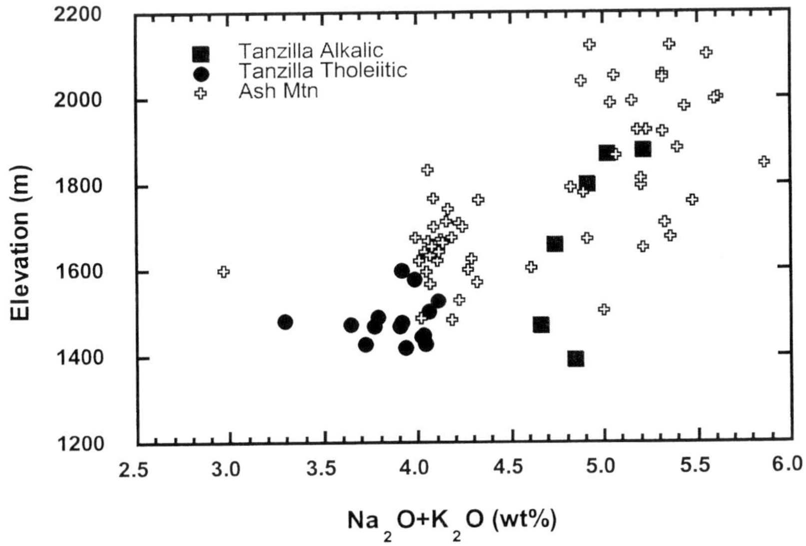

Fig. 7. Variation of total alkalis with elevation for Tanzilla Mountain (this study) and Ash Mountain (Moore *et al.* 1995) showing that alkalic lavas occur at the summit and near the base of Tanzilla Mountain.

glasses (Figs 5 & 6) are similar to those of enriched mid-ocean ridge basalts (EMORB) from NE Pacific, including the Endeavour segment of the Juan de Fuca Ridge (Karsten *et al.* 1990) and the Explorer Ridge (Michael *et al.* 1989). This suggests that the mantle source region has not been strongly modified by addition of subduction-related fluids, which would preferentially increase the K_2O concentrations. The higher incompatible element concentrations in Tanzilla Mountain alkalic basalts are higher than those in moderately enriched Endeavour segment EMORB but lower than those in strongly enriched EMORB from Tuzo Wilson Volcanic Field (Allan *et al.* 1993; Cousens *et al.* 1995). Relative to the tholeiites, the Tanzilla Mountain alkalic basalts can be generated by lower extents of melting of a source similar to that for Endeavour segment EMORB followed by greater extents of differentiation (Moore *et al.* 1995).

Table 1. *Major, minor and volatile element compositions for glasses from Tanzilla Mountain, British Columbia, Canada*

Sample	95C-057	95C-056	95C-058	95C-069	95C-076
Elevation (m)	1390	1420	1430	1430	1445
Rock type	1-alkalic	1-tholeiitic	1-tholeiitic	tholeiitic	1-tholeiitic
SiO_2	49.20	50.87	50.99	50.85	50.95
Al_2O_3	15.66	15.33	15.39	16.06	15.45
FeO	11.87	11.17		11.18	11.53
MgO	4.94	5.85	5.66	5.94	5.89
CaO	9.84	9.70	9.68	9.53	9.75
Na_2O	3.60	3.26	3.34	3.22	3.37
K_2O	1.25	0.68	0.70	0.50	0.65
TiO_2	2.66	1.86	1.98	1.86	1.94
P_2O_5	0.47	0.31	0.31	0.28	0.30
MnO	0.18	0.17	0.16	0.16	0.16
Cl	0.028	0.021	0.022	0.019	0.020
S	0.107	0.100	0.102	0.096	0.105
Total	99.950	99.450	99.660	100.20	99.950
H_2O (wt%)	0.684 (36)	0.594 (6)	0.558 (3)	0.489 (29)	0.559 (3)
H_2O mol (wt%)[1]	0.074 (14)	0.033 (7)	0.037 (4)	0.026 (2)	0.031 (7)
CO_2 (ppm)[1]	<30	<30	<30	<30	<30
P_{equil}[2]	42.5	64.0	60.0	53.5	60.0
Ice thickness[2]	472	711	667	594	667

Sample	95C-060	95C-072	95C-73	95C-068	95C-075
Elevation (m)	1450	1470	1470	1470	1475
Rock type	1-tholeiitic	1-alkalic	3-tholeiitic	1-tholeiitic	1-tholeiitic
SiO_2	50.77	48.94	51.64	50.87	51.12
Al_2O_3	50.37	15.25	15.77	15.44	15.84
FeO	11.30	11.92	11.28	11.36	11.32
MgO	5.85	4.87	5.67	6.12	5.85
CaO	9.68	10.00	9.53	9.59	9.68
Na_2O	3.34	3.53	3.25	3.22	3.14
K_2O	0.69	1.13	0.52	0.68	0.50
TiO_2	1.92	2.84	1.91	1.94	1.84
P_2O_5	0.30	0.57	0.29	0.31	0.29
MnO	0.16	0.16	0.17	0.17	0.15
Cl	0.018	0.041	0.017	0.020	0.018
S	0.099	0.111	0.095	0.095	0.095
Total	99.814	99.510	100.30	99.940	99.980
H_2O (wt%)	0.519 (3)	0.922 (32)	0.550 (13)	0.535 (40)	0.551 (25)
H_2O mol (wt%)[1]	0.03 (6)	0.102 (10)	0.036 (3)	0.030 (1)	0.046 (0)
CO_2 (ppm)[1]	<30	<30	<30	<30	<30
P_{equil}[2]	56.0	80.5	59.5	57.5	59.5
Ice thickness[2]	622	894	661	639	661

Table 1. *(continued)*

Sample	95C-104	95C-079	95C-067	95C-74	95C-106
Elevation (m)	1480	1485	1490	1505	1530
Rock type	1-tholeiitic	1-tholeiitic	1-tholeiitic	1-tholeiitic	3-tholeiitic
SiO_2	50.82	51.58	50.72	51.45	51.56
Al_2O_3	15.38	15.31	15.67	15.42	15.39
FeO	11.50	11.37	11.53	10.59	11.08
MgO	6.03	6.06	6.20	5.57	5.63
CaO	9.59	9.74	9.36	9.67	9.71
Na_2O	3.25	2.63	3.26	3.44	3.43
K_2O	0.66	0.66	0.52	0.62	0.68
TiO_2	1.93	1.88	1.87	1.92	1.91
P_2O_5	0.28	0.29	0.31	0.30	0.30
MnO	0.15	0.16	0.15	0.16	0.15
Cl	0.018	0.017	0.024	0.018	0.020
S	0.095	0.099	0.095	0.098	0.099
Total	99.840	100.10	99.840	99.390	100.09
H_2O (wt%)	0.514 (4)	0.534 (4)	0.551 (15)	0.512 (16)	0.531 (19)
H_2O mol (wt%)[1]	0.034 (7)	0.038 (6)	0.034 (9)	0.037 (2)	0.038 (8)
CO_2 (ppm)[1]	<30	<30	<30	<30	<30
P_{equil}[2]	55.5	57.5	59.5	55.5	57.5
Ice thickness[2]	617	661	661	617	639

Sample	HHB-610	HHB-609	HHB-606	95C-101	HHB-608
Elevation (m)	1550	1570	1580	1580	1595
Rock type	2-tholeiitic	1-tholeiitic	3-tholeiitic	3-tholeiitic	6-tholeiitic
SiO_2	–	–	–	51.20	–
Al_2O_3	–	–	–	15.48	–
FeO	–	–	–	11.10	–
MgO	–	–	–	5.89	–
CaO	–	–	–	9.42	–
Na_2O	–	–	–	3.32	–
K_2O	–	–	–	0.67	–
TiO_2	–	–	–	1.91	–
P_2O_5	–	–	–	0.29	–
MnO	–	–	–	0.15	–
Cl	–	–	–	0.020	–
S	–	–	–	0.094	–
Total	–	–	–	99.670	–
H_2O (wt%)	0.627 (104)	0.520 (20)	0.690 (60)	0.538 (2)	0.550 (20)
H_2O mol (wt%)[1]	0.036 (10)	0.034 (5)	0.072 (2)	0.035 (1)	0.037 (0)
CO_2 (ppm)[1]	<30	<30	<30	<30	<30
P_{equil}[2]	68.0	56.0	76.0	58.0	59.5
Ice thickness[2]	756	622	844	644	661

Studies of northern Canadian Cordilleran lavas (e.g. Francis & Ludden 1990, 1995; Charland *et al.* 1993) have shown that amphibole is a residual phase during melting and production of olivine nephelinitic to basanitic lavas. These lavas show depletions in elements that partition strongly into amphibole (K_2O, Rb and Ba) at a given concentration of another strongly incompatible element (e.g. La or P_2O_5). In Figure 6 the distribution of K_2O concentrations in Canadian Cordilleran basanites to nephelinites flattens out at *c.* 2 wt% as P_2O_5 increases above *c.* 0.7 wt% (Francis & Ludden 1990, 1995). The fact that both tholeiitic and alkalic Tanzilla Mountain lavas lie on the linear trend for samples with <0.7 wt% P_2O_5, as defined by tholeiites and alkalic basalts from the northeastern Pacific basin (Michael *et al.* 1989; Karsten *et al.* 1990; Allan *et al.* 1993; Cousens *et al.* 1995), implies that amphibole was not a residual phase during generation of Tanzilla Mountain magmas.

In contrast to the other subglacial volcanoes, alkalic basalt at Tanzilla Mountain occurs at both the summit and near the base (Fig. 7). The

Table 1. (*continued*)

Sample	95C-100	95C-108	HHB-503	HHB-507	HHB-511
Elevation (m)	1600	1660	1800	1880	1870
Rock type	1-tholeiitic	1-alkalic	1-alkalic	6-alkalic	4-alkalic
SiO_2	51.47	49.84	48.45	48.91	48.72
Al_2O_3	15.43	15.13	15.04	15.13	15.18
FeO	10.90	11.30	11.81	12.19	12.00
MgO	5.83	4.98	4.80	4.57	4.74
CaO	9.60	9.89	9.90	9.54	9.66
Na_2O	3.24	3.64	3.71	3.88	3.82
K_2O	0.67	1.10	1.20	1.33	1.21
TiO_2	1.85	2.54	2.90	2.94	2.88
P_2O_5	0.29	0.51	0.55	0.63	0.61
MnO	0.17	0.18	0.18	0.19	0.18
Cl	0.021	0.039	0.035	0.034	0.034
S	0.095	0.100	0.099	0.096	0.087
Total	99.690	99.380	98.807	99.580	99.250
H_2O (wt%)	0.515 (20)	0.803 (35)	0.768 (66)	0.658 (17)	0.624 (67)
H_2O mol (wt%)[1]	0.031 (4)	0.087 (20)	0.076 (8)	0.060 (11)	0.050 (3)
CO_2 (ppm)[1]	<30	<30	<30	<30	<30
P_{equil}[2]	55.5	60.0	56.0	40.0	35.5
Ice thickness[2]	617	667	622	445	394

Rock types are: 1, coherent pillows; 2, pillow breccia; 3, hyaloclastite tuff; 4, scoria, bombs; 5, lava flow; 6, dyke. Major elements were measured using electron microprobe at USGS/Menlo Park.

[1] H_2O and CO_2 concentrations measured using FTIR at the University of Miami/RSMAS. Molar absorptivities used were 63 ± 5 l/mol-cm for total dissolved water at 3535 cm^{-1}, 375 ± 20 l/mol-cm for the carbonate bands at 1515 and 1430 cm^{-1}, and 25 ± 5 l/mol-cm for the molecular water band at 1630 cm^{-1} (Dixon *et al.* 1995). Values in parentheses are 1s standard deviations in the last or last two decimal places.

[2] Ice thickness calculated from vapour saturation pressure, P_{equil} (Dixon 1997), assuming an ice density of 0.9 g cm^{-3}, 15 ± 15 ppm CO_2 in tholeiitic glasses and 0 ppm CO_2 in alkalic glasses.

lowest sample (95C-057) has higher K_2O (Fig. 6) at a given P_2O_5 content than the overlying alkalic basalts and may have been derived from a distinct parental magma composition.

Volatiles

All CO_2 concentrations are less than the detection limit of 30 ppm. Variations in H_2O, S and Cl with elevation are shown in Figures 8a, b and c. Tholeiites have roughly constant H_2O (*c.* 0.56 ± 0.07 wt%), S (980 ± 30 ppm) and Cl (200 ± 20 ppm) concentrations that do not vary systematically with elevation. Alkalic basalts have higher and more variable volatile concentrations that decrease with increasing elevation (0.62–0.92 wt% H_2O, 870–1110 ppm S, and 280–410 ppm Cl). These results contrast with S data for Ash Mountain, South Tuya and Tuya Butte, where the alkalic basalts contain intermediate to low S concentrations relative to the underlying high S tholeiitic basalts (Moore *et al.* 1995). The topographically lowest sample (95C-057), however, has lower water and Cl (Fig. 8a, c), in addition to higher K_2O (Fig. 6), than the overlying alkalic basalts.

Water behaves as an incompatible element during melting and crystallization processes (Dixon *et al.* 1988, 1997; Michael 1988; Danyushevsky *et al.* 2000). Therefore, water concentrations increase as extents of melting decrease or as extents of crystallization increase. When H_2O is plotted against another incompatible element (e.g. K_2O or P_2O_5), it is possible to distinguish trends produced by melting and crystallization processes (H_2O correlates linearly with non-volatile incompatible elements) from trends produced by degassing (H_2O falls below melting–crystallization trends). On a plot of H_2O against K_2O (Fig. 9a), Tanzilla Mountain tholeiitic glasses fall on a linear array with data from NE Pacific mid-ocean ridges (Juan de Fuca, Endeavour and Explorer ridges; Dixon *et al.* 1988; Michael 1988; Karsten *et al.* 1990). Magmatic water concentrations are slightly higher than, but similar to, those in basalts from the enriched Endeavour Ridge (Dixon *et al.* 1988). The higher water concentrations are consistent with greater extents of differentiation (lower MgO concentrations; Fig. 9b) of magma generated from a mantle source region similar to

Table 2. *Trace elements in Tanzilla Mountain whole-rock samples*

Sample	95C-070	95C-056	95C-078	95C-072	95C-055	95C-075	95C-079
Ba	300	270	315	290	265	265	265
Rb	12	14	18	14	14	14	14
Sr	388	374	410	386	368	358	334
Nb	16	16	20	16	16	16	14
Zr	117	111	129	117	111	114	102
U	18	16	18	20	18	16	14
Ni	143	169	139	131	200	194	235
Co	44	46	43	42	50	48	55
Sample	95C-067	95C-103	95C-105	95C-066	95C-065	95C-108	
Ba	265	460	535	355	365	445	
Rb	14	20	16	16	12	18	
Sr	354	548	638	434	444	468	
Nb	14	26	34	20	20	28	
Zr	111	150	153	132	135	117	
U	18	18	18	18	16	16	
Ni	220	88	27	104	132	215	
Co	53	32	35	38	42	54	

Trace elements (in ppm) measured using XRF at Geological Survey of Canada laboratories.

the enriched MORB source for the Endeavour Ridge. Because the magmatic water concentrations are similar to what is expected based on other mantle-derived melts, it is unlikely that these tholeiitic glasses have lost significant water by degassing.

In contrast, Tanzilla Mountain alkalic glasses fall on or below the line projected through the linear array of water and K_2O data (Fig. 9a). Thus, even though the concentrations of water are higher in the alkalic glasses, the H_2O/K_2O or H_2O/P_2O_5 are significantly lower, suggesting that the alkalic glasses have lost water due to degassing. The coherent trend of decreasing volatile species (H_2O, Cl and S) concentration with elevation (with the exception of 95C-057) suggests that all three volatile species were degassing during eruption of the alkalic basalts.

Vapour saturation pressure calculations

Dissolved water and carbon dioxide concentrations in glasses can be used to calculate vapour saturation pressures and the composition of the vapour phase in equilibrium with the melt at that pressure (Dixon & Stolper 1995; Dixon 1997). In this paper the vapour saturation pressures and assumptions about ice density are used to estimate ice thickness overlying the lavas during eruption. Because the solubility of CO_2 is much lower than that of H_2O, CO_2 will be mostly to completely lost from a magma before H_2O begins to degas (Dixon & Stolper 1995). Therefore, since the tholeiitic samples have not lost significant H_2O, it is unlikely that the CO_2 concentrations are zero. A concentration of 15 ± 15 ppm is assumed to calculate the pressure of vapour saturation. The vapour saturation pressure using CO_2 of 15 ppm is 60 ± 5 bars. Assuming an ice density of 0.9 g cm^{-3}, this is equivalent to 660 ± 60 m of ice. The full uncertainty using ± 15 ppm is ± 33 bars or ± 370 m of ice. Therefore, vapour saturation calculations for the tholeiites only allow us to say that they were erupted under less than *c.* 1 km of ice or water, but more than 300 m of ice or water (Fig. 10).

Because water has degassed from the alkalic lavas, it is reasonable to assume that the dissolved CO_2 contents are truly zero (Dixon 1997). Vapour saturation pressures calculated from dissolved water contents for the alkalic basalts range from 36 to 81 bars with alkalic glasses collected near the base and at the summit of the volcano yielding the highest and lowest estimated equilibration pressures, respectively. Assuming an ice density of 0.9 g cm^{-3}, a maximum pressure of 81 bars corresponds to an ice thickness of *c.* 0.9 km overlying the nascent Tanzilla Mountain (Fig. 10). The pressure differential from base to summit using the alkalic basalt data (45 bars) represents an ice thickness change of 500 m identical to the volcano height of 500 m implying that there was minor or no collapse of the overlying ice (cauldron formation) or catastrophic release of melt water (jökulhlaup) during eruption of Tanzilla Mountain.

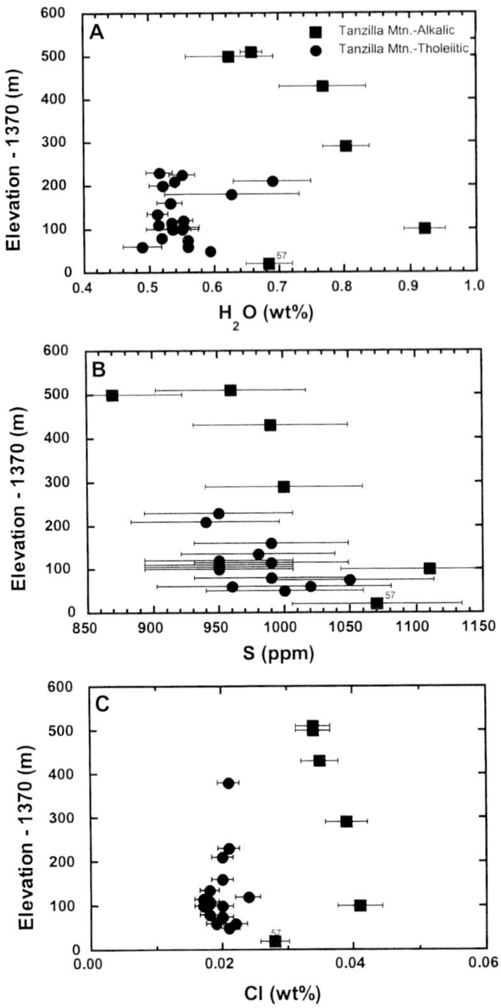

Fig. 8. Diagrams showing variation of (**A**) H_2O, (**B**) S and (**C**) Cl with elevation. Volatiles in tholeiitic glasses do not vary systematically with elevation consistent with quenching of pre-eruptive magmatic values. Volatile concentrations in alkalic glasses are initially higher than those in tholeiitic glasses because of lesser extents of melting and greater extents of differentiation. Systematic decrease in volatile contents with elevation (except 95C-057) in alkalic glasses is consistent with loss of H_2O, S and Cl during eruptive degassing.

Sample 95C-057, at the base of the volcano, has a low equilibration pressure of 42.5 bar (470 m ice thickness) for its location. If the minor element geochemistry of sample 95C-057 was identical to the samples erupted at the summit of Tanzilla Mountain, it would be reasonable to infer that the lava initially erupted at the summit, reached vapour–melt equilibrium for the lower summit pressure, and then flowed downhill. However, because of its distinct geochemistry, it is proposed that this unit may represent a separate and later eruption formed as the ice receded.

Discussion

Implications for timing of volcanism and deglaciation

It is possible to compare the estimate of maximum ice thickness based on magmatic volatiles at the initiation of subglacial volcanism with other estimates of ice thickness during the last glacial maximum. The most relevant work on ice-flow and ice thickness is a recent study of central British Columbia (late Fraser glaciation), *c.* 450 km south of this study area (Stumpf *et al.* 2000). Stumpf *et al.* (2000) categorized ice flow into three phases: (1) ice expansion (25–16 ka); (2) glacial maximum (16–15 ka); and (3) late glacial (15–13 ka). At its maximum, the ice sheet was 900 km wide, reached 2 to 3 km above present sea level (asl), and was roughly 1–2 km thick. This result is similar to the earlier Ice-4G model (Peltier 1994), that gave an ice thickness of approximately 2 km. During the late glacial period the ice sheet reached a maximum of 1.5 km asl and was <1 km thick. In this study area, the average elevation of the surrounding country rock is 1.3 km. Therefore, the maximum ice thickness is predicted to have been *c.* 1–2 km.

The volcanic morphology and volatile contents of the Tanzilla Mountain basalts can be used to infer timing of volcanism relative to glaciation. The eruption of Tanzilla Mountain was completely subglacial beneath 300 to 900 m of ice. The ice thickness at the initiation of volcanism is within uncertainty of the lower limit of maximum ice thickness; therefore Tanzilla Mountain erupted during or near the end of the maximum phase. Other subglacial volcanoes in the region, of similar height and built on countryrock with comparable elevations (*c.* 1.3 km), breached through the surface of the ice to erupt phreatomagmatically and subaerially, suggesting that they were erupted under thinner ice, probably during glacial retreat. Our estimates of vapour saturation pressures, combined with the observation that Tanzilla Mountain never reached the phreatomagmatic stage, suggest that the magmatic–phreatomagmatic transition probably occurs at an ice thickness

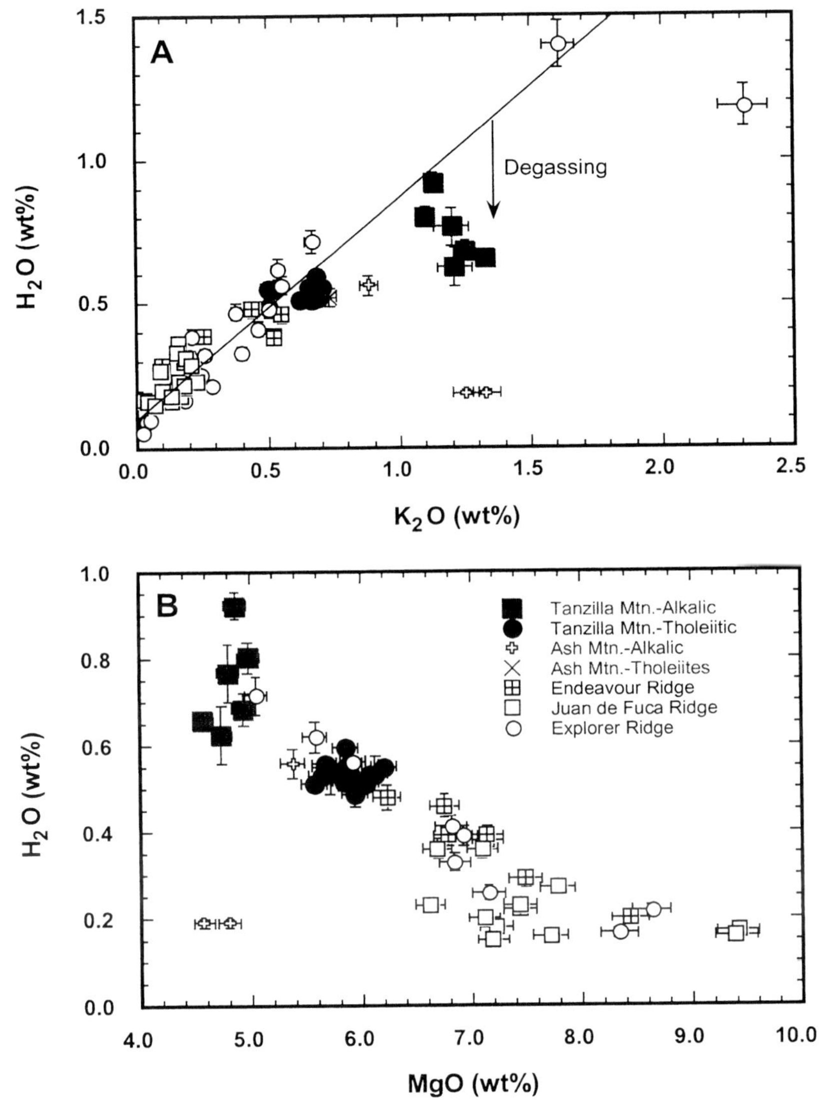

Fig. 9. (**A**) H_2O v. K_2O for Tanzilla Mountain (this study), Ash Mountain (Moore *et al.* 1995), Endeavour Ridge (Dixon *et al.* 1988), Explorer Ridge (Michael 1995) and Juan de Fuca Ridge (Dixon *et al.* 1988). Tanzilla tholeiites have water contents similar to enriched MORB from the NE Pacific. Tanzilla alkalic basalts have higher water concentrations, but lower H_2O/K_2O consistent with water loss during eruptive degassing. Ash Mountain basalts also have lower H_2O/K_2O consistent with degassing. (**B**) H_2O v. MgO for same data set.

of 200–300 m. Volcanism ceased before the disappearance of the glaciers, shown by the fact that none of the volcanoes are subaerial cinder cones. Thus, the subglacial volcanoes were most likely erupted during a period of waning glaciation and glacial retreat.

Volcanism associated with glacial retreat and subsequent isostatic rebound also has been described for Iceland (Hardarson & Fitton 1991; Sigvaldason *et al.* 1991; Jull & McKenzie 1996; Mary Gee *et al.* 1998; Slater *et al.* 1998). In central Iceland, magma production was 20–30 times higher from 10 000 to 4500 yr BP than during the period since 2900 yr BP. In the Reykjanes Peninsula, magma production is estimated to have been 20 times higher during postglacial

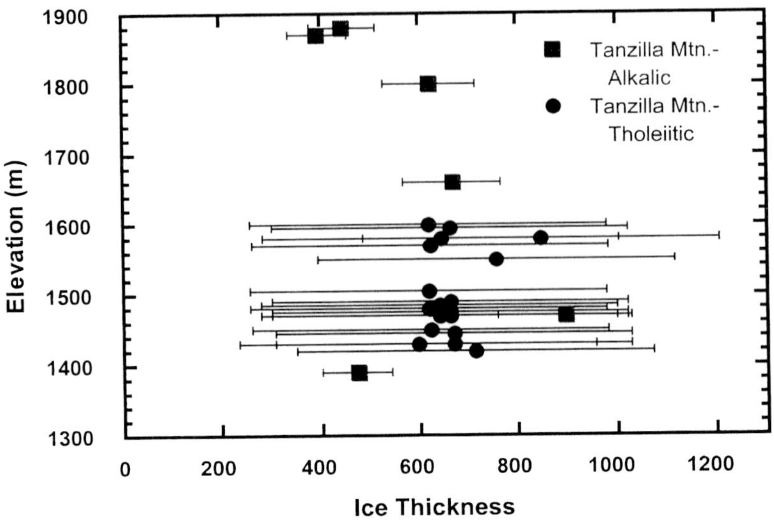

Fig. 10. Diagram showing calculated ice thickness v. elevation. Ice thicknesses calculated from vapour saturation pressures (Dixon 1997) and assuming an ice density of 0.9 g cm^{-3}, 15 ± 15 ppm CO_2 for the tholeiitic glasses and 0 ppm CO_2 for the alkalic glasses.

times than the period since 1100 yr BP. (Jull & McKenzie 1996). The lag time between glaciation and increased eruption rates in Iceland is about 1–3 ka, consistent with melt transport times from measurements of $^{226}Ra/^{230}Th$ from ocean island basalts and mid ocean ridge basalts (Slater *et al.* 1998). Jull & McKenzie (1996) concluded that removal of a 2 km-thick ice sheet is equivalent to moving a melt column up 0.6 km, resulting in an increase in melt fraction by about 0.2%. Increased mantle melting accounts for both increased magma supply and more depleted magma compositions erupted just after deglaciation.

Compared to Iceland, the subglacial volcanoes in northern British Columbia probably erupted through thicker crust and thinner ice. Intrusion into thicker and cooler crust leads to more differentiated compositions of the Canadian subglacial lavas compared to the Icelandic subglacial lavas. The more alkalic compositions (tholeiites to alkalic basalts) of the subglacial volcanoes compared with olivine tholeiites in Iceland are consistent with a smaller pressure differential from glacial to post-glacial time, resulting in less uplift and melting. It is also possible that this mechanism could produce a range of magma compositions, with extents of melting decreasing with depth in the melting column. Such a mechanism could produce the observed stratigraphy, with the shallow tholeiitic melts erupting first followed by the deeper alkalic melts.

Implications for deep v. shallow degassing models

Moore *et al.* (1995) suggested that the transition from tholeiitic to alkalic compositions was triggered by the transition from subglacial to shallow water or subaerial eruptions, as manifested by change to more explosive activity and then to subaerial lava flows (and by a marked reduction of sulphur in volcanic glass). They proposed that the mechanism for the compositional change was related to degassing within a stratified magma chamber. Prior to eruption, the lower density tholeiitic melts overlay higher density alkalic melts in shallow chambers underlying each of the volcanoes. During subglacial conditions, the lower density tholeiitic melts were the first to erupt. As the volcano grew through the ice (or ice-impounded water), the volcanic conduit vented to the atmosphere, producing a partial depressurization of the conduit and the subsurface chamber. This sudden reduction in confining pressure would cause enhanced vesiculation of volatile saturated melts, particularly of the more volatile-rich alkalic melts, causing them to rise to the top of the chamber and erupt.

The new data, in this paper, on Tanzilla Mountain are inconsistent with the stratified magma chamber model. Firstly, the volcanic edifice at Tanzilla Mountain never made the transition to shallow water or subaerial eruptions, and yet it displays the same bimodal tholeiitic–alkalic compositions as the other subglacial volcanoes.

Secondly, the alkalic lavas are compositionally heterogeneous, inconsistent with residence and homogenization within a magma reservoir. Thirdly, the decreasing volatile concentrations with elevation (with the exception of 95C-057), combined with the fact that the alkalic basalts at the base of Tanzilla Mountain fall on the Pacific oceanic basalt trend, suggest that most of the degassing occurred during the eruption in response to small changes in hydrostatic (cryostatic?) pressure and did not occur deep in the magma chamber.

Implication for volatile content of mantle source region

The similarity of H_2O/K_2O and Cl/K_2O for Tanzilla Mountain tholeiitic basalts and other enriched Pacific basalts implies that the mantle source region for the subglacial volcanoes is enriched in incompatible elements, but not preferentially enriched in volatiles. Edwards & Russell (2000) also concluded that lavas from the northern Cordilleran volcanic province have trace element abundance patterns similar to the average composition of oceanic island basalt and that the source region is likely asthenospheric. If the source region has undergone recent metasomatic enrichment, our data suggest the enriching agent was low degree partial melts, not a hydrous fluid. Note that high volatile contents in the Tanzilla Mountain glasses do not require the presence of hydrous minerals in the source region as proposed by Cousens *et al.* (1995) for the NE Pacific EMORB. Dixon *et al.* (1988) determined that the mantle source for the Endeavour Ridge contained *c.* 330 ppm H_2O. This amount of water can be easily accommodated in nominally anhydrous minerals (Bell & Rossman 1992), in particular clinopyroxene. If amphibole was present in the mantle source, it occurred in small quantities and was exhausted during partial melting.

Conclusions

Based on measurements of volatiles in glasses, lavas from Tanzilla Mountain, British Columbia, probably erupted under 300 to 900 m of ice. The absence of a flat-topped subaerial lava cap and the dominance of pillows and pillow breccias imply that Tanzilla Mountain did not reach a subaerial eruptive phase. Therefore the magmatic to phreatomagmatic transition for these subglacial volcanoes was at ice thicknesses <300 m. Maximum calculated ice thickness (*c.* 1 km) is at the lower end of the range predicted during the maximum of the Fraser glaciation (*c.* 1–2 km), and may indicate initiation of volcanism during the waning stages of glaciation. Temporal evolution from tholeiitic to alkalic compositions is not related to degassing within a stratified magma chamber during transition to the phreatomagmatic eruptive phase, but may reflect compositional gradients within a melting column with melting enhanced by lithostatic rebound following glacial retreat.

Tholeiites have roughly constant H_2O (*c.* 0.56 $\pm$ 0.07 wt%), CO_2 (<30 ppm), S (980 $\pm$ 30 ppm) and Cl (200 $\pm$ 20 ppm) concentrations. Their major, minor, and volatile element concentrations are similar to those found in basalts from the Endeavour Ridge, an enriched portion of the NE Pacific mid-oceanic ridge system. It is proposed that these tholeiites have degassed primarily a CO_2-rich vapour with insignificant loss of H_2O, S and Cl. Alkalic basalts have higher and more variable volatile concentrations that decrease with increasing elevation (0.62–0.92 wt% H_2O, <30 ppm CO_2, 870–1110 ppm S and 280–410 ppm Cl). It is suggested that the higher volatile concentrations are related to lower extents of melting followed by greater amounts of differentiation. The correlation with elevation implies that CO_2, H_2O, Cl and S were lost during eruptive degassing. The mantle source regions for the subglacial volcanoes is similar to that for the Endeavour Ridge, with *c.* 330 ppm H_2O in the mantle source. Amphibole is not a residual phase during generation of these magmas.

Electron microprobe analyses of major elements and sulphur in glasses were made by Lewis C. Calk in the laboratories of the US Geological Survey in Menlo Park, California. Some fieldwork funds were paid by the Geological Survey of Canada Project No. 92008 to C. J. Hickson. The manuscript benefited from thorough reviews by J. K. Russell, J. Allan, and S. Newman. This research was supported by NSF OCE-9702795 Early Career Award to J. E. Dixon.

References

ALLAN, J. F., CHASE, R. L., COUSENS, B., MICHAEL, P. J., GORTON, M. P. & SCOTT, S. D. 1993. The Tuzo Wilson Volcanic Field, NE Pacific: Alkaline volcanism at a complex, diffuse, transform-trench-ridge triple junction. *Journal of Geophysical Research*, **98**, 22 367–22 387.

ALLEN, C. C. 1990. Tuya Butte. *In*: WOOD, C. A. & KIENLE, J. (eds) *Volcanoes of North America*. Cambridge University Press, Cambridge, 119–121.

ALLEN, C. C., JERCINOVIC, M. J. & ALLEN, J. S. B. 1982. Subglacial volcanism in north-central British Columbia and Iceland. *Journal of Geology*, **90**, 699–715.

ATWATER, T. & STOCK, J. 1998. Pacific–North America plate tectonics in the Neogene southwestern United States: An update. *International Geology Review*, **40**, 375–402.

BELL, D. R. & ROSSMAN, G. R. 1992. Water in the earth's mantle: the role of nominally anhydrous minerals. *Science*, **255**, 1391–1397.

CHARLAND, A., FRANCIS, D. & LUDDEN, J. 1993. Stratigraphy and geochemistry of the Itcha Volcanic Complex, central British Columbia. *Canadian Journal of Earth Science*, **30**, 132–144.

COUSENS, B. L., ALLAN, J. F., LEYBOURNE, M. I., CHASE, R. L. & VAN WAGONER, N. 1995. Mixing of magmas from enriched and depleted mantle sources in the northeast Pacific: West Valley segment, Juan de Fuca Ridge. *Contributions to Mineralogy and Petrology*, **120**, 337–357.

DANYUSHEVSKY, L. V., EGGINS, S. M., FALLOON, T. J. & CHRISTIE, D. M. 2000. H_2O abundance in depleted to moderately enriched mid-ocean ridge magmas; Part I: incompatible behavior, implications for mantle storage, and origin of regional variations. *Journal of Petrology*, **41**, 1329–1364.

DEMETS, C. & DIXON, T. H. 1999. New kinematic models for Pacific-North America motion from 3 Ma to present, 1: evidence for steady motion and biases in the NUVEL-1A model. *Geophysical Research Letters*, **26**, 1921–1924.

DIXON, J. E. 1997. Degassing of alkalic basalts. *American Mineralogist*, **82**, 368–378.

DIXON, J. E., STOLPER, E. & DELANEY, J. R. 1988. Infrared spectroscopic measurements of CO_2 and H_2O in Juan de Fuca Ridge basaltic glasses. *Earth and Planetary Science Letters*, **90**, 87–104.

DIXON, J. E., CLAGUE, D. A. & STOLPER, E. M. 1991. Degassing history of water, sulfur, and carbon in submarine lavas from Kilauea Volcano, Hawaii. *Journal of Geology*, **99**, 371–394.

DIXON, J. E. & STOLPER, E. M. 1995. An experimental study of water and carbon dioxide solubilities in mid-ocean ridge basaltic liquids. Part II: Applications to degassing. *Journal of Petrology*, **36**, 1633–1646.

DIXON, J. E., STOLPER, E. M. & HOLLOWAY , J. R. 1995. An experimental study of water and carbon dioxide solubilities in mid-ocean ridge basaltic liquids. Part I: Calibration and solubility models. *Journal of Petrology*, **36**, 1607–1631.

DIXON, J. E., CLAGUE, D. E., WALLACE, P. & POREDA, R. 1997. Volatiles in alkalic basalts from the North Arch volcanic field, Hawaii: Extensive degassing of deep submarine-erupted alkalic series lavas. *Journal of Petrology*, **38**, 911–939.

EDWARDS, B. R. & RUSSELL, J. K. 1999. Northern Cordilleran volcanic province: A northern Basin and Range? *Geology*, **27**, 243–246.

EDWARDS, B. R. & RUSSELL, J. K. 2000. Distribution, nature and origin of Neogene-Quaternary magmatism in the northern Cordilleran volcanic province, Canada, *Geological Society of America, Bulletin*, **112**, 1280–1295.

FRANCIS, D. & LUDDEN, J. 1990. The mantle source for olivine nephelinite, basanite, and alkaline olivine basalt at Fort Selkirk, Yukon, Canada. *Journal of Petrology*, **31**, 371–400.

FRANCIS, D. & LUDDEN, J. 1995. The signature of amphibole in mafic alkaline lavas, a study in the northern Canadian Cordillera. *Journal of Petrology*, **36**, 1171–1191.

GABRIELSE, H. 1970. Geology of Jennings River map-area, B.C. (104-0). *Geological Survey of Canada, Paper*, **68-55**.

HARDARSON, B. S., FITTON, J. G. 1991. Increased mantle melting beneath Snaefellsjökull volcano during Late Pleistocene deglaciation. *Nature*, **353**, 62–64.

HICKSON, C. J. 1990. Canadian Cordillera: Volcano vent map and table. *In*: WOOD, C. A. & KIENLE, J. (eds) *Volcanoes of North America*. Cambridge University Press, Cambridge, 116–117.

HICKSON, C. J. 2000. Physical controls and resulting morphological forms of Quaternary ice-contact volcanoes in western Canada. *Geomorphology*, **32**, 239–261.

HICKSON, C. J., MOORE, J. G., CALK, L. C. & METCALFE, P. 1995. Intraglacial volcanism in the Wells Gray-Clearwater volcanic field, east-central British Columbia, Canada. *Canadian Journal of Science*, **32**, 838–851.

JULL, M. & MCKENZIE, D. 1996. The effect of deglaciation on mantle melting beneath Iceland. *Journal of Geophysical Research*, **101**, 21 815–21 828.

KARSTEN, J. L., DELANEY, J. R., RHODES, J. M. & LIIAS, R. A. 1990. Spatial and temporal evolution of magmatic systems beneath the Endeavour segment, Juan de Fuca Ridge: Tectonic and petrologic constraints. *Journal of Geophysical Research*, **95**, 19 235–19 256.

MACDONALD, G. A. & KATSURA, T. 1964. Chemical composition of Hawaiian lavas. *Journal of Petrology*, **5**, 82–133.

MARY GEE, M. A., TAYLOR, R. N., THIRLWALL, M. F. & MURTON, B. J. 1998. Glacioisostacy controls chemical and isotopic characteristics of tholeiites from the Reykjanes Peninsula, SW Iceland. *Earth and Planetary Science Letters*, **164**, 1–5.

MATHEWS, W. H. 1947. 'Tuyas' flat-topped volcanoes in northern British Columbia. *American Journal of Science*, **245**, 560–570.

MICHAEL, P. J. 1988. The concentration, behavior and storage of H_2O in the suboceanic upper mantle: Implications for mantle metasomatism. *Geochimica et Cosmochimica Acta*, **52**, 555–566.

MICHAEL, P. J. 1995. Regionally distinctive sources of depleted MORB: Evidence from trace elements and H_2O. *Earth and Planetary Science Letters*, **131**, 301–320.

MICHAEL, P. J., CHASE, R. L. & ALLAN, J. F. 1989. Petrologic and geologic variations along the Southern Explorer Ridge, Northeast Pacific Ocean. *Journal of Geophysical Research*, **94**, 13 895–13 918.

MOORE, J. G., HICKSON, C. J. & CALK, L. C. 1995. Tholeiitic-alkalic transition at subglacial volcanoes, Tuya region, British Columbia, Canada. *Journal of Geophysical Research*, **100**, 24 577–24 592.

PELTIER, W. R. 1994. Ice age paleotopography. *Science*, **265**, 195–201.

SIGVALDASON, G. E., ANNERTZ, K. & NILSSON, M. 1991. Effect of glacier loading/deloading on volcanism: postglacial volcanic production rate of the Dyngjufjöll area, central Iceland. *Bulletin of Volcanology*, **54**, 385–392.

SLATER, L., JULL, M., MCKENZIE, D. & GRONVÖLD, K. 1998. Deglaciation effects on mantle melting under Iceland: results from the northern volcanic zone. *Earth and Planetary Science Letters*, **164**, 151–164.

SOUTHER, J. G. 1977. Volcanism and tectonic environments in the Canadian Cordillera – a second look. *In*: BARAGAR, W. R. A., COLEMAN, L. C. & HALLS, J. M. (eds) *Volcanic Regimes in Canada*. Geological Association of Canada, Toronto, Special Paper, **16**, 3–24.

SOUTHER, J. G. 1990. Volcano tectonics of Canada. *In*: WOOD, C. A. & KIENLE, J. (eds) *Volcanoes of North America*. Cambridge University Press, Cambridge, 111–116.

SOUTHER, J. G., CREW, D. A. & OKULITCH, A. V. 1979. *Iskut River-Map 1418A, Sheet 104, 114* (scale 1:1 000 000). Geological Survey of Canada, Ottawa.

STUMPF, A. J., BROSTER, B. E. & LEVSON, V. M. 2000. Multiphase flow of the late Wisconsinan Cordilleran ice sheet in western Canada. *Geological Society of America, Bulletin*, **112**, 1850–1863.

WALLACE, P. & CARMICHAEL, I. S. E. 1992. Sulfur in basaltic magmas. *Geochimica et Cosmochimica Acta*, **56**, 1863–1874.

Layered, massive and thin sediments on Mars: possible Late Noachian to Late Amazonian tephra?

MARY G. CHAPMAN

US Geological Survey, 2255 N. Gemini Drive, Flagstaff, Arizona 86001, USA (e-mail: mchapman@usgs.gov)

Abstract: Data from instruments on the currently orbiting Mars Global Surveyor (MGS) suggest that as an alternative interpretation to lacustrine deposits, widespread sediments on Mars may be tephra deposits of variable age, formed in part by volcano–ice interactions. The materials are often associated with outcrops of mapped geological units that have each been previously interpreted as volcanic ash deposits with identified, but unconfirmed possible volcanic vents. Spectral investigation indicates that although some outcrops are basaltic, many show moderate to high concentrations of andesite, a composition at which large explosive eruptions may be possible. In addition, many outcrops are in areas suspected to be water/ice rich. On Earth, magma and groundwater can react to create violent explosive eruptions. Observations from MGS support a pyroclastic mechanism of deposition and show some morphologies consistent with volcano–ice interactions, including subaqueous eruptions. Perhaps MGS data are finally producing more definitive evidence of the widespread tephra that were predicted to be likely in the reduced atmospheric pressure of Mars.

Data from instruments on the currently orbiting Mars Global Surveyor (MGS) spacecraft are inspiring hypotheses about the planet's surface. In particular, the high-resolution (2–5 m per pixel) images from the narrow-angle Mars Orbiter Camera (MGS) have generated a new suggestion of possible widespread, sedimentary deposits, some as thick as 4 km, at nearly equatorial latitudes, that may be likely lacustrine deposits (Malin & Edgett 2000). Because standing bodies of water provide sites for exobiological evolution, this speculative interpretation of lacustrine origin may have a bearing on the studies of Martian terrain and therefore targeting of future missions. This paper presents evidence that may suggest an alternative origin for the deposits. Using the spatial associations of the material with geological units, suggested spectral compositions, possible palaeowater–ice locales, and geomorphologic attributes observed on MGS and Viking data, it appears that these outcrops may just as likely be tephra layers formed in part by volcano–ice interactions. Information presented in this paper is not meant to imply that lacustrine deposits do not exist on Mars; closed basins marked by channels on their peripheries are compelling evidence of their local presence. However the variably indurated layered mantles discussed in this paper can be several kilometres thick, frequently overtop the confines of basins, are in some cases relatively young, and may not have been deposited by fluvial channels.

Similar to the Martian layered materials, pyroclastic rocks can form massive deposits or bedded units that range from very thick beds (measured in metres) to thin laminations (<1 cm; Fisher & Schmincke 1984). Bed forms may be planar or show dune structures; internally the beds may be graded, cross-bedded, massive, or show alignment and orientation. Relative brightness variations may be due to grading, composition or alteration. Clast size ranges from fine-grained dust to large bombs. Deposits may be non-indurated to well-indurated depending on degree of welding and alteration.

Plotted locations of the many widespread sediment outcrops in MOC images obtained up to 31 October 2000 (Malin & Edgett 2000) show clear to close association with three mapped geological units, composing or plotting spatially close to outcrops of the (1) Medusae Fossae Formation, (2) the interior layered deposits (ILDs) of Valles Marineris, or (3) the subdued unit, a mantle on Noachian plains (Fig. 1; Scott & Tanaka 1986; Greeley & Guest 1987). (The relation between widespread outcrops and units mapped as the ILDs and subdued unit was noted by Malin & Edgett (2000).) Outlying deposits may be erosional remnants of these units.

From: SMELLIE, J. L. & CHAPMAN, M. G. (eds) 2002. *Volcano–Ice Interaction on Earth and Mars*. Geological Society, London, Special Publications, **202**, 273–293. 0305-8719/02/$15.00 © The Geological Society of London 2002.

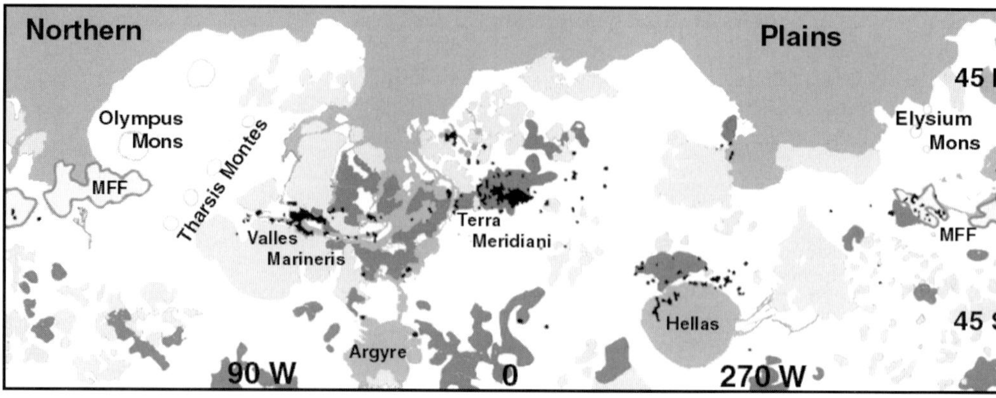

Fig. 1. Plotted locations of layered, massive and thin mesa outcrops in MOC images obtained up to 31 October 2000, shown in black (adapted from Malin & Edgett 2000); also shown are outcrops of the Medusae Fossae Formation (light with dark outline; labeled MFF), Hesperian lavas (light grey), suggested localities of water/ice (medium grey), the subdued unit (dark grey) and selected feature names.

Malin & Edgett (2000) grouped the light, intermediate, and dark-toned deposits in question into three classes of units: layered, massive, and thin (or LMT) mesa. From ancient to relatively young, the geological record of Mars is divided into three age systems: Noachian, Hesperian, and Amazonian (Condit 1978; Scott & Carr 1978) and, although Malin & Edgett noted that their grouping was not intended to imply stratigraphic continuity, their conclusion was that all of the LMT units are ancient materials (Early to Middle Noachian). The temptation to bracket these similar appearing materials into roughly one age range is not new. From careful observation of Viking data, Schultz & Lutz (1988) previously noted the similarities of these widespread materials, such as the fine-grained aspects and layering, and suggested that they may be palaeopolar deposits, roughly Middle Hesperian in age, predating the formation of the large Tharsis volcanoes, but following Lunae Planum ridged plains volcanism. The Hesperian palaeopolar origin is generally doubted because (1) the lack of tectonic features that would form due to reorientation of the spin axis (Grimm & Solomon 1986), (2) Tharsis-induced polar wandering would have to predate the Late Noachian (Tanaka 2000), and (3) the timing of polar wander would require some LMT materials (such as the Medusae Fossae Formation) to be much older than the stratigraphy indicates (Scott & Tanaka, 1982). Based on the ages of the mapped geological units and the stratigraphic relations discussed in this paper, it appears that the LMT units have a range of ages from Latest Noachian/Early Hesperian to Late Amazonian.

Possible composition: the alternative tephra hypothesis

From Viking data, many workers have noted the indurated and fine-grained nature of LMT deposits, based on outcrops that show steep layered cliffs, yardangs and preservation of faults and contacts (Ward 1979; Peterson 1981; Lucchitta 1982; Scott & Tanaka 1982; Nedell *et al.* 1987; McKay & Nedell 1988; Schultz & Lutz 1988). Logically, Malin & Edgett (2000) suggest that these attributes and the lack of boulders in talus of MOC images, the widespread occurrence of LMT deposits, and their pervasive location in impact basins constrain them to be sediments. Eolian, subaqueous or volcanic processes could generate widespread fine-grained sediments. As noted by Malin & Edgett (2000), it is unknown if eolian deposition could create regular, repeated beds of variable albedo. Compositionally, Mars tends to be mostly basaltic and the Thermal Emission Spectrometer (TES) on board MGS has not indicated the presence of carbonates or evaporites (Christensen *et al.* 1998; Bandfield *et al.* 2000); a potential problem for the lacustrine hypothesis of origin. Even 'White Rock', another LMT outcrop and a famous suspected crater-fill evaporite at 7°S, 335°, lacks spectral evidence for an aqueous origin (Ruff *et al.* 2000). However, with the exception of

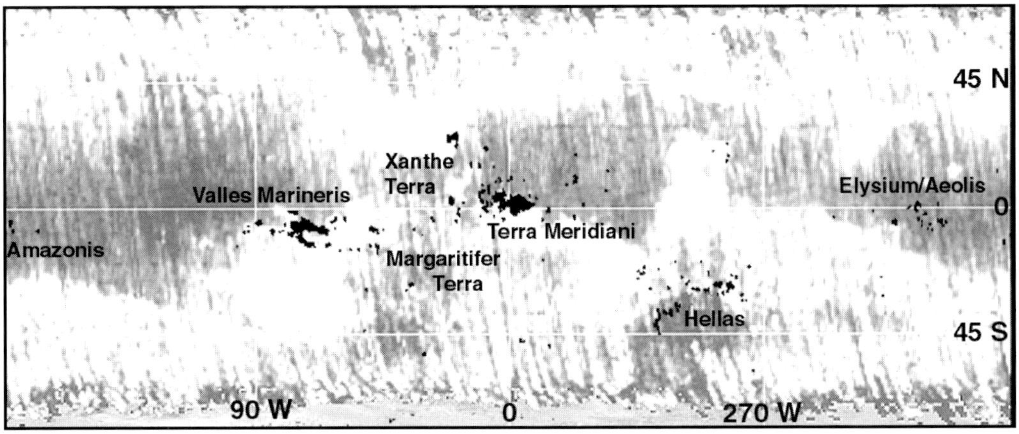

Fig. 2. Plotted locations of layered, massive and thin mesa outcrops in MOC images obtained through 31 October 2000, shown in black (adapted from Malin & Edgett 2000); shown in white are TES designated locations of moderate to high concentrations of andesite; grey indicates low to no andesite (adapted from Bandfield *et al.* 2000).

Medusae Fossae Formation outcrops, which emit basaltic spectra, many plotted LMT outcrops (Malin & Edgett 2000) appear to show weak to strong correlation with TES designated locations of moderate to high concentrations of andesite (Bandfield *et al.* 2000; Fig. 2).

On Earth, the basalt–andesite transition has been suggested to be a viscosity barrier in magma, above which large explosive eruptions are possible (Smith 1979; Francis & Wood 1982), but it is not entirely clear if TES spectral interpretations of Martian andesite are accurate (Noble & Pieters 2001). However, an andesite composition is not required to form widespread tephra deposits, especially on Mars where basalts can generate large explosive eruptions (Wilson & Head 1981, 1983). Interpretations of TES spectra have also noted a large concentration of crystalline hematite within the LMT material of north Terra Meridiani (Christensen *et al.* 1998) and other locations in Aram Chaos, Ares Valles, and Margaritifer Chaos (Christensen *et al.* 2000*a*, *b*; Noreen *et al.* 2000). In addition, local areas of concentrated crystalline hematite have been reported on or near the ILDs in Central Valles Marineris, Hebes Chasma and Eos Chasma (Noreen *et al.* 2000). Concentrated crystalline hematite is thought to be a result of hydrothermal processes, including precipitation from Fe-rich hydrothermal fluids (Christensen *et al.* 1998). Such fluids often emanate from cooling of ignimbrites and tuffs. Magmatic heat from below may also generate hydrothermal fluids (Noreen *et al.* 2000). In many areas, the LMT outcrops are noted to fill topographic lows including impact crater interiors (Malin & Edgett 2000). Ignimbrites fill topographic depressions and are known for their ability to surmount topographic barriers (i.e. impact crater rims). Voluminous terrestrial ignimbrites are largely confined to regions where the granitic crust exceeds 25 km in thickness (Coulon & Thorpe 1981). The MOLA (Mars Orbiter Laser Altimeter) Team estimated that below Valles Marineris and the subdued terrain a crust of unknown composition (but likely not silicic) ranges from 25 to 60 km thick (Zuber *et al.* 2000). The lower atmospheric pressure on Mars yields higher eruption velocities and lower final densities of decompressed gas, which would cause plinian ash columns to rise much higher, have near-vent clasts much larger, and distribute fine particles much farther than those on Earth (Wilson & Head 1983, 1994; Wilson & Heslop 1990). The expected mixture of larger clast size and widespread indurated fines is perhaps supported by MGS TES mapping data that places the LMT material in complex higher thermal-inertia areas (Mellon *et al.* 2000). Wilson & Head (1994) also suggested that ignimbrite formation should be more likely on Mars than Earth because the reduced atmospheric pressure enhances the collapse of eruption columns, but identifying large source areas for these pyroclastic flows from Viking data has been difficult. Francis & Wood (1982) made the point that identification of large ignimbrite sources from satellite data is often difficult on Earth as well due to many structures having low or negative relief, and obscure linear or arcuate fissures forming incomplete 'trapdoor' calderas, such as Pastos Grandes, Bolivia (Baker 1981).

The plotted LMT outcrops also correlate well with localities suggested to be somewhat water/ice rich, such as highland–lowland boundary areas, Hellas and Argyre Planitia, Valles Marineris, and areas along some channel paths (Fig. 1). Adding very small amounts of water (just 1 wt% H_2O) to a melt created from a parental SNC magma can create a 'sulphur-free' andesite (Minitti & Rutherford 2000), such as that suggested at the Pathfinder Site (McSween *et al.* 1999). In their paper, Minitti & Rutherford (2000) suggested that groundwater could interact with the magma on Mars before it gets to the surface by partial melting of water-bearing mantle, stoping or assimilation of hydrothermally altered crust, or direct introduction of hydrothermal fluids into a magma during caldera collapse. Besides possibly playing a role in generating andesite, water may have contributed to the formation of tephra. The Late Noachian–Early Hesperian age of the subdued material suggests that deposition of this material may have largely preceeded the accumulation of water from the mostly Late Hesperian outflow channels, but not that from Noachian valley networks. The overlapping ages of other units and outflow channel materials, and distinctive morphologies, imply that water–ice interactions may have influenced or triggered the development of chaos and the deposition of the ILDs and possibly the Medusae Fossae Formation. On Earth, rising magma may intersect surface water bodies or aquifers, resulting in violent explosive eruptions and magma fragmented into very fine tephra. These types of eruptions usually produce maars or tuff cones, involve basaltic magmas and are volumetrically insignificant (Francis & Wood 1982). Although, low volume is usually the case, some widespread high volume terrestrial ignmibrites exist, such as the Icelandic basaltic Ytri flow (Van Bemmelen & Rutten 1955). Eruption involving magma–ice interactions may have enlarged chasmata, formed areas of chaos, and melted ground-ice forming outflow channels that cut the subdued material (Chapman & Tanaka 2000, 2001*b*). Volcanic interaction with ice in Valles Marineris may have taken the form of subaqueous volcanoes (Chapman & Tanaka 2001*a*). Finally, fire fountaining (Keszthelyi *et al.* 2000) or basaltic ignimbrites may be responsible for the Medusae Fossae Formation.

Geological association

The Medusae Fossae Formation, the ILDs and the subdued unit were all previously acknowledged (from Viking data) to be fine-grained materials that range in age from Late Amazonian to Early Hesperian (perhaps Late Noachian). Each of the units has at one time or another been interpreted as some type of volcanic ash deposit. In addition, possible volcanic vents have been identified which may have sourced each deposit.

Medusae Fossae Formation

Medusae Fossae Formation forms 0.5 to about 2 km-thick piles of material that straddle the highland/lowland boundary in Amazonis and Elysium Planitia for a distance of about 5500 km (Tanaka 2000). The upper layers of the material are friable, lower layers are indurated to the degree that they form cliffs and yardangs. The formation has been stratigraphically dated as Late Hesperian to Early Amazonian in age (Scott & Tanaka 1986; Greeley & Guest 1987). Proposed origins for the formation include ignimbrites and pyroclastic rocks (Malin 1979; Ward 1979; Scott & Tanaka 1982), palaeopolar deposits (Schultz & Lutz 1988), carbonate platforms (Parker 1991) and rafted pumice deposits along an ancient shoreline (Mouginis-Mark 1993). Eolian deposits (Scott & Tanaka 1986; Greeley & Guest 1987), volcanic dust (Tanaka 2000) or other friable material (Malin 1979; Ward 1979) have also been suggested.

Using data from the MOLA on the MGS spacecraft, Sakimoto *et al.* (1999) investigated the proposed origins of the Medusae Fossae Formation. The topography, roughness, layering and slope properties suggest that the formation is a layered mantle deposit hundreds of metres thick with substantial internal relief. Their analysis is consistent with layers locally draped over the underlying terrain, a conclusion that does not support either lacustrine or carbonate deposition. As discussed, palaeopolar deposits are unlikely. Friable materials would require induration by some means that did not affect the upper layers. Alternatively, unwelded zones are common in ignimbrites.

Although Scott & Tanaka (1982) noted that ignimbrites from local fissures can bury their sources, they identified surface depressions on the formation that may indicate source vents. Tanaka (2000) suggested that the nearby, large volcanoes of Elysium and Olympus Mons and those of the Tharsis Rise could have emanated putative volcanic ash of the Medusae Fossae Formation. Finally, slightly welded or altered ash related to fire fountains from voluminous basaltic Cerberus lava flows, south of Elysium Mons, has been suggested as the type and source of the formation (Keszthelyi *et al.* 2000). Basaltic eruptions are often initiated by volatile-charged fire fountaining. Keszthelyi *et al.* (2000) did not discuss water–ice interaction and it is not clear how groundwater or ice may have played a role in deposition of the basaltic formation without inducing large maar type vents.

Interior layered deposits

LMT outcrops compose the interior layered deposits (ILDs) that form mounds and mesas within Valles Marineris chasmata (Fig. 3). The troughs or chasmata that compose the Valles Marineris range from 50 to 600 km wide and are interconnected, except for the entirely enclosed Hebes Chasma. The chasmata have been interpreted as grabens and/or collapse structures (Lucchitta *et al.* 1992). Many processes to form the chasmata have been suggested, including dynamic upwarping and rising magma caused by a local mantle plume (Hartmann 1973; Carr 1974; Wise *et al.* 1979; Chapman & Tanaka 2001*b*). The interior layered deposits have been mapped as Late Hesperian to Early Amazonian in age (Scott & Tanaka 1986; Greeley & Guest 1987; Witbeck *et al.* 1991). This is a stratigraphic age relative to the possible Early Hesperian opening of the chasmata (Lucchitta *et al.* 1992).

Mapping indicates that the ILDs formed at different times (Lucchitta 1990) and the deposits have distinct morphologies, suggesting different depositional histories (Komatsu *et al.* 1993). Origin of the interior deposits as eroded wall rock is not considered a viable hypothesis by most investigators because of the vastly different outcrop patterns of wall rock and interior deposits (Peterson 1981; Nedell *et al.* 1987; Lucchitta 1990; Komatsu *et al.* 1993). Although some young eolian deposits are present, great uniformity in thickness and sequence of deposits across several troughs is required if the deposits were largely eolian, because they would have to reflect global events (Komatsu *et al.* 1993) and this uniformity is not observed. The ILDs were suggested to be lacustrine on the basis of their apparent horizontal continuity, similarity in connected troughs, and local fine layering (McCauley 1978; Carr 1981; Peterson 1981; Lucchitta 1982; Nedell *et al.* 1987; McKay & Nedell 1988; Malin & Edgett 2000; Weitz & Parker 2000). The developing chasmata may have become partly filled with water, freed from a confined aquifer (Carr 1979) or accumulated via runoff from early, pre-outflow channel flow from Hydraotes Chaos to the north (Ori & Mosangini 1998) or overflow of Argyre Planitia to the south (Parker *et al.* 2000). Several chasmata lead to Late Hesperian–Early Amazonian outflow channels, which debouch into the northern plains, supporting the notion that the chasmata held water/ice at that time. The chasmata occur at the source of the catastrophic channels, on a topographic high, with no inlet channels. The lacustrine hypothesis is deficient in that a lake in the connected chasmata cannot explain the moats that separate the deposits from the trough walls, nor the angled beds observed on some ILD mounds, nor the lack of inlet channels that might have supplied the massive amount of sediments to confined chasmata (Lucchitta *et al.* 1992; Chapman & Tanaka 2001*a*). A volcanic origin for the ILDs is supported by the volcano-tectonic setting, layer diversity, low albedo and high competence of some layers, tuff-like weathering, location of dark materials (basaltic ash?), and

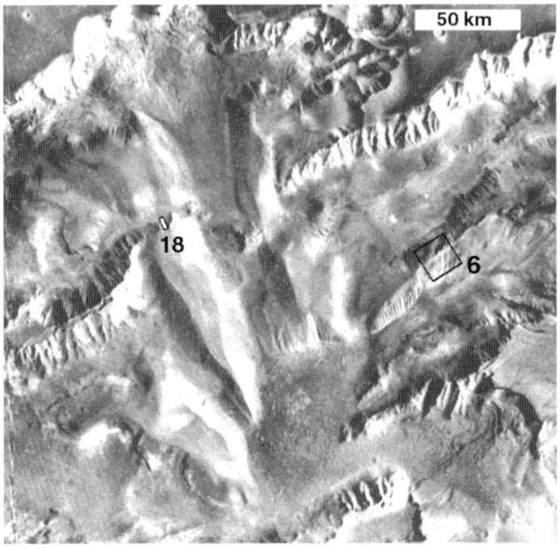

Fig. 3. Viking Orbiter image 297B42 (256 m per pixel) showing mounds and mesas in Ophir and central Candor Chasmata; boxes denote figure locations.

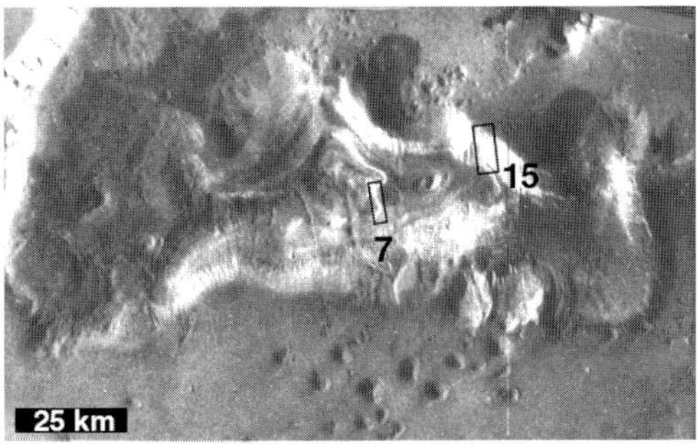

Fig. 4. Part of Viking Orbiter 897A40 (159 m/pixel) showing largest Gangis Chasma mound; boxes denote figure locations.

diversity between adjacent mounds (Peterson 1981; Lucchitta 1990; Witbeck *et al.* 1991; Weitz 1999).

More recently, workers have suggested that the ILDs are locally subaerial ash (Chapman & Tanaka 2001*a*), but mostly hyalotuffs produced by magma–water interaction and related to sub-ice volcanism based on erosional resemblance to Icelandic tuyas (table mountains) like Sellandafjall, topography and morphology (Fig. 4; Croft 1990; Lucchitta *et al.* 1994; Chapman & Tanaka 2001*a*). Tuya and ridge components are formed in three sequential stages: (1) pillow volcano, (2) tuff cone, and (3) hyaloclastite delta/subaerial cap lava; hyaloclastic ridges are linear forms that may or may not have a cap of subaerial lava (Fig. 5). Within their meltwater lakes, basaltic tuyas rapidly alter to palagonite. Similar to tuyas, the ILDs appear to have palagonitic compositions (Murchie *et al.* 2000) and display mound and ridge forms. Viking images show many mesas with caprock (Fig. 6), as do MOC images (Fig. 7). At least one of these distinctive mesas is associated with possible lava flows. For example, MOC image M3-00945, of ILDs in Melas Chasma, shows possible flows that are tentatively identified as lavas, which are traceable upslope to a rimmed pit containing a central flat-topped mesa (Fig. 8; Chapman & Tanaka 2001*a*). Coincidentally, the most favoured explanation for the outflow channels is catastrophic floods (Baker *et al.* 1992), which, on Earth, are formed only by dam bursts of surface water, some of which are caused by subglacial volcanic eruptions (Baker & Milton 1974). Tuyas occur at the source of catastrophic flood channels.

Although subaqueous volcanoes may bury their source, some possible vents for the ILDs have been identified in Viking and MOC images. These forms have been suggested to resemble regular feeder dykes, calderas, vents atop a tuya, fissures and volcanic pit chains (Lucchitta 1990; Lucchitta *et al.* 1994; Chapman & Tanaka 2001*a*). Within the ILDs of Melas Chasma, there is a partly buried, north-trending, large (55 km-long) open-ended trough (Fig. 9a). The trough has layers that are upturned on its sides, bent into a curve at the north nose, and stand

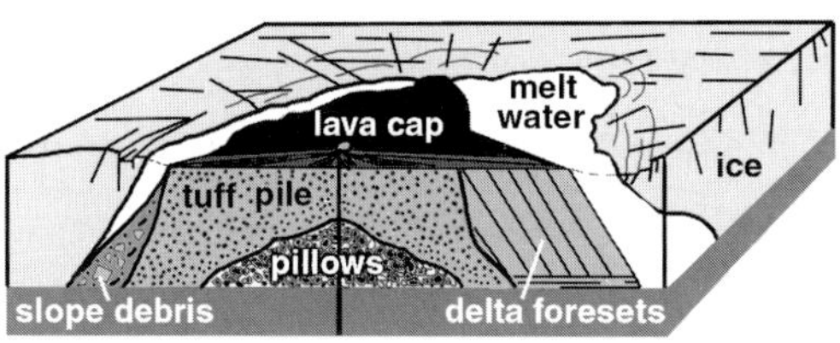

Fig. 5. Schematic view of a terrestrial tuya.

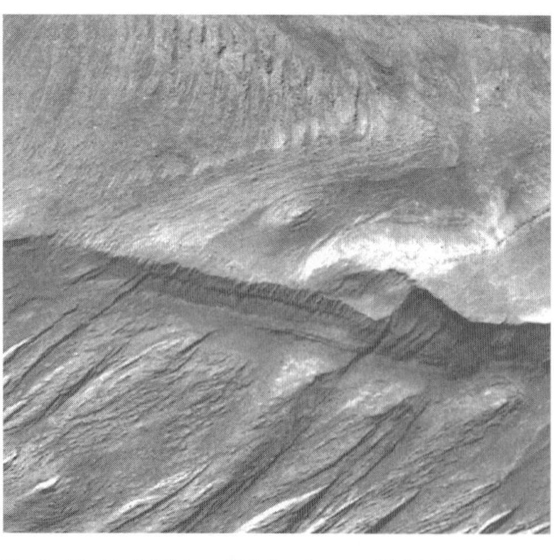

Fig 6. Part of Viking Orbiter image 815A58 (24 m per pixel) showing contacts between 'massive' material with grooves and ridges oriented up and down slope (flutes) and overlying layered resistant caprock; location shown in Figure 3.

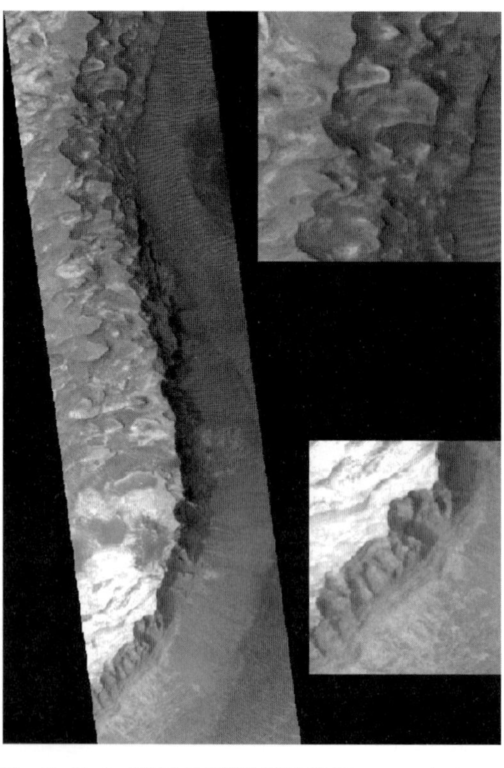

Fig. 7. Part of MOC M0804332 (2.86 m per pixel, 1.46 km wide) showing layered resistant caprock on Gangis Chasma mound; insets are enlarged; location shown in Figure 4.

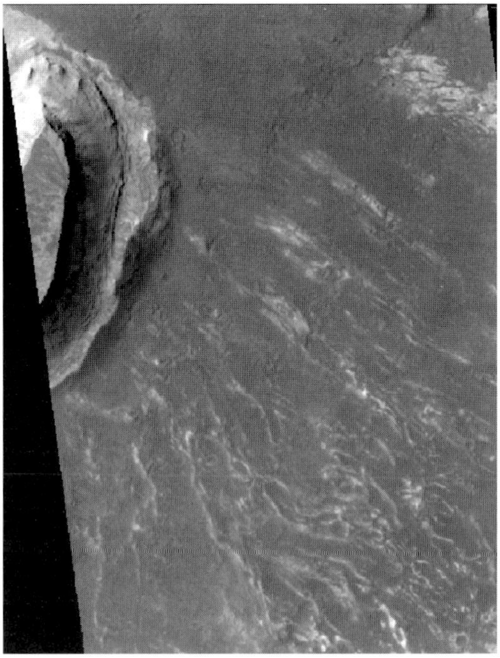

Fig. 8. Part of mound in Melas Chasma (at about lat. 13.2°S., long. 70.97°) on MOC M0300945; (5.66 m per pixel; 2.9 km image width) showing possible flows that are tentatively identified as lavas, which are traceable upslope to a rimmed pit with a central flat topped mesa (tuya?).

vertically in its interior. Although the structure could be an anticline, no structural disruptions of floor or wall rocks are observed along the trend of the trough. Therefore, it is possible that the feature is a partly buried, linear volcanic cone, as it resembles terrestrial open-ended tephra cones (Fig. 9b, c). The trough beds that are linear along its length and semi-circular at its end are similar in position to beds of some linear terrestrial tephra cones (Fig. 9c).

Subdued plains

In local areas of the highlands, the subdued material has an intermediate brightness and a friable surface appearance, buries large craters and was mapped as a Late Noachian-age unit (Scott & Tanaka 1986; Greeley & Guest 1987). The unit has a superimposed crater density similar to that of Early Hesperian ridged plains, which is depicted as possibly a younger unit and mapped as having a gradational or uncertain contact with the subdued cratered unit in south Lunae Planum (Scott & Tanaka 1986; Witbeck *et al.* 1991). Crater counts show the unit's surface to be comparable in age to the Early Hesperian ridged plains of Hesperia, Syrtis Major, Lunae Planum and a region south of Hellas (Schultz & Lutz 1988). Sand-sized material may cover the unit based on fine-component thermal inertia data (221–423 $J\,m^{-2}\,s^{-0.5}K^{-1}$) acquired by Viking (Christensen 1986; Edgett & Parker 1997; Christensen *et al.* 2000*b*), indicating an average surface particle size equivalent to coarse sand. The unit is layered (Christensen *et al.* 2000*b*) and contains the initial hematite area spotted by the TES instrument (Christensen *et al.* 1998), which is associated with the areas of fine ridges (Edgett & Malin 2000). The layers range from high to intermediate in relative brightness. Lower layers appear to be indurated forming cliffs; the uppermost hematite-rich material is friable.

Many workers have suggested that the subdued unit is some type of friable material, perhaps eroded by wind (Presley & Arvidson 1988; Schultz & Lutz 1988; Edgett & Parker 1997; Chapman 1999; Christensen *et al.* 2000*b*; Tanaka 2000). The subdued cratered unit has been interpreted to be a mantle of thin, interbedded lava flows and eolian material (Scott & Tanaka 1986; Greeley & Guest 1987); however flow lobes are rare. Schultz & Lutz (1988) suggested the material might be palaeopolar deposits. As discussed above, this origin is doubtful. Pyroclastic airfall (Moore 1990) and volcanic dust (Tanaka 2000) have also been suggested. The subdued material in Terra Meridiani was also interpreted to be water-laid sediments based on (1) the relatively low and flat topography of the area, (2) possible evaporite deposits noted by Lee (1993), but not confirmed by TES (see above) within large, old craters, (3) dunes and Viking rock abundance and thermal inertia that indicate sand-sized material, (4) few valley networks, (5) smooth surface of the subdued crater unit, and (6) lack of mantling relationships that would support volcanic or eolian airfall (Edgett & Parker 1997). Based on new MGS data, such as compositions indicated by TES, and MOC images showing subdued unit layers with strikingly different relative brightness and lithification states, and local regularly-spaced mounds that resemble fumorolic mounds (also observable on MOC images of the Medusae Fossae Formation), the material in Terra Meridiani has also been interpreted as likely layered ignimbrites (Chapman 1999; Chapman & Tanaka 2000, 2001*b*). Ignimbrites (1) follow pre-existing topography and flow into low areas (Fisher & Schminke 1984), (2) commonly contain sand-sized and larger grain sizes (Murai 1961; Fisher

& Schminke 1984) and (3) lack valley networks. Furthermore, contrary to Viking observations, MOC images show that the unit is not smooth, but contains numerous fine scale ridges (Edgett & Malin 2000). Mantling is an undiagnostic criterion as airfall, ash flow and water-laid sediments all form overburden materials that mantle/cover/subdue underlying units. Finally, as previously noted, differences in composition, grain size and alteration can form variable albedoes in terrestrial ash deposits and ignimbrites may be welded or unwelded.

On Earth, some of the largest-volume ignimbrites are found in continental rift zones (Francis & Wood 1982). The largest area of the subdued material is found in Xanthe, Margaritifer and Meridiani Terrae, associated with and downslope of Valles Marineris, a huge trough/rift system on Mars. Most of this subdued material may be related to initial fissuring of the Valles Marineris rifts (Chapman & Tanaka 2000). No obvious vents are located in Terra Meridiani, but the ability of volcanic ash to bury its source has been mentioned. A large tract of the subdued unit in Xanthe and Margaritifer Terrae, west and adjacent to Terra Meridiani, surrounds and is contained within areas of chaos (Edgett & Malin 2000). Chaos consists of jumbled and subsided blocks of material within subcircular and amorphous depressions. Chaos may be explained by collapse of overlying rocks due to the melting of ground ice and expulsion of ground water and rock debris (Sharp 1973; Komatsu *et al.* 2000). Magmatic intrusion could have provided the energy for the melting (Masursky *et al.* 1977; Max & Clifford 2000). Eruption and drainage of magma (Sharp 1973) or subsurface CO_2 ice (Hoffman 2000) may have contributed to chaos subsidence.

Edgett & Malin (2000) suggested that the roughly circular outlines of some chaos indicate that they are all impact craters filled with likely lacustrine sediments. However resurgent calderas also have roughly circular outlines and can erupt

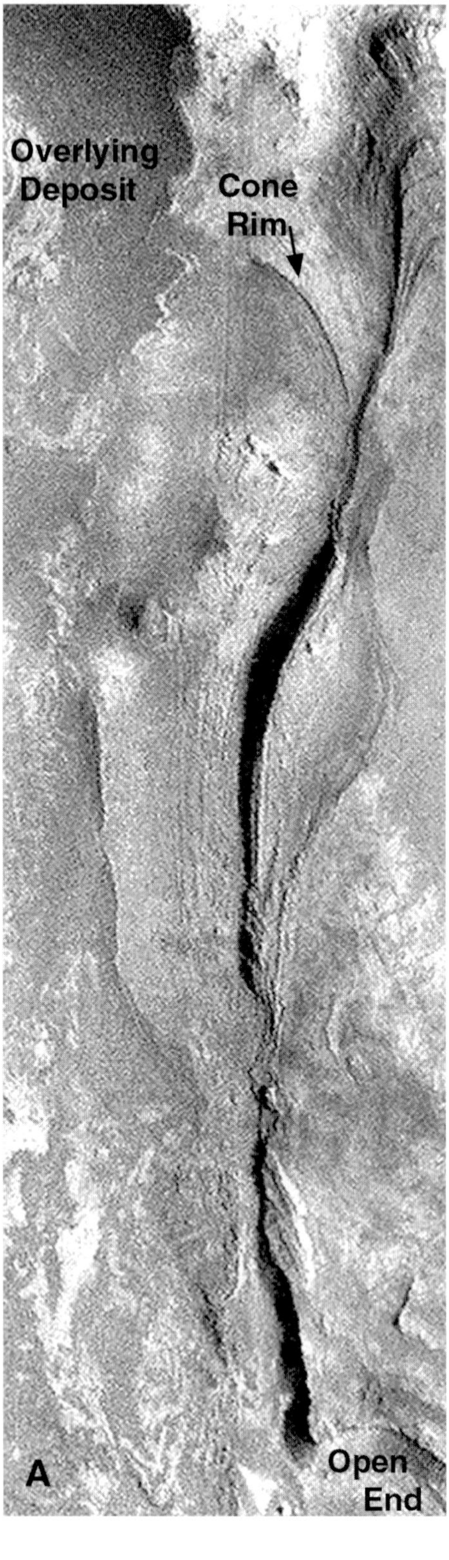

Fig. 9. (**a**) Part of Viking Orbiter image 915A22 (40 m per pixel; centered at lat. 11°S., long. 73.7) showing possible volcanic vent in Melas mound; note 55 km-long trough, open-ended to south, overlain by layered beds. (**b**) Linear pyroclastic cone (1.6 km long and 165 m high) in NW part of San Francisco Volcanic Field near Flagstaff, Arizona (Photo by J. F. McCauley, US Geological Survey). (**c**) The Sproul, an elliptical tephra cone (1 km long) with an open-ended, breached rim located in west part of San Francisco Volcanic Field near Flagstaff, Arizona (Photo by E. W. Wolfe, US Geological Survey); note semi-circularity of beds at end of cone and their linearity at the sides of the cone.

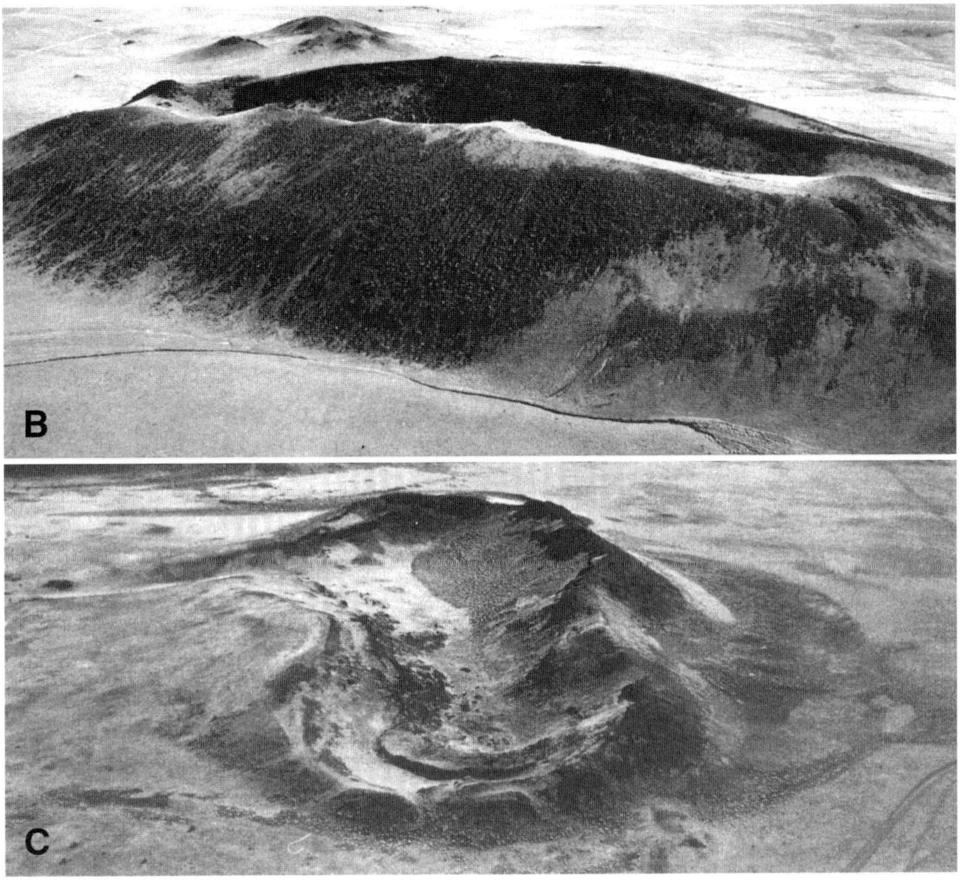

Fig. 9. (*continued*)

repetitively. Perhaps it is time to re-examine the origins of chaotic terrains with MGS data. Aram Chaos is the most circular chaotic area; other chaotic areas are amorphous or much less circular in outline. We knew from Viking data that Aram Chaos was the source of a small outflow channel and, unlike most impact craters and possibly due to burial, the rim is obscured and no ejecta deposits are observed. Within Aram Chaos, TES has detected concentrations of crystalline hematite (Christensen *et al.* 2000*b*; Noreen *et al.* 2000) and MOC images show possible small volcanic edifices (Lanz & Jaumann 2001). MOLA-derived topography of Aram Chaos, indicates that like the Valles Caldera, Aram Chaos is on a regional topographic high and contains interior mounds positioned asymmetric to its centre (Fig. 10). Unlike the Valles Caldera, Aram Chaos is very circular in outline and does not have precipitation-induced runoff channels eroded into its flank. The circularity of Aram may suggest an impact origin, but does not preclude later volcanic eruptions that may have utilized fractures formed by a previous impact. Late-stage volcanism may have generated tephra deposits that buried the impact ejecta and rim. Volcanic or subvolcanic activity may have melted ground ice and generated fluids that eroded the small outflow channel into the east rim of Aram (Lanz & Jaumann 2001).

It is possible that early eruptions from chasmata, and buried sites to the east, may have formed the subdued material in Xanthe and Margaritifer Terrae. The lower Martian gravity and atmospheric pressures and the likelihood of finer grained material suggest highly fluidized, more mobile and more dispersed ash flows (Wilson & Head 1983, 1994). Therefore, ash flows may have had the ability to move downslope a great distance perhaps travelling to Terra Meridiani. Late-stage eruption may have formed amorphously-shaped chaos and utilized buried

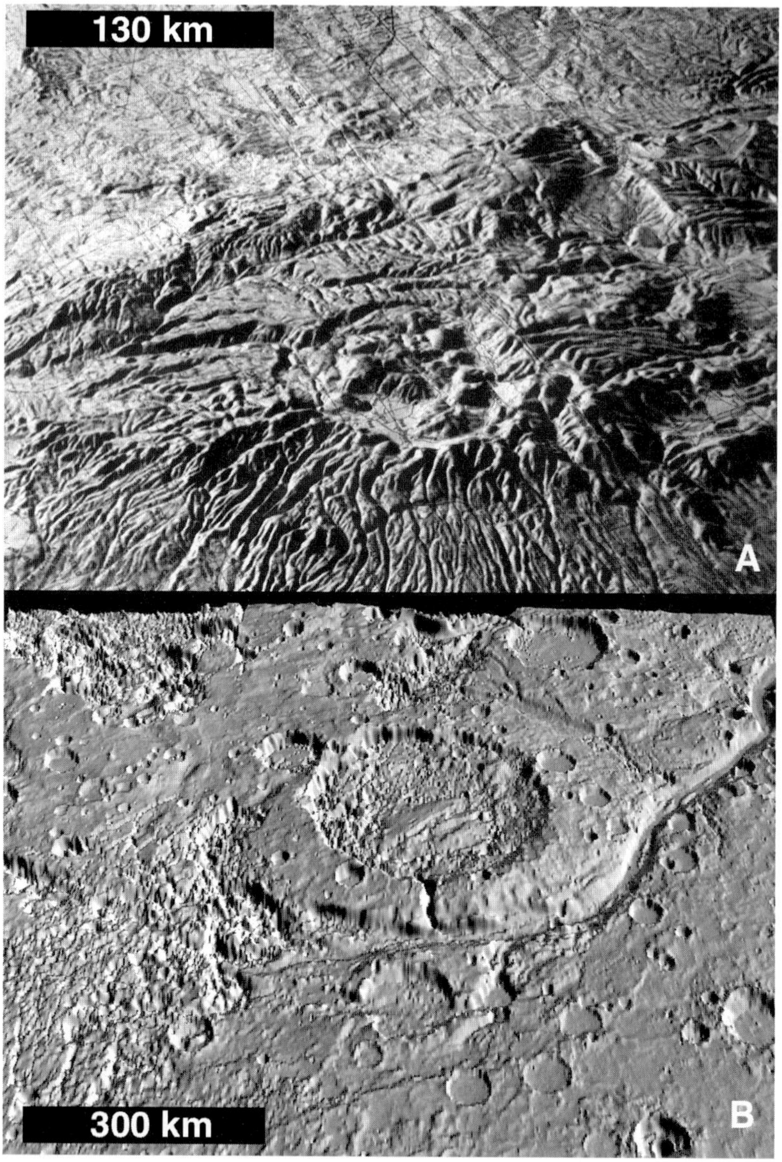

Fig. 10. Oblique views of (**a**) a relief map of the Jemez Mountains, New Mexico, showing the resurgent Valles Caldera (illumination from upper left; R. Bailey, US Geological Survey), and (**b**) Aram Chaos (illumination from right; courtesy of the MOLA Team); north is toward the right.

impact structures like Aram Chaos. Eruptions of fluid, gas (etc.) from beneath chaotic areas, cut previously deposited subdued materials in Xanthe and Margaritifer Terrae, may have emitted more ash, and melted ice to form floods that carved the associated outflow channels. Although other scenarios for the formation of chaos are possible, an eruptive origin for the chaotic terrains is consistent with nested collapse pits having tentatively identified maar rims (MOC AB108805) on the NE wall of Hydraotes Chaos (Chapman & Tanaka 2001*b*). Alternatively, some of the LMT deposits in the adjoining terrae were perhaps derived from eruptions associated with flood volcanism to the north. There is emerging evidence that a significant Early Hesperian volcanic deposit exists in the northern lowlands of Mars, which could have added a significant contribution to volcanic tephra in this time period (Head *et al.* 2001).

Stratigraphic age

The LMT material was suggested to represent ancient units based on MOC images that appear to show ILDs within and under walls of Valles Marineris and the subdued unit within large craters, and within circular depressions, such as Aram Chaos (Malin & Edgett 2000). If one draws a cross section between central Candor Chasma and a point just east of Juventae, it can be demonstrated that the ILDs and subdued material are (1) not particularly 'ancient' and (2) could not have formed at the same time (Fig. 11). On the plateau, subdued material and Hesperian ridged plains have a gradational contact and both blanket Noachian cratered material, i.e. both units appear to occupy the same stratigraphic horizon (Scott & Tanaka 1986; Greeley & Guest 1987; Witbeck *et al.* 1991). These units and the underlying Noachian rocks are cut by chasmata and Juventae's upper Hesperian outflow channel, Maja Valles (Scott & Tanaka 1986; Witbeck *et al.* 1991). In contrast, the ILDs overlie Valles Marineris wall rock at the breach between Candor and Ophir chasmata (Fig. 3; Lucchitta 1999) and upper Hesperian chaotic material on the floor of Juventae chasma (Fig. 12; Komatsu *et al.* 1993). It is clear from these stratigraphic relations that the subdued unit likely formed some time in the Early Hesperian, whereas the ILDs formed later during Late Hesperian to Early Amazonian time. On the opposite side of the planet, the Medusae Fossae Formation is superimposed on many highland and lowland units and has been stratigraphically dated as Late Hesperian to Early Amazonian in age (Scott & Tanaka 1986; Greeley & Guest 1987). If this formation is related to the Cerberus flood lavas (Keszthelyi *et al.* 2000), it may be as young as Late Amazonian in age.

Outcrop observations

The 'layered' units of the LMTs are high- to intermediate in relative brightness, thin (<200 m)

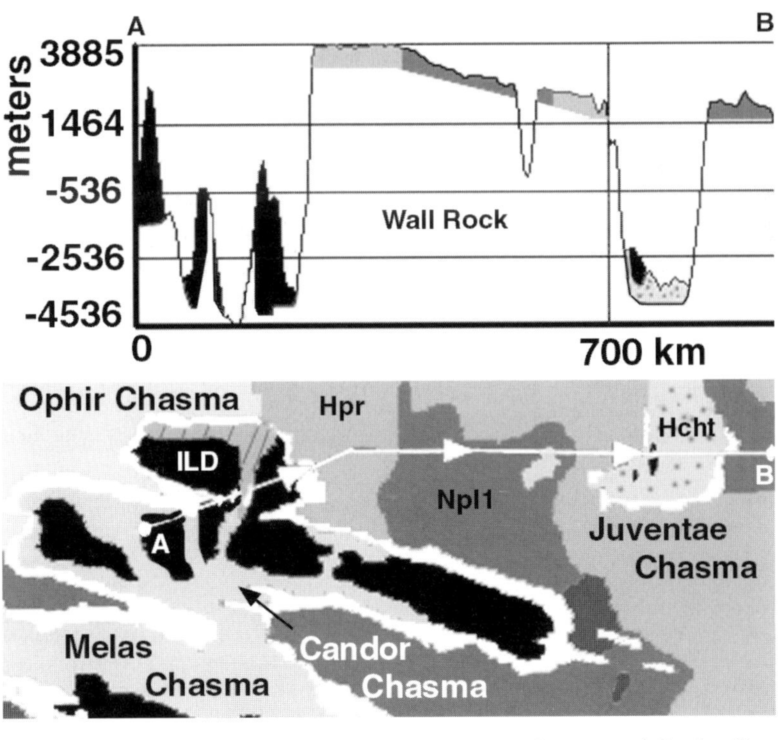

Fig. 11. Schematic geological cross section along MOLA profile A–B; from central Candor Chasma, to a point just east of Juventae Chasma (geological contacts taken from Scott & Tanaka 1986); interior layered deposits (ILD in black), Lower Hesperian intercrater plains (Hpr, light gray), subdued unit (Npl1, dark gray), Upper Hesperian chaotic material (Hcht, speckled).

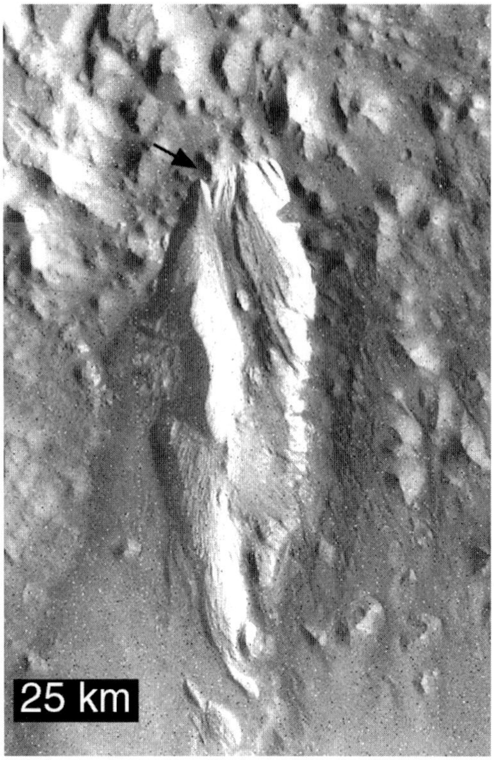

Fig. 12. Part of Viking Orbiter image 906A06 (72.5 m per pixel; centered at lat. 3.39°S., long. 61.54°) showing largest Juventae Chasma mound; mound eroded into yardangs that fork around a pre-existing smaller mound of chaotic material (arrow).

to thick (<2000 m), contain hundreds of sub units (beds); all LMTs were deemed ancient on the partial basis of this unit seemingly cropping out from the walls of Valles Marineris (Malin & Edgett 2000). However, these outcrops were previously noted on Viking images to occur only in flatter wall gullies; thus they were interpreted as fill, pasted on or abutting against wall rock (Lucchitta 1999). A detailed geological map of AB106306 (Fig. 13) does not support the interpretation that the ILDs are ancient. The map shows the deposit (1) to occur only on one side of the wall rock spur, which could be interpreted as younger deposits onlapping the walls and (2) some layers that dip as steeply as the talus slope indicating that they were deposited on top of pre-existing (older) talus. However, this map also shows an ILD unit with horizontal beds overlying the steeply bedded unit, possibly indicating a ramp (Fig. 14). Ramping of horizontal beds over steeply dipping beds may indicate a nearby up-slope source. If the material were volcanic, in addition to being erupted from the chasma floors, local eruption may have come from dykes within the wall rock. Sloping lava flows from wall dykes could support the later horizontal ramps. This relationship could explain the Malin & Edgett (2000) observation that in places LMTs are cropping out from the wall rock.

'Massive' units are light- to intermediate-toned, hundreds of metres to a few kilometres thick, not layered or poorly bedded, and have a distinct surface morphology of ridges and furrows oriented up- and down-slope (Malin & Edgett 2000). Although these authors noted that the unit is poorly bedded, their interpretation that the massive beds are everywhere likely lacustrine, shows a bias toward horizontal bedding. In fact, Viking and MOC images show the massive unit to have local layers that dip very steeply on ILDs mesas and mounds in the Valles Marineris chasmata (Chapman & Tanaka 2001*a*). Examination of the oriented ridges on MOC images shows that they are eroded in aligned wedges that point down or up slope, an indication that even crudely bedded materials are steeply dipping (Fig. 15). The evidence of steeply dipping beds on mesa flanks is indicative of deposition in an original mound form, such as a volcano. The massive units show a gradational or abrupt transition with often underlying, layered units (Lucchitta 1999; Malin & Edgett 2000) and, in Valles Marineris, commonly show an abrupt transition to resistant, overlying layered caprock units (Fig. 6; Lucchitta *et al.* 1992, Lucchitta 1999; Chapman & Tanaka 2001). The local caprock form resembles sub-ice volcanic tuyas; high relative brightness may be due to andesitic compositions and/or clay alteration (Chapman & Tanaka 2001*a*).

Also, in Valles Marineris, the layered and massive beds on the flanks of some ILDs in Melas and Gangis Chasmata locally exhibit abrupt mass termination of material lobes along a linear trend (Fig. 16a, b). This termination suggests terrace erosion or damming/constraint against some type of material that has been subsequently removed. This type of constrained morphology occurs on the flanks of terrestrial silicic tuyas, as these lavas are not hot enough to immediately melt the ice (Fig. 17; Tuffen *et al.* 2001). An identical situation might be possible on Mars if the ILDs showing this morphology were somewhat silicic, perhaps like the suggested andesitic compositions (Bandfield *et al.* 2000) and/or if the constraining material were a type of ice with a higher melting point, possibly a clathrate.

Malin & Edgett (2000) suggest that 'thin' mesa units unconformably overlie layered and

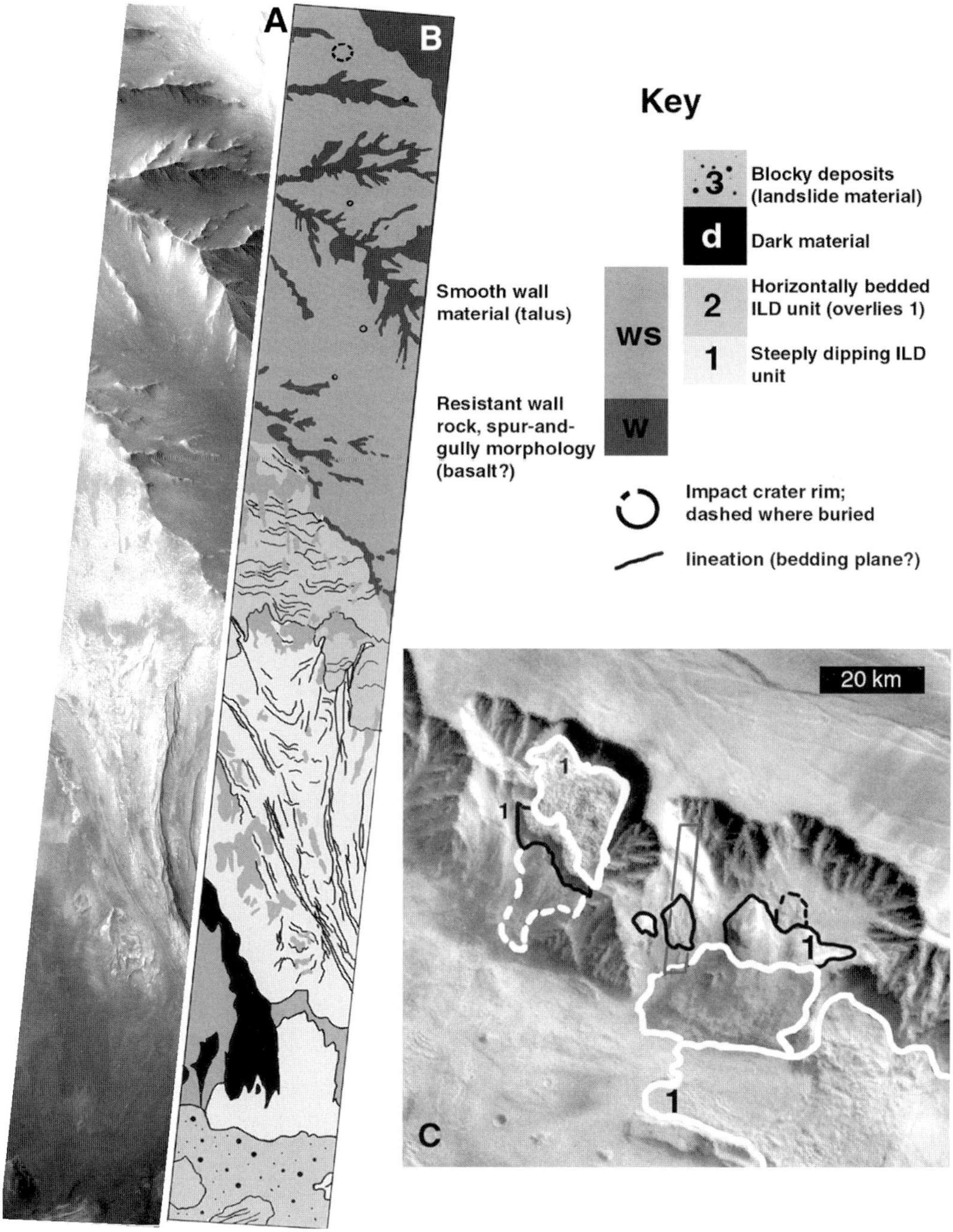

Fig. 13. (**a**) MOC AB106306 (4.52 m per pixel, 4.62 km wide) showing relatively bright 'layered' deposit on wall of NW Coprates Chasma in Valles Marineris; (**b**) geological map; (**c**) Viking context image showing MOC footprint; black contacts mark interior layered deposit outcrops on wall; white contacts mark landslide deposits ('1' denotes contacts taken from Lucchitta *et al.* 1994).

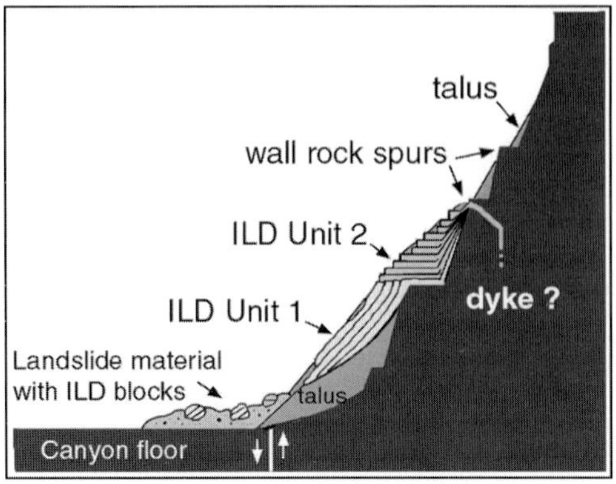

Fig. 14. Schematic cross-section and interpretation of geological units and relations shown in Figure 13b.

massive units and consist of mostly dark-toned, thin, mesa-forming or resistant caprock-forming materials with smooth, pitted, or intensely ridged and grooved surfaces. This classification groups several materials into one unit, i.e. pitted dark surfaces may be lava flows and not related to the ridged and grooved surfaces, and ridged and grooved surfaces might not form caprock. For example, by careful observation Malin & Edgett (2000) correctly noted that the ridged and grooved textures on the unit are not modern eolian dunes. This is demonstrated in MOC M1003164, where the ridges and grooves can be seen brittlely fracturing along the margin of the unit (Fig. 18). However this material is not a mesa- or caprock-forming unit as it is not resistant, like the layered caprock of the ILDs

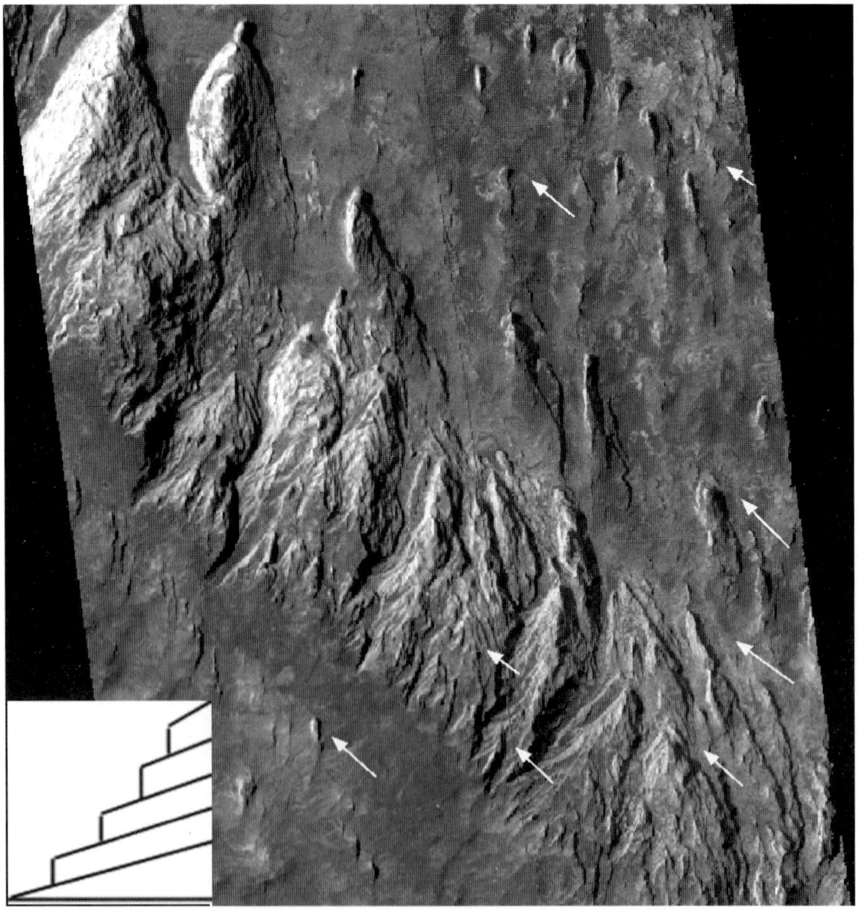

Fig. 15. Mosaic of MOC images 304405 and 401737 (about 5.7 m per pixel, combined 5.1 km width); arrows point to alignment of wedge points on flutes (grooves and ridges oriented up and downs slope); inset shows layers stacked in like manner to fluted material; location shown in Figure 4.

(a) (b)

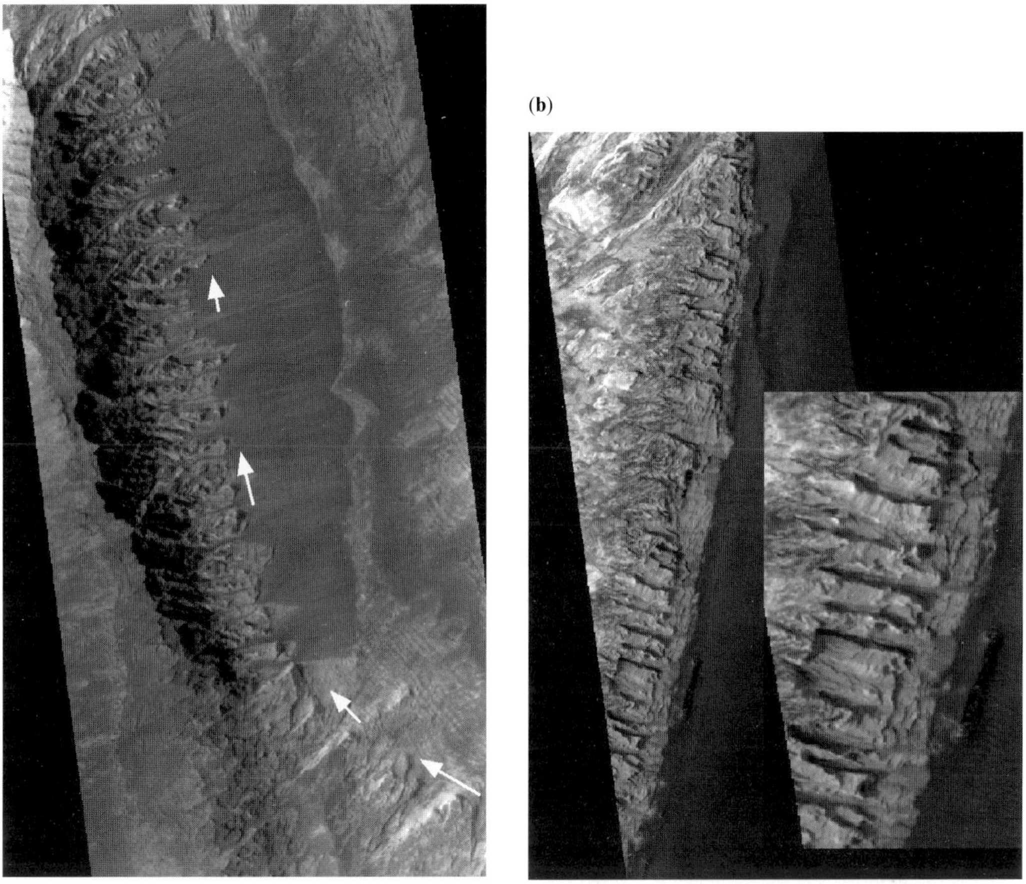

Fig. 16. Flank of two Valles Marineris mounds showing lobes that terminate along a lineation (trend indicated by white arrows) suggesting terrace erosion or confinement against subsequently removed material. (**a**) MOC M0804981 (5.67 m per pixel, 2.9 km wide) in Melas Chasma. (**b**) MOC M0705587 (4.28 m per pixel, 2.87 km wide) in Gangis Chasma.

mesas (Figs 6 & 7). More likely it is an indurated mantle on top of a caprock surface. Malin & Edgett (2000) interpret the ridges and grooves as an erosional expression, perhaps caused by exhumation of palaeodunes. However, the ridges and grooved texture may be relatively young and not exhumed. MOC image M0806284 (Fig. 19) shows dark material forming rimmed cones and dunes apparently associated with a dyke-like structure cutting the ILDs of western Candor Chasma, indicating young volcanism (B. K. Lucchitta, pers. comm., 2001). Pyroclastic ash from these types of vents may have formed much of the wind eroded dark mantles on mesas and large dunes on the Chasmata floors (Lucchitta 2001). The composition of the dark material is mafic (Geissler *et al.* 1990; Murchie *et al.* 2000) and the low relief of the cone rims suggest that they are maars or tuff cones composed of fine ash. Phreatomagmatic eruptions may have lower plume heights due to cooling of the magma, and thus ash erupted in the chasmata may have been confined to the canyons (Weitz 1999). The ash falling on caprock could form a mantle susceptible to eolian modification; induration might be due to freezing of volatiles or clay alteration. Ash falling on steep slopes may have been incorporated within massive slope materials or eroded to form dark floor deposits. This dark material on the chasmata floors is locally suggested to contain concentrations of crystalline hematite (Noreen *et al.* 2000).

Finally, layered and massive deposits within impact crater Gale were previously mapped from

Fig. 17. Icelandic rhyolitic tuya Kirkjufell showing ice constrained lava flow lobes on flank.

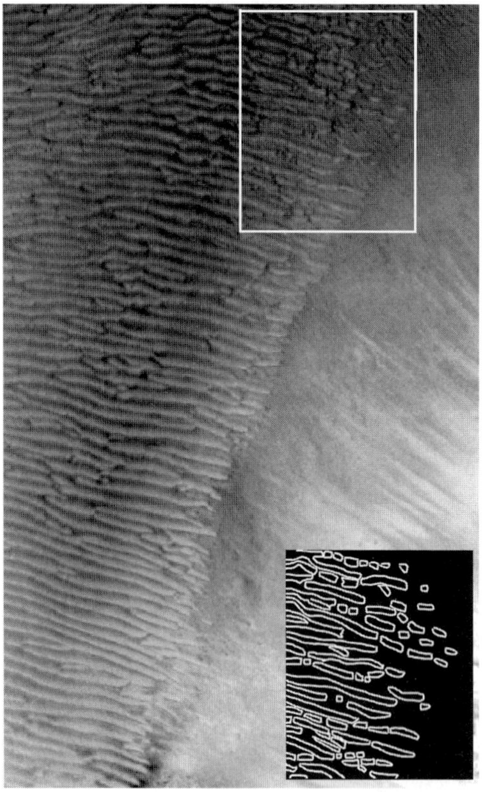

Fig. 18. Part of MOC M1003164 (2.86 m per pixel, 2.92 km wide) showing brittle fracturing of dunes (illustrated in sketch map of white-outlined area) on edge of deposit that mantles Candor mesa between Ophir and Candor Chasmata; dark dunes appear lighter due to image processing; location in Figure 3.

Viking data to be Amazonian in age and part of the middle member of the Medusae Fossae Formation (Scott & Chapman 1995). MOLA profiles over the Gale deposit, not the crater's central peak, show that the elevation of the material exceeds that of the rim of the impact basin (Fig. 20), an impossible situation for lacustrine rocks.

Discussion

Rather than ancient lacustrine deposits, the LMT sediments may be widespread ash deposits of variable age. Perhaps, MGS data are finally producing more definitive evidence of the pyroclastic flows that were predicted to be likely on Mars (Wilson & Head 1983). However, much more analysis of the data is needed, particularly of the subdued unit. Some of the possible pyroclastic materials seem to be related in space and time with water- or ice-rich localities and these fluids may have interacted. Volcano–ice interaction areas have as much or more exobiologic importance as do lacustrine materials. If the sediments are pyroclastic deposits, as suggested in this paper, they may have formed during the following generalized history:

During the Late Noachian to Early Hesperian, basaltic to andesitic magma, erupted along the initial Valles Marineris rift and other fissures in the highlands, may have encountered ground-ice and erupted violently, depositing voluminous widespread ignimbrites that locally blanketed the highlands and formed the layered subdued plains unit. Later faulting or collapse, related to growth

Fig. 19. Part of MOC M0806284 (2.86 m per pixel, 1.46 km wide) showing possible dyke (lineation) and young volcanic deposits in western Candor Chasma; note dark low-rimmed circular craters and dark dunes along indurated lineation.

of the Tharsis rise, began to form the Valles Marineris chasmata (Lucchitta *et al.* 1992). Eruptions east of Valles Marineris, beneath now ice-rich materials and ash, encountered water from subsurface ice and exploded violently, producing chaos and flooding that eroded subdued material. Late Hesperian to Early Amazonian eruptions within the deepening chasmata may have occurred beneath confined ice, forming interior-deposit tuyas and wallrock outcrops, and generating more floods downstream. Basaltic andesite material in the northern plains (Bandfield *et al.* 2000) may partly be a result of the floods that eroded and removed large amounts of Xanthe Terra and redistributed it into the Chryse basin and perhaps beyond (Chapman & Tanaka 2001*b*). Crystalline hematite deposits may have resulted from Fe-rich hydrothermal fluids generated by post-magmatic cooling alteration of subdued material ash flows and dark ash deposits on the ILDs. About this time in Martian history, basaltic plinian eruptions may have formed the Medusae Fossae Formation, related to late stage volatile-rich volcanism near Elysium and Olympus Mons.

The observational evidence suggested to support the generalized history is by no means unambiguous. Thus, the history is merely a hypothesis to be evaluated. None of the ideas in this paper are necessary to support the main tenet that the LMT sediments may be tephra. There are voluminous flood lavas of all ages on Mars near each of the major regions of LMT deposits, and voluminous tephra should have been produced from these eruptions, based on our best understanding of how they erupt (Wilson & Head 1983) and our best terrestrial analogues (Thordarson & Self 1993; Keszthelyi *et al.* 2000). If the widespread LMTs are not formed of tephra, then where are other candidate deposits for widespread tephra on Mars?

Future research will likely determine the origin of the LMT units. At this date, the Mars Global Surveyor spacecraft and its onboard instruments continue to orbit the planet and acquire data. Additionally, in January of 2003, NASA's goal is to place on Mars two landers equipped with Mars Exploration Rovers (MERs), to learn more about the ancient climate and putative water-rich environments through rock examination via MER instruments. These instruments include a PanCam, a Rock Abrasion Tool, a Miniature Thermal Emission Spectrometer, a Mössbauer Spectrometer, an Alpha Proton X-Ray Spectrometer and a Microscopic Imager (Squyres 1998). The orbital and future rover missions might provide evidence to determine if the LMT units are tephra deposits. For example, more detailed

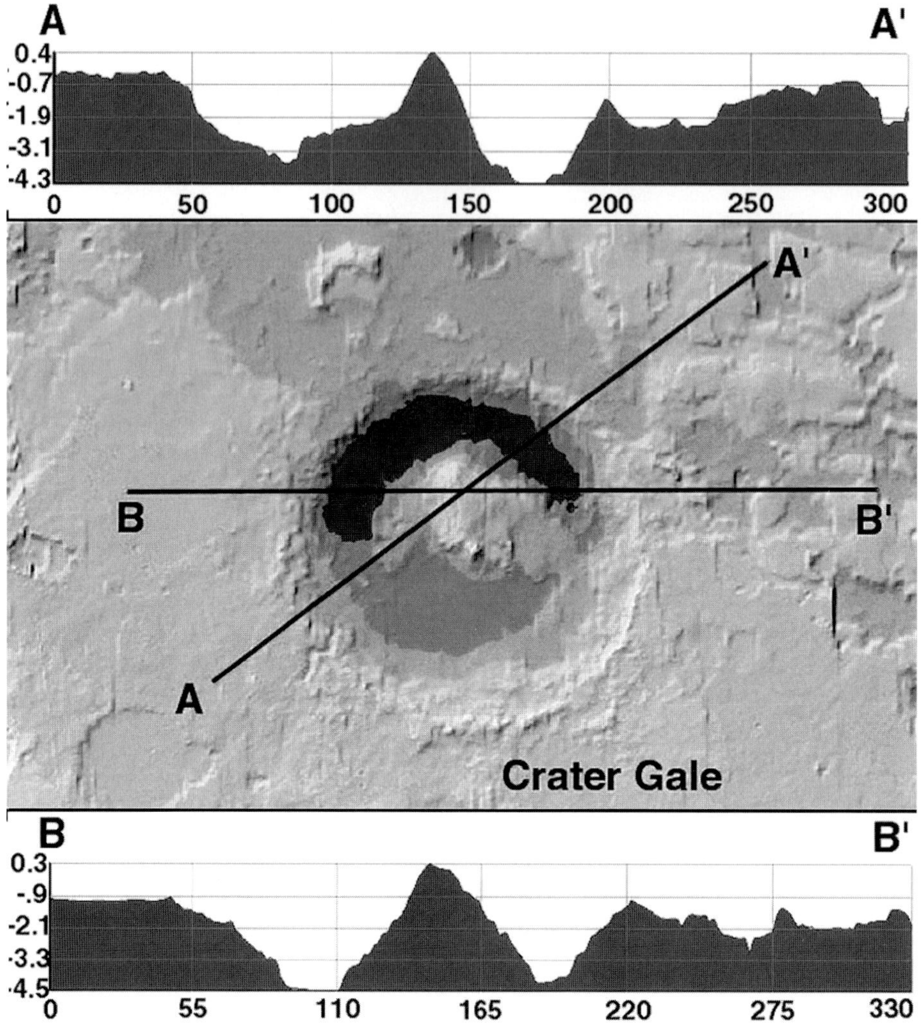

Fig. 20. MOLA topographic view of impact crater Gale (at about latitude 5°S., longitude 222°) showing profile locations (grey colours indicate topographic slices; local vertical lines mark errors; courtesy of MOLA Team). Topographic profiles A–A′ and B–B′ are both located well north of crater central peak (scale in km; vertical exaggeration is 10×).

MOLA digital elevation models could show topographic anomalies associated with possible calderas and subsurface magma chambers. Additional MOC images may show evidence of explosive volcanism such as edifice mantling material and smooth deposits associated with lobe-shaped features (Head & Wilson 1998) or more convincing examples of volcanic vents, fumarolic mounds, tent rocks or large gas cavities associated with terrestrial ash deposits. Interpretation of additional TES, mini TES and Alpha Proton X-Ray data may be able to better discriminate volcanic compositions, glasses and glass alteration products, such as palagonite, tachylite, chlorite, montmorillonite, bentonites, etc. The PanCam may take panoramic images that show surface boulders/rocks with fiamme and vesicular textures. There is also the possibility that this camera could view gullies eroded into lithic-rich tuffs or tuffs with ballistically emplaced clasts that distort impacted layers of ash. Finally, the Microscopic Imager has the capability to reveal ash shards within clasts broken by the Rock Abrasion Tool.

The author wishes to thank K. Tanaka and J. Kargel of the United States Geological Survey, J. Head of Brown University, and A. McEwen at the University of Arizona for their thoughtful reviews and timely suggestions.

References

BAKER, M. C. W. 1981. The nature and distribution of Upper Cenozoic ignimbrite centers in the Central Andes. *Journal of Volcanology and Geothermal Research*, **11**, 293–315.

BAKER, V. R., CARR, M. H., GULICK, V. C., WILLIAMS, C. R. & MARLEY, M. S. 1992. Channel and valley networks. *In*: KIEFFER, H. H., JAKOSKY, B. M., SNYDER, C. W. & MATTHEWS, M. S. (eds) *Mars*. University of Arizona Press, Tucson, 493–522.

BAKER, V. R. & MILTON, D. J. 1974. Erosion by catastrophic floods on Mars and Earth. *Icarus* **23**, 27–41.

BANDFIELD, J. L., HAMILTON, V. E. & CHRISTENSEN, P. R. 2000. A global view of Martian surface compositions from MGS-TES. *Science*, **287**, 1626–1630.

CARR, M. H. 1974. Tectonism and volcanism of the Tharsis region of Mars. *Journal of Geophysical Research*, **79**, 3943–3949.

CARR, M. H. 1979. Formation of Martian flood features by release of water from confined aquifers. *Journal of Geophysical Research*, **84**, 2995–3007.

CARR, M. H. 1981. *The Surface of Mars*. Yale University Press, New Haven.

CHAPMAN, M. G. 1999. Enigmatic terrain of north Terra Meridiani, Mars (abs.) *In*: *30th Lunar & Planetary Science Conference*, Lunar and Planetary Institute, March 15–18, Houston, Texas, LPSC CD, No. 1294.

CHAPMAN, M. G. & TANAKA, K. L. 2000. Chasmata, chaos, outflow channels, and interior deposits on Mars: produced by sub-ice eruptions? (abs.) *In*: *31st Lunar & Planetary Science Conference*, Lunar and Planetary Institute, March 15–18, Houston, Texas, LPSC CD, No. 1256.

CHAPMAN, M. G. & TANAKA, K. L. 2001*a*. Interior trough deposits on Mars: Sub-ice volcanoes? *Journal of Geophysical Research*, **106**, 10 087–10 100.

CHAPMAN, M. G. & TANAKA, K. L. 2001*b*. Related magma-ice interactions: possible origins of chasmata, chaos and surface materials in Xanthe, Margaritifer, and Meridiani Terrae, Mars. *Icarus*, in press.

CHRISTENSEN, P. R. 1986. The spatial distribution of rocks on Mars. *Icarus*, **68**, 217–238.

CHRISTENSEN, P. R., ANDERSON, D. L., CHASE, S. C., ET AL. 1998. Results from the Mars Global Surveyor Thermal Emission Spectrometer. *Science*, **279**, 1692–1698.

CHRISTENSEN, P. R., MALIN, M., MORRIS, D., BANDFIELD, J., LANE, M. & EDGETT, K. S. 2000*a*. The distribution of crystalline hematite on Mars from the Thermal Emission Spectrometer: Evidence for liquid water (abs.). *In*: *31st Lunar & Planetary Science Conference*, Lunar and Planetary Institute, March 15–18, Houston, Texas, LPSC CD, No. 1627.

CHRISTENSEN, P. R. & 15 OTHERS. 2000*b*. Detection of crystalline hematite mineralization on Mars by the Thermal Emission Spectrometer: Evidence for near-surface water. *Journal of Geophysical Research*, **105**, 9623–9642.

CONDIT, C. D. 1978. Distribution and relations of 4- to 10-km-diameter craters to global geologic units of Mars. *Icarus*, **34**, 465–478.

COULON, C. & THORPE, R. S. 1981. Role of continental crust in petrogenesis of orogenic volcanic associations. *Tectonophysics*, **77**, 79–93.

CROFT, S. K. 1990. Geologic map of the Hebes Chasma quadrangle, VM 500K 00077 (abs.). *National Aeronautics & Space Administration Technical Memorandum*, **4210**, 539–541.

EDGETT, K. S. & MALIN, M. 2000. The new Mars of MGS MOC: Ridged layered geologic unit (they're not dunes; abs.). *In*: *31st Lunar & Planetary Science Conference*, Lunar and Planetary Institute, March 15–18, Houston, Texas, LPSC CD, No. 1057.

EDGETT, K. S. & PARKER, T. J. 1997. Water on early Mars: Possible subaqueous sedimentary deposits covering ancient cratered terrain in western Arabia and Sinus Meridiani. *Geophysical Research Letters*, **24**, 2897–2900.

FISHER, R. V. & SCHMINCKE, H. U. 1984. *Pyroclasitic Rocks*. Springer-Verlag, New York.

FRANCIS, P. W. & WOOD, C. A. 1982. Absence of silicic volcanism on Mars: Implications for crustal composition and volatile abundance. *Journal of Geophysical Research*, **87**, 9881–9889.

GEISSLER, P. E., SINGER, R. B. & LUCCHITTA, B. K. 1990. Dark materials in Valles Marineris: Indications of the style of volcanism and magmatism on Mars. *Journal of Geophysical Research*, **95**, 14 399–14 413.

GREELEY, R. & GUEST, J. E. 1987. *Geologic map of the eastern equatorial region of Mars*. United States Geological Survey Miscellaneous Investigation Series Map **I-1802-B**, 1:15 000 000 scale.

GRIMM, R. E. & SOLOMON, S. C. 1986. Tectonic tests of proposed polar wander paths for Mars and the Moon. *Icarus*, **65**, 110–121.

HARTMANN, W. K. 1973. Martian surface and crust: Review and synthesis. *Icarus*, **19**, 550–575.

HEAD, J. W. & WILSON, L. 1998. Tharsis Montes as composite volcanoes?: 3. Lines of evidence for explosive volcanism in edifice construction (abs.). *In*: *29th Lunar & Planetary Science Conference*, Lunar and Planetary Institute, March 15–18, Houston, Texas, LPSC CD 29, No. 1124.

HEAD, J. W., KRESLAVSKY, M. A. & PRATT, S. 2001. Northern lowlands on Mars: Evidence for widespread volcanic flooding and tectonic deformation in the Early Hesperian (abs.). *In*: *32nd Lunar & Planetary Science Conference*, Lunar and Planetary Institute, March 15–18, Houston, Texas, LPSC CD 32, No. 1063.

HOFFMAN, N. 2000. White Mars: A new model for Mars' surface and atmosphere based on CO_2. *Icarus*, **146**, 326–342.

KESZTHELYI, L., MCEWEN, A. S. & THORDARSON, T. 2000. Terrestrial analogs and thermal models for Martian flood lavas. *Journal of Geophysical Research*, **105**, 15 027–15 049.

KOMATSU, G., GEISSLER, P. E., STROM, R. G. & SINGER, R. B. 1993. Stratigraphy and erosional landforms of layered deposits in Valles Marineris, Mars. *Journal of Geophysical Research*, **98**, 11 105–11 121.

KOMATSU, G., KARGEL, J. S., BAKER, V. R. STROM, R. G. ORI, G. G. MOSANGINI, C. & TANAKA, K. L. 2000. A chaotic terrain formation hypothesis: explosive outgas and outlow by dissociation of clathrate on Mars (abs.). *In*: *31st Lunar & Planetary Science Conference*, Lunar and Planetary Institute, March 15–18, Houston, Texas, LPSC CD, No. 1434.

LANZ, J. K. & JAUMANN, R. 2001. Possible volcanic constructs in Aram Chaos revealed by MOC and their impact on outflow channel genesis (abs.). *In*: *32nd Lunar & Planetary Science Conference*, Lunar and Planetary Institute, March 15–18, Houston, Texas, LPSC CD 32, No. 1574.

LEE, P. 1993. Briny lakes on early Mars? Terrestrial intracrater playas and Martian candidates. Abstract for the Workshop on Early Mars: How warm and how wet? *Lunar & Planetary Science Institute Technical Report* **93–03**, Lunar & Planetary Institute, Houston, Texas.

LUCCHITTA, B. K. 2001. Young dark mantles and light flows in Valles Marineris, Mars (abs.). *In*: *32nd Lunar & Planetary Science Conference*, Lunar and Planetary Institute, March 15–18, Houston, Texas, LPSC CD, No. 2059.

LUCCHITTA, B. K. 1999. *Geologic map of Ophir and Central Candor Chasmata (MTM -05072) of Mars*. United States Geological Survey Miscellaneous Investigation Series Map **I-2568**, scale 1:500 000.

LUCCHITTA, B. K. 1990. Young volcanic deposits in the Valles Marineris, Mars. *Icarus*, **86**, 476–509.

LUCCHITTA, B. K. 1982. Lakes or Playas in Valles Marineris. *National Aeronautics & Space Administration Technical Memorandum*, **85127**, 233–234.

LUCCHITTA, B. K., CLOW, G. D., GEISSLER, P. E., MCEWEN, A. S., SCHULTZ, R. A., SINGER, R. B. & SQUYRES, S. W. 1992. The canyon system on Mars. *In*: KIEFFER, H. H., JAKOSKY, B. M., SNYDER, C. W. & MATTHEWS, M. S. (eds) *Mars*. University of Arizona Press, Tucson, 453–492.

LUCCHITTA, B. K., ISBELL, N. K. & HOWINGTON-KRAUS, A. 1994. Topography of Valles Marineris: Implications for erosional and structural history. *Journal of Geophysical Research*, **99**, 3783–3798.

MALIN, M. C. 1979. Evidence of indurated deposits of fine materials (abstract). *National Aeronautics & Space Administration Conference Publication*, **2072**, 54.

MALIN, M. C. & EDGETT, K. S. 2000. Sedimentary rocks of Mars. *Science*, **290**, 1927–1937.

MASURSKY, H., BOYCE, J. M., DIAL, A. L. JR., SCHABER, G. G. & STROBELL, M. E. 1977. Classification and time of formation of Martian channels based on Viking data. *Journal of Geophysical Research*, **82**, 4016–4038.

MAX, M. D. & CLIFFORD, S. M. 2000. The initiation of Martian outflow channels through the catastrophic decomposition of methane hydrate (abs.). *In*: *31st Lunar & Planetary Science Conference*, Lunar and Planetary Institute, March 15–18, Houston, Texas, LPSC CD, No. 2094.

MCCAULEY, J. F. 1978. *Geological map of the Coprates quadrangle of Mars*. United States Geological Survey Miscellaneous Investigation Series Map **I-897**, Scale 1:5 000 000.

MCKAY, C. P. & NEDELL, S. S. 1988. Are there carbonate deposits in the Valles Marineris, Mars? *Icarus*, **73**, 142–148.

MELLON, M. T., JAKOSKY, B. M., KIEFFER, H. H. & CHRISTENSEN, P. R. 2000. High-resolution thermal inertia mapping from the Mars Global Surveyor Thermal Emission Spectrometer. *Icarus*, **198**, 437–455.

MCSWEEN, H. Y. JR. & 19 OTHERS. 1999. Chemical, multispectral, and textural constraints on the composition and origin of rocks at the Mars Pathfinder landing site. *Journal of Geophysical Research*, **104**, 8679–8715.

MINITTI, M. E. & RUTHERFORD, M. J. 2000. Genesis of the Mars Pathfinder 'sulfur-free' rock from SNC parental liquids. *Geochimica et Cosmochimica Acta*, **64**, 2535–2547.

MOORE, J. M. 1990, Nature of the mantling deposit in the heavily cratered terrain of northeastern Arabia, Mars. *Journal of Geophysical Research*, **95**, 14 279–14 289.

MOUGINIS-MARK, P. 1993. The influence of oceans on Martian volcanism. *In*: 24th *Lunar & Planetary Science Conference Abstracts*, Lunar and Planetary Institute, March 15–18, Houston, Texas, 1021–1022.

MURAI, I. 1961. A study of the textural characteristics of pyroclastic flow deposits in Japan. *Tokyo University Earthquake Research Institute, Bulletin*, **39**, 133–248.

MURCHIE, S., KIRKLAND, L., ERARD, S., MUSTARD, J. & ROBINSON, M. 2000. Near-infrared spectral variations of Martian surface materials from ISM imaging spectrometer data. *Icarus*, **147**, 444–471.

NEDELL, S. S., SQUYRES, S. W. & ANDERSEN, D. W. 1987. Origin and evolution of the layered deposits in the Valles Marineris, Mars. *Icarus*, **70**, 409–441.

NOBLE, S. K. & PIETERS, C. M. 2001. Type 2 terrain: Compositional constraints on the Martian lowlands (abs.). *In*: *32nd Lunar & Planetary Science Conference*, Lunar and Planetary Institute, March 15–18, Houston, Texas, LPSC CD, 32 No. 1230.

NOREEN, E., CHAPMAN, M. G., JOHNSON, J., TANAKA, K. L. & TITUS, T. A. 2000. Examinations of igneous alternatives to Martian hematite using terrestrial analogues. *Geological Society of America, Abstracts with Programs*, **32,** No. 7, A303.

ORI, G. G. & MOSANGINI, C. 1998. Complex depositional systems in Hydraotes Chaos, Mars: An example of sedimentary process interactions in the Martian hydrological cycle. *Journal of Geophysical Research*, **103**, 22 713–22 723.

PARKER, T. J. 1991. A comparison of the Martian Medusae Fossae Formation with terrestrial carbonate platforms. *In*: 27th *Lunar & Planetary Science Conference Abstracts,* Lunar and Planetary Institute, March 15–18, Houston, Texas, 1003–1004.

PARKER, T. J., CLIFFORD, S. M. & BANERDT, W. B. 2000. Argyre Planitia and the Mars global hydrologic cycle (abs.). *In*: *31st Lunar & Planetary Science Conference,* Lunar and Planetary Institute, March 15–18, Houston, Texas, LPSC CD, No. 2033.

PETERSON, C. 1981. A secondary origin for the central plateau of Hebes Chasma. *In*: *12th Lunar & Planetary Science Conference Proceedings*, Lunar and Planetary Institute, March 15–18, 1981, Houston, Texas, 1459–1471.

PRESLEY, M. A. & ARVIDSON, R. E. 1988. Nature and origin of materials exposed in the Oxia Palus–Western Arabia–Sinus Meridiani region, Mars. *Icarus,* **75**, 499–517.

RUFF, S. W., CHRISTENSEN, P. R., CLARK, R. N., KIEFFER, H. H., MALIN, M. C., BANDFIELD, J. L., JAKOSKY, B. M., LANE, M. D., MELLON, M. T. & PRESLEY, M. A. 2000. Mars 'White Rock' feature lacks evidence of an aqueous origin (abs.). *In*: *31st Lunar & Planetary Science Conference*, Lunar and Planetary Institute, March 15–18, Houston, Texas, LPSC 31th CD, No. 1945.

SAKIMOTO S. E. H., FREY, H. V., GARVIN, J. B. & ROARK, J. H. 1999. Topography, roughness, layering, and slope properties of the Medusae Fossae Formation form Mars Orbiter Laser Altimeter (MOLA) and Mars Orbiter Camera (MOC) data. *Journal of Geophysical Research*, **104**, 24 141–24 154.

SCHULTZ, P. H. & LUTZ, A. B. 1988. Polar wandering on Mars. *Icarus,* **73**, 91–141.

SCOTT, D. H. & CARR, M. H. 1978. *Geologic map of Mars.* United States Geological Survey Miscellaneous Investigation Series Map **I-1083**, 1 : 25 000 000 scale.

SCOTT, D. H. & CHAPMAN, M. G. 1995. *Geologic and topographic maps of the Elysium Paleolake basin, Mars.* United States Geological Survey Miscellaneous Investigation Series Map **I-2397**, 1 : 500 000 scale.

SCOTT, D. H. & TANAKA, K. L. 1982. Ignimbrites of the Amazonis Planitia region of Mars. *Journal of Geophysical Research*, **87**, 1179–1190.

SCOTT, D. H. & TANAKA, K. L. 1986. *Geologic map of the western equatorial region of Mars.* United States Geological Survey Miscellaneous Investigation Series Map **I-1802-A**, 1 : 15 000 000 scale.

SHARP, R. P. 1973. Mars–troughed terrain. *Journal of Geophysical Research*, **78**, 4063–4072.

SMITH, R. L. 1979. Ash flow magmatism. *Geological Society of America,* Special Paper, **180**, 5–28.

SQUYRES, S. W. 1998. The Athena Mars Rover science payload (abs.). Mars Surveyor 2001 Landing Site Workshop. NASA-Ames Research Center, January 26–27, Mountain View, California (http://cmex.arc.nasa.gov/Mars_2001/Squyres_abs.html).

TANAKA, K. L. 2000. Dust and ice deposition in the Martian geologic record. *Icarus,* **144,** 254–266.

THORDARSON, TH. & SELF, S. 1993. The Laki (Skaftar Fires) and Grimsvötn eruptions in 1783–1785. *Bulletin of Volcanology*, **55**, 233–263.

TUFFEN, H., GILBERT, J. & MCGARVIE, D. 2001. Products of an effusive subglacial rhyolite eruption: Blahnukur, Torfajokull, Iceland. *Bulletin of Volcanology*, **63**, 179–190.

VAN BEMMELEN, R. W. & RUTTEN, M. G. 1955. *Tablemountains of northern Iceland.*, E. J. Brill, Leiden, Netherlands.

WARD, A. W. 1979. Yardangs on Mars: evidence of recent wind erosion. *Journal of Geophysical Research*, **84**, 8147–8166.

WEITZ, C. M. 1999. A volcanic origin for the interior layered deposits in Hebes Chasma, Mars (abs.). *In*: *30st Lunar & Planetary Science Conference*, Lunar and Planetary Institute, March 15–18, Houston, Texas, LPSC CD, No. 1279.

WEITZ, C. M. & PARKER, T. J. 2000. New Evidence that the Valles Marineris interior layered deposits formed in standing bodies of water (abs.). *In*: *31st Lunar & Planetary Science Conference*, Lunar and Planetary Institute, March 15–18, Houston, Texas, LPSC CD, No. 1693.

WILSON, L. & HEAD, J. W. 1981. Ascent and eruption of basaltic magma on the Earth and Moon. *Journal of Geophysical Research*, **86**, 2971–3001.

WILSON, L. & HEAD, J. W. 1983. A comparison of volcanic eruption processes on Earth, Moon, Mars, Io, and Venus. *Nature,* **302,** 663–669.

WILSON, L. & HEAD, J. W. 1994. Mars: Review and analysis of volcanic eruption theory and relationships to observed landforms. *Reviews of Geophysics,* **32,** 221–264.

WILSON, L. & HESLOP, S. E. 1990. Clast sizes in terrestrial and Martian ignimbrite lag deposits. *Journal of Geophysical Research*, **95**, 17 309–17 314.

WISE, D. U., GOLOMBEK, M. P. & MCGILL, G. E. 1979. Tharsis province of Mars: Geologic sequence, geometry, and a deformation mechanism. *Icarus*, **38**, 456–472.

WITBECK, N. E., TANAKA, K. L. & SCOTT, D. H. 1991. *The geologic map of the Valles Marineris region, Mars.* United States Geological Survey Miscellaneous Investigation Series Map **I-2010**, scale 1 : 2 000 000.

ZUBER, M. T. & 14 OTHERS 2000. Internal structure and early thermal evolution of Mars from Mars Global Surveyor topography and gravity. *Science*, **287**, 1788–1793.

Rootless cones on Mars: a consequence of lava–ground ice interaction

S. A. FAGENTS[1,3], P. LANAGAN[2] & R. GREELEY[1]

[1] *Department of Geological Sciences, Box 871404, Arizona State University, Tempe, AZ 85287-1404, USA (e-mail: fagents@asu.edu)*
[2] *Lunar and Planetary Laboratory, University of Arizona, Tucson, AZ 85721, USA*
[3] *Present address: Hawaii Institute of Geophysics and Planetology/SOEST, University of Hawaii at Manoa, 2525 Correa Road, Honolulu, HI 96822, USA (e-mail: fagents@higp.hawaii.edu)*

Abstract: Fields of small cratered cones on Mars are interpreted to have formed by rootless eruptions due to explosive interaction of lava with ground ice contained within the regolith beneath the flow. Melting and vaporization of the ice, and subsequent explosive expansion of the vapour, act to excavate the lava and construct a rootless cone around the explosion site. Similar features are found in Iceland, where flowing lavas encountered water-saturated substrates. The martian cones have basal diameters of *c.* 30–1000 m and are located predominantly in the northern volcanic plains. High-resolution Mars Orbiter Camera images offer significant improvements over Viking data for interpretation of cone origins. A new model of the dynamics of cone formation indicates that very modest amounts of water ice are required to initiate and sustain the explosive interactions that produced the observed features. This is consistent with the likely low availability of water ice in the martian regolith. The scarcity of impact craters on many of the host lava flows indicates very young ages, suggesting that ground ice was present as recently as <10–100 Ma, and may persist today. Rootless cones therefore act as a spatial and temporal probe of the distribution of ground ice on Mars, which is of key significance in understanding the evolution of the martian climate. The location of water in liquid or solid form is of great importance to future robotic and human exploration strategies, and to the search for extraterrestrial life.

Rootless cone groups, also known as pseudocraters in Iceland, consist of small, hydrovolcanic cones located atop lava flows that have moved over a substrate containing water at the surface or within pore spaces (Thorarinsson 1951, 1953; Morrissey & Thordarson 1991; Thordarson *et al.* 1992). Similar to the formation of littoral cones at lava–ocean entries (Thordarson & Self 1991; Jurado-Chichay *et al.* 1996; Mattox & Mangan 1997), vaporization of the water by the hot lava leads to explosive excavation of the flow, distributing fragmented lava around the explosion site to form a rootless cone. These features are therefore quite distinct from primary cones constructed over volcanic conduits rooted deeper in the crust. Rootless cone material consists of scoria, spatter and a lesser proportion of lithic material derived from the substrate (Jónsson 1990; Morrissey & Thordarson 1991; Thordarson *et al.* 1992, 1998). Icelandic examples are found where lavas flowed over marshy ground, stream sediments, glacial outwash plains or into lake basins (Thorarinsson 1951, 1953). They exhibit a wide range of morphologies, from large, broad-cratered cones having convex-up slopes, to smaller, steeper cones with small craters and concave slopes.

On Mars, fields of cones, domes and mounds were identified in Viking Orbiter image data and interpreted to be rootless volcanic cones, formed by explosive interaction of lava and ice contained in the martian regolith (Allen 1979; Frey *et al.* 1979). The resolution and coverage of Viking data were not sufficient to ascribe confidently a rootless volcanic origin, however. Alternative origins include exhumed impact craters, rampart craters, lava-capped diatremes and pingoes (Hodges & Moore 1992).

More recent data from the Mars Orbiter Camera (MOC) on board the Mars Global Surveyor (MGS) spacecraft have re-imaged many of the same locations at much higher resolutions (1.5 to 10 metres per pixel; Malin *et al.* 1992). Therefore, these preliminary and tentative interpretations can now be examined much more closely. Many examples of apparent volcanic cone morphologies are seen, although in some cases their origins remain equivocal.

From: SMELLIE, J. L. & CHAPMAN, M. G. (eds) 2002. *Volcano–Ice Interaction on Earth and Mars*. Geological Society, London, Special Publications, **202**, 295–317. 0305-8719/02/$15.00

The presence of candidate rootless cones on Mars offers important clues as to the evolution of the martian surface and climate (Allen 1979; Frey & Jarosewich 1982; Greeley & Fagents 2001; Lanagan *et al.* 2001). Present climatic conditions promote dehydration of the near-surface regolith over much of Mars (Clifford & Hillel 1983; Fanale *et al.* 1986; Kuzmin 1988). The locations and ages of rootless cones can be used as indicators of past climatic fluctuations, representing times when water was more stable at the surface, and able to diffuse from the atmosphere into the regolith (Mellon & Jakosky 1995). Although CO_2 may also be present in the martian regolith, it is more volatile than H_2O, and is consequently less stable under the low atmospheric pressure conditions. We therefore consider H_2O to be the most likely near-surface regolith ice species.

Variations in cone morphology and spatial density may provide insights into regional and global variations in thickness and depth of burial of regolith ice. The very youthful appearance of some of the lava flows on which candidate rootless cones reside suggests that martian ground ice was present very recently (<10 to 100 Ma; Hartmann 1999; Hartmann *et al.* 1999; Hartmann & Berman 2000), and therefore may persist today (Mellon *et al.* 1997). The combination of water and volcanic heat provides a key precursory condition for biotic development (e.g. Sagan & Lederberg 1976). Therefore, understanding the distribution, ages and formation conditions of these features is of critical significance in guiding future exploration strategies, in terms of both the search for extraterrestrial life and a future human presence on Mars.

In this contribution the Viking and MOC evidence for rootless cones on Mars is considered, and their morphologies analysed in comparison to terrestrial analogues. Previous models for their formation and recent advances in understanding their emplacement mechanisms are discussed. Finally, the implications for the spatial and temporal distribution of water ice in the shallow crust are detailed.

Observations

Icelandic rootless cones

Rootless cone fields occur on many lava flows in Iceland, some of which have been documented in some detail (Thorarinsson 1951, 1953; Jónsson 1990; Morrissey & Thordarson 1991; Thordarson *et al.* 1992, 1998; Thordarson & Self 1993). The Myvatn District in north central Iceland is considered the terrestrial type locality for Icelandic rootless cones. Other cone fields include those located at Raudholar in south-western Iceland, in Thjorsardalur and on the Laki lava flow in south central Iceland, and at Álftaver and Landbrot in south Iceland. In each case, the cones formed in lavas that flowed into marshland, shallow lake basins (Myvatn, Raudholar) or shallow stream environments (e.g. the glacial outwash plains at Álftaver and Landbrot). Formation of rootless cones is commonly closely associated with preferred internal lava pathways or tubes (Morrissey & Thordarson 1991; Thordarson *et al.* 1992; Jurado-Chichay *et al.* 1996; Mattox & Mangan 1997; Greeley & Fagents 2001). The Icelandic rootless features generally consist of a well defined cone of scoria or spatter ranging from a few metres to >300 m in diameter, and 1 to 40 m in height (Thorarinsson 1953; Frey & Jarosewich 1982; Greeley & Fagents 2001), which may be surrounded by an outer apron of finer pyroclastic material. A lesser proportion of lithic material excavated from the substrate is also typically present (Jónsson 1990; Morrissey & Thordarson 1991; Thordarson *et al.* 1992, 1998).

Distinct differences in cone morphology and clast characteristics are evident both between and within the different cone fields (e.g. Thorarinsson 1951, 1953), despite the fact that some areas have undergone reworking and erosion (e.g. Larsen 2001). Large-diameter, broad-cratered cones resemble tuff cones or rings, produced by vigorous explosive phreatomagmatic activity at primary vents. Smaller, steeper features are reminiscent of magmatic cinder cones, spatter mounds, or hornitos (Figs 1 & 2). Summit craters vary widely in size (Fig. 2).

Pyroclast characteristics are also indicative of the vigour of the fragmentation process. Broad-cratered cones generally consist of a greater proportion of angular, lapilli-sized scoria having relatively high densities (2000–3000 kg m^{-3}; Thorarinsson 1951, 1953); these characteristics are consistent with thorough disruption of the lava by an external water source (Wohletz 1983). Steeper cones are predominantly composed of coarse spatter, suggesting only weak fragmentation by relatively small amounts of vapour. The cones typically exhibit crude bedding indicative of multiple explosive pulses of varying intensity, and a coarsening of clasts upwards reflecting a decline in eruptive vigour with time (Thordarson *et al.* 1992).

The cones of the Álftaver District formed on the *c.* AD 934 Eldgjá flow (Jóhannesson *et al.* 1990). These cones have been somewhat reworked by jökulhlaups from eruptions of Katla volcano (Larsen 2001), so their morphologies

(a)

(b)

Fig. 1. Icelandic rootless cones. (**a**) Rootless cones at Lake Myvatn have large craters and steep, convex-up slopes. (**b**) Rootless cones of the Álftaver District have small craters and steep 'witches hat' profiles.

are not pristine. However, field inspection suggests that those in the eastern part of the cone field retain many of their original features (e.g. summit craters intact, late stage spatter casts remaining *in situ* on the cone flanks), despite the fact that the basal diameters may be somewhat modified due to burial by eolian material or flood deposits. The measurable diameters are subject to some uncertainties therefore, but range from a few metres to >80 m with a modal diameter between 10 and 20 m (Greeley & Fagents 2001). These cones are rather more spatter-rich than the

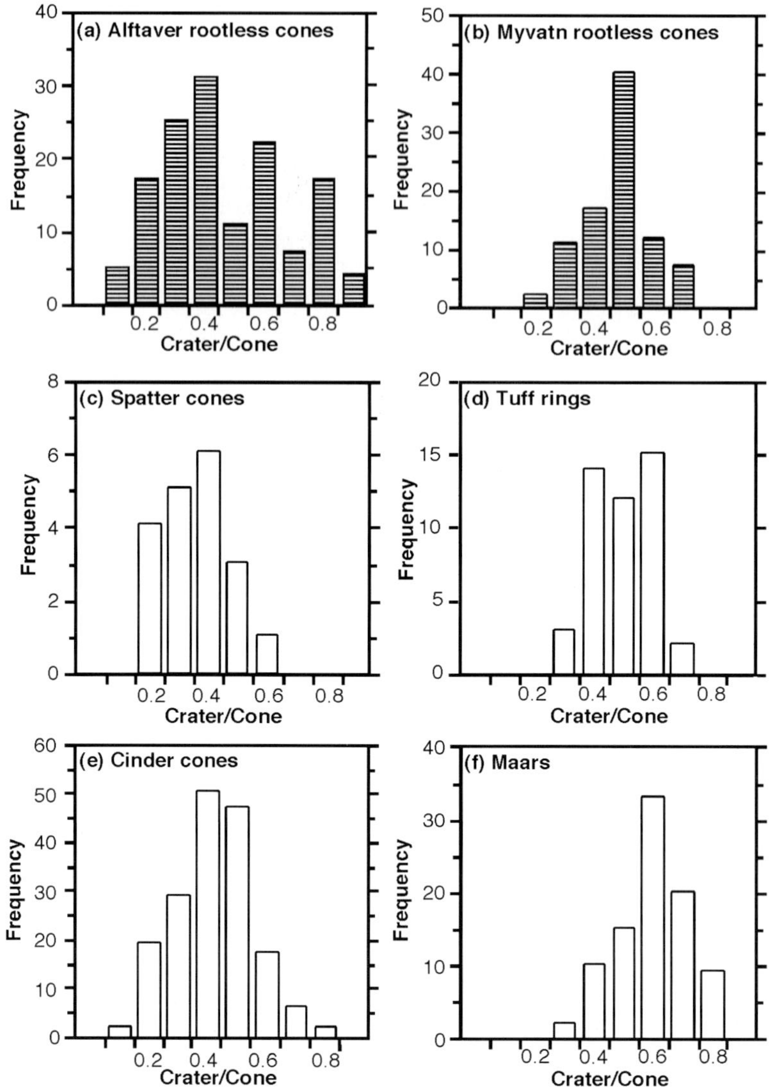

Fig. 2. Histograms of the ratio of crater diameter to cone diameter for rootless cones, in comparison to different types of primary volcanic cone. There is some uncertainty associated with Álftaver cone morphology because of erosion/deposition due to historic jökulhlaups (Larsen 2001). Data taken from Greeley & Fagents (2001), Frey & Jarosewich (1982), Pike & Clow (1981), and Wood (1979).

cones of the Myvatn District, which have basal diameters up to *c.* 320 m (Thorarinsson 1951, 1953). The Myvatn cones typically have larger crater/cone diameter ratios than those at Álftaver (Fig. 2).

The differences in cone morphology and clast characteristics are measures of the intensity of the explosive interactions involved in their genesis, whereby broad-cratered cones consisting predominantly of scoria were formed by more energetic interactions than small, steep spatter cones (Fig. 3). It is therefore inferred that the Myvatn cone-forming explosions were more vigorous than those at Álftaver. The explosivity is primarily dependent on the ability of the lava to interact with water (represented by the ratio of water to lava), which in turn is related to the rate of supply of lava to the explosion site, the geometry of the internal lava feeder pathways, and the accessibility of substrate water (Wohletz

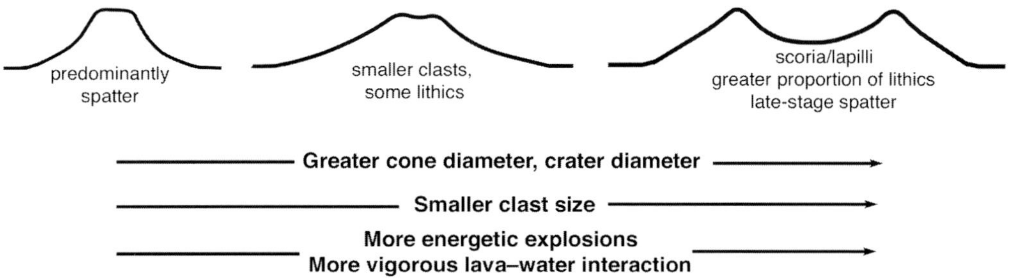

Fig. 3. Generalized interpretation of explosive intensity based on different cone morphologies.

1983, 1986; Mattox & Mangan 1997). However, the exact controls on rootless cone formation, and the role of the environment of formation are far from fully resolved, and require more work in the field.

Viking observations

Fields of cratered cones, mounds and domes were identified in Viking Orbiter images of Acidalia, Utopia, Elysium and Isidis Planitiae, Hephaestus Fossae, the north Olympus Mons aureole and Arrhenius province (Fig. 4; Frey *et al.* 1979; Frey & Jarosewich 1982; Mouginis-Mark 1985; Hodges & Moore 1992). The typical diameters of the cones means that they are only well resolved in images of less than *c.* 50 m per pixel. Images containing cones at 20–50 m per pixel represent only *c.* 2.7% of more than 12 000 images examined in a Viking-based study (Frey & Jarosewich 1982). Figure 5 shows a cone field in Eastern Acidalia, imaged by Viking Orbiter 1 at *c.* 48 m per pixel. These edifices were interpreted as rootless cones on the basis of their morphology, apparent superimposition on lava flows, typical large summit pits (where resolved), random distribution of cone clusters, and lack of association with eruptive fissures (Allen 1979; Frey *et al.* 1979; Frey & Jarosewich 1982).

The martian cones identified in the Viking Orbiter data are systematically larger than the terrestrial analogues (*c.* 200 to 1500 m; Frey *et al.* 1979; Frey & Jarosewich 1982; Hodges & Moore 1992), although the inconsistency in coverage and resolution of the Viking data probably precluded identification of smaller features. The size discrepancy between terrestrial and martian cones may be due, at least in part, to the differing environmental influences on volcanic explosion mechanisms. The low martian gravity and atmospheric pressure should lead to enhanced gas expansion, greater explosion velocities and more widely dispersed material than for the same initial conditions on Earth (Wilson & Head 1983, 1994; Fagents & Wilson 1996).

The cones are located in northern plains between latitudes of 10 and 48°N; an exception being the Arrhenius cones around 45°S (Fig. 4). This distribution reflects the split between ancient southern highlands and younger northern lowlands. Although there is good evidence of explosive volcanism in the southern hemisphere, which may itself be an indication of larger-scale magma–water interactions (Reimers & Komar 1979; Greeley & Spudis 1981; Mouginis-Mark *et al.* 1982, 1992; Scott 1982; Greeley & Crown 1990; Crown & Greeley 1993; Robinson *et al.* 1993; Greeley *et al.* 2000), it is the widespread presence of lava flows and effusive plains in the northern hemisphere (Greeley & Spudis 1981; Scott 1982), together with the greater atmospheric pressures at lower altitudes promoting ground ice retention (Fanale *et al.* 1986; Mellon & Jakosky 1995), that would commonly create conditions favourable for cone-forming events. Furthermore, given the relative youth of these plains, cones are more likely to exist in a state of preservation adequate for detection in spacecraft imagery.

MOC observations

Small cones may represent a category of volcanic edifice that has been somewhat overlooked due to the difficulty of distinguishing their detailed morphology in the Viking data. However, with the acquisition of a large amount of image data at resolutions of 3–10 m per pixel from the Mars Orbiter Camera (Malin *et al.* 1992), many such features are being clearly revealed for the first time. Small-scale explosive volcanism and lava–ice interactions may therefore be of greater significance to Mars' volcanic, geological and climatic evolution than previously thought. The increased detail afforded by the MOC data allows morphological and stratigraphic criteria

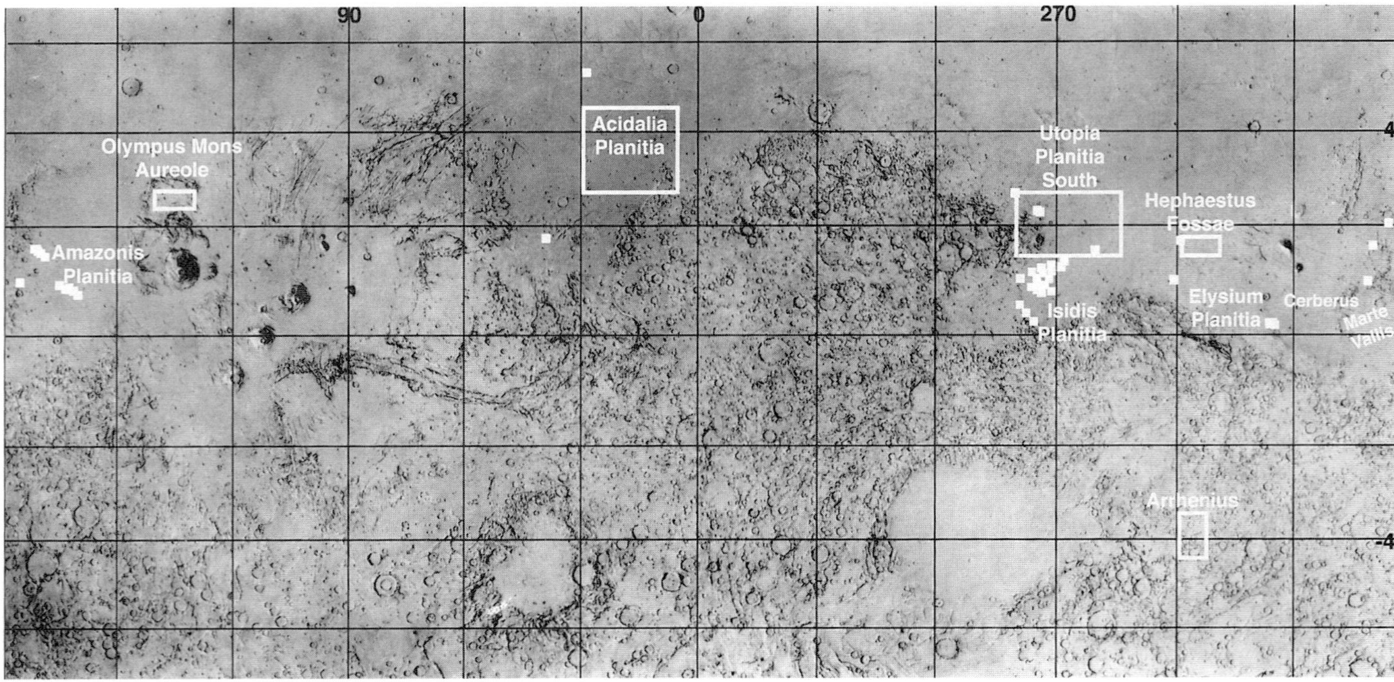

Fig. 4. Map showing locations of candidate martian rootless cones. Open boxes indicate areas identified in Viking coverage (Hodges & Moore 1992), filled squares show locations of MOC images in which cones are found. This map is a work in progress and will be updated as further MOC data are released. Scale: 10° equals 600 km at equator.

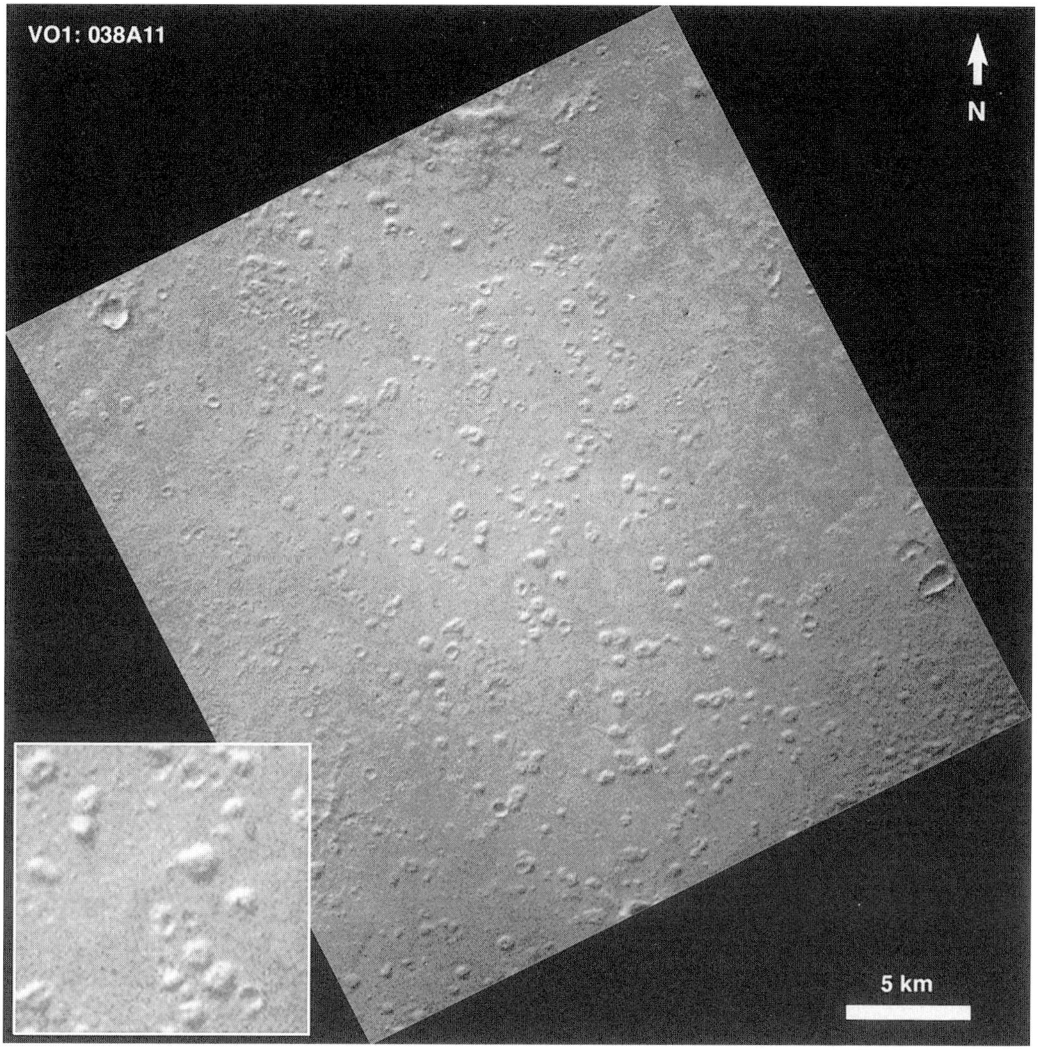

Fig. 5. Viking Orbiter image 038A11 (48 m per pixel) located in eastern Acidalia Planitia. Inset box shows a magnified portion of the centre of the image. The mounds are indistinct and poorly resolved, leading to ambiguity in interpretations of their origins.

designed to assess cone origin to be applied with a greater degree of confidence. Criteria suggestive of rootless origins include:

(a) positive-relief conical feature; crater floor not lying below the level of the surrounding terrain;
(b) clear superimposition on top of lavas or volcanic plains, down-flow from lava source (Thorarinsson 1951, 1953);
(c) no volumetrically significant lava flows emanating from or breaching the cone (although rare, small-volume rheomorphic flows are possible (von Komorowicz 1912; Thordarson *et al.* 1998));
(d) random clustering; no alignment along eruptive fissures (Thorarinsson 1951, 1953), although association with linear concentrations of water or ground ice is not precluded;
(e) possible association with lava tubes which could feed lava into the explosion site (Morrissey & Thordarson 1991; Thordarson *et al.* 1992; Jurado-Chichay *et al.*

1996; Mattox & Mangan 1997; Greeley & Fagents 2001);
(f) low regional slope.

Although none of these criteria alone is definitive evidence of rootless cone origins, the combination of several lines of evidence permits a more confident assessment to be made.

Alternative origins proposed for martian cones, mounds and domes include: (i) pedestal craters, which have ejecta blankets that armour a surface undergoing deflation, producing a plateau (McCauley 1973; Arvidson *et al.* 1976); (ii) rampart craters, which have been interpreted to form by impacts into a volatile-rich target (e.g. Carr *et al.* 1977; Gault & Greeley 1978; Mouginis-Mark 1987), and are characterized by a lobate ejecta blanket; and (iii) pingoes, which are ice-cored sediment mounds formed in periglacial lake or permafrost environments (Theilig & Greeley 1979; Lucchitta 1981; Rossbacher & Judson 1981; Cabrol *et al.* 2000). In many cases, the morphological and stratigraphic characteristics of candidate martian rootless cones rule out a non-volcanic origin (although in others, image resolution precludes a confident interpretation). For example, the cones observed in MOC images have morphologies quite different from those of impact craters: they are distinctly conical, and plateaux or lobate deposits are absent. Furthermore, their locations on lava flows do not provide the sedimentary basin setting required for pingo formation. The main

(a)

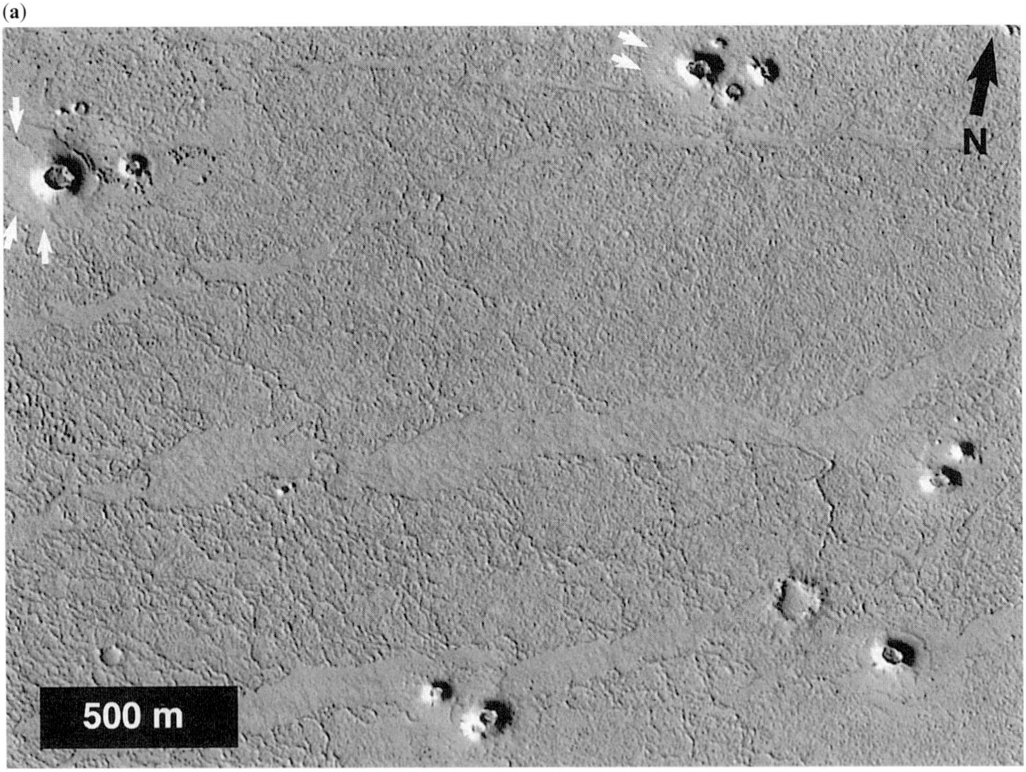

Fig. 6. MOC images of candidate rootless cones. Scene is illuminated from left except where otherwise stated. (**a**) Portion of image M0303958 of cone field near 24.8°N, 171.4°W (western Amazonis Planitia) showing overlapping cones (bottom of image) and smooth aprons surrounding cones suggestive of fine pyroclastic mantle (arrows). (**b**) Portion of image M0801962 of cone field near 26.0°N, 189.7°W (Amazonis Planitia). Note dense distribution of cones. (**c**) Portion of MOC image M0307175 of cone field near 12.8°N, 162.9°W (southern Amazonis Planitia). Note the overlapping cones. Scene is illuminated from the right. (**d**) Portion of image M0800090 of cone field near 2.6°N 215.8°W (southwestern Cerberus Plains). Note the presence of multiple overlapping cones. (**e**) Portion of image sp234405 (Isidis Planitia) showing multiple clustered cones. (**f**) Portion of image m0203503 (Isidis Planitia) with MOLA track superimposed. The topographic information yields a lower limit on the height of this cone of *c.* 25 m. Image is rotated 90 to enhance interpretation of MOLA track. (**g**) Oblique air photo of rootless cones at Lake Myvatn, Iceland, for comparison.

difficulty, therefore, lies in distinguishing rootless cones from primary vent edifices.

Figure 6 shows several examples of cratered cone fields imaged by MOC (locations shown in Fig. 4), which have been tentatively identified as rootless cones based on the criteria above. The cones shown here share many characteristics with the Icelandic cones. In many cases, the cones consist of a well formed edifice with a distinct summit crater. Some are surrounded by a smooth, thin, flat-lying halo which subdues the surface texture of the underlying lava flow (e.g. Fig. 6a), perhaps analogous to the apron of fine pyroclastic material associated with some Icelandic cones. The cone spatial distribution varies from widely spaced single cones or small clusters (Fig. 6a) to densely packed, multiple overlapping cones (Figs 6b, d, e) reminiscent of the Myvatn cones (Fig. 6g). Figures 6a through 6c show cones located on top of lavas exhibiting a distinctive surface texture termed platy-ridged morphology by Keszthelyi *et al.* (2000). This is interpreted to represent a style of sheet flow emplacement characterized by fluctuations in the effusion rate, leading to rafted plates of solidified crust which initially formed on relatively stagnant flow and were later disrupted by a surge in flow rate (Keszthelyi *et al.* 2000). This could provide a mechanism for continued supply of lava to interact with the ground ice.

The better resolution of the MOC data allows identification of cones down to much smaller sizes than was possible with Viking Orbiter images. Analysis of the morphometry of the cones in Amazonis Planitia indicates basal diameters in the range from *c.* 30 to 200 m with a modal diameter of *c.* 100 m an order of magnitude larger than the cones at Álftaver, Iceland (Fig. 7a), but more similar to the Myvatn cones. Where it can be discerned, the outer apron of

(**b**)

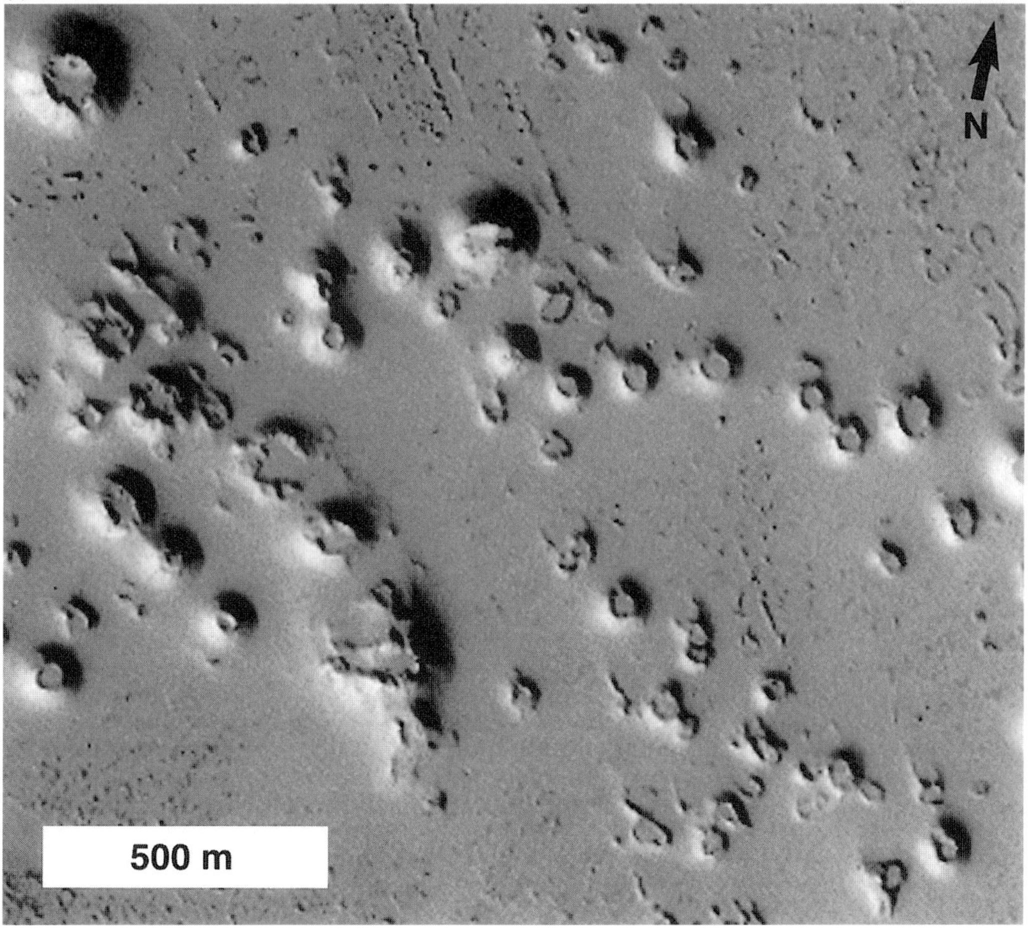

Fig. 6. (*continued*).

(c)

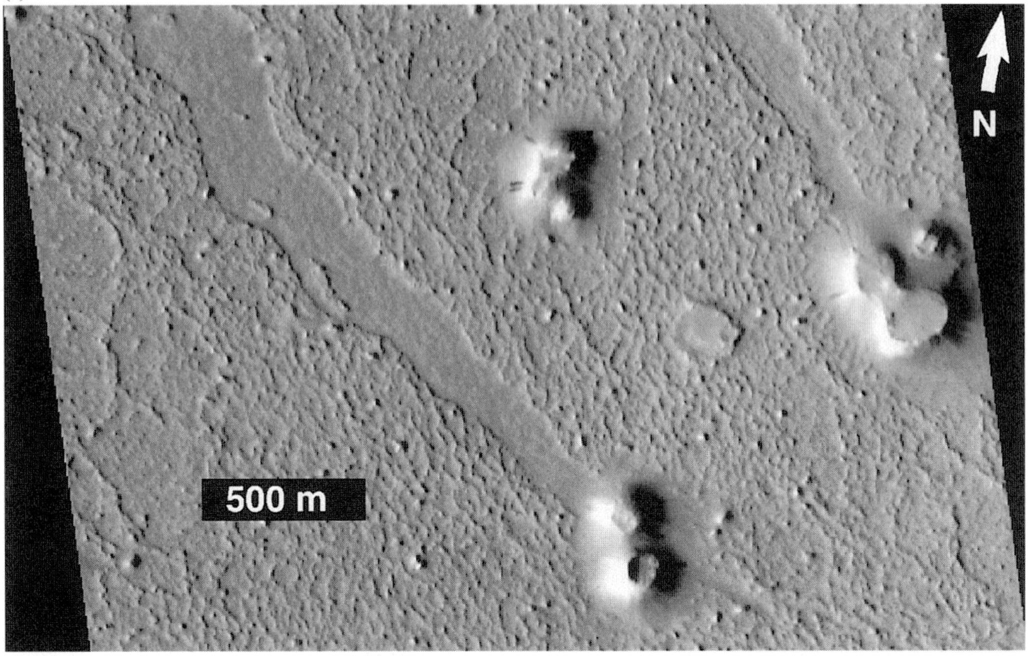

(d)

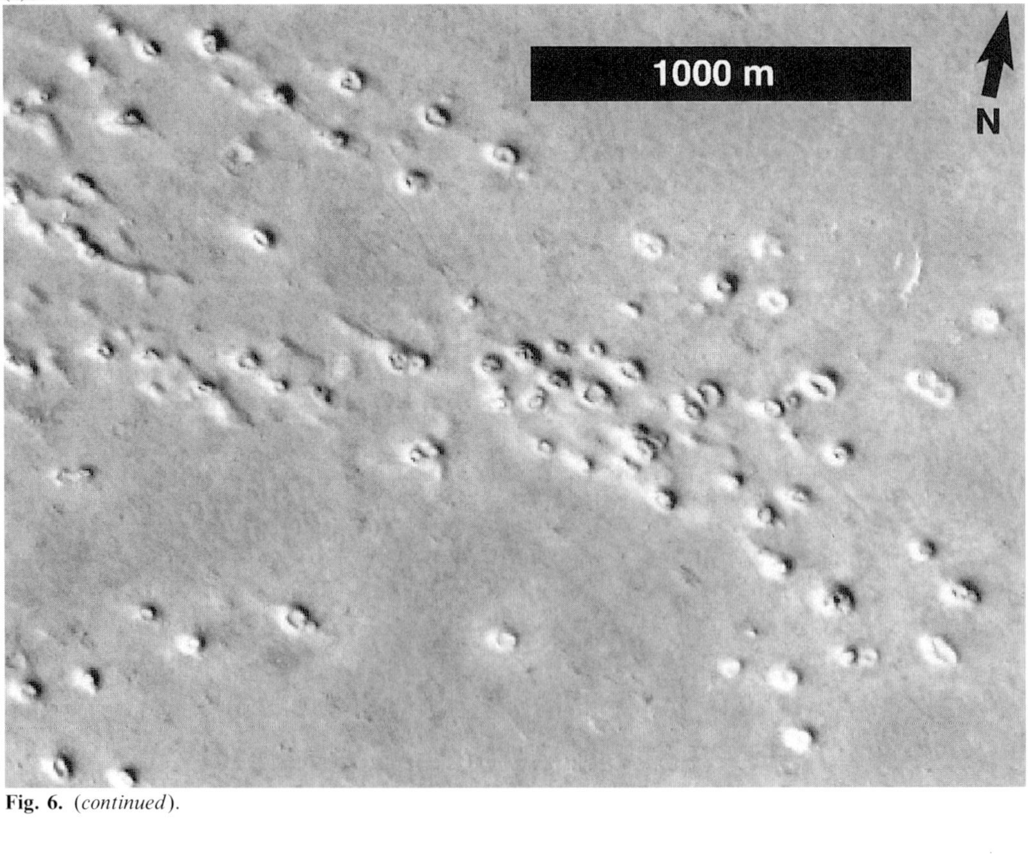

Fig. 6. (*continued*).

(e)

(f)

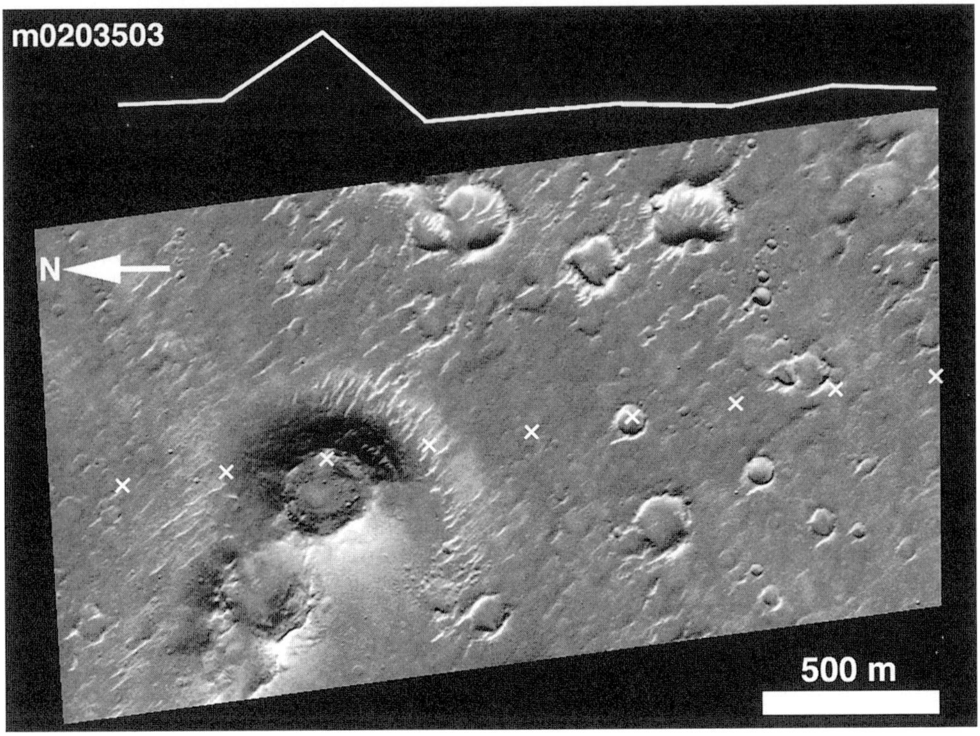

Fig. 6. (*continued*).

(g)

Fig. 6. (*continued*).

fine material extends to roughly twice the cone diameter. Elsewhere, in Isidis Planitia, cones ranging from 200 to 1000 m in basal diameter are found, but most lie in the range 300–600 m (Figs 6e, f & 7b). These cones are less pristine, and they may have undergone significant morphological modification by mass wasting processes, which renders their genesis less clear.

Precise measurements of the slopes of the cones' flanks are currently unavailable because they are too small to be adequately resolved by MOLA data. However, as these cones do not project true shadows in MOC images at incidence angles around 40, their slopes can be no steeper than 50. Rarely, a fortuitously placed MOLA track directly crosses a cone (Fig. 6f). Although typically only one or two shots land on the cone itself, these can be sufficient to provide a lower bound on the cone height. The example in Figure 6f yields a height of 25 m; others range up to *c.* 60 m. These are minimum estimates only; first, the spread of the MOLA footprint over such a small feature acts to average the topography somewhat, and second, the shots rarely land directly on the summits of the cones.

The size distributions of the Icelandic and martian cones indicate a larger modal diameter for the latter. The modal diameter for the Icelandic cones is on the order of 10–100 m whereas the martian cones have typical diameters of 100–500 m (Fig. 7). Plotting the distribution of the ratio of crater diameter to cone diameter for the Amazonis and Isidis cones (Fig. 8) indicates that their morphologies are generally similar to the broad-cratered Myvatn cones and phreatomagmatic tuff cones (Fig. 2). This, together with the larger cone diameters, suggests that energetic explosions were responsible for forming the martian constructs. This is likely to be related in large part to the low atmospheric pressure and gravity conditions, which act to promote explosive eruptions in the martian environment (Wilson & Head 1983, 1994; Fagents & Wilson 1996; Greeley *et al.* 2000), rather than being indicative of optimal lavawater interactions, as discussed further below.

Mechanisms of formation

Magma–water interaction

Although the formation of Icelandic rootless cones has never been witnessed at close quarters (Thordarson *et al.* 1998), extensive fieldwork has determined their origin to be related to explosive interaction of the lava with water trapped

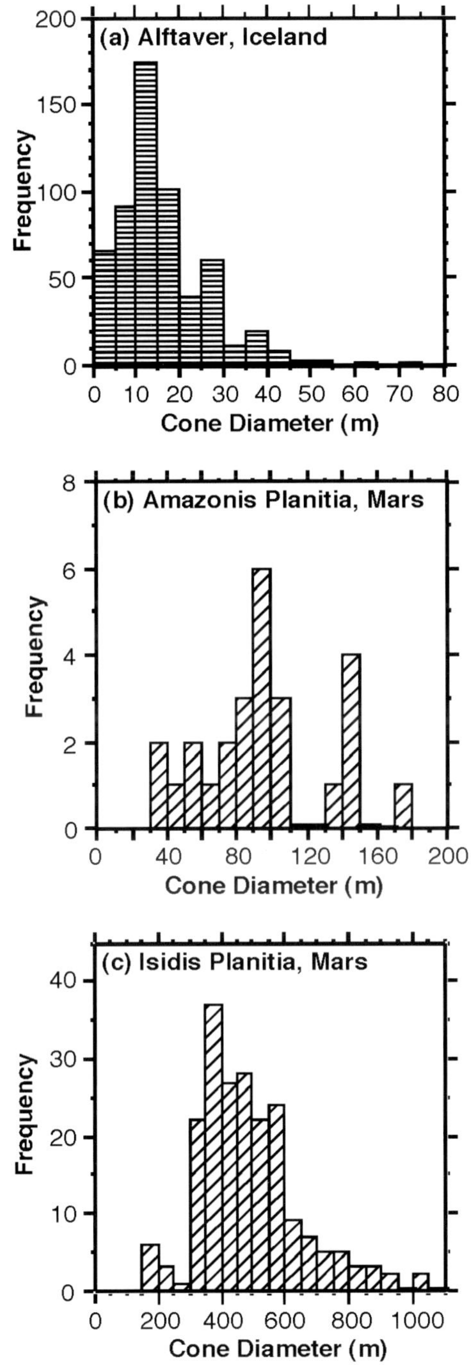

Fig. 7. Size–frequency plots of basal diameters for cones in (**a**) Álftaver, Iceland, (**b**) Amazonis Planitia, Mars, and (**c**) Isidis Planitia, Mars. There is some uncertainty associated with Álftaver cone diameters because of erosion/deposition due to historic jökullhlaups (Larsen 2001).

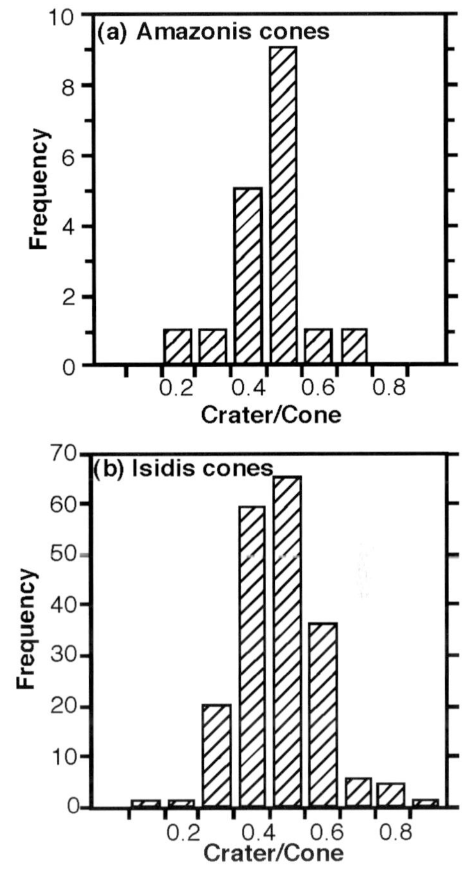

Fig. 8. Histograms of the ratio of crater diameter to cone diameter for cones in (**a**) Amazonis and (**b**) Isidis Planitiae.

beneath the flow (Thorarinsson 1953; Morrissey & Thordarson 1991; Thordarson *et al.* 1992). The best analogy for Icelandic rootless cones is perhaps found in littoral cones, which form as lava enters the ocean. Mattox & Mangan (1997) described observations of actively forming littoral cones on the pahoehoe flows of Kilauea volcano, Hawaii. They identified four mechanisms of littoral explosions, classified as either open mixing or confined mixing events. In the former case, collapse of a lava bench into the ocean exposes an incandescent scarp or active lava tube to the action of the surf, producing lithic blasts and/or tephra jets. The impact of waves is a key factor in inducing lava fragmentation and efficient lava–water mixing to generate the pressures required to eject the lava (Mattox & Mangan 1997). Confined mixing takes place when fractures allow water to enter a submerged lava tube, leading to sustained littoral fountains or intermittent bubble bursts, in which pockets of vapour displace the lava in an expanding bubble, until it is weakly fragmented into spatter and deposited around the explosion site. The fountains were interpreted to result from steady access to seawater, whereas the bubble bursts require lower, unsteady infiltration of the water (Mattox & Mangan 1997).

A considerable body of experimental work has investigated the key factors in controlling the style and intensity of explosive magma–water interactions (e.g. Wohletz 1983; Wohletz & McQueen 1984; Zimanowksi *et al.* 1991; Fröhlich *et al.* 1993; Büttner *et al.* 2000). Early studies were driven by the necessity of understanding hazards posed by industrial accidents involving spills of hot materials contacting cooler liquids (Lipsett 1966; Witte *et al.* 1970), but were found to be broadly applicable to phreatomagmatic phenomena (Colgate & Sigurgeirsson 1973; Peckover *et al.* 1973). The more vigorous events are termed explosive fuel–coolant interactions (FCI), and proceed as follows: (1) initial contact of water and melt produces an insulating vapour film; (2) instabilities at the melt–vapour interface cause collapse of the vapour film and direct contact of the melt and water, as well as initial melt fragmentation; (3) the subsequent mixing and increase in hot melt surface area promotes efficient heat transfer to the water, generating further vapour and promoting further fragmentation; (4) escalating cycles of water–melt mixing, vapour production and energy release rapidly result in explosive expansion of the mixture of vapour and finely fragmented melt (Colgate & Sigurgeirsson 1973; Sheridan & Wohletz 1981, 1983; Wohletz 1983, 1986). The initial contact geometry, the mass ratio of interacting coolant and fuel (in this case water and magma, respectively), and the degree and rate of mixing were found to be critical factors controlling the explosive intensity (Wohletz 1983; Wohletz & McQueen 1984; Zimanowksi *et al.* 1991; Fröhlich *et al.* 1993).

Experimental investigations of the influence of water–melt mass ratios showed that at low (<0.1) and high (>3.0) ratios the efficiency of transfer of thermal to mechanical (explosive) energy is very low, and explosive activity is weak or suppressed entirely. At intermediate values, the efficiency increases dramatically, leading to finely fragmented melt and highly energetic explosions (Sheridan & Wohletz 1981; Wohletz 1983, 1986; Wohletz & McQueen 1984). However, White (1996) pointed out that hydrovolcanic activity rarely involves pure water. The coolant is typically a mixture of water, rock particles and dissolved compounds. The properties of impure coolants have competing effects on explosive intensity: more vapour nucleation sites and smaller wetting angles promote vaporization and FCI initiation, but the increased coolant density and viscosity and reduced vapour production per unit volume of coolant probably tend to suppress explosivity (White 1996). White urged caution in the use of the experimentally derived energy release curves, suggesting instead that natural irregularities in vent conditions (e.g. the rate of lava supply or the degree of interpenetration of the melt and coolant) may instead dominate the explosion process.

The variety of cone morphologies and clast characteristics observed in our field studies indicate a continuum of explosive intensity related to the degree of lava–water interaction (Fig. 3), which probably depends on factors such as availability of water, rate of lava influx and explosion site geometry (Wohletz 1983, 1986; White 1996; Mattox & Mangan 1997). These factors all influence the ability of the lava and water to interact. The field relationships and geometric configuration of Icelandic rootless cones (cones overlying the host lava flow, which in turn overlies the water source) suggests variable, but generally relatively low degrees of mixing in a confined setting. For example, some cone and clast characteristics indicate significant mixing with sediment-laden coolant (finely fragmented lava, sedimentary lithics, broad-cratered large-diameter cones), at least early in the activity (Morrissey & Thordarson 1991; Thordarson *et al.* 1992); others indicate milder activity (coarse spatter, occasional lithics, steep, small diameter cones). However, since the typical cone size and pyroclast characteristics are not consistent with the very violent mixing and high pressures of large-scale FCI phreatomagmatic explosions, we infer that relatively mild explosive interactions take place. For cones consisting predominantly of spatter, the style of formation might be similar to the 'bubble burst' mechanism (Mattox & Mangan 1997), in which intermittent, transient explosion would result as the vapour pressure overcomes the confining pressure, provided the substrate was permeable and allowed water to flow back in towards the explosion site after each ejection. This relatively low-energy explosion mechanism explains the metre-scale sizes of some Icelandic cones.

Substrate heating models: predictions for Mars

On Mars, the possible presence of ground ice might provide the driving volatile for rootless cone-forming explosions (Allen 1979; Frey *et al.* 1979). However, some thermal models of the martian surface suggest that, at mid- to low-latitudes, the martian regolith should have become extensively or completely dehydrated due to the instability of water under low atmospheric pressure conditions and relatively warm temperatures (Clifford & Hillel 1983; Fanale *et al.* 1986; Clifford 1993). This obviously presents a problem for the formation of rootless cones in young volcanic plains. However, models accounting for heterogeneities in regolith properties (Paige 1992), variations in orbital parameters (Mellon & Jakosky 1995), and recondensation as vapour ascends through the regolith (Mellon *et al.* 1997), suggest that mechanisms do exist for the retention of ice at shallow levels and low latitudes. The identification of rootless cones and computation of depths of vaporization are important tests of these hypotheses on ground ice distribution.

Theoretical studies of rootless cone formation are remarkably few (Allen 1979; Mattox & Mangan 1997; Greeley & Fagents 2001). Although magma–water and magma–ice interactions are very important processes on Earth, and are likely to have taken place on Mars, attention has mainly been focused on large-scale phenomena such as maar-forming explosions, phreatoplinian activity and surtseyan eruptions (e.g. Sheridan & Wohletz 1983; Squyres *et al.* 1987). The formation of terrestrial rootless cones has therefore not been addressed in a comprehensive quantitative manner, thereby hampering the understanding of the requirements for their formation on Mars. However, some investigations into the thermal influences of lava flows and intrusions on saturated country rock are relevant and instructive.

Allen (1979) examined the implications of lava–water and lava–ice interaction on Earth and Mars, based on analytical expressions describing the temperature distribution of basaltic lava in contact with wet sediments (Jaeger 1957, 1959). A vapour layer with a maximum thickness of half the lava body was derived. Calculating the pressure in the vapour, Allen (1979) determined that the energetics of vapour expansion could adequately excavate a 5 m thick lava flow to produce a rootless cone.

Squyres *et al.* (1987) proposed a more detailed treatment of substrate heating due to a surface lava flow by solving numerically the one-dimensional heat conduction equation, accounting for surface radiative and convective cooling of basaltic lava flows on Mars. Having computed the variation of temperature with depth in the lava and substrate for a variety of lava thicknesses, the relative amounts of vapour and water were calculated. Although not specifically applied to rootless cone formation, this study found that the lava flow could only produce vapour to a depth equivalent to *c.* 0.3 times the lava flow thickness before the flow cooled too much to produce deeper heating.

One implication drawn from these studies is that, for rootless cones to form on Mars, regolith ice must be present at depths below the surface less than 0.3 to 0.5 times the flow thickness; the vaporization front cannot reach deeper layers before the thermal energy of the basaltic lava is exhausted. This implies that, for a plausible range of flow thicknesses between 1 and 50 m (Keszthelyi *et al.* 2000) ice must lie within 0.3 to 25 m of the surface. If the lavas were more silicic in composition, the depth of heated substrate would be even less, as a result of the lower eruption temperatures.

However, there are several oversimplifications in these substrate-heating models, which may significantly have influenced the validity of the results. Perhaps the most problematic aspect is that none of these treatments of substrate heating account for the fact that vapour is an extremely poor thermal conductor (e.g. Wohletz 1986), such that in a passive heating scenario the formation of a vapour layer beneath the lava flow would effectively insulate deeper water/ice from direct thermal interaction with the lava. However, field evidence suggests that, in the natural situation, some mixing takes place to ensure better thermal contact between the lava and water (Jónsson 1990; Morrissey & Thordarson 1991; Thordarson *et al.* 1992, 1998).

Further inaccuracies arise from the fact that the lava is treated as a uniform stationary (solid) cooling body with constant thermophysical properties (thermal conductivity, specific heat capacity). The release of latent heat on solidification is not adequately addressed, nor is the influence on the thermal profile of a solid surface crust over a molten flow core and cooler flow base. Properties such as specific heat and thermal conductivity vary greatly with vesicularity, olivine concentration, temperature, and from liquid to solid state (Murase & McBirney 1973; Robertson & Peck 1974; Roy *et al.* 1981), and can significantly influence heat transfer calculations (Fagents & Greeley 2001).

Additionally, the concept that the cones are formed by a single explosion (Allen 1979) is not supported by field and photogeologic evidence. For example, the bedding characteristics of Icelandic cones are not consistent with a single, transient explosion, instead reflecting multiple explosive pulses (Morrissey & Thordarson 1991; Thordarson *et al.* 1992). Furthermore, the volumes of both terrestrial and martian cones exceed the volume of lava that would be excavated by a single explosion, even if the lower bulk density of the cone is taken into account. The preferred concept of rootless cone formation involves repeated explosions, facilitated by continued flow of molten lava beneath a solid carapace. This would ensure delivery of a sufficient volume of lava to the explosion site to create the observed cones. The morphological characteristics of both Icelandic and martian rootless cones indicate that the lava must have developed a stationary, competent, solid crust, since cones are not rafted or deformed by subsequent flow motion. Repeated explosions would be facilitated by continued flow in a lava tube (Morrissey & Thordarson 1991; Thordarson *et al.* 1992; Jurado-Chichay *et al.* 1996; Mattox & Mangan 1997; Greeley & Fagents 2001), in the mobile core of an insulated sheet flow, or in the stagnating-surging emplacement interpreted to produce the platy-ridged lava surfaces on which the Amazonis Planitia cones reside (Keszthelyi *et al.* 2000).

As well as providing a continued lava supply, the additional advection of heat due to sustained flow of molten material would strongly influence both the heat transfer rate and total heating of the substrate (Fagents & Greeley 2001). A continually replenished source of heat will yield very different cooling profiles (and much greater heated substrate depths) than those calculated by Squyres *et al.* (1987).

Refinement of the concepts of rootless cone formation has led to the following model (Fig. 9):

(1) Initial emplacement of lava heats the substrate, vaporizing water on the surface and/or in the pores of substrate material (Iceland), or resulting from melting of interstitial ice (Mars); the lava surface and base cool and solidify while interior remains molten and mobile; vapour pressure builds due to confinement by country rock and overlying lava.

(2) Once a threshold pressure is exceeded (defined by a combination of the weight and strength of the overlying lava, together with the atmospheric pressure) the gas expands rapidly into the atmosphere, excavating and entraining the overlying lava and some proportion of substrate material; ejected lava fragments follow trajectories influenced by gas motions, and are distributed around the explosion site.

(3) Inflow of lava from the mobile flow core and accumulation of other debris fills the cavity and allows continued thermal interaction with water/ice at greater depths;

(4) Repeated vaporization, pressurization, excavation, and inflow produce intermittent ejections of material to build the cone until the supply of either lava or water diminishes.

Explosion dynamics

In light of this reassessment of rootless cone formation, and as an alternative to modelling substrate heating, another approach is to consider the dynamics of the phreatic explosions (Greeley & Fagents 2001). A treatment of the

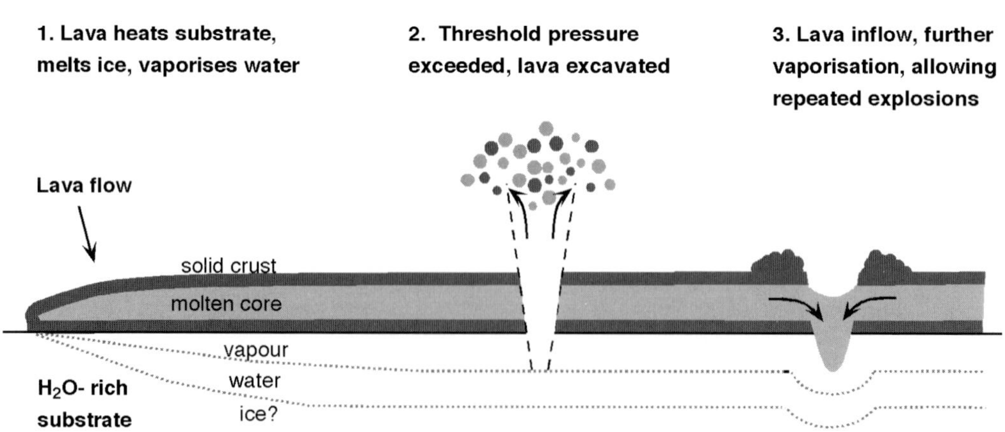

Fig. 9. Diagram depicting stages of rootless cone formation.

energetics of vapour decompression, lava ejection and displacement of the atmosphere surrounding the explosion site, yields the following equation of motion:

$$\left[P_{g0}\left(\frac{r_g}{r}\right)-P_a\right]r^2 = \frac{1}{2}\left(\frac{\mathrm{d}^2r}{\mathrm{d}t^2}\right)\{\rho_l[(r_g+r_l)^3-r_g^3] + \rho_a[(r+r_l)^3-(r_g+r_l)^3]\} \quad (1)$$

which describes the force imparted due to vapour expansion (left-hand side) in terms of the resulting acceleration of the masses of ejected lava, vapour and displaced atmosphere (Greeley & Fagents 2001). In this expression P_{g0} is the initial (pre-explosion) pressure in the gas phase, P_a is the atmospheric pressure, r is the instantaneous distance from the explosion source of the accelerating plug of ejecta (assumed to be accelerated initially en masse), r_g is the size of the region in which the vapour collects, r_l is the lava thickness, ρ_l and ρ_a are the density of the lava and atmospheric gas, and γ is the ratio of the specific heats at constant pressure and constant volume of the water vapour. Table 1 gives typical values for these parameters. The lava thickness, r_l, might initially be quite large (>25 m; e.g. Thordarson & Self 1993; Keszthelyi *et al.* 2000), but subsequent explosions might involve lesser thicknesses of lava from the flow core. The size of the gas region, r_g, is difficult to constrain, and depends on the details of the water–lava interaction (vigorous mixing *versus* trapped vapour pockets), and whether the vapour is distributed throughout substrate pore spaces. This study adopts a wide range of values (Table 1) to accommodate all possibilities. The initial gas pressure must overcome a threshold pressure determined by the mass of overlying lava, the yield strength of any molten lava, the tensile strength of the viscoelastic and/or brittle carapace (Hon *et al.* 1994; Rossi & Gudmundsson 1996), and the ambient atmospheric pressure. Thus a lower limit on P_{g0} might range from 1.2×10^5 to 2×10^6 Pa for Earth, and 2.6×10^4 to 1.3×10^6 Pa for Mars, based on plausible ranges of these parameters (Table 1). The simplicity of this model, specifically the fact that it does not 'care' how the vapour pressure is generated (whether by a relatively mild bubble burst mechanism or a more vigorous, overpressured FCI), allows a range of initial pressures applicable to both end member scenarios to be supplied as input parameters.

Numerical integration of equation (1) yields the ejection velocity of the fragmented lava slug, and subsequent computation of the trajectories of individual lava fragments subject to the aerodynamic influences of the atmospheric and volcanic gases (Wilson 1972; Fagents & Wilson 1993, 1996), allows the ejection distances to be derived. Synthesis of the ejection distances of a range of typical pyroclast sizes (1–30 cm) yields the diameter of the resulting cone. Thus the cone diameter can be related to the pre-explosion vapour pressure and the amount of water vapour involved. Figure 10 shows clast ejection distance plotted as a function of water vapour mass and pressure for 10 cm clasts ejected at an angle of 70° from horizontal. An

Table 1. *Ranges of values taken by model parameters*

Parameter	Units	Description	Value
P_{g0}	Pa	Initial (threshold) gas pressure	$=\rho_l g r_l+\sigma_T+\sigma_y+P_a$ 1.2×10^5–2×10^6 (Earth); 2.6×10^4–1.3×10^6 (Mars)
P_a	Pa	Atmospheric pressure	10^5 (Earth); 600 (Mars)
r	m	Distance from explosion source	Calculated within model
r_g	m	Size of gas region	0.3–30
r_l	m	Lava thickness	<150
ρ_a	$kg\,m^{-3}$	Atmospheric density	0.75 (Earth); 0.014 (Mars)
ρ_g	$kg\,m^{-3}$	Vapour density	Variable, calculated within model
ρ_l	$kg\,m^{-3}$	Lava/clast density*	2000–3000
σ_T	Pa	Lava tensile strength†	0–10^6; includes components for solid and viscous crust
σ_y	Pa	Lava yield strength‡	0–10^4

* Thorarinsson 1951, 1953.
† Touloukian *et al.* 1981; Hon *et al.* 1994; Rossi & Gudmundsson 1996.
‡ Shaw *et al.* 1968; Shaw 1969; Gauthier 1973; Pinkerton & Sparks 1978; McBirney & Noyes 1979; Murase 1981; Pinkerton & Stevenson 1992.

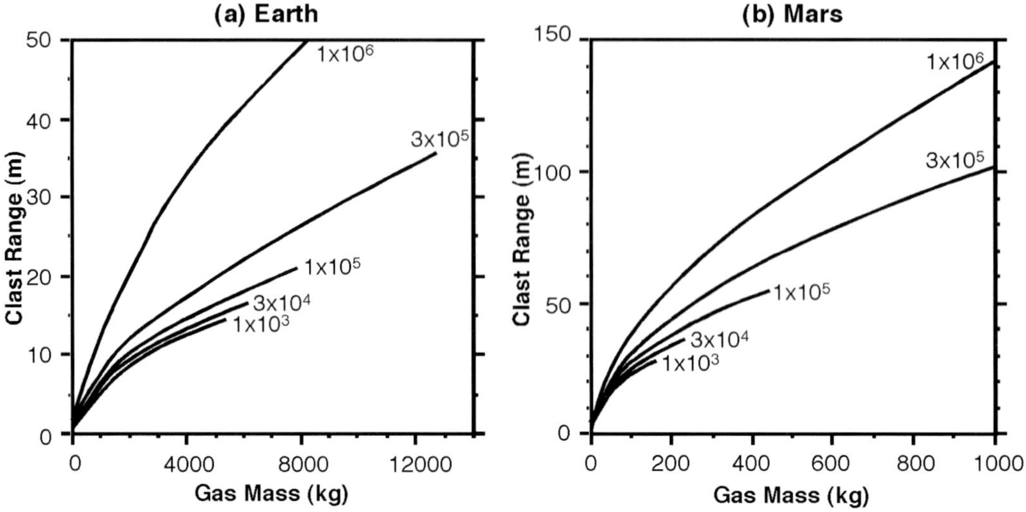

Fig. 10. Plots of ejection range as a function of gas mass for 10-cm clasts for (**a**) terrestrial and (**b**) martian rootless cone explosions. Curves are labelled with the pre-explosion gas pressure (in Pa). Lava density is 2500 kg m^{-3}, consistent with measurements of degassed lava and rootless cone material (Thorarinsson 1951, 1953). The ejection angle is taken as 70° from horizontal, following the finding by Chouet *et al.* (1974) that >90% of pyroclasts in transient strombolian-style explosions are ejected within 20° of vertical.

angle of 70° is chosen based on photoballistic evidence that >90% of ejecta in transient strombolian style explosions are emitted within 20° of vertical (Chouet *et al.* 1974). The pre-explosion lava thickness is taken to be 5 m. It is anticipated that only the initial explosion would eject a large thickness of lava (the initial flow depth); subsequent explosions would eject smaller volumes of material that had flowed into the resulting cavity (Fig. 9). Material ejected at angles greater than 70° falls closer to the vent and contributes to the upper flanks of the cone, so that the clast ejection ranges shown in Figure 10 represent the outer limit of the cone; i.e. doubling the clast ejection ranges yields the cone diameter. Table 2 summarizes the results of applying the model to different-sized cones on Earth and Mars. The ranges in water (vapour) mass shown in Table 2 are a function of the range of initial pressures used as input to the model.

It is clear from Figure 10 and Table 2 that substantially less water vapour is involved in cone-forming explosions on Mars relative to Iceland. This is attributed to the different environmental conditions on Mars. The low atmospheric pressure enhances the energy release due to gas expansion (the driving pressure term on the left-hand side of equation (1) is relatively large), producing greater accelerations and greater ejection velocities. Once the fragmental ejecta are launched, the low gravity enables them to remain aloft for longer, and the low density atmospheric gas imposes only modest aerodynamic drag decelerations. These two factors conspire to produce much greater clast ejection distances. Hence, the ultimate result is significantly wider edifices on Mars compared to Iceland.

Key conclusions of this approach include (Table 2):

Table 2. *Summary of model results*

Cone diameter (m)	Lava thickness (m)	Mass of H_2O required per explosion (kg)
Iceland		
20 m	1	14–300
	5	700–2400
	10	2200–3000
100 m	1	500
	5	3000
	10	–
Mars		
100 m	1	2–18
	5	170–350
	10	700–1150
500 m	1	40–165
	5	2 900
	10	15 000

Clast size and density are taken as 10 cm and 2500 kg m^{-3}, respectively.

(a) All pre-eruption conditions being equal, the current martian environment acts to produce cones with basal diameters 5 to 10 times greater than on Earth (Greeley & Fagents 2001);
(b) The typical 100 m-wide Amazonis cone requires only 14–38% of the water vapour per explosion that is required to form the typical 20 m-wide Álftaver cone, and only 0.4–11% of that required to form the typical 100 m-wide Myvatn cone.
(c) Martian cones as large as 500 m in diameter (such as those in Isidis) still require less water vapour (8–96%) than 100 m-wide Myvatn cones.
(d) Values of the mass ratio of water to lava derived from the model results are typically <0.01 for Icelandic cones and an order of magnitude lower for the (larger) martian cones, consistent with weakly explosive lava–water interactions (Sheridan & Wohletz 1981; Wohletz 1983, 1986; Wohletz & McQueen 1984).

These results are consistent with the lower availability of water on Mars, which is likely to have been locked up as interstitial ice in the regolith.

Discussion and conclusions

It is clear that the origins of martian volcanic cones cannot be determined solely on the basis of direct morphological comparison with terrestrial analogues. First, the significant differences in eruptive environment between the Earth and Mars (especially atmospheric pressure and gravity) critically influence eruptive style and the resulting volcanic feature (Wilson & Head 1983, 1994; Fagents & Wilson 1996; Greeley *et al.* 2000), such that inferences regarding water amounts based on morphology alone are invalid.

Furthermore, significant variations in the morphologies of terrestrial rootless cones are observed, from hornito-like accumulations of spatter to broad-cratered scoria cones. These differences are related to the ability of the lava to physically interact with the water, which in turn is probably influenced by such factors as the lava feeder system (Morrissey & Thordarson 1991; Thordarson *et al.* 1992), rate and constancy of lava supply (Mattox & Mangan 1997), explosion site geometry, and local hydrology, in ways that are poorly understood. The model predictions of the water–lava mass ratio support the notion of variable degrees of lava–water interaction; the explosions forming the larger Icelandic cones require a greater amount of vapour. However, all calculated ratios lie in the low efficiency domain of the energy release curves (Sheridan & Wohletz 1981; Wohletz 1983, 1986; Wohletz & McQueen 1984), indicating that relatively weak explosive interactions form Icelandic rootless cones (compared with highly explosive phreatoplinian interactions, for example), even despite the uncertainties associated with impure coolants (White 1996).

However, very low water–lava ratios are found for the martian cones, despite the fact that they tend to resemble broad-cratered Icelandic examples. The planetary environmental conditions, and not the amount of water involved, dominate the eruptive intensity and the resultant landform. It is possible that erroneous interpretations could be made based on photogeologic studies of martian volcanic cones. Therefore, a combination of careful data analysis, field analogue work and numerical modelling holds the best promise for determining cone origins and understanding the role and distribution of martian volatiles.

To that end, our modelling of rootless cone explosion dynamics builds on earlier treatments of substrate heating to investigate the variables controlling cone formation under martian conditions. The results of this work suggest that martian cones on the order of 100 to 500 m in diameter are very plausibly formed with modest amounts of water vapour derived from regolith ice. For example, if the vapour driving the explosions forming a 100 m cone is derived from a volume of substrate material equivalent to a cylinder of radius and depth 1 m (a rather conservative estimate of vent width and depth of ice-rich regolith), then the 2 to 1000 kg of ice required would occupy <0.1 to 30% of this volume. This is consistent with ice contained within interstices of a particulate regolith. If the explosions draw vapour from a larger volume of substrate, then even smaller volumes of ice-filled voids are required. Larger cones, of course, require correspondingly larger amounts of water vapour in their genesis, suggesting that ice stores are more extensive. However, this should also be readily accommodated without the need for ice to be present at depths of tens to hundreds of metres. We conclude that steam explosions and cone formation are likely consequences of lava flowing over substrates containing small amounts of ice, and that the broad-cratered martian cones are a result of energetic explosions promoted by the low gravity, low atmospheric pressure martian environment.

The timing of rootless cone field emplacement is of critical importance to understanding Mars' climatic history. Crater counting analyses performed on high-resolution MOC images yield age estimates as low as 10 to 100 Ma for surfaces in Elysium and Amazonis Planitiae (Hartmann 1999; Hartmann *et al.* 1999; Hartmann & Berman 2000). In particular, portions of lava surfaces in Marte Vallis and the western Amazonis Planitia cone field may be significantly younger than 10 Ma, even given a factor of 2 to 4 uncertainty in Hartmann's isochrons on the plots of crater spatial density versus crater diameter. The very young lava ages suggest that volcanism was active much more recently than previously suggested (Plescia & Saunders 1979; Neukum & Hiller 1981; McSween 1994), and may in fact be ongoing.

In terms of rootless cone formation, the implication that ground ice was present within the past 10 Ma contradicts past assertions that the martian regolith should essentially be devoid of water ice under recent climatic conditions (Clifford & Hillel 1983; Fanale *et al.* 1986; Clifford 1993). However, there are currently multiple viable hypotheses for recharge of ground ice. For instance, Mellon & Jakosky (1995) contended that variations in Mars' orbital parameters promote exchange of water between the atmosphere and regolith. The key factor is the oscillation in orbital obliquity (the tilt of Mars' spin axis relative to the orbital plane), which varies from 15° to 35° over periods on the order of 10^5 years (Mellon & Jakosky 1995). This influences the insolation at the surface and the atmospheric water abundance, allowing water to diffuse into and freeze in the first metre or two of substrate at low latitudes during periods of high obliquity. Clearly therefore, there have been many opportunities for regolith ice to contribute to cone formation at relatively low latitudes during the past 10 Ma.

Another possibility for replenishing of ground ice is transient fluvial activity. Geomorphological and mapping studies of the Elysium–Cerberus–Amazonis region have revealed many fluvial features, produced by water that may have been derived from subsurface aquifers or released due to large-scale melting of ground ice by extensive volcanism (Burr *et al.* 2000). During the late Amazonian Period, a large volume of water would have moved east from the Elysium Basin through Marte Vallis, debouching into the Amazonis Basin and providing abundant water to infiltrate the near-surface regolith. This is consistent with the distribution of the Cerberus–Amazonis cones, which lie in topographically low, flat areas, and are commonly found in proximity to fluvial features. Further studies will elucidate if fluvial activity is a viable mechanism for emplacing water and supplying ground ice stores in other areas in which rootless cones are found.

There is a clear need for further work on martian rootless cones. Figure 4 represents only a partial survey of the MOC images, which number in the tens of thousands. Acquisition of high-resolution images will continue into 2002, so there is plenty of scope for further identification of cone fields and detailed studies of their morphology and stratigraphy.

Further detailed fieldwork will also help to deconvolve the effects of lava supply rates, preferred lava pathways or tubes, explosion site geometry and substrate hydrology on the degree of lava–water interaction and the intensity of the ensuing explosion, and hence guide the development of realistic numerical treatments of rootless cone-forming activity.

On the modelling front, the explosion model (Greeley & Fagents 2001) provides fairly loose constraints on water requirements at this time, and does not adequately address the depth of ice-rich regolith from which the vapour is drawn to produce the observed features. The assumption is simply that there is sufficient water available. What is lacking, therefore, is a realistic and comprehensive model of both substrate heating/volatile vaporization and the explosion dynamics. Correcting the deficiencies of previous substrate-heating models in the light of our refined concept of rootless cone formation will be a significant contribution, and allow for much tighter constraints on water amounts and the depth of the regolith ice. In combination with a global search of the MOC data and assessment of the relative ages of the volcanic regions, this will permit determination of the spatial and temporal variations in ice depths and amounts across the locations in which rootless cones occur. Only then can we hope to deconvolve the influences of planetary environmental conditions, lava thickness and regolith ice content on cone size, and fully understand the implications for the history of Mars' volatile inventory.

The authors are grateful to T. Thordarson and T. Gregg for their thorough reviews, which resulted in significant improvements on the initial draft of this manuscript.

References

ALLEN, C. C. 1979. *Volcano–ice interactions on the Earth and Mars*. PhD thesis, University of Arizona, Tucson.

ARVIDSON, R. E., CORADINI, M., CARUSI, A., CORADINI, A., FULCHIGNONI, M., FEDERICO, C., FUNICIELLO, R. & SALOMONE, M. 1976. Latitude variations of wind erosion of crater ejecta deposits on Mars. *Icarus*, **27**, 503–516.

BURR, D. M., MCEWEN, A. S. & LANAGAN, P. D. 2000. Recent fluvial activity in and near Marte Valles, Mars. *Lunar and Planetary Science*, **XXXI**, No. 1951.

BÜTTNER, R., ZIMANOWSKI, B. & RÖDER, H. 2000. Short-time electrical effects during volcanic eruption: experiments and field measurements. *Journal of Geophysical Research*, **105**, 2819–2827.

CABROL, N. A., GRIN, E. A. & POLLARD, W. H. 2000. Possible frost mounds in an ancient martian lake bed. *Icarus*, **145**, 91–107.

CARR, M. H., CRUMPLER, L. S., CUTTS, J. A., GREELEY, R., GUEST, J. E. & MASURSKY, H. 1977. Martian impact craters and emplacement of ejecta by surface flow. *Journal of Geophysical Research*, **82**, 4055–4065.

CHOUET, B., HAMISEVICZ, N. & MCGETCHIN, T. R. 1974. Photoballistics of volcanic jet activity at Stromboli, Italy. *Journal of Geophysical Research*, **79**, 4961–4976.

CLIFFORD, S. M. 1993. A model for the hydrologic and climatic behavior of water on Mars. *Journal of Geophysical Research*, **98**, 10 973–11 016.

CLIFFORD, S. M. & HILLEL, D. 1983. The stability of ground ice in the equatorial region of Mars. *Journal of Geophysical Research*, **88**, 2456–2474.

COLGATE, S. A. & SIGURGEIRSSON, T. 1973. Dynamic mixing of water and lava. *Nature*, **244**, 552–555.

CROWN, D. A. & GREELEY, R. 1993. Volcanic geology of Hadriaca Patera and the Eastern Hellas Region of Mars. *Journal of Geophysical Research*, **98**, 3431–3451.

FAGENTS, S. A. & GREELEY, R. 2001. Factors influencing lava-substrate heat transfer and implications for thermomechanical erosion. *Bulletin of Volcanology*, **62**, 519–532.

FAGENTS, S. A. & WILSON, L. 1993. Explosive volcanic eruptions – VII. The ranges of pyroclasts ejected in transient volcanic explosions. *Geophysical Journal International*, **113**, 359–370.

FAGENTS, S. A. & WILSON, L. 1996. Numerical modeling of ejecta dispersal around the sites of volcanic explosions on Mars. *Icarus*, **123**, 284–295.

FANALE, F. P., SALVAIL, J. R., ZENT, A. P. & POSTAWKO, S. E. 1986. Global distribution and migration of subsurface ice on Mars. *Icarus*, **67**, 1–18.

FREY, H. & JAROSEWICH, M. 1982. Subkilometer martian volcanoes: Properties and possible terrestrial analogs. *Journal of Geophysical Research*, **87**, 9867–9879.

FREY, H., LOWRY, B. L. & CHASE, S. A. 1979. Pseudocraters on Mars. *Journal of Geophysical Research*, **84**, 8075–8068.

FRÖHLICH, G., ZIMANOWSKI, B. & LORENZ, V. 1993. Explosive thermal interactions between molten lava and water. *Experimental Thermal and Fluid Science*, **7**, 319–332.

GAULT, D. E. & GREELEY, R. 1978. Exploratory experiments of impact craters formed in viscous-liquid targets: Analogs for martian rampart craters? *Icarus*, **34**, 486–495.

GAUTHIER, F. 1973. Field and laboratory studies of the rheology of Mount Etna lava. *Philosophical Transactions of the Royal Society of London*, **A274**, 83–98.

GREELEY, R. & CROWN, D. A. 1990. Volcanic geology of Tyrrhena Patera, Mars. *Journal of Geophysical Research*, **95**, 7133–7149.

GREELEY, R. & FAGENTS, S. A. 2001. Icelandic pseudocraters as analogs to some volcanic cones on Mars. *Journal of Geophysical Research*, **106**, 20 527–20 546.

GREELEY, R. & SPUDIS, P. D. 1981. Volcanism on Mars. *Reviews of Geophysics and Space Physics*, **19**, 13–41.

GREELEY, R., BRIDGES, N. T., CROWN, D. A., CRUMPLER, L., FAGENTS, S. A., MOUGINIS-MARK, P. J. & ZIMBELMAN, J. R. 2000. Volcanism on the red planet: Mars. *In*: ZIMBELMAN, J. R. & GREGG, T. K. P. (eds) *Environmental Effects on Volcanic Eruptions: From Deep Oceans to Deep Space*. Kluwer Academic/Plenum Press, New York, 75–112.

HARTMANN, W. K. 1999. Martian cratering VI: Crater count isochrons and evidence for recent volcanism from Mars Global Surveyor. *Meteoritics and Planetary Science*, **34**, 167–178.

HARTMANN, W. K. & BERMAN, D. C. 2000. Elysium Planitia lava flows; crater count chronology and geological implications. *Journal of Geophysical Research*, **105**, 15 011–15 025.

HARTMANN, W. K., MALIN, M. C., MCEWEN, A. S., CARR, M. H., SODERBLOM, L. A., THOMAS, P. C., DANIELSON, E., JAMES, P. B. & VEVERKA, J. 1999. Evidence for recent volcanism on Mars from crater counts. *Nature*, **397**, 586–589.

HODGES, C. A. & MOORE, H. J. 1992. *Atlas of Volcanic Landforms on Mars*. United States Geological Survey Professional Paper **1534**.

HON, K., KAUAHIKAUA, J., DENLINGER, R. & MACKAY, K. 1994. Emplacement and inflation of pahoehoe sheet flows: Observations and measurements of active lava flows on Kilauea Volcano, Hawaii. *Geological Society of America, Bulletin*, **106**, 351–370.

JAEGER, J. C. 1957. The temperature in the neighborhood of a cooling intrusive sheet. *American Journal of Science*, **255**, 306–318.

JAEGER, J. C. 1959. Temperatures outside a cooling intrusive sheet. *American Journal of Science*, **257**, 44–54.

JÓHANNESSON, H., JAKOBSSON, S. P. & SAEMUNDSSON, K. 1990. *Geological Map of Iceland, Sheet 6 South Iceland* (Third edition). Icelandic Museum of Natural History and Icelandic Geodetic Survey, Reykjavík.

JÓNSSON, J. 1990. An interesting occurrence in a pseudocrater. *Náttúrufraedingurinn*, **60**, 69–73.

JURADO-CHICHAY, Z., ROWLAND, S. K. & WALKER, G. P. L. 1996. The formation of circular littoral cones from tube-fed pahoehoe; Mauna Loa, Hawaii. *Bulletin of Volcanology*, **57**, 471–482.

KESZTHELYI, L. P., MCEWEN, A. S. & THORDARSON, T. 2000. Terrestrial analogs and thermal models for martian flood lavas. *Journal of Geophysical Research*, **105**, 15 027–15 049.

KUZMIN, R. O. 1988. *The Cryolithosphere of Mars* (translated from Russian). NASA TT-20264.

LANAGAN, P. D., MCEWEN, A. S., KESZTHELYI, L. P. & THORDARSON, T. 2000. Rootless cones on Mars indicating the presence of shallow equatorial ground ice in recent times. *Geophysical Research Letters*, **28**, 2365–2368.

LARSEN, G. 2001. Holocene eruptions within the Katla volcanic system, south Iceland: Characteristics and environmental impact. *Jökull*, **49**, 1–28.

LIPSETT, S. G. 1966. Explosions from molten materials and water. *Fire Technology*, **2**, 118–126.

LUCCHITTA, B. K. 1981. Mars and Earth: Comparison of cold climate features. *Icarus*, **45**, 264–303.

MALIN, M. C., DANIELSON, G. E., INGERSOLL, A. P., MASURSKY, H., VEVERKA, J., RAVINE, M. A. & SOULANILLE, T. A. 1992. The Mars Observer Camera. *Journal of Geophysical Research*, **97**, 7699–7717.

MATTOX, T. N. & MANGAN, M. T. 1997. Littoral hydrovolcanic explosions: A case study of lava–seawater interaction at Kilauea Volcano. *Journal of Volcanology and Geothermal Research*, **75**, 1–17.

MCBIRNEY, A. R. & NOYES, R. M. 1979. Crystallisation and layering of the Skaergaard Intrusion. *Journal of Petrology*, **20**, 487–554.

MCCAULEY, J. F. 1973. Mariner 9 evidence for wind erosion in the equatorial and mid-latitude regions of Mars. *Journal of Geophysical Research*, **78**, 4123–4137.

MCSWEEN, H. Y. 1994. What we have learned about Mars from SNC meteorites. *Meteoritics*, **29**, 757–779.

MELLON, M. T. & JAKOSKY, B. M. 1995. The distribution and behavior of Martian ground ice during past and present epochs. *Journal of Geophysical Research*, **100**, 11 781–11 799.

MELLON, M. T., JAKOSKY, B. M. & POSTAWKO, S. E. 1997. The persistence of equatorial ground ice on Mars. *Journal of Geophysical Research*, **102**, 19 357–19 369.

MORRISSEY, M. M. & THORDARSON, T. 1991. Origins and occurrences of pseudocrater fields in S. Iceland. *Eos, Transactions of the American Geophysical Union*, **72**, 556.

MOUGINIS-MARK, P. J. 1985. Volcano/ground ice interactions in Elysium Planitia, Mars. *Icarus*, **64**, 265–284.

MOUGINIS-MARK, P. J. 1987. Water or ice in the martian regolith?: Clues from rampart craters seen at very high resolution. *Icarus*, **71**, 268–286.

MOUGINIS-MARK, P. J., WILSON, L. & HEAD, J. W. 1982. Explosive volcanism on Hecates Tholus, Mars: Investigation of eruption conditions. *Journal of Geophysical Research*, **87**, 9890–9904.

MOUGINIS-MARK, P. J., WILSON, L. & ZUBER, M. T. 1992. The physical volcanology of Mars. *In:* KIEFFER, H. H., JAKOSKY, B. M., SNYDER, C. W. & MATTHEWS, M. S. (eds) *Mars*. University of Arizona Press, Tucson, 424–452.

MURASE, R. 1981. Thermophysical properties of some magmatic silicate liquids. *Volcanological Society of Japan, Bulletin*, **26**, 161–185.

MURASE, R. & MCBIRNEY, A. T. 1973. Properties of some common igneous rocks and melts at high temperatures. *Geological Society of America, Bulletin*, **84**, 3562–3592.

NEUKUM, G. & HILLER, K. 1981. Martian ages. *Journal of Geophysical Research*, **86**, 3097–3121.

PAIGE, D. A. 1992. The thermal stability of near-surface ground ice on Mars. *Nature*, **356**, 43–45.

PECKOVER, R. S., BUCHANAN, D. J. & ASHBY, D. E. T. F. 1973. Fuel–coolant interactions in submarine vulcanism. *Nature*, **245**, 307–308.

PIKE, R. J. & CLOW, G. D. 1981. Revised classification of terrestrial volcanoes and catalog of topographic dimensions, with new results on edifice volume. *United States Geological Survey, Open File Report*, **81–1039**.

PINKERTON, H. & SPARKS, R. S. J. 1978. Field measurements of the rheology of lava. *Nature*, **276**, 383–385.

PINKERTON, H. & STEVENSON, R. J. 1992. Methods of determining the rheological properties of magmas at sub-liquidus temperatures. *Journal of Volcanology and Geothermal Research*, **53**, 47–66.

PLESCIA, J. B. & SAUNDERS, R. S. 1979. The chronology of the martian volcanoes. *In*: *Proceedings of the Tenth Lunar and Planetary Science Conference*, 2841–2859.

REIMERS, C. E. & KOMAR, P. D. 1979. Evidence for volcanic density currents on certain martian volcanoes. *Icarus*, **39**, 88–110.

ROBERTSON, J. B. & PECK, D. L. 1974. Thermal conductivity of vesicular basalt from Hawaii. *Journal of Geophysical Research*, **79**, 4875–4888.

ROBINSON, M. S., MOUGINIS-MARK, P. J., ZIMBELMAN, J. R., WU, S. S. C., ABLIN, K. K. & HOWINGTON-KRAUS, A. E. 1993. Chronology, eruption duration, and atmospheric contribution of the Martian volcano Apollinaris Patera. *Icarus*, **104**, 301–323.

ROSSBACHER, L. A. & JUDSON, S. 1981. Ground ice on Mars; inventory, distribution, and resulting landforms. *Icarus*, **45**, 39–59.

ROSSI, M. & GUDMUNDSSON, A. 1996. The morphology and formation of flow-lobe tumuli on Icelandic shield volcanoes. *Journal of Volcanology and Geothermal Research*, **72**, 291–308.

ROY, R. F., BECK, A. E. & TOULOUKIAN, Y. S. 1981. Thermophysical properties of rocks. *In:* Touloukian, Y. S., JUDD, W. R. & ROY, R. F. (eds) *Physical Properties of Rocks and Minerals*. McGraw-Hill, New York.

SAGAN, C. & LEDERBERG, J. 1976. The prospects for life on Mars; a pre-Viking assessment. *Icarus*, **28**, 291–300.

SCOTT, D. H. 1982. Volcanoes and volcanic provinces: Martian western hemisphere. *Journal of Geophysical Research*, **87**, 9839–9851.

SHAW, H. R. 1969. Rheology of basalt in the melting range. *Journal of Petrology*, **10**, 510–535.

SHAW, H. R., WRIGHT, T. L., PECK, D. L. & OKAMURA, R. 1968. The viscosity of basaltic magma: An analysis of field measurements in Makaopuhi lava lake, Hawaii. *American Journal of Science*, **266**, 255–264.

SHERIDAN, M. F. & WOHLETZ, K. H. 1981. Hydrovolcanic explosions: The systematics of water-pyroclast equilibrium. *Science*, **212**, 1387–1389.

SHERIDAN, M. F. & WOHLETZ, K. H. 1983. Hydrovolcanism: Basic considerations and review. *Journal of Volcanology and Geothermal Research*, **17**, 1–29.

SQUYRES, S. W., WILHELMS, D. E. & MOOSMAN, A. C. 1987. Large-scale volcano–ground ice interactions on Mars. *Icarus*, **70**, 385–408.

THEILIG, E. & GREELEY, R. 1979. Plains and channels in the Lunae Planum–Chryse Planitia region of Mars. *Journal of Geophysical Research*, **84**, 7994–8010.

THORARINSSON, S. 1951. Laxargljufur and Laxarhraun. A tephrochronological study. *Geografiska Annaler*, **H1–2**, 1–89.

THORARINSSON, S. 1953. The crater groups in Iceland. *Bulletin of Volcanology*, **14**, 3–44.

THORDARSON, T. & Self, S. 1991. Lava–seawater interactions at the Kupaianaha flow front, Kilauea volcano, Hawaii. *Eos, Transactions of the American Geophysical Union*, **72**, 566.

THORDARSON, T. & SELF, S. 1993. The Skaftár Fires (Laki) and Grímsvötn eruptions in 1783–1785. *Bulletin of Volcanology*, **55**, 233–263.

THORDARSON, T., MILLER, D. J. & LARSEN, G. 1998. New data on the Leidolfsfell cone group on south Iceland. *Jökull*, **46**, 3–15.

THORDARSON, T., MORRISSEY, M. M., LARSEN, G. & CYRUSSON, H. 1992. Origins of rootless cone complexes in S. Iceland. *In*: GEIRSDÓTTIR, A., NORDDAHL, H. & HELGADÓTTIR, G. (eds) *The 20th Nordic Geological Winter Meeting*. Icelandic Geological Society, Reykjavík.

TOULOUKIAN, Y. S., JUDD, W. R. & ROY, R. F. 1981. *Physical Properties of Rocks and Minerals*. McGraw-Hill, New York.

VON KOMOROWICZ, M. 1912. *Vulkanologische Studien auf Einigen Inseln des Atlantichen Oceans*. E. Schweizerbart'sche Verlagsbuchhandlung Nagele und Dr. Sprosser.

WHITE, J. D. L. 1996. Impure coolants and interaction dynamics of phreatomagmatic eruptions. *Journal of Volcanology and Geothermal Research*, **74**, 155–170.

WILSON, L. 1972. Explosive volcanic eruptions – II. The atmospheric trajectories of pyroclasts. *Geophysical Journal of the Royal Astronomical Society*, **30**, 381–392.

WILSON, L. & HEAD, J. W. 1983. A comparison of volcanic eruption processes on Earth, Moon, Mars, Io and Venus. *Nature*, **302**, 663–669.

WILSON, L. & HEAD, J. W. 1994. Mars: Review and analysis of volcanic eruption theory and relationships to observed landforms. *Reviews of Geophysics*, **32**, 221–264.

WITTE, L. C., COX, J. E. & BOUVIER, J. E. 1970. The vapor explosion. *Journal of Metals*, **22**, 39–44.

WOHLETZ, K. H. 1983. Mechanisms of hydrovolcanic pyroclast formation: Grain-size, scanning electron microscopy, and experimental studies. *Journal of Volcanology and Geothermal Research*, **17**, 31–63.

WOHLETZ, K. H. 1986. Explosive magma-water interactions: Thermodynamics, explosion mechanisms, and field studies. *Bulletin of Volcanology*, **48**, 245–264.

WOHLETZ, K. H. & MCQUEEN, R. G. 1984. Experimental studies of hydromagmatic volcanism. *In*: *Explosive Volcanism: Inception, Evolution, and Hazards*. National Academy of Sciences, Washington, DC, 158–169.

WOOD, C. A. 1979. Monogenetic volcanoes of the terrestrial planets. *In*: MERRILL, R. B. (ed.) *Proceedings of the Tenth Lunar and Planetary Science Conference*. Lunar Planetary Institute, Houston, Texas, 2815–2840.

ZIMANOWKSI, B., FRÖHLICH, G. & LORENZ, V. 1991. Quantitative experiments on phreatomagmatic explosions. *Journal of Volcanology and Geothermal Research*, **48**, 341–358.

The hyaloclastite ridge formed in the subglacial 1996 eruption in Gjálp, Vatnajökull, Iceland: present day shape and future preservation

M. T. GUDMUNDSSON, F. PÁLSSON, H. BJÖRNSSON & Þ. HÖGNADÓTTIR

Science Institute, University of Iceland, Hofsvallagötu 53, 107 Reykjavík, Iceland (e-mail: mtg@raunvis.hi.is)

Abstract: In the Gjálp eruption in 1996, a subglacial hyaloclastite ridge was formed over a volcanic fissure beneath the Vatnajökull ice cap in Iceland. The initial ice thickness along the 6 km-long fissure varied from 550 m to 750 m greatest in the northern part but least in the central part where a subaerial crater was active during the eruption. The shape of the subglacial ridge has been mapped, using direct observations of the top of the edifice in 1997, radio echo soundings and gravity surveying. The subglacial edifice is remarkably varied in shape and height. The southern part is low and narrow whereas the central part is the highest, rising 450 m above the pre-eruption bedrock. In the northern part the ridge is only 150–200 m high but up to 2 km wide, suggesting that lateral spreading of the erupted material occurred during the latter stages of the eruption. The total volume of erupted material in Gjálp was about 0.8 km^3, mainly volcanic glass. The edifice has a volume of about 0.7 km^3 and a volume of 0.07 km^3 was transported with the meltwater from Gjálp and accumulated in the Grímsvötn caldera, where the subglacial lake acted as a trap for the sediments. This meltwater-transported material was removed from the southern part of the edifice during the eruption. Variations in basal water pressure may explain differences in edifice form along the fissure. Partial floating of the overlying ice in the northern part is likely to have occurred due to high water pressures, reducing confinement by the ice and allowing lateral spreading of the edifice. The overall shape of the Gjálp ridge is similar to that of many Pleistocene hyaloclastite ridges in Iceland. Future preservation of the Gjálp ridge will depend on the rate of glacial erosion it will suffer. Besides being related to future ice flow velocities, the erosion rate will depend on the rate of consolidation due to palagonitization and shielding from glacial erosion while depressions in the ice are gradually filled by ice flow directed towards the Gjálp hyaloclastite ridge.

It has been known since the middle of the twentieth century that hyaloclastite ridges and tuyas can be formed in eruptions within glaciers and ice sheets (Kjartansson 1943; Mathews 1947; van Bemmelen & Rutten 1955). Progressively more refined models of the formation of hyaloclastite mountains have since been presented, together with detailed analysis of the composite units of these mountains (e.g. Jones 1969; Smellie & Skilling 1994; Werner *et al.* 1996). These studies have revealed how hyaloclastite mountains commonly consist of a succession of basal pillow lavas, overlain by hyaloclastites that are sometimes capped with subaerial lavas. The models have developed on the basis of field studies of exposed hyaloclastite mountains that were mostly formed within Pleistocene ice sheets in Iceland, British Columbia and Antarctica. The extent to which glacial erosion has modified the exposed hyaloclastite volcanoes is difficult to ascertain, but it has been suggested that removal of edifices occurs in regions of fast ice flow (Behrendt *et al.* 1995; Bourgeois *et al.* 1998).

The eruption at Gjálp in the western central part of the temperate Vatnajökull ice cap, Iceland, in October 1996 was observed in much greater detail than any previous subglacial eruption. The initial ice thickness was several hundred metres and conditions may have been similar to those that occurred under Pleistocene ice sheets in Iceland. Therefore the Gjálp eruption provided important data on various aspects of subglacial volcanic activity. These include observations of (1) melting rates in a subglacial eruption, (2) the response of the ice cap to rapid subglacial melting during the eruption, (3) drainage of meltwater from the eruption site, and (4) formation of a hyaloclastite ridge within a glacier. A first report on the main aspects of the eruption, touching on all the above points, was published by Gudmundsson *et al.* (1997). Alsdorf & Smith (1999), Björnsson *et al.* (2001)

From: SMELLIE, J. L. & CHAPMAN, M. G. (eds) 2002. *Volcano–Ice Interaction on Earth and Mars*. Geological Society, London, Special Publications, **202**, 319–335. 0305-8719/02/$15.00 © The Geological Society of London 2002.

and Gudmundsson *et al.* (2002) used InSAR radar interferometry coupled with other data to deduce the flow field of the ice depressions around Gjálp in the aftermath of the eruption. Steinthorsson *et al.* (2000) provided petrological analyses of the Gjálp magma, and isotopic analyses of Sigmarsson *et al.* (2000) suggested that the erupted magma belongs to the Grímsvötn volcanic system. This paper presents the results of geophysical surveying in 1997–2000. The form, volume and morphology of the subglacial ridge created in the eruption is presented and comparisons made with Pleistocene hyaloclastite ridges in Iceland. Finally, the preservation of subglacial hyaloclastite mountains is discussed with respect to possible effects of compaction due to palagonitization, and temporal diversions in the ice flow field because of disruption and melting of the glacier by the subglacial eruption.

The 1996 eruption

The site of the eruption was located between the central volcanoes Grímsvötn and Bárdarbunga (Fig. 1), in an area where the pre-eruption ice thickness was 550–750 m (Björnsson 1988; Björnsson *et al.* 1992). The pre-eruption bedrock topography was made of mounds and short ridges, with relief of 100–200 m rising from a plateau of elevation 950–1100 m above sea level. The eruption had been preceded by an increase in regional seismicity for several months (Einarsson *et al.* 1997). On 29 September, 1996, a 5.4 magnitude earthquake occurred in Bárarbunga, followed by an intense earthquake swarm. The onset of continuous tremors at about 22 hours GMT on 30 September is considered to mark the start of the eruption. The following morning, two depressions (cauldrons) in the ice surface were observed from the air (Fig. 2). These cauldrons grew bigger during the first day, while the meltwater drained along the glacier bed into the subglacial lake within the Grímsvötn caldera (Gudmundsson *et al.* 1997). After 31 hours, the subglacial eruption broke through the ice cover and a subaerial hydromagmatic eruption commenced. Also on the second day of the eruption, a new cauldron started to form in the ice surface to the north of the other two. At that time the subglacial eruption fissure had reached its maximum length, about 6 km; the location of the volcanic fissure has been defined as the line drawn along the bottom of the elongated ice depressions (Fig. 2). The eruption lasted 13 days, ending on October 13. By that time, about 3 km^3 of ice had melted and accumulated in Grímsvötn (Gudmundsson *et al.* 1997); the meltwater subsequently drained from Grímsvötn on 5 November three weeks after the end of the eruption. The volume of tephra dispersed over Vatnajökull in the subaerial eruption was about 10 million m^3, two orders of magnitude less than the total volume of erupted materials (Gudmundsson *et al.* 1997). A minor subglacial eruption may have taken place on the SE corner of Bárðarbunga at the same time as the Gjálp eruption, since two shallow depressions also formed in the ice surface there (Fig. 1c).

An important observation in the Gjálp eruption was the very high heat transfer rate that can occur between an erupting volcano and the overlying ice; the observed heat flux was 3×10^5 W m^{-2} (Gudmundsson *et al.* 1997). This heat flux is too high to be explained by pillow lava formation, calling for fragmentation of the magma to glass as the dominant mode of cooling and solidification (Gudmundsson *et al.* 1997). This has implications for the edifice since it suggests that it all should be predominantly made of fragmented material and that crystalline rocks can only make up a minor part of the volume. The heat released during the eruption and the first six weeks following it was equivalent to that released by fragmentation to pyroclastic glass and cooling of 0.4 km^3 of basic/intermediate magma (Gudmundsson *et al.* 1997). These findings were based on observations made during and immediately following the eruption. However, the distribution of volcanic material under the glacier and the shape of the new ridge was unknown, and also how much material was transported with the meltwater into Grímsvötn. In order to gather

Fig. 1. (**a**) Map of Vatnajökull and its surroundings, including the ice cap (white), and Pleistocene hyaloclastite formations (black). The area in Figure 1 (c) is indicated by the indexed box. (**b**) Map of Iceland with the volcanic zones (shaded) and Pleistocene basaltic hyaloclastite formations (black) (after Jóhannesson & Saemundsson 1998). (**c**) Ice surface map of the Gjálp area (summer 1998). The Gjálp volcanic fissure is shown (heavy black line),and the arrows indicate the subglacial path of the meltwater from Gjálp. The boundaries of the calderas of Bárðarbunga and Grímsvötn are indicated. The two black dots in the SE corner of Bárðarbunga are small depressions in the ice surface, considered to have been formed in a minor subglacial eruption at the same time as the Gjálp eruption (Gudmundsson *et al.* 1997). The boundary of the post-eruption Grímsvötn–Gjálp ice drainage basin is indicated (dashed line). The boundaries of the water drainage basin are approximately the same.

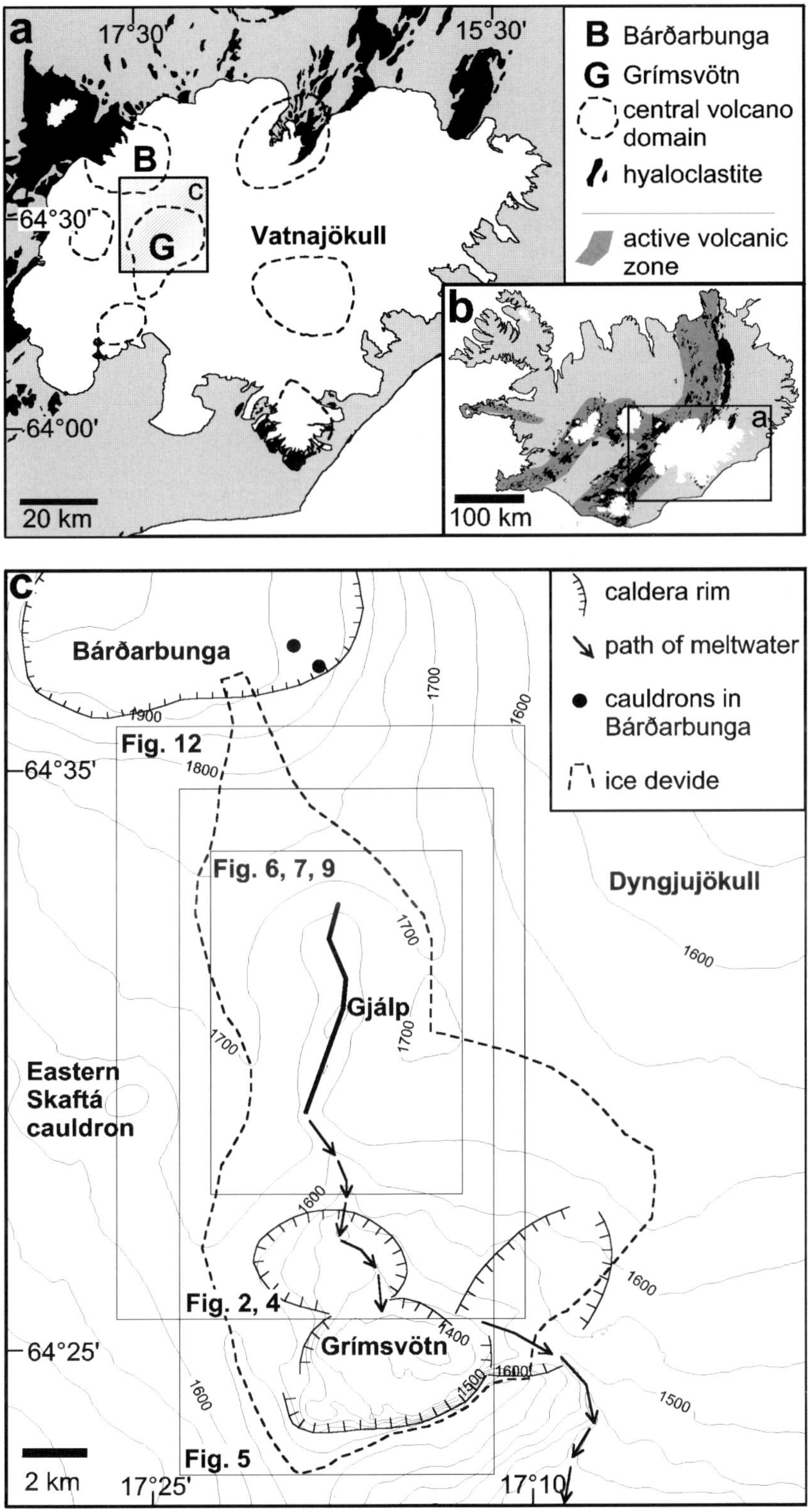
a
17°30'
15°30'
B Bárðarbunga
G Grímsvötn
central volcano domain
hyaloclastite
active volcanic zone
B
C
G
64°30'
Vatnajökull
64°00'
20 km
b
a
100 km
c
caldera rim
path of meltwater
cauldrons in Bárðarbunga
ice devide
Bárðarbunga
1900
1800
1700
1600
Fig. 12
64°35'
Fig. 6, 7, 9
Dyngjujökull
Gjálp
Eastern Skaftá cauldron
Fig. 2, 4
1400
1500
Grímsvötn
64°25'
Fig. 5
2 km
17°25'
17°10'

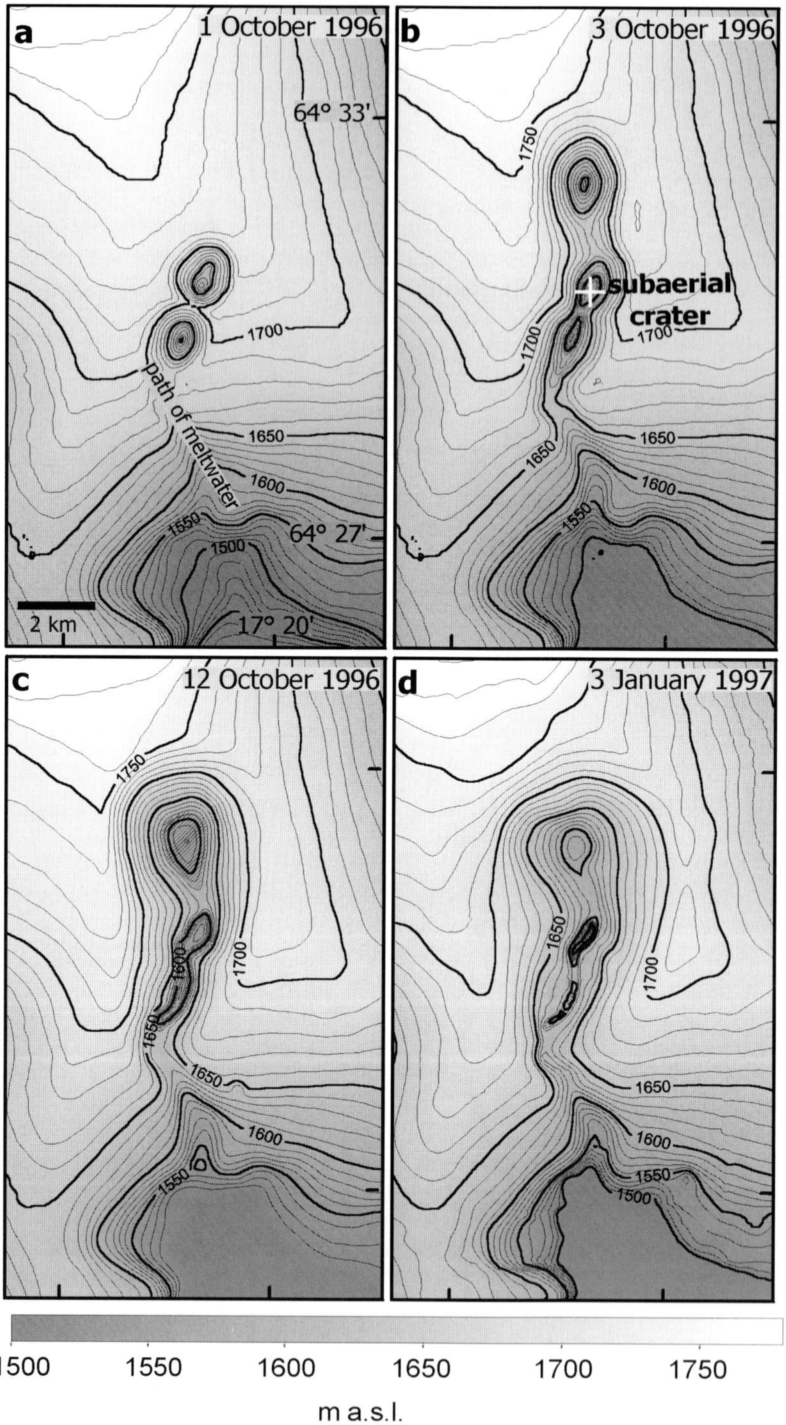

Fig. 2. Maps showing the development of ice cauldrons during and shortly after the Gjálp eruption. The deep trough extending for 3.5 km along the central-southern part of the depression on 12 October is the canyon formed in the ice surface. Water flowed southwards along the canyon during the latter part of the eruption and in the following months. The ice in the central part of the northern cauldron (hatched) on 12 October may have been floating due to high basal water pressure (see section on morphology of volcano).

information on the above questions, a programme of geophysical surveying was carried out in 1997–2000. The programme included:

(1) Radio echo soundings, to map the new subglacial volcano.
(2) Gravity surveying, to determine the density of the volcano.
(3) Repeated traverses of the ice surface to determine the shape and volume of the depressions in the ice surface at the eruption site. Initially this was done with airborne altimetry, but later by oversnow traverses with Differential Global Position System (DGPS) surveying of sub-metre accuracy.
(4) Evaluation of ice flow into the depressions, by measuring surface velocities in a network of stakes, with sub-metre DGPS.
(5) Seismic reflection, gravity profiling, radio echo soundings and DGPS surface profiling in Grímsvötn to determine the distribution and thickness of new sediments derived from Gjálp, by comparing the results with those of earlier surveys.
(6) Field observations of the exposed part of the volcano in 1997, before it was covered by the glacier.

As a result of subglacial melting and meltwater drainage during the eruption, a heavily crevassed depression formed in the glacier surface, enclosing the eruption site. The ice surface was also covered with a layer of tephra. The crevasses and the tephra cover have made traverses of the area difficult, resulting in irregular location of geophysical survey lines. Therefore, the results presented cannot be regarded as final, since in future when conditions become easier, more optimal surveys with denser line spacing will be possible. The present data nevertheless provide a clearer picture of the subglacial volcano than before, as well as new indications of its thermal state in the period following the eruption.

Direct observations

At the end of the eruption, the volcano itself was completely hidden by water and ice at the bottom of the ice canyon overlying the southern part of the fissure (Fig. 2). A flat tephra cone had been built on ice, encircling the opening in the glacier. This tephra cone disappeared as the ice forming its foundations was melted. Continued melting above the volcanic edifice removed the ice cover at the site of the subaerial crater, and by January 1997, the top of the edifice had become visible. On 5 April 1997 the exposed part was a sharp ridge, 250–300 m long, that rose about 40 m above the bottom of the ice canyon (Fig. 3). The ridge was inspected in June 1997. It was predominantly made of unconsolidated pyroclastic glass fragments of ash to lapilli size. Temperature in the pile rose fast with depth, and was about 70°C at 0.5 m below the surface. Further exploration of the edifice was not possible; later in the summer and autumn of 1997 the top of the edifice was under water, since the water level rose due to surface ablation. By spring 1998 the edifice was covered by inflow of ice.

Radio echo soundings

A radio echo sounder operating at 1–5 MHz frequency (Sverrisson *et al.* 1980) was used for imaging of the subglacial bedrock. Extensive surveys carried out in the 1980s mapped the surrounding area in considerable detail (Björnsson 1988; Björnsson *et al.* 1992). The main source of information on the pre-eruption bedrock on the eruption site were survey lines from 1991 (Björnsson *et al.* 1992) and 1993 (Fig. 4). The pre-eruption bedrock map (Fig. 5a) shows ridges and mounds rising 100–200 m above a gently northwards sloping bedrock. The elevation of the pre-eruption bed near the southern end of the Gjálp 1996 fissure was 1100–1150 m but about 1000 m at the northern end. The 1996 fissure was located 0.5–1 km to the west of a 4 km-long, about 1 km-wide and 100–150 m-high ridge (hatched on Fig. 5a) considered to have formed in a subglacial eruption in 1938 (Björnsson 1988; Gudmundsson & Björnsson 1991).

Radio echo surveys were conducted at the eruption site in 1997, 1998 and 2000 (Fig. 4). In 1997, only a limited area in the northernmost part of Gjálp could be mapped with the sounder due to heavy crevassing in other parts. In 1998 reasonable coverage was obtained for the northern part and further gaps were filled in 2000. In parts where the slopes of the edifice were very steep (>30°), its width was poorly constrained by the radio echo soundings; reflections are received from the top of the edifice but downward-travelling energy from the transmitter, as it encounters a steep slope, is reflected downwards and thus not detected by the radio echo receiver. These problems were partly resolved by using the results of gravity profiling to constrain the results (see following section).

Figure 6 shows some of the radio echo sections used in the mapping of the volcano. Heavy crevassing in the southern and central parts not only made traverses difficult but also degraded signal quality. Only the top part of the

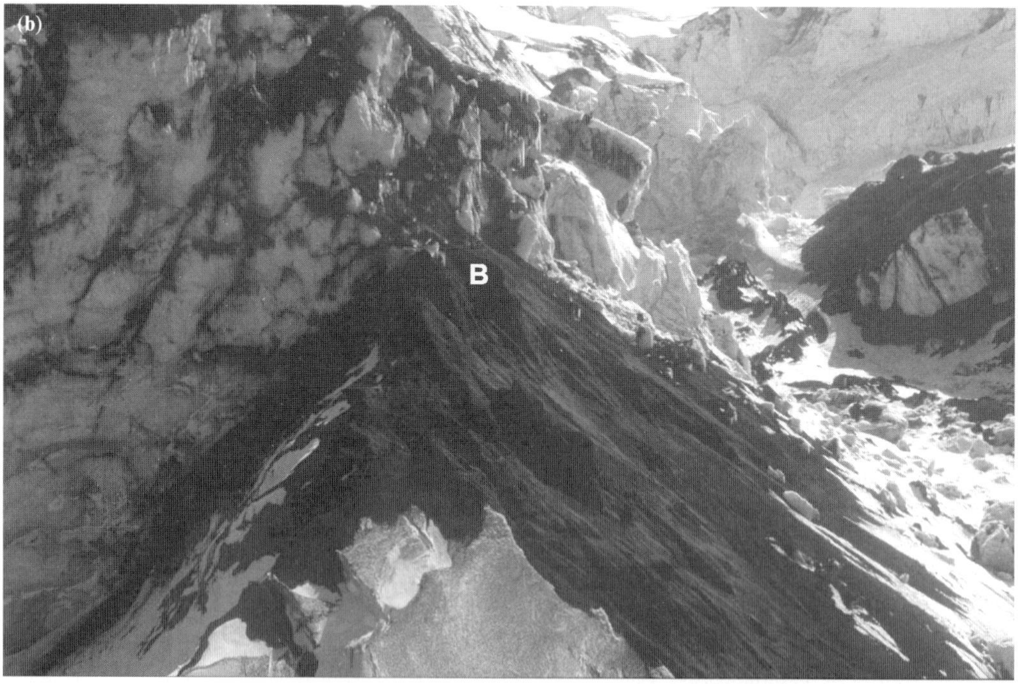

Fig. 3. (**a**) View (looking east) of the Gjálp ridge as seen from an aeroplane on 5 April 1997. Meltwater is present, flowing in the supraglacial ice canyon formed after the third day of the eruption (Fig. 2). The ridge is partly snow covered and ice flowing in from the eastern side is banked against the ridge flank. The canyon wall is about 100 m high. (**b**) View along the Gjálp ridge in June 1997. The photo is taken at the point marked as A on Figure 3a, looking towards point B. The ice cliff seen in the left foreground is 25–30 m high.

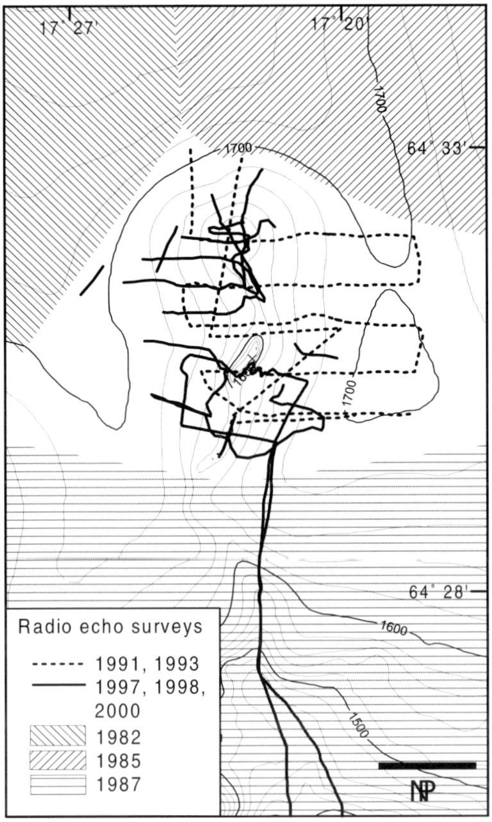

Fig. 4. Map showing the location of radio echo sounding lines used to define the pre- and post-eruption bedrock surface. The results of the surveys of 1982, 1985 and 1987 are summarized in Björnsson (1988).

edifice is detected in the soundings, but the data are consistent with a steep narrow ridge. In contrast, the wide northern part is well resolved by the radio echo soundings.

Gravity profiling

Gravity data were collected using a LaCoste-Romberg G-meter in 1997 and 1998, and a Scintrex CG-3M in 2000 (Fig. 7). Differential GPS was used for elevation control and positioning, yielding elevation accuracy of 1–2 m. For the two profiles surveyed in 2000, elevations were measured using higher precision GPS equipment providing elevation accuracy of about 0.1 m.

In order to remove effects of pre-eruption bedrock and ice surface topography from the data, the gravitational effects of ice and pre-eruption bedrock were calculated from digital elevation models (DEMs) (the method is outlined in Gudmundsson & Milsom 1997). A reduction density of 900 kg m^{-3} was used for the ice. For the pre-eruption bedrock a density of 2300 kg m^{-3} was used, the average value for the uppermost 0.5–1.0 km of the crust within the volcanic zones in Iceland (Pálmason 1971). Regional trends were removed and the residual anomaly was used for forward modelling along three profiles (Fig. 7). All profiles show positive anomalies, narrow and sharp in the south and central parts but wider in the northern part, in qualitative agreement with the radio echo data. The gravity models were constructed using the Gravmag software (Pedley *et al.* 1991) where polygons of finite length, striking at right angles to the survey line ($2\frac{1}{2}$-D), are created and their form and density adjusted until the calculated effects of the model fit with the observed data. The volume occupied by the Gjálp edifice has in the gravity data reductions been assigned an ice density of 900 kg m^{-3}. This implies that the anomalies in Figure 7 arise because of the density contrast between the ridge and the ice; it is this contrast that is used in the model calculations. Thus, if the density of the edifice is assumed to be twice the ice density, a 10% change in assumed density results in 20% change in density contrast and a corresponding 20% change in calculated anomaly amplitude. In the following discussion the density referred to is the assumed edifice density.

The model for the northern part is based on the radio echo soundings. A reasonable fit is obtained using an edifice density of 1900 kg m^{-3} (density contrast of 1000 kg m^{-3}) which would be consistent with a mass of water-saturated pyroclastic glass, but too low for any significant volume of pillow lavas or other crystalline rocks. However, separation of residual anomaly from the regional field is not as straightforward as for the central and southern parts. It is possible that the anomaly should have larger amplitude, thus a density of 2200–2300 kg m^{-3} (contrast 1300–1400 kg m^{-3}) cannot be ruled out. This implies that a significant fraction of the northern part may be pillow lavas, the present data are not conclusive. The model for the southern part is consistent with a 200–250 m-high narrow ridge of density 1900 kg m^{-3}, the same as in the preferred model for the northern part.

To account for the heavy crevassing in the central part, a 100 m-thick and 1 km-wide area of the surface ice is assigned a reduced density of 860 kg m^{-3}, assuming that the crevasses are partly filled by low density snow and firn. The model results shown on Figure 7 for the central part yield the very low density of 1500–1600 kg m^{-3}. This implies that any volume of

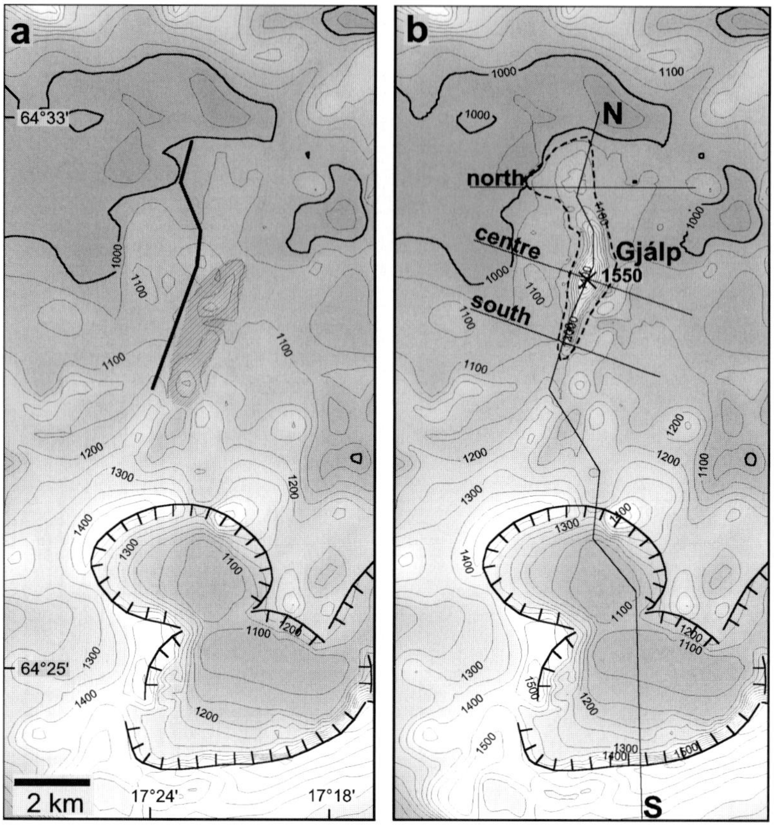

Fig. 5. (**a**) Map of pre-eruption bedrock topography in the Gjálp area, after Björnsson *et al.* (1992) and 1991 and 1993 data (survey line locations shown in Fig. 4). The bold gray line marks the volcanic fissure in the bedrock, and the hatched area marks the ridge that formed in the eruption in 1938 (Björnsson, 1988; Gudmundsson & Björnsson 1991). Bedrock contour interval is 50 m. (**b**) Map of bedrock in the Gjálp area after the 1996 eruption based on direct observations, radio echo data and gravity surveying. The 1996 ridge is shown with a dashed line. The cross sections shown on Figure 8 (north, centre, south) and Figure 11 (N–S) are also indicated. Contour interval is the same as in (**a**).

pillow lava present is insignificant since pillow lavas commonly have a density of 2300–2400 kg m^{-3} (Pálsson *et al.* 1984).

Several alternative models are possible for the central part but all have a very low mean density. Some of the possibilities are: (1) A steeper, narrower ridge of higher density would have the same mass as the one shown in the model. However, such a ridge would create a narrower anomaly with a higher peak value and provide a worse fit to the observed anomaly. (2) A layered model, with a somewhat higher density in the lower part while the upper part would have even lower density. This would be consistent with the lower part being water-saturated and the upper part dry, with a groundwater table at perhaps 1200–1250 m elevation. A low groundwater level is highly unlikely though, since it is to be expected that the level of the Grímsvötn lake imposes a lower bound on the basal water pressure within the Grímsvötn drainage basin (Fig. 1). Thus, groundwater level within the ridge should be equivalent or higher than 1350–1400 m asl, the level of the Grímsvötn lake in 1997–2000 (Science Institute, unpublished data). (3) A third possible interpretation would be that the central part is hot, with steam making up a large fraction of the pore space.

Morphology of volcano

By combining the information from direct observations in 1997, the radio echo soundings

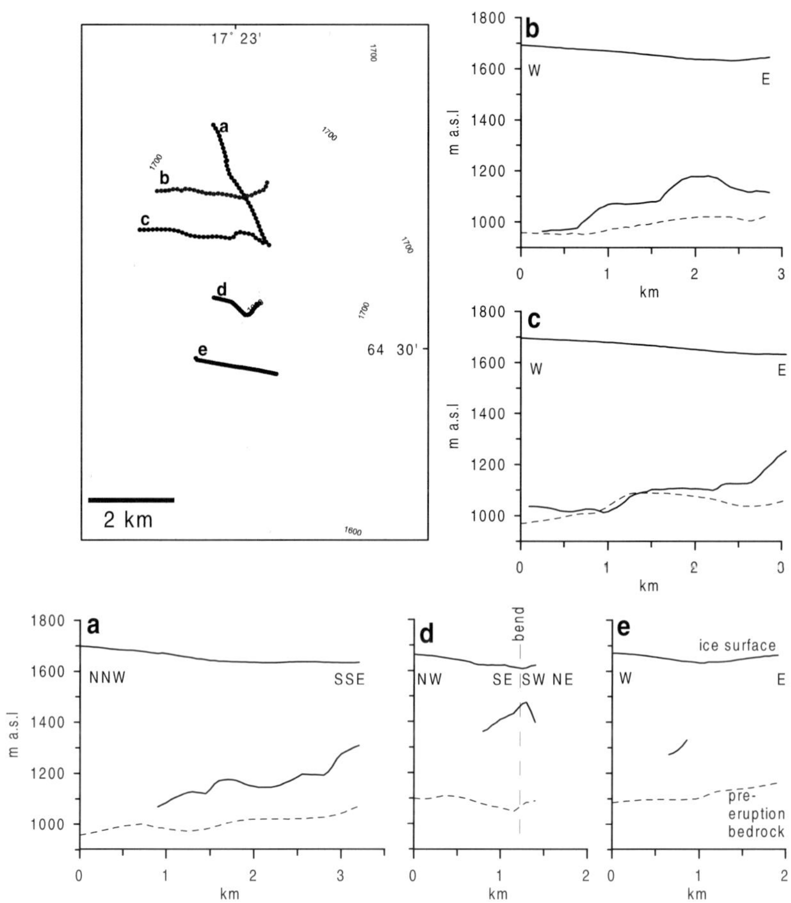

Fig. 6. Five of the radio echo survey lines used to define the morphology of the Gjálp ridge (map contour interval 20 m). The pre-eruption bedrock surface is shown as a dashed line. The northern part of the ridge is well defined (**a**, **b** and **c**), whereas difficult terrain, heavy crevassing giving rise to scatter and attenuation of the radar signal, and steep slopes of the edifice made the use of radio echo soundings more difficult in the central and southern parts (**d** and **e**), and the ridge is less well defined. (**d**) The survey line approaches the ridge from the NW at an angle, making the apparent slope of the ridge less steep than the actual slope.

and indications from the gravity data on the basal width of the edifice, a map of the edifice has been constructed (Fig. 5b). Given the uncertainties in the data, the map should be regarded as a smoothed image of the true form of the volcano. However, the overall picture is clear and there are considerable contrasts in the form of different parts of the volcano created in the eruption (Figs 8 & 9). In the southern and central parts the ridge is narrow, with a relief of 200–250 m in the southern part, rising steeply to 400–450 m in the central part, where the top reaches 1500–1550 m above sea level. The combined information from the radio echo soundings and the gravity show that in the southern and central parts the slopes are steep, 30°–35° (Fig. 8). The width of the southern part is about 500 m but that of the central part is close to 1 km. The shape of the northern part is different. The elevation drops northwards from the central part, to about 1150–1200 m. The total width of the northern part is about 2 km (Fig. 8). It is distinctively asymmetric with respect to the volcanic fissure, extending about 1.5 km to the west from the fissure. The eastern part of the edifice, centred over the volcanic fissure, is a broad ridge rising 150–200 m above the pre-eruption bedrock. The part extending to the west is about 100 m-thick, gently sloping towards west. This shoulder is located under the area

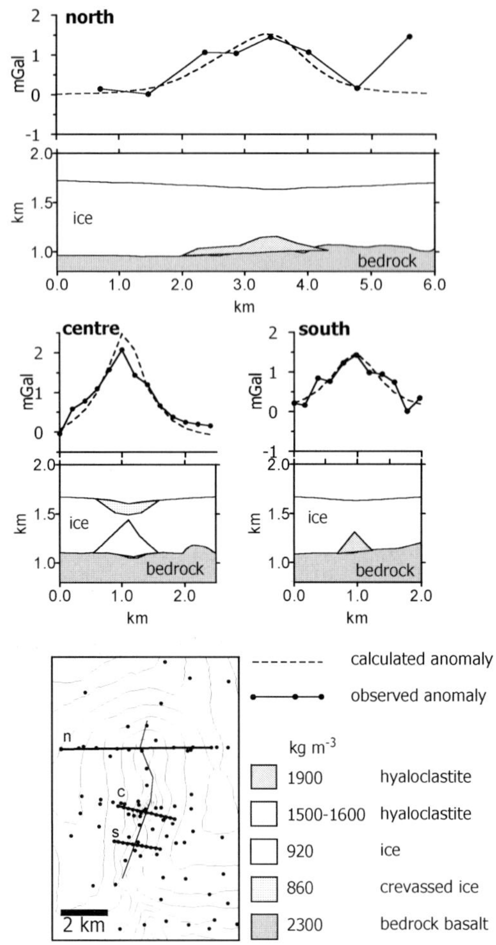

Fig. 7. Forward gravity models of the Gjálp ridge. The location of the three profiles is shown on the map (north, centre and south). The bodies corresponding to the ridge have finite half strike lengths (perpendicular to the profiles): north ±0.6 km, centre ±0.5 km, south ±1 km. These strike lengths were determined on the basis of the estimated dimensions of the ridge (Fig. 5) using the radio-echo soundings, the direct observations and the length of the ice cauldrons. The rise at the eastern end of the north-profile marks the start of a positive gravity anomaly that has a source in the shallow crust to the east of Gjálp; it is not be related to the Gjálp edifice and no attempt is made to explain it further here.

where a westwards widening of the ice depressions was observed late in the eruption (Fig. 2c). A possible explanation for the widening of the ridge is migration of the volcanic material westwards along the glacier bed. Another possibility is that an eruption occurred on a short fissure to the west of the main fissure late in the eruption.

The latter option is less likely for several reasons. Firstly, formation of such fissures to the side of a main fissure is uncommon. Secondly, observations of seismicity and course of events late in the eruption do not suggest any large changes, other than a decline in activity (Einarsson *et al.* 1997; Gudmundsson *et al.* 1997). In contrast the ice surface and bedrock data suggest that conditions for spreading may have been favourable, since confinement of the edifice by the surrounding ice was minimal or non-existent because of floating of the ice in the central part of the northern cauldron. The fact that water drained southwards shows that basal water pressure was highest at the northern end of the volcanic fissure, that is under the northern cauldron. Water level in the ice canyon late in the eruption was about 1560 m asl. This provides a lower limit on the basal water pressure under the northern cauldron as it must have been at least equivalent to the weight of a water column reaching from the bed up to 1560 m elevation.

The static ice overburden pressure at the base of the glacier is $P_i = \rho_i g(z_i - z_b)$ where ρ_i equals $900\,kg\,m^{-3}$, the average density of the overlying glacier; g is gravitational acceleration; and z_i and z_b are the elevations of the ice surface and the bedrock respectively. Similarly, the water pressure at the base of the glacier is $P_w \geq \rho_w g(z_w - z_b)$, where ρ_w equals $1000\,kg\,m^{-3}$ and denotes water density and z_w is the water level in the ice canyon. The ice floats where $P_w \geq P_i$. For z_w (1560 m asl) and z_b (1000 m asl; the average bedrock height under the northern cauldron around the edifice), the basal water pressure exceeds the ice overburden pressure ($P_w > P_i$) where the ice surface in the northern cauldron is below 1620 m asl ($z_i \leq 1620$ m). This area is shaded in Figure 2c. It is about 2 km-long (N–S) and 1.3 km-wide (E–W). The above results suggest that the edifice under the northern cauldron was not confined by the surrounding ice, leading to the formation of the shoulder by westwards flow or slumping and sliding of the volcanic material. The slight dip of the bedrock (Fig. 8) may explain why sliding occurred towards west but not towards east.

The overall shape of the ridge formed in Gjálp conforms to length/width statistics of a large number of Pleistocene hyaloclastite ridges in the volcanic zones in Iceland presented by Chapman *et al.* (2000). In Figure 10, Gjálp is compared with Skridutindar, a late Pleistocene hyaloclastite ridge of similar length and volume in SW Iceland. Several other ridges of similar length, height and volume exist in the volcanic zones and hyaloclastite mountains with low height/width ratios are present, analogous to the northern part of Gjálp.

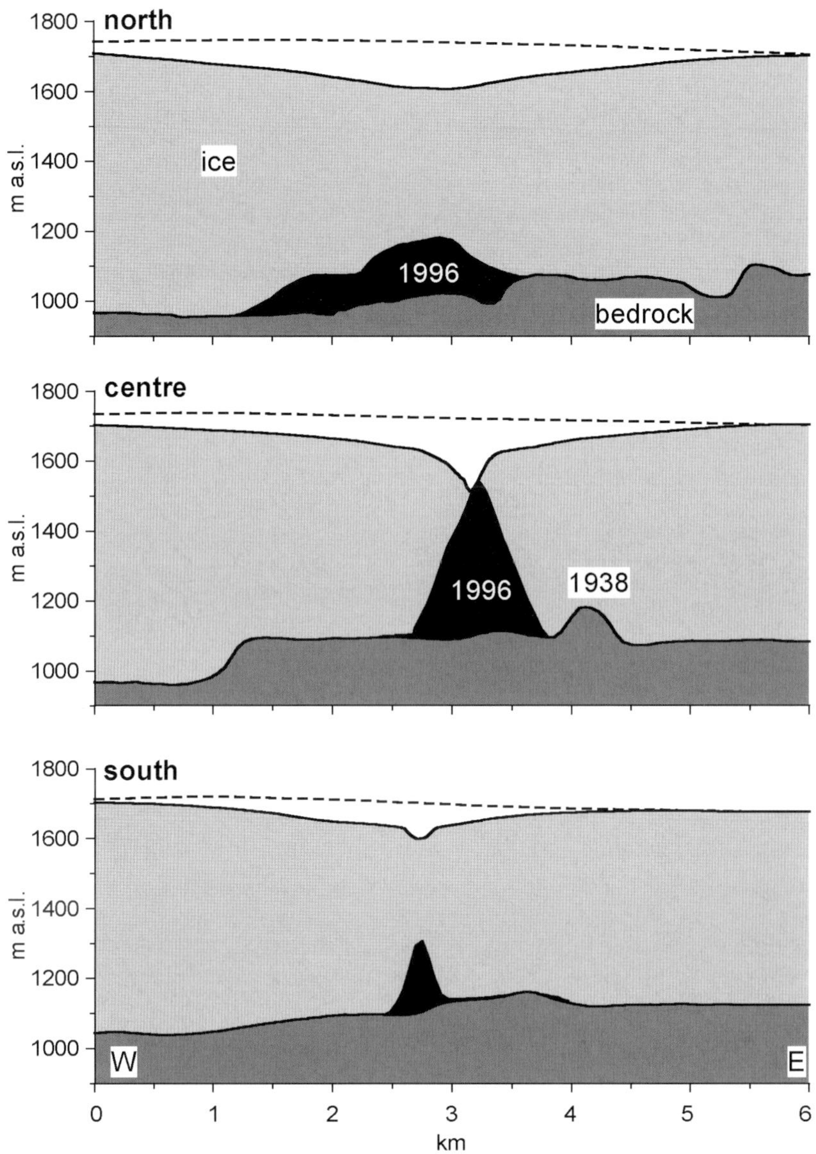

Fig. 8. Cross-sections of Gjálp based on direct observations, radio echo and gravity data. The northern end of the ridge formed in 1938 is situated about 1 km to the east of the 1996 ridge in the central section shown. The dashed lines show the pre-eruption ice surface.

Volume of erupted material

The edifice is about 6 km long and has a volume of about 0.70 km^3. The uncertainty in the volume is about 20% or 0.15 km^3. A density of 1900 $kg\,m^{-3}$ calls for a highly porous body. If the pore spaces are filled with water of density 1000 $kg\,m^{-3}$, and the grain density of volcanic glass is taken as 2750 $kg\,m^{-3}$ (Oddsson 1982), the average porosity is almost 50%. If the volcanic pile is saturated with water at the pressure boiling point down to several hundred metres depth, the mean water density may be 850–900 $kg\,m^{-3}$ (e.g. Ingebritsen & Sanford 1998), yielding a porosity of 45%.

The low mean density of the central part of the edifice is enigmatic. If it is caused by a high steam fraction in the edifice, a mean density of

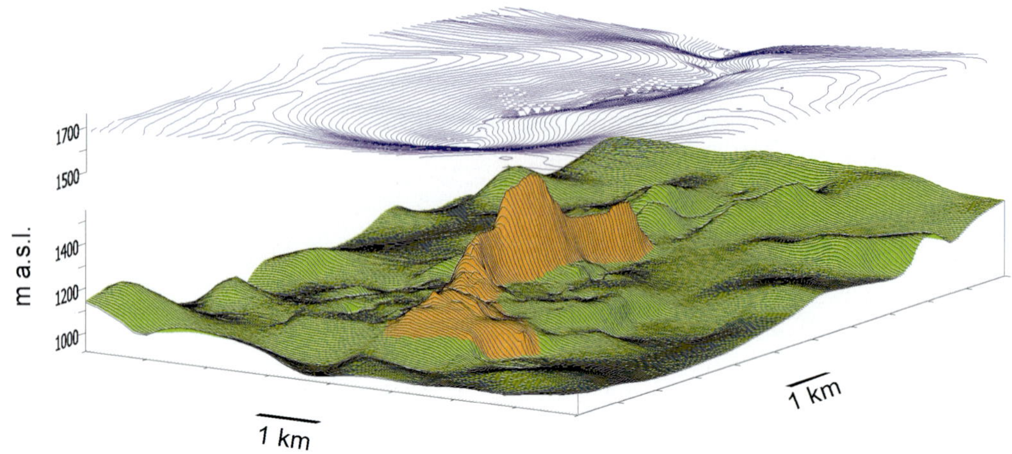

Fig. 9. A perspective plot of Gjálp, viewed from the NW. The vertical axis is in metres above sea level. The upper surface is that of the ice in 1997. The 1996 ridge is coloured orange brown whereas the surrounding bedrock is shown green. The image draws out the contrasting shape of the northern part (close to the observer) with that of the central and southern parts.

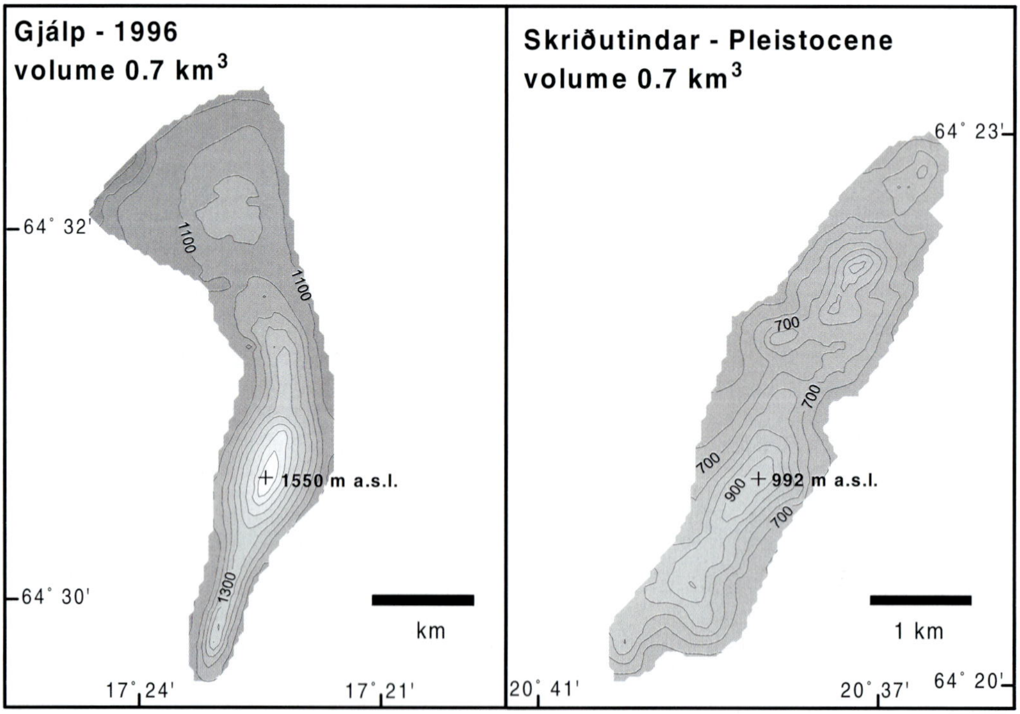

Fig. 10. Maps comparing the morphologies of the Gjálp ridge and the Pleistocene hyaloclastite ridge of Skridutindar in the Western Volcanic Zone in Iceland. The two ridges are morphologically similar and have approximately the same volume and length.

$1600\,kg\,m^{-3}$ could be explained by a 45% porosity with about three quarters of the pore space taken up by steam and one fourth by water. These would be the conditions found in vapour dominated geothermal systems that exist on some active volcanoes (Stefánsson & Björnsson 1982). Whether this may be the case for this ice covered volcano is a subject of further research.

A part of the volcanic material was transported subglacially by the meltwater into the Grímsvötn subglacial lake (Fig. 11). Surveys carried out in Grímsvötn show that this material makes up two piles or deltas were the meltwater issued into the subglacial lake, and that the total volume of these sediments is about $0.07\,km^{-3}$ (Gudmundsson *et al.* 2000). It is likely that this material was mainly derived from the southern part of the ridge, the section closest to Grímsvötn. The southernmost section of the edifice is small compared to the volume of ice melted over that part of the fissure, which supports the suggestion of removal of material by the meltwater. Analyses of the sediments carried by the jokulhlaup from Grímsvötn in November 1996 revealed that no material formed in the Gjálp eruption escaped, showing that Grímsvötn was an effective trap for the sediments (Maria *et al.* 2000). In addition to the edifice itself and material transported into Grímsvötn, a minor part of the volcanic material was airborne tephra; its volume was about $0.01\,km^3$ (Gudmundsson *et al.* 1997). Thus, the best estimate of the total volume of erupted material at Gjálp in 1996, is $0.8\,km^3$.

Possible occurrence of pillow lava

An important question is whether pillow lavas make up a significant fraction of the total volume of the edifice. The low density of the southern and central parts argues against the existence of pillows. However, it does not rule out that pillows at the base of the edifice could account for up to 10% of the total volume.

For the northern part the result is more uncertain. A mean density of $2200–2300\,km\,m^{-3}$ is possible, which would be consistent with the northern part being predominantly made of pillow lava. A magnetic survey of the area might help resolving this question, since pillow lavas often have high remanent magnetization while hyaloclastite tuffs have virtually none (e.g. Kristjánsson 1970). An aeromagnetic survey of Gjálp was carried out in May 1997; no magnetic anomalies that could be associated with the Gjálp ridge were observed (Science Insitute, University of Iceland, unpublished data). The significance of this result is limited however. The titanomagnetite that is often the main source of magnetization in unaltered pillow lavas has a low Curie point, commonly 100–200°C (Kristjánsson 1970) and it is likely that substantial parts of the edifice were above this temperature in June 1997, as the heat transfer rate was still very high (Gudmundsson *et al.* 1999, 2002). Future magnetic surveys will reveal whether changes in magnetization have occurred since May 1997, and may clarify whether pillows

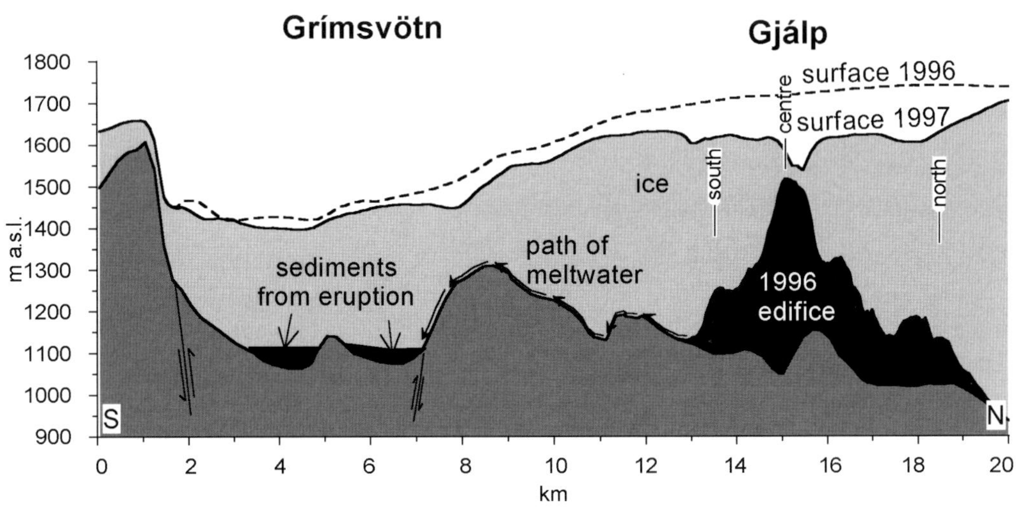

Fig. 11. Section through Grímsvötn and Gjálp (see Fig. 5b for location). The path of meltwater from Gjálp to Grímsvötn is shown (arrows) and the location of sediments derived from Gjálp and deposited within Grímsvötn (Gudmundsson *et al.* 2000).

make up a significant fraction of the volume in the northern part of Gjálp.

In summary, the present data provide the following constraints on pillow lava existence: The minimum is that no pillows exist, consistent with the gravity models in Figure 7. The maximum possible pillow lava content would occur if 10% of the central and southern parts and the whole northern part is made of pillows; this would mean that one third of the volume of the edifice was pillow lavas. The high heat flow observed during the eruption (Gudmundsson *et al.* 1997) does not contradict the possibility of one third of the volume being pillow lava, since the major part of the erupted magma would still fragment into glass, enough to account for the high heat flow.

Volume of magma erupted

With the total volume of erupted material as $0.8\,km^3$, an estimate of the volume of magma erupted can be obtained. The appropriate magma density for the basaltic andesite erupted in Gjálp (Maria *et al.* 2000; Steinthorsson *et al.* 2000) is close to $2600\,kg\,m^{-3}$ (Murase & McBirney 1973). If it is assumed that the volume of pillow lava is insignificant and by using the same assumptions as above on porosity and grain density, the resulting volume is $0.45\,km^{-3}$. If the maximum possible volume of $0.25\,km^{-3}$ for pillows (porosity *c.* 0.25) and the minimum possible volume of $0.55\,km^{-3}$ for glass/tephra are used, a magma volume of $0.50\,km^{-3}$ is calculated. These values are slightly higher but not significantly different from the initial estimate of $0.4\,km^3$ which was based on melted ice volumes and calorimetry (Gudmundsson *et al.* 1997).

Future preservation of the edifice

When the Gjálp ridge is compared with ridges formed within the Pleistocene ice sheet, the question arises of how well have these ridges been preserved and what may be the preservation potential of the newly formed Gjálp ridge. The data provide information on the initial shape of the Gjálp ridge, and the evidence suggests that at the end of the eruption it was composed mainly of an unconsolidated pile of volcanic glass/tephra. If such a pile was subjected to fast flow of ice overriding the edifice, it might suffer heavy subglacial erosion. In such a scenario a large part or the whole edifice might disappear in a short period of time, as may have been the case in West Antarctica (Behrendt *et al.* 1995) and has been suggested for certain areas of Iceland (Bourgeois *et al.* 1998). Bourgeois *et al.* (1998) pointed out that ridges formed at or close to ice divides should not be removed by glacial erosion. Two further factors that must greatly reduce the removal potential are, firstly, consolidation of the edifice by palagonitization and, secondly, the effect the eruption has on the ice flow field as ice flow is diverted into the depressions, towards the newly formed edifice.

Palagonitization

The pile of glass formed in the eruption may alter to palagonite, as has a large part of the hyaloclastite tuffs making up the Pleistocene ridges and tuyas (Jones 1969; Jakobsson 1979; Chapman *et al.* 2000). The rate of alteration in the subglacial environment is unknown; under subaerial conditions in Surtsey, basaltic tephra was palagonitized to dense tuff in only 1–2 years when subjected to mild hydrothermal activity at temperatures of 80–100°C (Jakobsson 1978). The observed temperatures in the top of the Gjálp edifice in June 1997 and a powerful geothermal heat flux revealed by continued subglacial melting (Gudmundsson *et al.* 1999) from the end of the eruption through 1999, indicates high temperatures in at least some parts of the edifice. This suggests that considerable palagonitization of the edifice may already have taken place. However, in the absence of samples from the edifice, no proof exists of palagonitization actually occurring at Gjálp. This is a subject of further research.

Diversion of ice flow

The Gjálp eruption occurred near the ice divide between the drainage basins of Grímsvötn, Dyngjujökull and the Eastern Skaftá Cauldron (Figs 1 & 12) (Björnsson 1988; Gudmundsson *et al.* 1997). This location suggests favourable conditions for preservation compared to other areas. No direct measurements exist of ice flow velocities in the Gjálp area prior to the eruption. However, the location of ice divides and the direction of ice flow (Fig. 12) is known (Björnsson 1988; Björnsson *et al.* 1992). An estimate of ice flow velocities can be made on the basis of mass balance of this part of Vatnajökull (Björnsson *et al.* 1998) and the size of the ice drainage basin upstream of the Gjálp ridge. By assuming that the glacier was in steady state before the eruption, the balance velocity (Paterson 1994) can be calculated. The result for the pre-eruption balance velocity at the centre of the Gjálp fissure is about $10\,m\,a^{-1}$ (*c.* 3 cm per day) towards south. This can be compared to the post-eruption flow

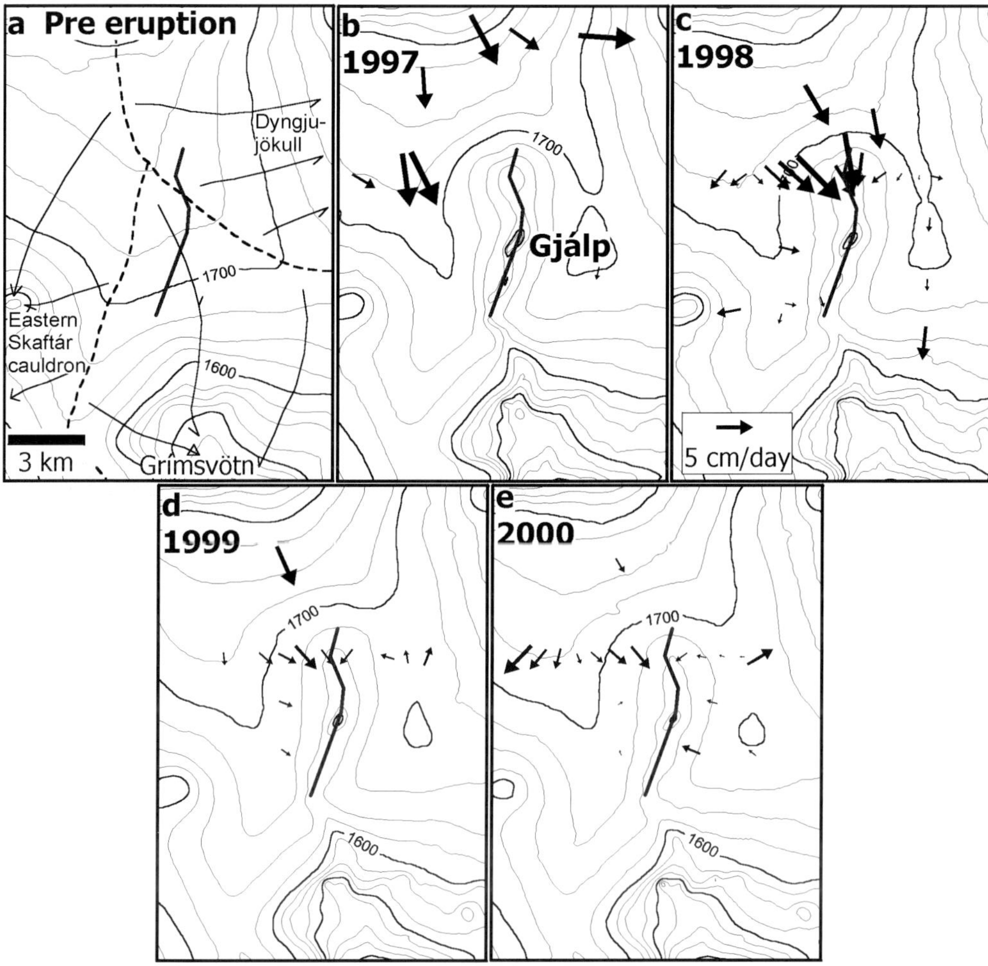

Fig. 12. Maps showing changes in the ice flow field in the Gjálp area in the first four years after the eruption. All maps have contour intervals of 25 m and show the Gjálp fissure (bold grey line). (**a**) The pre-eruption ice flow field with ice divides (dashed) and ice flow lines (solid with arrows). The ice flow lines are drawn perpendicular to surface contours. (**b**)–(**e**) Post-eruption horizontal ice flow velocities in 1997–2000, calculated from displacements of stakes measured with sub-metre DGPS during the period June–September each year. Ice flow velocities have declined with time but inside the depressions they are directed inwards, towards the Gjálp subglacial ridge.

field. During and after the eruption, ice flow inside the depressions was directed towards the centre (Fig. 12) with the observed velocities considerably higher than the pre-eruption balance velocities. Although the velocities have gradually been reduced, this flow field has persisted since the eruption. It will probably do so for several years, while the depression still exists. While flow of ice from all directions converging on the Gjálp ridge persists, no removal of volcanic material by glacial erosion should occur. If palagonitization occurs at Gjálp at a rate similar to that observed in Surtsey (i.e. a few years), the ridge should have consolidated to a large degree before ice starts flowing over it. This indicates that shielding from the regional ice flow for several years during infilling of depressions formed in subglacial eruptions may be an important factor in preserving mountains formed in these eruptions.

Volunteers of the Iceland Glaciological Society took part in the fieldwork and the Society together with the National Power Company of Iceland provided logistical assistance. E. Sturkell did GPS surveying for the gravity survey in June 2000. S. Gudbjornsson, the chief pilot of the Icelandic Aviation Authority is thanked for enthusiastic participation during radar altimetry surveys. This research was supported by a the University of Iceland Research Fund, a special grant from the Icelandic Government and grants from the Icelandic Road Authority. Constructive reviews by O. Bourgeois, J. Behrendt and J. Smellie improved the quality of this paper.

References

ALSDORF, D. E. & SMITH, L. C. 1999. Interferometric SAR observations of ice topography and velocity changes related to the 1996, Gjálp subglacial eruption, Iceland. *International Journal of Remote Sensing*, **20**, 3031–3050.

BEHRENDT, J. C., BLANKENSHIP, D. D., DAMASKE, D. & COOPER, A. K. 1995. Glacial removal of late Cenozoic subglacially emplaced volcanic edifices by the West Antarctic ice sheet. *Geology*, **23**, 1111–1114.

VAN BEMMELEN, R. W. & RUTTEN, M. G. 1955. *Tablemountains of northern Iceland.* E. J. Brill, Leiden.

BJÖRNSSON, H. 1988. Hydrology of ice caps in volcanic regions. *Societas Scientarium Islandica*, **45**.

BJÖRNSSON, H., PÁLSSON, F. & GUDMUNDSSON, M. T. 1992. *Vatnajökull, Northwestern part, 1: 100 000, Subglacial Surface Map.* National Power Company and Science Institute, Reykjavík.

BJÖRNSSON, H., PÁLSSON, F., GUðMUNDSSON, M. T. & HARALDSSON, H. H. 1998. Mass balance of western and northern Vatnajökull, Iceland 1991–1995. *Jökull*, **45**, 35–58.

BJÖRNSSON, H., ROTT, H., GUDMUNDSSON, S., FISCHER, A., SIEGEL, A. & GUDMUNDSSON, M. T. 2001. Glacier–volcano interactions deduced by SAR interferometry. *Journal of Glaciology*, **47**, 58–70.

BOURGEOIS, O., DAUTEUIL, O. & VAN VLIET-LANOË, B. 1998. Pleistocene subglacial volcanism in Iceland: tectonic implications. *Earth and Planetary Science Letters*, **164**, 165–178.

CHAPMAN, M. G., ALLEN, C. C., GUDMUNDSSON, M. T., GULICK, V. C., JAKOBSSON, S. P., LUCCITTA, B. K., SKILLING, I. P. & WAITT, R. B. 2000. Volcanism and ice interactions on Earth and Mars. *In:* ZIMBELMAN, J. R., & GREGG, T. K. P. (ed.) *Environmental effects on volcanic eruptions.* Kluwer Academic/Plenum Publishers, New York, 39–74.

EINARSSON, P., BRANDSDÓTTIR, B., GUDMUNDSSON, M. T., BJÖRNSSON, H., SIGMUNDSSON, F. & GRÖNVOLD, K. 1997. Center of the Iceland hotspot experiences volcanic unrest. *Eos*, **78**, **369**, 374–375.

GUDMUNDSSON, M. T. & BJÖRNSSON, H. 1991. Eruptions in Grímsvötn 1934–1991. *Jökull*, **41**, 21–46.

GUDMUNDSSON, M. T. & MILSOM, J. 1997. Gravity and magnetic studies of the subglacial Grímsvötn volcano, Iceland: Implications for crustal and thermal structure. *Journal of Geophysical Research*, **102**, 7691–7704.

GUDMUNDSSON, M. T., SIGMUNDSSON, F., & BJÖRNSSON, H. 1997. Ice–volcano interaction of the 1996 Gjálp subglacial eruption, Vatnajökull, Iceland. *Nature*, **389**, 954–957.

GUDMUNDSSON, M. T., HÖGNADÓTTIR, Þ., BJÖRNSSON, H., PÁLSSON, F. & SIGMUNDSSON, F. 1999. Heat flow and ice cap response after the 1996 Gjálp eruption in Vatnajökull, Iceland. *Eos*, **80**, 333.

GUDMUNDSSON, M. T., HÖGNADÓTTIR, Þ., PÁLSSON, F. & BJÖRNSSON, H. 2000. Grímsvötn: Eldgosi 1998 og breytingar á botni, rúmmáli og jarhita 1996–1999 [The 1998 eruption in Grímsvötn and changes in bedrock topography, lake volume and geothermal activity 1996–1999]. *University of Iceland, Science Institute Report*, **RH-03–2000**.

GUDMUNDSSON, S., GUDMUNDSSON, M. T., BJÖRNSSON, H., SIGMUNDSSON, F., ROTT, H. & CARTENSEN, J. M. 2002. Three-dimensional glacier sur-face motion maps at the Gjálp eruption site, Iceland, inferred from combining InSAR and other ice displacement data. *Annals of Glaciology*, **34**, 315–322.

INGEBRITSEN, S. E. & SANFORD, W. E. 1998. *Groundwater in geologic processes.* Cambridge University Press, Cambridge.

JAKOBSSON, S. P. 1978. Environmental factors controlling the palagonitization of the Surtsey tephra, Iceland. *Geological Society of Danmark, Bulletin*, **27**, Special Issue, 97–105.

JAKOBSSON, S. P. 1979. Outline of the petrology of Iceland. *Jökull*, **29**, 57–73.

JÓHANNESSON, H. & SAEMUNDSSON, K. 1998. *Geological map of Iceland. 1: 500 000. Bedrock geology* (2nd edition). Icelandic Institute of Natural History, Reykjavík.

JONES, J. G. 1969. Intraglacial volcanoes of the Laugarvatn region, southwest Iceland – I. *Quarterly Journal of the Geological Society, London*, **124**, 197–211.

KJARTANSSON, G. 1943. *Árnesingasaga, I. Náttúrulsing Árnessýslu.* [The geology of Árnessýsla]. Árnesingafélagið, Reykjavík.

KRISTJÁNSSON, L. 1970. Paleomagnetism and magnetic surveys in Iceland. *Earth and Planetary Science Letters*, **8**, 101–108.

MARIA, A., CAREY, S., SIGURDSSON, H., KINCAID, H. & HELGADÓTTIR, G. 2000. Source and dispersal of jökulhlaup sediments discharged to the sea following the 1996 Vatnajökull eruption. *Geological Society of America, Bulletin*, **112**, 1507–1521.

MATHEWS, W. H. 1947. 'Tuyas', flat topped volcanoes in northern British Columbia. *American Journal of Science*, **245**, 560–570.

MURASE, T. & MCBIRNEY, A. R. 1973. Properties of some common igneous rocks and their melts at high temperatures. *Geological Society of America Bulletin*, **84**, 3563–3592.

ODDSSON, B. 1982. Rock quality designation and drilling rate correlated with lithology and degree of alteration in volcanic rocks from the 1979 Surstey drill hole. *Surtsey Research Progress Report*, **IX**, 94–97.

PÁLMASON, G. 1971. Crustal structure of Iceland from explosion seismology. *Societas Scientarium Islandica*, **40**.

PÁLSSON, S HARALDSSON, G. I. & SIGFÚSSON, G. H. 1984. Elismassi og poruhluti bergs [Density and porosity of rocks]. *National Energy Authority Report*, **OS-84048/VOD-18B**.

PATERSON, W. S. B. 1994. *The physics of glaciers*. Pergamon/Elsevier, Kidlington.

PEDLEY, R. C., BUSBY, J. B. & DABEK, Z. K. 1991. *Gravmag user manual*. British Geological Survey, Keyworth.

SIGMARSSON, O., KARLSSON, H. R. & LARSEN, G. 2000. The 1996 and 1998 subglacial eruptions beneath the Vatnajökull ice sheet in Iceland: contrasting geochemical and geophysical inferences on magma migration. *Bulletin of Volcanology*, **61**, 468–476.

SMELLIE, J. L. & SKILLING, I. P. 1994. Products of subglacial eruptions under different ice thicknesses: two examples from Antarctica. *Sedimentary Geology*, **91**, 115–129.

STEFÁNSSON, V. & BJÖRNSSON, S. 1982. Physical aspects of hydrothermal systems. *American Geophysical Union, Geodynamic Series*, **8**, 123–145.

STEINTHORSSON, S., HARDARSON, B. S., ELLAM, R. M. & LARSEN, G. 2000. Petrochemistry of the Gjálp 1996 subglacial eruption, Vatnajökull, SE Iceland. *Journal of Volcanology and Geothermal Research*, **98**, 79–90.

SVERRISSON, M., JÓHANNESSON, Æ. & BJÖRNSSON, H. 1980. Radio-echo equipment for depth sounding of temperate glaciers. *Journal of Glaciology*, **93**, 477–485.

WERNER, R., SCHMINKE, H. U. & SIGVALDASON, G. 1996. A new model for the evolution of table mountains: volcanological and petrological evidence from Herdubreid and Herdubreidartögl volcanoes (Iceland). *Geologische Rundschau*, **85**, 390–397.

Subglacial volcanic features beneath the West Antarctic Ice Sheet interpreted from aeromagnetic and radar ice sounding

JOHN C. BEHRENDT[1,3], D. D. BLANKENSHIP[2], D. L. MORSE[2], C. A. FINN[3] & R. E. BELL[4]

[1] *Institute of Arctic and Alpine Research, University of Colorado, Boulder, CO 80309–0450 USA (e-mail: behrendj@stripe.colorado.edu)*
[2] *University of Texas Institute of Geophysics, Austin, TX 78759–8345, USA*
[3] *United States Geological Survey, Denver, CO 80225, USA*
[4] *Lamont Doherty Earth Observatory, Columbia University, Palisades, N.Y. 10964, USA*

Abstract: The West Antarctic Ice Sheet (WAIS) flows through the volcanically active, late Cenozoic West Antarctic rift system. Active subglacial volcanism and a vast (>10^6 km^3) extent of subglacial volcanic structures have been interpreted from aerogeophysical surveys over central West Antarctica in the past decade, combined with results from 1960s and 1970s aeromagnetic profiles over the WAIS. Modelling of magnetic anomalies constrained by radar ice sounding shows volcanic sources at the base of the ice throughout large areas, whose subglacially erupted hyaloclastite edifices have been eroded by moving ice, as in Iceland. The 1800 m-high divide of the WAIS is underlain by the 400 km-long volcanic Sinuous Ridge, which rises above sea level; most hyaloclastite edifices there have also been glacially removed, indicating migration of the ice divide through time. Northeast of the divide of the WAIS there is a 400-nT positive magnetic anomaly over the shallowest, most rugged bedrock topography (elevation +380 m above sea level), probably comprising sub-aerially erupted flows erupted when the Sinuous Ridge area was deglaciated. Uplift of the Sinuous Ridge may have forced the advance of the WAIS. Other aspects of the subglacial volcanism in Antarctica can be observed in Iceland and have a direct bearing on our understanding of the subglacial conditions of the WAIS and its dynamics.

The West Antarctic Ice Sheet (WAIS) flows through the immense enigmatic West Antarctic rift system (Fig. 1). The rift is characterized by exposures of alkaline volcanic rocks (LeMasurier & Thomson 1990) and by >10^6 km^3 of subglacial volcanic rocks interpreted from aeromagnetic data (Behrendt *et al.* 1994). A few exposures are as old as *c.* 30 Ma, but most ages range from <15 Ma to the present; active volcanism can be expected throughout the rift system. The WAIS and the late Cenozoic volcanic activity in the West Antarctic rift system have been coeval since at least Miocene time, although the area has been deglaciated at times during this period (e.g. Scherer 1991, Scherer, *et al.* 1998). The source of the extensive volcanism is probably a mantle plume (Behrendt *et al.* 1992; Hole & LeMasurier 1994). The 1800 m-high divide of the WAIS is underlain by the poorly defined, 400 km-long Sinuous Ridge (Fig. 1). Jankowski *et al.* (1983) interpreted the Sinuous Ridge to comprise volcanic rocks based on >1000-nT anomalies observed on several widely spaced aeromagnetic profiles.

The presence of volcanic activity (present or past) beneath the WAIS can only be inferred from geophysical measurements (e.g. Blankenship *et al.* 1993, 2000; Behrendt *et al.* 1995). However, as pointed out by Bourgeois *et al.* (2000), in Iceland, tectonic activity and volcanism occurred during the last glaciation beneath an ice sheet about 1000 m thick. Therefore, comparisons with the WAIS are useful.

In this paper we discuss aeromagnetic results (Fig. 2) constrained by bedrock elevation from radar ice sounding to examine whether the hypothesis for glacial removal of volcanic edifices (Behrendt *et al.* 1995) is valid for the Sinuous Ridge area beneath the WAIS divide, and compare these interpretations with observations in Iceland. The Antarctic results are also discussed in Behrendt *et al.* (2002).

From: SMELLIE, J. L. & CHAPMAN, M. G. (eds) 2002. *Volcano–Ice Interaction on Earth and Mars*. Geological Society, London, Special Publications, **202**, 337–355. 0305-8719/02/$15.00

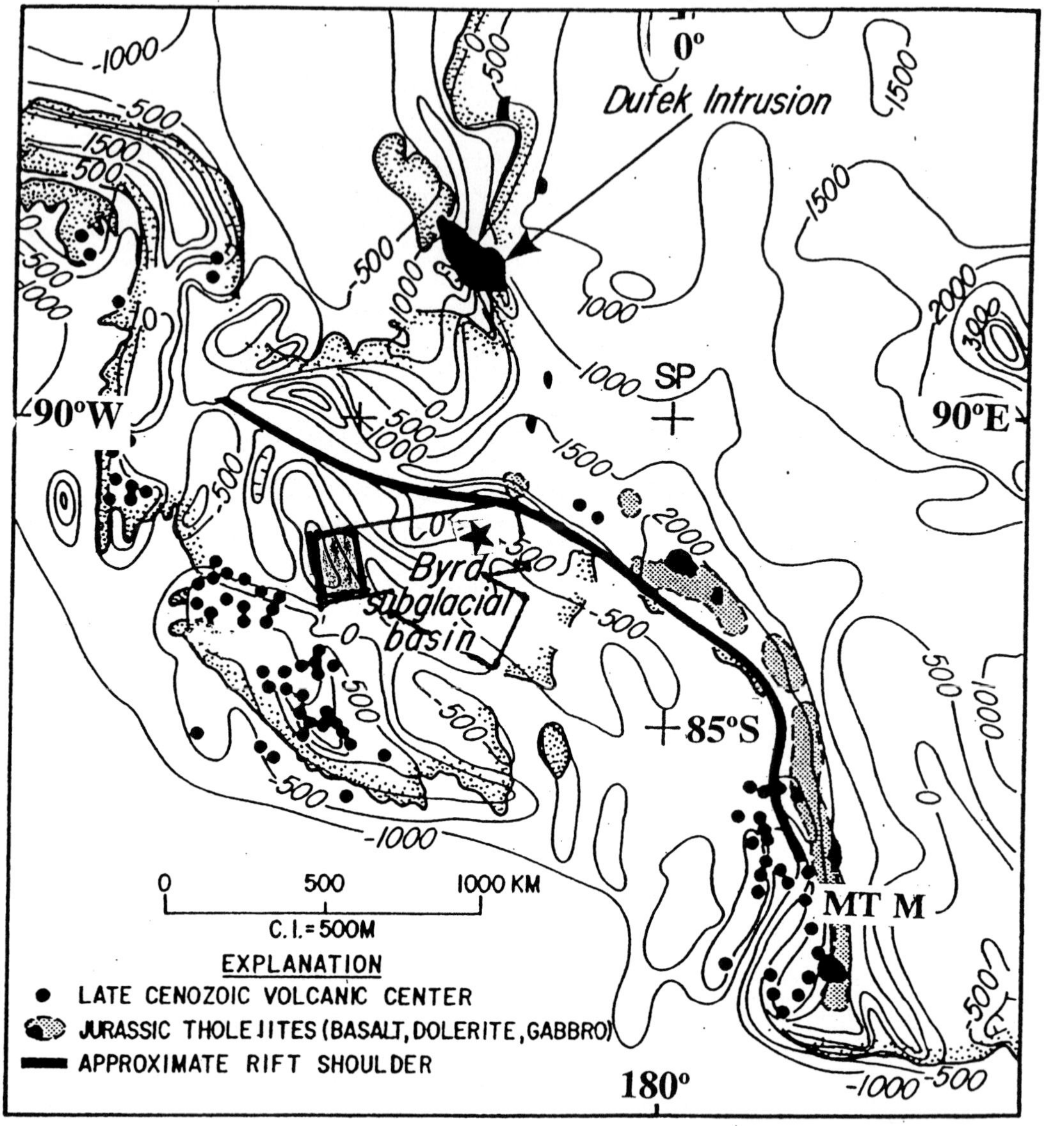

Fig. 1. Generalized isostatically compensated (after ice removal) bedrock elevation map of part of Antarctica (from Behrendt *et al.* 1994; modified from Drewry 1983). Coarse stippled margin shows edge of present grounded ice. Irregular box centred over sub sea level area indicates the location of the aeromagnetic survey in central West Antarctica (Fig. 2). Shaded rectangular box in centre of figure indicates part of aeromagnetic and radar surveys (Figs 4 & 5) and high topography of Sinuous Ridge (the area which rises above sea level at the rectangular box). Bentley Subglacial Trench (Fig. 5) is deep area at the south of shaded box. The CASERTZ survey surrounds active 'Mt. CASERTZ', indicated by a star. Cross marks South Pole (SP). The Meridian at 0° (grid north) is at top of the map following usual convention for small-scale maps of Antarctica; in contrast, true north is at top of Figures 4–8, 10 and 11). Mt. Melbourne (Figs 11 & 12) is also indicated (MT M).

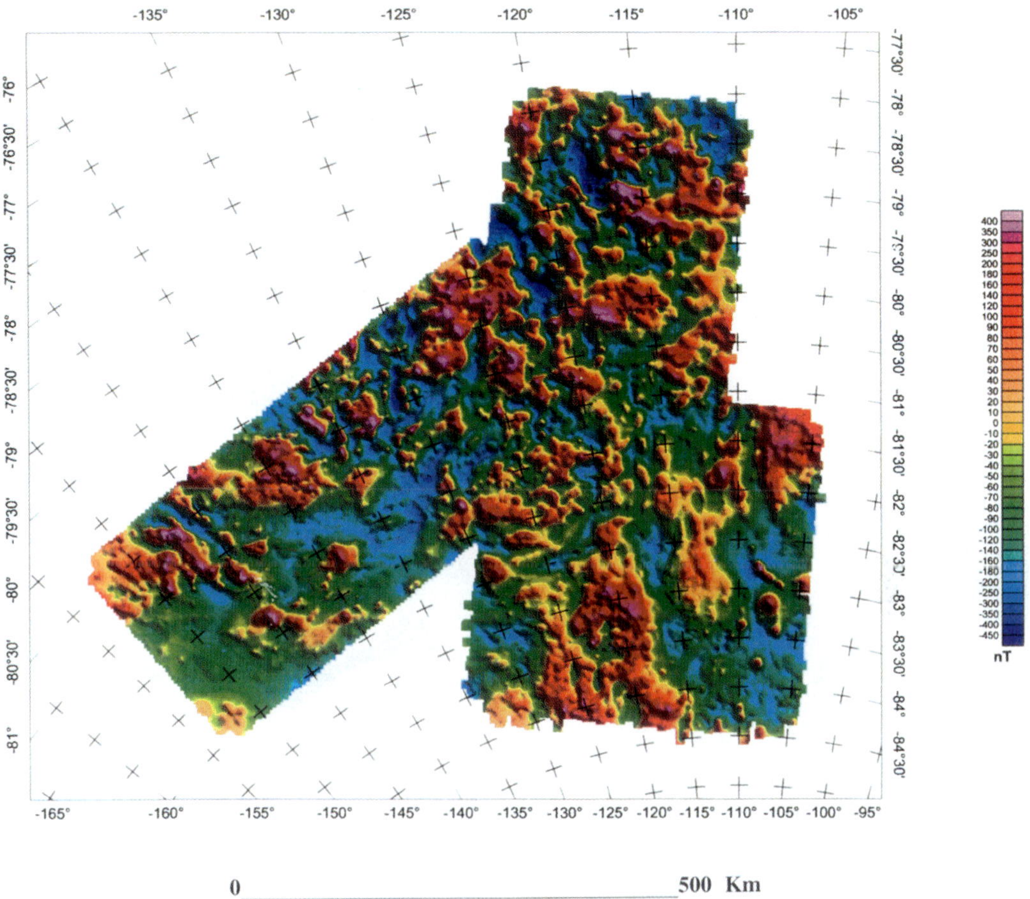

Fig. 2. Central west Antarctica aeromagnetic anomaly map (after Sweeney *et al.* 1999). The rugged, high-amplitude magnetic anomaly field is indicative of the shallow source volcanic centres interpreted to underlie this area of the WAIS (Behrendt *et al.* 1991*b*, 1994, 1998).

Glacial removal of unconsolidated volcanic edifices erupted beneath the West Antarctic Ice Sheet

LeMasurier (1990) interpreted observed hyaloclastites associated with volcanic exposures (primarily flows; Fig. 1) throughout the West Antarctic rift system as evidence of subglacial eruptions. High-amplitude magnetic anomalies have been long known to mark volcanoes and exposures of volcanic rocks in the rift and provided initial evidence of extensive deposits of inferred volcanic rocks beneath the WAIS (e.g. Behrendt 1964). The extent of submarine and subglacial late Cenozoic volcanism is inferred from geophysical data to be much larger than the mapped volcanic exposures (Fig. 1). Several hundred volcanic centres are interpreted from shallow-source magnetic anomalies over the sea and ice covered areas of the rift system (about 100 in the Ross Sea survey, Behrendt *et al.* 1991*a* and 30–40 in the 1991–1992 CASERTZ survey (the part of the survey block surrounding 'Mt. CASERTZ' in Figure 1, Behrendt *et al.* 1994)). Many volcanic centres are apparent in the steep gradient, short wavelength magnetic anomalies over central West Antarctica (Fig. 2). Blankenship *et al.* (1993) reported geophysical evidence of an active volcano (Fig. 1) in the 1991–1992 CASERTZ survey.

Magnetic sources of the high-amplitude short-wavelength anomalies observed over the WAIS (e.g. Fig. 2) could be any age, but we would not expect to observe (e.g. Behrendt *et al.* 1991*b*) significant anomalies at the flight elevation and

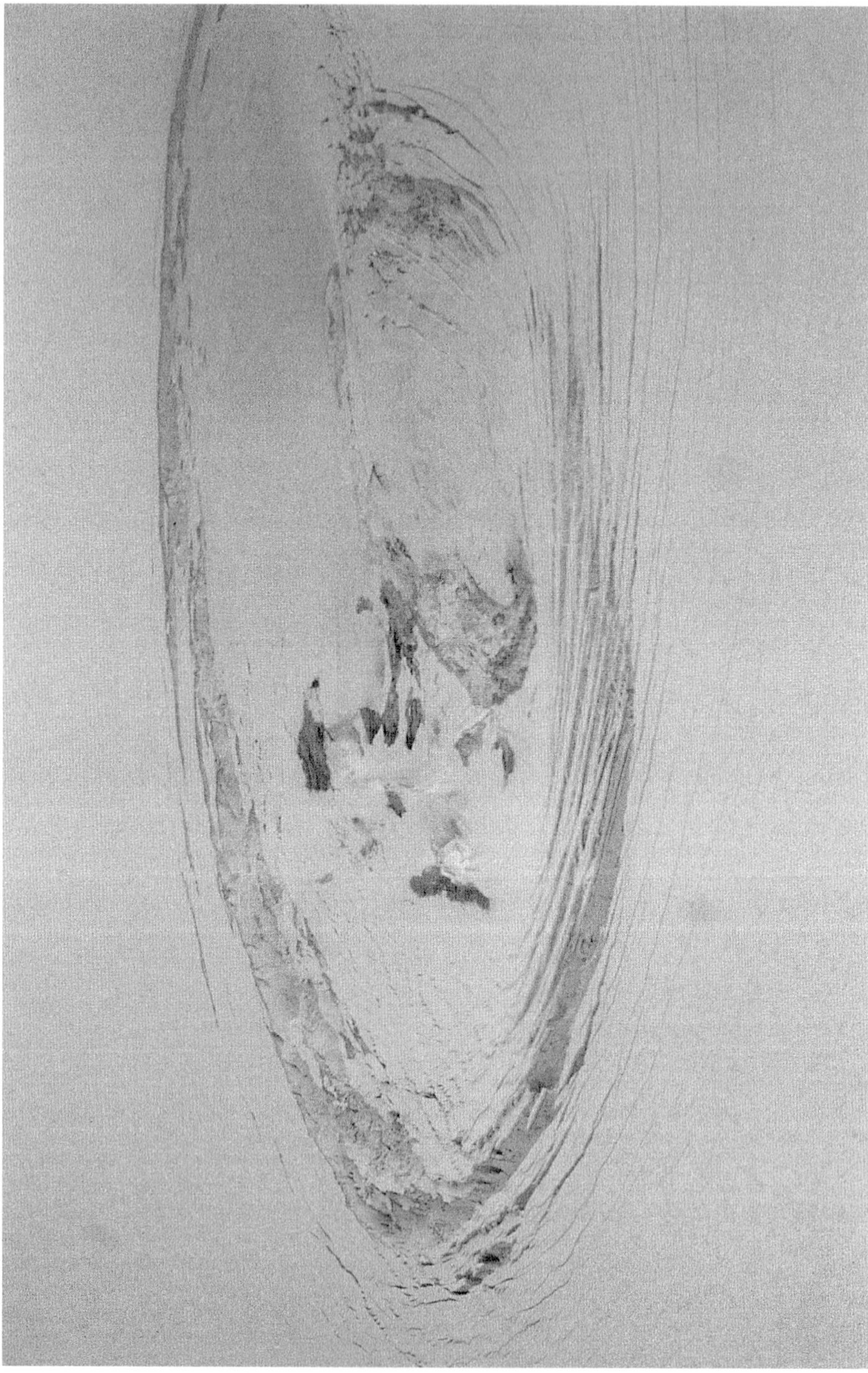

Fig. 3. Photograph of approximately 2 km-diameter ice surface depression marking subglacial melting caused by volcanic activity in Iceland. Collapse is the result of a jökulhlaup about a week earlier (M. T. Gudmundsson, pers. comm., 2001).

ice thickness from the thin Jurassic Ferrar dolerite sills which may be present beneath the ice. High-amplitude magnetic anomalies may also result from several-kilometre-thick Jurassic intrusions such as the Dufek intrusion (Fig. 1), magnetic granitoids (Maslanjy *et al.* 1991) or a thick pile of Kirkpatrick basalt flows (Behrendt *et al.* 1991*b*). A significant component of remanent magnetization, which would be likely for volcanic rocks and some plutonic rocks (Q in the Dufek intrusion ranges from 1.3 to 5; Behrendt *et al.* 1991*b*, 1996) would not produce anomalies characteristic of the present field direction as we observe (Behrendt *et al.* 1994). Because the Antarctic plate has been essentially stationary for the past 100 Ma, Late Cretaceous magnetic granitoids, for example, cannot be ruled out based on magnetization directions; however, we would not expect magnetizations or susceptibilities as high as are required to produce the observed anomalies over the 1–3 km-thick ice sheet at 1 km flight elevations above the ice.

The great volume ($>10^6$ km^3) of volcanic rocks and subvolcanic intrusions calculated from magnetic anomalies to exist beneath the ice sheet and ice shelf (Behrendt *et al.* 1994) raises the question as to why there are such limited exposures (Fig. 1). Behrendt *et al.* (1995, 1997) suggested that the large exposed volcanoes in the West Antarctic rift system (e.g. in the Transantarctic Mountains and Marie Byrd land areas, Fig. 1) may be the exceptions. The ages of these high peaks range as far back in time as several million years (LeMasurier & Thomson, 1990) and most were likely erupted through the WAIS. Because the exposed flows were erupted subaerially they are quite erosion resistant. In contrast, Behrendt *et al.* (1995) concluded that glacial 'removal' of volcanic edifices, originally consisting of subglacially erupted debris (probably mostly pillow breccias and hyaloclastites) underlain by volcanic centres (i.e. subvolcanic intrusions >1 km thick), and now marked by prominent magnetic anomalies over the West Antarctic Ice Sheet (Fig. 2), Ross Ice Shelf and Ross Sea may be the general case. It is important to note that if, for example, very magnetic, <50–400 m-thick volcanic flows were also present, as is quite likely, these would not produce the observed (>50–100 nT) magnetic anomalies at the survey flight elevations (*c.* 1 km) and ice thicknesses (*c.* 1–3 km) present. Behrendt *et al.* (1995) inferred that small volcanic peaks must be episodically erupted from active volcanoes erupting beneath the moving ice. They cited the example of a volcanically active peak ('Mt. CASERTZ', Fig. 1) reported beneath *c.* 1800–1900 m-thick moving ice (Blankenship *et al.* 1993). There is no magnetic anomaly over the apparently non-magnetic edifice ('Mt. CASERTZ'), which sits on the edge of the source of a 40 km-wide circular *c.* 500 nT anomaly caused by the subvolcanic intrusion associated with this active volcano (Behrendt *et al.* 1994). Behrendt *et al.* (1995) suggested that this small active peak is a transitory feature. Although there is a broad depression indicated by the laser altimeter and AVHRR satellite imagery (Blankenship *et al.* 1993), the snow surface directly over the subglacial peak is inconspicuous, unlike some subglacial eruptions in Iceland (e.g. Figure 3 which is similar in appearance to the Gjalp eruption of 1996; MT Gudmundsson, pers. comm., 2001).

Behrendt *et al.* (1995) further suggested that other small volcanic edifices (probably made up of unconsolidated, easily eroded hyaloclastite and pillow breccia) have been continuously removed by the ice during and after their eruption. Even quite thick deposits of such mixed volcanic material would not produce observable magnetic anomalies as is the case for the 'Mt. CASERTZ' edifice. Any high remanent magnetizations would cancel in a randomly oriented pile of debris; altered volcanic material would have low susceptibilities. See discussion of the models below.

Gudmundsson *et al.* (1997) reported, from the 1996 subglacial eruption beneath the 500 to 750-m thick ice of Vatnajökull, Iceland, observations of a 200–300 m-high hyaloclastite ridge which is probably also transitory. Bourgeois *et al.* (1998) concluded that ice streams in Iceland remove hyaloclastite deposits as they are erupted, just as suggested by Behrendt *et al.* (1995).

The hypothesis of glacial removal as discussed by Behrendt *et al.* (1995) does not assume anything about either a wet- or cold-based ice sheet. Because of the generally high present accumulation rate (Vaughan *et al.* 1999) we would expect a cold-based WAIS at present and quite likely in the past. Volcanic debris might be nearly as much connected to the overriding ice sheet as to the bedrock, as pointed out by Behrendt *et al.* (1996). Coincident with subglacial eruption, magma would melt ice; this water would chill the magma. The erupted material would likely be very poorly crystallized, and extensively fragmented. The fragmented material could be readily removed ('removal' is probably more appropriate than 'erosion') by the ice.

The exposed peaks of >3 km-high volcanoes in the West Antarctic rift system (e.g. many in Figure 1 including Mt. Melbourne as discussed below) are capped by lava flows which were probably erupted subaerially (e.g. McIntosh &

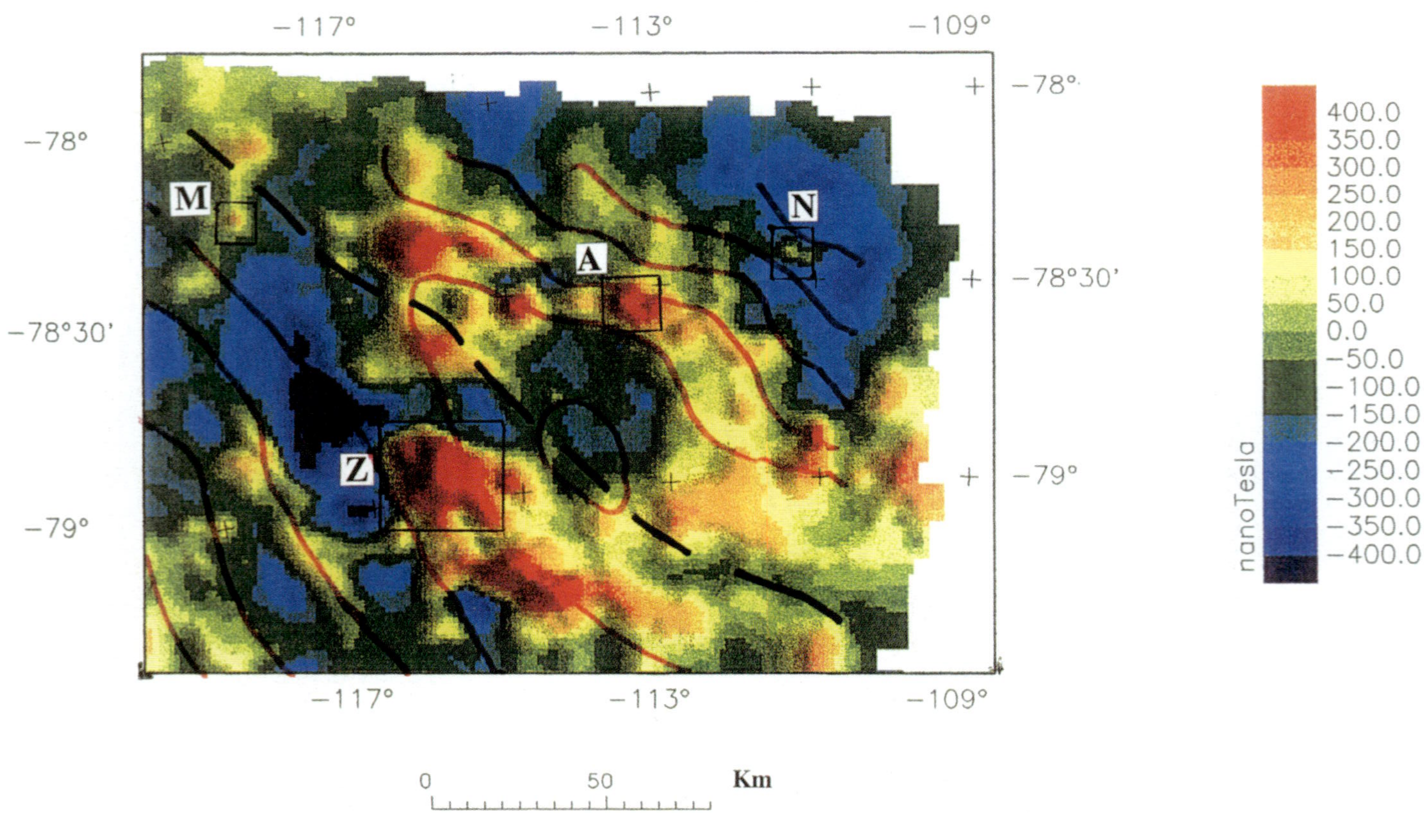

Fig. 4. Shaded aeromagnetic map of northernmost area of Figure 2 marked by rectangular box in Figure 1. The circular magnetic anomaly pattern is interpreted as caused by volcanic rocks associated with an inferred caldera complex. Anomalies A Z, M and N are modelled as shown in Figures 6–8 and 10, respectively. Generalized WAIS snow surface contours from SOAR data (WAISCORES 2000) are superimposed at 50 m contour interval; the dome elevation is 1800 m. The WAIS divide is indicated by a heavy dashed line.

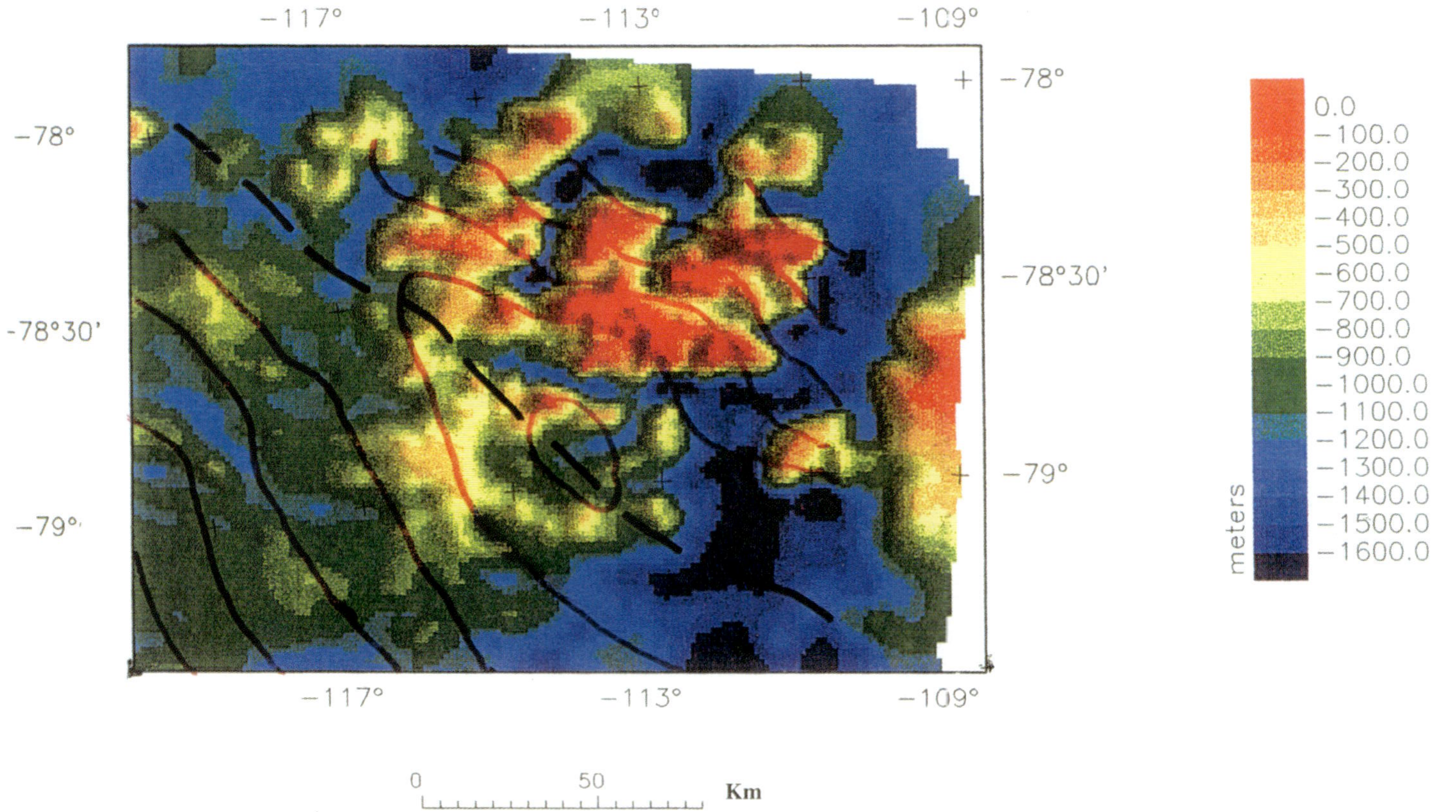

Fig. 5. Bedrock elevation map compiled from radar ice sounding of area of Figure 4. The lowest elevations at the south are the Bentley Subglacial Trench. Generalized WAIS snow surface contours from SOAR data (WAISCORES 2000) are superimposed at 50 m contour interval; the dome elevation is 1800 m. The WAIS divide is indicated by a heavy dashed line.

Gamble 1991), and which protected these peaks from subsequent glacial erosion as is apparent from the old ages reported (LeMasurier & Thomson 1990). Marine seismic reflection and magnetic gradiometer profiles over volcanic structures penetrating the sedimentary section in the Victoria Land basin provide evidence to support the glacial removal hypothesis in the formerly glaciated Ross Sea continental shelf (Behrendt *et al.* 1995). Smellie & Skilling (1994) and Skilling (1994) have discussed subglacial eruptions beneath much thinner Antarctic ice than the WAIS and noted that 'subglacial volcanoes and their products, which are typically poorly consolidated, have a low preservation potential owing to easy removal by the surrounding ice' (Smellie & Skilling 1994).

Gudmundsson *et al.* (1997), based on numerous classic Icelandic papers, noted the tripartite character of volcanic eruption through ice sheets in Icelandic exposures (i.e. pillow basalts at the base, hyaloclastite beneath thinner ice, and subaerial flows at the top). If this were the case in the WAIS area, we would see mostly subaerially erupted peaks (cf. Fig.1), the hyaloclastites and other volcanic debris (e.g. pillow breccia) would have been removed, and thin basalt flows would not be observable at flight elevations and ice thicknesses of the various surveys as discussed in the magnetic model section below. Only volcanic centres (i.e. subvolcanic intrusions) would produce the high amplitude magnetic anomalies observed over the WAIS.

Aeromagnetic and radar ice sounding survey

From December 1991 to January 1997 this study acquired an orthogonal gridded aeromagnetic survey at a 5 km line spacing (Blankenship *et al.* 1993, 2000; Behrendt *et al.* 1994) over central West Antarctica (Fig. 2) combined with radar ice sounding, aerogravity and laser surface altimeter measurements. The magnetic data are available in Sweeney *et al.* (1999). The amplitudes of the magnetic anomalies range from >1000 to −500 nT. These amplitudes are very high considering that the shallowest possible sources lie at a minimum of 1.5 km below the survey aircraft; the vast majority are >3 km. (See the discussion of anomaly M (Fig. 4) in the model section below; a 300 nT anomaly would increase in amplitude to about 1300 nT if the sensor were only 200 m above the source.) The sources of most of the short wavelength anomalies dominating the magnetic field are essentially at the base of the ice and are interpreted to be caused mostly by late Cenozoic volcanic centres (subvolcanic intrusions) as discussed by Behrendt *et al.* (1964, 1991*b*, 1994, 1998) and Jankowski *et al.* (1983). Jonsson *et al.* (1991) showed and discussed the magnetic anomaly map of Iceland, which is dominated by the volcanic terrain; the similarity this map and other aeromagnetic surveys over volcanic terrains to Figure 2 is striking.

Results obtained for the aeromagnetic and bedrock elevations over the west end of the Sinuous Ridge are shown in Figures 4 and 5. The prominent positive and negative anomalies range from 1200 (anomaly Z) to −500 nT (Fig. 4). The magnetic field is dominated by a 70 km-diameter circular pattern of positive anomalies whose amplitudes range from 400–1200 nT surrounding a low amplitude *c.* −150 nT central area. Behrendt *et al.* (1998) interpreted this pattern of anomalies to mark a volcanic caldera(?) complex associated with late Cenozoic volcanism, which is probably related to tectonic uplift of the Sinuous Ridge. Similar circular anomaly patterns surrounding central 'lows' (with similar amplitudes and diameters) mark other active or young known calderas such as Yellowstone (Bhattacharyya & Leu 1975; Smith & Braille 1994), Jemez, and San Juan (Committee for the Magnetic Map of North America 1987; USGS, unpublished data) in the tectonically active Rocky Mountains area of the USA

Behrendt *et al.* (1998) interpreted a shallow Curie isotherm to be the cause of the regional negative magnetic anomaly surrounding the caldera positive anomalies. We would not particularly expect much topographic evidence to remain for a subglacially erupted caldera considering that the 700 ka Yellowstone caldera has been 'scoured by glaciers – so there is little to see there today' (Francis 1993). Possibly a caldera could form subglacially via magma withdrawal leading to surface collapse, as suggested by J. L. Smellie (pers comm., 2001). Were this the case we would expect the collapsed volume to be chaotic and therefore essentially non magnetic because the likely high remanent magnetization would have random directions which would cancel the effect. The induced magnetization alone would still produce low amplitude, shallow source anomalies in the center of the circular pattern. This is observed (Fig. 4). The shallow subglacial sources of the very high-amplitude anomalies that compose the circular structure in Figure 4 are volcanic, although caldera origin cannot be proven. Interpretation of a meteorite impact structure in a volcanic terrain can also not be ruled out.

The WAIS surface contours overlain on the magnetic map (Fig. 4) indicate that the ice divide

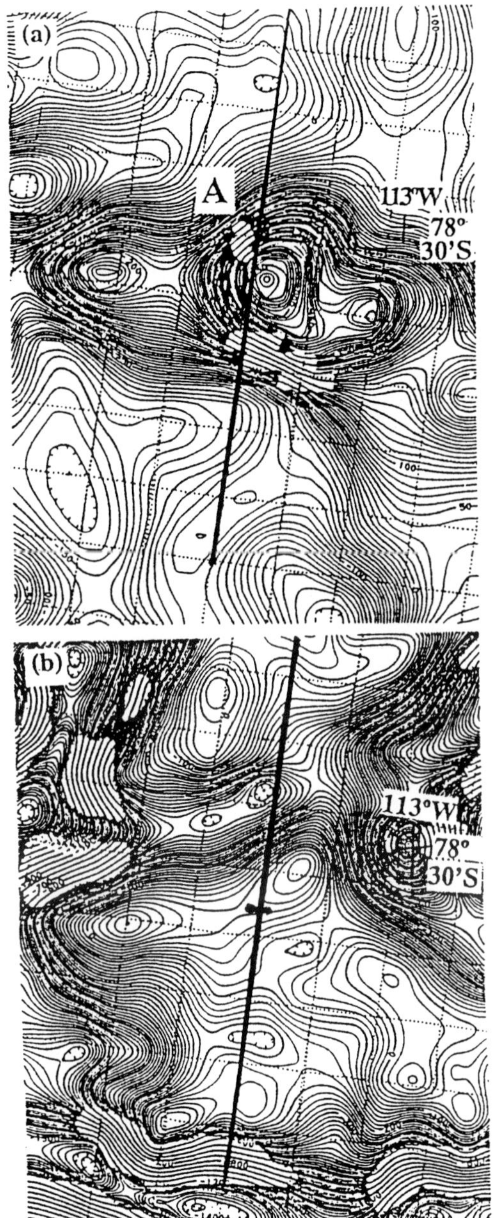

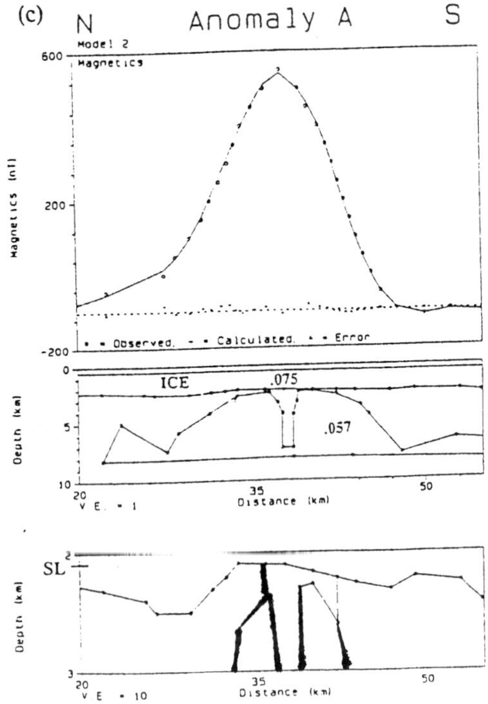

Fig. 6. (**a**) Detailed aeromagnetic map of area of anomaly A (located in Fig. 4). Contour interval 10 nT. Grid survey lines are spaced at 5 km and the long edges of the map trend true north. Location of modelled profile is indicated. (**b**) Bedrock elevation in area of anomaly A. Contour interval 20 m. Grid lines are the same as in (a). Contour interval 20 m. Location of modelled profile is shown, with tic indicating location of maximum magnetic anomaly amplitude.
(**c**) Theoretical two $\frac{1}{2}$-D (two and one half dimensional, a special case of three dimensional model in common usage) model fit to aeromagnetic profile for anomaly A. Susceptibilities indicated in SI. Central body has strike length of 3 km to west and 5 km to east; outside body has strike length of 17 km to west and 10 km to east. SL indicates sea level. V.E., vertical exaggeration.

overlies the inferred caldera marked by the circular magnetic pattern, but we consider that this is a coincidence, because the location of the WAIS divide is probably transitory, as discussed below. Comparison of the magnetic map (Fig. 4) and bedrock elevation map (Fig. 5) shows only minimal correlation of subglacial topography with the caldera magnetic pattern, indicating to us that the probable volcanic debris edifices have been removed by glacial action. The highest bedrock surface does not underlie the WAIS divide (Fig. 5), but is 80 km NE of the 1800 m elevation of the snow surface dome. Several specific anomalies are examined in detail below to demonstrate our interpretation of glacial removal of subglacially erupted volcanic edifices.

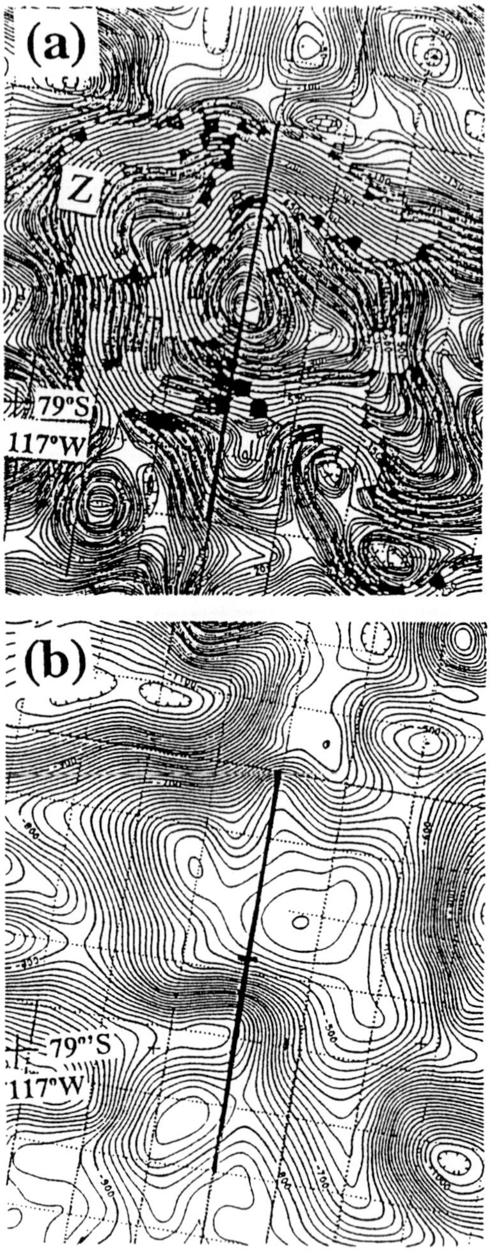

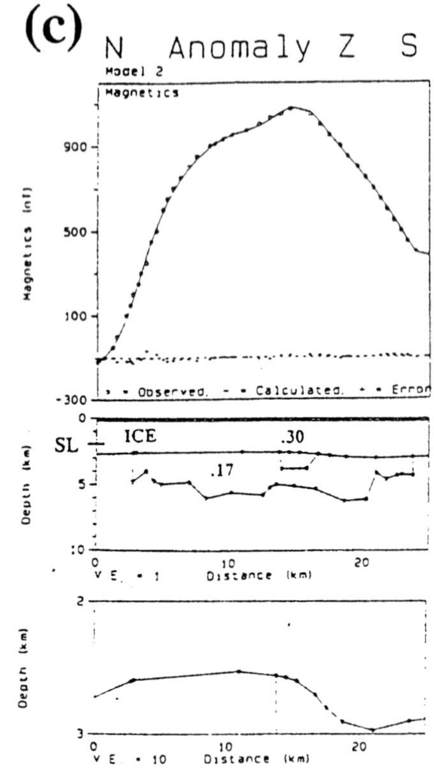

Fig. 7. Aeromagnetic (**a**) and bedrock elevation (**b**) maps of area of anomaly Z (located in Fig. 4); parameters as in Figure 6(**c**). Theoretical two $\frac{1}{2}$-D model fit to aeromagnetic profile for anomaly Z. Susceptibilities indicated in S.I. Central body has strike length of 3 km to east and 3 km to west; outside body has strike length of 10 km to east and 10 km to west.

Magnetic models

Figures 6 and 7 illustrate two of the anomalies composing the circular pattern interpreted as marking a caldera (Fig. 2). The topographic correlation with the anomalies is slight. The two $\frac{1}{2}$-D (two and one half dimensional, a special case of three dimensional model in common usage) magnetic models illustrate the very high apparent susceptibilities in S.I. dimensionless units required to produce the amplitudes observed at the aircraft elevation over the thick ice.

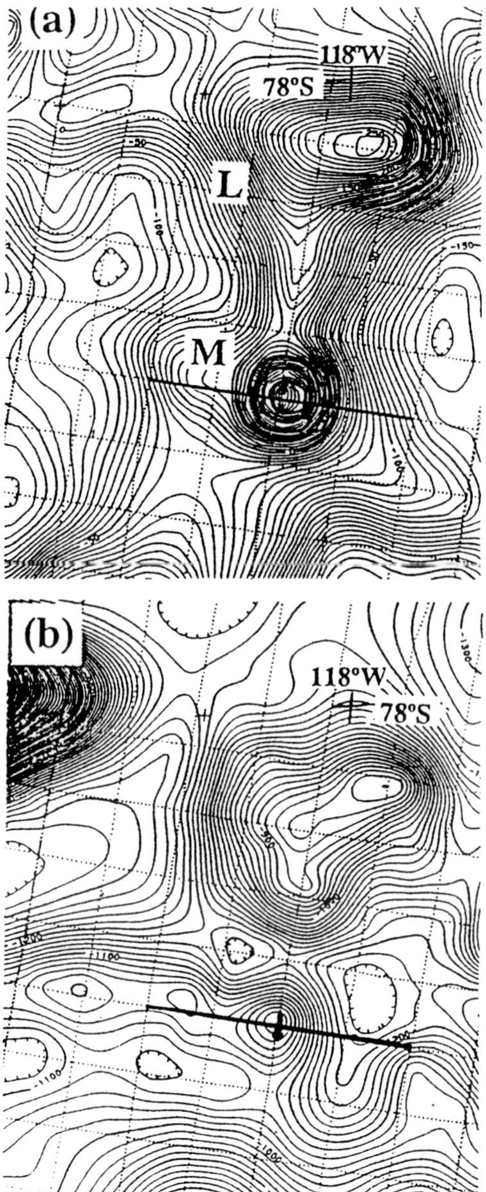

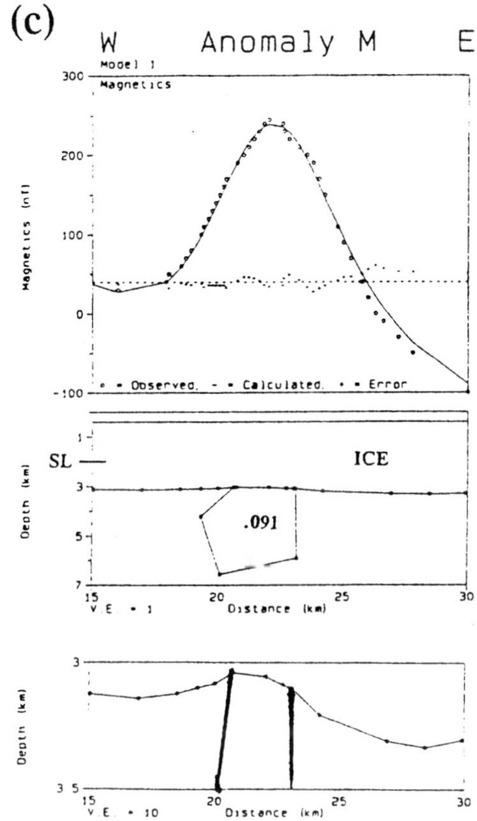

Fig. 8. Aeromagnetic (**a**) and bedrock elevation (**b**) maps of area of anomaly M (see Fig. 4); parameters as in Figure 6(**c**). Theoretical two $\frac{1}{2}$-D model fit to aeromagnetic profile for anomaly M. Susceptibilities indicated in S.I. Body has strike length of 2.5 km to north and 2.5 km to south.

We interpret these high values to indicate that the primary magnetization is a high but unknown remanent magnetization in the present field direction, consistent with that reported for other late Cenozoic volcanic rocks in the West Antarctic rift system as discussed by Behrendt *et al.* (1996). The more magnetic cores of these models (and Figs 6–8, discussed below) were required for the two $\frac{1}{2}$-D models as can be seen from the magnetic contour maps in Figures 6a and 7a. These cores geologically are considered to be reasonable.

The sources of anomalies A and Z (Figs 6 & 7) are interpreted to be primarily the subvolcanic intrusions remaining after the volcanic edifices were removed by the ice. Anomaly Z with an associated amplitude about 1200 nT is the highest in the area shown in Figure 4. The two calculated alternate models (not shown) were fit to anomaly Z, having half and twice the

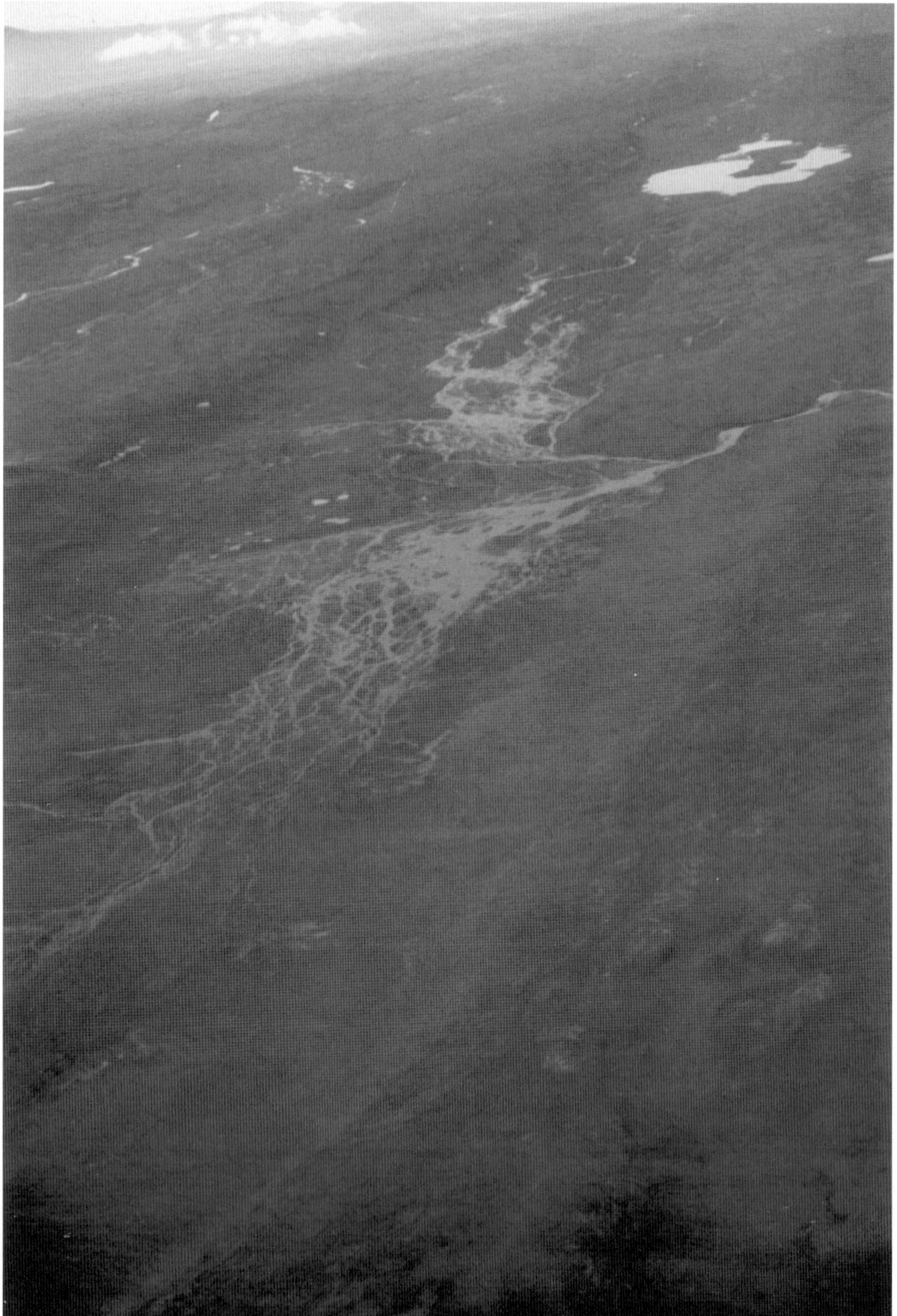

Fig. 9. Aerial view of glacially smoothed hyaloclastite ridges in Iceland with topographic relief comparable to relief of the base of the WAIS (cf. Figs 6–8b). The topography has also been modified by later fluvial erosion and possible tectonic activity (M. T. Gudmundsson, pers. comm., 2001).

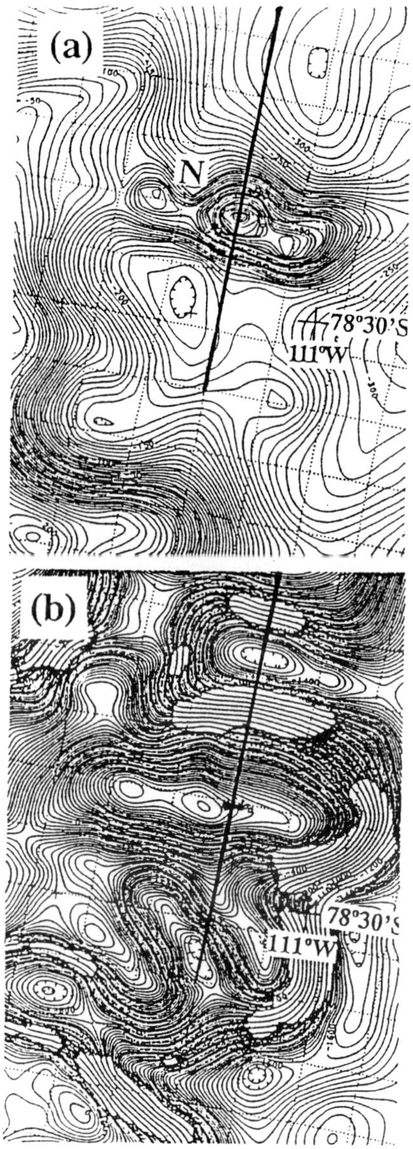

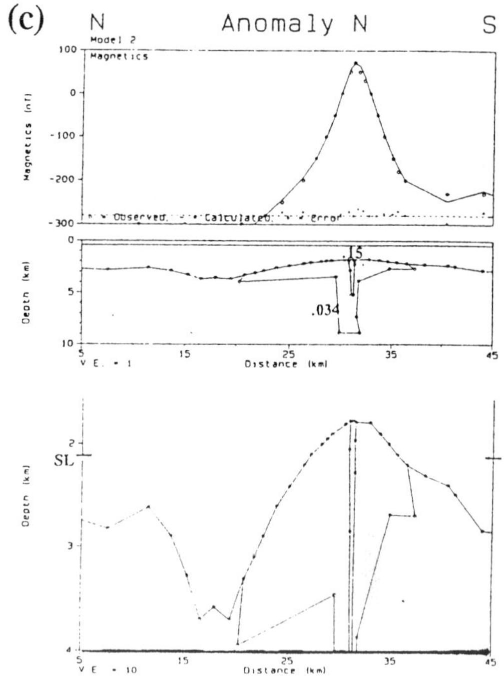

Fig. 10. Aeromagnetic (**a**) and bedrock elevation (**b**) maps of area of anomaly N (see Fig. 4); parameters as in Figure 6. (**c**). Theoretical two $\frac{1}{2}$-D model fit to aeromagnetic profile for anomaly N. Susceptibilities indicated in S.I. Central body has strike length of 2.5 km to west and 2 km to east; outside body has strike length of 8 km to west and 8 km to east.

thickness shown in Figure 7; associated susceptibilities are respectively about twice and half of that indicated in the figure. The higher susceptibility is too high to be likely and the base of the body of twice the thickness seems unreasonably deep. The irregular base of the model was required by the smooth bedrock topography as measured by the radar ice sounding. Alternatively, a geologically reasonable, more complex number of variable-susceptibility bodies could have been used as well with a smoother base.

Several very thin bodies were also calculated (not shown) to examine the possibility of volcanic flows as thin as 200–400 m causing this anomaly; the apparent susceptibilities of 2.0 and 1.0 S.I. respectively, seem unreasonably high even for volcanic rocks. A subvolcanic intrusion is the preferred interpretation in this study.

A common feature of the models in Figures 6 and 7 as well as of other models that were fitted to the circular caldera(?)-like pattern of anomalies (Fig. 4), is the slight topographic expression in the radar-defined bedrock surface. The

10 times vertical exaggeration in the bottom panel of Figure 4 displays this correlation best. The former volcanic edifices, probably composed of incompetent volcanic debris, appear to have been removed.

In similar fashion, models were fitted to other typical volcanic-appearing anomalies shown in Figure 4 away from the caldera pattern, which also support the removal hypothesis. The small-area magnetic anomaly M (Figs 4 & 8) is one such example. As was the case shown in Figures 6 and 7 any former volcanic debris edifice appears to have been largely removed. Again as with anomaly Z above, several very thin bodies were calculated to examine the possibility of volcanic flows causing this anomaly. As was the case with anomaly Z the apparent susceptibility of 2.1 seems unreasonably high for a 100 m-thick flow and 0.44 S.I. remotely possible for a 350 m-thick flow based on comparisons of measured susceptibilities for the McMurdo Volcanics (e.g. Behrendt *et al.* 1987, 1996) even taking high Q into account. Therefore, as above, we interpret that a subvolcanic intrusion is the most likely explanation even for this relatively low amplitude Anomaly M.

The effect of the high flight elevation was calculated over the source of anomaly M to illustrate its influence on the anomalies over the WAIS (e.g. Fig. 2). Decreasing the flight elevation to 200 m a common survey elevation in non-mountainous, basement terrains in other parts of the world, increased the amplitude of the calculated anomaly from about 300 nT (Fig. 8) to about 1300 nT.

Another example (not shown), anomaly L (Fig. 8a) to the north of anomaly M was also modelled with the same result again suggesting glacial removal of most of a possible former edifice associated with the *c.* 20-km wide, 200-m high residual topographic feature.

Figure 9 shows a glacially smoothed hyaloclastite deposit >800 ka) in Iceland that has relief similar to the tops of the sources of the magnetic anomalies beneath the WAIS (e.g. Figs 6–8). The topography has apparently been modified by later fluvial erosion and possible tectonic activity.

In contrast to the above examples is anomaly N, over the highest, most rugged topography in the surveyed area (Figs 4, 5 & 10). The approximately 400-nT positive residual anomaly is fitted well by a peak that resembles a subaerially-erupted volcano composed of magnetic flows. For comparison in this study, the magnetic anomaly was modelled (Figs 1, 11) over exposed, active Mt. Melbourne (Fig. 12) in the Transantarctic Mountains (Behrendt *et al.* 1991*b*). The similarity between anomaly N and Mt. Melbourne is apparent (Figs 10 & 11).

Anomaly N is interpreted as caused by a subaerially-erupted volcano constructed at a time when the WAIS was significantly lower or

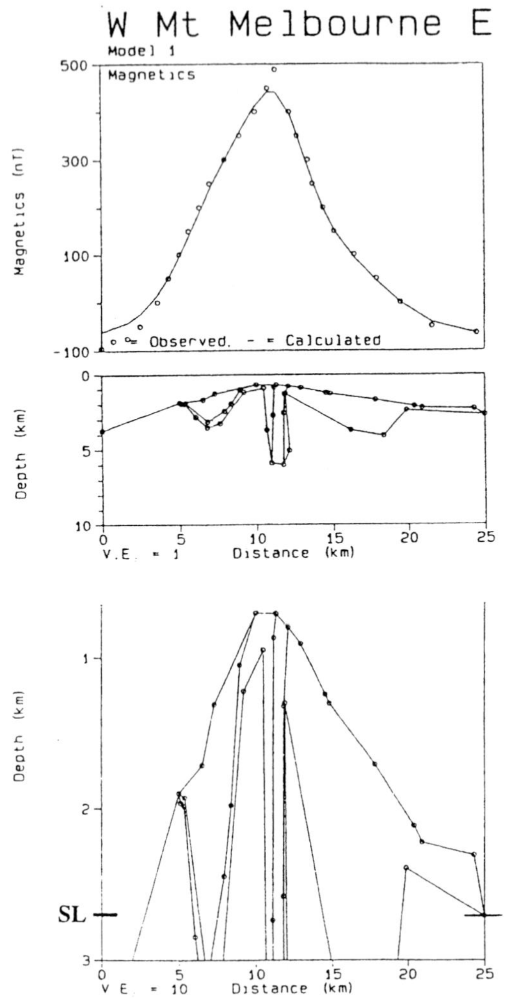

Fig. 11. Theoretical two $\frac{1}{2}$-D model fit to aeromagnetic profile for anomaly over Mt. Melbourne (Fig. 1). Topography from USGS Mt.Melbourne 1:250 000 topographic map. Aeromagnetic data from Behrendt *et al.* 1991*b*. Susceptibilities are not indicated in figure to avoid confusion, but range from 0.04 to 0.30 S.I. from outside to central bodies respectively. Strike lengths bodies range from 2.5 km to 7 km to north and to the south. The relatively complex model was the simplest possible fit to the topography from the map. The complex bodies were required by the two $\frac{1}{2}$-D computation that requires variable strike lengths out of the plane of the profile. A more simple 2-D model could probably be fit to the anomaly by a simple magnetic linear ridge.

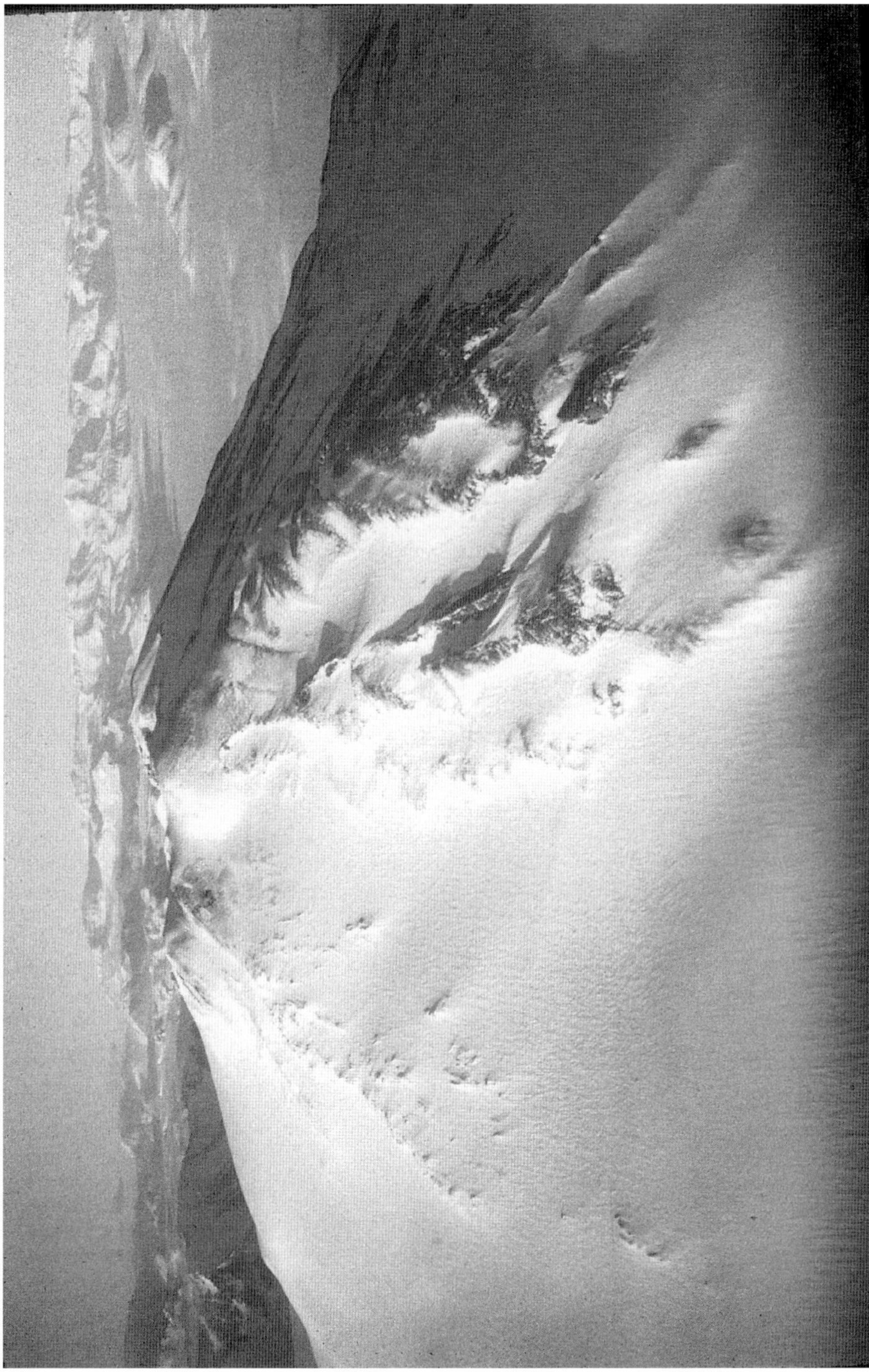

Fig. 12. Aerial view of Mt. Melbourne looking to northwest (Fig. 1). The subaerially erupted edifice (elevation 2732 m) is similar in relief and magnetization to the inferred subaerially erupted source of anomaly N (Fig. 10).

absent. If the WAIS were present but lower at the time of eruption, it is suggested that the volcanic source rose up through the ice sheet and subaerial flows were erupted as at Mt.Melbourne. Figure 5 shows that the present divide of the WAIS lies south of the highest bedrock topography. Scherer *et al.* (1998) have shown that the WAIS collapsed as recently as 400 ka (and presumably at other times). Drewry (1983, Sheet 6) showed between 250 and 1000 m of isostatic rebound in the area of the Sinuous Ridge if the WAIS were removed. The high topography associated with the Sinuous Ridge (Figs 1 & 5) would stand well above sea level after rebound. There would also be an isostatic depression of the Sinuous Ridge area by its own mass, as is well known for other volcanic areas.

Anomaly N and the associated bedrock topography (Fig. 10) also provide evidence (assuming approximately coeval eruption) that the subglacially erupted former edifices (Figs 6–8) if erupted subaerially, would have had edifices similar to that in Figures 10 and 11 and different from subglacial 'Mt CASERTZ' volcano (Behrendt *et al.* 1994, 1995) interpreted as comprised of debris with any possible high remanent magnetization cancelled by mixing. In our interpretation, any subglacially erupted edifices would have been transitory (like 'Mt. CASERTZ') and a function of ice flow velocity and rate of eruption.

Glaciological implications

The greatest uncertainty in any speculation of glaciological effects is the unknown age of the various volcanic structures beneath the WAIS. The shallow Curie isotherm inferred to be the cause of the regional negative magnetic anomaly (Behrendt *et al.* 1998), including the centre of the circular caldera(?) pattern and of the magnetic low surrounding anomaly N (Figs 4 & 10) suggests a young age, possibly Holocene, but this is speculative. Blankenship *et al.* (1993) reported geophysical evidence of active subglacial volcanism beneath the WAIS and active subaerial volcanism throughout the area is likely (LeMasurier 1990).

LeMasurier (1990) stated that volcanic eruptions began at about 28–30 Ma, as measured by K–Ar dates from exposures in Marie Byrd Land and have been increasingly frequent toward the present. LeMasurier & Rex (1983) reported the presence of hyaloclastites deposited at 25 Ma in Marie Byrd Land as evidence of initial glaciation. More recently, Wilch & McIntosh (2000) re-dated and reinterpreted the same locality and reported the oldest evidence for thin local ice or snow at 29–27 Ma (at Mt. Petras). Therefore the earliest evidence of late Cenozoic volcanism and ice in the present area of the WAIS have approximately the same age. However, the presence of glaciation associated with high altitude volcanoes is not evidence of a grounded ice sheet filling the Byrd Subglacial Basin.

The earliest evidence of the WAIS reported from marine research over the Ross Ice Shelf is Late Miocene, as summarized by Anderson & Shipp (2000). It is suggested that uplift of the Sinuous Ridge may have forced the advance of the WAIS at a time when the present Byrd Subglacial Basin area may have only been covered by an ice shelf or pack ice. Subaerial eruption forming the highest topography (e.g. anomaly N, Fig. 10) might have provided a nucleus for glaciation there.

Bourgeois *et al.* (1998) have suggested that moving ice has removed hyaloclastite edifices erupted beneath small Pleistocene ice caps in Iceland (e.g. Fig. 9) and that in other places underlying former ice divides, steep hyaloclastite ridges remain after ice cap collapse. For example Figure 13 shows part of the Pleistocene hyaloclastite ridge Skridutindar north of Laugarvatn in southwestern Iceland. Bourgeois *et al.* (2000) used the location of preserved volcanic structures (such as our Fig. 13) as proxies for the location of former ice divides during the last glacial maximum.

Behrendt *et al.* (1997) noted that 'the Sinuous Ridge underlies the ice divide in West Antarctica so we might expect stagnant ice and negligible shear stress at its base and therefore minimal erosion or removal of associated volcanic edifices.' The results presented here show that this is not the case. Therefore, this study interprets the evidence of glacial removal of probable volcanic debris volcanic edifices (e.g. Figs 6–8) erupted beneath the ice to suggest that the divide of the WAIS has migrated from its present location. Considering the approximately five times greater accumulation on the ocean side of the WAIS contrasted to the Ross Ice Shelf side at present (Vaughan *et al.* 1999) and that this condition has probably varied greatly during the history of the WAIS, considerable migration of the divide through time is to be expected.

Were ice coring operations to proceed as proposed for the area (Figs 4 & 5; WAISCORES 2000), sampling of subglacial material over the sources of the magnetic anomalies such as A Z, and N (Figs 6, 7 & 10) would help test the interpretations within this study.

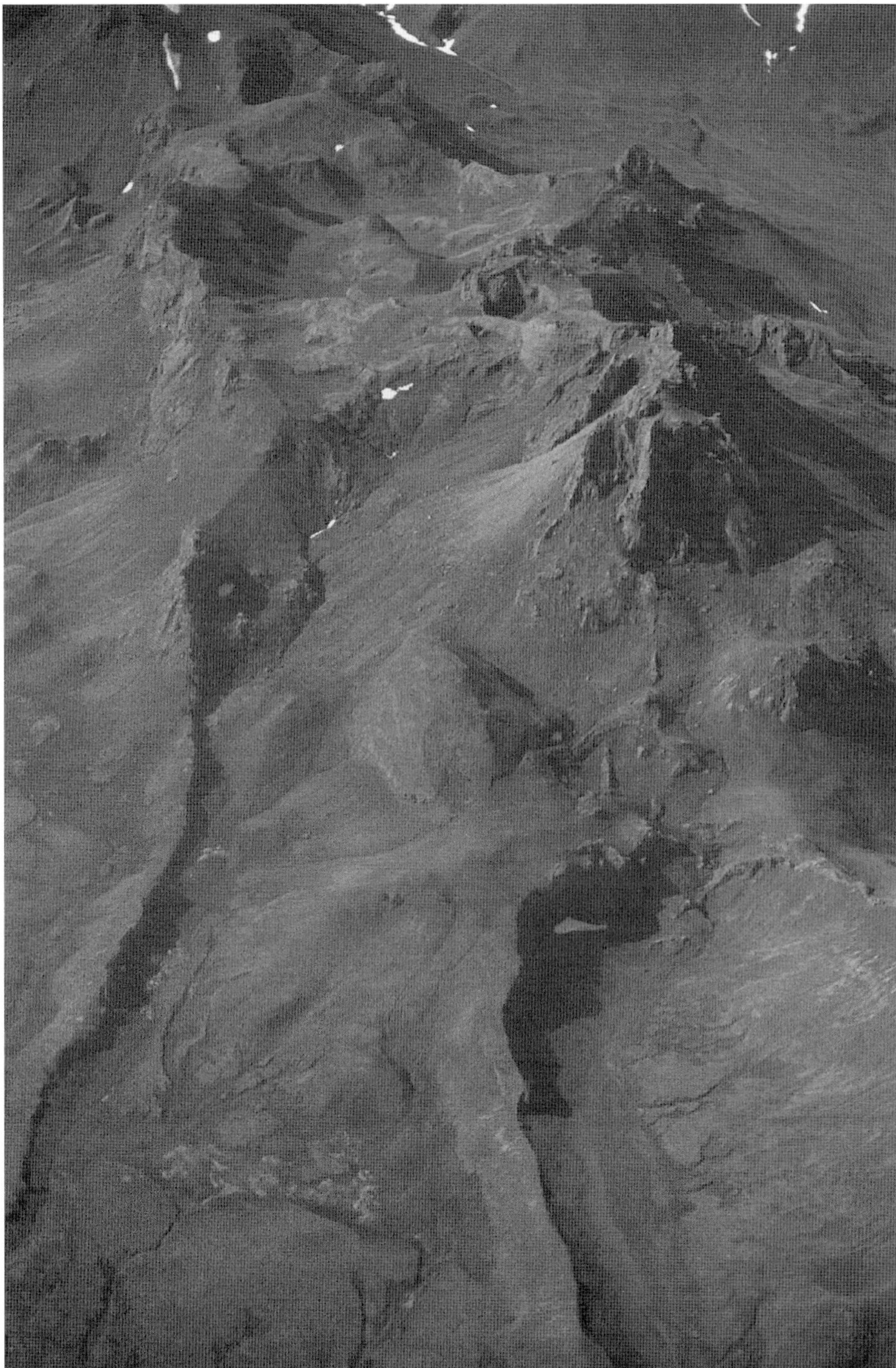

Fig. 13. View of Skridutindar, a steep Pleistocene hyaloclastite ridge north of Laugarvatn, southwestern Iceland, interpreted as having been erupted beneath a former ice divide (Bourgeois *et al.* 1998, 2000).

The aerogeophysical surveys were made by the Support Office for Aerogeophysical Research (SOAR). We thank the field operations team and the personnel of Kenn Borek Air Ltd. for their long term technical support. The work was supported by National Science Foundation Grant OPP-9319877 and the U.S. Geological Survey. The aeromagnetic survey was compiled by R. Sweeney at the US Geological Survey and the radar ice sounding survey was compiled at the University of Texas. M. Siders, J. Hollin, and M. Gudmundsson provided helpful reviews. We also thank organisers of the 2000 Iceland Field Excursion and overflight, in particular S. Jakobsson and M. Gudmundsson.

References

ANDERSON, J. B. & SHIPP, S. S. 2000. Evolution of the West Antarctic Ice Sheet. *In*: ALLEY, R. B. & BINDSHADLER, R. A. (eds) *The West Antarctic Ice Sheet: behavior and environment*. American Geophysical Union, Antarctic Research Series, **77**, 45–57.

BEHRENDT, J. C. 1964. Distribution of narrow-width magnetic anomalies in Antarctica. *Science*, **144**, 995–999.

BEHRENDT, J. C., COOPER, A. K. & YUAN, A. 1987. Interpretation of marine magnetic gradiometer and multichannel seismic-reflection observations over the western Ross Sea shelf, Antarctica. *In*: COOPER, A. K. & DAVEY, F. J. (eds) *The Antarctic Continental Margin Geology and Geophysics of the Western Ross Sea*. Circum-Pacific Council for Energy and Natural Resources Earth Science Series, **5B**, 155–178.

BEHRENDT, J. C., BLANKENSHIP, D. D., FINN, C. A., BELL, R. E., SWEENEY, R. E., HODGE, S. R. & BROZENA, J. M. 1994. Evidence for late Cenozoic flood basalts(?) in the West Antarctic rift system revealed by the CASERTZ Aeromagnetic Survey. *Geology*, **22**, 527–530.

BEHRENDT, J. C., BLANKENSHIP, D. D., DAMASKE, D., COOPER, A. C., FINN, C. A., & BELL, R. E. 1997. Geophysical Evidence for late Cenozoic Subglacial Volcanism Beneath the West Antarctic Ice Sheet and Additional Speculation as to its Origin. *In*: RICCI, C. A. (ed.) *The Antarctic region: geological evolution and processes*. Terra Antarctica Publication, Siena, 539–546.

BEHRENDT, J. C., BLANKENSHIP, D. D., DAMASKE, D. & COOPER, A. K. 1995. Removal of late Cenozoic subglacially emplaced volcanic edifices by the West Antarctic Ice Sheet. *Geology*, **23**, 527–530.

BEHRENDT, J. C., DUERBAUM, H. J., DAMASKE, D., SALTUS, R., BOSUM, W. & COOPER, A. K. 1991*a*. Extensive volcanism and related tectonism beneath the western Ross Sea continental shelf, Antarctica: interpretation of an aeromagnetic survey. *In*: THOMSON, M. R. A., CRAME, J. A. & THOMSON, J. W. (eds) *Geological evolution of Antarctica*. Cambridge University Press, Cambridge, 299–304.

BEHRENDT, J. C., LEMASURIER, W. E., COOPER, A. K., TESSENSOHN, F. & DAMASKE, D. 1991*b*. The West Antarctic rift system – a review of geophysical investigations. *In*: ELLIOT, D. H. (ed.) *Contributions to Antarctic Research II*. American Geophysical Union, Antarctic Research Series, **53**, 67–112.

BEHRENDT, J. C., LEMASURIER, W. E. & COOPER, A. K. 1992. The West Antarctic rift systema propagating rift 'captured' by a mantle plume. *In*: KAMINUMA, K. & YOSHIDA, Y. (eds) *Recent progress in Antarctic earth science*. Terra Publishing, Tokyo, 315–322.

BEHRENDT, J. C., SALTUS, R., DAMASKE, D., MCCAFFERTY, A., FINN, C. A., BLANKENSHIP, D. D. & BELL, R. E. 1996. Patterns of late Cenozoic volcanic and tectonic activity in the West Antarctic Rift System revealed by aeromagnetic surveys. *Tectonics*, **15**, 660–676.

BEHRENDT, J. C., FINN, C. A., BLANKENSHIP, D. D. & BELL, R. E. 1998. Aeromagnetic evidence for a volcanic caldera(?) complex beneath the divide of the West Antarctic Ice Sheet. *Geophysical Research Letters*, **25**, 4385–4388.

BEHRENDT, J. C., BLANKENSHIP, D. D., MORSE, D. L., FINN, C. A. & BELL, R. E. 2002. Removal of subglacially erupted volcanic edifices beneath the divide of the West Antarctic Ice Sheet interpreted from aeromagnetic and radar ice sounding surveys. *In*: GAMBLE, J. A., SKINNER, D. N. B., HENRYS, S. & LYNCH, R. (eds) *Antarctica at the close of a Millennium*. Royal Society of New Zealand, Bulletin, **35**.

BHATTACHARYYA, B. K. & LEU, L.-K. 1975. Analysis of magnetic anomalies over Yellowstone National Park: mapping of the Curie point isothermal surface for geothermal reconnaissance. *Journal of Geophysical Research*, **80**, 4461–4465.

BLANKENSHIP, D. D., BELL, R. E., HODGE, S. M., BROZENA, J. M. BEHRENDT, J. C. & FINN, C. A. 1993. Active volcanism beneath the West Antarctic ice sheet. *Nature*, **361**, 526–529.

BLANKENSHIP, D. D., MORSE, D. L., FINN, C. A., BELL, R. E., PETERS, M. E., KEMPF, S. D., HODGE, S. M., STUDINGER, M., BEHRENDT, J. C. & BROZENA, J. M. 2000. Geologic controls on the initiation of rapid basal motion for the ice streams of the southeastern Ross Embayment: a geophysical perspective including new airborne radar sounding and laser altimetry results. *In*: ALLEY, R. B. & BINDSCHADLER, R. A. (eds) *The West Antarctic Ice Sheet: behavior and environment*. American Geophysical Union, Antarctic Research Series, **77**, 105–121.

BOURGEOIS, O., DAUTEUIL, O. & VAN VLEIT-LANOE, B. 1998. Pleistocene subglacial volcanism in Iceland: tectonic implications. *Earth and Planetary Science Letters*, **164**, 165–178.

BOURGEOIS, O., DAUTEUIL, O. & VAN VLIET-LANOE, B. 2000. Geothermal control on flow patterns in the last glacial maximum ice sheet of Iceland. *Earth Surface Processes and Landforms*, **25**, 59–76.

COMMITTEE FOR THE MAGNETIC MAP OF NORTH AMERICA. 1987. *Magnetic Map of North America. 1:5 000 000 scale*. Geological Society of North America.

DREWRY, D. J. 1983. *Antarctica: Glaciological and geophysical folio*. Cambridge University Press, Cambridge.

FRANCIS, P. 1993. *Volcanoes a Planetary Perspective*. Oxford University Press, New York.

GUDMUNDSSON, M. T., SIGMUNDSSON, F. & BJÖRNSSON, H. 1997. Ice–volcano interaction of the 1996 Gjálp subglacial eruption, Vatnajökull, Iceland. *Nature*, **389**, 994–957.

HOLE, M. J. & LEMASURIER, W. E. 1994. Tectonic controls on the geochemical composition of Cenozoic alkali basalts from West Antarctica. *Contributions to Mineralogy & Petrology*, **117**, 187–202.

JANKOWSKI, E. J., DREWRY, D. J. & BEHRENDT, J. C. 1983. Magnetic studies of upper crustal structure. *In*: OLIVER, R. L., JAMES, J. B., & JAGO, J. B. (eds) *Antarctic earth science*. Australian Academy of Science, Canberra, 197–203.

JONSSON, G., KRISTJANSSON, L. & SVERRISSON, M. 1991. Magnetic surveys of Iceland. *Tectonophysics*, **189**, 229–247.

LEMASURIER, W. E. & REX, D. C. 1983. Rates of uplift and the scale of ice level instabilities recorded by volcanic rocks in Marie Byrd Land, West Antarctica. *In*: OLIVER, R. L., JAMES, J. B. & JAGO, J. B. (eds) *Antarctic earth science*. Australian Academy of Science, Canberra, 663–670.

LEMASURIER, W. E. 1990. Late Cenozoic volcanism on the Antarctic plate: an overview. *In*: LEMASURIER, W. E. & THOMSON, J. W. (eds) *Volcanoes of the Antarctic plate & southern oceans*. American Geophysical Union, Antarctic Research Series, **48**, 1–19.

LEMASURIER, W. E. & THOMSON, J. W. (eds) 1990. *Volcanoes of the Antarctic plate and southern oceans*. American Geophysical Union Antarctic Research Series, **48**.

MASLANYJ, M. P., GARRETT, S. W., JOHNSON, A. C., RENNER, R. G. B., & SMITH, A. M. 1991. Aeromagnetic map of West Antarctica (Weddell Sea Sector). *BAS GEOMAP Series*, Sheet 2 Scale 1:2 500 000, British Antarctic Survey, Cambridge.

MCINTOSH, W. C. & GAMBLE, J. A. 1991. A subaerial environment for the Hallett Coast volcanoes. *In*: THOMSON, M. R. A., CRAME, J. A. & THOMSON, J. W. (eds) *Geological evolution of Antarctica*. Cambridge University Press, Cambridge, 657–661.

SCHERER, R. P. 1991. Quaternary and Tertiary microfossils from beneath Ice Stream B: Evidence for a dynamic West Antarctic ice sheet history. *Palaeogeography Palaeoclimatology, Palaeoecology*, **90**, 359–412.

SCHERER, R. P., ALDAHAN, A., TULACZYK, S., POSSNERT, G., ENGELHARDT, H. & KAMB, B. 1998. Pleistocene collapse of the West Antarctic Ice Sheet. *Science*, **281**, 82–85.

SKILLING, I. P. 1994. Evolution of an englacial volcano: Brown Bluff, Antarctica. *Bulletin Volcanology*, **56**, 573–591.

SMELLIE, J. L. & SKILLING, I. P. 1994. Products of subglacial volcanic eruptions under different ice thicknesses: two examples from Antarctica. *Sedimentary Geology*, **91**, 115–129.

SMITH, R. B. & BRAILLE, L. W. 1994. The Yellowstone hotspot. *Journal of Volcanology and Geothermal Research*, **61**, 121–187.

SWEENEY, R. E., FINN, C. A., BLANKENSHIP, D. D., BELL, R. E. & BEHRENDT, J. C. 1999. Central West Antarctica aeromagnetic data: a web site for distribution of data and maps (on-line edition). *US Geological Survey, Open-File-Report*, **99–420**. World Wide Web Address: http://green wood cr.usgs.gov/pub/open-file-reports/ofr-99-0420/cwantarctica.html.

VAUGHAN, D. G., BAMBER, J. L., GIOVINETTO, M. & COOPER, A. P. R. 1999. Reassessment of net surface mass balance in Antarctica. *Journal of Climate*, **12**, 933–946.

WAISCORES. 2000. World Wide Web Address: http:/ www.ig.utexas.edu/research/projects/wais/inland /inland.html.

WILCH, T. I. & MCINTOSH, W. C. 2000. Eocene and Oligocene volcanism at Mt. Petras Marie Byrd Land: implications for middle Cenozoic ice sheet reconstructions in West Antarctica. *Antarctic Science*, **12**, 477–491.

Spectroscopic and geochemical analyses of ferrihydrite from springs in Iceland and applications to Mars

J. L. BISHOP[1] & E. MURAD[2]

[1] *SETI Institute/NASA-Ames Research Center, MS 239-4, Moffett Field, CA 94035, USA (e-mail: jbishop@mail.arc.nasa.gov)*

[2] *Bayerisches Geologisches Landesamt, Aussenstelle Marktredwitz, Leopoldstrasse 30, Postfach 389, D-95603 Marktredwitz, Germany*

Abstract: Ferrihydrite samples were collected from a thermal spring and a cold stream in the Landmannalaugar region of Iceland. Chemical and spectroscopic analyses have been performed on the air-dried and fine-grained fractions of these samples. The ferrihydrite from the cold stream is a pure sample, containing small amounts of Ca, P and Si. The ferrihydrite from the thermal spring is a less pure sample, containing larger amounts of amorphous Si and P with some of the Si incorporated in the ferrihydrite structure. The spectral character of these Icelandic ferrihydrites is compared with those of synthetic ferrihydrites and other iron oxide/oxyhydroxide minerals. Ferrihydrite is characterized by a broad Fe^{3+} excitation band near $10\,900\,cm^{-1}$ (*c.* 0.92 μm), a strong Fe–O vibrational feature near $475\,cm^{-1}$ (*c.* 21 μm), and multiple bands due to H_2O and OH. Highly pure ferrihydrite has a pair of spectral bands near 1400 and $1500\,cm^{-1}$ (*c.* 7 μm). Natural ferrihydrites frequently exhibit an extra band near $950–1050\,cm^{-1}$ (*c.* 10 μm) that is attributed to Si–O bonds. Hydrothermal springs may have been present at one time on Mars in association with volcanic activity. Ferrihydrite formation in such an environment may have contributed to the ferric oxide-rich surface material on Mars.

This study encompasses spectroscopic analyses of the ferric oxyhydroxide mineral ferrihydrite, characterization of two ferrihydrite deposits associated with volcanic activity on Iceland, and speculation on the formation of ferrihydrite in similar geological settings on Mars when volcanism was active there. Chemical and mineralogical analyses of naturally occurring iron oxides–oxyhydroxide minerals are important for understanding the surface processes on Mars because of the abundant ferric oxides–oxyhydroxides there. Two iron-rich samples were collected from the Landmannalaugar region of Iceland during August, 2000. Landmannalaugar is located near the centre of the southern part of the island, on the northern flanks of the Myrdalsjökull glacier, near 19°W, 64°N. Several thermal springs are present in this area as well as many cold water run-offs that colour this region with orange–red pigments. Previous studies of ferrihydrite or other iron oxides from Landmannalaugar have not been found in the literature.

Ferrihydrite structure and spectral properties

Early studies of the mineral ferrihydrite presented formulae including Fe, O, OH and H_2O and a hexagonal close packing structure similar to that of hematite (Towe & Bradley 1967; Chukhrov *et al.* 1973) for what was previously considered to be amorphous iron oxide. Although Chukhrov *et al.* (1973) initially suggested the formula $Fe_5HO_8 \cdot H_2O$ for ferrihydrite, a number of others have been suggested as well, and Stanjek & Weidler (1992) found that none of the currently proposed formulae for this mineral were helpful in understanding its structure. Russell (1979) used infrared (IR) spectroscopy and D_2O exchange to show that OH is an essential component of the ferrihydrite structure. Murad & Schwertmann (1980) used Mössbauer spectroscopy to compare the structure of ferrihydrite with those of other iron oxides and Murad *et al.* (1988) evaluated the influence of crystallinity on the magnetic ordering of ferrihydrites using Mössbauer spectroscopy. The structure of ferrihydrite has been hotly debated since then. Eggleton & Fitzpatrick (1988) first proposed that some of the Fe in ferrihydrite is in tetrahedral coordination and more recent studies by Zhao *et al.* (1994) and Janney *et al.* (2000) supported the presence of some tetrahedral Fe. Other studies by Manceau *et al.* (1990), Pankhurst & Pollard (1992) and Manceau & Gates

From: SMELLIE, J. L. & CHAPMAN, M. G. (eds) 2002. *Volcano–Ice Interaction on Earth and Mars*. Geological Society, London, Special Publications, **202**, 357–370. 0305-8719/02/$15.00

(1997) supported only octahedral coordination of Fe for ferrihydrite. Some discrepancy in the structure of ferrihydrite is certainly due to differences from one sample to another and may be linked to natural v. synthetic samples. Carlson & Schwertmann (1981) characterized the typical 2-line and 6-line ferrihydrite structures (named after the number of their X-ray diffraction peaks) and observed that natural ferrihydrites exhibit a range of crystallinities between the common 2-line and 6-line forms.

Reflectance spectroscopy in the visible and IR has been reported for synthetic ferrihydrites and ferrihydrite–montmorillonite aggregates in studies related to remote sensing and surface mineralogy of Mars (Bishop *et al.* 1993, 1995). Reflectance spectroscopy has also been applied to detection of ferrihydrite and other iron oxides in soils using second derivatives of the crystal field theory bands (Scheinost *et al.* 1998). The current study provides more detail about the spectral bands observed for ferrihydrite and includes both natural and synthetic samples.

Aqueous processes on Mars

Direct evidence for liquid water on Mars has not been found. However, surface features in orbital images have led to a variety of theories about aqueous processes on Mars. Carr (1981) suggested that abundant flowing water was needed to account for the many runoff and outflow channels depicted in Viking images, and climatic models describing a warm and wet early Mars were developed in order to explain these features (Pollack *et al.* 1987; Fanale *et al.* 1992). Analyses by Baker *et al.* (1991), Clifford (1993) and Parker *et al.* (1993) supported the presence of aqueous processes, and perhaps sedimentary deposits and an ancient ocean. The possibility of recent network formation, glaciation events, mass wasting and groundwater seepage have also been considered (Carr 1995). Analysis of Martian surface topography from the Mars Orbiter Laser Altimeter (MOLA) on the Mars Global Surveyor (MGS) led Head *et al.* (1999) to support the idea that oceans reigned on Mars during a previous era. The higher-resolution (*c.* 3 metre pixel size) images that the Mars Observer Camera (MOC) on MGS is continuing to return to Earth are providing new information for speculation and modelling of martian surface activity, including aqueous processes. Recently observed erosional features on Mars have been explained by groundwater seepage and surface runoff (Malin & Edgett 2000*b*) and sedimentary deposits (Malin & Edgett 2000*a*). Although these are not unique explanations for the surface features observed on Mars (e.g. Hoffman 2000; Musselwhite *et al.* 2001), a number of scenarios involving aqueous and subaqueous processes on Mars are possible, and consideration of potential aqueous mineral formation and chemical alteration mechanisms are warranted.

Methods

Natural ferrihydrite samples

Moist samples were collected using a plastic trowel from two sites in the Landmannalaugar region. They were enclosed in water-tight plastic bags and returned to the lab. Images of the sample collection sites are shown in Figure 1. One sample (No. 498) was collected near a thermal spring just off the walkway nearest the enclosed

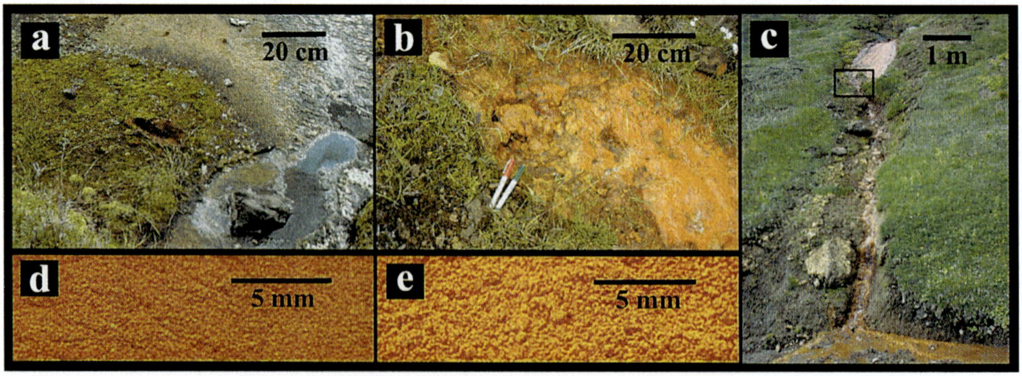

Fig. 1. Images of sample collection sites at Landmannalaugar for the hot springs sample (**a**), and cold springs sample (**b**, **c**). Images of the dried and sieved samples: No. 498 from the hot springs site (**d**) and No. 499 from the cold stream site (**e**).

structures at the Landmannalaugar campsite. Material rich in iron oxides was evident by its colour in deposits on the banks above the spring. The top centimetre of material was removed from an area *c.* 20 cm long and the material below this moss layer was collected. The water in the thermal pool was warm to the touch at the time of collection and temperatures of 60–76°C have been recorded (J. Kristjansson, pers. comm.). A second sample (No. 499) was collected from the mouth of one of many small, cold streams that emanated from the side of the volcanic ridge bordering the roadway on the west upon entering the Landmannalaugar site from the north. The water is presumed to be partly meltwater derived from a nearby glacier that percolated through the volcanic ridge. Iron oxides were readily identified by the bright orange–red colours pigmenting the cold springs. Fine-grained material was deposited and the rocks covered by water were heavily altered. Vegetation was abundant on the sides of the small stream, but not directly in the water's path. Fine-grained material was scooped up from the region shown in Figure 1b and the water became murky with loosened material following collection. Figure 1c shows the cold stream against the volcanic ridge.

Laboratory pH measurements were performed on water in the sample bags. The pH of the water from both sites was about 7.5–8. This is consistent with independent measurements by J. Kristjansson (pers. comm.) who studied aerobic thermophiles in the Landmannalaugar hot springs. The moist samples were air dried in covered glass dishes, gently shaken, then dry sieved to <125 µm and <45 µm for analysis. Both size fractions are similar in appearance. As individual ferrihydrite grains are several nm in diameter, the sieved particles are composites of multiple grain aggregates. Images of the <45 µm samples are shown in Figure 1d for the hot springs ferrihydrite and in Figure 1e for the cold stream ferrihydrite in order to show the colour of the samples and the texture of the powder. The cold springs sample (499) in Figure 1e is more pure than the hydrothermal springs sample (498) in Figure 1d which may explain why the particles of sample 499 clump together more than in the other one.

Other samples

Synthetic iron oxide–oxyhydroxide minerals prepared for other studies are included here for comparison with the natural ferrihydrites. The synthetic 2-line ferrihydrite No. 46 and the goethite No. 54 were prepared according to procedures outlined by Schwertmann & Cornell (2000) and were included in the study by Bishop *et al.* (1993). The synthetic 2-line ferrihydrite No. 254 was prepared by A. Scheinost and D. G. Schulze, whereas the synthetic <5 µm grain size hematite was purchased from Aldrich. All samples were dry sieved to <45 µm particle size. The natural goethite No. 802.2 was characterized using Mössbauer spectroscopy in a previous study (Murad 1979).

Analyses

Major element compositions of the <125 µm fractions of the samples were determined using X-ray fluorescence (XRF) at the University of Massachusetts at Amherst as in a previous study (Bishop *et al.* 1995). These data are summarized in Table 1. In addition to iron, the presence of silicon, calcium and phosphorous is notable. Samples were also sent to ACME analytical labs in Vancouver for Fe and S abundance measurements. The sulphur levels were below detection limits of 0.01% and the iron abundances were consistent with those measured using XRF.

X-ray diffraction (XRD) was carried out using a Bruker D8-Advance θ/2θ diffractometer equipped with a Cu tube (operated at 40 kV and 40 mA), a variable divergence slit system, a rotating sample stage, a graphite diffracted-beam monochromator and a NaI(Tl) scintillation counter. Samples were gently pressed into a top-loading sample holder that was rotated at 10 rpm during measurement, and scans were continuous from 3 to 70° 2θ at a scanning rate of 60 s per 0.02° 2θ.

Visible-infrared reflectance spectra were measured as described in detail previously (Bishop

Table 1. *Major elements for Icelandic springs samples*

wt% oxides	Hot spring sample 498	Cold spring sample 499
SiO_2	24.3	12.6
Al_2O_3	0.6	0.7
Fe_2O_3	62.0	77.5
MgO	0.6	0.5
CaO	2.2	3.0
Na_2O	0.7	0.8
K_2O	0.2	0.1
P_2O_5	7.0	1.8
MnO	0.4	0.4
LOI	20.0	29.2
Total	98.1	97.6

et al. 1995). Bidirectional visible/near-infrared spectra were measured relative to Halon under ambient conditions at the Reflectance Experiment Laboratory (RELAB) at Brown University. Biconical infrared reflectance spectra were measured relative to a rough gold surface using a Nicolet 740 Fourier transform interferometer (FTIR) in a H_2O- and CO_2-purged environment. Composite, absolute reflectance spectra were prepared by scaling the FTIR data to the bidirectional data near 1.2 μm.

Transmittance spectra were measured on KBr pellets using a Nicolet Magna 550 Fourier-transform spectrometer as in a previous study (Murad & Bishop 2000). Spectra were recorded in the range 4000–400 cm^{-1} at a resolution of 4 cm^{-1} and 200 scans were averaged for each sample. The 13 mm KBr disks were prepared by mixing 1 mg sample with 300 mg dehydrated KBr in a hand mortar, and pressing the mixture for several minutes at 10 tonnes in an evacuated die. The background was recorded on an empty sample holder. In order to minimize the effects of atmospheric H_2O and CO_2 on the spectra, the sample and background measurements were carried out after simultaneously enclosing neighbouring sample holders in the measurement chamber, so that both are measured under the same environmental conditions.

Results of X-ray diffraction and chemical analyses

X-ray diffraction (XRD) curves of the natural ferrihydrites from Iceland are shown in Figure 2. The XRD pattern of sample 499 is characteristic of pure ferrihydrite, with a broad major peak at *c.* 2.67 Å and a subordinate broad peak near 1.51 Å. Sample 498 has similar broad peaks

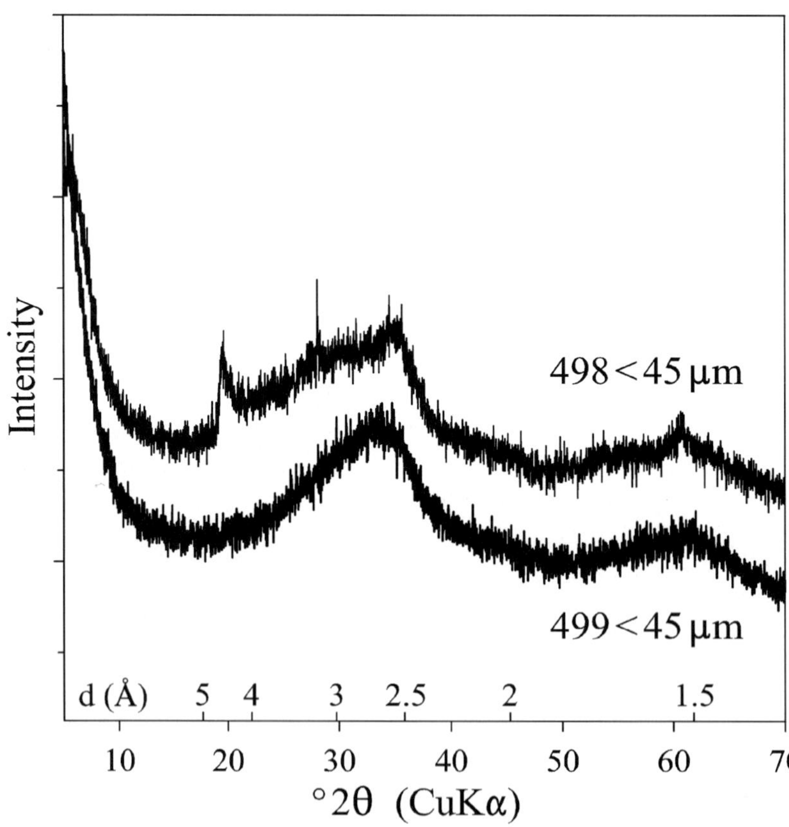

Fig. 2. X-ray diffraction curves for Icelandic ferrihydrites collected at Landmannalaugar. Both the hot springs sample (498) and the cold springs sample (499) are shown. The XRD curve for sample 499 indicates this is pure ferrihydrite, while the XRD curve for sample 498 shows that some impurities are present. A few d-spacing values have been added to facilitate identification of the peaks.

near 2.57 and 1.52 Å plus an asymmetric peak at c. 4.50 (corresponding to the 02,11 diffraction of smectite) and a sharp peak at 3.16 Å (possibly resulting from a trace of detrital feldspar). Because the count rates are extremely low for ferrihydrite, even minor admixtures of better-crystallized minerals become over-emphasized (this may be one reason why ferrihydrite has only been identified as a unique mineral fairly recently).

Barrón *et al.* (1997) found that small amounts of phosphate in solution retard the crystallization of synthetic ferrihydrite at alkaline pH. The phosphate found in our Icelandic samples may be influencing the formation of ferrihydrite and subsequent transformation to hematite or goethite as well. Gálves *et al.* (1999) studied the influence of phosphate levels (P/Fe 0 to 1.5), pH and temperature on ferrihydrite formation and found that increasing the P/Fe levels in solution decreased the degree of ferrihydrite transformation to hematite or goethite upon ageing. Gálves *et al.* (1999) performed lab experiments that encompass the formation conditions of our Icelandic samples. For pH c. 7–8 and 25°C they found that a mixture of hematite and goethite formed upon ageing of ferrihydrite in aqueous solution with 0 to 0.16 P/Fe. For pH c. 7–8 and 50°C they found that only goethite formed upon ageing of ferrihydrite in aqueous solution with 0.16–0.5 P/Fe. Silica and organic matter also retard crystallization and transformation of ferrihydrite to other iron oxide minerals (Schwertmann 1988; Schwertmann & Murad 1988; Childs 1992).

Visible/near-infrared reflectance spectra

Reflectance spectra are shown from 0.4 to 2.6 μm in Figure 2 for the Icelandic ferrihydrites, and synthetic iron oxide–oxyhydroxide minerals including two ferrihydrites (Fh46, Fh254), a goethite (Gt54) and a hematite (Hm129). These data were measured under ambient conditions using the bidirectional spectrometer and contain strong hydration bands due to adsorbed water. As described in Bishop *et al.* (1993) the extended visible region spectral character of ferrihydrite includes a gently sloping reflectance maximum near 0.8 μm and a broad absorption band centred near 0.92 μm (or c. 10 900 cm^{-1}). This band is observed for both the natural and synthetic 2-line ferrihydrites (Fig. 3) and is listed in Table 2 along with other spectral features.

The crystal field theory bands for iron oxides are described in detail in Morris *et al.* (1985) and Burns (1993). The goethite spectrum included in Figure 3 exhibits the typical absorption bands observed at 0.65 and 0.91 μm which are much narrower than those observed for ferrihydrite. Hematite exhibits a band centre near 0.85–0.86 μm and a shoulder near 0.65 μm. As the grain size of iron oxide minerals is reduced, the crystal field theory bands are also weakened and spectral features in this region are broadened (Morris *et al.* 1989). There is no evidence in the visible region spectra for nontronite or Fe-rich smectite absorption bands at 0.62 and 0.95 μm; however, a small amount of these smectites cannot be precluded.

Shown in Figure 4a are reflectance spectra from 4000 to 7500 cm^{-1} (c. 1.33–2.5 μm) of the Icelandic ferrihydrites and synthetic iron oxide–oxyhydroxide minerals measured under purged H_2O and CO_2 conditions, where most of the adsorbed water is removed. This gives a better indication of the OH and H_2O bands due to the mineral structure. Fundamental stretching and bending vibrations occur in the mid-IR region for H_2O and OH and overtones and combinations of these vibrations occur in the near infrared (NIR) region. These figures are plotted in terms of wave number in order to facilitate comparison of the NIR and mid-IR features. The band assignments given for the NIR features in ferrihydrite spectra are based on analysis of the spectral properties of H_2O and OH in smectite clays (Bishop *et al.* (1994) and references therein). Figure 4b includes normalized reflectance spectra across the region of the OH combination bands. A combination of the water stretching and bending absorptions is centred near 5180 cm^{-1} (1.93 μm) in spectra of the Icelandic ferrihydrites. This feature is similar in the spectrum of Fh46, but is broadened and shifted to 5130 cm^{-1} (1.95 μm) for sample Fh254. These bands are listed in Table 2.

Combinations of the OH stretching and bending vibrational absorptions occur in the range 4300 to 4600 cm^{-1}, and are weaker than the corresponding water combination bands. The Icelandic ferrihydrite spectra have multiple bands here, supporting the idea that the OH sites in the structure are non-unique. Bands are observed at 4375, 4440 and 4550 cm^{-1} (2.20, 2.25 and 2.29 μm) for both the <45 μm and <125 μm sieved fractions of the samples collected from the thermal spring (498) and the cold stream (499). The relative strengths of these bands vary such that the 4375 cm^{-1} band is much stronger in the impure thermal sample (498). The synthetic ferrihydrites exhibit related bands at 4320 and 4440 cm^{-1} (2.31 and 2.25 μm) for Fh46 and at 4340 and 4440 cm^{-1} (2.30 and 2.25 μm) for Fh 254. These synthetic samples also have a weak, broad band centred near

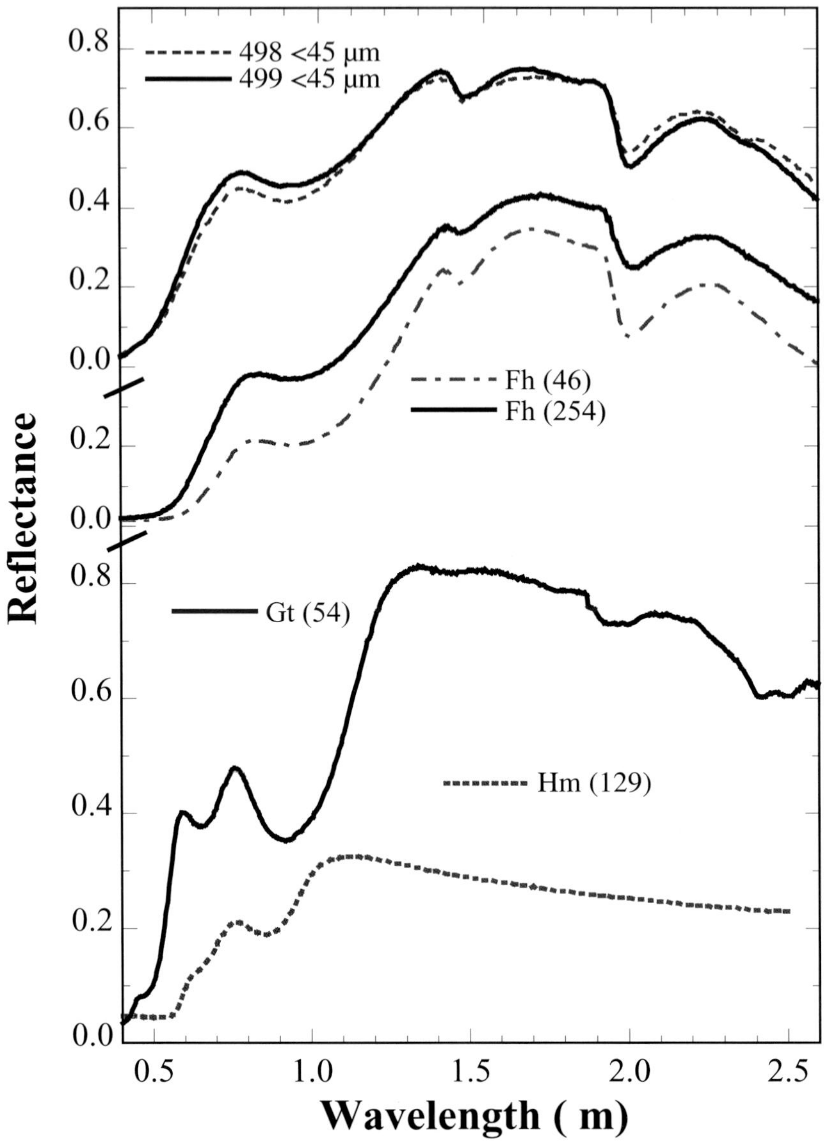

Fig. 3. Reflectance spectra from 0.4 to 2.5 µm of the Icelandic springs ferrihydrite (498, 499) and synthetic iron oxide–oxyhydroxide minerals.

4140 cm^{-1} (2.42 µm), which is also observed for goethite (Fig. 4b). This indicates that some of the OH sites in the synthetic ferrihydrite samples are similar to the OH sites in goethite.

Smectites also exhibit OH combination bands in this region in reflectance spectra. The 4370 cm^{-1} band observed in nontronite spectra is due to OH groups octahedrally bound to two Fe atoms. This may be similar to the octahedrally coordinated Fe and OH bonds in ferrihydrite. Montmorillonite exhibits a band due to OH bonded to two Al atoms at 4530 cm^{-1} and ferruginous smectite exhibits a band at 4470 cm^{-1} due to OH bonded to one Fe and one Al, as well as the nontronite band (Bishop *et al.* 1999). A minor amount of smectite, indicated by XRD of the impure thermal sample (498), may be contributing to the spectral features in this region, especially the band at 4375 cm^{-1} which is stronger for sample 498 than for sample 499. Perhaps

Table 2. *Spectral features for natural and synthetic ferrihydrites*

Sample	Transmittance, cm^{-1}	Reflectance, cm^{-1} [μm]	Band assignment
499 (pure, nat.)	475, 680 sh	*c.* 680 [*c.* 14.7]	Fe–O
498 (nat.)	475, 670 sh	*c.* 600, *c.* 680 [*c.* 16.7, *c.* 14.7]	Fe–O
46 (synth.)	455, 580	760 [13.2]	Fe–O
254 (synth.)	455, 580	750 [13.3]	Fe–O
499 (pure, nat.)	*c.* 850 sh	(850–900?) [*c.* 11]	δOH
46, 254 (synth.)	*c.* 840 sh		δOH
499 (pure, nat.)	965	*c.* 1030 [*c.* 9.7]	Si–Fe–O?
498 (nat.)	1010	960 sh, 1140 [10.4, 8.8]	Si–Fe–O?
499 (pure, nat.)	1385, 1495	1385, 1495 [7.2, 6.7]	
498 (nat.)		1395, 1520 [7.2, 6.6]	
46, 254 (synth.)	1360, 1495	1385, 1470 [7.2, 6.8]	
499, 498, 46, 254	1630	1620–1630 [6.1–6.2]	δH_2O
499 (pure, nat.)	3400	*c.* 3360 broad [*c.* 3.0]	νH_2O
498 (nat.)	3400	*c.* 3380 broad [*c.* 3.0]	νH_2O
498, 498 (heated)	3380		νH_2O
46 (synth.)	3380	3400 broad [*c.* 2.9]	νH_2O
254 (synth.)	3380	3400 broad [*c.* 2.9]	νH_2O
499 (pure, nat.)		3730 sh [*c.* 2.7]	νOH
498 (pure, nat.)		3560 sh, 3735 [*c.* 2.7–2.8]	νOH
46 (synth.)		3645 [2.7]	νOH
254 (synth.)		3640 sh [2.7]	νOH
499 (pure, nat.)		4375, 4440, 4550 [2.29, 2.25, 2.20]	νOH + δOH
498 (nat.)		4375, 4440, 4550 [2.29, 2.25, 2.20]	νOH + δOH
46 (synth.)		4340, 4440 sh [2.30, 2.25]	νOH + δOH
254 (synth.)		4320, 4440 sh [2.31, 2.25]	νOH + δOH
499, 498		5180 [1.93]	$\nu H_2O + \delta H_2O$
46 (synth.)		5170 [1.93]	$\nu H_2O + \delta H_2O$
254 (synth.)		5130 [1.95]	$\nu H_2O + \delta H_2O$
499 (pure, nat.)		7060 [1.42]	$\nu H_2O + 2\delta H_2O$
498 (nat.)		6990 [1.43]	$\nu H_2O + 2\delta H_2O$
46, 254 (synth.)		7160 [1.40]	$\nu H_2O + 2\delta H_2O$
499 (pure, nat.)		*c.* 7300 sh [*c.* 1.37]	2νOH
498 (nat.)		7305 [1.37]	2νOH
499, 498, 46, 254		*c.* 10 900 broad [0.92]	Fe CFT

Band assignments are for features observed in Figures 2–6 and are consistent with data presented by Farmer (1974), Ryskin (1974), Russell (1979), and Kodama (1985). sh, shoulder; ν, stretching vibration; δ, bending vibration; CFT, crystal field theory band.

some Al and/or Si incorporated into the natural ferrihydrite structure produces stronger bands at higher wave numbers in spectra of these samples than are observed in spectra of synthetic ferrihydrites because of AlOH or SiOH bonds in the natural samples.

Mid-infrared reflectance and transmittance spectra

The optical constants of minerals include the real (n) and imaginery (k) indices of refraction and provide information about how a mineral surface interacts with radiation. Reflectance and transmittance spectra were both measured in the mid-IR region and exhibit differences in band shape and intensity because transmittance spectra show only the absorbance bands due to molecular vibrations, whereas reflectance spectra in this region depend on both optical constants and are therefore influenced by vibrational absorptions and radiation scattering among grains. The transmittance spectra can be more simply related to mineral structure and the reflectance spectra are more useful for remote sensing. Emission spectra recorded for planetary surfaces can be approximated by 1-reflectance spectra when Kirchhoff's Law holds (e.g. Salisbury 1993), but it is much more difficult to compare emission and transmittance spectra.

Mid-IR reflectance and transmittance spectra are shown from 400 to 4000 cm^{-1} in Figure 5 for the thermal springs ferrihydrite sample (498) and

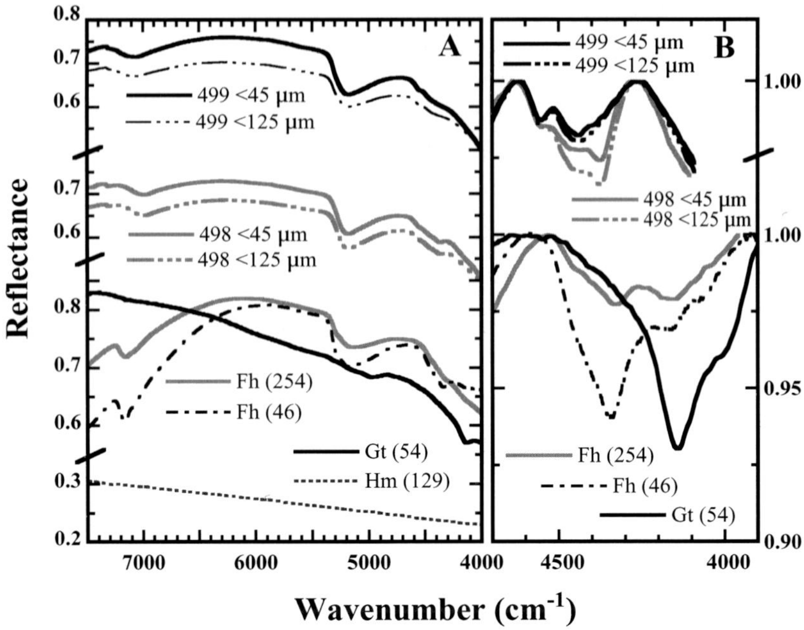

Fig. 4. Spectra of the Icelandic springs ferrihydrite (498, 499) and synthetic iron oxide–oxyhydroxide minerals: (**a**) reflectance spectra from 4000 to 7500 cm^{-1} and (**b**) normalized reflectance spectra of the OH combination region. The normalized spectra were set to 1 near 4620 and 4260 cm^{-1} for the Icelandic ferrihydrites and near 4500–4600 and 3850–3950 cm^{-1} for the others.

in Figure 6 for the cold stream ferrihydrite sample (499). A spectral band due to Fe–O bonds in the Icelandic ferrihydrites is observed at 475 cm^{-1} (21.1 µm) with a shoulder near 670–680 cm^{-1} in transmittance spectra. A broader band centred near 680 cm^{-1} (14.7 µm) is observed for the Fe–O vibrations in the reflectance spectra of these samples. Another strong band is observed at 965 cm^{-1} (10.4 µm) for the pure ferrihydrite sample (499) and at 1010 cm^{-1} (9.9 µm) for the less pure ferrihydrite sample (498) in transmittance spectra. These bands are thought to involve Si bonds because they only occur in natural ferrihydrites when Si is also present. Corresponding bands are observed at *c.* 1030 cm^{-1} (9.7 µm) for sample 499 and at 1140 cm^{-1} (8.8 µm) with a shoulder at 960 cm^{-1} (10.4 µm) for sample 498 in the reflectance spectra (Figs 5 & 6). A similar band is also present in transmittance spectra of a pure natural ferrihydrite sample collected in a cold spring in the Czech Republic (Murad, unpublished spectra). The presence of this band at higher energies for sample 498 is consistent with the presence of some smectite and a contribution of an Si–O stretching band, which occurs near 1000–1050 cm^{-1} in transmittance spectra and near 1200–1250 cm^{-1} in reflectance spectra (e.g. Farmer 1974; Salisbury *et al.* 1991). A smectite Si–O–Si bending vibration may also be contributing to the *c.* 475 cm^{-1} band in transmittance spectra of samples 498 (observed at longer wavenumbers in reflectance spectra).

The strong 965 cm^{-1} band in transmittance for ferrihydrites might be explained by Fe–O–Si stretching, Fe–O stretching in FeOH groups, and Si–O stretching in SiOH groups. Spectral analyses of silica gel found that the Si–O stretching band observed near 960 cm^{-1} shifted toward higher energies and weakened with dehydration as the SiOH groups are removed (Hino & Sato 1971). OH deformation bands are also observed in this region for some oxyhydroxides. Ryskin (1974) described a strong OH bending band near 1000–1100 cm^{-1} for $M(OH)_2$ groups where the cation forms non-identical bonds to the oxygen atom, forms some tetrahedral bonding and may be associated with sp-hybridization of the O valence electrons. Unique assignment for this band cannot be given at this time.

Water bands are observed in transmittance and reflectance spectra of all ferrihydrite samples near 1630 and 3400 cm^{-1}. The band at 1630 cm^{-1} (6.1 µm) is due to the bending vibrations of H_2O. Bending vibrations for OH in ferrihydrite were shown by Russell (1979) to occur

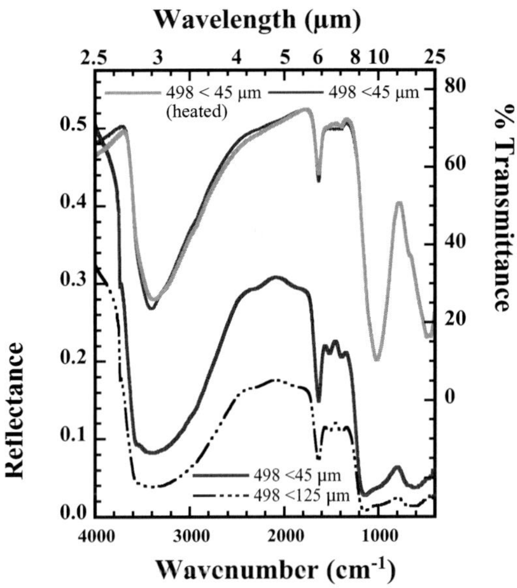

Fig. 5. Mid-IR reflectance and transmittance spectra from 400 to $4000\,cm^{-1}$ of Icelandic ferrihydrite from the Landmannalaugar hot spring.

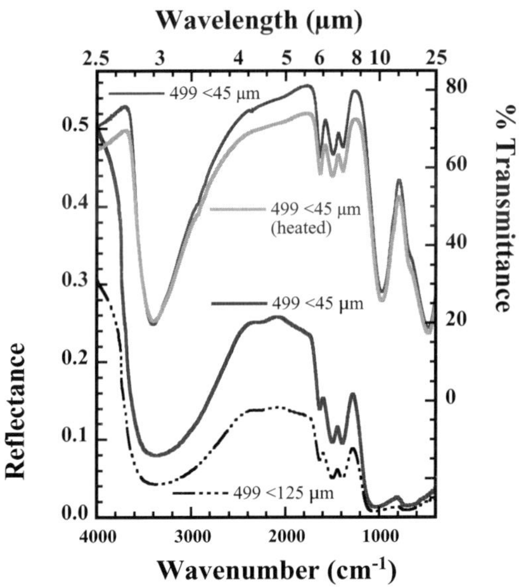

Fig. 6. Mid-IR reflectance and transmittance spectra from 400 to $4000\,cm^{-1}$ of Icelandic ferrihydrite from the Landmannalaugar cold spring.

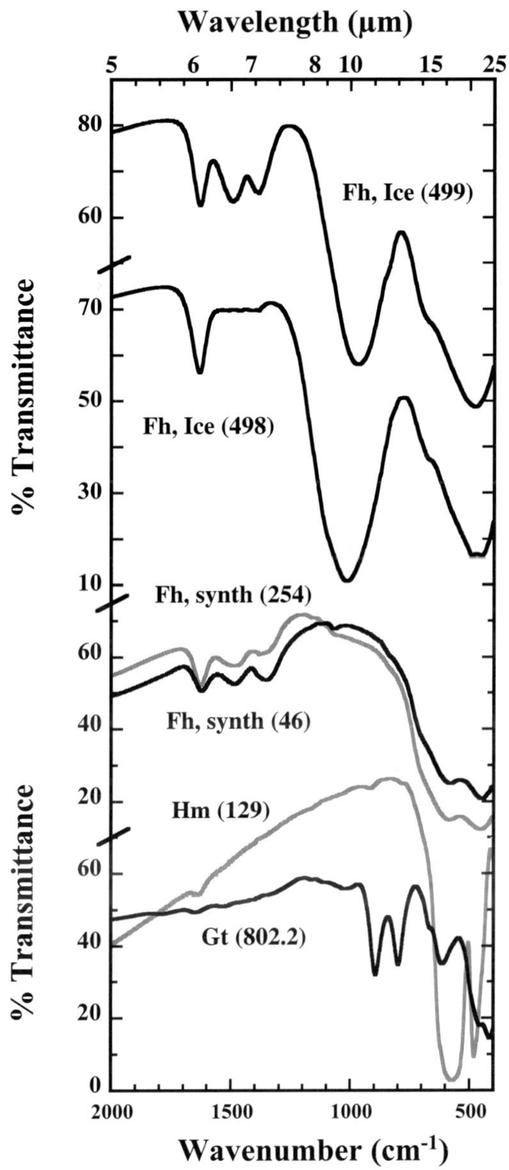

Fig. 7. Transmittance spectra from 400 to $2000\,cm^{-1}$ of Icelandic ferrihydrites and synthetic iron oxide–oxyhydroxide minerals.

as weak, broad features near $800\,cm^{-1}$ and are observed as a shoulder near $850\,cm^{-1}$ (11.8 µm) in transmittance spectra of sample 499. Stretching vibrations for OH and asymmetric plus symmetric stretching vibrations for H_2O all contribute to the broad band centered near $3400\,cm^{-1}$. This band is shifted slightly toward lower energies for the spectra of sample 498 after dehydrating the samples by heating at 100°C for *c.* 1 hour; however, even less change is observed in the spectra of sample 499. The small changes in sample 498 upon dehydration could be explained by the presence of a small amount of smectite in that sample. Weak features are observed near 3560 and 3735 cm^{-1} (2.8 and 2.7 μm) in reflectance spectra of sample 498 that are characteristic of OH stretching vibrations. Weak shoulders are present at these energies in the reflectance spectra of sample 499 as well. These bands indicate that hydroxyl groups are present in these natural ferrihydrites in multiple forms. Reflectance spectra of nontronite and montmorillonite exhibit OH stretching bands at 3570 and 3630 cm^{-1}, respectively. There is no evidence for the montmorillonite band; however, some nontronite or ferruginous smectite could be contributing to the OH stretching bands in the spectrum of sample 498.

Additional features are observed at 1385 and 1495 cm^{-1} (7.2 and 6.7 μm) in the transmittance and reflectance spectra of sample 499. Related, weaker features are observed at 1395 and 1520 cm^{-1} (7.2 and 6.6 μm) in the reflectance spectra of sample 498, but not in the transmittance spectra. Although carbonates have a strong band here, no other bands characteristic of carbonates are observed and XRD showed no evidence of carbonates. Ryskin (1974) attributed weak bands in this region to OH bending vibrations of acid O_3SiOH groups. He stated that these bands are also observed for other acid salts and that they can be split into a doublet by resonance interactions. However, this assignment may be problematic because he found that this band is generally not observed above 1400 cm^{-1}. Unique assignment of this band is not possible at this time.

Transmittance spectra are shown in Figure 7 from 400 to 2000 cm^{-1} for the Icelandic ferrihydrites (498, 499) and selected iron oxide–oxyhydroxide minerals. Spectra of the synthetic ferrihydrites (Fh 46, Fh 254) contain Fe–O bands near 455 and 580 cm^{-1} (22.0 and 17.2 μm) that are near the bands observed for hematite (Hm 129) and goethite (Gt 802.2) in this region (Fig. 7); however both the Fh 46 and Fh 254 spectra exhibit the characteristic ferrihydrite bands at 1360 and 1495 (7.4 and 6.7 μm). The strong bands near *c.* 1000 cm for the natural ferrihydrites are not observed for the synthetic samples studied here.

Analysis of OH vibrations using NIR and Mid-IR spectra

In order to confirm the presence of OH combination bands in ferrihydrite near 4500 cm^{-1}, the mid-IR fundamental bands for ferrihydrite and other iron oxides are evaluated here. Stretching and bending vibrations of the same group can combine together giving rise to bands at roughly the vibrational energy of the sum of the stretching and bending energies (e.g. Gaffey *et al.* 1993). The synthetic hematite spectrum includes Fe–O bands at 480 and 570 cm^{-1}, which are characteristic of hematite, and also contains weak water bands near 1635 and 3400–3500 cm^{-1}, probably due to adsorbed water on the grain surfaces. The hematite spectrum contains no features in the 4500 cm^{-1} region (Fig. 4) and contains only a hint of weak features near 800 and 900 cm^{-1} that could be due to minor OH contributions on a few altered grains. The goethite spectrum exhibits features at 370, 410 and 610 cm^{-1} with shoulders at 450 and 655 cm^{-1} due to Fe–O vibrations and at 790 and 890 cm^{-1} due to OH bending vibrations as reported earlier for this sample by Murad (1979) and for goethite in general by Ryskin (1974). This goethite spectrum also contains strong OH stretching vibrations at 3180 and *c.* 3400 cm^{-1} as well as weaker features due to water. Averaging the sums of the OH stretching and bending vibrations in goethite for the two OH sites gives 4130 cm^{-1}, which is very close to the OH combination band observed at 4140 cm^{-1} for goethite in Figure 4.

Adding together the approximate energies of the OH bending (850 cm^{-1}) and stretching (3560 and 3730 cm^{-1}) bands from shoulder features in the mid-IR spectra for the Icelandic ferrihydrites gives values near 4410 and 4580 cm^{-1} for the OH combination bands. These are consistent with the two bands observed at 4440 and 4550 cm^{-1} in Figure 4. Because a third band was observed in the NIR spectra near 4375 cm^{-1}, this implies the presence of another OH stretching band at lower energies in the ferrihydrite spectra. (A small amount of smectite may also be contributing to these features for sample 498). Given an OH bending vibration at *c.* 850 cm^{-1}, this OH stretching vibration would be near 3525 cm^{-1}. Because of the multiple, overlapping OH and water bands in this region, it is difficult to identify this band without D_2O exchange or dehydration experiments.

Weak shoulders are also observed at *c.* 840 cm^{-1} for ferrihydrites 46 and 254 in Figure 7 and are assigned to OH bending vibrations. Adding together the OH bending (840 cm^{-1}) and stretching (3645 cm^{-1}) vibrations for ferrihydrite 46 gives 4485 cm^{-1}, which is close to the value of the OH combination shoulder near 4440 cm^{-1} (Fig. 4). The strongest OH combination band occurs at 4340 and 4320 cm^{-1} for ferrihydrites 46 and 254, respectively, and a weaker band is also present near 4140 cm^{-1}. This implies that the dominant OH stretching bands occur for these synthetic ferrihydrite samples near 3500 cm^{-1}

and that a minor component OH stretching band occurs near 3300 cm^{-1}.

The NIR spectra of the synthetic ferrihydrites show a feature near 6900–7200 cm^{-1} that contains both the overtones of the OH stretching bands and a combination of the H_2O stretching plus an overtone of the H_2O bending vibrations (see Fig. 4 and Table 2). For the Icelandic ferrihydrites these features are extended from *c.* 6900 to *c.* 7300 cm^{-1}. The water combination band occurs near 6900–7000 cm^{-1} and the OH overtone occurs near 7100–7300 cm^{-1} (see Fig. 4 and Table 2). This is consistent with the trends observed for the OH combination features, where the bands for the Icelandic ferrihydrites are observed at higher energies than those for the synthetic ferrihydrites. The OH overtone band is a sharp, weak band in reflectance spectra of smectites and is observed at *c.* 7100 cm^{-1} (*c.* 1.41 μm) for montmorillonite and *c.* 7000 cm^{-1} (*c.* 1.43 μm) for nontronite.

Applications to Mars

The exciting discoveries of possible oceans (Head *et al.* 1999), water seepage (Malin & Edgett 2000*a*), sedimentary layers (Malin & Edgett 2000*b*), and a gray (specular) hematite deposit (Christensen *et al.* 2000*a*) on Mars support earlier suggestions of aqueous processes (Baker *et al.* 1991) and open many new questions about surface water and the possibility of alteration minerals. The ferrihydrite samples associated with hydrothermal volcanic activity on Iceland may have implications for Mars. Volcanism is thought to have been prevalent in the early history of Mars and alteration minerals associated with hydrothermal fluids have been suggested as a source of the Martian soil (Newsom *et al.* 1999). Ferrihydrite may have formed in thermal pools or meltwater run-off streams during the time of volcanism on Mars. Work by Gálvez *et al.* (1999) suggested that ferrihydrite left undisturbed at our collection sites (i.e. at the edges of the water) would form hematite and goethite in the cold stream over time and would form primarily goethite in the thermal spring over time. Ferrihydrite deposited at banks or stream beds and allowed to dry would be more stable.

Low-temperature alteration studies of ferrihydrite are needed in order to determine the stability of this mineral on the surface of Mars and how its spectral and other properties might change upon dehydration. Possibly this ferrihydrite is converted to other nanophase iron oxides such as hematite. However, the effects of dry heating on synthetic 2-line and 6-line ferrihydrites showed that ferrihydrite can be extensively dehydrated without breaking down (Stanjek & Weidler 1992), so a dehydrated version of ferrihydrite on Mars might be stable. Upon dehydration, the nanophase iron oxide grains are likely to become disaggregated particles whatever their mineral structure, and are likely to become dispersed by the dust storms on Mars. If the bands near 1400 and 1500 cm^{-1} are still present in dehydrated ferrihydrite, it should be possible to use these bands to identify this nanophase iron oxyhydroxide mineral remotely if it is sufficiently abundant in the pixel measured. In the case that these bands are not retained in the spectra of the dehydrated ferrihydrite, then spectral identification on Mars would be difficult.

Current thermal IR measurements by TES are recorded from 200–2000 cm^{-1} for surface spots of 3–5 km (Christensen *et al.* 2000*b*), which is much improved over the 130 km spatial resolution achieved by the infrared spectrometers on Mariner 6 and 7 (e.g. Pimentel *et al.* 1974). Still, the spatial resolution of TES would not be expected to allow detection of the *c.* 1400 and 1500 cm^{-1} bands. Carbonates also exhibit spectral features near 1400 cm^{-1} but have not been observed on Mars to date; Christensen *et al.* (2000*b*) interpreted this to indicate that carbonates comprise less than 10% of any 3–5 km spot measured on the surface of Mars. Places to look for dehydrated ferrihydrite on Mars would include in or near the regions where TES has identified gray hematite, features that looks like flow channels or potential hydrothermal regions near volcanic features. Another thermal infrared spectrometer, Mini-TES, will be on the 2003 Mars Exploration Rovers (MERs) and will have a spectral range similar to TES (see http://athena.cornell.edu/the_mission/ins_minites.html). This instrument will be able to measure samples at much closer range and may have a better chance of detecting features near 1400 and 1500 cm^{-1} for dehydrated ferrihydrite, if it is present where these rovers are deployed. The 10 wave number resolution of mini-TES should give 5–7 datapoints for each of these bands if observed.

Dehydrated ferrihydrite formed in Mars' early history may be contributing to the current surface soils and atmospheric dust on Mars. Analysis of the martian soils by the instruments on Mars Pathfinder showed that they are chemically and mineralogically distinct from the rocks, and that they contain iron oxyhydroxides, but do not show the expected evidence for crystalline hematite (Bell *et al.* 2000). The magnetic tests indicated the presence of maghemite and/or magnetite in the dust and soils and suggested that the magnetic component must be intimately mixed with the silicates and other components (Hviid *et al.* 1997; Madsen *et al.* 1999). Ferrihydrite could be a precursor to these minerals. Studies of ferrihydrite–silicate aggregates have also shown that a nanophase iron oxide mineral like ferrihydrite, intimately mixed with smectite grains, is consistent with some, but not all, of the spectral and chemical observations for Mars (Bishop *et al.* 1995).

The compositional heterogeneity of the martian surface has been described recently through detailed analyses of the Imaging Spectrometer for Mars (ISM) data (Bibring *et al.* 1990; Erard *et al.* 1994; Murchie *et al.* 2000). Earlier studies of ISM data found regional variation in the 3 μm water band on Mars (Bibring *et al.* 1990; Erard *et al.* 1994); however the Murchie *et al.* (2000) work evaluated variations in the 3 μm band, together with a number of other spectral factors, and was able to define several 'anomalous soils' in the high albedo layered deposits in western Candor Chasma and in the lower albedo soils in south-eastern Lunae Planum and Sinus Merdiani. Murchie *et al.* (2000) suggested hydrated ferric oxyhydroxides, such as ferrihydrite, as a possible mineralogical explanation for the stronger water band in these regions. The Omega spectrometer scheduled to fly on Mars Express in 2003 is a modified version of ISM; it has a spectral range of 0.5 to 5.2 μm and is expected to have a spatial resolution of *c.* 400 m (http://sci.esa.int/content/doc/90/22160_.htm). If there are regions containing ferrihydrite on Mars, they should exhibit a weak ferric band near 0.92 μm and a strong *c.* 3 μm water band, which could be identified with this instrument. Unfortunately, these bands are not exclusive to ferrihydrite.

Summary

Ferrihydrite spectra are characterized by IR features due to Fe–O, OH and H_2O and a broad crystal field theory band due to excitations of the ferric iron. A broad Fe–O doublet is observed for the synthetic ferrihydrite spectra in this study near 455 and 580 cm^{-1}, which is similar in general to the Fe–O bands observed for goethite and hematite spectra. The natural ferrihydrite spectra have Fe–O bands near 475 and 1000 cm^{-1}, where the higher energy band is strongly influenced by the amount of Si present in the sample and might be due to some Si–O bonds. Analysis of the OH spectral features for ferrihydrites and other iron oxides indicates that natural and synthetic ferrihydrites contain multiple OH sites in their structures. For the samples studied here, the synthetic ferrihydrites contained a minor contribution of OH sites similar to those observed for goethite.

Spectral identification of ferrihydrite in the lab would be simplest using the transmittance bands at *c.* 475 and 1000 cm^{-1}. These bands are broadened in reflectance spectra and would be more difficult to use for ferrihydrite identification in remote sensing. The most characteristic spectral feature for ferrihydrite is a doublet near 1400 and 1500 cm^{-1}, which occurs for pure samples whether they are natural or synthetic. Because these are not strong bands, high spatial resolution and an adequate abundance of ferrihydrite would be required for identification.

Ferrihydrite may have formed on Mars in association with hydrothermal activity as in the case of the Icelandic ferrihydrites studied here. This mineral could have become partly dehydrated on the surface of Mars and it is unknown how loss of water would affect the mineral stability and spectral features. Greater abundances of nanophase ferric oxides near volcanic features or signs of earlier aqueous processes may indicate that ferrihydrite was formed initially in these regions, whether or not it is the iron oxide mineral found today. Many features of the volcano–ice interactions in Iceland serve as examples for what is observed on Mars. Ferrihydrite formation may be another example.

Support for JLB from NASA's Mars Data Analysis Program is much appreciated. Reflectance spectra were measured at RELAB, a multi-user, NASA-supported facility (NAG5-3871). Assistance from C. Allen with sample collection, from J. Kristjansson with information about the Icelandic springs, from T. Hiroi with the bi-directional spectra, from J. Johnson, T. Roush, P. Schiffman and A. Zent with helpful editorial comments and from M. Chapman and J. Smellie with editorial handling is much appreciated.

References

Baker, V. R., Strom, R. G., Gulick, V. C., Kargel, J. S., Komatsu, G. & Kale, V. S. 1991. Ancient oceans, ice sheets and the hydrological cycle on Mars. *Nature*, **352**, 589–594.

Barrón, V., Gálves, N., Hochella, M. F. & Torrent, J. 1997. Epitaxial overgrowth of goethite on hematite synthesized in phosphate media: A scanning force and transmission electron microscopy study. *American Mineralogist*, **82**, 1091–1100.

Bell III, J. F., McSween Jr, H. Y., Murchie, S. L. *et al.* 2000. Mineralogic and Compositional Properties of Martian Soil and Dust: Results from Mars Pathfinder. *Journal of Geophysical Research*, **105**, 1721–1755.

Bibring, J.-P., Combes, M., Langevin, Y. *et al.* 1990. ISM observations of Mars and Phobos:

First Results. *In*: *Proceedings of the 20th Lunar and Planetary Science Conference*, LPI, Houston, 461–471.

BISHOP, J. L., PIETERS, C. M. & BURNS, R. G. 1993. Reflectance and Mössbauer spectroscopy of ferrihydrite-montmorillonite assemblages as Mars soil analog materials. *Geochimimica et Cosmochimica Acta*, **57**, 4583–4595.

BISHOP, J. L., PIETERS, C. M. & EDWARDS, J. O. 1994. Infrared spectroscopic analyses on the nature of water in montmorillonite. *Clays and Clay Minerals*, **42**, 701–715.

BISHOP, J. L., PIETERS, C. M., BURNS, R. G., EDWARDS, J. O., MANCINELLI, R. L. & FROESCHL, H. 1995. Reflectance spectroscopy of ferric sulfate-bearing montmorillonites as Mars soil analog materials. *Icarus*, **117**, 101–119.

BISHOP, J. L., MURAD, E., MADEJOVÁ, J., KOMADEL, P., WAGNER, U. & SCHEINOST, A. 1999. Visible, Mössbauer and infrared spectroscopy of dioctahedral smectites: Structural analyses of the Fe-bearing smectites Sampor, SWy-1 and SWa-1. *In*: *11th International Clay Conference, June 1997*, Ottawa, 413–419.

BURNS, R. G. 1993. *Mineralogical Applications of Crystal Field Theory*. Cambridge University Press, Cambridge.

CARLSON, L. & SCHWERTMANN, U. 1981. Natural ferrihydrites in surface deposits from Finland and their association with silica. *Geochimica et Cosmochimica Acta*, **45**, 421–429.

CARR, M. H. 1981. *The Surface of Mars*. Yale University Press, New Haven.

CARR, M. H. 1995. The martian drainage system and the origin of valley networks and fretted channels. *Journal of Geophysical Research*, **100**, 7479–7507.

CHILDS, C. W. 1992. Ferrihydrite: A review of structure, properties and occurrence in relation to soils. *Zeitschrift fur Pflanzenernährung und Bodenkunde*, **155**, 441–448.

CHRISTENSEN, P. R., BANDFIELD, J. L. & CLARK, R. N. ET AL. 2000*a*. Detection of crystalline hematite mineralization on Mars by the Thermal Emission Spectrometer: Evidence for near-surface water. *Journal of Geophysical Research*, **105**, 9623–9642.

CHRISTENSEN, P. R., BANDFIELD, J. L., SMITH, M. D., HAMILTON, V. E. & CLARK, R. N. 2000*b*. Identification of basaltic component on the Martian surface from Thermal Emission Spectrometer Data. *Journal of Geophysical Research*, **105**, 9609–9621.

CHUKHROV, F. V., ZVYAGIN, B. B., ERMILOVA, L. P. & GORSHKOV, A. I. 1973. New data on iron oxides in the weathering zone. *In*: *Proceedings of the International Clay Conference, Madrid*, 1972, 333–341.

CLIFFORD, S. M. 1993. A model for the hydrological and climatic behavior of water on Mars. *Journal of Geophysical Research*, **98**, 10 973–11 016.

EGGLETON, R. A. & FITZPATRICK, R. W. 1988. New data and a revised structural model for ferrihydrite. *Clays and Clay Minerals*, **36**, 111–124.

ERARD, S., MUSTARD, J., MURCHIE, S., BIBRING, J.-P., CERRONI, P. & CORADINI, A. 1994. Martian aerosols: Near-infrared spectral properties and effects on the observation of the surface. *Icarus*, **111**, 317–337.

FANALE, F. P., POSTAWKO, S. E., POLLACK, J. B., CARR, M. H. & PEPIN, R. O. 1992. Mars: Epochal climate change and volatile history. *In*: KIEFFER, H. H., JAKOSKY, B. M., SNYDER, C. W. & MATTHEWS, M. S. (eds) *Mars*. University Arizona Press, Tucson, 1135–1179.

FARMER, V. C. 1974. The layer silicates. *In*: FARMER, V. C. (ed.) *The infrared spectra of minerals*. The Mineralogical Society, London, 331–363.

GAFFEY, S. J., MCFADDEN, L. A., NASH, D. & PIETERS, C. M. 1993. Ultraviolet, visible, and near-infrared reflectance spectroscopy: Laboratory spectra of geologic materials. *In*: PIETERS, C. M. & ENGLERT, P. A. J. (eds) *Remote geochemical analysis: elemental and mineralogical composition*. Cambridge University Press, Cambridge, 43–77.

GÁLVES, N., BARRÓN, V. & TORRENT, J. 1999. Effect of phosphate on the crystallization of hematite, goethite, and lepidocrocite from ferrihydrite. *Clays and Clay Minerals*, **47**, 304–311.

HEAD III, J. W., HIESINGER, H., IVANOV, M. A., KRESLAVSKY, M. A., PRATT, S. & THOMSON, B. J. 1999. Possible ancient oceans on Mars: Evidence from Mars Orbiter Laser Altimeter Data. *Science*, **286**, 2134–2137.

HINO, M. & SATO, T. 1971. Infrared absorption spectra of silica gel-water, water-D_2, and water-^{18}O systems. *Chemical Society of Japan Bulletin*, **44**, 33–37.

HOFFMAN, N. 2000. White Mars: A new model for Mars' surface and atmosphere based on CO_2. *Icarus*, **146**, 326–342.

HVIID, S. F., MADSEN, M. B., GUNNLAUGSSON, H. P. ET AL. 1997. Magnetic Properties Experiments on the Mars Pathfinder Lander: Preliminary Results. *Science*, **278**, 1768–1771.

JANNEY, D. E., COWLEY, J. M. & BUSECK, P. R. 2000. Structure of synthetic 2-line ferrihydrite by electron nanodiffraction. *American Mineralogist*, **85**, 1180–1187.

KODAMA, H. 1985. *Infrared Spectra of Minerals. Reference Guide to Identification and Characterization of Minerals for the Study of Soils*. Agriculture Canada, Ottawa.

MADSEN, M. B., HVIID, S. F., GUNNLAUGSSON, H. P., KNUDSEN, J. M., GOETZ, W., PEDERSEN, C. T., DINESEN, A. R., MOGENSEN, C. T. & OLSEN, M. 1999. The magnetic properties experiment on Mars Pathfinder. *Journal of Geophysical Research*, **104**, 8761–8779.

MALIN, M. & EDGETT, K. S. 2000*a*. Sedimentary rocks of early Mars. *Science*, **290**, 1927–1937.

MALIN, M. C. & EDGETT, K. S. 2000*b*. Evidence for recent groundwater seepage and surface runoff on Mars. *Science*, **288**, 2330–2335.

MANCEAU, A. & GATES, W. P. 1997. Surface structural model for ferrihydrite. *Clays and Clay Minerals*, **45**, 448–460.

MANCEAU, A., COMBES, J. M. & CALAS, G. 1990. New data and a revised structural model for ferrihydrite: Comment. *Clays and Clay Minerals*, **38**, 331–334.

MORRIS, R. V., LAUER JR, H. V., LAWSON, C. A., GIBSON JR, E. K., NACE, G. A. & STEWART, C. 1985. Spectral and other physicochemical properties of submicron powders of hematite (α-Fe2O3), maghemite (γ-Fe2O3), magnetite (Fe3O4), goethite (α-FeOOH), and lepidocrocite (γ-FeOOH). *Journal of Geophysical Research*, **90**, 3126–3144.

MORRIS, R. V., AGRESTI, D. G., LAUER JR, H. V., NEWCOMB, J. A., SHELFER, T. D. & MURALI, A. V. 1989. Evidence for pigmentary hematite on Mars based on optical, magnetic and Mössbauer studies of superparamagnetic (nanocrystalline) hematite. *Journal of Geophysical Research*, **94**, 2760–2778.

MURAD, E. 1979. Mössbauer spectra of goethite: Evidence for structural imperfections. *Mineralogical Magazine*, **43**, 355–361.

MURAD, E. & SCHWERTMANN, U. 1980. The Mössbauer spectrum of ferrihydrite and its relations to those of the other iron oxides. *American Mineralogist*, **65**, 1044–1049.

MURAD, E. & BISHOP, J. L. 2000. The infrared spectrum of synthetic akaganéite, β-FeOOH. *American Mineralogist*, **85**, 716–721.

MURAD, E., BOWEN, L. H., LONG, G. J. & QUIN, T. G. 1988. The influence of crystallinity on magnetic ordering in natural ferrihydrites. *Clay Minerals*, **23**, 161–173.

MURCHIE, S., KIRKLAND, L., ERARD, S., MUSTARD, J. & ROBINSON, M. 2000. Near-infrared spectral variations of martian surface materials from ISM imaging spectrometer data. *Icarus*, **147**, 444–471.

MUSSELWHITE, D. S., SWINDLE, T. D. & LUNINE, J. I. 2001. Liquid CO_2 breakout and the formation of recent small gullies on Mars. *Lunar Planetary Science XXXII*, CD-ROM No. 1030 (abstr.).

NEWSOM, H. E., HAGERTY, J. J. & GOFF, F. 1999. Mixed hydrothermal fluids and the origin of the Martian soil. *Journal of Geophysical Research*, **104**, 8717–8728.

PANKHURST, Q. A. & POLLARD, R. J. 1992. Structural and magnetic properties of ferrihydrite. *Clays and Clay Minerals*, **40**, 268–272.

PARKER, T. J., GORSLINE, D. S., SAUNDERS, R. S., PIERI, D. C. & SCHNEEBERGER, D. M. 1993. Coastal Geomorphology of the Martian Northern Plains. *Journal of Geophysical Research*, **98**, 11 061–11 078.

PIMENTEL, G. C., FORNEY, P. B. & HERR, K. C. 1974. Evidence about hydrate and solid water in the martian surface from the 1969 Mariner infrared spectrometer. *Journal of Geophysical Research*, **79**, 1623–1634.

POLLACK, J. B., KASTING, J. F., RICHARDSON, S. M. & POLIACKOFF, K. 1987. The case for a warm, wet climate on early Mars. *Icarus*, **71**, 203–224.

RUSSELL, J. D. 1979. Infrared spectroscopy of ferrihydrite: Evidence for the presence of structural hydroxyl groups. *Clay Minerals*, **14**, 109–114.

RYSKIN, Y. I. 1974. The vibrations of protons in minerals: Hydroxyl, water and ammonium. *In*: FARMER, V. C. (ed.) *The infrared spectra of minerals*. The Mineralogical Society, London, 137–181.

SALISBURY, J. W. 1993. Mid-infrared spectroscopy: Laboratory data. *In*: PIETERS, C. M. & ENGLERT, P. A. J. (eds) *Remote geochemical analysis: elemental and mineralogical composition*. Cambridge University Press, Cambridge, 79–98.

SALISBURY, J. W., WALTER, L. S., VERGO, N. & D'ARIA, D. M. 1991. *Infrared (2.1–25 μm) spectra of minerals*. Johns Hopkins University Press, Baltimore.

SCHEINOST, A. C., CHAVERNAS, A., BARRÓN, V. & TORRENT, J. 1998. Use and limitations of second-derivative diffuse reflectance spectroscopy in the visible to near-infrared range to identify and quantify Fe oxide minerals in soils. *Clays and Clay Minerals*, **46**, 528–536.

SCHWERTMANN, U. 1988. Goethite and hematite formation in the presence of clay minerals and gibbsite at 25°C. *Soil Science Society of America Journal*, **52**, 288–29.

SCHWERTMANN, U. & MURAD, E. 1988. The nature of an iron oxide – organic iron association in a peaty environment. *Clay Minerals*, **23**, 291–299.

SCHWERTMANN, U. & CORNELL, R. M. 2000. *Iron oxides in the laboratory. Preparation and characterization*. Wiley-VCH, Weinheim (Germany).

STANJEK, H. & WEIDLER, P. G. 1992. The effect of dry heating on the chemistry, surface area, and oxalate solubility of synthetic 2-line and 6-line ferrihydrites. *Clay Minerals*, **27**, 397–412.

TOWE, K. M. & BRADLEY, W. F. 1967. Mineralogical constitution of colloidal 'hydrous ferric oxides'. *Journal of Colloid Interface Science*, **24**, 384–392.

ZHAO, J., HUGGINS, F. E., FENG, Z. & HUFFMAN, G. P. 1994. Ferrihydrite: Surface structure and its effects on phase transformation. *Clays and Clay Minerals*, **42**, 737–746.

Geochemical and mineralogical analyses of palagonitic tuffs and altered rinds of pillow basalts in Iceland and applications to Mars

JANICE L. BISHOP[1], P. SCHIFFMAN[2] & R. SOUTHARD[3]

[1] *SETI Institute/NASA-Ames Research Center, MS 239-4, Moffett Field, CA 94035, USA (e-mail: jbishop@mail.arc.nasa.gov)*

[2] *Department of Geology, University of California, Davis, CA 95616, USA*

[3] *Department of Land, Air and Water Resources, University of California, Davis, CA 95616, USA*

Abstract: Samples of altered pillow basalts and hyalotuffs were collected from a volcanic tuya and hyaloclastite ridge in western Iceland. Altered basaltic material from regions such as Hlöðufell tuya and Thórólfsfell ridge may be similar to the altered basaltic surface fines on Mars. Geochemical and mineralogical analyses have been performed on the Icelandic samples in order to characterize the properties that distinguish palagonitization from other forms of low temperature alteration in this environment. Major elements were measured using an electron microprobe and mineralogy was determined through X-ray diffraction and visible-infrared reflectance spectroscopy. The primary focus in this study was on the <2 μm size fractions of the Hlöðufell altered pillow basalt and Thórólfsfell palagonitic tuff samples. Both volcanic alteration products contain at least some smectite and serpentine clay minerals, as well as poorly crystalline layer silicates. The palagonitic tuff contains more crystalline clay minerals, fewer nanophase iron oxides/oxyhydroxides, and has a higher Al/Fe ratio compared to the altered pillow basalt. Spectra of the <2 μm fractions of both Icelandic samples share similarities with the extended visible region spectra of the bright martian soils measured by Pathfinder and the infrared spectra of the martian dust measured by spectrometers on the Mariner missions.

Altered basaltic rocks, including pillow lavas and palagonitized hyalotuffs, are a significant component of the lithofacies of subglacial volcanoes on Earth, and perhaps on Mars as well. Von Waltershausen (1846) first recognized palagonitized tuffs from Palagonia, Sicily, and Honnorez (1981) subsequently noted their widespread occurrence in Iceland. Since that time, many geochemical and mineralogical studies of weathering and palagonitization have been used to characterize alteration trends in glassy basalts and andesites (e.g. Colman 1982; Staudigel & Hart 1983; Fisher & Schminke 1984; Nesbitt & Young 1984). Collectively, these studies have shown that volcanic glass and olivine are the least stable components of basaltic deposits, whereas iron–titanium oxide minerals are the most resistant to alteration. According to these and other studies, the typical low temperature alteration products of mafic volcanic rocks are a mixture of short-range ordered aluminosilicates (e.g. allophane and imogolite), amorphous iron oxides/oxyhydroxides, and poorly crystalline and crystalline clay minerals (including smectite and kaolinite). Palagonitization and formation of weathering rinds on basalts and andesites are also accompanied by large reductions in Ca, Mg, Na and K depletion in Si, oxidation of the Fe, and incorporation of water (e.g. Fisher & Schminke 1984).

The palagonitization and alteration of the eruptive products of subglacial basaltic volcanism on Iceland have long been studied in order to characterize alteration rates, composition of the altered glass and the influence of hydrothermal and environmental factors on primary mineralogy (Furnes 1978; Jakobsson 1978; Jakobsson & Moore 1986; Crovisier *et al.* 1987; Le Gal *et al.* 1999). Palagonitization of basaltic tephra on Surtsey occurred rapidly in response to hydrothermal activity. When subjected to temperatures ranging from 40 to 100°C large portions of the tephra cone were palagonitized within 2–4 years (Jakobsson 1978). Conversely, palagonitization at ambient surface temperatures on Iceland requires a few to several thousand years (Furnes 1978). Low temperature alteration, including palagonitization, has been shown to

From: SMELLIE, J. L. & CHAPMAN, M. G. (eds) 2002. *Volcano–Ice Interaction on Earth and Mars*. Geological Society, London, Special Publications, **202**, 371–392. 0305-8719/02/$15.00

modify significantly the primary magnetic mineralogy of Icelandic basaltic deposits. A study of hematite in basaltic Icelandic lava flows by Beske-Diehl & Huiling (1993) showed that the magnetic properties of basalts could be used as alteration indicators for these basalts. A more recent study identified fine-grained (<2 μm) secondary iron oxides that were associated with palagonitization of hyaloclastite glass fragments in samples collected from western Iceland (Goguitchaichvili *et al.* 1999).

In this study, we present new geochemical and mineralogical data on altered pillow basalts and hyalotuffs from subglacial volcanic remnants in the Laugarvatn mountains of southwestern Iceland, the site of seminal studies on the origin of intraglacial volcanoes (Jones 1969, 1970). The pillow lava and hyalotuff sequences in these Quaternary volcanic systems are believed to represent temporarily eruptive products of magma–water interactions within shallowing glacial meltwater vaults (Jones 1969). The approximate sample collection sites and images of the samples are shown in Figures 1 and 2. Palagonitized tuff samples for this study were collected from a prominent outcrop at *c.* 50–100 m elevation above the base, on the western face of the Thórólfsfell hyaloclastite ridge. Samples of

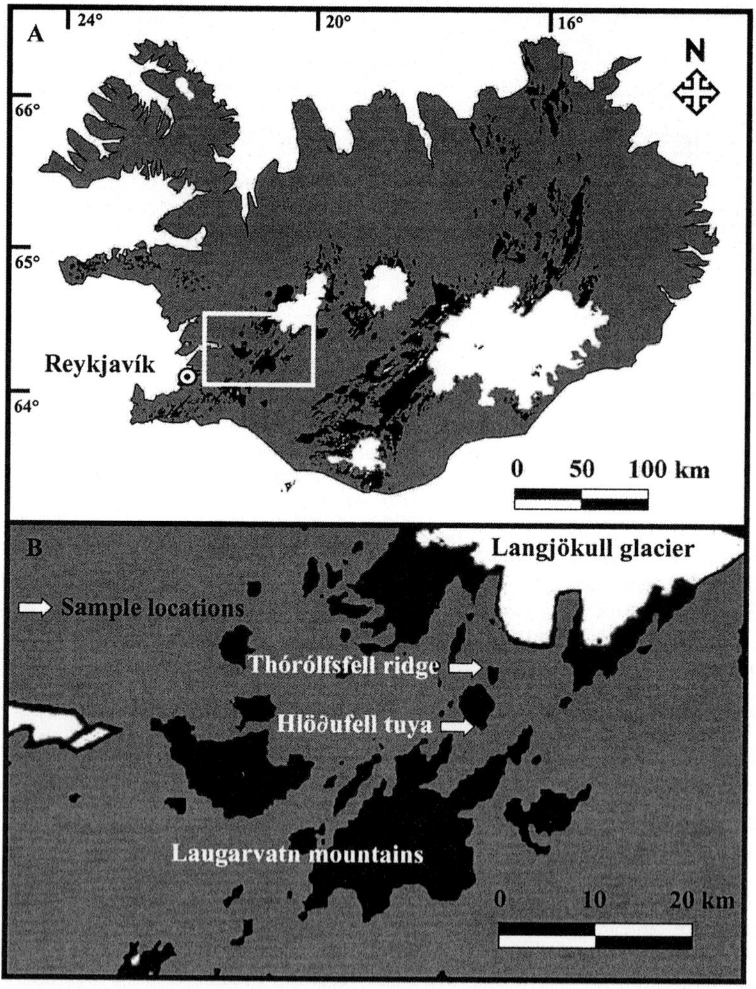

Fig. 1. Map indicating locations of sample collection sites: (**a**) Iceland, and (**b**) Laugarvatn mountains, SW Iceland. Samples were collected on the southern flank of the Hlöðufell tuya and on the NW face of the Thórólfsfell ridge which are located at the northern edge of the Laugarvatn mountains, just south of the Langjökull glacier (map after Jóhannesson & Saemundsson 1998).

Fig. 2. Images of the sample collection sites: (**a**) Thórólfsfell ridge, (**b**) palagonitic tuff in Thórólfsfell ridge, (**c**) altered surface of pillow basalt on Hlöðufell tuya, and (**d**) an altered glassy selvage of pillow basalt on Hlöðufell tuya. The red bar represents *c.* 1 metre in (**a**) and *c.* 5 cm in (**b**, **c**, and **d**).

altered pillow basalt were collected from *c.* 200 to 500 m elevation above base on the southern face of the Hlöðufell tuya. Geochemical and mineralogical analyses have been performed on these samples in order to characterize the similarities and differences in the alteration products of basaltic volcanic material formed via palagonitization and surface alteration. In addition to providing mineralogical information about the samples, analysis of the reflectance spectra measured here provide insights into the spectral bands that can be used for remote detection of fresh and altered volcanic material. This study has implications for basaltic alteration in terrestrial settings, as well as on Mars, as discussed below.

Composition of Mars

The surface mineralogy on Mars holds information about the environmental and geochemical record on that planet and may even provide clues to when water may have been present there. Visible/near-infrared (NIR) reflectance spectra, thermal-region IR emission spectra, chemistry and magnetic parameters are all combined in efforts to characterize the mineralogy of the surface rocks and soils on Mars. The rock compositions are basaltic to andesitic (McSween *et al.* 1999; Bandfield *et al.* 2000; Christensen *et al.* 2000) and many of the rocks near the Pathfinder lander contain surface coatings (McSween *et al.* 1999). The Martian soils contain elevated Fe and S abundances compared to the rocks (Bell *et al.* 2000; Foley *et al.* 2001) and unique identification of specific minerals has been difficult in the soils because they are thought to be a composite of multiple poorly-crystalline phases. The magnetic properties of the Martian soils near the Viking (Hargraves *et al.* 1977) and Pathfinder (Madsen *et al.* 1999) landing sites are consistent with tiny (nanometer-sized) maghemite and/or magnetite grains embedded in 1–2 μm silicate particles. The alteration processes that control which silicate and iron oxide minerals form from volcanic material are directly linked to environmental conditions and physical factors. Spectroscopic analyses of terrestrial soils and alteration products with known weathering histories and/or formation conditions are essential in order to interpret the spectroscopic data from Mars. Studies of the geochemistry, mineralogy and spectral character of altered tephra have been performed on samples from the Hawaiian islands (e.g. Singer 1982; Morris *et al.* 1990; Bell *et al.* 1993; Bishop *et al.* 1998*a*) in order to use altered Hawaiian tephra as possible analogue materials for the Martian surface soils.

The composition of Mars has historically been determined through telescopic observations, and more recently through landers and orbiters. The spectral properties of bright and dark regoliths on Mars have been summarized by Soderblom (1992) and Roush *et al.* (1993). Typical bright region Mars soil spectra exhibit a weak visible feature due to iron oxides (Singer *et al.* 1979; McCord *et al.* 1982; Bell *et al.* 1990); this band has been observed near 0.86 µm in many regions and is consistent with fine-grained, red hematite (Bell *et al.* 1990). Spectra of less common bright-region Mars soils exhibit a weak visible feature at longer wavelengths, which is more consistent with nanophase ferric oxyhydroxides (e.g. ferrihydrite, schwertmannite) than hematite. For Lunae Planum and the Oxia region of western Arabia this ferric absorption feature is observed near 0.88 and 0.92 µm, respectively (McCord *et al.* 1982; Murchie *et al.* 1993, 2000). Recent analyses of the soils measured by the imager on Mars Pathfinder are consistent with these earlier telescopic and orbital measurements; however, the spectra of the soils near the Pathfinder lander exhibit weaker ferric bands in general and indicate variation in the soil mineralogy on the scale of a metre or less (Bell *et al.* 2000).

The compositional heterogeneity of the Martian surface has been described recently through detailed analyses of the Imaging Spectrometer for Mars (ISM) data (Bibring *et al.* 1990; Erard *et al.* 1994; Murchie *et al.* 2000). Differences in the bands near 2 and 3 µm cannot be attributed to simple albedo changes or rock–soil mixing; they are best explained by differences in mineralogy and are especially evident in Lunae Planum, Syrtis Major and Valles Marineris. The spectral units exhibiting stronger than usual 3 µm H_2O bands are termed ‘anomalous soils’. These anomalous soils are found in a variety of regions and exhibit variations in their visible/NIR spectral properties, implying that multiple processes are responsible for formation of these hydrated soil units on Mars. Murchie *et al.* (2000) observed the unusually strong 3 µm band in spectra of the high-albedo layered deposits in western Candor Chasma and in the lower-albedo soils in southeastern Lunae Planum and Sinus Merdiani. They suggested four possible compositional sources for this stronger 3 µm band based on work done by others: ferric sulfate minerals such as schwertmannite or hydrated ferric oxyhydroxides such as ferrihydrite (Burns 1994; Bishop & Pieters 1995; Bishop & Murad 1996), hydrovolcanic glass (Farrand & Singer 1992) or hydrous carbonates (Calvin *et al.* 1994).

Chemical mixing models have shown that simple soil-rock mixing scenarios are inadequate to explain the chemical composition of the martian rocks and soils (McLennan 2000; McSween & Keil 2000) and that the global dust on Mars resulted from weathering of basaltic rather than felsic rocks (McLennan 2000; McSween & Keil 2000). McSween & Keil (2000) further suggested that palagonitization of basalt better explains the chemical trends observed for Martian soil, than do other alteration processes. The presence of a diagenetic mineral (e.g. an evaporite) is also required in order to account for some or all of the S, Cl, Mg and part of the Fe (McLennan 2000; McSween & Keil 2000). Mineral fractionation is also required to explain the Fe_2O_3/TiO_2 trends observed for soils near Mars Pathfinder. Volcano–ground ice interactions have been invoked as contributors to the martian mineralogy (Squyres *et al.* 1987) following early studies of the spectral properties of material resulting from palagonitization of basaltic tephra (Allen *et al.* 1981; Singer 1982). More recent laboratory analyses of the mineralogy and spectroscopic properties of the alteration or palagonitization products of basaltic ash and tephra have been performed (Bell *et al.* 1993; Morris *et al.* 1993, 1996; Roush & Bell 1995; Bishop *et al.* 1998*a*; Schiffman *et al.* 2002) and indicate that a variety of minerals and spectroscopic features form depending on the alteration conditions.

Methods

Samples

The primary focus in this study is on the <2 µm size fractions of the two Icelandic samples collected from the palagonitic tuff and the altered pillow basalt as described earlier. Bulk samples were reduced to powders using a tungsten carbide mill. Separation of the <2 µm fraction was performed on sub-samples of the powders that were dispersed in dilute Na_2CO_3 (0.11 g l^{-1}) and centrifuged; the supernatant containing the <2 µm material was decanted (Jackson 1979). This process was repeated several times to ensure nearly complete removal of the <2 µm fraction. The suspensions containing the Na-saturated material were dialyzed in cellulose tubes against distilled water to remove excess Na. These products were air dried, then the fine-grained particles were partially disaggregated through dry sieving. The dry <2 µm particles that passed through a 45 μm sieve and those that clustered in larger than 125 μm aggregates were collected for spectroscopic analysis.

For comparison with the clay-size material, the original palagonitic tuff sample was also dry sieved to <45 and 45–125 μm for spectroscopic analysis. The altered material collected at the base of a large pillow lava on Hlöðufell was dry sieved only to <125 μm because insufficient fines were present in this sample for measurement of a <45 μm fraction.

Analyses

Wavelength dispersive electron microprobe analyses were performed on the <2 μm size fractions of the palagonitic tuff and altered pillow basalt, as well as several grains in the <125 μm dry sieved altered pillow basalt and several spots in a transect across an altered rim of the pillow basalt. This technique was used as in past studies (Schiffman *et al.* 2000, 2002) to determine the major element composition of these altered basalt samples. Back-scattered electron (BSE) images were collected from epoxide-impregnated, polished sections. The sample powders were smear-mounted on double-sided sticky carbon tape and then carbon coated.

X-ray diffraction (XRD) analyses were performed on oriented aggregates of the <2 μm fraction mounted on glass slides with a Diano 8000 diffractometer (Cu Kα radiation). Two diffractograms were produced for each sample: the first was run following Mg saturation of the clay minerals (to replace exchangeable Na and other cations) and another was performed following glycerol solvation of the Mg-saturated clay minerals in the <2 μm fractions (Whittig & Allardice 1982).

Visible-infrared reflectance spectra were measured as described in detail previously (Bishop *et al.* 1995). Bidirectional visible/near-infrared spectra were measured relative to Halon (a calibrated, white reflectance standard) under ambient conditions at the Reflectance Experiment Laboratory (RELAB) at Brown University. Infrared biconical reflectance spectra were measured relative to a rough gold surface using a Nicolet 740 Fourier transform interferometer (FTIR) in a H_2O- and CO_2-purged environment. Composite, absolute reflectance spectra were prepared by scaling the FTIR data to the bidirectional data near 1.2 μm.

Oxyhydroxide and short-range-ordered layer silicate phases of the bulk materials were selectively extracted with ammonium oxalate and with citrate-dithionite (Soil Survey Laboratory Staff 1996). Ammonium oxalate (0.2 M, pH 3) extracts a suite of short-range-ordered compounds including aluminosilicates (e.g. allophane and imogolite), iron oxyhydroxides (e.g. ferrihydrite), and organically-complexed Fe and Al. Citrate-dithionite reduces and complexes (hence, solubilizes) ferric iron in crystalline and short-range-order Fe-oxyhydroxides, probably dissolves some 'allophane-like' aluminosilicates, and extracts Fe and Al from humus-metal complexes (Wada 1989). Extracted Fe, Al, and Si were analysed by ICP or atomic absorbance spectrometry.

Results: chemistry, mineralogy, reflectance spectra

The average chemical composition of samples in this study are shown in Table 1 along with an average of soil compositions for sites A-2, A-4, A-5, A-9, A-10 and A-15 near the Mars Pathfinder lander reported by Foley *et al.* (2001). Although expressing these data as oxides may be problematic because of the extremely fine-grained and variable hydrated nature of the alteration materials, this form is used in order to facilitate comparison with the chemistry of martian soils and other geological studies. The iron is a combination of ferrous and ferric iron, but is expressed as FeO in Table 1. The <2 μm fraction of the palagonitic tuff contains relatively higher SiO_2 and lower FeO compared to the <2 μm fraction of the altered pillow basalt. The lower oxide totals for the altered portions of the polished samples in Table 1 reflect water in the epoxide mounts. Both water and an uneven surface can produce lower totals for the powdered samples. Relatively flat sample regions were selected in this study in order to reduce the surface effects. The analyses of the two <2 μm particle size powders have somewhat higher weight percent oxide totals with respect to that of the *in situ* alteration (Table 1). This difference may reflect the presence of fresh (i.e. anhydrous) glass particles as well as hydrous clays in the <2 μm particle size powders.

BSE imaging of a Thórólfsfell ridge hyalotuff sample demonstrates that grains of glassy basaltic ash are ubiquitously rimmed by alteration rinds characterized by lower mean-atomic-number (as reflected by the higher brightness of the fresh glass with respect to the rinds as seen in Fig. 3a, b.) These rinds, whose maximum thickness is <50 μm have physically melded together in places, partially consolidating the hyalotuff. Compositional analyses of the rinds (Table 1) indicate that they are significantly depleted in

Table 1. *Major elements for Thórólfsfell ridge and Hlöðufell tuya samples*

wt% oxides	Pillow basalt, Hlöðufell			Palagonitic tuff, Thórólfsfell			Mars soil at MPF average of 6 sites*
	<2 μm (500a)	*in situ* vein	*in situ* glass	<2 μm (501a)	*in situ* rind	*in situ* glass	
SiO_2	33.3	29.8	48.9	44.6	41.9	48.3	40.6 ± 0.7
Al_2O_3	24.0	24.3	14.6	22.0	14.3	16.1	9.6 ± 0.3
FeO	22.1	20.6	12.4	12.2	12.7	10.6	20.1 ± 0.2
MgO	1.2	0.6	7.6	5.9	3.4	8.4	8.3 ± 1.8
CaO	1.1	1.3	12.0	1.0	4.4	11.5	5.8 ± 0.2
Na_2O	2.1	0.1	2.2	3.0	0.0	2.1	3.5 ± 1.1
SO_3	.bdl	.bdl	0.04	0.4	.bdl	0.06	6.1 ± 1.2
K_2O	0.0	0.0	0.2	0.1	0.1	0.2	0.9 ± 0.05
P_2O_5	0.2	0.3	0.3	0.1	0.1	0.3	0.7 ± 0.2
MnO	0.6	0.0	0.2	0.2	0.1	0.1	0.3 ± 0.1
TiO_2	2.8	2.6	1.8	1.1	2.0	1.5	0.9 ± 0.1
Total	87.4	79.7	100.1	90.7	79.0	99.1	97.9

Note that all Fe is shown as FeO; this is likely to be present as both FeO and Fe_2O_3 in the samples. *averages of soil data from Foley *et al.* (2001) and Fe_2O_3 converted to FeO, which decreased the total oxide sum. bdl, below detection limits of 0.03 wt% S as SO_3.

MgO, CaO, and Na_2O (and presumably greatly enriched in H_2O) with respect to the adjacent fresh glass. The Ca-depletion is graphically evident in X-ray dot maps (Fig. 3c). Conversely, the enrichments or depletions in Fe, Al, and Si (Table 1; Fig. 3) accompanying the formation of the rinds have been less significant than the changes in Mg, Ca and Na.

BSE imaging of a glassy selvedge from a basaltic pillow lava at the base of Hlöðufell Tuya reveals that the alteration is restricted to thin – generally <100 μm thick-veins that cut through fresh glass, but not microphenocrysts of olivine and plagioclase (Fig. 4a). Compositional analyses of these veins (Table 1) indicate that they are also significantly depleted in MgO, CaO, and Na_2O (and presumably greatly enriched in H_2O) with respect to their adjacent fresh glass. The strong Ca-depletion in the veins is clearly seen in X-ray dot maps (Fig. 4c). However, unlike the compositional changes recorded in the rinds of the hyalotuff, the veins cutting the pillow basalts are also apparently depleted in SiO_2 and enriched in FeO and Al_2O_3 (Table 1 Fig. 4).

The magnitudes of the elemental enrichments and depletions discussed above are better evaluated by converting the reported concentrations from weight to molar percentages. This method yields a more rigorous comparison of the elemental fluxes that accompany glass hydration. For these calculations we have assumed that (1) the fresh glass contains approximately 0.5 wt% H_2O, and that (2) the weight percent of water in the alteration products is equivalent to the difference of 100% minus the sum of the oxides given in Table 1. The results of these calculations are presented in Table 2 and indicate that the fluxes of major and minor elements accompanying alteration have been somewhat different for the hyalotuff compared to the pillow basalt samples. Specifically, if we arbitrarily define an element as migratory, whose oxide flux is >30 mol% then the elements Al, P, Ti, and Fe have behaved conservatively in the pillows, whereas only Ti and Fe have been conserved in the hyalotuffs. The calculated molar abundances for selected elements in the fresh glass and altered rims or veins are shown graphically in Figure 5.

The X-ray diffractograms of the <2 μm fractions are shown in Figure 6. The altered pillow basalt diffractograms are shown in Figure 6a where the Mg-saturated and glycerol solvated sample is compared with the Mg-saturated version. The clay minerals in this sample are non-crystalline or have only short-range order and we suggest that they represent pedogenic weathering products. The hyalotuff diffractograms are shown in Figure 6b and indicate that this sample contains some poorly crystalline or 'nanocrystalline' smectite-like clay minerals. The interlayer spacings of the smectite component are 1.47 nm with Mg and expand with glycerol to higher, but variable d-spacings. The variability of the d-spacings in these expanded samples results in a weaker peak. Although this peak is weak, it unequivocally demonstrates that smectite is an alteration product of the hyalotuffs and not of the pillow lavas.

The XRD results are corroborated by data from *in situ* cation exchange experiments on polished mounts of the hyalotuff and pillow basalt samples. Details of this technique which

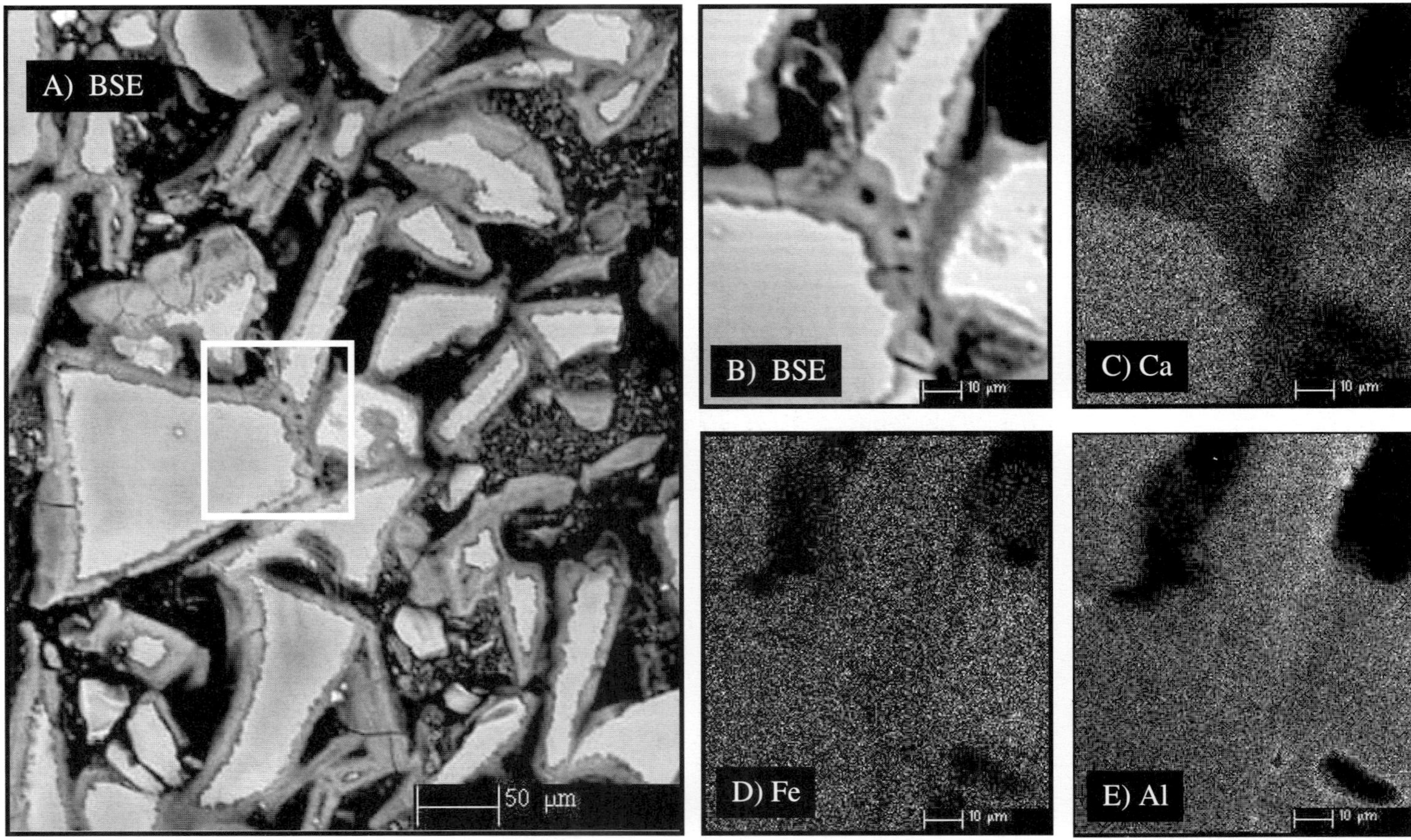

Fig. 3. BSE images of palagonitic tuff material from Thórólfsfell ridge. An image including several glassy grains and their altered rinds is shown in (**a**), followed by an enlarged view of the region in the white box in (**b**). Elemental dot maps of the region in (**b**) are shown for Ca in (**c**) for Fe in (**d**) and for Al in (**e**).

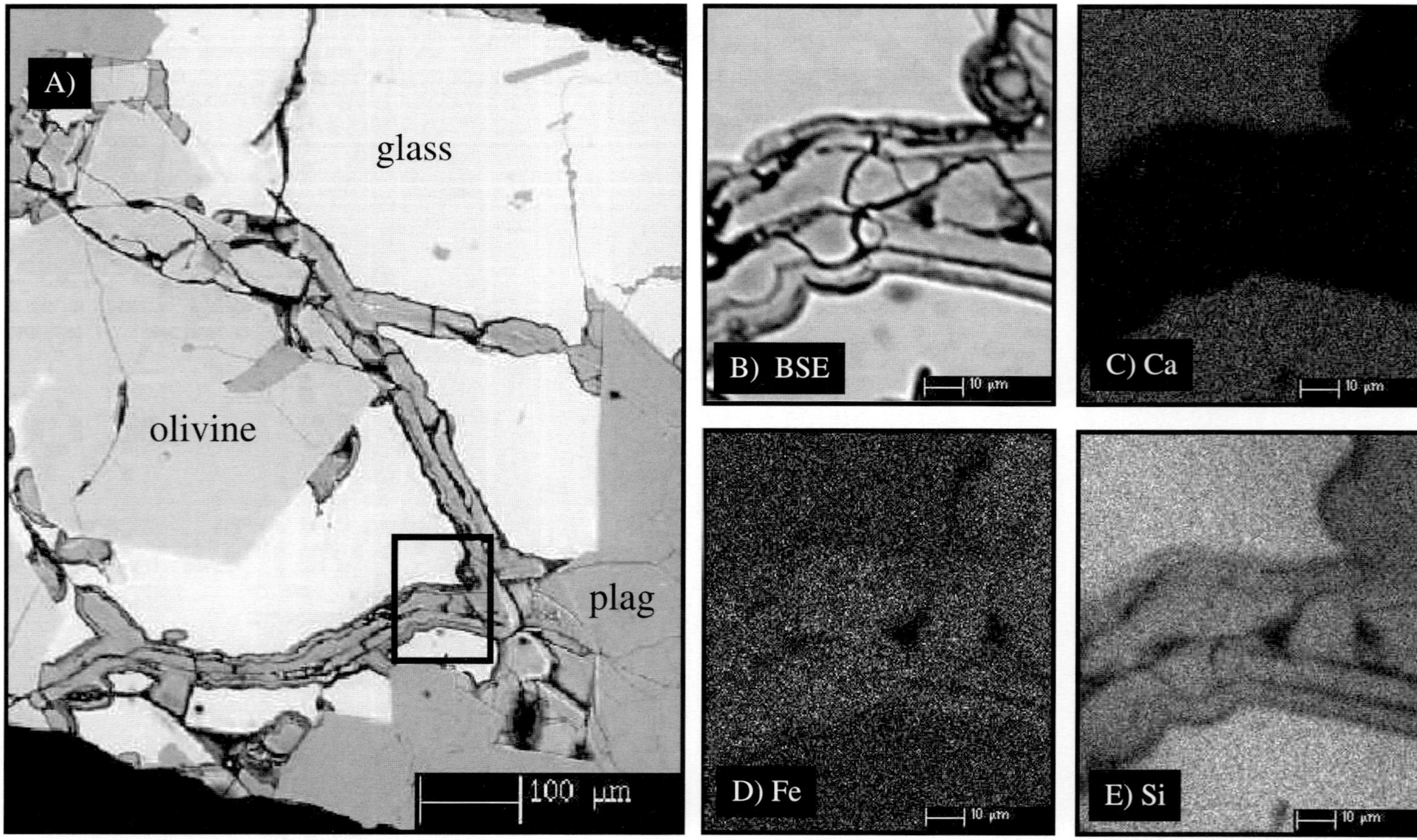

Fig. 4. BSE images of alteration rinds of pillow basalts from Hlöðufell tuya. An image showing glass, olivine and plagioclase grains, plus several altered veins is displayed in (**a**). An enlarged view of the region in the black box is shown in (**b**) with the corresponding elemental dot maps for Ca in (**c**) for Fe in (**d**) and for Si in (**e**).

Table 2. *Molar abundances of major elements for Thórólfsfell ridge and Hlöðufell tuya samples*

Mole % oxides	Pillow basalt, Hlöðufell					Palagonitic tuff, Thórólfsfell				
	in situ glass	*in situ* vein	rel.% Δ	<2 μm	rel.% Δ	*in situ* glass	*in situ* rind	rel.% Δ	<2 μm	rel.% Δ
SiO_2	50.1	22.3	−56	28.8	−42	50.0	29.4	−41	39.3	−21
Al_2O_3	8.8	10.7	21	12.2	39	9.8	5.9	−40	11.4	17
FeO	10.6	12.9	21	16.0	50	9.2	7.5	−19	9.0	−2
MgO	11.6	0.7	−94	1.6	−87	13.0	3.6	−73	7.8	−40
CaO	13.2	1.0	−92	1.0	−92	12.8	3.3	−74	1.0	−93
Na_2O	2.2	0.1	−97	1.8	−19	2.1	0	−100	2.6	22
SO_3									0.3	
K_2O	0.1	0.0	−100	0.0	−100	0.1	0.0	−66	0.1	−57
P_2O_5	0.1	0.1	−27	0.1	−44	0.1	0.0	−77	0.0	−72
MnO	0.2	0	−100	0.4	153	0.1	0.1	−32	0.2	71
TiO_2	1.4	1.5	5	1.8	31	1.2	1.1	−10	0.7	−37
H_2O	1.7	50.8	2880	36.3	2030	1.7	49.2	2750	27.6	1500
Total	100	100		100		100	100		100	

Note that all Fe is shown as FeO; this is likely to be present as both FeO and Fe_2O_3 in the samples. In converting the wt% abundances in Table 1 to mole % values, the fresh glass is assumed to contain 0.5 wt% H_2O. For the altered samples H_2O is assumed to be present in sufficient quantity to give a sum of 100 wt% in Table 1. The relative change in mole % (rel.% Δ) is the relative percent difference between the molar abundances of the altered glass and <2 μm fractions compared to the fresh glass.

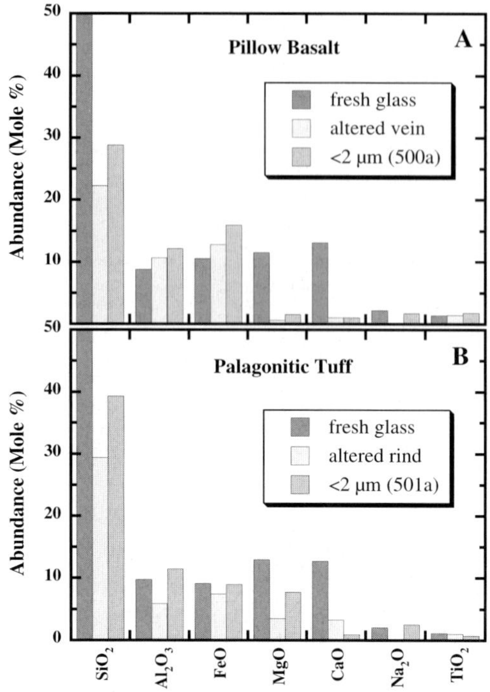

Fig. 5. Chemical variance in molar abundance of selected elements as oxides between the fresh glass, altered regions and the <2 μm fractions of (**a**) the altered pillow basalt and (**b**) the palagonitic tuff.

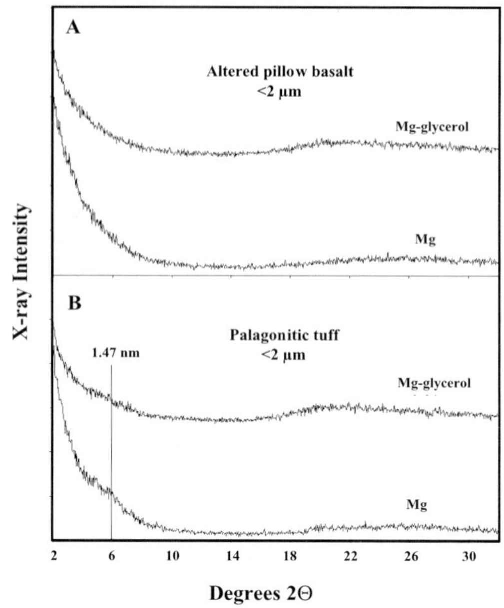

Fig. 6. X-ray diffractograms of the <2 μm fractions of (**a**) the altered pillow basalt and (**b**) the palagonitic tuff. Diffractograms are shown for the Mg-saturated and the Mg-saturated-glycerol-solvated treatments of each sample.

entails 'staining' of the samples with 1 molar CsCl solutions followed by X-ray dot mapping and compositional analysis, and which provides qualitative as well as quantitative estimation of cation exchange capacity (CEC), have been presented by Schiffman & Southard (1996). Figure 7 shows complementary BSE images and Cs–Lα X-ray dot maps recorded in the vicinity of alteration rinds in the hyalotuff, and alteration veins in the pillow basalts. The intensity of Cs–Lα X-rays is much greater in the palagonitized rinds of the hyalotuffs than in the veins that cut through glass in the pillow basalts. The fresh glass exhibits only background levels of Cs–Lα X-rays. The CEC's (calculated directly from the wt% of Cs in these regions) of the alteration materials in the hyalotuff rinds v. pillow veins are approximately 25 meq/100 g and 4 meq/100 g respectively. The former value presumably reflects the presence of exchangeable interlayer sites in smectite within the palagonitized rinds; the latter value presumably reflects the amorphous nature of the pedogenic alteration products in the pillow veins.

Table 3. *Elemental composition of oxalate and dithionite selective dissolution extracts of Thórólfsfell ridge and Hlöðufell tuya samples*

Sample	% Al_o	% Si_o	% Fe_o	% Al_d	% Si_d	% Fe_d	Al_o/Si_o	Fe_o/Fe_d
Palagonitic tuff, Thórólfsfell	4.87	8.34	4.53	0.16	0.25	3.28	0.61	1.38
Pillow basalt, Hlöðufell	4.34	6.28	5.00	0.72	0.38	6.81	0.72	0.73

Note: the % values are elemental mass % (not oxides), and the molar ratios are elemental molar ratios (not oxides). $_o$ denotes oxalate extraction; $_d$ denotes dithionite extraction.

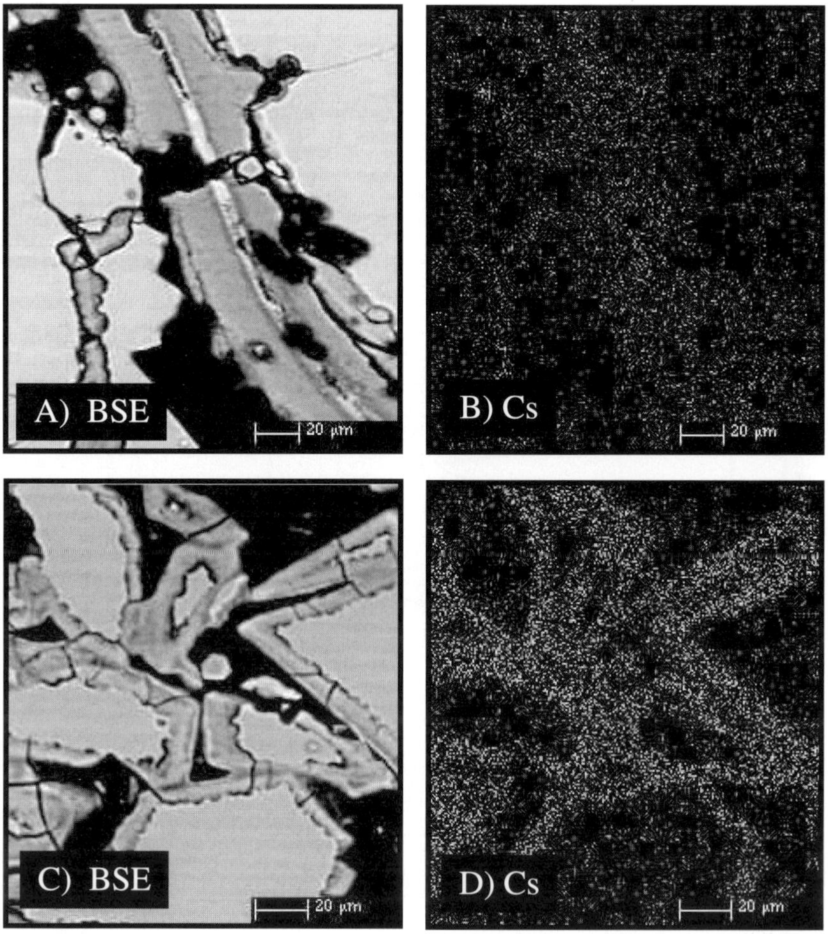

Fig. 7. Complementary BSE images and Cs–Lα X-ray dot maps are shown here for an altered vein in the Hlöðufell pillow basalt (**a**, **b**) and an alteration rind in the Thórólfsfell hyalotuff (**c**, **d**). The intensity of dots is directly correlated with the cation exchange capacity (CEC) in the minerals and glass. The fresh glass has essentially no CEC, and the faint intensity of dots in these regions are due to background X-radiation.

Results of the selective dissolution by oxalate and dithionite are shown in Table 3 and are consistent with the XRD, cation exchange and elemental composition results. The elements measured following oxalate extraction are denoted by subscript 'o' and the elements measured after dithionite extraction are denoted by subscript 'd'. Both the palagonitized tuff and the pillow basalt contain short-range-order layer silicates that are oxalate soluble. Oxalate also extracts Al from humus–metal complexes, but both samples probably have relatively little incorporated organic matter. Thus, the Al_o/Si_o molar ratios indicate the presence of allophane in both the tuff and the pillow basalt. The Fe values show that the pillow basalt has more total 'secondary Fe' (dithionite extractable) than the tuff. An Fe_o/Fe_d ratio near one can be interpreted to mean that all of the 'secondary Fe' (dithionite extractable) has short-range order. A ratio higher than one often indicates that very fine-grained magnetite was present in the sample. The results suggest that the tuff may contain some fine-grained magnetite (although not attracted to a magnet, see below) and that the secondary iron oxyhydroxides in the tuff have short-range order. In contrast, the Fe_o/Fe_d ratio in the pillow basalt indicates that some of the iron oxyhydroxides are crystalline (extractable by dithionite, but not by oxalate). Either goethite or hematite could be the crystalline phase, but neither was identifiable by XRD. The

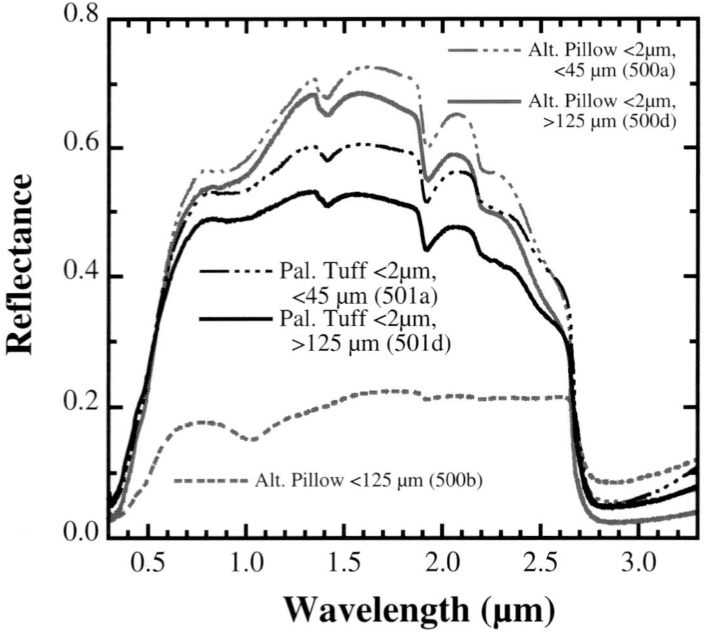

Fig. 8. Reflectance spectra from 0.3 to 3.3 µm of the Icelandic tuff and altered pillow basalt samples. Spectra are shown here for three separates of the altered pillow basalts: dry sieved to <125 µm and sediment fractionated to <2 µm; the clay-size fraction was further separated by dry sieving into <45 µm and >125 µm. Spectra are shown here for the <2 µm fraction of the palagonitic tuff sample that were also separated by dry sieving to <45 µm and >125 µm.

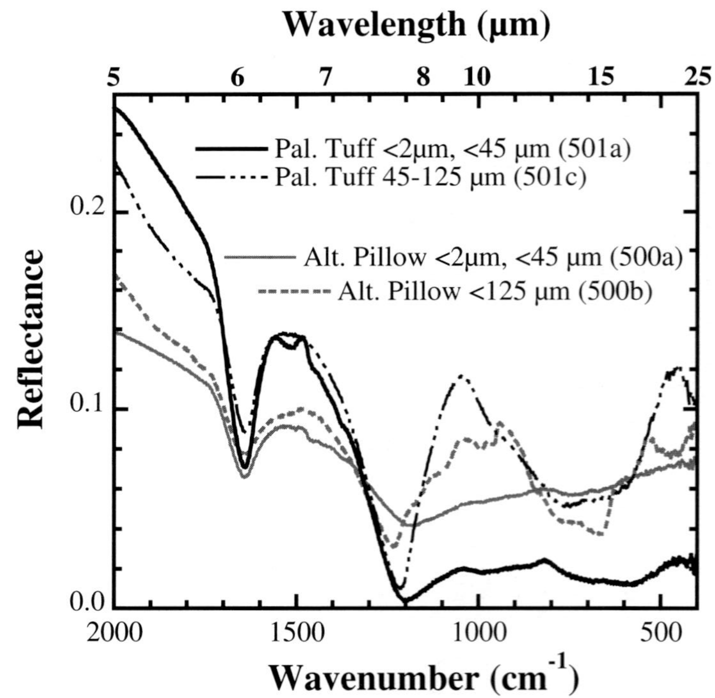

Fig. 9. Reflectance spectra from 2000 to 400 cm^{-1} (5–25 µm) of the Icelandic tuff and altered pillow basalt samples. Spectra of the finest clay fraction (sedimentation fractionated to <2 µm and dry sieved to <45 µm) and a coarser size fraction of the altered material are shown here. For the palagonitic tuff a spectrum is shown of the 45–125 µm size fraction and for the altered pillow a spectrum is shown for the <125 µm size fraction.

relatively high Al_d content of the pillow basalt may be the result of Al substitution in goethite produced by surface weathering.

The <125 μm particle size material collected near the base of the pillow basalt outcrops contains a variety of fresh glassy and crystalline basaltic grains, as well as some highly altered grains. The relative proportion of fresh material is much higher in this sample than in the <2 μm fraction of the pillow basalt (Table 1) or in the altered veins shown in Figure 4. The <125 μm fraction also responds to a strong magnet. A portion of this sample dry sieved to <45 μm gives a weaker magnetic response. Crushed and sieved fractions of the palagonitic tuff do not respond to the magnet. The <2 μm fractions of neither sample show a magnetic response. This suggests that a magnetic mineral, such as magnetite or titanomagnetite, is present in the pedogenically altered basalt, but not in the palagonitically altered tuff. Because the <125 μm sample is attracted more strongly than the <45 μm sample, the magnetic component appears to reside in the larger grains and may not be present in the alteration products.

Reflectance spectra of the altered pillow basalt samples and the palagonitic tuff samples are shown in Figure 8 from 0.3 to 3.3 μm. The spectrum of the <125μm altered material exhibits a band centred at 1.02 μm that is characteristic of high-Ca pyroxene, a broad 3 μm band characteristic of water in glass, and weak features near 1.4, 1.9 and 2.2 μm that are associated with alteration minerals. Spectra of the <2 μm altered pillow and palagonitic tuff samples all contain a weak, broad ferric oxide band centred near 0.9 μm followed by stronger alteration features near 1.4, 1.9 and 2.2–2.3 μm (due to combinations and overtones of OH and H_2O in layer silicates), and a strong *c.* 3 μm water band consistent with glassy and/or hydrated phases.

Reflectance spectra of selected altered pillow basalt samples and palagonitic tuff samples are shown in Figure 9 as a function of wave-number from 2000 to 400 cm^{-1} (5–25 μm). Each of these spectra contains a band near 1630 cm^{-1} due to water and near 1200–1250 cm^{-1} due to silicates. Past this strong silicate feature, surface scattering dominates the spectrum rather than volume scattering and the spectral features associated with vibrations in the mineral structure are manifested as upward 'Reststrahlen' features (e.g. Salisbury 1993). The weak features near 1450 and 1520 cm^{-1} in spectra of the <2 μm fractions are attributed to a small amount of carbonate residue from the sedimentation process. Both the spectrum of the <125 μm fraction of the altered pillow basalt and the spectrum of

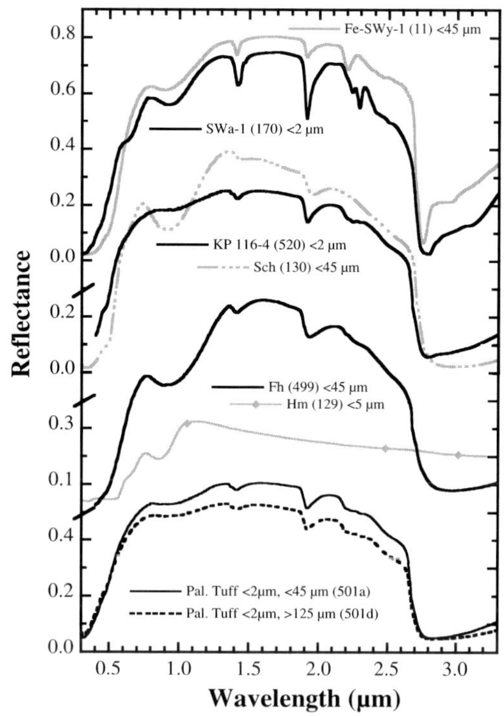

Fig. 10. Reflectance spectra from 0.3 to 3.3 μm of the Icelandic tuff and selected smectites and iron oxides. Spectra of the two versions of the <2 μm palagonitic tuff samples are shown here in the lower portion of the figure. In the upper portion of the figure are spectra of smectite-bearing samples including: an Fe-exchanged SWy-1 montmorillonite (11), a ferruginous smectite (170), and a palagonitic tuff sample from Kilauea, Hawaii (520). Spectra of fine-grained iron oxide-bearing minerals are also shown here for comparison; these are schwertmannite (130), ferrihydrite (499) and hematite (129). The particle sizes are given for each sample.

the 45–125 m fraction of the palagonitic tuff contain silicate features commonly observed for basalts. The spectra of the <2 μm fractions in Figure 9 are much weaker at lower energies (longer wavelengths) because of the small particle sizes.

Comparison with minerals and other alteration products – reflectance spectra

Comparison of the samples in the current study with minerals and other alteration products can be used in order to identify the components contributing to the spectral features observed in Figures 8 and 9. Shown in Figures 10 and 11 are

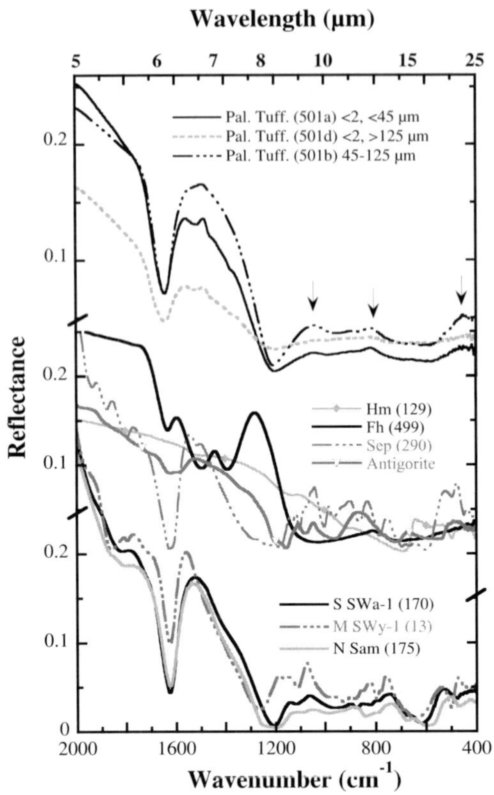

Fig. 11. Reflectance spectra from 2000 to 400 cm^{-1} (5–25 μm) of the Icelandic tuff and selected smectites and iron oxides. The samples shown here include two versions of the <2 μm palagonitic tuff samples and the 45–125 μm portion of this sample in the upper region. In the center of the figure are shown spectra of hematite (129), ferrihydrite (499), sepiolite (290) and antigorite. In the lower section are shown spectra of a ferruginous smectite, SWa-1 (170), a montmorillonite, SWy-1 (13) and a nontronite, Sampor (175). Each of these samples contains grains <5 μm that are clustered into larger aggregates. Weak bands in the Icelandic tuff spectra are indicated by arrows.

reflectance spectra of the palagonitic tuff samples as well as spectra of a ferruginous smectite (SWa-1), a montmorillonite (SWy-1), and a nontronite from Sampor (Slovakia; Bishop *et al.* 1999), an Fe-treated SWy-1 montmorillonite (Bishop *et al.* 1993), an antigorite sample (Salisbury *et al.* 1991), a palagonitic tuff collected near Kilauea, Hawaii (Schiffman *et al.* 2002), a natural schwertmannite (Bishop & Murad 1996), a natural Icelandic ferrihydrite (Bishop & Murad 2002), a sepiolite (Sep) from the Clay Minerals Society Source Clays Repository, and a synthetic hematite from Aldrich. The hematite and ferruginous smectite spectra exhibit features in the visible region (a shoulder near 0.6 and sharp maximum near 0.7–0.75 μm) that are characteristic of crystalline ferric oxides and are not observed in the spectra of the palagonitic tuff samples. The broader, gently sloping visible spectral character observed for the nanophase ferric oxide species in the Fe-exchanged montmorillonite, ferrihydrite and schwertmannite samples is more consistent (although not identical) with the spectral character in this region observed for the palagonitic tuff samples. This suggests that the ferric oxides that form in the alteration rinds are nanophase and/or poorly crystalline.

The steeper slope from 0.7 to 1.3 μm in the altered pillow spectra is due to a higher abundance of crystalline iron oxides/oxyhydroxides in these samples compared to the palagonitic tuffs. The presence of crystalline iron oxyhydroxides in the altered pillow spectra (Fig. 8) is consistent with the lower Fe_o/Fe_d ratio observed for the pillow basalt (Table 3). Thus, at least some, and perhaps much, of the iron expressed as FeO in Tables 1 and 2 for the <2 μm fractions of altered material is due to ferric oxides/oxyhydroxides. Differences in the layered silicate bands are observed for the NIR spectra of the altered pillow and the palagonitic tuff samples. Both contain a spectral feature at 1.42 μm which is typical for montmorillonite and some other Al-bearing clay minerals; however, there is another feature near 1.37–1.38 μm that appears as a strong shoulder for the altered pillow samples, but as only a weak shoulder for the palagonitic tuff samples. This band is attributed to Si–OH bonds and is described in more detail below. The water and OH in these hydrated minerals would account for the assumed water component in the chemical analyses of the altered phases (Table 1).

A comparison of the water and hydroxyl bands in the NIR spectral region shows that features in the palagonitic tuff samples correspond in general with the bands observed for smectites, but that the bands in the palagonitic tuff spectra are broadened, which implies the presence of hydrated ferric oxides such as the ferrihydrite and schwertmannite shown in Figure 10. The bands near 1.92 m in the <2 μm fractions of both Icelandic samples are broadened toward longer wavelengths and indicate that bound water is present. The palagonitic tuff samples exhibit multiple features in the range 2.20–2.33 μm while the altered pillow basalts contain a stronger feature near 2.2 μm.

The broad band extending from 2.7 to 3.1 μm or longer contains multiple, overlapping features due to stretching vibrations of structural OH, bound water, adsorbed water, and an overtone of the water-bending vibration (e.g. Bishop *et al.* 1994). The strong band near 2.7–3.3 μm in the palagonitic tuff spectra is consistent with the presence of both a sheet silicate near 2.7–2.8 μm and a hydrated ferric oxide species near 2.9–3.3 μm. The steep slope near 2.7–2.8 μm is characteristic of samples containing a small amount or poorly crystalline layer silicates. The stronger asymmetry of this band towards 2.8 μm in the palagonitic tuff spectra, compared to the altered pillow spectra, indicates that a higher abundance of layer silicates is present in these samples.

The mid-IR spectra of the palagonitic tuff spectra in Figure 11 show that spectrally dominant components of the 45–125 μm fraction and the <2 μm fraction are similar. Compacting of the <2 μm fraction of this sample into coarse aggregates (>125 μm) produces a broader water band near 1630 cm^{-1} (*c.* 6.1 μm) and weaker spectral contrast at lower energies (longer wavelengths). Comparison of the altered tuff spectra with the ferruginous smectite spectrum in Figure 11 shows that these samples are in general similar; however, the characteristic smectite doublet near 450 and 550 cm^{-1} (*c.* 22, 18 μm) is observed as a single broadened band in the altered tuff spectra. This broadening could be explained by the presence of poorly crystalline smectites and/or the presence of smectite plus another silicate mineral. A strong reflectance minimum is observed in the altered tuff spectra near 1200 cm^{-1} (*c.* 8.3 μm) and weak reflectance maxima are observed near 800 and 1050 cm^{-1} (*c.* 12.5, 9.5 μm; Figs 8 & 10). The band near 1200 cm^{-1} is characteristic of the Si–O stretching bands in Ca-rich plagioclase and many layer silicates (e.g. Salisbury I. 1991). The weak bands near 1050 and 800 cm^{-1} are attributed to layer silicates and iron oxyhydroxides, respectively. Ferrihydrite and goethite spectra exhibit a band at *c.* 800–850 cm^{-1} (e.g. Fig. 7; Bishop & Murad 2002) that is consistent with the palagonitic tuff spectra, whereas hematite spectra contain a band near 650 cm^{-1} (*c.* 15 μm) that is not observed in the palagonitic tuff spectra. The antigorite spectrum shown in Figure 10 (from Salisbury *et al.* 1991) includes broad features near 850 and 500 cm^{-1} that are also consistent with bands observed in the alteration products.

Magnetite exhibits one IR band near 550 cm^{-1}. This could be contributing to the spectrum of the <125 μm fraction of the altered pillow in Figures 9 and 11, although this magnetite band cannot be uniquely identified here because it overlaps with the Si–O bending bands due to silicate minerals in basalts. Because this sample is magnetic, it is likely that magnetite is present and contributing to this feature. The <2 μm portion of the altered pillow is not magnetic and has much weaker spectral features in the mid-IR region. The 550 cm^{-1} spectral band characteristic of magnetite is substantially weakened for fine-grained samples and may not be observed for the ultra-fine magnetite grains attributed to the high Fe_o/Fe_d ratio for the palagonitic tuff sample.

Normalized reflectance spectra of the <2 μm fractions of the palagonitic tuff and altered pillow samples, along with a number of other silicates and iron oxides, are shown in Figure 12 from 5500 to 4200 cm^{-1} (*c.* 1.8–2.4 μm). This spectral region spans the water combination band (fundamental stretching plus bending vibrations) near 5200 cm^{-1} and the structural OH combination band near 4250–4600 cm^{-1}. The altered pillow basalt spectrum contains a band due to water at 5200 cm^{-1} and a band characteristic of OH bound to two Al atoms at 4550 cm^{-1} (e.g. Bishop *et al.* 1999). Both of these features are broader than those observed for montmorillonite (SWy 13 in Fig. 12a) and the water band is at a lower energy than that of montmorillonite, which is consistent with the presence of a poorly crystalline aluminous smectite and nanophase iron oxyhydroxide, such as ferrihydrite. A shoulder near 4350 cm^{-1} could be due to sulphate minerals such as jarosite and schwertmannite, Fe-smectites such as SWa (170) and NG (26), or sepiolite (290) (Fig. 12b,c). As S was not detected for the altered pillow basalt sample, small amounts of Fe–smectite or sepiolite are more likely to be responsible for the spectral feature near 4350 cm^{-1}.

The spectrum of the palagonitic tuff contains a water band centered near 5220 cm^{-1}, which is still lower than the smectite band at 5240 cm^{-1}. The structural OH feature for this sample is very broad (4330–4550 cm^{-1}) implying that a mixture of OH sites is present. The palagonitic tuff spectrum is consistent with the presence of a mixture of Al-rich and Fe-rich smectite, a Si-bearing phase or mineral such as allophane, imogolite, serpentine clay minerals or sepiolite and nanophase-iron oxide/oxyhydroxide such as ferrihydrite. At least some of these components must be poorly crystalline in order to account for the broadened features.

Allophane, imogolite, serpentine clays, sepiolite and palygorskite all have a combination of Si and Al in sites bound to OH and could be responsible for broadening of some of the IR features observed in the <2 μm fractions of the altered hyalotuff and pillow basalt in this study. The serpentine clays lizardite, antigorite and

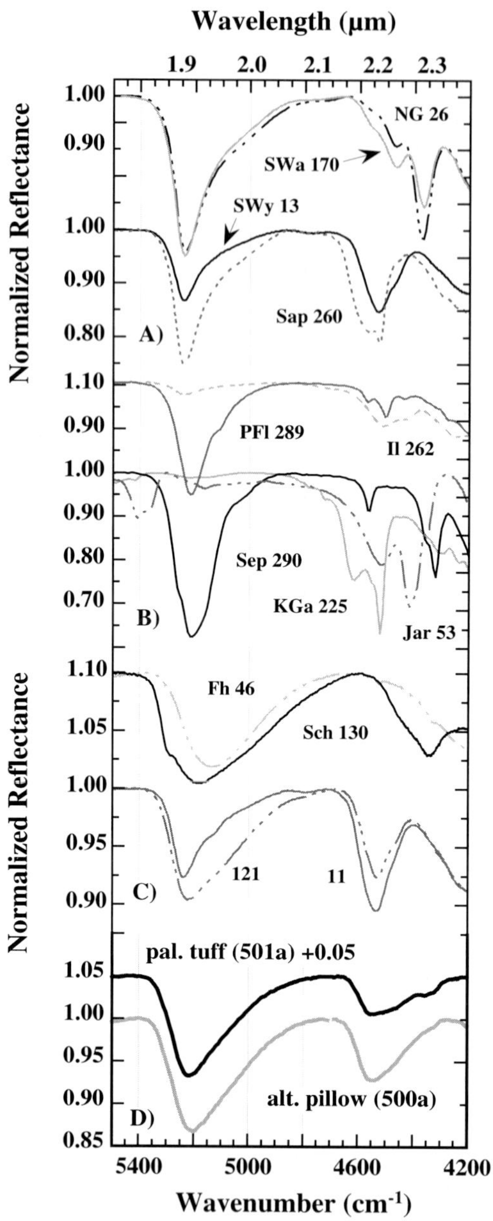

Fig. 12. Normalized reflectance spectra from 5500 to 4200 cm^{-1} (*c.* 1.8–2.4 μm): (**a**) smectites: nontronite (NG 26), ferruginous smectite (SWa-1 170), montmorillonite (SWy-1 13), saponite (Sap 260), (**b**) clay minerals: illite (Il 262), attapulgite (PFl 289), sepiolite (Sep 290), kaolinite (KGa 225), and jarosite (Jar 53) (**c**) nanophase ferric oxyhydroxides: schwertmannite (Sch 130), ferrihydrite (Fh 46) and smectite-ferric assemblages: ferrihydrite–montmorillonite (11), ferrihydrite–schwertmannite–montmorillonite (121) and (**d**) the Icelandic altered pillow basalt (500a) and palagonitic tuff (501a).

chrysotile are layered silicates containing tetrahedrally and octahedrally coordinated Si, Mg, Al and Fe (e.g. Bailey 1980). Serpentines typically form during weathering of olivine, but are relatively unstable and frequently alter to smectites, especially in the presence of water (Velde 1985). Attapulgite and sepiolite (PFl 289 and Sep 290 in Fig. 12b) are clay minerals with chain-like structures and are less common than sheet silicates (e.g. Deer *et al.* 1975). The terms sepiolite, attapulgite and palygorskite have been used for a variety of fibrous, hydrated silicate minerals and they may result from hydrothermal activity and alteration of igneous rocks, although they are not common (Jones & Galan 1988). Singer (1979) and Velde (1985) observed palygorskite and sepiolite in association with smectite in altered igneous material formed under low-temperature conditions.

Some features in spectra of both the altered tuff and altered pillow basalt samples are best explained by Si–OH groups in allophane, imogolite, serpentines or chain-structure clays in association with smectites. The NIR spectral bands for hydroxyl groups on silica surfaces were characterized by Anderson & Wickersheim (1964). They found a sharp OH stretching band at 2.67 μm (3740 cm^{-1}) for fully dehydrated silica gel containing only OH (no H_2O). They also observed the overtone for this band at 1.37 μm (7326 cm^{-1}) and a hydroxyl stretching and bending combination band near 2.2 μm. A NIR spectral study of magnesian serpentines showed bands at 1.38–1.40 and near 2.3 μm that attributed these to structural OH (King & Clark 1989). The serpentines in this study contain primarily Mg and Si (and some Fe) and exhibit multiple overlapping bands in the 1.38–1.40 μm region. These features were later attributed to structural OH stretching overtone bands (Clark *et al.* 1990*a*), and the bands at 1.38 μm in serpentine are attributed here to Si–OH overtones and the bands at 1.39–1.40 μm to overtones of the OH bound to Fe and Mg. The spectral features observed in the altered tuff and altered pillow samples attributed to Si–OH in these minerals include: (1) the shoulder at 1.37–1.38 μm (Fig. 7); (2) a shift in the 1.92 μm band toward longer wavelengths or lower energies (5220 cm^{-1} for the altered tuff and 5200 cm^{-1} for the altered pillow, Fig. 12); (3) the additional weak features contributing to the OH combination bands near 2.2–2.3 μm; and (4) the broadening of the smectite doublet near 20–25 μm (400–450 cm^{-1}) in Figures 9 and 11. These features are indicative of the presence of serpentine clays, short-range ordered Al/Si clay minerals (e.g. allophane/imogolite) and/or chain-structure clay minerals in the altered fractions of both Icelandic samples; however, the spectra indicate that this component is relatively more abundant in the altered pillow basalt than in the altered tuff material. The compositional data (Table 1) do not support a significant component of chain-structure clay minerals in the altered portions of either the pillow basalts or palagonitized hyalotuffs. Detailed XRD analyses in a study of altered palagonitic basalts from Kilauea showed the presence of some serpentine minerals (Schiffman *et al.* 2002) suggesting that serpentines are more likely to be the minerals responsible for these spectral features in the altered Icelandic basalts studied here.

Comparison with Mars – chemistry, mineralogy, reflectance spectra

Shown in Figure 13 are extended visible-region spectra of the <2 μm fractions of the Icelandic alteration products and selected soil spectra from Mars (Yingst *et al.* 1999). The visible spectra of the <2 μm fractions of both Icelandic samples exhibit similar spectral bands and slope to the spectra of the bright Martian soils measured by Pathfinder, but the <2 μm Icelandic samples are brighter. The coarser aggregates of the <2 μm alteration products have darker visible/NIR spectra than the <2 μm fractions that were dry sieved to <45 μm and may have implications for duricrust on Mars. The soils measured by Mars Pathfinder are characterized by a smooth upward slope from 500 to 700 nm and a broad maximum near 750–800 nm (Bell *et al.* 2000). Some soil units have a stronger shoulder feature near 600 nm, a reflectance maximum near 750 nm, and exhibit a weak, broad band near 900 nm. An example of these soils is shown in Figure 13 as an average of several 'disturbed' soil spectra (data from Yingst *et al.* 1999) and are similar in spectral slope and band positions to synthetic aggregates of

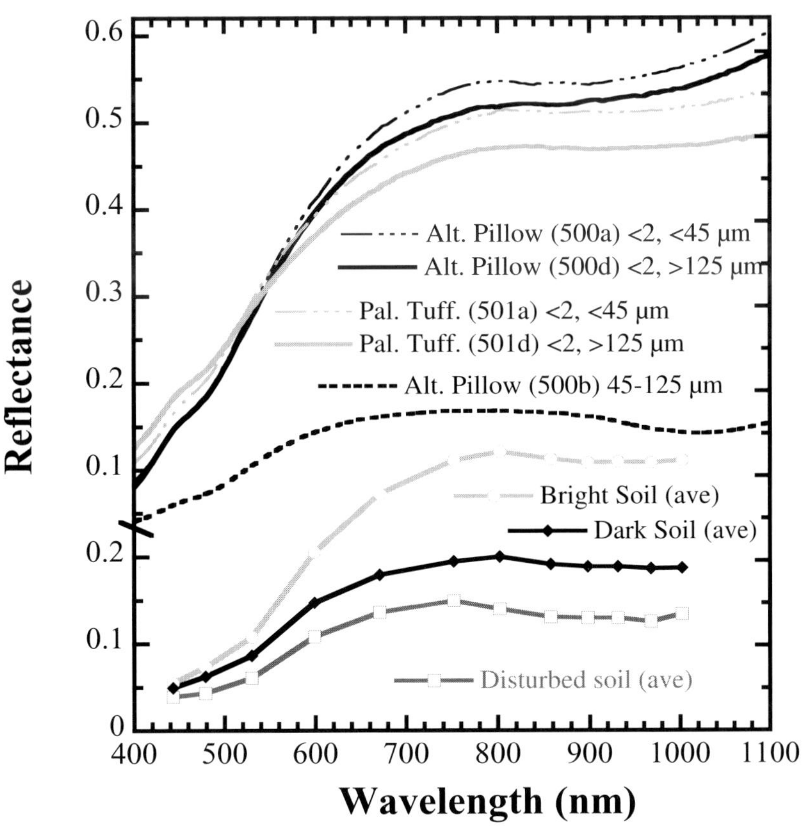

Fig. 13. Extended visible region spectra of Icelandic samples compared to IMP spectra of bright, dark and disturbed Mars soil units. The Mars soil spectra are averages of several point spectra from Yingst *et al.* (1999). Spectra are shown for multiple size fractions of the altered Icelandic palagonitic tuff and pillow basalt.

nanophase iron oxides, sulphates and silicates (Bishop *et al.* 1998*b*).

The mineralogy of the sample influences the spectral shape in this region, whereas the brightness of the reflectance is a factor of the sample mineralogy and texture. Smaller particle sizes and fluffy sample texture tend to give brighter reflectance as seen for the spectra of the altered pillow basalt and altered tuff in Figure 13. The $<125\,\mu m$ fraction of the altered pillow basalt is much darker than the other samples. Comparing the $<2\,\mu m$ fractions of the altered pillow basalt and the altered tuff also indicates a difference in brightness between these clay-size sedimentation fractions that were dry sieved to $<45\,\mu m$ and those that were compacted into aggregates greater than $125\,\mu m$. The mean size of the atmospheric dust on Mars has been calculated to be *c.* 2–3 μm in diameter in multiple studies using data from Viking and Mars Pathfinder (e.g. Pollack *et al.* 1995; Smith *et al.* 1997). One obvious question, then, is why the soil spectra on Mars are so dark relative to the fine-grained components of altered terrestrial basalts. Our data show that aggregation of the $<2\,\mu m$ diameter alteration material produces significant darkening of the spectra in this region; this textural effect may explain some of the patches of darker albedo observed in the Martian soils at the Pathfinder landing site (e.g. Smith *et al.* 1997).

The chemical compositions of the samples in this study follow expected trends with alteration. Decreases in the abundance of Ca, Mg and Si were observed with increasing alteration as were increases in Al and Fe (Fig. 5). However, differences were observed in the chemical trends with alteration for the palagonitically altered hyalotuff and the pedogenically weathered pillow basalt. The chemical patterns for the pillow basalt indicate a trend of increasing alteration from the fresh glass to the altered rims/veins to the $<2\,\mu m$ fraction of the altered material. This is observed in the increasing Al and Fe abundances and significant drops in Mg and Ca levels. Conversely, the chemical patterns for the hyalotuff show that the Al and Fe decrease in the altered rinds compared to the fresh glass, but that the Al and Fe increase again for the $<2\,\mu m$ size fraction of the altered rinds. Although the Mg and Ca levels do decrease in the palagonitically altered material, much higher levels of Mg and Ca are found in the palagonitic alteration products than in the altered pillow basalts. The chemical trends are much more complicated for the palagonitic tuff than for the altered pillow basalt.

Altered volcanic material formed in large-scale palagonitic tuffs or via surface alteration of ash, tephra or lava could be contributing to the 1–3 μm sized dust particles on Mars. Smectites, nanophase iron oxides, and layer silicates, such as allophane and imogolite (or perhaps small amounts of chain-structure clays), are present in the altered rinds of palagonitic glass and in altered veins through glass in a pillow basalt. Studies of palagonitic and pedogenic weathering of volcanic basalts collected near Kilauea, Hawaii, have shown that major differences in the chemistry and mineralogy of the products result for these two processes (Schiffman *et al.* 2002). In this study of samples from one hyalotuff and one pillow basalt from SW Iceland, we found that the alteration products formed from both of these processes produced nanophase and poorly crystalline iron oxides and silicates. Differences in the samples studied here include a higher abundance of smectites and/or more crystalline smectites occurring in the palagonitically altered tuff than in the pedogenically altered pillow basalt. Another difference is the presence of a magnetic mineral, such as magnetite, in the pedogenically altered pillow.

Differences in the chemical trends observed for alteration in the palagonitic and pedogenic samples may also have implications for Mars. Revised chemical analyses for the rocks and soils measured by Mars Pathfinder have been reported recently by Foley *et al.* (2001). The higher FeO level observed in the pedogenic weathering product of the pillow basalt in our study is more consistent with the high FeO abundance in the Mars soil, while the higher MgO levels in the palagonitically altered tuff are more consistent with that observed for the Mars soil. Both altered samples in our study have much higher Al_2O_3 abundance than has been observed for the soils on Mars. The Al_2O_3 abundance for the rocks near Mars Pathfinder are *c.* 10–12 wt% (Foley *et al.* 2001), which is lower than the Al_2O_3 levels in the fresh glass in our study, but not sufficiently lower to support local alteration of the martian rocks to produce the martian soils by either of the processes studied here.

If serpentine clays, poorly crystalline aluminosilicates and/or chain-like clay minerals, such as sepiolite, do occur together with smectites in the alteration products of Icelandic basalts, this may have important implications for the soil mineralogy on Mars. Chemical models performed by Toulmin *et al.* (1977), based on the Viking chemical measurements (Baird *et al.* 1977), indicated that smectite clays were a likely component of the martian soil; smectite abundance was estimated at *c.* 20–80%. Hanel *et al.* (1972) found silicate bands in the IR spectra of

Martian dust measured by Mariner 9 and spectral analyses by Hunt *et al.* (1973) suggested the presence of smectites on Mars because of similarities in the bands near 10 and 20 μm; however later analyses pointed out that the martian spectra are missing the characteristic smectite doublet near 400–600 cm^{-1} (20–25 μm) and the characteristic NIR band(s) at *c.* 2.2–2.3 μm (Roush 1989; Clark *et al.* 1990*b*; Roush *et al.* 1993; Roush & Bell 1995). Detailed analysis of NIR martian surface spectra has shown the presence of a weak band near 2.2–2.3 μm in some regions that are not associated with altitude and are attributed to the surface material (Bell & Crisp 1993; Murchie *et al.* 1993; Beinroth & Arnold 1996). Poorly crystalline and/or altered smectites may not have clearly resolvable features, and, thus, could be responsible for the silicate features observed on Mars (Bishop *et al.* 1993; Burns 1993). Spectral analyses of mid-IR spectral measurements of smectites and a spectrally bland altered volcanic material found that 15 wt% crystalline montmorillonite mixed with this neutral material should be detectable in IR spectra of Mars by features at 9.7 and 19.9 μm (Roush & Orenberg 1996). Our study shows that the presence of crystalline and poorly crystalline smectites plus serpentine clays in altered volcanic basalts gives weakened and broadened mid-IR silicate bands. Recent modelling of TES spectra of Mars include an estimate of *c.* 15% silicate dust (in the atmosphere and on the surface) in many regions, but does not provide details of the mineralogy of this dust (Bandfield *et al.* 2000). The Omega instrument on Mars Express (http://sci.esa.int/content/doc/90/22160_.htm) should enable detection of the *c.* 2.2–2.4 μm spectral bands characteristic of alteration minerals if regions of about 400 m (spot size on Mars) or larger of altered volcanic material are present on the surface. The Mini-TES spectrometer on the Mars Exploration Rovers (Squyres & Team 2001) will be able to characterize the mid-IR silicate and iron oxide bands in the surface materials near the rovers and Mini-TES is expected to be able to identify alteration minerals, if present.

Summary

Subglacial volcanic activity and the subsequent palagonitization and pedogenic weathering of its glassy eruptive products on Iceland have led to the formation of a variety of silicate and iron oxide-rich alteration products that may serve as a model for chemical alteration on Mars. Comparison of the chemistry and mineralogy of the <2 μm fractions of the altered palagonitic tuff and altered pillow basalt in our study showed differences in these two materials. The altered palagonitic tuff contains more crystalline layer silicates, including smectite clays, while the altered pillow basalt contains more crystalline iron oxide minerals. Both contain poorly crystalline and/or nanophase layer silicates and iron oxides/oxyhydroxides. The formation of alteration products in a similar manner on Mars may have contributed to the fine-grained dust and soil on Mars that is dominated by poorly crystalline and nanophase iron oxides/oxyhydroxides and silicates. The extended visible region spectral properties of both Icelandic alteration products studied here are similar in band position and slope to what was observed by Pathfinder for the Martian soils. Our study shows that the presence of smectites and serpentine clays in altered volcanic basalts gives weakened and broadened mid-IR silicate bands and may explain the mid-IR spectra of the dust and soil on Mars.

Support for JLB from NASA's Mars Data Analysis Program, for PS from the Committee on Research at the University of California, Davis, and for all authors from NASA's Mars Fundamental Research Program are much appreciated. Reflectance spectra were measured at RELAB, a multi-user, NASA-supported facility (NAG5-3871). Thanks are due to T. Hiroi for assistance with the bi-directional spectra, to L. Gruendler with the graphics, to W. Calvin, T. Roush, M. Schulte and M. Staid for helpful editorial comments and to M. Chapman and J. Smellie for editorial handling of the manuscript.

References

Allen, C. C., Gooding, J. L., Jercinovic, M. & Keil, K. 1981. Altered basaltic glass: A terrestrial analog to the soil of Mars. *Icarus*, **45**, 347–369.

Anderson, J. H. & Wickersheim, K. A. 1964. Near infrared characterization of water and hydroxyl groups on silica surfaces. *Surface Science*, **2**, 252–260.

Bailey, S. W. 1980. Structures of layer silicates. *In*: Brindley, G. W. & Brown, G. (eds) *Crystal structures of clay minerals and their X-ray identification*. Mineralogical Society, London.

Baird, A. K., Castro, A. J., Clark, B. C., Toulmin III, P., Rose Jr, H., Keil, K. & Gooding, J. L. 1977. The Viking X ray fluorescence experiment: Sampling strategies and laboratory simulations. *Journal of Geophysical Research*, **82**, 4595–4624.

Bandfield, J. L., Hamilton, V. E. & Christensen, P. R. 2000. A global view of martian surface compositions from MGS-TES. *Science*, **287**, 1626–1630.

BEINROTH, A. & ARNOLD, G. 1996. Analysis of weak surface absorption bands in the near-infrared spectra of Mars obtained by Phobos-2. *Vibrational Spectroscopy*, **11**, 115–121.

BELL III, J. F. & CRISP, D. 1993. Groundbased imaging spectroscopy of Mars in the near-infrared: Preliminary results. *Icarus*, **104**, 2–19.

BELL III, J. F., MCCORD, T. B. & OWENSBY, P. D. 1990. Observational evidence of crystalline iron oxides on Mars. *Journal of Geophysical Research*, **95**, 14 447–14 461.

BELL III, J. F., MORRIS, R. V. & ADAMS, J. B. 1993. Thermally altered palagonitic tephra: A spectral and process analog to the soil and dust of Mars. *Journal of Geophysical Research*, **98**, 3373–3385.

BELL III, J. F., MCSWEEN JR, H. Y., MURCHIE, S. L. *ET AL.* 2000. Mineralogic and Compositional Properties of Martian Soil and Dust: Results from Mars Pathfinder. *Journal of Geophysical Research*, **105**, 1721–1755.

BESKE-DIEHL, S. & HUILING, L. 1993. Magnetic properties of hematite in lava flows from Iceland: Response to hydrothermal alteration. *Journal of Geophysical Research*, **98**, 403–417.

BIBRING, J.-P., COMBES, M., LANGEVIN, Y. *ET AL.* 1990. ISM observations of Mars and Phobos: First Results. *In*: *Proceedings of the 20th Lunar and Planetary Science Conference*, LPI, Houston, 461–471.

BISHOP, J. L. & PIETERS, C. M. 1995. Low-temperature and low atmospheric pressure infrared reflectance spectroscopy of Mars soil analog materials. *Journal of Geophysical Research*, **100**, 5369–5379.

BISHOP, J. L. & MURAD, E. 1996. Schwertmannite on Mars? Spectroscopic analyses of schwertmannite, its relationship to other ferric minerals, and its possible presence in the surface material on Mars. *In*: DYAR, M. D., MCCAMMON, C. & SCHAEFER, M. W. (eds) *Mineral Spectroscopy: A tribute to Roger G. Burns*. The Geochemical Society, Houston, 337–358.

BISHOP, J. L. & MURAD, E. 2002. Spectroscopic and geochemical analyses of ferrihydrite from springs in Iceland and applications to Mars. *In*: SMELLIE, J. L. & CHAPMAN, M. G. (eds) *Volcano–Ice Interaction on Earth and Mars*. Geological Society, London, Special Publications, **202**, 357–370.

BISHOP, J. L., FRÖSCHL, H. & MANCINELLI, R. L. 1998*a*. Alteration processes in volcanic soils and identification of exobiologically important weathering products on Mars using remote sensing. *Journal of Geophysical Research*, **103**, 31 457–31 476.

BISHOP, J. L., PIETERS, C. M. & BURNS, R. G. 1993. Reflectance and Mössbauer spectroscopy of ferrihydrite-montmorillonite assemblages as Mars soil analog materials. *Geochimica et Cosmochimica Acta*, **57**, 4583–4595.

BISHOP, J. L., PIETERS, C. M. & EDWARDS, J. O. 1994. Infrared spectroscopic analyses on the nature of water in montmorillonite. *Clays and Clay Minerals*, **42**, 701–715.

BISHOP, J. L., PIETERS, C. M., BURNS, R. G., EDWARDS, J. O., MANCINELLI, R. L. & FROESCHL, H. 1995. Reflectance spectroscopy of ferric sulfate-bearing montmorillonites as Mars soil analog materials. *Icarus*, **117**, 101–119.

BISHOP, J. L., SCHEINOST, A., BELL, J., BRITT, D., JOHNSON, J. & MURCHIE, S. 1998*b*. Ferrihydrite-schwertmannite-silicate mixtures as a model of Martian soils measured by Pathfinder. *Proceedings of the 30th Lunar and Planetary Science Conference*, CD-ROM #1803.

BISHOP, J. L., MURAD, E., MADEJOVÁ, J., KOMADEL, P., WAGNER, U. & SCHEINOST, A. 1999. Visible, Mössbauer and infrared spectroscopy of dioctahedral smectites: Structural analyses of the Fe-bearing smectites Sampor, SWy-1 and SWa-1. *In*: KODAMA, H., MERMUT, A. R. & TORRANCE, J. K. (eds) *Clays for our future. Proceedings of the 11th International Clay Conference, Ottawa, June 1997*, 413–419.

BURNS, R. G. 1993. Rates and mechanisms of chemical weathering of ferromagnesian silicate minerals on Mars. *Geochimica et Cosmochimica Acta*, **57**, 4555–4574.

BURNS, R. G. 1994. Schwertmannite on Mars: Deposition of this ferric oxyhydroxysulfate mineral in acidic saline meltwaters. *In*: *Proceedings of the 25th Lunar and Planetary Science Conference*, LPI, Houston, 203–204.

CALVIN, W. M., KING, T. V. V. & CLARK, R. N. 1994. Hydrous carbonates on Mars? Evidence from Mariner 6/7 infrared spectrometer and ground-based telescopic spectra. *Journal of Geophysical Research*, **99**, 14 659–14 675.

CHRISTENSEN, P. R., BANDFIELD, J. L., SMITH, M. D., HAMILTON, V. E. & CLARK, R. N. 2000. Identification of basaltic component on the Martian surface from Thermal Emission Spectrometer Data. *Journal of Geophysical Research*, **105**, 9609–9621.

CLARK, R. N., KING, T. V. V., KLEJWA, M. & SWAYZE, G. A. 1990*a*. High spectral resolution reflectance spectroscopy of minerals. *Journal of Geophysical Research*, **95**, 12 653–12 680.

CLARK, R. N., SWAYZE, G. A., SINGER, R. B. & POLLACK, J. B. 1990*b*. High-resolution reflectance spectra of Mars in the 2.3 μm region: Evidence for the mineral scapolite. *Journal of Geophysical Research*, **95**, 14 463–14 480.

COLMAN, S. M. 1982. *Chemical weathering of basalts and andesites: Evidence from weathering rinds.* United States Geological Survey Professional Paper, **1246**.

CROVISIER, J. L., HONNOREZ, J., FRITZ, B. & PETIT, J. C. 1987. Dissolution of subglacial volcanic glasses from Iceland: Laboratory study and modelling. *Applied Geochemistry*, Special Issue **1**, 55–81.

DEER, W. A., HOWIE, R. A. & ZUSSMAN, J. 1975. *An introduction to the rock-forming minerals*. Longman, London.

ERARD, S., MUSTARD, J., MURCHIE, S., BIBRING, J.-P., CERRONI, P. & CORADINI, A. 1994. Martian aerosols: Near-infrared spectral properties and effects on the observation of the surface. *Icarus*, **111**, 317–337.

FARRAND, W. H. & SINGER, R. B. 1992. Alteration of hydrovolcanic basaltic ash: Observations with visible and near-infrared spectrometry. *Journal of Geophysical Research*, **97**, 17 393–17 408.

FISHER, R. V. & SCHMINKE, H.-U. 1984. *Pyroclastic rocks*. Springer-Verlag, Berlin.

FOLEY, C. N., ECONOMOU, T. E. & CLAYTON, R. N. 2001. Chemistry of Mars Pathfinder samples determined by the APXS. *In*: *Proceedings of the 32nd Lunar and Planetary Science Conference*, LPI, Houston, CD-ROM #1979.

FURNES, H. 1978. Element mobility during palagonitization of a subglacial hyaloclastite in Iceland. *Chemical Geology*, **22**, 249–264.

GOGUITCHAICHVILI, A., PRÉVOT, M., DAUTRIA, J.-M. & BACIA, M. 1999. Thermodetrital and crystallodetrital magnetization in an Icelandic hyaloclastite. *Journal of Geophysical Research*, **104**, 29 219–29 238.

HANEL, R. A., CONRATH, B. J., HOVIS, W. A., KUNDE, V. G., LOWMAN, P. D., PEARL, J. C., PRABHAKARA, C. & SCHLACHMAN, B. 1972. Infrared spectroscopy experiment on the Mariner 9 mission: Preliminary results. *Science*, **175**, 305–308.

HARGRAVES, R. B., COLLINSON, D. W., ARVIDSON, R. E. & SPITZER, C. R. 1977. The Viking magnetic properties experiment: Primary mission results. *Journal of Geophysical Research*, **82**, 4547–4558.

HONNOREZ, J. 1981. The aging of the oceanic crust at low temperature. *In*: EMILIANI, C. (eds.) *The Oceanic Lithosphere*. John Wiley & Sons Inc., New York, 525–588.

HUNT, G. R., LOGAN, L. M. & SALISBURY, J. W. 1973. Mars: Components of infrared spectra and the composition of the dust cloud. *Icarus*, **18**, 459–469.

JACKSON, M. L. 1979. *Soil chemical analysis–advanced course*, 2nd edition. M. L. Jackson, University of Wisconsin, Madison.

JAKOBSSON, S. P. 1978. Environmental factors controlling the palagonitization of the Surtsey tephra, Iceland. *Geological Society of Denmark Bulletin*, **27**, 91–105.

JAKOBSSON, S. P. & MOORE, J. G. 1986. Hydrothermal minerals and alteration rates at Surtsey volcano, Iceland. *Geological Society America Bulletin*, **97**, 648–659.

JÓHANNESSON, H. & SŒMUNDSSON, K. 1998. *Geologic map of Iceland. 1:50 000. Bedrock Geology*. Náttúrufraedistofnun Íslands.

JONES, B. F. & GALAN, E. 1988. Sepiolite and palygorskite. *In*: BAILEY, S. W. (ed.) *Hydrous Phyllosilicates, Reviews in Mineralogy*. Mineralogical Society of America, Washington, DC, 631–664.

JONES, J. G. 1969. Intraglacial volcanoes of the Laugarvatn region, southwest Iceland, I. *Quarterly Journal of the Geological Society, London*, **124**, 197–211.

JONES, J. G. 1970. Intraglacial volcanoes of the Laugarvatn region, southwest Iceland, II. *Journal of Geology*, **78**, 127–140.

KING, T. V. V. & CLARK, R. N. 1989. Spectral characteristics of chlorites and Mg-serpentines using high-resolution reflectance spectroscopy. *Journal of Geophysical Research*, **94**, 13 997–14 008.

LE GAL, X., CROVISIER, J.-L., GAUTHIER-LAFAYE, F., HONNOREZ, J. & GRAMBOW, B. 1999. Meteoric alteration of Icelandic volcanic glass: Long term changes in the mechanism. *Comptes Rendus de l'Académie des Sciences, Sciences de la Terre et des Planetes*, **329**, 175–181.

MADSEN, M. B., HVIID, S. F., GUNNLAUGSSON, H. P., KNUDSEN, J. M., GOETZ, W., PEDERSEN, C. T., DINESEN, A. R., MOGENSEN, C. T. & OLSEN, M. 1999. The magnetic properties experiment on Mars Pathfinder. *Journal of Geophysical Research*, **104**, 8761–8779.

MCCORD, T. B., CLARK, R. N. & SINGER, R. B. 1982. Mars: Near-infrared spectral reflectance of surface regions and compositional implications. *Journal of Geophysical Research*, **87**, 3021–3032.

MCLENNAN, S. M. 2000. Chemical composition of Martian soil and rocks: Complex mixing and sedimentary transport. *Geophysical Research Letters*, **27**, 1335–1338.

MCSWEEN JR, H. Y. & KEIL, K. 2000. Mixing relationships in the Martian regolith and the composition of the globally homogeneous dust. *Geochimica et Cosmochimica Acta*, **64**, 2155–2166.

MCSWEEN JR, H. Y., MURCHIE, S. L., CRISP, J. A. *ET AL*. 1999. Chemical, multispectral, and textural constraints on the composition and origin of rocks at the Mars Pathfinder landing site. *Journal of Geophysical Research*, **104**, 8679–8715.

MORRIS, R. V., GOODING, J. L., LAUER, JR, H. V. & SINGER, R. B. 1990. Origins of Marslike spectral and magnetic properties of a Hawaiian palagonitic soil. *Journal of Geophysical Research*, **95**, 14 427–14 434.

MORRIS, R. V., MING, D. W., GOLDEN, D. C. & BELL III, J. F. 1996. An occurrence of jarosite tephra on Mauna Kea, Hawaii: Implications for the ferric mineralogy of the Martian surface. *In*: DYAR, M. D., MCCAMMON, C. & SCHAEFER, M. W. (eds) *Mineral spectroscopy: A tribute to Roger G. Burns*. The Geochemical Society, Houston, 327–336.

MORRIS, R. V., GOLDEN, D. C., BELL III, J. F., LAUER JR, H. V. & ADAMS, J. B. 1993. Pigmenting agents in Martian soils: Inferences from spectral, Mössbauer, and magnetic properties of nanophase and other iron oxides in Hawaiian palagonitic soil PN-9. *Geochimica et Cosmochimica Acta*, **57**, 4597–4609.

MURCHIE, S., KIRKLAND, L., ERARD, S., MUSTARD, J. & ROBINSON, M. 2000. Near-infrared spectral variations of martian surface materials from ISM imaging spectrometer data. *Icarus*, **147**, 444–471.

MURCHIE, S., MUSTARD, J., BISHOP, J., HEAD, J., PIETERS, C. & ERARD, S. 1993. Spatial variations in the spectral properties of bright regions on Mars. *Icarus*, **105**, 454–468.

NESBITT, H. W. & YOUNG, G. M. 1984. Prediction of some weathering trends of plutonic and volcanic rocks based on thermodynamic and kinetic considerations. *Geochimica et Cosmochimica Acta*, **48**, 1523–1534.

POLLACK, J. B., OCKERT-BELL, M. E. & SHEPARD, M. K. 1995. Viking lander image anaylsis of Martian atmospheric dust. *Journal of Geophysical Research*, **100**, 5235–5250.

ROUSH, T. L. 1989. Infrared transmission measurements of martian soil analogs. *In*: LEE, S. (ed.) *MECA Workshop on Dust on Mars III*. LPI Technical Report **89-01**, 52–54.

ROUSH, T. L. & BELL III, J. F. 1995. Thermal emission measurements 2000–400 cm^{-1} (5–25 μm) of Hawaiian palagonitic soils and their implications for Mars. *Journal of Geophysical Research*, **100**, 5309–5317.

ROUSH, T. L. & ORENBERG, J. B. 1996. Estimated detectability limits of iron-substituted montmorillonite clay on Mars from thermal emission spectra of clay-palagonite physical mixtures. *Journal of Geophysical Research*, **101**, 26 111–26 118.

ROUSH, T. L., BLANEY, D. L. & SINGER, R. B. 1993. The surface composition of Mars as inferred from spectroscopic observations. *In*: PIETERS, C. M. & ENGLERT, P. A. J. (eds) *Remote geochemical analysis: Elemental and mineralogical composition*. Cambridge University Press, Cambridge, 367–393.

SALISBURY, J. W. 1993. Mid-infrared spectroscopy: Laboratory data. *In*: PIETERS, C. M. & ENGLERT, P. A. J. (eds) *Remote geochemical analysis: Elemental and mineralogical composition*. Cambridge University Press, Cambridge, 79–98.

SALISBURY, J. W., WALTER, L. S., VERGO, N. & D'ARIA, D. M. 1991. *Infrared (2.1–25 μm) Spectra of Minerals*. John Hopkins University Press, Baltimore, MD.

SCHIFFMAN, P. & SOUTHARD, R. J. 1996. Cation exchange capacity of layer silicates and palagonitized glass in matic volcanic rocks: a comparative study of bulk extraction and *in situ* techniques. *Clays Clay Mineralogy*, **44**, 624–634.

SCHIFFMAN, P., SPERO, H. J., SOUTHARD, R. J. & SWANSON, D. A. 2000. Controls on palagonitization versus pedogenic weathering of basaltic tephra: Evidence from the consolidation and geochemistry of the Keanakako'i ash member, Kilauea volcano. *Geochemistry Geophysics Geosystems*, **1**, 2000GC000068.

SCHIFFMAN, P., SOUTHARD, R. J., EBERL, D. D. & BISHOP, J. L. 2002. Distinguishing palagonitized from pedogenically-altered basaltic Hawaiian tephra: mineralogical and geochemical criteria. *In*: SMELLIE, J. L. & CHAPMAN, M. G. (eds) *Volcano–Ice Interaction on Earth and Mars*. Geological Society, London, Special Publications, **202**, 393–405.

SINGER, A. 1979. Palygorskite in sediments: Detrital, diagenetic or neoformed. A critical review. *Geologische Rundschau*, **68**, 996–1008.

SINGER, R. B. 1982. Spectral evidence for the mineralogy of high-albedo soils and dust on Mars. *Journal of Geophysical Research*, **87**, 10 159–10 168.

SINGER, R. B., MCCORD, T. B., CLARK, R. N., ADAMS, J. B. & HUGUENIN, R. L. 1979. Mars surface composition from reflectance spectroscopy: A summary. *Journal of Geophysical Research*, **84**, 8415–8426.

SMITH, P. H., BELL III, J. F., BRIDGES, N. T., BRITT, D. T., GADDIS, L., GREELEY, R., KELLER, H. U., HERKENHOFF, K. E., JAUMANN, R., JOHNSON, J. R., KIRK, R. L., LEMMON, M., MAKI, J. N., MALIN, M. C., MURCHIE, S. L., OBERST, J., PARKER, T. J., REID, R. J., SABLOTNY, R., SODERBLOM, L. A., STOKER, C., SULLIVAN, R., THOMAS, N., TOMASKO, M. G., WARD, W. & WEGRYN, E. 1997. Results from the Mars Pathfinder camera. *Science*, **278**, 1758–1765.

SODERBLOM, L. A. 1992. The composition and mineralogy of the Martian surface from spectroscopic observations: 0.3–50 m. *In*: KIEFFER, H. H., JAKOSKY, B. M., SNYDER, C. W. & MATTHEWS, M. S. (eds) *Mars*. University of Arizona Press, Tucson, 557–593.

SOIL SURVEY LABORATORY STAFF 1996. *Soil Survey Laboratory Methods Manual, Soil Survey Investigations*. USDA National Soil Survey Center, Lincoln, NE.

SQUYRES, S. W. & TEAM, AND THE ATHENA SCIENCE TEAM. 2001. The Athena science payload for the 2003 Mars Exploration rovers. *First Landing Site Workshop for 2003 Mars Exploration Rovers*, 65–66.

SQUYRES, S. W., WILHELMS, D. E. & MOOSMAN, A. C. 1987. Large-scale volcano-ground ice interactions on Mars. *Icarus*, **70**, 385–408.

STAUDIGEL, H. & Hart, S. R. 1983. Alteration of basaltic glass: Mechanisms and significance for the oceanic crust-seawater budget. *Geochimica et Cosmochimica Acta*, **47**, 337–350.

TOULMIN III, P., BAIRD, A. K., CLARK, B. C., KEIL, K., ROSE, J. H. J., CHRISTIAN, R. P., EVANS, P. H. & KELLIHER, W. C. 1977. Geochemical and mineralogial interpretation of the Viking inorganic chemical results. *Journal of Geophysical Research*, **82**, 4625–4634.

VELDE, B. 1985. *Clay minerals: A physico-chemical explanation of their occurrence*. Elsevier, New York.

VON WALTERSHAUSEN, S. 1846. Über die submarinen Ausbrüche in der tertiaren Formation des Val di Noto im Vergleiche mit verwandten Erscheinungen am Atna. *Göttinger Studien*, **1**, 371–431.

WADA, K. 1989. Allophane and imogolite. *In*: DIXON, J. B. & Weed, S. B. (eds) *Minerals in soil environments*. Soil Science Society of America, Madison, WI, 1051–1087.

WHITTIG, L. D. & ALLARDICE, W. R. 1982. X-ray diffraction analysis. *In*: KLUTE, A. (ed.) *Methods of soil analysis. Part 1: Physical and mineralogical methods*. American Society of Agronomy, Madison, WI, 331–362.

YINGST, R. A., REID, R. J. & SMITH, P. H. 1999. Preliminary science results from the recalibration of IMP images. *In*: *5th International Mars Conference*, Pasadena, CA, CD-ROM #6199.

Distinguishing palagonitized from pedogenically-altered basaltic Hawaiian tephra: mineralogical and geochemical criteria

P. SCHIFFMAN[1], R. J. SOUTHARD[2], D. D. EBERL[3] & J. L. BISHOP[4]

[1] *Department of Geology, University of California, Davis, CA 95616, USA (e-mail: PSchiffman@UCDavis.edu)*

[2] *Department of Land, Air, and Water Resources, University of California, Davis, CA 95616, USA*

[3] *US Geological Survey, 3215 Marine Street, Boulder, CO 80303-1066, USA*

[4] *SETI Institute, NASA-Ames Research Center, M.S. 239-4, Moffett Field, CA 94035, USA*

Abstract: Palagonitization is a common, but imperfectly defined process that greatly modifies the physical and chemical properties of glassy basaltic tephra deposited in subaquatic/subglacial environments on Earth and perhaps Mars. It also results in textures and mineralogies that are distinct from other forms of (mainly pedogenic) low temperature alteration. Specifically, the process of palagonitization (1) initially results in the formation of 'palagonitized glass', a quasi- or nano-crystalline, rind-like material that contains smectite, as well as lesser amounts of other clays (e.g. serpentine), and (2) eventually results in consolidation of tephra, mediated through the accretion of palagonitized glass and later-formed authigenic cements. Conversely, pedogenic weathering of glassy basaltic tephra is characterized by disaggregation of tephra, and formation of a wide range of pedogenic products, including layer silicates (although not primarily smectite), short-range-order aluminosilicates and oxyhydroxides, whose composition reflects the intensity of the weathering environment. These mineralogical and textural properties can be readily recognized through a variety of techniques including electron microscopy/microprobe analysis, reflectance spectroscopy, X-ray diffraction and soil chemistry. Analyses of samples collected from the summit regions of Kilauea and Mauna Kea volcanoes on the island of Hawaii are presented here in order to illustrate differences between palagonitization and pedogenic weathering of glassy basaltic tephra. In the young Hawaiian tephras studied, palagonitization has occurred in response to hydrothermal activity shortly after deposition. Although some, non-hydrothermally affected tephras may eventually become palagonitized, those that have been strongly desilicated by intense pedogenic weathering will probably never become palagonitized.

Glassy basaltic tephra, deposited through subglacial or other forms of hydrovolcanism, is an inherently unstable material under ambient surface conditions. On Earth and perhaps other 'blue' planets, these tephra deposits are readily altered through subsequent contact with meteoric water, seawater or hydrothermal fluids. Since basaltic glass (sideromelane) is black and altered equivalents are characterized by a range of colours including yellow, orange, red, brown and green, it is generally easy to distinguish fresh from altered vitric tephra in outcrop and by remote sensing techniques. But beyond this qualitative distinction, any systematic classification of altered tephra is rarely attempted. Many published lithological descriptions of glassy basaltic tephras collectively refer to these altered equivalents, which may range in consistency from soil to fully consolidated tuff, as 'palagonite'. For example, planetary scientists (e.g. Allen *et al.* 1981; Farrand & Singer 1991; Banin *et al.* 1992; Golden *et al.* 1993; Buemi *et al.* 1998) have noted for many years that the spectral properties of parts of the Martian regolith are a close match to that of 'palagonite', in reference to either soil and/or rock. 'Palagonite', formally neither a mineral nor a rock, is a material whose definition has been debated since it was first described by Sartorius von Waltershausen in the mid nineteenth century (Sigurdsson 1999). Honnorez (1981) comprehensively reviewed the nomenclatural nightmare confusion surrounding this

From: SMELLIE, J. L. & CHAPMAN, M. G. (eds) 2002. *Volcano–Ice Interaction on Earth and Mars*. Geological Society, London, Special Publications, **202**, 393–405. 0305-8719/02/$15.00

material, and suggested that the use of the term 'palagonite' should be discontinued. We concur and further suggest the use of more precise terms, specifically (1) 'palagonitized glass', the quasi/nano-crystalline alteration product of sideromelane, (2) 'palagonitized tuff', a consolidated rock comprising unaltered glass, palagonitized glass and authigenic mineral cements, and (3) 'palagonitization', the process that results in the formation of both palagonitized glass and palagonitized tuff.

The fundamental question we address below is whether palagonitized glass is chemically and mineralogically distinct from pedogenically weathered basaltic glass. Whereas some studies (e.g. Hay & Iiijima 1968) have considered palagonitization essentially a pedogenic or weathering process, and others (e.g. Jakobsson 1978) a purely hydrothermal one, it is generally accepted that palagonitization may occur by both these mechanisms (especially in submarine settings), albeit at vastly different rates (Fisher & Schminke 1984; Singer & Banin 1990). Most workers would agree that palagonitization is a process entailing hydrolysis, oxidation and significant compositional modification of sideromelane, in which leaching of silica and silica as well as alkaline cations may eventually lead to local precipitation of zeolitic and other cements. There is also a growing body of data that indicates that palagonitized glass has a smectite-like structure (Hay & Iijima 1968; Singer 1974; Eggleton & Keller 1982; Zhou *et al.* 1992), even in the initial stages of sideromelane alteration. Conversely, some workers (e.g. Gooding & Keil 1978; Bell *et al.* 2000; Morris *et al.* 2001) have suggested that smectite is not an essential component of 'palagonite'.

In this contribution, we present a comprehensive data set collected using a wide variety of analytical techniques (i.e. electron microprobe, X-ray diffraction, reflectance spectroscopy, soil chemistry) and with which we characterize various forms of low-temperature alteration affecting a suite of Holocene basaltic tephra from Kilauea volcano as well as a late Pleistocene to early Holocene tephra from Mauna Kea. Because weathering rates are so high in Hawaii, very recent (i.e. <500 year-old) tephras have already established clear alteration trends (Schiffman *et al.* 2000). The intent of the present study is to establish useful criteria, which can be readily applied in both the field and laboratory, to distinguish palagonitization from other forms of alteration, specifically pedogenic weathering. We also wish to stimulate further discussion regarding the nature and origin of palagonitization at a time when the planetary science community is focusing on terrestrial subglacial and subaquatic basaltic volcanism as analogues for early martian processes.

Geological setting

Samples examined in this study are vitric tholeiitic tephras from the Keanakako'i Ash Member of Kilauea volcano in Hawaii. These tephras, covering approximately 150 km^2 adjacent to the present caldera (McPhie *et al.* 1990), were deposited by phreatomagmatic pyroclastic surges originating within the summit region of Kilauea between 1500 and 1790 AD (Swanson *et al.* 1998), perhaps in response to magmas erupted into a summit crater lake (Mastin, 1997). The summit region of Kilauea volcano experiences marked gradients in rainfall, soil pH and soil–ground temperature that have profoundly affected weathering/alteration of the Keanakako'i tephras since their deposition (Hay & Jones 1972; Schiffman *et al.* 2000). The west side of the caldera, situated in the orographic rain shadow ($<150\,cm\,a^{-1}$) of both Kilauea and Mauna Loa, is severely affected by acid-aerosol fallout derived from summit fumarolic gases, which has locally reduced soil pH to ≤ 4.0. The east side of the caldera, although unaffected by acid-aerosol fallout, lies in a region of orographically-enhanced annual rainfall ($>300\,cm\,a^{-1}$) and average air temperatures of 17–18°C. The caldera rim itself is enclosed by a circumferential fault system along which steam vents (Cazadevall & Hazlett 1983) have locally elevated the temperature of tephra deposits to <80°C (Schiffman *et al.* 2000). Collectively, these environmental variables have had a marked control on the patterns of alteration affecting the Keanakako'i tephras (Schiffman *et al.* 2000). Specifically, (1) tephra deposited on the lee side of the caldera (in regions of low soil pH) is undergoing congruent dissolution; (2) tephra deposited on the windward side (in regions of high rainfall and near neutral soil pH) is undergoing weathering to a wide variety of pedogenic clay minerals (e.g. in samples 117-1 and 118-2, below); and (3) tephra deposited near active and fossil steam vents along circumferential faults that bound the caldera are palagonitized (e.g. samples 116-3 and 116-4, below). The locations of Keanakako'i Ash tephra samples examined for this study are shown in Figure 1.

For the sake of comparison, we also examined JSC Mars-1, a sample that NASA has recently made available for scientific and educational uses as a Martian soil simulant (Allen *et al.* 1998). JSC Mars-1 is the sieved (<1 mm) fraction of

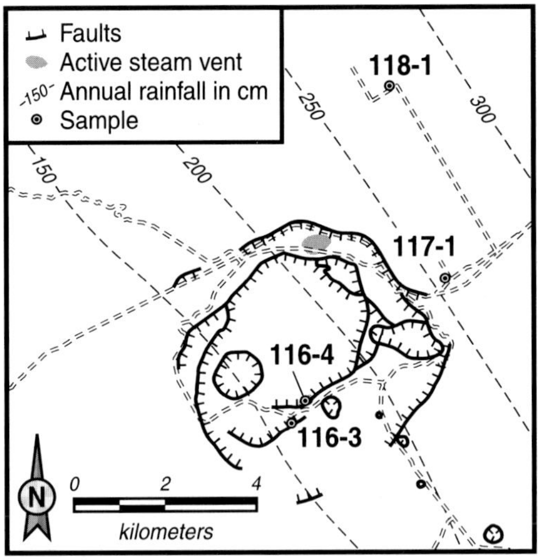

Fig. 1. Map showing the location of the four Keanakako'i Ash tephra samples examined in this study. The samples are located along the margins of Kilauea caldera as defined by a circumferential fault system indicated in the figure. The figure also shows contours for mean annual precipitation (in cm) taken from Giambelluca and Schroder (1998). The location of the JSC Mars-1 sample (not shown) is approximately 40 km to the NW of the map area.

pedogenically altered, hawaiite ash tephra from a late Pleistocene to early Holocene cinder cone on the flanks of Mauna Kea (Wolfe & Morris 1996). At 1850 m elevation in the saddle between Mauna Kea and Mauna Loa, the cinder cone is situated in a relatively dry and cool region on the Big Island with an annual rainfall of approximately 60 cm a^{-1} and an average air temperature of 12°C (Giambelluca & Schroder 1998).

Results

Outcrop characteristics

The samples used in this study vary greatly in their *in situ* field characteristics, specifically their colour and degree of consolidation. Dry tephra (Munsell) colours of the Keanakako'i samples are of orange and red hues: 5 YR 4/6 for the palagonitized samples 116-3 and 116-4; 7.5 YR 4/6 for the pedogenically altered samples 117-1 and 118-1. The JSC Mars-1 has a dry tephra colour of 7.5 YR 4/4. Samples 116-3 and 116-4 were taken from near-vertical, fault-scarp exposures of very well-consolidated tephra (i.e. vitric tuff). Samples 117-1, 118-1, and Mars JSC-1JSC Mars-1 were collected from unconsolidated tephra deposits (i.e. soils) exposed by excavation of a soil pit or road cut.

Petrography

The petrographic features of the *in situ* alteration textures of these samples are perhaps best displayed in back-scattered electron (BSE) images collected from epoxide-impregnated, polished sections. BSE images of representative samples from this study are presented in Figure 2. Figure 2a depicts the texture of unaltered Keanakako'i vitric ash. The degree of consolidation of altered samples observed in outcrop is readily evident in their micro-textures as well. Figure 2b depicts a consolidated, palagonitized sample (116-4) in which approximately 20 µm-thick alteration rinds coat both the exterior and interior (i.e. vesicle walls) of the vitric ash fragments. Although this sample has an appreciable porosity (as reflected by the black epoxide), the rinds around especially the smaller fragments have welded with one another to result in the consolidation observed in the outcrop. Conversely, Figure 2c depicts the micro-texture of a pedogenically altered but unconsolidated Keanakako'i tephra sample (118-1). In this sample, the tephra has largely weathered to a heterogeneous mixture of fine-grained alteration products, which, unlike the rinds of the consolidated tephra sample, are physically separated from the vitric ash fragments. Smaller vitric fragments have been completely altered, but the cores of larger ones (such as the one depicted in Figure 2c) retain their vitroclastic textures. Figure 2d depicts the microtexture of the Mars JSC-1JSC Mars-1 tephra. Although the ash in this tephra is much less vitric, its microtexture is quite similar to that of pedogenically altered, unconsolidated Keanakako'i tephra (cf. Fig. 2c, d).

X-ray diffraction

Initial X-ray diffraction (XRD) analyses were performed on oriented aggregates of the <2 µm fraction mounted on glass slides. Details of the technique are presented in Bishop *et al.* (2002). Results of X-ray diffraction analysis of fractionated samples from representative Keanakako'i tephras as well as the Mars JSC-1JSC Mars-1 tephra are shown in Figure 3. The <2 µm size fractions of these samples were analysed both after (1) $MgCl_2$-saturation, and (2) $MgCl_2$-saturation followed by glycerol-solvation. The patterns exhibit two essential features of the altered Hawaiian tephras; (1) they

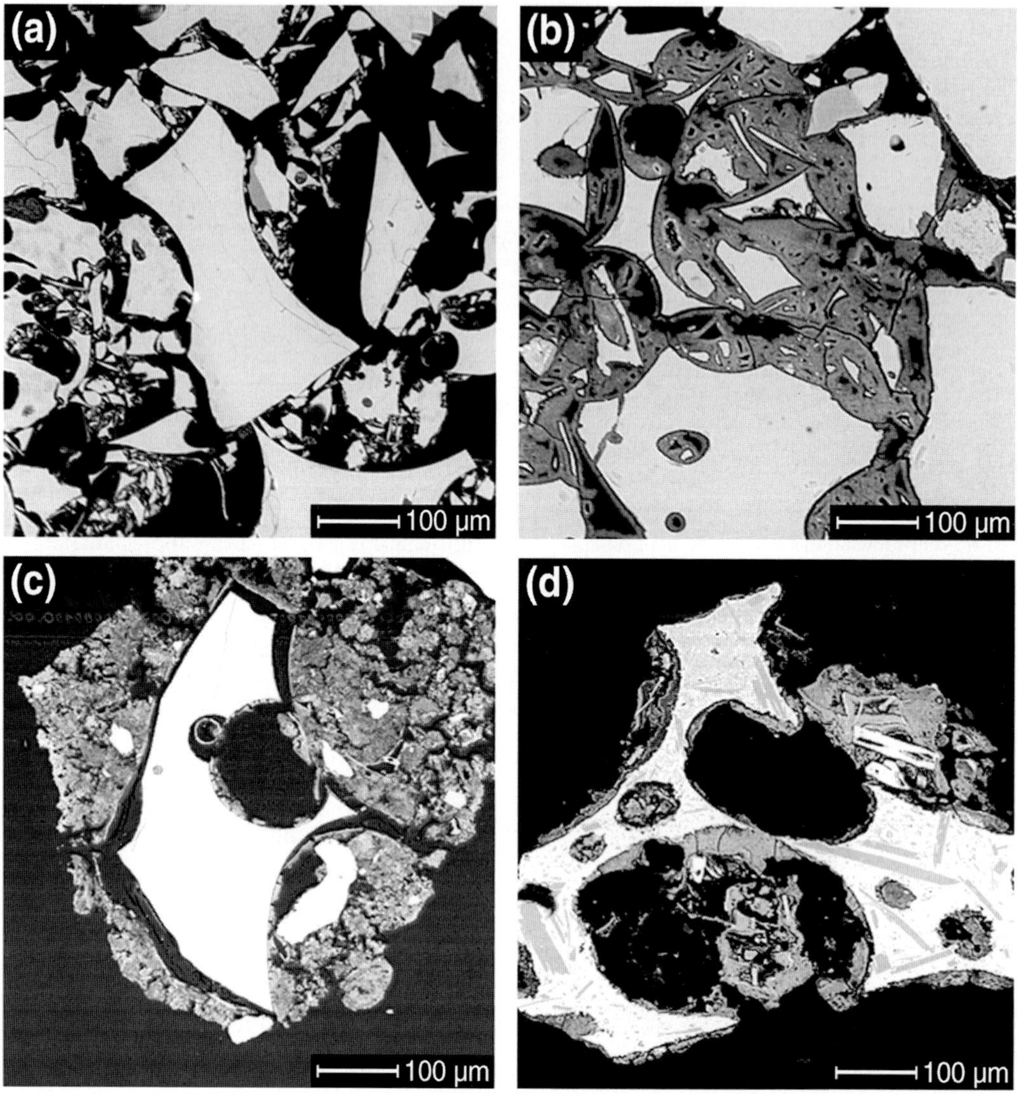

Fig. 2. Back-scattered electron images comparing consolidated and unconsolidated, altered tephras: (**a**) fresh Keanakako'i Ash tephra, (**b**) palagonitized Keanakakoi Ash (**c**) pedogenically weathered Keanakako'i, Ash, and (**d**) JSC Mars-1 tephra.

are uniformly poorly-crystalline and/or of extremely small crystallite size, and (2) the palagonitized tephas are mineralogically distinct from the other tephras. Specifically, palagonitized Keanakako'i tephras (i.e. 116-3 and 116-4, Fig. 3a, b) display smectite-like behaviour in their response to glycerol solvation by expansion of a weak (001) reflection from *c*. 1.55 nm to *c*. 1.8–1.9 nm. Conversely, pedogenically-affected tephras (e.g. 118-1 and the JSC Mars-1 samples, Fig. 3c, d) contain no obvious smectite, or any other crystalline clays with crystallite sizes capable of coherent X-ray diffraction.

The <2 μm size fraction from Keanakako'i tephra sample 116-4 is probably composed mainly of materials from the palagonitized rinds on vitric ash particles as can be seen on the BSE image (Fig. 2b). This sample's clay fraction was investigated in more detail by (1) the Bertaut-Warren-Averbach ('MudMaster') technique for calculating crystallite size distribution based on XRD peak shapes (Drits *et al.* 1998; Eberl *et al.*

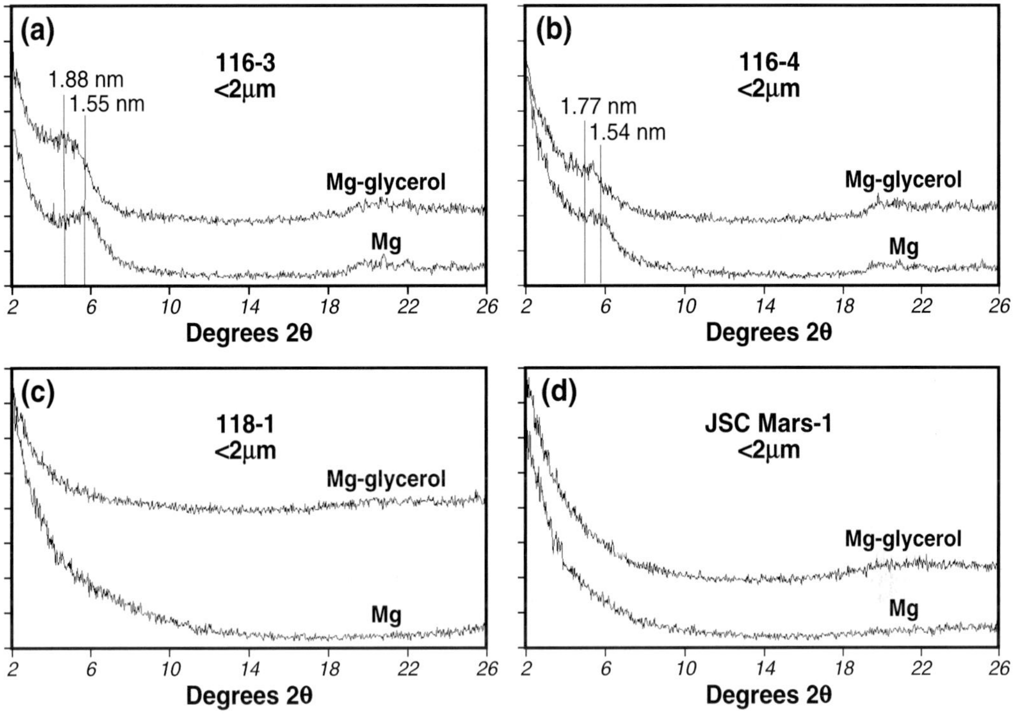

Fig. 3. X-ray diffraction traces of the <2 μm size fraction of representative altered tephra samples. (**a**) 116–3, (**b**) 116-4, (**c**) 118-1 and (**d**) JSC Mars-1. The former two are palagonitized samples that display clays with characteristics of smectite. The latter two are pedogenically altered tephras and are either amorphous or contain clays of crystallite sizes smaller than that required for coherent diffraction.

1998*a*), (2) the polyvinylpyrrolidone (PVP-10) intercalation technique for determining fundamental particle thickness (Eberl *et al.* 1998*b*), and (3) NewMod, a technique for modelling mixed-layered clays (Reynolds 1985). Upon K-saturation and glycolation, the 116-4 <2 μm size fraction expanded to only 14 Å, implying the presence of some high-charged smectite layers, which did not fully expand to 17 Å. The powder pattern, obtained from a K-saturated sample, dehydrated by heating to 300°C overnight and X-rayed in a dry atmosphere, was calculated by MudMaster. The resulting plot (Fig. 4a) reveals that smectite in the palagonitized sample has a 2.7 nm mean thickness (in the c-direction) for MacEwan crystallites (i.e. non-fundamental particles), with an asymptotic shape to the crystallite size distribution. The (060) reflection of this sample has a 1.517 Å d-spacing indicating that the smectite could be either di-octahedral (e.g. nontronite) or tri-octahedral (e.g. saponite). MudMaster calculations of the (060) implies that the smectite crystallites also have a very small mean crystallite size (3.2 nm) in the b-direction (Fig. 4b).

The XRD pattern (Fig. 5) of sample 116-4 after Na-saturation and PVP-10 intercalation reveals that the <2 μm fraction of the palagonitized tephra is not composed of purely fundamental (i.e. 1 nm thick) smectite particles. PVP-10 intercalation will completely swell and disperse all fundamental particles (Eberl *et al.* 1998*b*), and the resultant XRD pattern is indicative of only a non-swelling, mixed-layered component (if present). Figure 5 reveals that a small amount of a 7 Å phase (kaolinite or serpentine?) must be mixed-layered with smectite in the palagonitized tephra clay fraction. Similarly, the small shoulder at 16.5 Å indicates that some of the smectite crystallites (presumably those of high charge) did not swell completely upon PVP-10 intercalation.

The results of the MudMaster calculations and PVP-10 intercalation were incorporated into a NewMod (Reynolds 1985) model of the mixed-layering found in the palagonitized tephra clay fraction. The best-fitting model (shown in Fig. 6) is a mixed layer structure composed of 90% smectite and 10% serpentine layers, both Fe-rich. The model also assumes a crystallite size

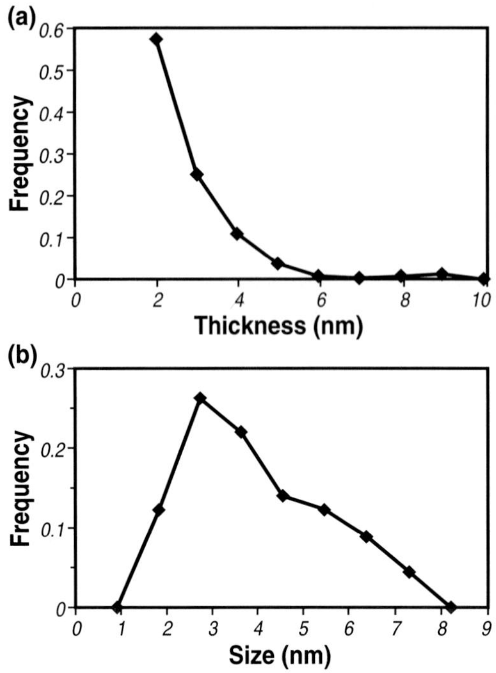

Fig. 4. Results of MudMaster calculations of crystallite size distribution of MacEwan particles of smectite in palagonitized Keanakako'i Ash tephra (116-4). (**a**) crystallite thickness distribution in the c-direction; (**b**) crystallite size distribution in the b-direction.

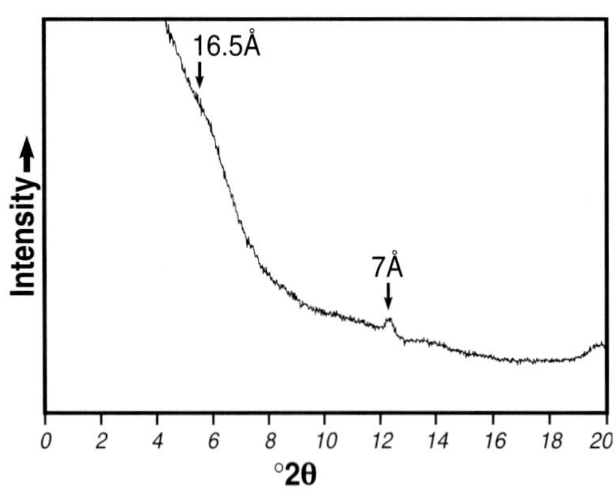

Fig. 5. X-ray diffraction pattern of Na-saturated and PVP-10 intercalated $<2\,\mu m$ size fraction of palagonitized Keanakako'i Ash (sample 116-4). The small peak at 12.2 degrees represents a 7 Å phase interlayered with smectite. Although the PVP-10 has dispersed most of the smectite fundamental particles, the small break in slope in the rising background just below 6° represents a small amount of high-charge smectite crystallites which were not dispersed by the PVP-10.

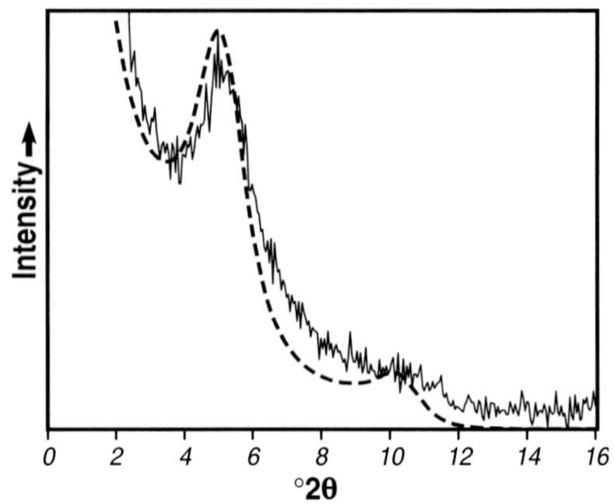

Fig. 6. NewMod model (dashed line) of Ca- and ethylene glycol-saturated <2 mm fraction of 116-4. The model assumes a mixed layer structure with 90% smectite and 10% serpentine. See text for more details.

distribution in which elementary (i.e. 1 nm) smectite particles constitute 2/3 of the total smectite layers. The remaining 33% of smectite layers have thicknesses (in the c-direction) between 2 and 6 nm, with the crystallite size distribution taken from the results of the MudMaster calculations (Fig. 4a). In summary, the results of the XRD analysis and subsequent modeling of the 116-4 $<2\,\mu m$ clay fraction indicate that the Keanakako'i palagonitized glass is: (1) a mixture of short-range-order and crystalline material; (2) that the crystalline material is a mixed-layered phase composed of mainly smectite and a small amount of a 7 Å phase (possibly serpentine); (3) that the intercalated smectite is poorly ordered with both low- and high-charge layers, and (4) that these crystallites are very small, with a mean crystallite size of <4 nm, which explains their weak X-ray diffraction patterns.

Selective dissolution

This method, used routinely by soil scientists, can help identify alteration products whose extremely fine-grained (i.e. nanophase) and/or amorphous nature renders them unidentifiable by standard X-ray diffraction techniques. For this study, alteration materials in the Hawaiian bulk samples were selectively extracted with ammonium oxalate and with citrate-dithionite. The oxalate extracts ferrihydrite, allophane, imogolite and humus metal complexes (if present), and dissolves fine-grained magnetite and fresh basaltic glass (if present). Similarly, the dithionite reduces and complexes ferric iron in crystalline and short-range-order Fe-oxides (also, if present). More

Table 1. *Composition of oxalate (o) and dithionite (d) selective dissolution extracts of bulk palagonitized and pedogenically altered Keanakako'i tephra and the pedogenically altered JSC Mars-1 soil simulant*

Sample	$\%Fe_o$	$\%Al_o$	$\%Si_o$	$\%Fe_d$	$\%Al_d$	$\%Si_d$	Fe_o/Fe_d	Al_o/Si_o	CEC
116-3 (palagonitized)	1.64	0.97	1.04	0.68	0.38	0.83	2.41	0.97	73–92
116-4 (palagonitized)	2.35	1.78	1.43	1.03	0.49	1.05	2.29	1.29	47–59
117-1 (pedogenic)	2.02	1.26	1.08	2.76	0.08	0.09	0.73	1.20	0
118-1 (pedogenic)	3.32	3.64	2.37	5.94	0.80	0.48	0.56	1.59	0
Fresh Keanakako'i glass	1.39	nd	nd	nd	nd	nd	nd	nd	0
JSC Mars-1 (pedogenic)	4.04	2.40	1.93	7.26	0.78	1.20	0.56	1.29	7–17

Note: The % values are reported on an elemental mass basis. The Al_o/Si_o ratio is reported on an elemental molar basis. CEC was measured using an *in situ* CsCl-staining technique (Schiffman and & Southard, 1996). nd, not determined. The range of measured cation exchange capacity (CEC in meq $100\,g^{-1}$) of these samples is also shown.

details of these techniques are presented in Bishop *et al.* (2002) and Soil Survey Laboratory Staff (1996). In Table 1 the selective dissolution results are presented as Fe_o, Al_o, and Si_o (oxalate-extractable Fe, Al, and Si) and as Fe_d, Al_d, and Si_d (dithionite-extractable Fe, Al, and Si).

These data emphasize significant differences between the palagonitized and pedogenically altered tephras. For example, Fe_o in the palagonitized Keanakako'i tephras averages 2.0% whereas Fe_o in the pedogenically altered Keanakako'i tephras averages 2.7%. Similarly Fe_d in the palagonitized Keanakako'i tephras averages 0.9% whereas Fe_d in the pedogenically altered Keanakako'i tephras averages 4.4%. The oxalate extractable Si and Al values are also generally higher in the pedogenically altered tephras, especially in the most strongly weathered soils (JSC-Mars 1 and 118-1). As indicated by these relatively lower total extractable Fe, Al and Si contents, palagonitized Keanakako'i tephras contain less ferrihydrite, goethite, allophane and imogolite than pedogenically altered equivalents. The unusually high Fe_o/ Fe_d ratios (i.e. >1) for the palagonitized tephras apparently reflect the relative abundance of fresh basaltic glass in these samples. Approximately 1.4% Fe was removed from fresh Keanakako'i vitric tephra by oxalate extraction.

Within the pedogenically-altered tephras, the variations in selective dissolution values indicate variable weathering intensity, either due to climate or duration of weathering. Specifically, the differences in oxalate values for the pedogenically-affected Keanakako'i tephras appear to reflect variations in intensity of weathering: the nearly 2:1 ratio of oxalate-extractable cations between samples 118-1 and 117-1 presumably reflects the increase in annual precipitation between these two sample locations (Fig. 1). Similarly, the selective dissolution data indicate that the JSC Mars-1 material is the most extensively weathered of all samples, presumably a reflection of this sample's much greater duration of exposure to pedogenic processes, but possibly also a function of the higher alkali cation content of this tephra relative to the Keanakako'i tephras. The weathering intensity is especially reflected by the Fe_d values because the Fe-oxyhydroxides soluble in citrate-dithionite tend to accumulate residually as more soluble constituents are dissolved and leached during weathering.

Electron microprobe analysis of tephra alteration products

Wavelength dispersive electron microprobe analyses were used to determine the composition of alteration products in the tephra samples. Analyses were performed on a Cameca SX-50 microprobe operated at 15 KeV and 5 nA beam current. To minimize specimen damage, the beam was rastered over the sample at a magnification of 60 000×. The average compositions of alteration products in samples 116-4, 117-1, 118-1, and Mars JSC Mars-1 are shown in Table 2. The composition of fresh Keanakako'i and JSC Mars-1 glasses is also shown. Because of the very fine-grained nature of the alteration products (especially pedogenic material), the oxide totals (which ranged from approximately 50 to 80 wt%) are not presented in Table 2. Instead, all the compositions (including the glasses) are presented as cation proportions/22 oxygens (as if the compositions were smectites) because the extremely fine-grained and variably hydrated nature of the alteration materials make meaningful comparison of oxide totals problematic at best. Figure 7 depicts compositional variations among the sample suite in terms of Al-content and total cation content

Table 2. *Electron microprobe analyses of Keanakako'i and JSC Mars-1 glass and their alteration products*

	Na	Mg	Al	Si	P	K	Ca	Ti	Mn	Fe	cation sum
Keanakako'i fresh glass	0.60	1.70	2.06	6.75	0.04	0.07	1.51	0.24	0.02	1.26	14.26
116-4 alteration	0.06	2.42	2.14	7.31	0.00	0.13	0.26	0.23	0.01	0.94	13.49
117-1 alteration	0.04	0.13	4.80	4.14	0.02	0.01	0.26	0.67	0.12	4.58	14.78
118-1 alteration	0.01	0.08	7.18	3.85	0.07	0.00	0.19	0.57	0.02	1.90	13.88
JSC Mars-1 fresh glass	0.94	0.77	2.53	6.85	0.14	0.37	0.87	0.30	0.03	1.25	14.04
JSC Mars-1 alteration	0.03	0.14	3.42	6.63	0.11	0.06	0.50	0.64	0.01	1.36	12.90

All analyses are presented as cation proportions based on 22 oxygens.

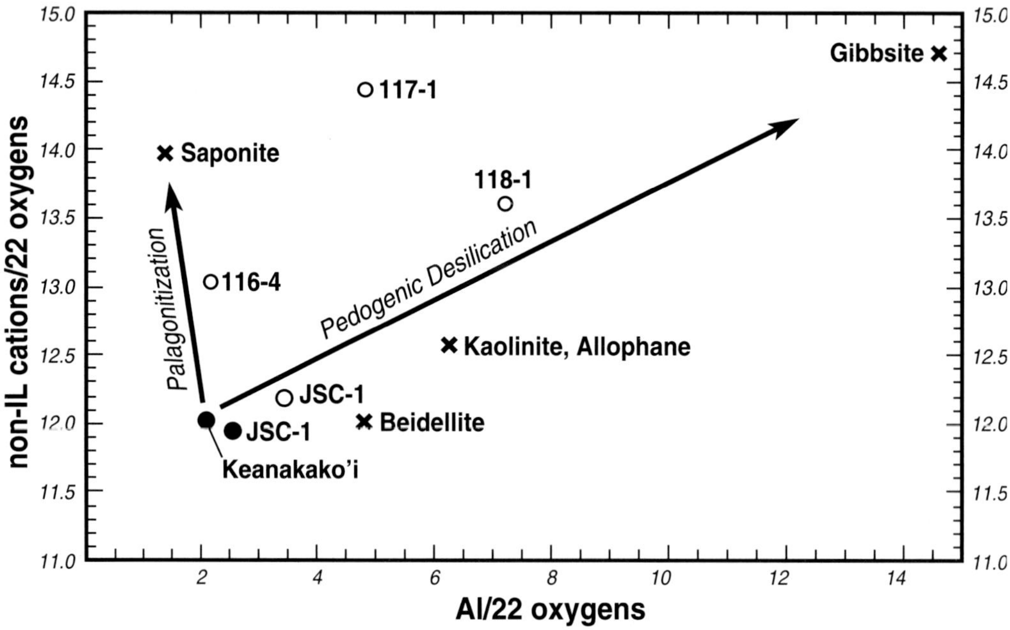

Fig. 7. Compositions of alteration products of tephra samples determined by electron microprobe analysis. All the analyses are recalculated on a 22 oxygen basis. Filled circles are microprobe compositions of fresh glasses; open circles are microprobe compositions of altered glasses; crosses are the end-member compositions of some common low-temperature clay minerals.

excluding interlayer (I.L.) cations (i.e. Na, K, Ca). Relative to the parent materials (i.e. the fresh Keanakako'i and JSC Mars-1 glasses, which plot in close proximity), the palagonitized glass (sample 116-4) is enriched in total non-I.L. cations, but with no accompanying increase in Al-content. Conversely, the pedogenically-affected Keanakako'i tephras (i.e. 117-1 and 118-1) have alteration products enriched in both total non-I.L. cations and aluminum with respect to the fresh glass. Figure 7 also shows idealized compositions of some common pedogenic and/or hydrothermal layer silicates: saponite, beidellite, kaolinite, allophane and gibbiste. The palagonitized rinds of sample 116-4, which have total non-I.L. cation contents of approximately 13/22 oxygens, plot close to midway on a trend between the parent material and saponite. The alteration products of tephra weathering in sample 118-1 plot close to the midway point of a line between their parent material's composition and that of gibbsite, the latter a hypothetical end-member weathering product of extreme desilication. The interpretation of where the pedogenically altered samples (e.g. sample 117-1) plot on this figure is probably complicated somewhat by the accumulation of Fe and suggests that loss of silica ('pedogenic desilication') is not the only major process in the pedogenic alteration of the tephras.

Cation exchange capacity

The cation exchange capacity (CEC) of the alteration minerals in these tephras was measured *in situ* with the electron microprobe using the CsCl staining technique described by Schiffman & Southard (1996). For this technique, polished sections are immersed overnight in a 1 M CsCl solution. After thorough rinsing in distilled water to remove any precipitated CsCl salt, the sections are carbon coated and analysed for Cs-content with the electron probe. The CEC of an individually analysed spot is directly related to its Cs-content (Schiffman & Southard 1996). Palagonitized rinds in the Keanakako'i tephras (samples 116-3 and 116-4) have CEC's that range from approximately 50 to 90 meq/100 g (Table 1), indicative of their high smectite content. Conversely, fine-grained pedogenically altered material in Keanakako'i tephra samples 117-1 and 118-1 have no detectable CEC using this method. The fine-grained alteration products in the JSC Mars-1 tephra have a CEC between approximately 10 and 20 meq 100 g^{-1}. These results imply that both the pedogenically altered Keanakako'i and JSC Mars-1 tephras have very low smectite contents.

Reflectance spectroscopy

Reflectance spectroscopy provides an independent assessment of the mineralogy in these samples and is especially useful for the identification of fine-grained layer silicates and various forms of iron hydroxides. Visible, near- and mid-infrared (IR) spectra were collected on selected samples using techniques described in Bishop *et al.* (1995). Figure 8 presents spectra on two palagonitized Keanakako'i samples (116-3, 116-4), two pedogenically-weathered Keanakako'i samples (117-1 and 118-1) as well as the Mars JSC-1JSC Mars-1 tephra. There are distinct differences between the spectral characteristics of the palagonitized v. pedogenically-altered Keanakako'i samples, especially in the near- and mid-infrared regions where the palagonitized samples have considerably brighter reflectances (in the 0.8 to 2.5 µm as well as the 3.5 to 5 µm range). These samples also exhibit strong vibrational absorption bands for structural OH (at 1.4 and 2.2–2.3 µm) and HO (at *c.* 1.9 and 3 µm and at 1640 cm^{-1} or *c.* 6.1 µm). In general, the visible and near-IR reflectance spectra of the two palagonitized samples contain features characteristic of smectite (Fig. 9). However, other spectral features, such as the multiple weak bands near 2.3 µm, imply that

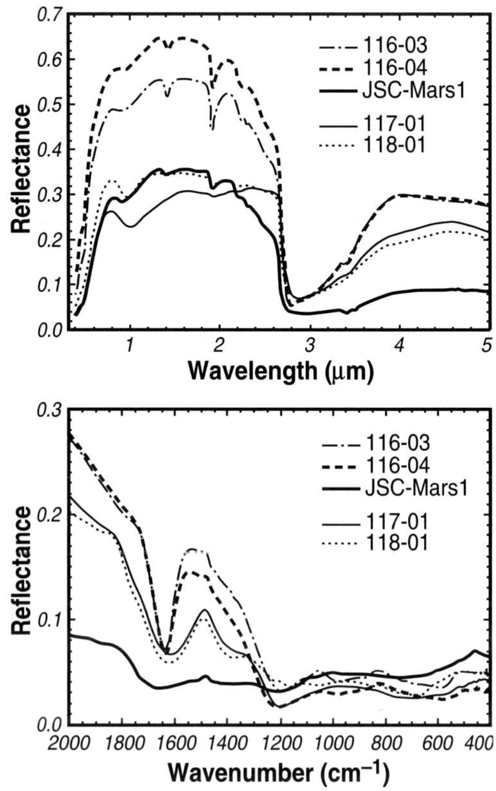

Fig. 8. Near- (top panel) and mid- (bottom panel) infrared reflectance spectra for palagonitized (samples 116-3 and 116-4) and pedogenically altered (samples 117-1 and 118-1) Keanakako'i tephras, as well as the JSC-Mars1 tephra. The two palagonitized samples differ markedly from the other three in total reflectance, as well as in smectite-like characteristics, such as strong bands at *c.* 1.4 and *c.* 1.9 µm due to OH and H_2O vibrations respectively.

another layer silicate may also be present in small quantities. The asymmetric shape of the broad 3 µm band also indicates the higher abundance of layer silicates in the two palagonitized samples. The sharp feature near 2.7 µm is due to the structural OH stretching, while the broad 3 µm band is due to water.

What other clays might be mineralogical components of 'palagonitized glass'? An overtone of the structural OH stretching band in many Al-bearing clays occurs near 1.4 µm (e.g. Clark *et al.* 1990; Bishop *et al.* 1994). The related band due to Si–OH is observed at 1.37 µm (Anderson & Wickersheim 1964). Serpentines exhibit bands from 1.38–1.40 µm depending on the cations bound to the structural OH (King & Clark 1989). Reflectance spectra of the 1.4 µm region are shown in Figure 10 for the palagonitized samples

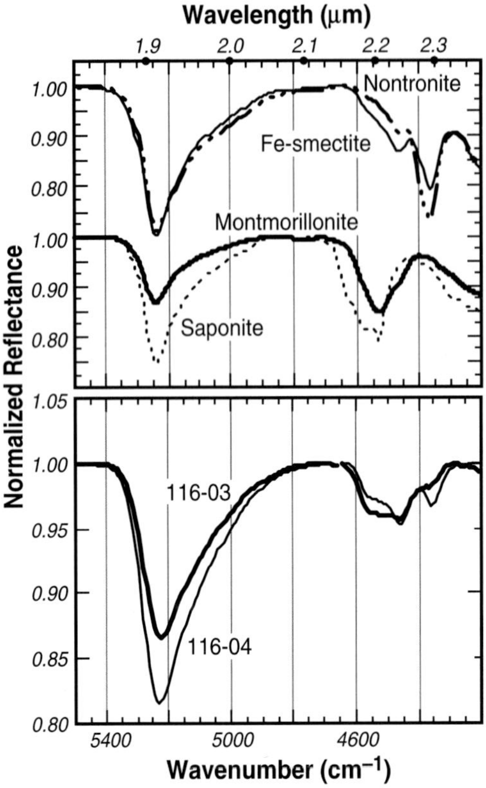

Fig. 9. Normalized near-infrared reflectance spectra of smectites (top panel) and the 2 palagonitized Keanakako'i tephras (bottom panel).

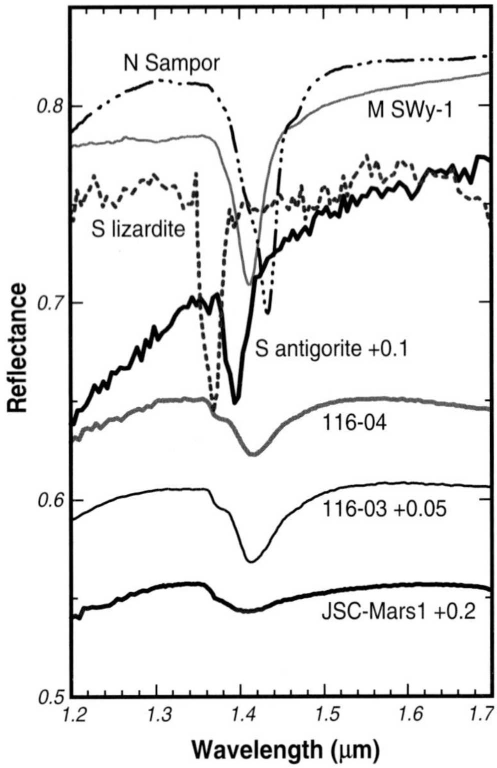

Fig. 10. Reflectance spectra of palagonitized Kilauea samples 116-03 (offset +0.05) and 116-4, JSC-Mars1 (offset +0.2), montmorillonite (M) SWy-1, nontronite (N) Sampor, and the serpentine (S) clays (from King & Clark 1989) lizardite and antigorite (offset +0.1).

(116-3 and 116-4), the JSCMars1 sample, and selected clay minerals, including serpentines. The chemistry and spectral properties of the montmorillonite (SWy-1) and nontronite (Sampor) samples were described in a recent study (Bishop *et al.* 1999). The lizardite sample contains primarily Si and Mg, with a small amount of Fe(III), whereas the antigorite sample contains Si, Mg, Fe(II), Fe(III) and Al cations (King & Clark 1989). The high-resolution spectra of these samples show multiple bands in this region, which are not resolved in the lower-resolution serpentine spectra included in Figure 10. The spectral features in this region observed for the palagonitization samples (116-3 and 116-4) are attributed primarily to montmorillonite with a small amount of serpentine. The broadened and weaker band observed for JSCMars-1 is probably due to some poorly crystalline clay minerals.

Unlike the palagonitized samples, the reflectance spectra for the two pedogenically-altered samples, as well as the JSC Mars-1 tephra, are characterized by much lower spectral contrast, as well as weaker and broader OH and H_2O bands. The JSCMars-1 may also have some layer silicates, indicated by the distinctive bands near 2.2 μm but the rounded 3 μm band is characteristic of water bound in hydrated minerals or glass. The two pedogenically-altered Keanakako'i samples (117-1 and 118-1) exhibit stronger bands at 1.0 μm due to Fe(II) in either pyroxene or glass, but very weak features representative of hydrous layered silicates (at 1.4, 1.9 and 2.2 μm.). The spectra of samples 117-1 and 118-1 are also consistent with the presence of a nanophase ferric oxyhydroxide such as ferrihydrite based on the broad reflectance maximum near 0.8, broad shoulder near 0.9 μm on the *c.* 1 μm band, and broad 3 μm band.

In summary, the two pedogenically altered samples contain the most crystalline and/or the most abundant quantities of iron oxides–oxyhydroxide minerals, followed by the palagonitized samples and JSCMars 1. Conversely, the spectra of the palagonitized samples exhibit the strongest layer silicate clay bands (characteristic of smectite), followed by the JSC Mars-1 sample, then the two pedogenically altered samples.

Discussion and conclusions

Is palagonitization distinct from pedogenic weathering?

Based on the data presented above, we believe that a strong case can be made that there is a clear distinction between palagonitization and pedogenic alteration of vitric basaltic tephras, at least under the weathering regime on the summit of Kilauea volcano. To summarize the criteria:

(1) *Textural*: As indicated by field observation and BSE micrography, palagonitization results in consolidation of tephra whereas soil formation results in degradation disaggregation of tephra.
(2) *Mineralogical*: As indicated by XRD and reflectance spectroscopy, palagonitization produces a mineral suite containing smectite with lesser amounts of other crystalline (e.g. serpentine) and short-range-order aluminosilicates and Fe–oxyhydroxides, whereas pedogenically altered tephras contain mainly the short-range-order materials (ferrihydrite, allophane/imogolite).
(3) *Chemical*: As indicated by the results of the selective dissolution, palagonitized samples have less extractable Al, Si, and Fe, particularly Fe extracted by citrate-dithionite, than the pedogenically altered tephra. The CEC of palagonitized samples is substantially higher than that of the soils; the chemical composition of fine-grained alteration materials in palagonitized samples is close to that of smectite, whereas the composition of fine grained alteration materials in the soils is generally more desilicated and therefore more aluminous.

A working definition of palagonitized tephra

Based on the data from the Keanakako'i tephras, it is suggested that palagonitized basaltic tephra must contain three essential features which distinguish it from other forms of alteration: (1) palagonitized tephra must contain vitric ash particles with rinds of palagonitized glass, and (2) palagonitized tephra must be partly to largely consolidated into tuff. Moreover, (3) the palagonitized glass should contain smectite-like crystallites.

Implications for palagonite as a product of sub-glacial volcanic processes

Palagonitization of Keanakako'i vitric tephras is limited to regions affected by (fossil) hydrothermal steam vents along the circumferential fault system that defines Kilauea Caldera (Schiffman *et al.* 2000). Presumably palagonitization occurred shortly after the deposition of the tephras approximately 500 years ago. Elsewhere on the summit of Kilauea, these same vitric tephras have been variably altered by soil forming processes since their deposition, with a distinct weathering gradient that covaries with annual rainfall (Hay & Jones 1972; Schiffman *et al.* 2000). However, there is no evidence that non-hydrothermal weathering processes on the summit of Kilauea have resulted in, or are currently resulting in, the palagonitization of these tephras.

How generally applicable are the observations made on weathering of basaltic tephra on the summit of Kilauea volcano? Specifically, are they useful for interpreting the weathering history of basaltic tephras in drier and cooler climates? It has long been recognized that subglacially-erupted hyalotuffs, e.g. in Iceland, are ubiquitously palagonitized (Jones 1969, 1970; Furnes 1978). Has this palagonitization been accomplished by a short hydrothermal process (e.g. as in Surtsey; Jakobsson 1978), or by a longer pedogenic process (as envisioned by Le Gal *et al.* 1999)? The mineralogic and geochemical data on the JSC Mars-1 tephra may provide some insight into these questions.

Unlike the warm and wet conditions on the windward side of Kilauea, the climate at the JSC Mars-1 locality is much cooler and is therefore somewhat similar to the climate of the interior of Iceland. The JSC Mars-1 tephra has been weathering in this climatic regime for several thousand years (Wolfe & Morris 1996), yet the textural, mineralogical and geochemical data presented above do not indicate that this tephra is undergoing palagonitization, at least as defined above. Rather, this tephra appears to be more pedogenically weathered (as indicated by Fe, Al, and Si extracted by selective dissolution) than any of the Keanakako'i tephras examined in this study.

In a related study, Bishop *et al.* (2002) examined the alteration products of some sub-glacially erupted, pillow basalts and hyalotuffs from the Laugarvatn region of Iceland. Both parent materials are of similar age and have been exposed to nearly identical climatic conditions since their deposition, yet the lavas display mineralogical and geochemical properties indicative of incipient pedogenic weathering, whereas the hyalotuffs are palagonitized. One hypothesis for these observations, and one that is identical to the Kilauea scenario, is that the Laugarvatn hyalotuffs were palagonitized by hydrothermal activity shortly after their deposition. Although no one has yet reported the active palagonitization of subglacially erupted tephras, the elevated heat flow from the site of the 1996 Gjálp subglacial eruption (Gudmundsson *et al.* 1997) implies that palagonitization may presently be affecting the underlying (buried) hyaloclastite ridge. This hydrothermal mechanism for palagonitization would be essentially identical to that which occurred at Surtsey (Jakobsson 1978).

Fieldwork in Hawaii by PS and Iceland was made possible by grants from the Committee on Research at the University of California, Davis. Reflectance spectra were measured at RELAB, a multi-user, NASA-supported facility (NAG5-3871). Support for JLB from NASA's Mars Data Analysis Program is much appreciated. Thanks are due to T. Hiroi for assistance with the bi-directional spectra, to C. Allen and S. Jakobsson for helpful editorial comments and to M. Chapman and J. Smellie for editorial handling of the manuscript.

References

ALLEN, C. C., JAGER, K. M., MORRIS, R. V., LINDSTROM, D. J., LINDSTROM, M. M. & LOCKWOOD, J. P. 1998. Martian soil simulant available for scientific, educational study. *EOS, Transactions*, **79**, 405–409.

ALLEN, C. C., GOODING, J. L., JERCINOVIC, M. J. & KEIL, K. 1981. Altered basaltic glass: a terrestrial analog to the soils of Mars. *Icarus*, **45**, 347–369.

ANDERSON, J. H. & & WICKERSHEIM, K. A. 1964. Near infrared characterization of water and hydroxyl groups on silica surfaces. *Surface Science*, **2**, 252–260.

BANIN, A., CLARK, B. C. & WANKE, H. 1992. Surface chemistry and mineralogy. *In*: KIEFFER, H. H., JAKOSKY, B. M., SNYDER, C. W. & MATTHEWS, M. S. (eds) *Mars*. University of Arizona Press, Tuscon, 594–625.

BELL, J. F., MCSWEEN, H. Y. JR, CRISP, J. A. *ET AL.* 2000. Mineralogic and compositional properties of Martian soil and dust: Results from Pathfinder. *Journal of Geophysical Research*, **105**, 1721–1755.

BISHOP, J. L., PIETERS, C. M. & & EDWARDS, J. O. 1994. Infrared spectroscopic analyses on the nature of water in montmorillonite. *Clays and Clay Minerals*, **42**, 701–715.

BISHOP, J. L., PIETERS, C. M., BURNS, R. G., EDWARDS, J. O., MANCINELLI, R. L. & FROESCHL, H. 1995. Reflectance spectroscopy of ferric sulfate-bearing montmorillonites as Mars soil analog materials. *Icarus*, **117**, 101–119.

BISHOP, J. L., FROSCHL, H. & & MANCINELLI, R. L. 1998. Alteration processes in volcanic soils and identification of exobiologically important weathering products on Mars using remote sensing, *Journal of Geophysical Research*, **103**, 31 457–31 476.

BISHOP, J. L., MURAD, E., MADEJOVÁ, J., KOMADEL, P., WAGNER, U. & SCHEINOST, A. 1999. Visible, Mössbauer and infrared spectroscopy of dioctahedral smectites: Structural analyses of the Fe-bearing smectites Sampor, SWy-1 and SWa-1. *In*: KODAMA, H., MERMUT, A. R. & TORRANCE, J. K. (eds) *Proceedings of the 11th International Clay Conference*, Ottawa, June 1997, 413–419.

BISHOP, J. L., SCHIFFMAN, P. & SOUTHARD, R. J. 2002. Geochemical and mineralogical analyses of palagonitic tuffs and altered rinds of pillow lavas on Iceland and applications to Mars. *In*: SMELLIE, J. & CHAPMAN, M. G. (eds) *Volcano–ice interactions on Earth and Mars*. Geological Society, London, Special Publications, **202**, 371–392.

BUEMI, A., CIMINO, G. & STRAZZULLA, G. 1998. Characterization of Etnean soils and application to Mars. *Journal of Geophysical Research*, **103**, 13 667–13 674.

CAZADEVALL, T. J. & HAZLETT, R. W. 1983. Thermal areas on Kilauea and Mauna Loa Volcanoes, Hawaii. *Journal of Volcanology and Geothermal Research*, **16**, 173–188.

CLARK, R. N., KING, T. V. V., KLEJWA, M. & SWAYZE, G. A. 1990. High spectral resolution reflectance spectroscopy of minerals. *Journal of Geophysical Research*, **95**, 12 653–12 680.

DRITS, V. A., EBERL, D. D. & SRODON, J. 1998. XRD measurement of mean thickness, thickness distribution and strain for illite and illite-smectite crystallites by the Bertaut-Warren-Averbach technique. *Clays and Clay Minerals*, **46**, 38–50.

EBERL, D. D., DRITTS, V. A. & SRODON, J. 1998*a*. Deducing growth mechanisms for minerals from the shapes of crystal size distributions. *American Journal of Science*, **298**, 499–533.

EBERL, D. D., NUESCH, R., SUCHA, V. & TSIPURSKY, S. 1998*b*. Measurement of fundamental illite particle thickness by X-ray diffraction using PVP-10 intercalation. *Clays and Clay Minerals*, **46**, 89–97.

EGGLETON, R. A. & KELLER, J. 1982. The palagonitization of limburgite glass – a TEM study. *Neues Jahrbuch fur Mineralogie Montashefte*, **7**, 321–336.

FARRAND, W. H. & SINGER, R. B. 1991. Spectral analysis and mapping of palagonite tuffs of Pavant Butte, Millard County, Utah. *Geophysical Research Letters*, **18**, 2237–2240.

FISHER, R. V. & SCHMINCKE, H.-U. 1984. *Pyroclastic rocks*. Springer-Verlag, New York.

FURNES, H. 1978. Element mobility during palagonitization of a subglacial hyaloclastite in Iceland. *Chemical Geology*, **22**, 249–264.

GIAMBELLUCA, T. W. & SCHRODER, T. A. 1998. Climate. *In*: JUVIK, S. P. & JUVIK, J. O. (eds) *Atlas of Hawaii*. University of Hawaii Press, Honolulu, 49–59.

GOLDEN, D. C., MORRIS, R. V., MING, D. W., LAUER, H. V. JR. & YANG, S. R. 1993. Mineralogy of three slightly palagonitized basaltic tephra samples from the summit of Mauna Kea, Hawaii. *Journal Geophysical Research*, **98**, 3401–3411.

GOODING, J. L. & KEIL, K. 1978. Alteration of glass as a possible source of clays on Mars. *Geophysics Research Letters*, **5**, 727–730.

GUDMUNDSSON, M. T., SIGMUNDSSON, F. & BJORNSSON, H. 1997. Ice-volcano interaction of the 1996 Gjálp subglacial eruption, Vatnajokull, Iceland. *Nature*, **389**, 954–957.

HAY, R. L. & IIJIMA, A. 1968. *Nature and origin of palagonitic tuffs of the Honolulu Group on Oahu, Hawaii*. Geological Society of America, Memoirs, **116**, 338–376.

HAY, R. L. & JONES, B. F. 1972. Weathering of basaltic tephra on the island of Hawaii. GSA *Geological Society of America, Bulletin*, **83**, 317–332.

HONNOREZ, J. 1981. The aging of the oceanic crust at low tempertature. *In*: EMILIANI, C. (ed.) *The Oceanic Lithosphere*, *The Sea, v.7*. John Wiley, New York, 525–588.

JAKOBSSON, S. P. 1978. Environmental factors controlling the palagonitization of the Surtsey tephra, Iceland. *Geological Society of Denmark, Bulletin*, **27** (Special Issue), 91–105.

JONES, J. G. 1969. Intraglacial volcanoes of the Laugarvatn region, south west Iceland – I. *Quaternary Journal Geologicalal Society, London*, **124**, 197–211.

JONES, J. G. 1970. Intraglacial volcanoes of the Laugarvatn region, south west Iceland. II. *Journal of Geology*, **78**, 127–140.

KING, T. V. V. & CLARK, R. N. 1989. Spectral characteristics of chlorites and Mg-serpentines using high-resolution reflectance spectroscopy. *Journal of Geophysical. Research*, **94**, 13 997–14 008.

LE GAL, X., CROVISIER, J.-L., GAUTHIER-LAFAYE, F., HONNOREZ, J. & GRAMBOW, B. 1999. Meteoric alteration of Icelandic volcanic glass: long term changes in the mechanism, *Compte Rendus, Académie des Sciences, Paris, Sciences de la Terre et des Planetes*, **329**, 175–181.

MASTIN, L. G. 1997. Evidence for water influx from a caldera lake during the explosive hydromagmatic eruption of 1790, Kilauea volcano, Hawaii. *Journal of Geophysical Research*, **102**, 20 093–20 109.

MCPHIE, J., WALKER, G. P. L. & CHRISTIANSEN, R. L. 1990. Phreatomagmatic and phreatic fall and surge deposits from Kilauea volcano, Hawaii, 1790 AD: Keanakakoi Ash Member. *Bulletin of Volcanology*, **52**, 334–354.

MORRIS, R. V., GOLDEN, D. C., MING, D. W., SHELFER, T. D., JORGENSEN, L. C., BELL, J. F., GRAFF, T. G. & MERTZMAN, S. A. 2001. Phyllosilicate-poor palagonitic dust from Mauna Kea Volcano (Hawaii): A mineralogical analogue for magnetic Martian dust? *Journal of Geophysical Research Planets*, **105**, 1757–1817.

REYNOLDS, R. C. JR. 1985. MEWMOD©, a computer program for the calculation of one-dimensional diffraction patterns of mixed-layered clays. R. C. Reynolds, Jr., 8 Brook Dr., Hanover, NH 03755, USA.

SCHIFFMAN, P., SPERO, H. J., SOUTHARD, R. J. & SWANSON, D. A. 2000. Controls on palagonitization versus pedogenic weathering of basaltic tephra: Evidence from the consolidation and geochemistry of the Keanakako'i ash member, Kilauea volcano. *Geochemistry Geophysics Geosystems*, **1**.

SCHIFFMAN, P. & SOUTHARD, R. J. 1996. Cation exchange capacity of layer silicates and palagonitized glass in mafic volcanic rocks: a comparative study of bulk extraction and in situ techniques. *Clays and Clay Minerals*, **44**, 624–624.

SIGURDSSON, H. 1999. *Melting the Earth*. Oxford University Press, Oxford.

SINGER, A. 1974. Mineralogy of palagonitic material from the Golan Heights, Israel. *Clays & Clay Minerals*, **22**, 231–240.

SINGER, A. & BANIN, A. 1990. Characteristics and mode of formation of palagonite – a review. *Proceedings of the International Clay Conference*, **9**, 173–181.

SOIL SURVEY LABORATORY STAFF. 1996. *Soil Survey Laboratory Methods Manual*. Soil Survey Investigations Report No. 42, version, 3.O. US Department of Agriculture, Natural Resources Conservation Service. National Soil Survey Center, Lincoln, Nebraska.

SWANSON, D. A., FISKE, R. S., ROSE, T. R. & KENEDI, C. L. 1998. Prolonged deposition of the Keanakako'i Ash Member, Kilauea. *EOS, Transactions*, **79**, 927.

WOLFE, E. W. & MORRIS, J. 1996. *Geological Map of the Island of Hawaii*. United States Geological Survey Miscellaneous Series Field Studies Map **I-2524-A**.

ZHOU, Z., FYFE, W. S., TAZAKI, K. & VAN DER GAAST, S. J. 1992 The structural characteristics of palagonite from DSDP Site 335. *Canadian Mineralogist*, **30**, 75–81.

Identifying bio-interaction with basaltic glass in oceanic crust and implications for estimating the depth of the oceanic biosphere: A review

H. FURNES[1], I. H. THORSETH[1], T. TORSVIK[2], K. MUEHLENBACHS[3], H. STAUDIGEL[4] & O. TUMYR[1]

[1] *Geological institute, University of Bergen, Allegt.41, 5007 Bergen, Norway (e-mail: Harald.Furnes@geol.uib.no)*
[2] *Department of Microbiology, University of Bergen, Jahnebk. 5 5007 Bergen, Norway*
[3] *Department of Geology, University of Alberta, Edmonton, Alta T6G 2E3, Canada*
[4] *Scripps Institution of Oceanography, University of California, La Jolla, CA 92093-0225, USA*

Abstract: The alteration of basaltic glass in the volcanic part of the oceanic crust is, to a substantial extent, biologically mediated. Evidence of microbial interaction with basaltic glass can be provided by a number of independent observations, such as: (1) Textures at the alteration front, generated by dissolution of the glass and subsequent precipitation. These bio-generated textures can be defined as a granular type (dominant) and a tubular type, and show size and form which are compatible with microbial etching. (2) Filament-like structures, representing organic remains, appear in connection with bio-generated textures. (3) Within areas of the bio-generated textures, particularly at the alteration front, DNA and ribosomal RNA have been demonstrated to be present in relatively young samples. (4) X-ray mapping shows that carbon and nitrogen invariably appear within the bio-generated textures, in young samples most strongly enriched at the alteration front. (5) Carbon isotopes ($\delta^{13}C$) in carbonates extracted from the glassy margin of pillows show highly variable values which can be explained in terms of bio-fractionation of the ^{12}C and ^{13}C isotopes. Estimates of the proportion of bio-genetic alteration products of basaltic glass, on the basis of textural relationships, suggest that bio-alteration is dominant compared to abiotic alteration in the upper 300 m of the oceanic crust, and declines to become insignificant at a depth of about 500 m.

Alteration of basaltic glass may occur as an abiotic, chemical–physical process, or as a result of biological interaction. Alteration of basaltic glass yields a pale yellow to dark brown material, commonly referred to as palagonite, and involves a complicated process of incongruent and congruent dissolution and contemporaneous precipitation, hydration and pronounced chemical exchange (Hay & Iijima 1968*a*, *b*; Eggleton & Keller 1982; Staudigel & Hart 1983; Furnes 1984; Crovisier *et al.* 1987; Bednarz & Schmincke 1989; Zhou & Fyfe 1989; Thorseth *et al.* 1991; Daux *et al.* 1994; Schiffman *et al.* 2000). The palagonitization process takes place at low to high temperatures (e.g. Jakobsson 1972, 1978; Jakobsson & Moore 1986; Zierenberg *et al.* 1995).

During the last decade much attention has been focused towards the existence of microbes attached to, and contributing to the alteration of the basaltic glass rim of pillow lavas in the upper part of the oceanic crust. The first suggestions of microbial pitting of volcanic glass were made by Jones & Goodbody (1982) and Ross & Fisher (1986), but no convincing mechanism of how microbes may actually facilitate dissolution of glass was provided. Later, in a study of Icelandic hyaloclastites the presence of bacteria hosted within alteration textures of the basaltic glass were found (Thorseth *et al.* 1992). Subsequently, similar textures were found in the glassy part of pillow lavas from several DSDP/ODP drill sites, at depths down to 550 m (the limit at which fresh glass has been found) into the volcanic basement where fresh glass can be found (Thorseth *et al.* 1995*a*; Furnes *et al.* 1996; Fisk *et al.* 1998; Torsvik *et al.* 1998; Furnes & Staudigel 1999). These typical alteration textures, attributed to bio-alteration of the glass, are common in volcanic rocks of the *in situ* ocean floor basement, ranging in age from Quaternary to 170 Ma (e.g. Fisk 1999; Furnes *et al.* 2001*a*). Thus, the presence of microbes at the glass–water interface seems

From: SMELLIE, J. L. & CHAPMAN, M. G. (eds) 2002. *Volcano–Ice Interaction on Earth and Mars*. Geological Society, London, Special Publications, **202**, 407–421. 0305-8719/02/$15.00

firmly established. However, as to the relationships between microbes and pits and channels in the glass, various possibilities may be considered: (1) The microbes get trapped (dead or alive) by the pore space at the interface; (2) the weathering goes on and the microbes live off the weathering but do not help it along much; and (3) the microbes greatly aid and benefit from the weathering zone which they colonize.

On the basis of the bacteria hosted in the Icelandic hyaloclastites, Thorseth *et al.* (1992) suggested that microbes may cause local variations in pH and/or ligands that allow them to chemically 'drill' into a silicate substrate. This process was subsequently verified in experiments (Thorseth *et al.* 1995*b*; Staudigel *et al.* 1995; 1998). Also, Welsh & Ullman (1996) demonstrated that organic acids derived by metabolic processes are more effective in dissolving some silicates than inorganic acids at the same pH. Further, as will be demonstrated below, the sizes and shapes of the corrosion pits and channels in the glass are similar to microbes. The microbial influence on the alteration process of basaltic glass has additionally been confirmed by documenting the presence of DNA and ribosomal RNA within the structures of anticipated biological origin (Thorseth *et al.* 1995*a*; Furnes *et al.* 1996; Giovannoni *et al.* 1996; Torsvik *et al.* 1998). Further evidence of microbial fingerprints within altered glass has been demonstrated by low and high $\delta^{13}C$ values (Furnes *et al.* 2001*a*), as well as the element distribution, particularly of C and N (Torsvik *et al.* 1998; Furnes *et al.* 1999). All of the above-mentioned features strongly suggest that the third alternative is the most likely, and will be discussed below.

Recent and ongoing studies of ophiolitic rocks of different ages have demonstrated textural and chemical evidence pointing towards bio-alteration. These ophiolites include, so far, the Troodos ophiolite in Cyprus (90 Ma; Staudigel & Gillis 1990), the Mirdita ophiolite in Albania (160 Ma; Robertson & Shallo 2000), and the Solund-Stavfjord ophiolite in west Norway (443 Ma; Dunning & Pedersen 1988). In particular, traces of bio-alteration have been found within the Troodos oceanic crust, which occurred during an early stage of ocean-floor alteration. This was demonstrated by alteration textures similar to those found in the *in situ* oceanic crust, the identification of organic remains and by carbon isotopes (Furnes *et al.* 2001*b*).

This paper provides a review of different approaches that have proved useful in the identification of bio-alteration of basaltic glass from a number of drill sites in the oceanic crust. A total of eleven drill holes from the oceanic crust of the slow-spreading Atlantic Ocean and the intermediate-spreading Costa Rica Rift and Lau Basin make the data base of information (Furnes *et al.* 2001*c*). Further, the implications these findings have had for evaluating the degree of bio-alteration at different levels in the volcanic pile of the oceanic crust are reviewed, thus providing a method for mapping the depth of the oceanic biosphere.

Identifying bio-alteration

For the identification of bio-alteration of basaltic glass, the following features were considered: (1) Textural relationships at the alteration front (i.e. the interface between fresh and altered glass), showing size and shape features comparable to microbes; (2) organic remains within bio-reminiscent textures; (3) DNA within bio-reminiscent textures; (4) elevated carbon and nitrogen concentration at alteration fronts; (5) carbon isotopic values of carbonate in anticipated bio-altered glassy pillow margins indicative of bio-fractionation.

Textures

Alteration textures, developed adjacent to fractures and vesicle walls in basaltic glass as a result of biotic or abiotic processes, have been studied in thin sections by scanning electron microscopy (SEM) and optical microscopy. In samples of different ages, or within the same thin section, various stages in the alteration process can be studied. The alteration front, i.e. the interface between altered and fresh glass, may be smooth or irregular. The abiotic alteration zones, consisting of pale yellow to dark brown amorphous to poorly crystalline palagonite, appear as banded material of approximately equal thickness on both sides of fractures, with smooth alteration fronts that are symmetrical with respect to the fracture around which it developed. Alteration studies of basaltic glass in pillow lavas of *in situ* oceanic crust show that most commonly the textures associated with irregular alteration fronts may be classified as either granular or tubular. These are taken to represent biotic mediation of the glass. During the initial stage in the development of the granular type, the alteration texture appears as isolated spherical structures along fractures (Fig. 1a), as small patches of isolated and/or coalesced bodies along fractures, or at the intersection between fractures (Fig. 1b). At the more advanced stage, the granular type defines thick,

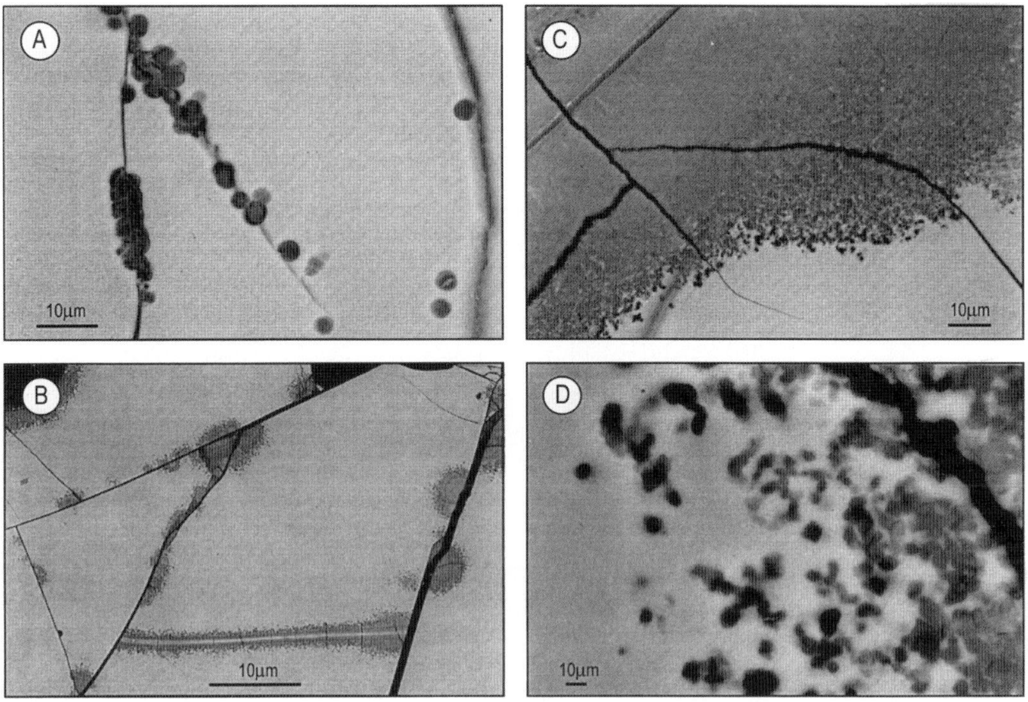

Fig. 1. SEM images showing various stages in the development of granular texture, taken to predominantly represent bio-generated alteration of basaltic glass. (**a**) Incipient alteration along fractures, seen as dark, spherical bodies. Light grey: fresh glass. ODP sample 106-648B, 1R-1, 37–38; depth into volcanic basement: 7 m; age: Quaternary; Detrick *et al.* (1988). (**b**) Fractures along which different degrees of granular type alteration (dark grey) have taken place. Light grey: fresh glass. DSDP sample 46-396B, 17R-3, 62–70; depth into volcanic basement: 151 m; age: 10 Ma; Dmitriev *et al.* (1978). (**c**) Typical granular texture (dark grey). Note that the texture becomes less distinct away from the alteration front (interface between fresh and altered glass). Light grey: fresh glass. ODP sample 148-896A, 6R-3, 30-34; depth into volcanic basement: 46 m; age: 5.9 Ma; Alt *et al.* (1996). (**d**) Typical granular texture (dark grey spherical bodies) at the alteration front. Light grey: fresh glass. DSDP sample 70-504B, 35-1, 106-113; depth into volcanic basement: 261 m; age: 5.9 Ma; Alt *et al.* (1996).

persistent zones consisting of numerous coalesced spherical bodies (Fig. 1c), and isolated and/or small groups of a few coalesced bodies at the alteration front (Fig. 1d). A study of the alteration of the glass rims of pillow lavas of the Troodos ophiolite show similar granular textures. The characteristic features of the granular texture, viewed by the optical microscope, are shown in Figure 2. The dominant size of the spherical bodies is between 0.2–0.6 μm, and only 7% exceed 1.5 μm. Due to the size and shape of these structures, they have been taken to represent microbially-generated etch pits, filled by precipitated, amorphous material (Thorseth *et al.* 1992, 1995*b*; Alt & Mata 2000). It should be kept in mind, however, that etch grooves of similar size as bacteria have also been produced during experimental alteration of basaltic glass in a sterile environment (Sæbø 1998). Thus, on the basis of the granular texture alone, it may be unsound to state categorically that this feature gives unequivocal evidence of bio-alteration. However, as will be demonstrated below, whenever areas of these structures are X-rayed, they show the presence of organic carbon and other bio-elements (N, P, S), and in some cases even organic remains and DNA have been identified. Hence, we take the granular texture as bio-generated. The granular type may be symmetrically arranged around fractures (e.g. Fig. 2a), but in many cases no symmetry exists. The granular texture is much more common than the tubular type (below), and can be found at any depth where fresh glass still is present (*c*. 550 m into the volcanic basement of the *in situ* oceanic crust). This type is also the dominant bio-alteration texture in pillow lava rims of relatively young, little-deformed or metamorphosed ophiolites (Fig. 2b) so far studied (Troodos (Cyprus) and Mirdita (Albania)).

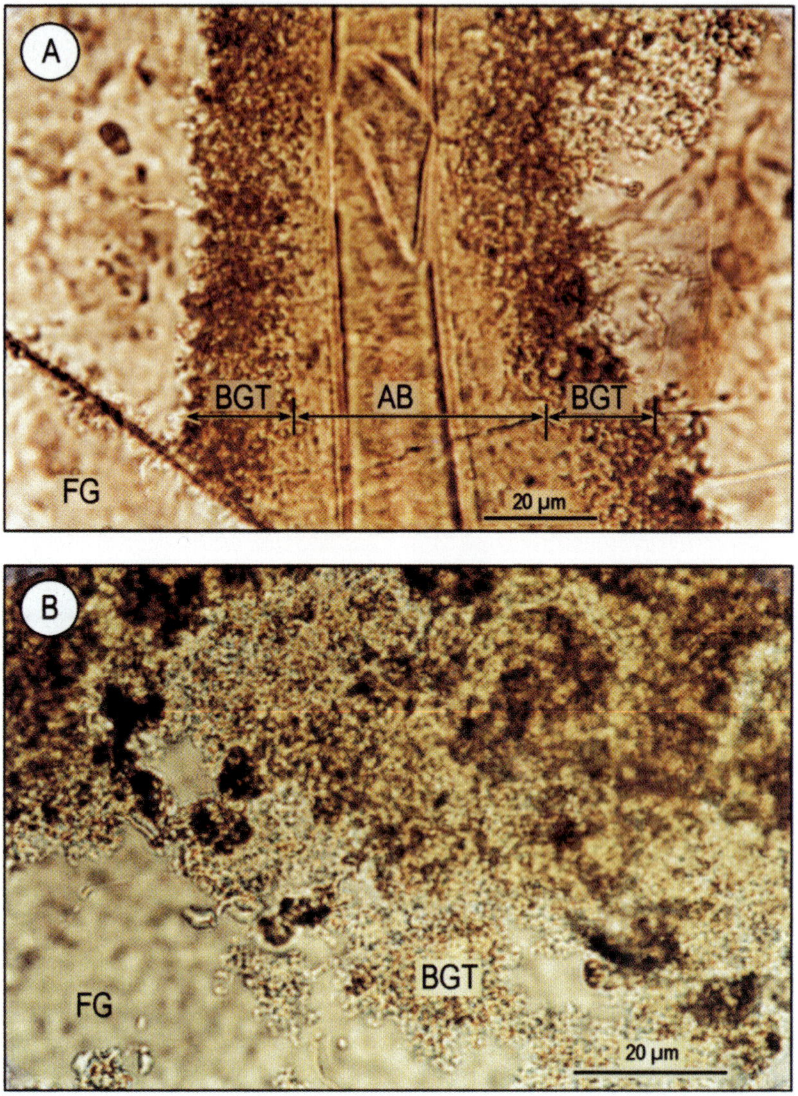

Fig. 2. Typical granular texture viewed through an optical microscope. (**a**) Granular texture of assumed biogenic origin (BGT) adjacent to assumed abiotic alteration product (AB). ODP sample 148-896A, 27R-1, 114–117; depth into volcanic basement: 237 m; age: 5.9 Ma; Alt *et al.* (1996). (**b**) Granular texture of assumed biogenic origin (BGT) in altered basaltic glass from pillow lava (sample 3-Al-00) of the Late Jurassic Mirdita ophiolite, Albania (Robertson & Shallo 2000). FG, fresh glass.

The tubular type is defined by the presence of straight or curved tubes, sometimes branching, or rarely defining a spiral shape. The tubes may attain lengths >100 µm, but are generally shorter (Fig. 3*a*, *b*). Most commonly the width of the tubes is *c*. 1–3 µm, also consistent with microbial etching. These structures seem to appear most commonly at depth levels 100–200 m into the volcanic basement. In the pillow lavas of the Troodos ophiolite, a larger type of tube was found than those so far reported from the *in-situ* oceanic crust. These tubes may reach lengths up to 500 µm, diameters up to 20 µm, and show well-defined segmentation into plates 5–10 µm thick (Fig. 4). The locally high population density of these large tubular, segmented bodies indicate that they may represent *Beggiatoa*-like organisms (Furnes *et al.* 2001*b*), sulphide-oxidizing

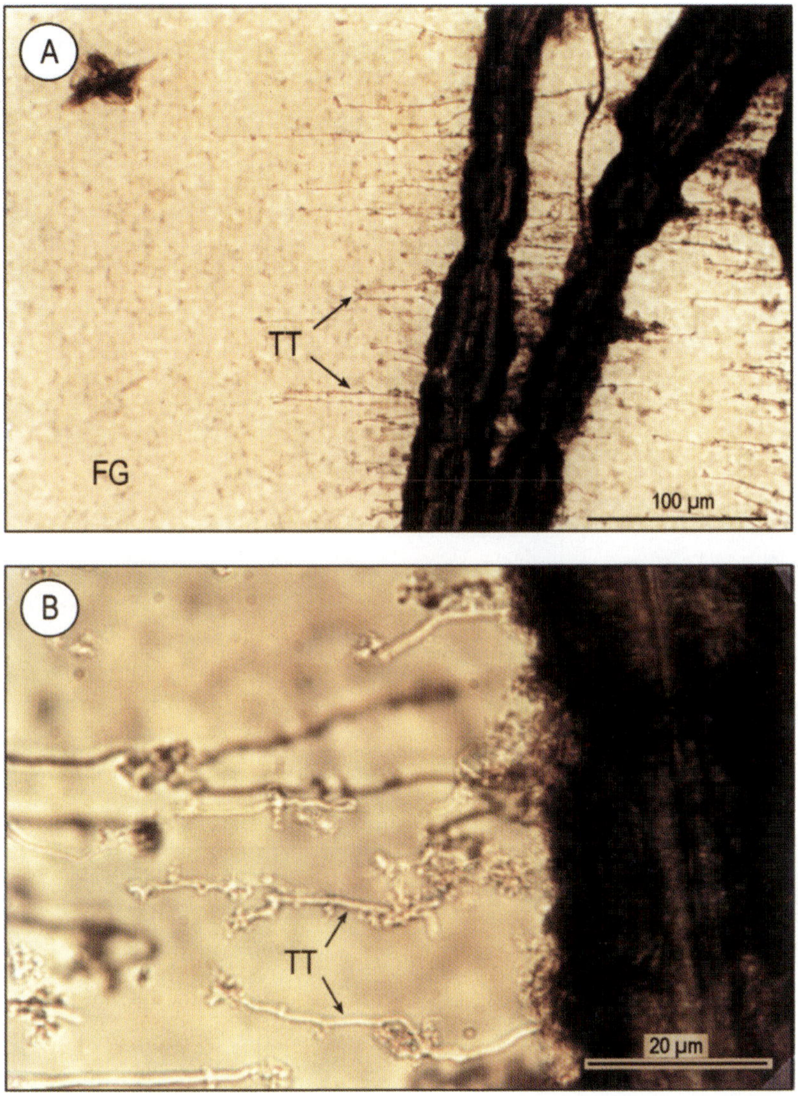

Fig. 3. Optical photomicrographs of tubular texture (TT) of assumed bio-generated origin. FG, fresh glass. (**a**) General view showing approximately parallel orientation of tubes. (**b**) Detailed picture from (a) showing bifurcation and buds on the tubes. DSDP sample 70-504B, 48-2, 121-126; depth into volcanic basement: 375 m; age: 5.9 Ma; Alt *et al.* (1996).

microbes known to form mats at the interface between anaerobic and aerobic environments, conditions that are frequently found at the oceanic floor (Jannasch 1995).

Organic remains

At the alteration front of the granular textures, features that appear to represent organic remains are found in the glassy rims of pillow lavas from *in situ* oceanic crust, as well as in the Troodos ophiolite. These features occur as long (at least 6 µm) and *c.* 50–300 nm wide, branching filaments, occasionally twisted, that in many cases can be seen to be attached to the fresh glass at the alteration front (Fig. 5a, b). The filaments, considered to represent microbial appendages, are partly buried by spherical and/or elliptical bodies *c.* 50–200 nm in diameter

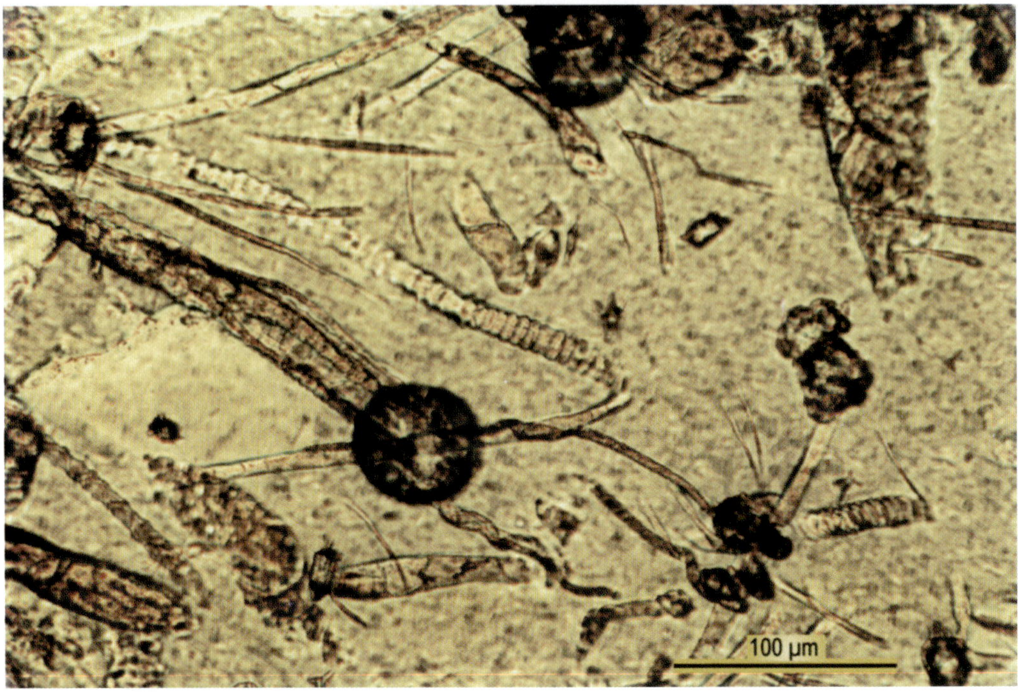

Fig. 4. Optical photomicrographs of tubular texture, comprising large segmented tubes surrounded by fresh basaltic glass. Dark, round structures are vesicles. Sample CY-1-30 from pillow lava of the Troodos ophiolite.

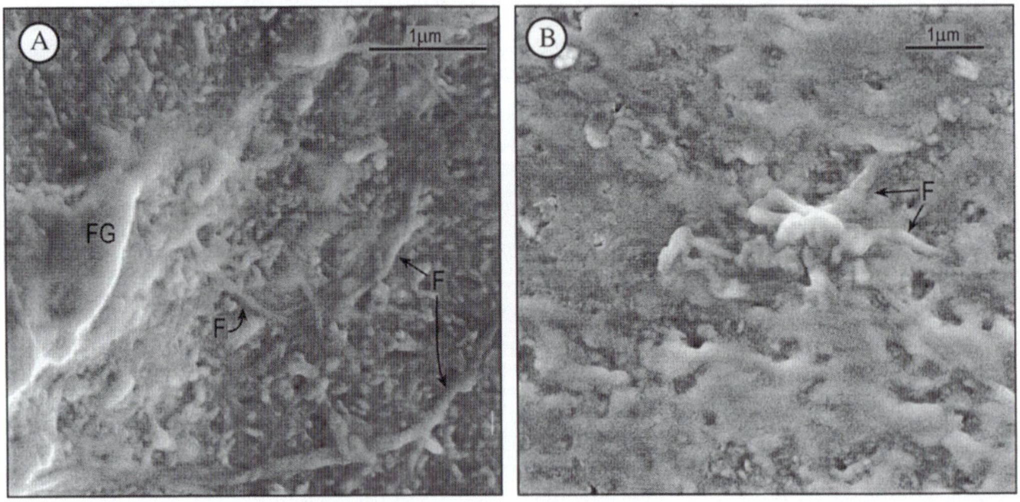

Fig. 5. SEM images of organic-like remains (biofilm and filaments) (F) within areas of typical granular textures with high carbon content. The organic remains are in places attached to the fresh glass (FG), and they are partly buried by bean- to spherical-shaped material, best seen in the upper left and lower right corners of image A. (**a**) Pillow lava sample CYP-99–11 from the Troodos ophiolite; Furnes *et al.* (2001*b*). (**b**) Pillow lava from DSDP Hole 418A (sample 418A-55-4, 112-114); depth into volcanic basement: 300 m; age: 110 Ma; Robinson *et al.* (1979).

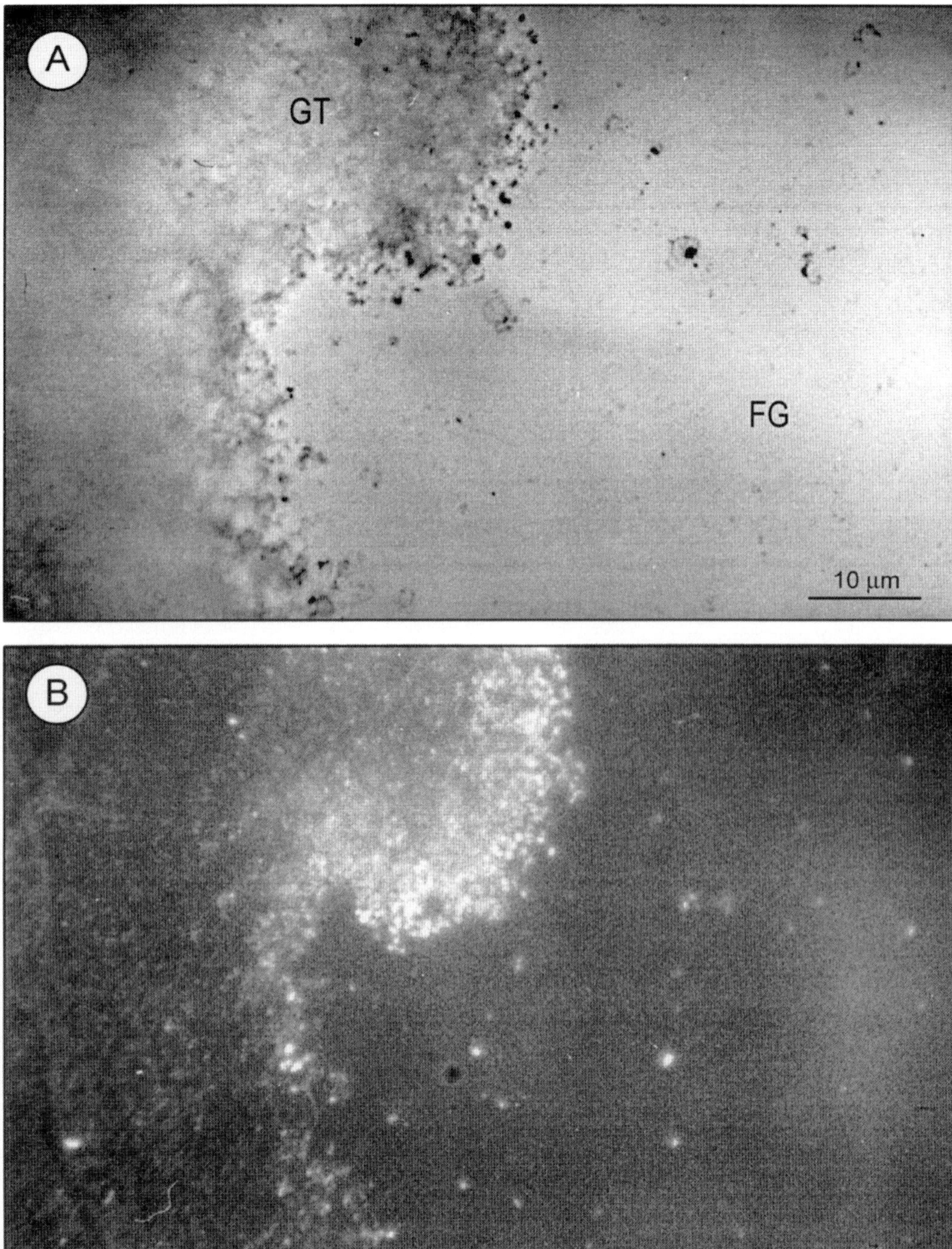

Fig. 6. Photomicrographs of an alteration front of the granular type (GT) from thin section of sample 148-896A, 11R-1, 72-73, viewed as (**a**) light transmission, and (**b**) epi-illumination for DAPI dye. FG, fresh glass.

(Fig. 5a). The size range for the structures indicated (50–300 nm) is, in general, smaller than that of bacteria, but is comparable to that of living nano-organisms found in Australian sandstones (Uwins *et al.* 1998).

DNA analyses

In order to verify the existence of microbes, independent of the textural evidence, samples were investigated, prior and subsequent to DNA-specific DAPI staining, by ordinary light and epifluorescence microscopy. Thin sections, embedded in epoxy, were stained with 10 μg/ml^{-1} DAPI (4,6 diamino-phenyl-indole, a fluorescent dye that specifically stains DNA) for 1 minute. Excess stain was removed by washing with distilled water. The method follows Porter & Feig (1980). Samples were examined in a Nikon Microphot microscope using light transmission and epifluorescence with excitation light wavelength of 365 nm, and emission wavelength of 420 nm. This method has been successfully applied by using different types of DNA-specific dyes (Thorseth *et al.* 1995*a*; Furnes *et al.* 1996; Giovannoni *et al.* 1996; Furnes *et al.* 2001*a*), and the result of DAPI staining is shown in Figure 6. The altered material that proved successful in the

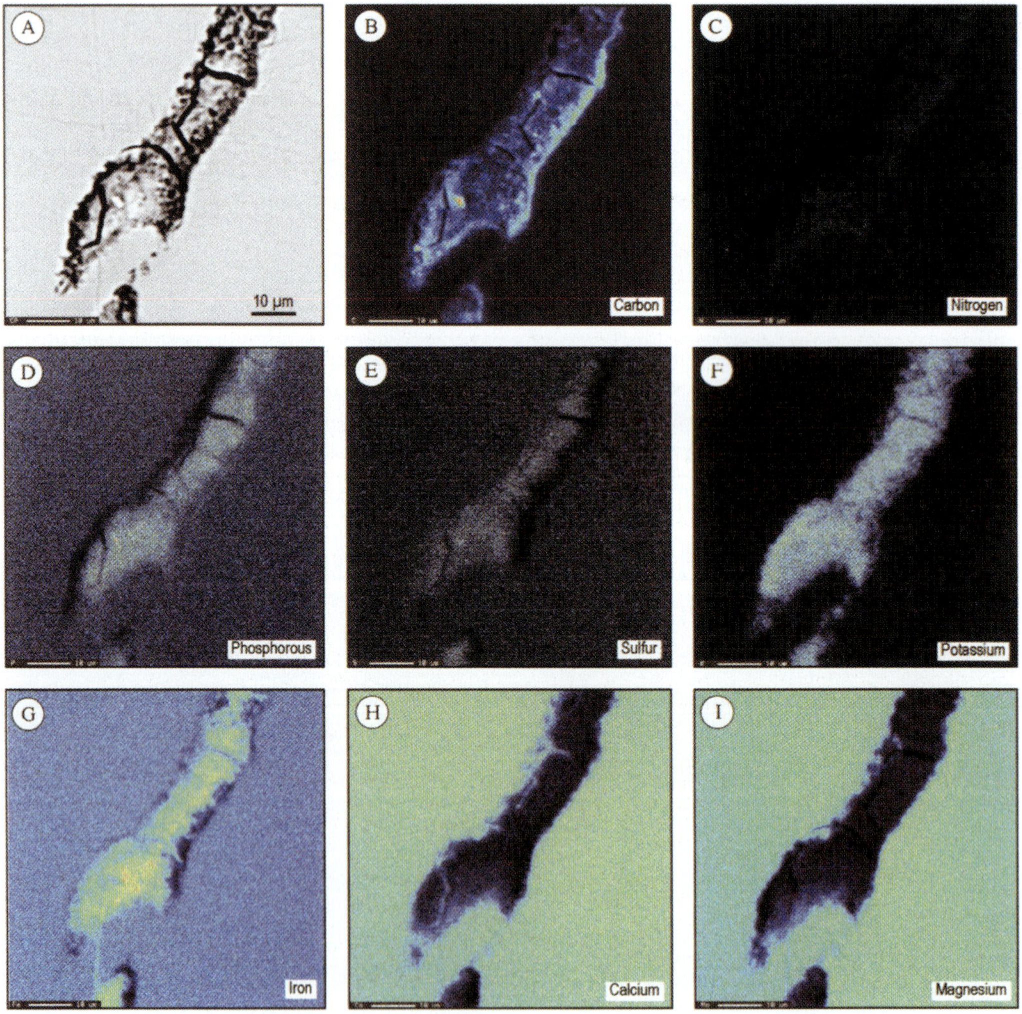

Fig. 7. SEM image (**a**) and X-ray maps showing the distribution of carbon (**b**), nitrogen (**c**), phosphorpous (**d**), sulphur (**e**), potassium (**f**), iron (**g**), calcium (**h**), and magnesium (**i**) of typical granular-type altered basaltic glass. Relative concentration scale (from lowest to highest): black–blue–green–yellow–red. DSDP sample 49-411, 2R-2, 19-27; depth into volcanic basement: 9 m; age: 1 Ma; Luydendyk *et al.* (1978).

staining experiments was obtained from Hole 896A (6 Ma; Alt *et al.* 1996) and Hole 396B (10 Ma; Dmitriev *et al.* 1978). It should be pointed out, however, that most staining experiments failed due to the difficulty in getting the dye into the altered areas of samples. This is because of the sealing effect during authigenic mineral formation. The spherical bodies of the granular type texture, particularly at the alteration front, fluoresce, demonstrating the presence of DNA (Fig. 6). Similarly, epifluorescent microscopy of the DAPI-stained tubular type of texture shows that DNA is localized as chains of spherical cells, particularly at the tips of the tubular bodies. The presence of archaeal as well as bacterial ribosomal RNA within micro-niches in DAPI-stained thin sections has also been demonstrated (see Torsvik *et al.* 1998, and references therein).

Element mapping

Additional strong evidence that the above-described textures are associated with, or caused by micro-organisms is provided by X-ray mapping, particularly with respect to carbon. A large number of samples have been X-ray mapped, and invariably where SEM images indicate the presence of bio-generated textures (Fig. 7a), the carbon content is high (Fig. 7b). The highest carbon concentrations appear most commonly at the alteration front (Torsvik *et al.* 1998; Furnes *et al.* 1999). This is in good agreement with the presence of filaments and DNA (Figs 5 & 6). Filaments are interpreted to be organic remains, which are occasionally observed at the alteration front. Within samples of the oldest crust, carbon is mostly equally distributed throughout the granular alteration zone (Furnes *et al.* 1999). It is assumed that carbon, represented by organic remains, is generally encrusted by amorphous material of the bio-genetic structures. Bio-mediated precipitation of elements, leached from the sub-stratum to micro-organisms, is a well-known phenomenon (e.g. Ferris *et al.* 1987). A recent study by McKinley *et al.* (2000) on secondary minerals within fractures in the Columbia River basalts showed that they were intimately intermixed with carbonaceous matter, an occurrence taken to represent degradation products of microbes. Nitrogen, like carbon, is also generally enriched within the granular alteration zones, but not to the same extent (Fig. 7c). This is, however, not to be expected since N:C values of marine bacteria have been reported to range from 0.07 to 0.35 (Fagerbakke *et al.* 1996). Phosphorous may, in shallow and young samples, show a patchy distribution to a general enrichment throughout the granular alteration zones (Fig. 7d). In old samples, however, phosphorous is invariably depleted. Sulphur shows a pattern similar to that of phosphorous (Fig. 7e), but may also appear as high-concentration spots about 1 μm in diameter at the alteration front in old samples at great depth. Phosphorous and sulphur are well-known bio-elements (e.g. Ehrlich 1999), and their high-concentration appearance, together with carbon and nitrogen, indicate that they are bio-concentrated. Potassium is always enriched within the zones of bio-alteration (Fig. 7f). This may be due to two phenomena; (1) microbial uptake of potassium, and (2) crystallization of potassium-bearing authigenic minerals (Torsvik *et al.* 1998; Furnes *et al.* 1999; Alt & Mata 2000). Iron, probably an important element for the microbial colonization and growth on the glass surfaces, is generally enriched within the alteration zone (Fig. 7g). Calcium and magnesium are nearly always depleted (Fig. 7h, i), demonstrating that carbon is not bound in carbonates, but occurs as organic compounds.

Carbon isotopes

Based on the strong indications that the alteration of glass is microbially mediated, we predict that this should be reflected in the carbon isotope ratio. The stable isotopes of carbon (^{12}C and ^{13}C) may be fractionated by either biotic or abiotic processes (Oremland 1988). The largest carbon reservoirs, marine carbonates and biogenic organic matter, show isotopic compositions with mean $\delta^{13}C$ values of approximately 0‰ and −25‰, respectively (Hoefs 1997). The $\delta^{13}C$ values of carbonates extracted from fresh basalts, representing mantle values, range from approximately −5 to 7‰ (Alt *et al.* 1996; Hoefs 1997). Most micro-organisms obtain carbon and energy from the oxidation of organic matter. During respiration, organic material is oxidized to CO_2, which may be subsequently precipitated as carbonate minerals. During this process biogenic fractionation will lead to ^{12}C-enriched CO_2, and hence carbonate depleted in $\delta^{13}C$. Positive $\delta^{13}C$ signatures, on the other hand, may result from lithotrophic utilisation of CO_2 when methanogenic *Archaea* produce methane from H_2 and CO_2. As a result of loss of the ^{12}C-enriched methane the remaining carbonate will be enriched in ^{13}C. Diagenetic dolomite with $\delta^{13}C$ as high as +14‰ has been reported from sediments in DSDP Hole 479 (Gulf of California), indicating a biogenic CO_2 reservoir related to active methanogenesis (Kelts & McKenzie 1982). Similarly, pore water in anoxic sediments

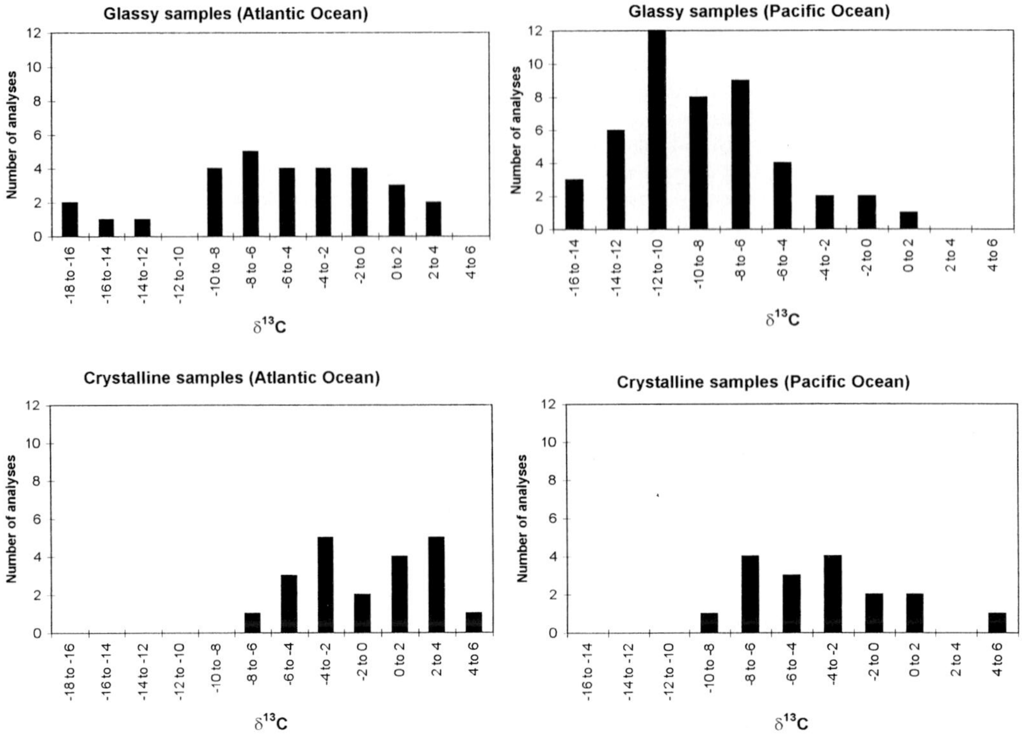

Fig. 8. Distribution of $\delta^{13}C$ in glassy and crystalline pillow lavas from DSDP/ODP drilling sites in the Atlantic Ocean and Pacific Ocean (Costa Rica Rift and Lau Basin). Data from Furnes *et al.* (2001*b*).

from a number of DSDP holes also showed $\delta^{13}C$ values in the range +1 to +14‰ (Hoefs 1997).

An extensive collection of pillow lava rims have been analysed for carbon isotopes. Since textural evidence of microbial influence upon the rocks has only been found in glass and not in crystalline rock, samples of pillow rims have been split in two, i.e. a glassy part and a crystalline part. Glassy samples from a number of drill sites from the Atlantic Ocean show a large range in the $\delta^{13}C$ values (from −16.3 to +2.69‰), a range which differs from that of the crystalline samples (from −6.4 to +4.6‰; Fig. 8). Analyses of samples from the Pacific Ocean (Costa Rica Rift and Lau Basin) show the same relationship between glassy and crystalline samples as those from the Atlantic Ocean, but the $\delta^{13}C$ values of the glassy samples are more abundant in the −14 to −6‰ range, and hardly any positive $\delta^{13}C$ values are found (Fig. 8). These results are further independent evidence that the glassy samples have been bio-processed, and that the crystalline samples in general show values which are more akin to the original magmatic values for basalts, i.e. in the range −5 to −7‰ (Alt *et al.* 1996; Hoefs 1997). The predominance of very low $\delta^{13}C$ values can be attributed to alteration influenced by bacteria and the oxidation of organic matter. The difference in the $\delta^{13}C$ values of the glassy samples from Atlantic and Pacific ocean crust has been ascribed to higher H_2 production in the slow-spreading, fault-dominated Atlantic crust, due to more extensive serpentinisation. Thus, positive $\delta^{13}C$ values in the glass from pillow lavas of the Atlantic crust suggest lithotrophic utilisation of CO_2, in which methanogenic *Archaea* produced CH_4 from H_2 and CO_2 (Furnes *et al.* 2001*a*; Fig. 9). However, several crystalline samples from the Atlantic Ocean also show positive $\delta^{13}C$ values. This feature may be explained in a manner similar to the glassy samples, though we do not see textural evidence for bio-alteration.

Implications for estimating bio-alteration of the oceanic crust and depth of the oceanic biosphere

Bio-generated features such as alteration textures (Figs 1–4), filament-like structures (Fig. 5),

DNA (Fig. 6), the presence of carbon and nitrogen (Fig. 7), and specific $\delta^{13}C$ values that can be attributed to microbial fractionation (Figs 8 & 9) in the glassy part of pillow lavas, have been abundantly reported from altered basaltic glass in both modern oceanic crust and ancient ophiolites. Individually, the above-mentioned features may not prove extensive bio-alteration of the volcanic glass in oceanic crust. Taken together, however, we consider them to provide a strong case for bio-alteration. Thus they provide good markers for mapping the distribution of the subsurface biosphere in oceanic crust and where bio-alteration processes are presently active.

Investigations of the glassy rims of pillow lavas from the deepest drill holes in the oceanic crust have demonstrated that bio-generated granular textures can be found throughout the volcanic pile, down to a depth of 550 m (Furnes & Staudigel 1999). Quantification of biotic and abiotic alteration of the glassy rims of basaltic pillows, based on the abundance of anticipated bio-generated textures (granular and tubular), can be achieved by point counting thin sections viewed through the optical microscope. In such quantification, several problems should be kept in mind when evaluating the results: (1) the method of identification and classification of biotic and abiotic alteration textures relies on petrographic recognition; (2) we assume that all granular textures are bio-generated; (3) for samples recovered by drilling oceanic crust, only small amounts of material are usually available for investigation; and (4) poor recovery rates and underestimation of glassy rocks (e.g. hyaloclastites) may bias sampling. These problems in collecting biogenic and abiotic alteration data in altered basaltic glass are discussed in Furnes & Staudigel (1999) and Furnes *et al.* (2001*c*).

Furnes & Staudigel (1999) estimated the relative proportion of alteration products in the glass rims of pillow lavas from some of the deepest drill holes using point-counting methods. Their results indicated that bio-alteration plays a major to dominant role in the upper 300 m of oceanic crust, and progressively becomes less important at deeper levels (Fig. 10).

The depth of active bio-alteration in the oceanic crust has been discussed by Furnes *et al.* (1999). It was proposed that, where gradients exist in carbon concentration across bio-generated textures (e.g. at an alteration front),

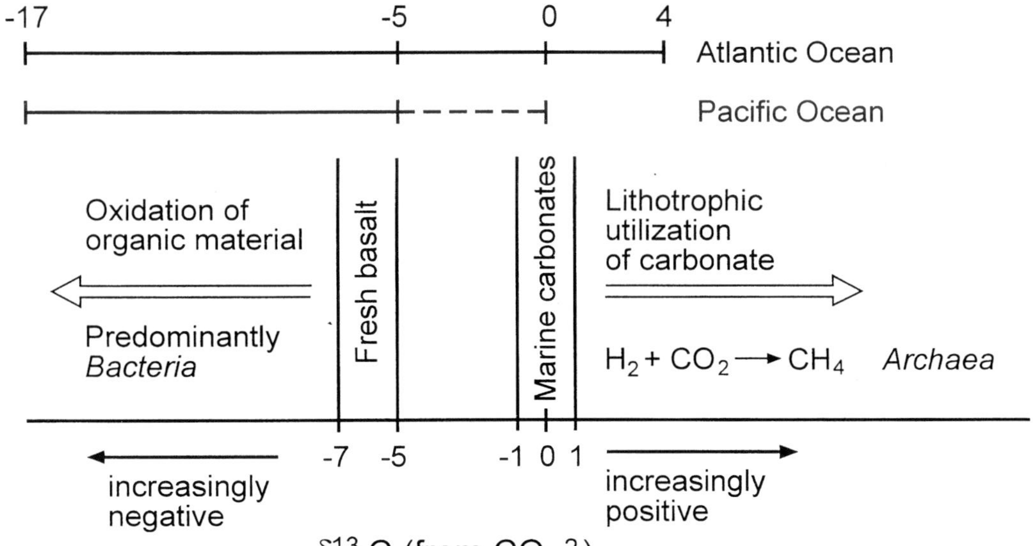

Fig. 9. Sketch indicating the opposite $\delta^{13}C$ paths generated by different microbial metabolisms, which may offer an explanation for the $\delta^{13}C$ data for pillow lavas from the Atlantic Ocean and Pacific Ocean (Lau Basin and Costa Rica Rift). The ranges of $\delta^{13}C$ values from the Atlantic Ocean and the Pacific oceans are shown. Solid line from $\delta^{13}C$ values of -17 to $+4$‰ in the Atlantic Ocean indicates equal distribution of collected data. For the data collection of the Pacific Ocean, equal distribution of $\delta^{13}C$ values is represented in the range -17 to 5‰. The broken line between -5 and 0‰ indicates fewer data points. Data from Torsvik *et al.* (1998) and Furnes *et al.* (2001*a*).

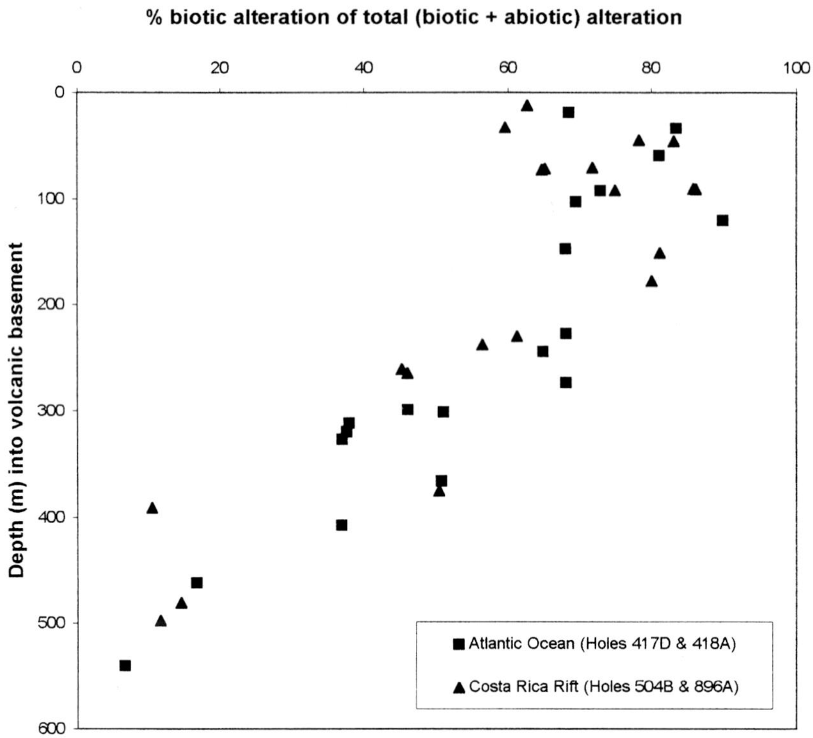

Fig. 10. Estimated percentage biotic alteration to total (biotic + abiotic) alteration relative to depth (m) into the volcanic basement of samples from the Atlantic Ocean (Holes 417D and 418A) and the Costa Rica Rift (Holes 504B and 896A). Modified from Furnes & Staudigel (1999).

the regions of highest carbon concentration would indicate active bio-alteration. In the case of fossil bio-activity, it was proposed that carbon gradient would gradually vanish. By considering carbon concentrations at, and inside the alteration front in progressively deeper samples from Hole 504B, microbes appear to interact with fresh glass down to a maximum depth of *c.* 380 m into the volcanic basement, at an ambient temperature of *c.* 100°C. The maximum and minimum values of the alteration data plotted in Figure 10 suggest that optimum bio-alteration takes place at a depth around 100 m. This corresponds to a present-day sub-seafloor temperature in Hole 504B of about 90°C (Guerin *et al.* 1996). However, Hole 504 B is today *c.* 200 km away from the active spreading ridge. At this distance from the spreading axis, the hydrothermal activity is strongly reduced and the crust undergoes conductive re-heating (Alt *et al.* 1986). At a near-axis position, where effective cooling of the crust takes place by hydrothermal flow (Sleep 1991), the temperature at a given depth would be considerably lower than the present-day temperature. The authigenic mineral history of Hole 504B, studied in detail by Alt *et al.* (1986), shows a complicated multi-stage hydrothermal history, in which the temperatures even at some distance into the sheeted dyke complex may have been as low as 100°C. It was further suggested that most of the bio-alteration occurred during the most hydrothermally active stage (Furnes *et al.* 2001*c*), and thus at lower temperatures than those determined for Hole 504B (Guerin *et al.* 1996).

Concluding remarks

Alteration textures attributed to microbial activity, along with the presence of DNA, carbon, nitrogen, organic remains and $\delta^{13}C$ values of carbonates extracted from altered glass are out of the range for magmatic and other abiotic processes, and together provide compelling evidence that microbes played an important role in the alteration of basaltic glass. The alteration apparently takes place in oceanic crust at any depth where temperatures permit life to exist. Also, the bio-produced features seem to survive even in the oldest known (*c.* 170 Ma) *in situ* oceanic crust, as well as in ophiolites. However, it is still largely unknown what resources are provided to microbes by basaltic glass, and further, what microbial taxa are involved. These are future problems to be resolved.

Other important issues related to the bio-alteration of the upper oceanic crust is the quantification of its rate and extent, and the role of the process in element interchange between oceanic crust and the hydrosphere. This can only be achieved by carefully studying sections through the crust, accounting for the volcanic construction and factors such as the local hydrogeology, temperature, the density of fractures, and authigenic mineral formation. Poor recovery when drilling *in situ* oceanic crust severely limits the full range of information. However, bio-alteration features are also preserved in the volcanic rocks of ophiolites, and ophiolites are known to extend back to Archaean time. Alteration studies of such complete volcanic sequences may thus contribute significantly to a better and more complete picture of bio-alteration in oceanic crust.

Studies over the last 5–6 years, leading to the present knowledge of bio-alteration of the *in-situ* oceanic crust and ophiolites, have been funded by grants (to the Bergen Group) from the Research Council of Norway (Grant no. 110833/40 and the Strategic University Program SUBMAR). K. Muehlenbachs acknowledges support from NSERC of Canada. We thank the reviewers Norman Sleep and Jack Farmer for constructive reviews of the manuscript. Jane Ellingsen kindly helped with the illustrations.

References

ALT, J. C. & MATA, P. 2000. On the role of microbes in the alteration of submarine basaltic glass: a TEM study. *Earth and Planetary Science Letters*, **181**, 301–313.

ALT, J. C., HONNOREZ, J., LAVERNE, C. & EMMERMANN, R. 1986. Hydrothermal alteration of a 1 km section through the upper oceanic crust, Deep Sea Drilling Project hole 504B: Mineralogy, chemistry, and evolution of seawater-basalt interactions. *Journal of Geophysical Research*, **91**, 10 309–10 335.

ALT, J. C., LAVERNE, C., VANKO, D. A. *ET AL.* 1996. Hydrothermal alteration of a section of upper oceanic crust in the eastern equatorial Pacific: A synthesis of results from site 504 (DSDP Legs 69, **70**, and 83, and ODP Legs 111, **137**, 140 and 148). *In*: *Proceedings of the Ocean Drilling Program, Scientific Results*. College Station, Texas, **148**, 417–434.

BEDNARZ, U. & SCHMINCKE, H.-U. 1989. Mass transfer during sub-seafloor alteration of the upper Troodos crust (Cyprus). *Contributions to Mineralogy and Petrology*, **102**, 93–101.

CROVISIER, J. L., HONNOREZ, J. & EBERHART, J. P. 1987. Dissolution of basaltic glass in seawater: Mechanism and rate. *Geochimica et Cosmochimica Acta*, **51**, 2977–2990.

DAUX, V., CROVISIER, J. L., HEMOND, C. & PETIT, J. C. 1994. Geochemical evolution of basaltic rocks subjected to weathering: fate of the major elements, rare earth elements, and thorium. *Geochimica et Cosmochimica Acta*, **58**, 4941–4954.

DETRICK, R., HONNOREZ, J., BRYAN, W. B., JUTEAU, T. *ET AL.* 1988. *Proceedings of the Ocean Drilling Project, Initial Reports*, **106/109**. US Government Printing Office, Washington DC.

DMITRIEV, L., HEIRTZLER, J. *ET AL.* 1978. *Initial Reports of the Deep Sea Drilling Project*, **46**, Washington DC, US Government Printing Office.

DUNNING, G. R. & PEDERSEN, R. B. 1988. U/Pb ages of ophiolites and arc-related plutons of the Norwegian Caledonides: implications for the development of Iapetus. *Contributions to Mineralogy and Petrology*, **98**, 13–23.

EGGLETON, R. A. & KELLER, J. 1982. The palagonitization of limburgite glass – a TEM study, *Neues Jahrbuch für Mineralogie*, **7**, 321–336.

EHRLICH, H. E. 1999. Microbes as geologic agents: their role in mineral formation. *Geomicrobiology Journal*, **16**, 135–153.

FAGERBAKKE, K. M., Heldal, M. & NORLAND, S. 1996. Content of carbon, nitrogen, oxygen, sulfur and phosphorous in native aquatic and cultured bacteria. *Aquatic Microbial Ecology*, **10**, 15–27.

FERRIS, F. G., FYFE, W. S. & BEVERIDGE, T. J. 1987. Bacteria as nucleation sites for authigenic minerals in a metal-contaminated lake sediment. *Chemical Geology*, **63**, 225–232.

FISK, M. R. 1999. New shipboard laboratory may answer questions about deep biosphere. *EOS, Transactions, American Geophysical Union*, **30**, 580.

FISK, M. R., GIOVANNONI, S. J. & THORSETH, I. H. 1998. Alteration of oceanic volcanic glass. textural evidence of microbial activity. *Science*, **281**, 978–979.

FURNES, H. 1984. Chemical changes during progressive subaerial palagonitization of a subglacial olivine tholeiitic hyaloclastite: A microprobe study. *Chemical Geology*, **43**, 271–264.

FURNES, H. & STAUDIGEL, H. 1999. Biological mediation in ocean crust alteration: how deep is the deep biosphere ? *Earth and Planetary Science Letters*, **166**, 97–103.

FURNES, H., THORSETH, I. H., TUMYR, O., TORSVIK, T. & FISK, M. R. 1996. Microbial activity in the alteration of glass from pillow lavas from Hole 896A. *In*: *Proceedings of the Ocean Drilling Program, Scientific Results*. College Station, Texas, **148**, 191–206.

FURNES, H., MUEHLENBACHS, K., TUMYR, O., TORSVIK, T. & THORSETH, I. H. 1999. Depth of active bio-alteration in the ocean crust: Costa Rica Rift (Hole 504B). *Terra Nova*, **11**, 228–233.

FURNES, H., MUEHLENBACHS, K., TORSVIK, T., THORSETH, I. H. & TUMYR, O. 2001*a*. Microbial fractionation of carbon isotopes in altered basaltic glass from the Atlantic Ocean, Lau Basin and Costa Rica Rift. *Chemical Geology*, **173**, 313–330.

FURNES, H., MUEHLENBACHS, K., TUMYR, O., TORSVIK, T. & XENOPHONTOS, C. 2001*b*. Biogenic alteration of volcanic glass from the Troodos ophiolite, Cyprus. *Journal of the Geological Society, London*, **158**, 75–84.

FURNES, H., STAUDIGEL, H., THORSETH, I. H., TORSVIK, T., MUEHLENBACHS, K. & TUMYR, O. 2001*c*. Bio-alteration of basaltic glass in the oceanic crust: Examples from the Atlantic Ocean, Lau basin and Costa Rica Rift. *Geochemistry, Geophysics, Geosystems*, **2**, Paper number 2000GC000150.

GIOVANNONI, S. J., FISK, M. R., MULLINS, T. D. & FURNES, H. 1996. Genetic evidence for endolithic microbial life colonizing basaltic glass/seawater interfaces. *In*: *Proceedings of the Ocean Drilling Program, Scientific Results*, College Station, Texas, **148**, 207–214.

GUERIN, G., BECKER, K., GABLE, R. & PEZARD, P. A. 1996. Temperature measurements and heat-flow analysis in hole 504B. *In*: *Proceedings of the Ocean Drilling Program, Scientific Results*. College Station, Texas, **148**, 291–305.

HAY, R. L. & IIJIMA, A. 1968*a*. *Nature and origin of palagonite tuffs of the Honolulu Group on Oahu, Hawaii.* Geological Society of America, Memoirs, **116**, 338–376.

HAY, R. L. & IIJIMA, A. 1968*b*. Petrology of palagonite tuffs of the Koko Craters, Oahu, Hawaii. *Contributions to Mineralogy and Petrology*, **17**, 141–154.

HOEFS, J. 1997. *Stable Isotope Geochemistry*. Springer, Berlin.

JAKOBSSON, S. P. 1972. On the consolidation and palagonitization of the tephra of the Surtsey volcanic island. *Surtsey Progress Report*, **VI**, 1–8.

JAKOBSSON, S. P. 1978. Environmental factors controlling the palagonitization of the Surtsey tephra, Iceland. *Geological Society of Denmark, Bulletin*, **27**, 91–105.

JAKOBSSON, S. P. & MOORE, J. G. 1986. Hydrothermal minerals and alteration rates at Surtsey volcano, Iceland. *Geological Society of America, Bulletin*, **97**, 648–659.

JANNASCH, H. W. 1995. Microbial interactions with hydrothermal fluids. *In*: HUMPHRIS, S. E., ZIERENBERG, R. A., MULLINEAUX, L. S. & THOMSON, R. E. (eds) *Seafloor Hydrothermal Systems. Physical, Chemical, Biological, and Geological Interactions*. Geophysical Monograph, **91**, 273–296.

JONES, B. & GOODBODY, Q. H. 1982. The geological significance of endolithic algae in glass. *Canadian Journal of Earth Sciences*, **19**, 671–678.

KELTS, K. & MCKENZIE, J. A. 1982. Diagenetic dolomite formation in Quaternary anoxic diatomaceous muds of Deep Sea Drilling Project Leg 64, Gulf of California. *Initial Reports of the Deep Sea Drilling Project*, **64**, Washington, DC, US Government Printing Office, 553–569.

LUYENDYK, B. P., CANN, J. R. *ET AL.* 1978. *Initial Reports of the Deep Sea Drilling Project*, **49**. US Government Printing Office, Washington.

MCKINLEY, J. P., STEVENS, T. O. & WESTALL, F. 2000. Microfossils and paleoenvironments in deep subsurface basalt samples. *Geomicrobiology Journal*, **17**, 43–54.

OREMLAND, R. S. 1988. Biochemistry of methanogenic bacteria. *In*: ZEHNDER, A. J. B. (ed.) *Biology of Anaerobic Microorganisms*. John Wiley & Sons, New York, 641–705.

PORTER, K. & FEIG, Y. S. 1980. The use of DAPI for identifying and counting aquaric microorganisms. *Limnology and Oceanography*, **25**, 943–948.

ROBERTSON, A. & SHALLO, M. 2000. Mesozoic-Tertiary tectonic evolution of Albania in its regional Eastern Mediterranean context. *Tectonophysics*, **316**, 197–254.

ROBINSON, P. T., FLOWER, M. F. J., SWANSON, D. A. & STAUDIGEL, H. 1979. Lithology and eruptive stratigraphy of Cretaceous oceanic crust, western Atlantic Ocean. *Initial Reports of the Deep Sea Drilling Project,* **LI**, **LII**, **LIII**. Washington DC, US Government Printing Office.

ROSS, K. A. & FISHER, R. V. 1986. Biogenic grooving on glass shards, *Geology*, **14**, 571–573.

SÆBØ, L. 1998. *Eksperimentell omdaning av basaltisk glas: Effekt av pH, oksalsyre og temperatur*. Candidatus Scientiarium (thesis), University of Bergen.

SCHIFFMAN, P., SPERO, H. J., SOUTHARD, R. J. & SWANSON, D. A. 2000. Controls on palagonitization versus pedogenic weathering of basaltic tephra: Evidence from the consolidation and geochemistry of the Keanakako'i Ash Member, Kilauea Volcano. *Geochemistry, Geophysics, Geosystems*, **1**, Paper number 2000GC000068.

STAUDIGEL, H. & GILLIS, K. 1990. The timing of hydrothermal alteration in the Troodos ophiolite. *In*: MALPAS, J., MOORES, E. M., PANAYIOTOU, A. & XENOPHONTOS, C. (eds) *Ophiolites Oceanic Crustal Analogues, Proceedings of the Symposium 'Troodos 1987'.* The Geological Survey Department, Ministry of Agriculture and Natural Resources, Nicosia, Cyprus, 665–671.

STAUDIGEL, H. & HART, S. R. 1983. Alteration of basaltic glass: processes and significance for the oceanic crust-seawater budget. *Geochimica et Cosmochimica Acta*, **47**, 337–350.

STAUDIGEL, H., CHASTAIN, R. A., YAYANOS, A. & BOURCIER, R. 1995. Biologically mediated dissolution of glass. *Chemical Geology*, **126**, 119–135.

STAUDIGEL, H., YAYANOS, A., CHASTAIN, R., DAVIES, G., VERDURMEN, E. A. Th., SCHIFFMAN, P., BOURCIER, R. & DE BAAR, H. 1998. Biologically mediated dissolution of volcanic glass in seawater. *Earth and Planetary Science Letters*, **164**, 233–244.

THORSETH, I. H., FURNES, H. & TUMYR, O. 1991. A textural and chemical study of Icelandic palagonite of varied composition and its bearing on the mechanism of the glass-palagonite transformation. *Geochimica et Cosmochimica Acta*, **55**, 731–749.

THORSETH, I. H., FURNES, H. & HELDAL, M. 1992. The importance of microbiological activity in the alteration of natural basaltic glass. *Geochimica et Cosmochimica Acta*, **56**, 845–850.

THORSETH, I. H., TORSVIK, T., FURNES, H. & MUEHLENBACHS, K. 1995*a*. Microbes play an important role in the alteration of oceanic crust. *Chemical Geology*, **126**, 137–146.

THORSETH, I. H., FURNES, H. & TUMYR, O. 1995*b*. Textural and chemical effects of bacterial activity on basaltic glass: an experimental approach. *Chemical Geology*, **119**, 139–160.

TORSVIK, T., FURNES, H., MUEHLENBACHS, K., THORSETH, I. H. & Tumyr, O. 1998. Evidence for microbial activity at the glass-alteration interface in oceanic basalts. *Earth and Planetary Science Letters*, **162**, 165–176.

UWINS, P. J. R., WEBB, R. I. & TAYLOR, A. P. 1998. Novel nano-organisms from Austalian sandstones. *American Mineralogist*, **83**, 1541–1550.

WELCH, S. A. & ULLMAN, W. J. 1996. Feldspar dissolution in acidic and organic solutions: Compositional and pH dependence of dissolution rate. *Geochimica et Cosmochimica Acta*, **60**, 2939–2948.

ZIERENBERG, R. A., SCHIFFMAN, P., JONASSON, I. R., TOSDAL, R., PICKTHORN, W. & MCCLAIN, J. 1995. Alteration of basalt hyaloclastite at the off-axis Sea Cliff hydrothermal field, Gorda Ridge. *Chemical Geology*, **126**, 77–99.

ZHOU, Z. & FYFE, W. S. 1989. Palagonitization of basaltic glass from DSDP Site-335, Leg-37 – Textures, chemical composition, and mechanism of formation. *American Mineralogist*, **74**, 1045–1053.

Index

Page numbers in *italics* refer to figures; those in **bold** type refer to figures